The Genetics of Human Populations

A Series of Books in Biology

The Genetics of Human Populations

L. L. Cavalli-Sforza
STANFORD UNIVERSITY

W. F. Bodmer
OXFORD UNIVERSITY

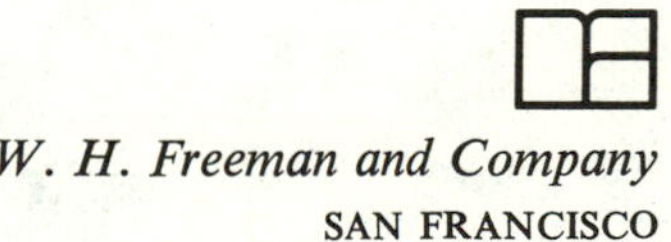

W. H. Freeman and Company
SAN FRANCISCO

to the memory of R. A. Fisher

Printed in the United States of America

International Standard Book Number: 0-7167-1018-8

Library of Congress Catalog Card Number: 72-120303

10 9 8 7 6 5 4

Contents

Preface

Individuals belonging to the same species differ with respect to a multitude of inherited characteristics. Population genetics is largely concerned with understanding the nature and source of these inherited differences, with predicting the changes in the relative frequencies of the different types in a population, and with determining the conditions when equilibrium between the forces affecting their frequencies is reached. The theory underlying these predictions is effectively the quantitative theory of evolution.

Following the rediscovery of Mendelism at the turn of the century, the mathematical theory of evolution based on Mendelian genetics was developed, mostly by R. A. Fisher, J. B. S. Haldane, and Sewall Wright. Much further theoretical work has been based on the foundations laid down by these three great scientists. Experimental population geneticists—starting with S. S. Chetverikov, who, working with *Drosophila*, was the first to identify large sources of genetic variability in natural populations, and continuing with the classic work of Theodosius Dobzhansky, E. B. Ford, and others—have provided basic observations on natural and artificial populations that test evolutionary theories and predictions. Although studies of experimental population genetics are necessarily restricted to short periods of time and usually to limited geographic areas, such microevolutionary studies have, nevertheless, helped greatly in understanding the general phenomena of evolution on a larger time scale (namely, macroevolution). Data from evolutionary comparisons on a molecular level are adding important new information to challenge the validity of our evolutionary theories.

Since the pioneering work of W. Weinberg it has been clear that the analysis of human genetic data must rely heavily on concepts of population genetics. Because it is not possible to carry out controlled experimental breeding with humans, the study of inheritance depends on observations of those matings that happen to take place in the human population. Human population genetics is needed, therefore, to predict the frequencies with which possible types of matings will be found.

Medical genetics, the study of inherited diseases, depends on a knowledge of human population genetics. This dependence has been emphasized by our growing awareness that individual reactions to drugs and other agents, and also susceptibility to some chronic diseases, may often be, at least in part, genetically controlled.

Several disciplines, including public health, require more knowledge of the range of individual variation and how environmental and genetic factors determine it. For example, no longer can we ignore the need for different educational responses to individual differences. Sociologists cannot afford to ignore the sources of differences when studying interactions between individuals. Physical anthropologists are becoming increasingly aware of the need to provide a theoretical background for their observations. Psychologists must take into account genetic differences in their studies of human behavior. Last, but by no means least, it is important that the demographer, whose aim is the prediction of the future course of population changes, also realize the need to take account of population genetic principles and, especially, of the family as the unit of information in making demographic predictions.

So far no comprehensive treatment of the genetics of human populations emphasizes the interpretation of data in relation to the theoretical models. This book is an attempt to fill the need for such a treatment. We have included some original work, which we were led to in the course of writing this book; for example, we looked into the effects of ascertainment biases on the estimation of mutation rates, the treatment of the hemoglobin ACS polymorphism including the effects of migration, a new method for calculating effective population size taking into account age structure, estimates of the selective effects of the association between ABO and duodenal ulcer, some of the treatment of the effects of selection on quantitative characters, and some of the results on molecular evolution.

For those readers with a more extensive knowledge of mathematics, we have included, in worked examples at the end of the chapters, more extensive mathematical treatment of some of the problems. We have also felt it appropriate to include an introductory first chapter that summarizes basic concepts of genetics in a form that is meant to be little more than an annotated glossary of terms and concepts most important for an understanding of human population genetics. In the same vein we have included in an appendix a brief summary of some basic concepts of statistics; this also should be considered as little more than an annotated list of definitions and formulas.

We have not attempted to be comprehensive in our review of genetic differences, though we have tried to present at least a brief description of all those which we use for the sake of illustration. The most important source of information on genetic differences in humans is the catalogue published by V. A. McKusick, *Mendelian Inheritance in Man.* We have provided at the end of individual chapters bibliographies of general references relevant to the chapters and have collected all the references cited together at the end of the book.

Many of our colleagues were helpful to us in the course of writing this book. In particular, we would like to express our gratitude to E. Anderson, M. Feldman, D. Kennedy, I. Gottesman, I. M. Lerner (deceased), R. Lewontin, and C. Stern for making many constructive comments and criticisms on the typescript. Margaret Muller transformed our crude sketches into clear and attractive illustrations, the late Jackie Dale typed what turned out to be a very long manuscript, while Judy Kidd helped with the proofs and index, and our wives were patient throughout our ordeal.

This book had its origin in courses given mostly at Stanford by the two authors. The manuscript was finished in the summer of 1969, and few additions or modifications were made after that date.

May 1971
and
March 1977

L. L. Cavalli-Sforza, Stanford
W. F. Bodmer, Oxford

Introduction

Several advances have been made in human population genetics since our book was completed in 1969, though most of the fundamentals of the subject as we presented it then have not changed. Here we shall highlight briefly some of the major relevant developments and give appropriate references so readers can pursue these topics further.

At the time our book was first published there were major developments in techniques for distinguishing different chromosomes by patterns of banding; these have revolutionized human chromosome studies. Each chromosome has its own highly characteristic banding pattern and can be easily distinguished from all the others. Furthermore, the banding patterns are sufficiently detailed that translocations, inversions, and deletions frequently can be defined precisely with respect to their position on the chromosome in relation to these banding patterns. Two major techniques are used: First, Caspersson and colleagues developed techniques dependent on fluorescent intensity staining differences; second, a number of laboratories throughout the world simultaneously developed techniques for simple staining with Giemsa under mild conditions of incubation. As a result of these developments, there has been much more research on the incidence and effects of a large variety of human chromosome changes and even documentation of polymorphisms for small inversions around the centromere. For an elementary account of these developments, see *Genetics, Evolution and*

Man by Bodmer and Cavalli-Sforza, 1976, especially Chapters 4 and 8; whereas detailed original reports can be found in journals, such as *Cell and Cytogenetics*, and also in *Population Cytogenetics: Studies in Humans*, edited by E. B. Hook and I. H. Porter (Academic Press, New York, 1977).

Another major advance in human genetics has been the development of somatic cell genetics through cell fusion techniques; this advance has extraordinarily increased our knowledge of the human gene linkage map. Well over a hundred genes are now mapped, many of them down to the level of specific chromosome bands, and all chromosomes have at least one gene assigned to them. An elementary account of these developments can be found in Chapter 5 of *Genetics, Evolution and Man*, while up-to-date accounts are published periodically in Proceedings of the Human Gene Linkage Workshops of which Volume 3 was published in 1977, edited by McKusick et al. *Birth Defects Original Article Series*, the National Foundation, Vol. XII, No. 7, 1976. Enzyme polymorphisms have been detected and, of course, associated studies of heterozygosity levels have advanced. A new edition of the book *Human Biochemical Genetics* by H. Harris was published in 1975 by North-Holland, and is a valuable general reference.

We have furthered our understanding of the HLA system. There are now three loci—*HLA-A*, *B*, and *C*—with approximately twenty alleles at the *HLA-A* and *B* loci, and seven at the *HLA-C* locus, each determining serological specificities detected on most tissues. In addition, there is a fourth locus, *HLA-D*, first determined by mixed lymphocyte culture reactions and recently correlated with serological determinants identified specifically on B lymphocytes. Three international histocompatibility workshops have been given since the publication of our book, namely, *Histocompatibility Testing 1972*, edited by Dausset and Colombani; *Histocompatibility Testing 1975*, edited by Kissmeyer-Nielsen; and *Histocompatibility 1977*, edited by W. F. Bodmer et al. (all published by Munksgaard, Copenhagen). These workshops describe in detail further advances in the HLA field: In 1972 they gave the world-wide distributions of HLA gene frequencies; in 1975 they established the *HLA-C* locus and, by mixed lymphocyte culture typing, the *HLA-D* locus; and in 1977 they defined clearly the serological specificities associated with the *HLA-D* locus and made major associations of HLA with disease.

The relationship of HLA with disease has been a rapidly expanding field of research and is well reviewed in the book *HLA and Disease*, edited by Dausset and Svejgaard, published in 1977 by Munksgaard, Copenhagen. Now many diseases, especially some of the autoimmune and chronic diseases, including rheumatoid arthritis, show striking HLA associations. In some instances we can probably explain these by the presence of immune response genes in the HLA region and linkage disequilibrium between them and the genes controlling the detected antigens. Linkage disequilibrium is now clearly a key phenomenon for the HLA system and helps explain many HLA and disease associations. Persistence of linkage disequilibrium can be surmised for some combinations of alleles of the HLA region loci and provides suggestive evidence for the action of natural selection on this system. The chemistry of the gene products is well developed, and it is clear that those of the *HLA-A*, *B*, and *C* loci are closely homologous and probably represent the products of duplicate genes, whereas the product of the *HLA-D* locus is quite different. In addition, now we know that three components of the complement system are coded by genes in the HLA region.

Very recently our understanding of gene organization at the molecular level has expanded through the application of the now famous recombinant DNA technique. Many surprises are in store, including even genes that overlap or their protein products that are made from noncontiguous DNA segments, perhaps in different combinations. The clustering of duplicated genes with perhaps quite complex patterns of control of expression may represent a major feature of the genetic organization of higher organisms.

The estimation of fitness in populations with overlapping generations has been the subject of extensive research by Charlesworth (*Theoretical Population Biology*, **1**:352–70, 1970; **3**:377–95, 1972; and **6**:108–33, 1974). To B. and D. Charlesworth we also owe the English translation, with some additions (in particular on demographic aspects) of the excellent book by A. Jacquard, *The Genetic Structure of Populations* (Springer-Verlag, Berlin, Heidelberg, and New York, 1974). After an introduction to the standard statistical properties of Mendelian populations, this book gives the theory of the genetic relationships between individuals of a population, following closely the elegant approaches started by G. Malecot and further developed by French scholars. There follows an exposition of the theories of mutation and selection and a chapter on migration (by D. Courgeau). Also included is an analysis of genetic distances and an exposition of principal components analysis. The book is almost entirely theoretical but does contain a few examples and a chapter at the end on selected human applications.

Considerable discussion of quantitative inheritance and continuous variation has been stimulated mostly by the problems of individual, race, and social class differences in IQ. Interpretations of existing data have ranged widely. L. J. Kamin, in *The Science and Politics of IQ* (Erlsbaum, Hillsdale, N.J., 1977), has challenged the contention that the available data prove the existence of a genetic component of IQ. Among the several inconsistencies and flaws in the data analyzed by Kamin were those concerning the contributions of the late Sir Cyril Burt. Later developments of this story have made headlines in the newspapers. At the opposite end of the spectrum, A. R. Jensen (*Genetics and Education*, Methuen, London, 1972; and *Educability and Group Differences*, Harper & Row, New York, 1973) has maintained a staunch hereditarian position. In spite of minor flaws in the analysis, C. Jencks et al. (*Inequality*, Basic Books, New York, 1972; gave what is perhaps the most balanced account of the inheritance of IQ. (See also *Race Differences in Intelligence*, Loehlin, Lindzey and Spuhler, W. H. Freeman, San Francisco, 1975). Jencks and coworkers used path coefficients, a method developed by S. Wright in the 1920s; it became popular in sociology and econometrics but had few applications in genetics. D. C. Rao and colleagues have also applied path analysis extensively to data on IQ and other continuous variates (*American Journal of Human Genetics*, **26**:331–59, 1974; **26**:767–72, 1974; **27**:509–20, 1975; **28**:228–42, 1976; *Behaviour Genetics*, **7**:147–59, 1977). Different fitting techniques and slightly different models have been used by L. J. Eaves and others (for a summary see J. Royal, Stat. Soc. Series, A General, **140**:324–55, 1977).

It would be impossible to review here these and many other contributions from other sources. In general the heterogeneity of the available data and the assumptions implicit in the models used for fitting make the results of the analyses unconvincing. Among the problems is the complexity of expectations of cultural transmission, theoretical models and probes for which were initiated by one of us in collaboration with M. Feldman (*Theoretical Population Biology*, **4**:42–55, 1973; **9**:238–59, 1976; **11**:161–81, 1977; *American Journal of Human Genetics*, **25**:618–37, 1973; *Annals of Human Biology*, **2**:215–26, 1975; *Proceedings of the N.A.S.*, **73**:1689–92, 1976; *Genetics*, 1978, in press). Some discussion of current ideas and controversies can be found in a symposium on genetic epidemiology held at the University of Hawaii in 1977 (now in press). This symposium will also contain information on recent developments in the techniques of segregation analysis and in the now developing art of pedigree analysis.

One area that was dealt with only briefly in our book has subsequently become notorious under the name of "sociobiology," and in particular through the book of that title by E. O. Wilson. In Chapter 10 we discuss natural selection and the sex ratio, and the evolution of sexual dimorphism, both key topics in discussions of group and kin selection. We do not, however, discuss the important work of Hamilton on the theory of kin selection (see especially *Journal of Theoretical Biology*, **7**:1–52, 1964; *Journal of Theoretical Biology*, **12**:12–45, 1966; *Science*, **156**:477–88, 1967; *Nature*, **228**:1218–20, 1970; also *Reviews of Ecology and Systematics*, **3**:193–232, 1972) and several further developments, including models for the interaction of biological and cultural evolution (Cavalli-Sforza and M. Feldman, as mentioned above).

Molecular evolution, human evolution and racial differentiation were the subject of three chapters in our book *Genetics, Evolution and Man*, which contains a more up-to-date account of these topics than the present book. With respect to the study of evolution, one development worth mentioning is a novel method for the analysis of evolutionary trees (A. Piazza et al., *Tissue Antigens*, **5**:445–63, 1975; and L. Cavalli-Sforza and A. Piazza, *Theoretical Population Biology*, **8**:127–65, 1975).

Problems at the interface of genetics and medicine, as well as of genetics and society, dealt with rather briefly in the last chapter of the present book, were the focus of two long chapters in *Genetics, Evolution and Man.*

1

The Basic Concepts of Genetics

Darwin formulated his theory of evolution by natural selection without knowledge of Mendelian genetics. The most widely held theory of inheritance in Darwin's time was the blending theory, according to which the heritable potentialities of the parents were simply mixed, or blended, like liquids in the offspring. Once blended, the parental attributes could never again be separated. Darwin realized that such a theory would rapidly eliminate the heritable variation that was an essential ingredient of his theory, unless there was a very high rate of spontaneous production of new variability. As a result he sought, unsuccessfully, to find a more suitable theory of inheritance than the blending theory. He never, apparently, came in contact with Mendel's great work. Had he been aware of it and recognized its importance, he might have changed, even further, the course of modern biology.

Those scientists who rediscovered Mendelism at the turn of the century did not at first realize that it was entirely compatible with Darwin's theory of evolution by natural selection. It remained for R. A. Fisher, J. B. S. Haldane, and Sewall Wright to show how many of the phenomena of evolution could be readily explained in terms of Mendelian inheritance. An understanding of the theory of inheritance is a prerequisite for the study of population genetics, and thus for the study of the genetics of human populations.

Almost all our readers will have had, at one time or another, some instruction in Mendelism and the concepts of modern biology; some may welcome a review of these fundamental subjects. Knowledge of the concepts of molecular biology has developed over the last 10–20 years, and only recently have they been introduced into high school and college courses. Appreciation of these concepts can help greatly in understanding basic genetics. The following short introduction to basic genetics with reference to molecular biology is intended to serve as a refresher and also as a compact guide or annotated glossary to the genetics needed for the study of the human population genetics.

The real subject of this book starts in the second chapter, to which readers not wishing to review the general concepts of genetics can proceed directly.

1.1 Cells, DNA, and Protein

THE NUMBER AND VARIETY OF CELLS THAT FORM ONE HUMAN INDIVIDUAL

A man can, theoretically, be resolved into his constituent cells which number about 10^{14} (one hundred thousand billion) in an adult individual. Many of these cells can be cultured *in vitro*. All of them have a number of common features. They all have an outer envelope whose contents, the **cytoplasm**, include various organelles, which are common to most cell types. Every cell potentially able to reproduce has a **nucleus**, that is, a spherical body usually centrally located with a membrane around it, containing the material chemically known as **deoxyribonucleic acid (DNA)** organized in bodies known as the **chromosomes.** In other respects, the shape, size, internal structure, composition, and function of various cell types may differ greatly. Almost all cells are highly differentiated and have specific functions. In spite of their differences, almost all cells of an organism contain the same amount and type of DNA.

STRUCTURE AND FUNCTION OF DNA

It is known from experiments in microorganisms, the essentials of which have also been shown to be true for higher organisms, that DNA contains the information needed to make new cells essentially identical to their parents.

DNA also contains the information for producing the various specific types of cells and for making sure the right ones are available at the right time. Thus every cell contains all of the same information but uses only a fraction of it to develop its specific activities.

The amount of DNA per human cell is of the order of 6×10^{-12} grams. This is a very small quantity, but the information contained is nevertheless enormous and sufficient for the purpose of directing the synthesis of a human individual. DNA molecules constitute one class of large molecules. A molecule of DNA is usually a very long double strand. Each strand is made of a series of elements attached one to the other in a linear sequence. These elements are called **nucleotides**, of which there are four types. They are distinguished by the presence of the substances adenine (whose symbol is A), guanine (G), thymine (T), and cytosine (C). The two strands carry substantially the same information in complementary forms and are strictly paired along their length. Structural reasons restrict which nucleotides can be opposite each other in the complementary strands. The rules of complementarity are that if in one strand there is an A at a given position, in the other strand there can only be a T at that position, and vice versa; and that if in one strand there is G at a given position, there can only be C in the other, and vice versa. These rules make the total quantity of G equal to that of C, and that of A equal to that of T. The nucleotide pairs AT, TA, GC, and CG are generally the only allowable combinations in double-stranded DNA. We may consider, as an example,

a sample of a strand of DNA

containing just nine nucleotides:	A T T A G A C A A
its complementary strand:	T A A T C T G T T
the double strand formed by their pairing:	A T T A G A C A A | | | | | | | | | T A A T C T G T T

Though DNA is almost always in the double-stranded form, it sometimes occurs as a single strand. The double-stranded form has different physical properties from those of unpaired single strands. A double strand is extremely thin, having a diameter of 2 millimicrons (2×10^{-9} meters or 2 nanometers) and can be almost endlessly long. A nucleotide pair occupies 0.34 millimicrons of the length of a double strand. The DNA in a single human cell comprises some six billion nucleotide pairs. It is present mainly in the chromosomes. We know little about the fine structure of a human chromosome but it is possible that its DNA is simply a very long unbroken thread. For the longest chromosome, this thread completely uncoiled would be 6 centimeters long, that is about 500,000 times the average diameter of a cell. A nucleus that contains all the 46 chromosomes present in each human cell, is only about ten-thousandths of a millimeter in diameter and therefore it is clear that DNA, in spite of a certain rigidity due to its double strandedness, must usually be folded many, many times within the chromosomes.

PROTEINS, THEIR STRUCTURE AND FUNCTION

The cytoplasm and the nucleus contain other important chemical constituents, many of which belong to another class of large molecules, the ***proteins****.*

Proteins are made of folded filaments called **polypeptides**. Some protein molecules consist of a single filament, others are made of two or more identical or different ones. Some proteins also have other substances associated with them. A polypeptide is a linear series of equally spaced elements called **amino acids**. There are twenty main types of amino acids, whose individual names are given at the bottom of Table 1.1. Each amino acid is made of one short common element that has one end with acidic properties, and another with alkaline properties. In addition, the element has a differentiating side group. Amino acids attach to one another to form polypeptide chains, an acidic end uniting with the alkaline end of another amino acid, and so on.

Polypeptides found in the human body can be formed by as few as a dozen or as many as a few hundred amino acids. The number, types, and sequences of the amino acids in a polypeptide chain determine its properties. A polypeptide tends to assume a given shape by folding in a three-dimensional way determined by the physicochemical properties of its constituent amino acids. The shapes of protein molecules differ widely. The number of different protein molecules in existence is enormous, probably in the hundreds of thousands for an organism like man. Some are very abundant in certain tissues. Thus, for example, there is about one kilogram of the protein hemoglobin in the blood, and several kilograms of the muscle protein actomyosin in the adult human body. One kilogram of hemoglobin is roughly ten billion billion molecules. At the other extreme, some proteins (for example regulatory proteins) are represented by a few molecules found in relatively few cells of the body.

As a consequence of their different physicochemical properties and shape, every protein has a different function. Perhaps the most common-type of protein activity is that of an enzyme.

Enzymes are proteins capable of increasing enormously the rate of, that is **catalyzing**, some special reaction. For example, the attachment of phosphoric acid to glucose, which is the first step in the utilization of this sugar, can be facilitated by a particular enzyme, which is called a kinase. Other enzymes mediate later steps in the utilization of glucose. In general there is a specific enzyme for each step of each of the many biochemical processes that go on in the body. The reactions catalyzed by enzymes might also occur in their absence, but at reaction rates so slow as to be practically meaningless. Enzymes are thus responsible for the **metabolism** of all small molecules, namely the degradative reactions that free the chemical energy contained in various compounds and make it available for other processes, and the synthetic reactions needed to transform molecules available in the environment into those necessary for growth and reproduction of cells. There are many thousands of different enzymes in a cell, each with some specific metabolic function.

Other protein functions involve transfer or storage of a variety of chemical substances.

Hemoglobin and myoglobin take up, store, transport, and release oxygen; ferredoxin has a similar function with iron; haptoglobin binds hemoglobin, and so on. Some proteins have mechanical functions such as the role of collagen in connective tissue or that of elastin in elastic fibers. Still others have special protective functions. For example, there is a large class, probably numbering at least several thousnds, of slightly different proteins, the gamma globulins, which have the property of binding with specific foreign substances, antigens, and serve to protect the body against them and to facilitate their capture and elimination. Antigens can, for instance, be substances present on the surface of bacteria or other parasites. Gamma globulins thus participate in a major way in the body's defense against infections.

1.2 The Synthesis of Proteins and the Genetic Code

Proteins cannot reproduce themselves. The information for making a given protein is coded in a DNA nucleotide sequence.

A given sequence of nucleotides in DNA, starting from a given point, is converted into a unique corresponding sequence of amino acids according to rules that are summarized by a simple dictionary called the genetic code (see Table 1.1). Three adjacent nucleotides in a DNA strand correspond to one amino acid, which is thus determined by a word of three letters, a triplet, in the DNA code. These letter triplets are " read " in sequence and converted by the machinery of protein synthesis into the corresponding amino acids. The letters are the four possible nucleotides, A, G, C, and T, corresponding to adenine, guanine, cytosine, and thymine. There are thus $4 \times 4 \times 4 = 64$ possible different triplets, 61 of which actually determine amino acids. There are, however, only twenty major amino acids, and it is found that there are from one to six triplets, coding for each of these. The remaining three triplets out of the 64 determine termination of the polypeptide chain.

The genetic code given in Table 1.1 is that of the bacterium *Escherichia coli.* It probably differs very little, if at all, from that of man, though there is no direct evidence for this. We have a unique perspective from which to view the unity of life when we consider the close similarities, even identities, of many of the basic phenomena common to all living organisms.

TABLE 1.1
The Genetic Code

1st letter	2nd letter: a	b	A	B	3rd letter
a	Phe	Ser	Tyr	Cys	a
	Phe	Ser	Tyr	Cys	b
	Leu	Ser	chain end	chain end	A
	Leu	Ser	chain end	Try	B
b	Leu	Pro	His	Arg	a
	Leu	Pro	His	Arg	b
	Leu	Pro	Gln	Arg	A
	Leu	Pro	Gln	Arg	B
A	Ile	Thr	Asn	Ser	a
	Ile	Thr	Asn	Ser	b
	Ile	Thr	Lys	Arg	A
	Met	Thr	Lys	Arg	B
B	Val	Ala	Asp	Gly	a
	Val	Ala	Asp	Gly	b
	Val	Ala	Glu	Gly	A
	Val	Ala	Glu	Gly	B

Note: Each amino acid is coded by a triplet of three bases, as shown in the table, which is a compact way of setting out the sixty-four possible triplets.

The four bases are denoted by the letters a, b, A, and B. In DNA the four bases are:

a = Adenine　　A = Thymine
b = Guanine　　B = Cytosine

In messenger-RNA they are:

a = Uracil　　A = Adenine
b = Cytosine　　B = Guanine

The twenty amino acids are identified as follows:

Ala	= Alanine (A)	Lys	= Lysine (K)
Arg	= Arginine (R)	Met	= Methionine (M)
Asn	= Asparagine (N)	Phe	= Phenylalanine (F)
Asp	= Aspartic acid (D)	Pro	= Proline (P)
Cys	= Cysteine (C)	Ser	= Serine (S)
Glu	= Glutamic acid (E)	Thr	= Threonine (T)
Gln	= Glutamine (Q)	Try	= Tryptophan (W)
Gly	= Glycine (G)	Tyr	= Tyrosine (Y)
His	= Histidine (H)	Val	= Valine (V)
Ile	= Isoleucine (I)	Chain End.	
Leu	= Leucine (L)		

For example the triplet bAB stands for Gln = Glutamine. This implies that Guanine-Thymine-Cytosine codes for glutamine in DNA and Cytosine-Adenine-Guanine codes for glutamine in RNA. The letters in parentheses are a shorthand single-letter code often used for the amino acids.

Source: Based on Crick, 1966.

Protein synthesis has been reproduced in the test tube and been shown to depend on three kinds of ***ribonucleic acids (RNA).***

Ribonucleic acids are similar to DNA. They differ mainly in the sugar component of the nucleotide, ribose, which is analogous to the deoxyribose in DNA; in the fact that T (thymine) is replaced by U (uracil), which has the same pairing properties (U pairs with A); and in that they may occur as relatively short *single* strands. There are three forms of RNA.

Messenger RNA (*m-RNA*). DNA does not act directly, but through an intermediary form of ribonucleic acid called messenger RNA. Messenger RNA is formed by the action of an enzyme, RNA polymerase, which copies the sequence from one strand of DNA and forms an RNA strand of the complementary sequence. Only one of the two DNA strands is generally active as a "template" for m-RNA synthesis, the other DNA strand remaining uncopied. The m-RNA sequence is identical to that in the uncopied strand of DNA except that T is replaced by U. Messenger RNA remains single stranded. One molecule or strand can generally be used for the manufacture of a large number of protein molecules, each with the amino acid sequence corresponding to the m-RNA's nucleotide sequence. The number of nucleotides in an m-RNA strand is three times the number of amino acids in the polypeptide chain copied from it plus three nucleotides to indicate the termination of the chain, and possibly, another triplet for its initiation. The mechanism for chain initiation is still not entirely clear.

Transfer RNA (*t-RNA*). Before amino acids are joined to form a polypeptide chain, they must be "activated" by the attachment of a special phosphoric acid group. They are then attached to another type of RNA called transfer RNA. There are as many varieties of t-RNA molecules as there are triplets that can determine amino acids in the code. The attachment of an amino acid to its corresponding t-RNA is directed by a specific enzyme. The t-RNA carries at one end the activated amino acid and at a special position has a triplet of nucleotides complementary to the triplet code for its amino acid. This latter triplet, which is part of the m-RNA, is often called a **codon** and that on the t-RNA the **anticodon**. The principle of pairing complementarity is used for the recognition by the t-RNA of its appropriate triplet on the m-RNA.

Ribosomal RNA (*r-RNA*). Alignment of the t-RNA and the m-RNA and the progression of polypeptide synthesis along the m-RNA, which is "read" from beginning to end, is mediated by special particles, the **ribosomes**. These are present in the cell in large numbers. A ribosome is made from a third form of ribonucleic acid called ribosomal RNA (whose structure is not yet well determined) together with a large number of protein molecules, twenty or more, whose detailed function is also not yet clear.

Figure 1.1 gives a schematic representation of the whole process of protein synthesis. The amounts of all the basic materials, including enzymes, that are needed for the process are carefully controlled by the cell in relation to its environment. The first step of protein synthesis, the transfer of information from DNA to m-RNA through the synthesis of m-RNA on the DNA template, is called **transcription**; the second step, the synthesis of proteins from m-RNA, is called **translation**.

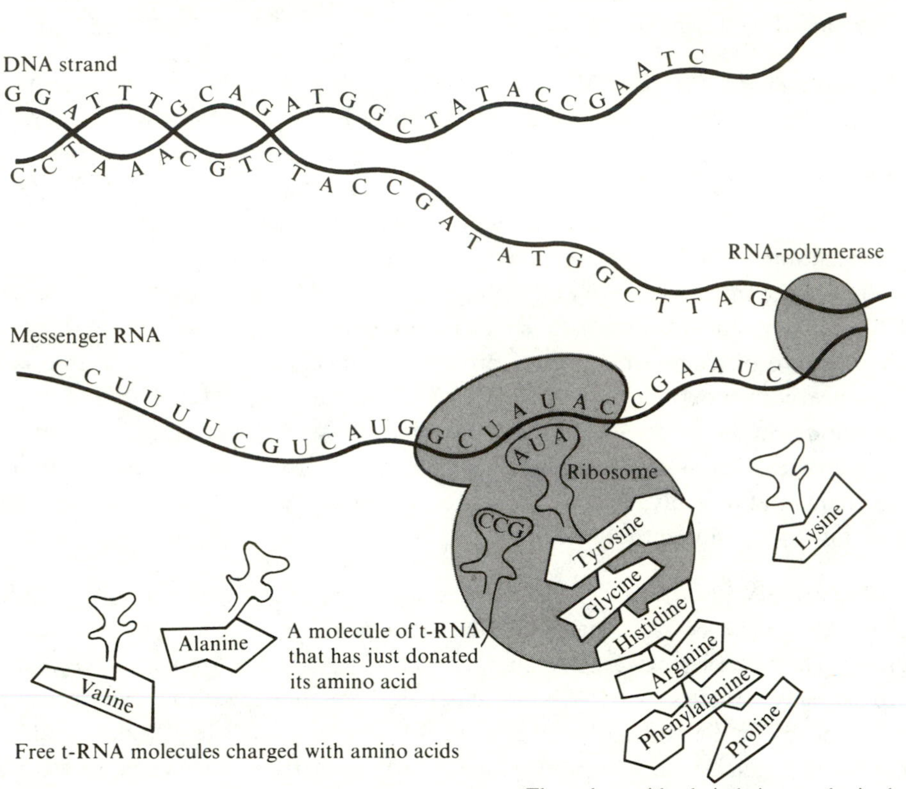

FIGURE 1.1
A scheme of protein synthesis.

Some mechanisms of regulation of protein synthesis act during transcription.

Regulation of protein synthesis allows for adaptation to the living environment. Only when a certain substance, for example, a sugar, is present in the environment is it worthwhile for a cell to develop the enzymes needed to utilize the substance (enzyme induction). At other times, when a certain end product of a reaction chain, necessary for growth (for example, some amino acid), is present in the environment, cells can cease making it and the enzymes necessary for its synthesis (enzyme repression). Mechanisms have so developed that most proteins are made only when they are useful. Some such mechanisms have been shown to operate in microorganisms during transcription, that is, by stopping the production of the specific m-RNA when the proteins coded by it are not needed. The information about the presence or absence of the chemical substances in the environment is conveyed to the DNA by other specific "regulatory" proteins coded for by other DNA segments, as is shown diagrammatically in Figure 1.2.

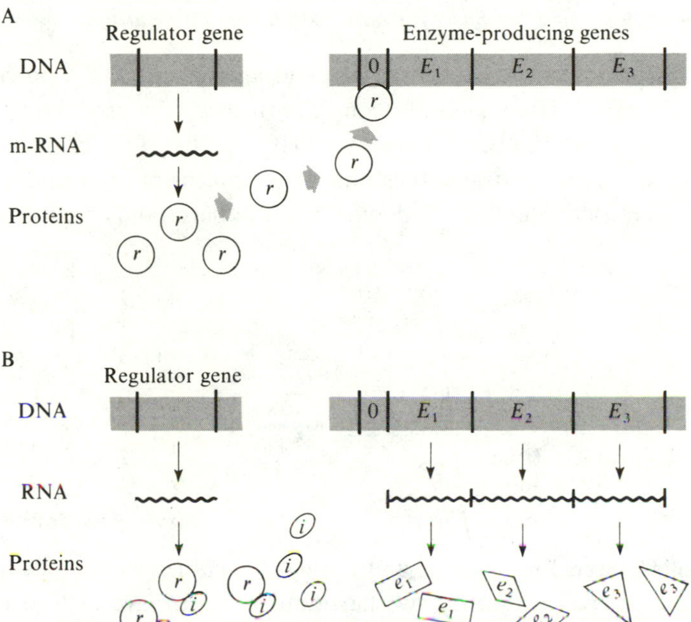

FIGURE 1.2
Enzyme induction, a phenomenon investigated in bacteria, may help the understanding of the regulation of protein synthesis in higher organisms, which is largely unknown. A: The substance *i*, utilized in the reaction chain promoted by enzymes e_1, e_2, e_3 is absent. The regulatory protein *r* produced by the regulator gene *R*, in that absence of *i*, combines with an end (0) of the DNA region that produces enzymes e_1, e_2, e_3 and inhibits formation of the m-RNA required for their synthesis. Therefore genes E_1, E_2, E_3 making these enzymes are inactive and the enzymes are not formed. B: Substance *i* is present and combines specifically with *r*, which is now unable to combine with the region 0. Messenger RNA from E_1, E_2, E_3, and the corresponding enzymes e_1, e_2, e_3 are produced.

Other mechanisms of regulation have been described. Those of the cells of higher organisms are largely unknown, but are believed to be similar in some respects to those operating in microorganisms. Undoubtedly, the regulation of protein synthesis must be at the basis of **differentiation**, which is the development of different cells and tissues. Evidence from various sources shows that various proteins are formed, and correspondingly, that various portions of the DNA are active in different tissues and cells at different times during development.

1.3 Duplication of DNA and Cell Reproduction

DNA has a central role in the life process in both supplying the information for the synthesis of proteins and the basis for duplicating this information.

In the presence of nucleotide precursors and an enzyme, DNA-polymerase, an identical copy of the DNA molecule can be formed by the production of strands complementary to those in the original "template" DNA. If the process takes place in the presence of just one strand, then only the complementary strand is produced. When, however, a double strand is copied, two double strands are formed, as outlined below.

```
                                                              A  T  G  A  C  G
                                                              |  |  |  |  |  |
A T G A C G    A*, A*, A*, T*                                 T* A* C* T* G* C*
| | | | | |  + T*, T*, C*, C*,     replication
T A C T G C    C*, G*, G*, G*    --------------->             A* T* G* A* C* G*
(old double strand) (free        (DNA polymerase)             |  |  |  |  |  |
               nucleotide                                     T  A  C  T  G  C
               precursors)                                    (two new double strands)
```

The nucleotides used for synthesis and the newly synthesized strands are marked with asterisks. It is clear that the two new double strands are each formed of one old strand and one completely new complementary strand. Two new cells can thus be formed from an old one, and each will have exactly the same information for building the basic cellular materials.

It is clearly essential that each daughter cell receives a complete complement of DNA. This is assured in higher organisms, such as man, by a mechanism called **mitosis**.

The DNA is contained in the nucleus of a cell in parcels consisting of long filaments wrapped in a protein matrix. The parcels are called chromosomes and a human cell has 46.

In a cell that is not actively reproducing, the chromosomes are probably uncoiled and indistinguishable under a light microscope. When a cell is preparing for reproduction, the chromosomes undergo coiling and structural reorganization, as a consequence of which each becomes much shorter and thicker. It is then visible under the light microscope. The name chromosome, meaning "colored body," comes from its staining properties. Each human chromosome has a characteristic shape and size, generally similar in every cell and individual, although it is not always possible to distinguish, morphologically, each of the chromosomes in a given cell.

Mitosis, the process of cell reproduction, assures an equal distribution of chromosomes to each daughter cell.

In the preparation for cell division, each individual chromosome has its DNA split internally into two separable threads called chromatids, each presumably constituted from one of the two new DNA double strands. The division is, however, not initially apparent and a human cell starting mitosis still has just 46 visible chromosomes. Then follows a complex process of preparation within the cell for the precise division of each chromosome into two equal parts. Each daughter cell must receive one copy of each chromosome, and thus a complete set of the 46 chromosomes. A specific part of the chromosome that contains visibly less DNA, the **centromere**, is attached to the double thread of DNA contained in each chromosome at this stage. The centromere splits in two with each of the resulting new centromeres attached to a different one of the two chromatids.

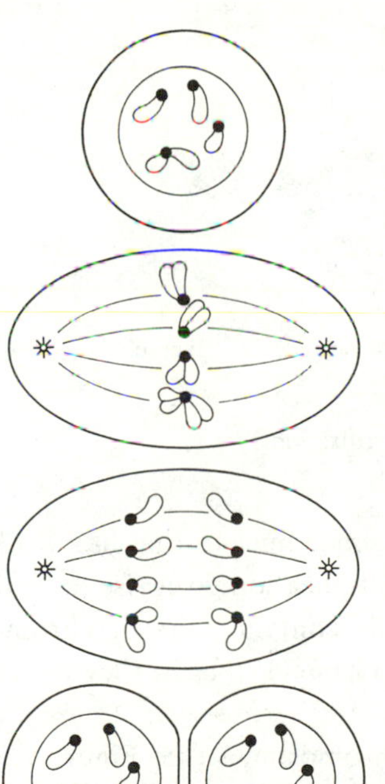

Prophase: the cell at the beginning of mitosis. The chromosomes are represented as oval structures with the centromeres indicated as dots. The DNA in the chromosome occupies half of the total volume and is highly folded. It duplicates before mitosis. The chromosomes can only be visualized individually during mitosis, when they are highly contracted.

Metaphase: the new cell poles are shown on the left and on the right. The nuclear membrane has dissolved; the centromeres, attached to the equator of the spindle defining the major axis of cell division, have not divided, but the rest of each chromosome has.

Anaphase: migration toward the poles. Each half has followed the centromeres, which are migrating to the opposite poles.

Telophase: separation. Two new cells identical to the original one have been formed.

FIGURE 1.3
Mitosis in a cell of a hypothetical organism that has one pair of identical chromosomes and one pair whose members are different.

At this stage a spindle has formed in the cytoplasm that extends along the major axis of division and defines the two poles of the cell. The nuclear membrane has disappeared so that the chromosomes can move freely through the cytoplasm. They attach by their centromere to the central or equatorial region of the spindle. Each chromosome is separated visibly into two moieties. One of the two centromeres of each dividing chromosome moves to one pole and the other to the other pole. This assures that the daughter chromosomes separate in a balanced way to form the two daughter cells (Figure 1.3).

The chromosome set is double in diploid organisms.

Man's 46 chromosomes are made up of 22 pairs, each of which comprises two members that are alike and an additional pair, the sex chromosomes, whose members are alike in females but quite dissimilar in males. This scheme is the consequence of sexual reproduction, which depends on the fusion of two cells, the **gametes** (a male gamete, or sperm, and a female gamete, or egg), to form a new individual. Each gamete carries 23 chromosomes, one member of each pair. When the two gametes unite by the process known as **fertilization** (Figure 1.4), the nucleus

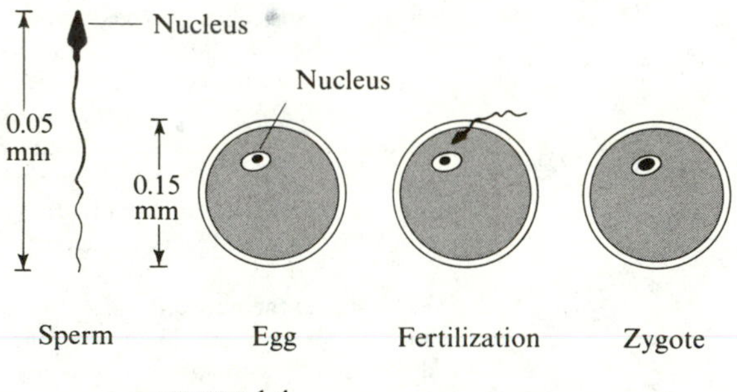

FIGURE 1.4
Human gametes and fertilization.

of the sperm and that of the egg fuse to form a single nucleus that has 46 chromosomes. The new cell is called a **zygote**, and its multiplication gives rise to all the cells of the new organism. An organism whose cells contain pairs of chromosomes formed in this way is called **diploid**. Gametes are **haploid**: that is, they have one set of chromosomes instead of two.

Both a man and a woman have 46 paired chromosomes (see Figure 1.5), with one pair in the female being different from the corresponding pair in the male, as already mentioned. All the other 22 pairs are made of two members that are like each other in shape and size. There are, however, practical limitations to recognition of chromosomes due to the fact that, at least in man, some pairs are so similar to others that it is usually impossible to tell them apart. In practice, one can easily

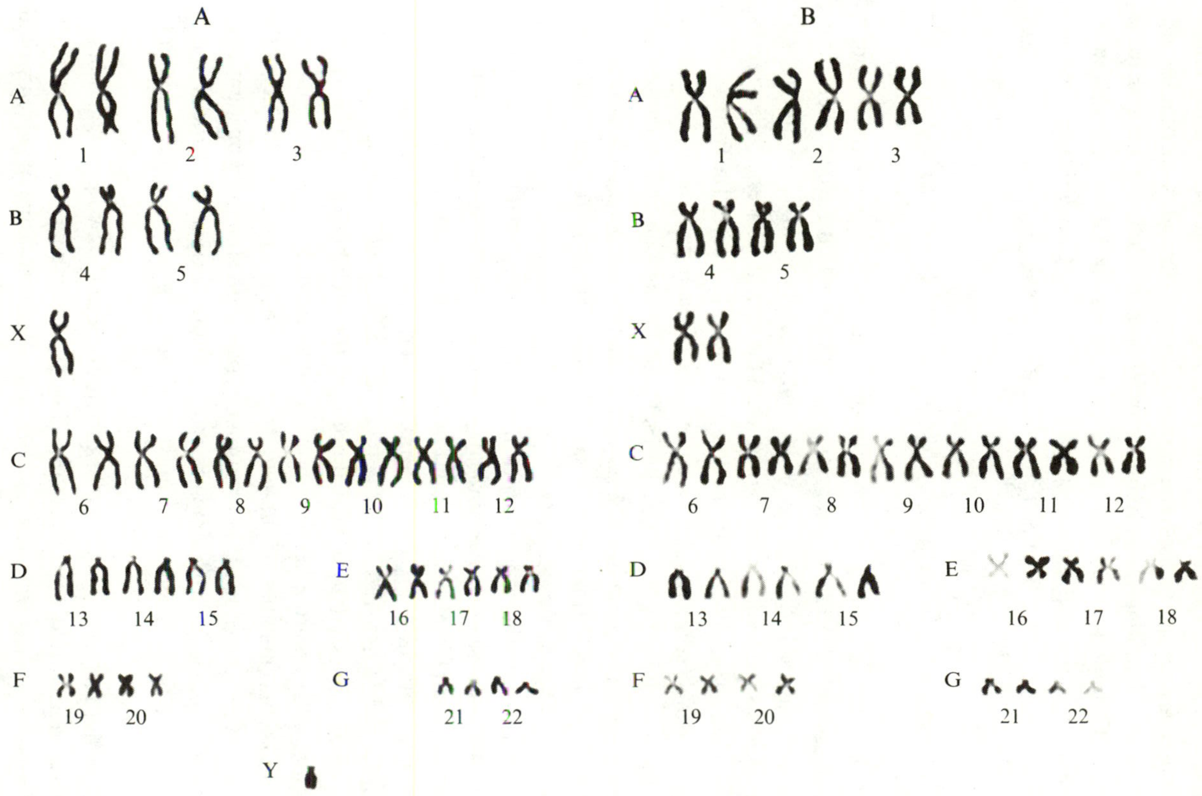

FIGURE 1.5

A: The chromosomes of a male, arranged in pairs according to their size, indicating the seven groups that are conventionally distinguished. B: The chromosomes of a female, arranged in the same way. The chromosomes are often referred to by their numbers, assigned by decreasing size as indicated underneath the groups. (Photos by L. Razavi.)

recognize in man the seven groups indicated in the figure by letters A–G, using the relative position of the centromere and possible other peculiarities as distinguishing features. The pair that differs between males and females is responsible for the sex of the individual, and the members of this pair are called the **sex chromosomes** The sex chromosomes present in the female, conventionally called X chromosomes, are intermediate in size among the human chromosomes, while in the male there is one X and one differentiated sex chromosome responsible for maleness, the Y chromosome, which is one of the smallest of the human set. Chromosome arrangements, such as those shown in Figure 1.5, are called **karyotypes**.

1.4 The Formation of Gametes: Meiosis, or Reduction

The human zygote and almost all the cells of an adult that are derived from it have 46 chromosomes. There must, therefore, be a process that reduces their number to 23 during gamete formation in order that the union of two gametes will produce a new individual with 46 chromosomes. This special process, called reduction, or **meiosis**, is a successive pair of modified mitoses. It assures the formation of gametes each of which contains one member only of each pair of chromosomes. One chromosome from each pair must be included in each gamete because each pair of chromosomes has a unique DNA sequence and thus a special set of functions. The complete set of instructions is necessary for the formation of new cells and organisms that are like their precursors.

The meiosis represented in Figure 1.6 is that of the same hypothetical organism used in Figure 1.3; the organism has one pair whose members are identical and one whose members are different, which could be the sex chromosomes. At the beginning of the first meiotic division, DNA double strands divide, but centromeres do not. This is followed by separation of the members of each homologous pair. The phenomenon that assures an equal distribution of the members of each pair, one to each daughter cell is **pairing** between homologues. Pairing serves as a guide for the centromere of one member of the pair to move to one pole, and that of the other member to the other pole. It is presumably based on similarity or identity in nucleotide sequence. In pairs whose members are not identical, such as the X and Y at least a part of the two chromosomes must be similar, or homologous, enough to assure pairing.

If pairing did not take place, the regular distribution to the two poles, and hence to the daughter cells would not be assured. The two members of one pair might migrate to the same pole, giving rise to unbalanced gametes, one of which would contain both members of the pair, and the other none. This event, called **nondisjunction**, happens very rarely. It tends to happen more often for the sex chromosomes probably because of their more limited pairing capacity and also because of the small size of the Y chromosome.

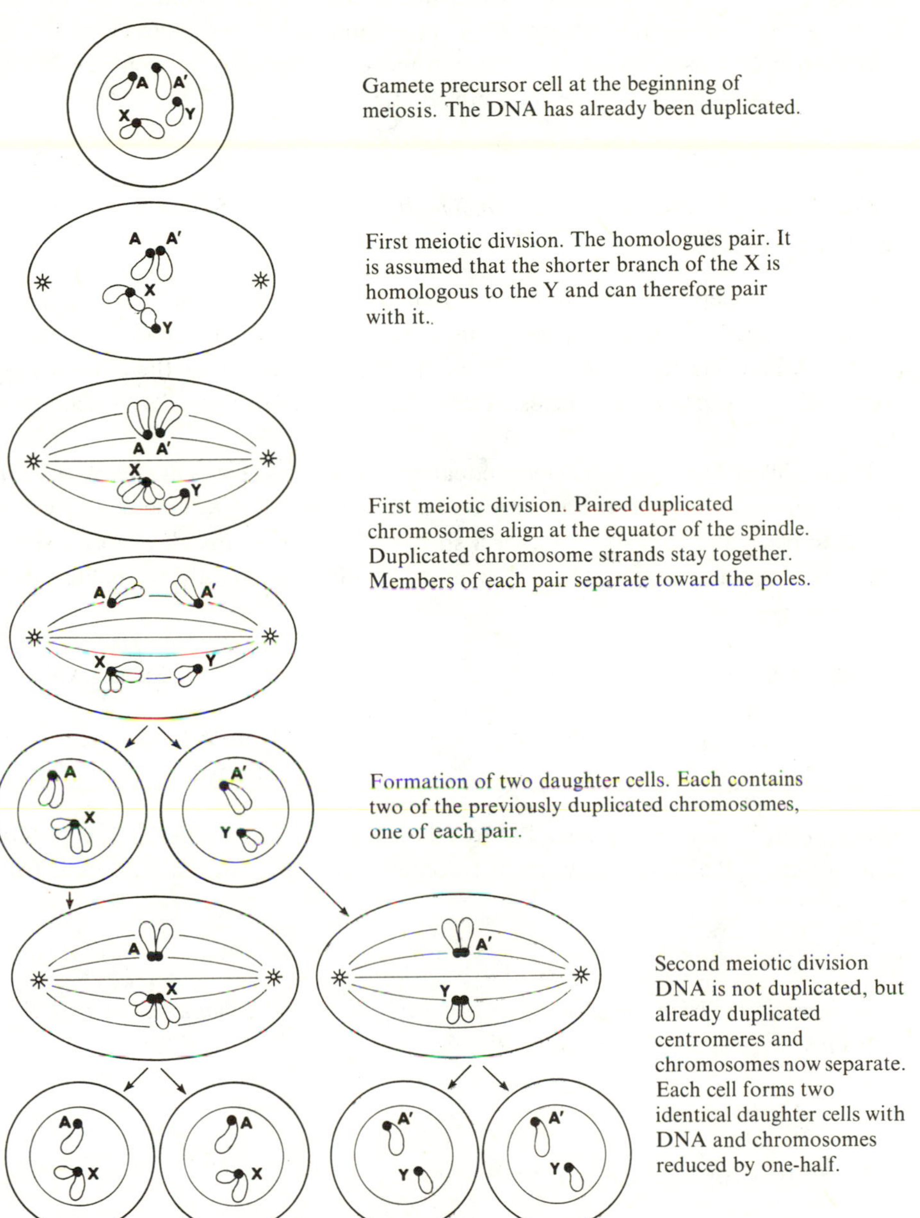

FIGURE 1.6
Meiosis in a hypothetical male who has two chromosome pairs, one of which is the X, Y sex-chromosome pair, the other (labeled A, A′), a pair of homologous chromosomes.

The first meiotic division assures the separation of members of pairs and thus the reduction of chromosome numbers by one-half. The amount of DNA has not halved. This happens in a second meiotic division. At this division, DNA does not duplicate, but the centromeres do. Thus the second division produces two pairs of cells with both DNA and number of chromosomes reduced in each new cell to one-half the amounts in the original premeiotic cell.

Inheritance of sex is determined in mammals by the distribution of the sex chromosomes among the sperm.

The meiosis depicted in Figure 1.6 is that of a male cell. It has produced four gametes, or sperm cells, two of them containing one A and one Y and two containing one A and one X. A female of this hypothetical species has the chromosome constitution AA XX so that meiosis forms four gametes or egg cells, which are all AX.

Thus, while male meiosis produces equal numbers of two types of sperm, AX and AY, female meiosis gives rise to only one type of egg cell, AX. One of two events can take place at fertilization. Either a sperm AY fertilizes an AX egg or a sperm AX fertilizes an AX egg. These two events should be equally probable. Their outcomes are

Egg AX ↘	Sperm AY ↙		Egg AX ↘	Sperm AX ↙
Zygote AA XY (male)			Zygote AA XX (female)	

Thus, an equal number of males and females should be formed in the next generation. A male will always receive its Y chromosome from the father and its X from the mother.

In general every individual receives one chromosome of a homologous pair from the father and the other from the mother.

1.5 Mutation and Selection

Mitosis, fertilization, and meiosis assure the constancy of chromosome number and type in every cell of an individual and in all individuals of the same species. An understanding of mitosis and meiosis provides the basis for predicting the laws of inheritance. In order for inheritance to be accessible for study, however, there must exist genetic differences between individuals. If everybody had black hair there would be no way of studying the inheritance of hair color. We know that there are individual differences for many characters. This book will be almost entirely concerned with the maintenance and changes in frequency of such individual differences in human populations. We shall now outline what is known concerning the nature and origin of these differences.

DNA is the carrier and transmitter of the information needed to make cells and individuals. Differences between individuals that are transmissible to their progeny —or, in other words, heritable—must therefore be due to differences in DNA. It is clear that if a change can occur in DNA and the changed DNA is copied, all the descendants made by the new DNA will carry and transmit the new DNA form. A substance having the capacity to be copied and also to carry information for making other substances, as does DNA, has all the basic properties required of the "hereditary substance." Changes in it will be inherited. If the change in DNA affects the types of proteins that can be made, the effect of the change also will be inherited. A heritable change is called a **mutation**. If a mutation takes place in a cell destined to become a gamete then it may be transmitted to the progeny. If it happens in a cell not destined to form a gamete, a **somatic cell**, the effects are limited to the individual, and the change is called a **somatic mutation**.

What kinds of changes can occur in DNA and what are their probable consequences? Knowledge of DNA structure and function leads to the following considerations:

1. The simplest change that can occur in DNA is the substitution of one nucleotide for another. If, in the sequence of DNA,

TTTACGTAG

(considering one strand only for simplicity), the second nucleotide is changed to *G*, the new sequence will read

T*G*TACGTAG,

which we can call DNA* to distinguish it from the first. We can, on the basis of the genetic code (Table 1.1) predict the change in the polypeptide synthesized by this DNA. The original sequence of DNA will specify the tripeptide

lysine–cysteine–isoleucine;

DNA* will lead to the synthesis of

threonine–cysteine–isoleucine.

A nucleotide substitution does not always lead to an amino acid substitution because many triplets may correspond to the same amino acid. An amino acid change, however, must involve the substitution of at least one nucleotide. Practically all mutations that have been observed in proteins, for example, in the hemoglobins of man, involve only one nucleotide substitution. The first mutation ever described in terms of a specific protein change is the mutation to hemoglobin S, which is responsible for an anemia widespread among Africans (see Chapter 4). Hemoglobin S differs from the standard hemoglobin in having a sequence of amino acids that is different at just one of 146 positions, the glutamic acid at position 6 of the beta polypeptide chain being replaced in hemoglobin S by valine. This change can be achieved by only one nucleotide substitution.

The substitution of a single nucleotide is the most common kind of change observed.

2. Another possible change is the insertion or deletion of one or more nucleotides. Because of the mechanism of translation from m-RNA into protein, this has much more drastic consequences than a substitution. A polypeptide is usually made up of a sequence of at least a hundred amino acids. If there is a deletion or insertion of one nucleotide in a DNA or RNA sequence, the message will be changed completely from that point onwards. This is because nucleotides are read successively in groups of three, so that the omission or insertion of one will change completely the translation into amino acids. Suppose, for example, that the original DNA is

AAA ACG AAA CCG AAG CAT CTT ...

and that the fourth nucleotide is deleted. The message becomes

AAA CGT TTC CGA AGC ATC TT ...,

which we may label as DNA*. Assuming that the beginning of the polypeptide corresponds to the left end of the sequence as written, DNA will be read in protein language (see Table 1.1) as

phenylalanine–cysteine–phenylalanine–glycine–
phenylalanine–valine–glutamic acid;

the DNA* will be read

phenylalanine–alanine–leucine–alanine–serine–chain termination.

After the phenylalanine, which is not affected, all subsequent amino acids are changed. In addition the sixth triplet now reads "chain termination," meaning that a peptide of only five amino acids will be produced instead of the presumably much longer one originally specified. This type of change is called a **frameshift mutation.** If a deletion or insertion affects three nucleotides in a row, frameshift does not occur and there will only be a gap, with perhaps a replacement of an amino acid next to the gap. Deletion or insertion of three nucleotides not in sequence leads to a profound alteration between the first and the last changes, but to no alteration on either side of them. Such mutations have been artificially produced and have helped in elucidating the genetic code.

Even a single amino acid substitution, as in hemoglobins, can substantially alter the function of a protein and so be observable as an effect on the whole organism.

As a first approximation, we call a **gene** *that segment of DNA that codes for a given protein or polypeptide chain; however, not all DNA segments code for protein.*

Changes of the type just described, especially the substitution of one nucleotide for another, are often called **point mutations** or **gene mutations**. Grosser changes affecting more than one gene may, however, occur. Some of these are referred to as **chromosomal aberrations**. They involve the loss, addition, or displacement of a major part of a chromosome or chromosomes. They are often large enough to be visible under the light microscope as a change in the shape or size of the affected chromosome.

There are four major types of chromosomal aberrations.

1. *Deletion*: A section of a chromosome is missing. Using letters to stand for genes we may diagram, as an example,

abcdefghijklmn ⟶ *abcdklmn*,
↑ ↑

in which the arrows indicate the points of breakage and rejoining after elimination of the intermediate segment. Such points can also be within a gene.

2. *Duplication*: A section of chromosome is duplicated.

abcdefghijkl ⟶ *abcdefcdefghijkl*
↑ ↑

The sort of duplication shown in this diagram is sometimes called a tandem duplication because the duplicated piece is immediately adjacent to the original piece. A duplicated segment may also be elsewhere on the same chromosome or on another chromosome.

3. *Inversion*: A section of chromosome is rotated 180°.

abcdefghijklm *abcgfedhijklm*
↑ ↑

4. *Translocation:* A chromosome segment is transferred from one chromosome to another. Translocation usually is reciprocal, as in the example below.

Chromosome A: *abcdefghijkl* ⟶ *abcdefghLMN*
↑
Chromosome B: *ABCDEFGHIJKLMN* ⟶ *ABCDEFGHIJKijkl*
↑

Deletions usually have the most drastic effects on the individuals carrying them, especially if the deleted piece is long. A number of translocations have been described in man; the possibility for such description is particularly good for those in which the segments translocated from one chromosome to another are long enough to be visible in cytological preparations of mitotic chromosomes. Pairing at meiosis offers advantages for the cytological study of chromosomal aberrations because it leads to characteristic configurations of the altered chromosomes. The famous little fruit fly *Drosophila*, which has been widely used in studies that have led to our present understanding of genetics, is very useful for this type of observation because it has cells in the salivary glands of its larvae with very big chromosomes that are permanently paired. The size of the chromosomes and their banding pattern makes it possible to identify even quite small deletions and other chromosomal aberrations. Furthermore, chromosomal aberrations in the fruit fly can be readily identified because they lead to characteristic loops in the pairing pattern of the salivary-gland chromosomes. In man, conditions for observation are not so favorable. The investigation of meiosis in the human male has to be done with testicular biopsy material obtained by puncturing the testes, in which meiosis takes place. It is almost impossible to observe meiosis in females owing to the difficulty of obtaining sample material from the human ovary.

Occasional irregularities at meiosis, such as nondisjunction, give rise to the formation of gametes with an abnormal number of chromosomes.

Gametes may be formed in which there are either two chromosomes of one homologous pair, or none. In the former case, human zygotes will have 47 chromosomes, and in the latter, 45 chromosomes. These abnormalities usually have fairly serious consequences. The best known example is Down's syndrome, formerly called mongolism, in which one of the small chromosomes, number 21, is represented three times. This cytological condition is called 21-trisomy.

Down's syndrome can be caused by translocation of part of chromosome 21 *to another chromosome as well as by nondisjunction.*

The original carrier of the translocation may show no effect. Let us assume that the translocation is with chromosome 15. This individual will then have a normal 21, a normal 15, a long 15 to which a piece of 21 is attached (called 15-21), and a short 21 (called 21′), which may be diagrammed as follows, with the chromosome 21 material shown by a thicker line,

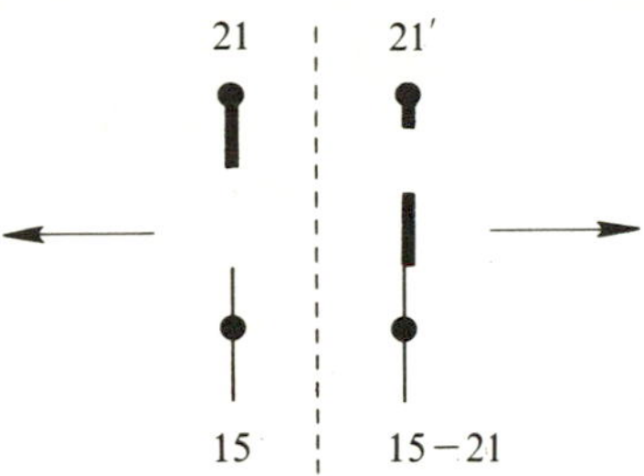

The arrows represent movement to the poles at anaphase. The 21′ fragment is, in practice, often missing or, at least, not visible. At meiosis each gamete will receive one 15 and one 21 centromere. The separation occurs along the dotted line as shown above giving rise to gametes having either 15 and 21 (which may be designated as type I) and to gametes having 15-21 and 21′ (type II). When fertilized with normal gametes, which have 15 and 21, the type I gametes will give rise to individuals that are fully normal, and the type II gametes will give rise to phenotyphically normal individuals that carry the translocation.

The configuration at meiosis, however, can also be

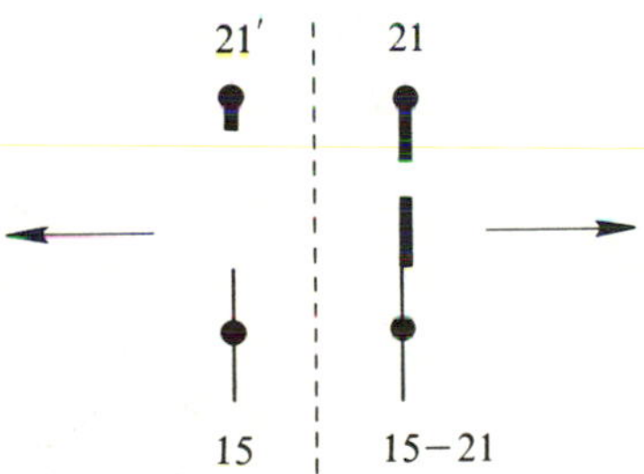

The gametes formed that have 15 and 21′ may be designated type III; those that have 15-21 and 21, type IV. Gametes of type IV when united with a normal gamete will produce an individual with 46 chromosomes—but three of these are effectively chromosomes 21. He will, therefore, be affected by Down's syndrome. The individual formed by the union of a type III and a normal gamete will have a deletion for one 21 chromosome, which is probably lethal at an early age.

The fact that chromosome 21, or a part of it, is in a new position does not apparently affect its action on the individual. Only the way in which it is transmitted to the progeny is affected.

The presence of extra chromosomes or the absence of chromosomes is called **aneuploidy**. When one of a chromosome pair is lost in a diploid organism, the condition is called **monosomy**, and when one is added, **trisomy**. **Euploidy** refers to a normal chromosome complement. (Eu for good, ploidy, for multiplicity; similarly, we have haploid for one, diploid for two, and so on.) Very rare gametes may arise in which there was no reduction of the diploid chromosome number: in man they have the full set of 46 chromosomes. When united with a normal gamete with 23 chromosomes, they produce individuals that have three chromosome sets (69 chromosomes). These individuals are called **triploids** and usually die at a very early age.

Chromosomal aberrations, especially aneuploidy, cause a considerable unbalance of the organism's make up.

Aneuploidy is often either lethal or gives rise to serious pathology. Some of the most extreme congenital malformations are a consequence of chromosomal change. It is known that an appreciable fraction of early abortions are due to this type of change (see Chapter 10). However, even gross alterations in the constitution of a chromosome are not incompatible with the survival and reproduction of single cells. Many human cells cultivated for a long time *in vitro* have a modified chromosome set, usually aneuploid, often with as many as 60 to 70 chromosomes. There is a fraction of normal cells in the human body that are tetraploid—that is, have their chromosomes in sets of four. They arise from a nuclear division that is not followed by cell division. Their functional significance is not known. Even octoploid cells in normal tissues have been described.

If an organism of a new type is so handicapped in some way that it's capacity to contribute progeny to the next generation is limited, the chance that the new type will be represented in future generations is decreased. The reverse may also be true: an organism of a new type may somehow be more able to produce progeny than the old type. Once transmissible changes occur, they will *automatically* be exposed to the forces of change that are called **natural selection**. If a new type leaves more progeny than the "normal" type in the environment in which both happened to live, it will multiply more rapidly than the normal type, at least to start with, and may eventually supplant it. If a new type is not as capable of reproducing itself as is the normal type, it will eventually become extinct or be maintained at a low frequency by mutation. Natural selection is caused by the differential capacity of the various hereditary types present in a population to leave progeny in a given environment. It can be identified with the popular phrase "survival of the fittest" provided one defines the word "fit" as the capacity to leave fertile progeny. The concept of fitness will be discussed in much more detail later, especially in Chapter 6.

Most mutations are deleterious.

There are two related reasons why the majority of mutations in any organism are detrimental. One is that mutation is a random change in a functional structure that is exceedingly complex. It is therefore likely to make that structure less efficient or even totally inadequate. Changing a connection in a computer at random is hardly likely to improve its performance. Even if, owing to a mutation, only one of the many thousands of proteins necessary for the life of a normal person is not present, is nonfunctional, or does not function adequately, the change may be lethal. The second reason why most mutations are detrimental is that we have a long evolutionary history before us, during which many of the possible changes that turned out to be advantageous have already been incorporated. Thus most mutations are deleterious, though many have little effect and some may turn out to be advantageous.

The frequency of mutations is low. There are, however, mutagenic agents that can increase it.

DNA is a very stable substance and its replication is a very precise process, meaning that very few mistakes are made. There are, however, conditions that affect the frequency with which mistakes happen during DNA duplication. There are also conditions that stimulate mutation in "resting"—that is, nonreproducing—DNA.

Under normal conditions, the frequency with which the average nucleotide changes is of the order of one in a billion or more per generation. This frequency, however, varies considerably among organisms and among genes and depends to some extent on the prevailing environmental conditions. Temperature and ionizing radiations increase mutation rate. More generally, any radiation of sufficient energy to determine chemical changes in DNA can give rise to mutations. Ultraviolet light is a potent mutagen, especially at a wavelength of around 260 millimicrons, which is that maximally absorbed by nucleic acids. This wavelength is however, almost absent in solar radiation by the time it reaches the earth's surface. In any case, our germ cells are generally screened from light of this type.

There are many chemical agents that are now known to have mutagenic action. They usually either react chemically with DNA, or are "analogues" of essential parts of the nucleotides. This means that they are sufficiently similar to the nucleotides to be incorporated into DNA in place of the ordinary nucleotides. These chemical agents then have, however, abnormal pairing relationships with other nucleotides and thus lead to errors during duplication. In order to be effective, mutagens must be able to reach the gonads, the organs in which the formation of germ cells takes place. Many mutagens active *in vitro*, for example caffein, are probably not dangerous for man because they do not reach the gonads in sufficient concentration. Ionizing radiation, for example, X-rays if sufficiently hard can penetrate the gonads and increase mutation rates over their spontaneous level.

Mutagenic agents are relatively unspecific in the sense that they do not affect just one gene, but tend to affect all genes almost equally. There are, however, some mutagens that attack only certain nucleotides, changing them into others. As the number of sites in the total DNA of an organism that such specialized mutagens can attack is large, this cannot lead to any great specificity of action in terms of which genes are affected or the way in which they are affected.

1.6 Dominance and Recessiveness: Phenotype and Genotype

If we examine individual organisms of the same species and find they are different, then the question arises: is the difference genetic—that is, heritable—or not? Obtaining the answer would be simple for an experimental geneticist. He would breed from the organisms and observe their descendants. If the progeny maintained the difference or it reappeared among later descendants then it would most probably be genetically determined. The test would have, of course, to be done under controlled environmental conditions, to ensure that the descendants of the two types shared the same environment.

The observation of the pattern of inheritance among the progeny of a certain organism tells us what the **genotype**, or genetic (DNA) constitution, of the organism really is. The outward appearance of the organism, or his **phenotype**, can be misleading. Similar phenotypes can arise from different genotypes. Similar genotypes may also give rise to different phenotypes if the organisms live in different environments.

The fact that our chromosome set is normally double, man being a diploid organism, means that almost all of our genes are represented twice. If one of two homologous genes is normal but the other has been affected by a mutation, providing that one normal gene is sufficient for normal function, the individual is protected against the possible deleterious effects of the mutation. This is perhaps the main reason why it is advantageous for an organism to be diploid (see Chapter 10).

Diploidy introduces a special problem regarding the identification of genotype from observation of the phenotype. Suppose that a given gene can exist in two different states, A and A'. The difference between the two may be at only one nucleotide position or, possibly, at many. Assume that A specifies the production of an active enzyme, while that specified by A' is either inactive or less functional. A diploid individual has two genes and can thus be one of three genotypes: AA, having inherited from both his father and his mother the same type of gene, the one that specifies the active enzyme; $A'A'$, having inherited from both parents the gene that specifies the inactive enzyme; or AA' having inherited one gene of each type. If the presence of the gene that specifies the active enzyme is necessary for survival, individuals of genotype $A'A'$ may be severely diseased and perhaps even die at an early age. This is the case, for example, for the gene that specifies an enzyme necessary for the normal utilization of the sugar galactose, which is contained in milk. Unless treated, persons with two of these genes, both of which are nonfunctional genes, have the disease named galactosemia. The genotype AA', which has only one gene making the normal enzyme, generally makes only half the amount of enzyme made by AA, but this is usually enough to allow the person to digest milk. The biochemist measuring the exact enzyme activity of the two types, can tell the difference between AA and AA'. Until this became possible, however, phenotypes of the individuals AA and AA' were apparently identical with respect to milk digestion.

We now must introduce some basic genetic terminology. Individuals who carry two genes of the same type, such as AA or $A'A'$, are said to be **homozygotes**, meaning that as zygotes they were formed by the union of "same" gametes. Individuals who carry a pair of different genes, such as AA' are called **heterozygotes**, meaning formed by the union of "different" gametes. The definition may be validly applied only by considering one particular pair of genes at a time. Individuals may be homozygous for one gene pair and heterozygous for another. Different types of the same gene arising by mutation are called **alleles**, short for allelomorphs, meaning different forms. If the heterozygote looks like one of the two homozygotes, for example if AA is indistinguishable from AA' in its capacity to digest milk, then the allele A is said to be **dominant** to the allele A', and A' is said to be **recessive**. The word recessive is derived from a Latin root meaning yielding, retiring, or hiding. We can refer to AA as the dominant homozygote and to $A'A'$ as the recessive homozygote. The way to distinguish AA from AA' when A is dominant with respect to A' is by a test of genotype; that is, by looking at the progeny obtained from suitable crosses. The analysis of parents and other relatives may also be useful for distinguishing the genotype. Sometimes, of course, as with galactosemia, more refined tests can differentiate clearly between homozygotes and heterozygotes for an outwardly dominant gene. If we could actually examine the DNA we would always be able to distinguish homozygotes from heterozygotes. Upper case letters are often used for dominant alleles, and lower case for recessives.

1.7 The Laws of Mendelian Inheritance

We are now in a position to understand fully the laws of inheritance as they were discovered and described by Mendel in 1865. They can, in fact, be deduced on the basis of the information that has been summarized so far in this chapter, none of which was available in Mendel's time. Mendel derived his laws from careful, planned observations of the results of breeding experiments with garden peas. His laws remain valid for all diploid organisms, including man.

From the facts of meiosis and fertilization we know that, apart from rare exceptions, every individual derives one chromosome of a pair, and hence one gene of each pair, from one parent and the other chromosome, or gene, from the other parent. All chromosomes that are not different in the two sexes, called **autosomes**, behave in this way. The genes on the sex chromosome have special rules of inheritance that will be considered a little later. Which particular chromosome enters a gamete is determined at random and therefore either chromosome may enter a particular gamete with equal probability. This rule has almost no exceptions. Possible exceptions called meiotic drive have been described in some experimental organisms but they are rare, and claims to their existence in man are, in general, not well substantiated.

An assumption that Mendel made, and that is substantiated in practice, is that male gametes fertilize female gametes independently of the particular chromosomes or genes that each of them contains.

The expected outcome of a cross can be predicted as a function of the genotypes of the mates as follows.

1. Given the genotype of each parent, establish what gametes each will form and in what proportions. For example, if the mates are *GG* and *Gg*, the first will form only *G* gametes while the second will form $\frac{1}{2}$ *G* gametes and $\frac{1}{2}$ *g*.

2. Work out all combinations of male and female gametes. The genotype of a zygote is derived by combining the genes of the two gametes. The proportion of zygotes having each combination will be the product of the relative frequencies of each type of gamete in each mate. Thus, if we assume that the male is *GG* and the female *Gg*, all sperm will be *G* but they may fertilize either a *G* egg or a *g* egg, each of which occurs with a frequency of $\frac{1}{2}$. The resulting zygotes *GG* and *Gg* therefore also each occur with a frequency of $\frac{1}{2}$. A slightly more complicated example, that of a cross between two heterozygotes is given in the diagram that follows. In this example, we can pool identical classes of zygotes and sum their proportions.

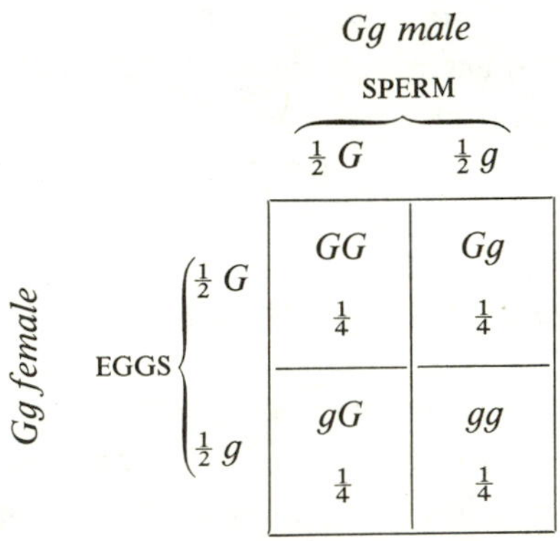

There are four possible combinations of gametes. The genotypes of the zygotes thus formed are simply the combinations of the gametes that gave rise to them. The proportions of each zygote are the products of the proportions of the corresponding gametes. Summing over identical genotypes, since it is immaterial whether G or g comes from the father or the mother, we have

$$\begin{array}{ccc} GG & gG + Gg & gg \\ \frac{1}{4} & \frac{1}{2} & \frac{1}{4} \end{array}$$

3. Determine the proportions of phenotypes. If G is fully dominant over g, the dominant phenotype, which is displayed by all individuals of genotypes GG, Gg, and gG, occurs among the progeny of the cross $Gg \times Gg$ with a frequency $\frac{1}{4} + \frac{1}{4} + \frac{1}{4} = \frac{3}{4}$, and the recessive phenotype, displayed by gg individuals, occurs with an expected frequency of $\frac{1}{4}$. There is, therefore, no **segregation** of distinct phenotypic classes among the progeny of the $Gg \times Gg$ cross shown in the diagram.

In all cases, random sampling effects will, of course, generally lead to observed proportions that differ somewhat from the expected. A brief survey of the statistical techniques which are used to assess the agreement between observed and expected is given in the appendices.

The expected results of all possible crosses between two individuals with respect to one locus having two alleles are given in Table 1.2. A cross between heterozygotes is often called an **intercross** and that between a heterozygote and a homozygote a **backcross**. These terms arose from the application of genetics to plant breeding.

TABLE 1.2
The Possible Types of Crosses Involving One Locus with Two Alleles G and g

Mates		*Progeny*				
Male	*Female*	*Genotype proportions*			*Phenotypes if G is dominant*	
		GG	*Gg*	*gg*	*G*	*g*
GG	*GG*	all	—	—	all	—
GG	*Gg*	1/2	1/2	—	all	—
Gg	*GG*	1/2	1/2	—	all	—
GG	*gg*	—	all	—	all	—
gg	*GG*	—	all	—	all	—
Gg	*Gg*	1/4	1/2	1/4	3/4	1/4
Gg	*gg*	—	1/2	1/2	1/2	1/2
gg	*Gg*	—	1/2	1/2	1/2	1/2
gg	*gg*	—	—	all	—	all

The mere fact of a gene's being on a sex chromosome means that it is governed by special patterns of inheritance that are related to sex (sex linkage).

The Y chromosome in man seems to carry very few genes. There is no gene that is known to be present both on the Y and the X and therefore all X chromosome genes are effectively haploid in the male (see Chapter 10). Consider first an **X-linked gene**—that is, a gene on the X chromosome—which is recessive in females. The presence of the recessive gene on the single X of the male almost always produces the phenotype that corresponds to that of the recessive homozygote in the female. This in itself indicates that the corresponding gene is not present on the Y, and such a male is said to be **hemizygous**. The patterns of inheritance expected for such an X-linked recessive are summarized in Table 1.3, using hemophilia (a

TABLE 1.3

Patterns of X-linked Inheritance. X^h *represents an* X *chromosome carrying the recessive gene for Hemophilia* A. *Affected individuals are boxed.*

Mating Type[a]	*Parents*		*Offspring*			
	Female	*Male*	*Female*		*Male*	
1	XX	XY	XX		XY	
2	XX^h	XY	$\frac{1}{2}$XX	$\frac{1}{2}XX^h$	$\frac{1}{2}$XY	[$\frac{1}{2}X^hY$]
3	[X^hX^h]	XY	XX^h		[X^hY]	
4	XX	[X^hY]	XX^h		XY	
5	XX^h	[X^hY]	$\frac{1}{2}XX^h$	[$\frac{1}{2}X^hX^h$]	$\frac{1}{2}$XY	[$\frac{1}{2}X^hY$]
6	[X^hX^h]	[X^hY]	[X^hX^h]		[X^hY]	

[a]Since the X^h chromosome is very rare (see Chapter 4) mating types 3, 5 and 6 are rarely, if ever, observed. Mating types 2 and 4 are those most characteristic for an X-linked recessive.

hereditary disease in which blood coagulation is severely delayed) as an example. The second, third, and fourth mating types show the associations of a character with sex that are generally the diagnostic for X-linkage. In mating type two, unaffected heterozygous XX^h "carrier" females give rise only to normal females, half of which are, however, carriers, while half of the males are affected. Type 3 gives only affected males as sons of affected females. For a rare defect, the frequency of the mutant chromosome X^h is so low that mating types 3, 5 and 6 rarely, if ever, occur. Mating types 2 (normal carrier female and normal male) and 4 (normal female and affected male) are then those that are commonly used to diagnose X-linkage in human pedigrees. The male-associated familial pattern of inheritance of hemophilia was perhaps already recognized at the time the Talmud was written.

A complete study of the genetic transmission of a trait requires the analysis of how it is transmitted in all possible matings.

This, however cannot be done for many traits that are rare. For these traits, it may in fact be very difficult to find enough homozygotes for matings between them to be studied. Often they are so rare that even heterozygotes are very difficult to find. The homozygote for a rare gene almost always has both parents carrying it. Thus, if heterozygotes are rare, the probability of observing a mating between them is exceedingly low. For many rare human traits that show up in heterozygotes, the homozygotes have never been seen.

For instance chondrodystrophy, a congenital defect determining irregular growth of bones, especially of the long ones, causes a peculiar type of dwarfism, determined by a dominant gene. Since its frequency in the population is of the order of 1 in 10,000, marriages between heterozygotes are very rare. If mating were at random, such marriages would occur only once in every hundred million marriages. This is about the total number of marriages that occur in the human species in five years. Genes such as that for chondrodystrophy are called dominants because they show up in the heterozygote, though it is often not clear what the homozygote would look like. Homozygosity for some such rare genes is known to be lethal.

The study of the genetic transmission of a trait is carried out by an analysis of enough pedigrees to provide information on its mode of inheritance. A rigorous analysis is complicated and its essential principles will be given in Appendix II. Here we shall simply indicate some rules of thumb that allow the geneticist to carry out a superficial analysis.

When a certain character or group of characters can be classified into three nonoverlapping categories, it is possible that the three categories correspond to two homozygotes and a heterozygote for a pair of alleles. If there were a pair of alleles that gave three such classes for a human character, marriages between homozygotes of one class should produce as progeny only homozygotes of that class; those between the different classes of homozygotes should give as progeny only heterozygotes, which are expected to be somewhat intermediate in appearance between the two homozygotes. The other three possible marriages should give the expected segregations. Deviations from this scheme could be explained as being due to errors of classification, or to illegitimate paternity. In theory, mutation might also account for deviations but this would be a very rare event.

Examples from human genetics of characters for which a pair of alleles produces three observable classes are some hereditary anemias, like the sickle-cell anemia and C hemoglobin anemia, common among Africans, and thalassemia, common in some European and Oriental populations. The person who is heterozygous for any of these diseases has a very mild anemia, which can be detected by special laboratory tests. The homozygote has severe anemia and, especially in the case of thalassemia, does not usually survive to reproductive age. Such genes are called **homozygous lethals**. When these anemias were first discovered, laboratory tests were not able to detect all heterozygotes with certainty, thus giving rise to errors of classification. Improvements in techniques have made possible the identification in the laboratory of the three genotypes for each of these traits.

When only two phenotypic classes are clearly distinguishable, the first problem is to test whether the inheritance corresponds to a Mendelian scheme with dominance. When one type is rare and the other, by definition the "normal," is common, the following rules are useful in making a prediction. In a dominant defect, one of the two parents of the affected individual is also usually affected, and if this is so, the frequency of affected sibs is about $\frac{1}{2}$. With a rare recessive defect, on the other hand the parents of the affected are usually both unaffected. In this case both parents are usually heterozygous and the expected proportion of affected progeny is $\frac{1}{4}$. Quite frequently parents of a person having a rare recessive defect are related. These rules are subject to several exceptions. Particularly with very rare dominants, instances may be reported of an affected individual with unaffected parents. The affected individual is then probably the result of a mutation in one of the two gametes that formed it.

There are also other difficulties in the interpretation of pedigrees. Some genes do not express themselves in all individuals that carry them. They are referred to as genes of incomplete or low **penetrance**. This may happen in either recessive or dominant states. It is also possible that the **viability** (which is partially synonymous with fitness as defined for natural selection) of some genotypes may be low. These and other complications related to the sampling of human pedigrees make their rigorous analysis an involved procedure.

Sex linkage is often discovered because of a different frequency of a trait in males and females, but this rule is far from foolproof. Many traits have a different incidence in the two sexes without being sex linked. Thus, pattern baldness is much more frequent among males but the study of pedigrees shows that it is not sex linked. Such characters are called sex conditioned, or **sex limited**. True sex linkage of a gene, whether dominant or recessive, is ascertained by transmission studies. A sex linked recessive will very often show up in sons of unaffected parents. Affected females must have an affected father. Other features of sex linkage will be apparent from the study of Table 1.3. Sex linkage also gives rise to a clear difference between the progeny of reciprocal matings, for example those in which the father is A and mother a compared with those in which the father is a and the mother A.

Such reciprocal differences can also be found in so-called **maternal inheritance** and **cytoplasmic inheritance**. In maternal inheritance the phenotype of the progeny is determined by the phenotype of the mother and is independent of that of the father. In cytoplasmic inheritance, the trait is transmitted by components of the cytoplasm of the germ cell, and therefore usually only through the maternal parent because egg cells have much more cytoplasm than sperm and so contribute most of the cytoplasm of the fertilized zygote.

Genes located on nonhomologous chromosome pairs are transmitted independently of each other.

Migration of homologous chromosome pairs to the cell poles at meiosis occurs at random. The only restriction is that if one member goes to one pole, the other goes to the opposite pole. When two chromosome pairs are considered, the mode of separation of the members of one pair does not in general influence that of any other. Thus, if an individual is heterozygous for two genes located on different chromosome pairs, the two pairs of genes, which may we refer to as *Aa* and *Bb*, will behave independently of each other during segregation at meiosis. Because of the rules we have already described, we expect an equal number of gametes to receive *A* and *a* and an equal number *B* and *b*. We can now add, because of the independence rule, that of those gametes receiving *A*, we expect equal numbers to receive *B* and *b* and the same will be true for those receiving *a*. The only restriction is that each gamete must have one allele of the *Aa* pair and one allele of the *Bb* pair. There will therefore be four types of gametes formed by the individual under consideration, all expected with equal frequency: $\frac{1}{4}AB$, $\frac{1}{4}Ab$, $\frac{1}{4}aB$, $\frac{1}{4}ab$. The progeny of a cross between double heterozygotes are expected to be as given in Table 1.4.

TABLE 1.4

The Cross Between Two Double Heterozygotes AaBb. The two gene pairs A, a and B, b are located on different chromosomes. Since each gamete has a frequency of $\frac{1}{4}$, all combinations in the body of the table occur with a probability of $\frac{1}{4} \times \frac{1}{4} = \frac{1}{16}$. Equivalent genotypic classes are indicated by boxes.

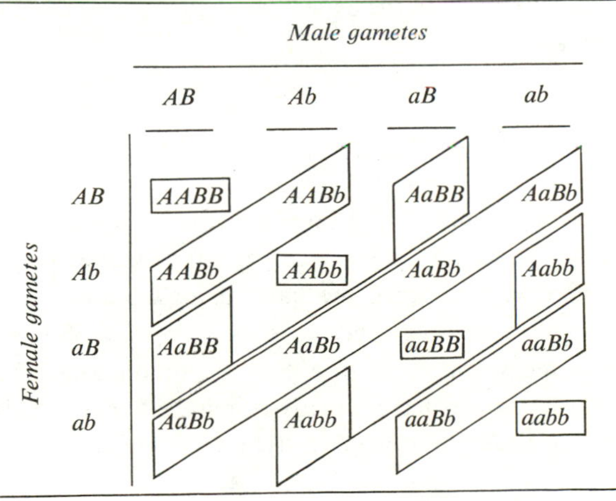

Female gametes	Male gametes: *AB*	*Ab*	*aB*	*ab*
AB	*AABB*	*AABb*	*AaBB*	*AaBb*
Ab	*AABb*	*AAbb*	*AaBb*	*Aabb*
aB	*AaBB*	*AaBb*	*aaBB*	*aaBb*
ab	*AaBb*	*Aabb*	*aaBb*	*aabb*

Pooling equivalent genotypic classes in the table, the segregation is

AABB	*AaBB*	*AABb*	*AaBb*	*AAbb*	*Aabb*	*aaBB*	*aaBb*	*aabb*	*Total*
1	2	2	4	1	2	1	2	1	16

Assuming dominance of *A* over *a* and *B* over *b*, the expected phenotype proportions are

AB	*Ab*	*aB*	*ab*	*Total*
9	3	3	1	16

These results could have been predicted in a simpler way by taking the phenotypic segregation for two pairs of genes and noting that they are independent. Simple multiplication of proportions then gives

Phenotype		*A* $\frac{3}{4}$	*a* $\frac{1}{4}$
Phenotype	*B* $\frac{3}{4}$	*AB* $\frac{9}{16}$	*aB* $\frac{3}{16}$
	b $\frac{1}{4}$	*Ab* $\frac{3}{16}$	*ab* $\frac{1}{16}$

The expected results of other types of crosses can be computed similarly. For example, a cross between a double heterozygote and a double recessive homozygote, *AaBb* × *aabb* gives a segregation of 1 : 1 : 1 : 1 for the four genotypes *AB*/*ab*, *Ab*/*ab*, *aB*/*ab*, and *ab*/*ab*, which correspond to the four phenotypes if *A* and *B* are dominant. In such crosses the expected frequencies of the gametes formed by the double heterozygote, are equal to the expected frequencies of the phenotypes obtained among the progeny.

1.8 Linkage and Recombination

Genes on the same chromosome do not generally obey the rule of independent transmission. In fact, genes on the same chromosome would always be transmitted together unless there were a special mechanism for separating them. This mechanism exists and is called **crossing-over.**

Crossing-over, a reciprocal event in a pair of homologous chromosomes, takes place when the chromosomes have already duplicated at the beginning of the first meiotic division. For detecting it genetically, a chromosome must be marked with two pairs of genes: *Aa* and *Bb*, say. Crossing-over can thus be observed by analyzing the gametes formed by an individual who is heterozygous for two pairs of genes located on the same pair of chromosomes, as diagrammed below.

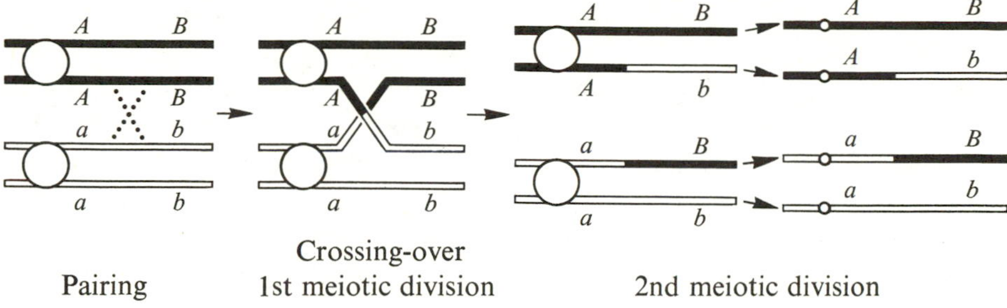

Of the four types of gametes formed, *AB* and *ab* carry the genes in the same combination as in the chromosomes of the double heterozygote, and therefore in the same combination in which they were in the parents of the double heterozygote. The other two, *Ab* and *aB* carry them in new combinations. *AB* and *ab* are called the **parental**, or **nonrecombinant**, types and *Ab* and *aB* the **recombinant** types.

A few crossovers are expected to occur at meiosis in every chromosome pair approximately in proportion to the length of the chromosomes. A given pair of chromosomes in one gamete-forming cell may have zero, one, or more crossovers which are distributed roughly at random along the length of the chromosome. If two genes that are being considered are fairly close together the probability of a crossover between them is low, and thus, on average, few recombinant types will be formed for these genes. If they are far apart, the probability is higher. The observed frequency of recombinants will, therefore, depend on the distance between the gene pairs being studied, and can be used as a measure of this distance.

The recombination fraction r is computed from the frequency of recombinant types ($Ab + aB$ in the example just discussed) among all the gametes tested from the double heterozygote. The expected frequency of *Ab* is equal to that of *aB*, and therefore each of them is equal to $r/2$. Conversely, the nonrecombinant types are each expected to occur with a frequency of $(1 - r)/2$.

The double heterozygote we have been discussing received *A* and *B* from one parent, *a* and *b* from the other. This combination, or **linkage phase**, with the two dominants from one parent, is called **coupling**. The other possible combination, one dominant and one recessive from each parent gives rise to the double heterozygote in **repulsion**, as summarized below:

Parents	*Double heterozygote*	*Phase*	*Parental gamete types*	*Recombinant gamete types*
AABB × *aabb*	*AB/ab*	coupling	*AB* *ab*	*Ab* *aB*
AAbb × *aaBB*	*Ab/aB*	repulsion	*Ab* *aB*	*AB* *ab*

From a repulsion double heterozygote, the expected frequency of *Ab* and *aB* types is $(1 - r)/2$ and that of *AB* and *ab* is $r/2$, exactly the reverse as for coupling.

The two phases, coupling and repulsion, are also sometimes called **cis** and **trans** respectively.

If two gene pairs are far apart on the same chromosome, the recombination fraction *r* approaches $\frac{1}{2}$ (see Appendix II, Section II.16) and the results are then indistinguishable from those obtained for two gene pairs on different chromosomes.

Linkage is the tendency of genes on the same chromosome to segregate together. It is close when the two genes are near each other and recombination between them is low. It is loose when they are far apart, and linkage may then be difficult to demonstrate directly. It may, however, be shown by finding that the two genes are linked with others that lie between them on the chromosome. All genes on the same chromosome, therefore, form one **linkage group**. There are expected to be as many linkage groups as the haploid chromosome number.

Exchanges, or more probably their consequences, can be observed in meiotic chromosomes when paired homologues separate during the first meiotic division. Points of probable exchange become visible as so-called **chiasmata** following pairing during meiosis, and can be counted. This gives an idea of the average number of exchanges expected per chromosome, which increases with chromosome length. Crossing-over in this context is thought to be mediated by actual breakage and rejoining of homologous chromosomes in corresponding positions, leading to a reciprocal exchange of chromosome segments. Some pedigrees illustrating linkage and recombination in man are shown in Appendix II.

1.9 The Gene Concept

The demonstration that a character shows Mendelian inheritance implies a simple determination of the character at a physiological level.

"Mendelizing" characters can often be reduced to a difference in a specific protein. Many factors can, however, obscure Mendelian inheritance. The environment may lead to phenotypic variations that may make it difficult or impossible to distinguish the phenotypes corresponding to different genotypes. Another serious disturbance is that more than one gene may affect the same character. Mendelian analysis may then still be possible, but usually becomes exceedingly complicated even in an experimental organism, in which any desired cross can be made. It is obviously more difficult in man, where the geneticist has to rely on the matings that happen to exist in the population. The refinement of phenotypic analysis usually helps to elucidate cases in which the transmission is obscure. Morphological characters especially are subject to the action of a great number of genes and are therefore less easily analyzed. The inheritance of such complex traits is dealt with specifically in Chapter 9. When analysis can be carried out at the biochemical level, particularly at the level of the proteins or of their direct products, genetic problems are usually greatly simplified. Conversely, the resolution of differences in terms of single genes, and therefore usually single proteins, can greatly help in solving the problem of understanding the functions of the relevant proteins.

We have provisionally used a definition of a gene as the segment of DNA that directs the synthesis of a given polypeptide or protein. This is not general enough, because some DNA regions probably do not synthesize protein. Some, for instance, are believed to synthesize only r-RNA or t-RNA. Moreover there are various kinds of genes. There is evidence that some DNA regions coding for several proteins can be activated or inactivated in a block. When this happens there must be some physiological connection between the DNA segments that form the functional unit. A "gene" as revealed by some types of analysis may thus be a more complex chromosome region than that synthesizing a given polypeptide.

A gene is thus a DNA segment recognizable by its specific function.

Cistron is a word commonly used for that segment of the DNA that codes for a particular polypeptide chain. This word derives from a combination of the terms cis and trans. The so called cis-trans test compares the phenotypes of the trans (or repulsion) and cis (or coupling) heterozygotes. If two functionally similar mutants belong in the same cistron the *cis* combination will usually be found to be normal while the *trans* usually shows the mutant phenotype.

It is worth adding that the name **locus** (plural: loci) is usually used to indicate the position of a gene on a chromosome. The word gene is often used to designate both locus and allele, but it is better to use one of the two distinguishing terms when confusion may arise. Locus and cistron are sometimes used interchangeably. Recombination can occur between mutant alleles affecting the same cistron, and may even occur if the mutations affect neighboring nucleotide pairs. Thus, the nucleotide pair is the minimal unit of mutation and recombination, while the cistron is the minimal *functional* unit.

The total number of genes in man was estimated in the past to be of the order of tens of thousands. This approximation was arrived at by indirect methods and most likely was an underestimate. Knowledge of the average size of polypeptide chains, or of proteins, and of the total DNA content of a human cell should provide a better estimate, assuming most of the DNA codes for proteins. Taking the average size of a polypeptide to be about 100 amino acids, the number of nucleotides coding for the average polypeptide will be 300, and therefore 600 in double-stranded DNA. The number of nucleotides in a haploid human cell is about 3×10^9; thus the number of cistrons should be $3 \times 10^9/600$ or approximately five million. This may be an overestimate, because an appreciable fraction of genes exists in multiple copies or because there may be, as has sometimes been suggested, a fraction of the DNA, that has no obvious function (see Chapter 11).

General References

All the references cited in the body of the text are collected together at the end of the book. A few selected papers and books are, in addition, listed at the end of each chapter under the heading General References. These have been chosen because of their special relevance to the contents of the particular chapter and are intended as a guide to further reading. The general references given at the end of this chapter, however, relate to the overall contents of the book and have been divided into those concerned specifically with population genetics and evolution, those concerned with human genetics, and, finally, some general references in genetics and molecular biology. Many of these references are especially relevant to later chapters of the book, though they are listed only at the end of Chapter 1.

POPULATION GENETICS AND EVOLUTIONARY THEORY

Cold Spring Harbor Symposia on Quantitative Biology, Vol. 20, "Population Genetics: The Nature and Causes of Genetic Variability in Populations," Cold Spring Harbor, New York: The Biological Laboratory, 1958. (Contains a number of important papers on the theory of population genetics.)

Crow, J. F., and M. Kimura, *An Introduction to Population Genetics Theory*. New York: Harper and Row, 1970. (The most recent and comprehensive account of the mathematical theory of population genetics.)

Dobzhansky, T., *Genetics and the Origin of Species*. New York; Columbia University Press, 1951. (A classical book on the processes of species formation.)

Ewens, W. J., *Population Genetics*. London: Methuen, 1969. (A compendium of chosen subjects of mathematical genetics.)

Falconer, D. S., *Introduction to Quantitative Genetics*, Edinburgh; Oliver and Boyd, 1960. (A good general text on quantitative genetics.)

Fisher, R. A., *The Genetical Theory of Natural Selection*. New York: Dover, 1958. (A classic.)

Ford, E. B., *Ecological Genetics*. New York: John Wiley and Sons, 1964.

Haldane, J. B. S., *The Causes of Evolution*. New York: Harper, 1932. (A classic.)

Kempthorne, O., *An Introduction to Genetic Statistics*. New York: John Wiley and Sons, 1957. (A detailed mathematical treatment dealing mainly with problems on quantitative genetics.)

Kimura, M., *Population Genetics*. Tokyo: Baifukan, 1960. (An excellent summary (in Japanese) of the theory of population genetics—useful because of the mathematics even to those who cannot read Japanese).

Li, C. C., *Population Genetics*. Chicago: University of Chicago Press, 1955. (An elementary introduction to the mathematics of population genetics.)

Malécot, G., *Les mathematiques de l'hérédité*. Paris: Masson, 1948. (A summary of many of Malécot's major contributions to population genetics. An English translation of this book has been published—*The Mathematics of Heredity*. San Francisco: W. H. Freeman and Company, 1970.)

Mayr, E., *Animal Species and Evolution*. Cambridge, Mass.: The Belknap Press of Harvard University Press, 1963. (A comprehensive description of the processes of evolution.)

Moran, P. A. P., *The Statistical Processes of Evolutionary Theory*. Oxford: The Clarendon Press, 1962. (An advanced book on several mathematical aspects of population genetics.)

Simpson, G. G., *The Major Features of Evolution*. New York, Columbia University Press, 1953. (Another well-known discussion on the overall evolutionary process with special reference to the paeleontological record.)

Wright, S., *Evolution and the Genetics of Populations, Vols. I and II*. Chicago: Genetic and Biometric Foundation, 1967. (Collects rare contributions to mathematical genetics of this author.)

HUMAN AND MEDICAL GENETICS

Boyer, S. H., ed. *Papers on Human Genetics*. Englewood Cliffs, N.J.: Prentice-Hall, 1963. (A collection of important papers in human genetics originally published between 1908 and 1963.)

British Medical Bulletin, Vol. 25(1), *New Aspects of Human Genetics* (edited by C. E. Ford, and Harry Harris). London: Medical Department, the British Council, 1969. (A collection of papers concerning recent progress in various aspects of human genetics.)

Burdette, W. J., ed., *Methodology in Human Genetics*. San Francisco: Holden-Day, 1962.

Cold Spring Harbor Symposia on Quantitative Biology, Vol. 24, 1964, *Human Genetics*, Cold Spring Harbor, Long Island, New York, Cold Spring Harbor Laboratory of Quantitative Biology.

Dobzhansky, T., *Mankind Evolving*. New Haven: Yale University Press, 1962. (An elementary exposition of human inheritance and evolution.)

Harris, Harry, *Human Biochemical Genetics*. Cambridge: Cambridge University Press, 1959.

Huron, R., and J. Ruffie, *Les méthodes en génétique générale et en génétique humaine*. Paris: Masson, 1959. (A useful textbook—in French.)

Lerner, I. M., *Heredity, Evolution, and Society*. San Francisco: W. H. Freeman and Company, 1968. (A stimulating elementary discussion of genetics and evolution with special reference to current problems of human society.)

McKusick, V. A., *Human Genetics.* Englewood Cliffs, N.J.: Prentice Hall, 1964. (A good concise elementary text.)

McKusick, V. A., *Mendelian Inheritance in Man.* Baltimore: The Johns Hopkins Press, 1968. (A comprehensive catalogue of human genetic variants.)

Neel, J. V., and Schull, W. J., *Human Heredity.* Chicago: The University of Chicago Press, 1954.

Stern, C., *Principles of Human Genetics* (second edition). San Francisco: W. H. Freeman and Company, 1960. (A well-known comprehensive handbook of human genetics.)

Vogel, F., *Lehrbuch der Allgemeinen Humangenetik*, Berlin: Springer-Verlag, 1961. (A useful comprehensive text—in German.)

GENERAL GENETICS AND MOLECULAR BIOLOGY

Crow, J. F., *Genetics Notes.* Minneapolis: Burgess, 1950. (An excellent concise, but elementary, text.)

Hayes, W., *The Genetics of Bacteria and their Viruses.* New York: John Wiley and Sons, 1964. (The genetic basis for molecular biology.)

Peters, J. A., ed., *Classical Papers in Genetics.* Englewood Cliffs, N. J.: Prentice Hall, 1959. (A collection of many significant contributions to the development of genetics.)

Rieger, R., A. Michaelis, and M. M. Green, *A Glossary of Genetics and Cytogenetics.* New York: Springer-Verlag, 1968. (Comprehensive and accurate—a first rate handbook.)

Srb, A. M., R. D. Owen, and R. S. Edgar, *General Genetics* (second edition). San Francisco: W. H. Freeman and Company, 1965. (A general textbook of genetics.)

Stent, G. S., *Molecular Genetics*, San Francisco: W. H. Freeman and Company, 1970. (A general textbook.)

Watson, J. D., *Molecular Biology of the Gene.* New York: W. A. Benjamin, 1965. (An introduction to those concepts of molecular biology that are fundamental for genetics.)

Strickberger, M. W., *Genetics.* New York: Macmillan, 1968. (Another general textbook.)

2

Mendelian Populations

2.1 Some Basic Factors of Population Genetics

According to a widely accepted definition, a species is the most inclusive group of individuals that can potentially interbreed and produce fertile offspring. By this criterion, all human beings of all races belong to a single species, *Homo sapiens*. Interbreeding among human beings is, however, in practice limited by geographic, social, religious, and psychological barriers. As a result of these restrictions a great variety of partially differentiated local groups has evolved. The study of a sample of humans must therefore be accompanied, to be meaningful, by a careful description of the population from which it originates. This requires knowledge of the procedures by which the sample was obtained.

Our discussions of evolution will often refer to **Mendelian populations**—that is, populations of interbreeding individuals who share a common pool of genes, which are, of course, transmitted from one generation to the next according to Mendel's laws (see Dobzhansky, 1951). Simple Mendelian segregation may, of course, be disturbed by a variety of factors. These will not, however, in general, affect our definition of the Mendelian population. We will take this to mean a population that is interbreeding without major internal restrictions, that is sufficiently well-defined geographically and in other ways for meaningful samples to be taken from it, and within which we study characters that obey the Mendelian rules of inheritance.

An important property of Mendelian inheritance is that the genetic composition of an infinitely large population, undisturbed by differential selection, mutation, or environmental and other stimuli, will not change. Natural populations, however, reproduce under the influence of forces, in particular mutation and selection, that do mediate changes in their genetic composition. Mutation and selection are the two basic driving forces of evolution. In the first part of this chapter we will survey briefly some of the general basic factors of population genetics as a prelude to the more detailed treatment they will receive in later chapters.

Mutation produces new alleles, or, more generally, new heritable changes, in the genetic material but is a rare event.

The mutation rate per gene per generation is of the order 10^{-4} to 10^{-8}, and sometimes even lower. It should be emphasized that mutation rates in higher organisms including man are often defined somewhat loosely. In practice, we do not estimate the *rate* of *mutation* but the *frequency* of new *mutant genes*. More precisely, we estimate the frequency of mutant genes among the gametes that have joined with a gamete of the opposite sex to form a zygote.

Mutation is such a rare event that it can easily be overlooked. If we are interested in observing mutation we must be prepared to test a large enough number of individuals to identify with some certainty even the rarest "sport." This is especially true for the study of rare human malformations, congenital anomalies, and inborn errors of metabolism, which are often only the result of recent mutations. There is one fact that can assist us in our study of mutation: a mutant gene, although carried initially in a single individual, may be favored by selection (or by chance) and thus increase in frequency until it may be more easily detected. A mutant gene is favored by selection if individuals carrying it tend to leave more progeny than other individuals, either because they have a higher fertility, or because their offspring have a greater capacity to survive in their environment.

Gametes containing new mutations for any given gene are rare. If one of them unites with another gamete to form a zygote, it is thus almost certain that the other gamete will carry the normal gene, meaning that the new individual will be heterozygous. In other words, most new mutant genes are carried in the heterozygous state by the individuals who bear them. So long as it is rare, a mutant gene will only very occasionally occur in the homozygous form necessary for its expression in double dose.

Most mutant genes are deleterious. They are, in general, less deleterious in the heterozygous than in the homozygous state.

We have already noted the reasons why mutant genes are usually deleterious: a mutation is a random change in a highly complex structure that has evolved to meet the challenge of a complicated, variable, and often hostile environment.

Most organisms are especially adapted to the environment in which they live, an adaptation which may be almost unbelievably sophisticated. It is not surprising therefore that *random* genetic changes are largely deleterious.

***Polymorphism**, the condition in which a mutant gene is frequent, can be transient or stable.*

When the environment changes, new solutions to new problems are needed, which creates the possibility that some previously deleterious mutant genes may become beneficial. Even if the environment does not change, however, a random genetic change may occur that is advantageous. In either case, the mutant gene that confers some benefit will increase in frequency and will do so more quickly the greater the advantage. During the initial period of selection for a new mutation most individuals carrying it will be heterozygotes. Only when the frequency of the mutant gene has increased enough that there may be a reasonable number of matings between heterozygotes will homozygous progeny be produced. From then on, the mutant gene will be exposed to selection in the homozygous state as well as in the heterozygous state. If individuals who are homozygous for the mutant gene are as fit, or more fit than the heterozygotes, selection will go to completion and the "normal" allele will disappear and be replaced by what was originally the new mutant allele. However, it may happen, for example, that a mutant gene which is advantageous in the heterozygous state is not an asset in the homozygous state. Selection will then slow down as the frequency of the mutant gene increases and eventually an equilibrium state may be reached in which the normal and new alleles are both present. Such a state is called a **balanced polymorphism**. A polymorphism is a condition in which the population contains at least two phenotypes (and presumably at least two genotypes) neither of which is rare—that is, neither of which occurs with a frequency less than, say, one percent. There may be, of course, and there often are, more than two alleles, and therefore, more than two phenotypes, for a single locus.

If we examined individuals of a population at any given time we should probably find that for some loci there are no heterozygotes at all, while for others there are two or more allelles segregating (that is, present) in the population. At many of the loci having more than one allele, one may be very common and the others extremely rare. Some loci, however, are truly polymorphic. Unless data collected at different times are available, which are at present very difficult to obtain for human populations, it is usually impossible to decide whether a population is really in equilibrium with respect to the alleles at any of these loci. Indirect evidence may sometimes be available. It may be possible, for example, to assume that, unless significant changes in the environment have taken place in the recent past, the population is at or near equilibrium. However, this is an assumption that may be almost unverifiable.

A few concrete examples of different types of mutant genes and knowledge about them may help to clarify the discussion.

Rarely, persons are found who have unusual substances or abnormal concentrations of normal substances in their urine. Well-known examples are phenylketonuria and galactosemia. Many other abnormalities in amino-acid and sugar metabolism are also known. Persons affected by these traits often exhibit complex and debilitating syndromes that are a consequence of the metabolic disturbance. The "inborn error of metabolism" is frequently the result of an enzymatic deficiency. The heterozygote, even though it usually has half the amount of enzyme that the normal has, very often has enough of the protein to be normal or almost so. Thus, the defect is recessive. These deleterious recessive mutant genes remain at a low frequency in the population. In fact, the heterozygous carriers also often have a slightly lower reproductivity than the normal homozygotes, a fact that contributes in an important way to keeping the abnormal allele rare. Albinism is a more conspicuous abnormality also determined by rare recessive genes. It is the result of an inability to synthesize pigment in skin, hair, and eyes. Chondrodystrophy, a dominant and conspicuous defect, is an alteration of the bone forming cartilages that results in abnormal growth of the limbs. This causes a considerable reduction in stature and gives rise to a characteristic pattern of dwarfism.

Polymorphism occurs for several diseases, anomalies, and traits.

A defect such as color blindness occurs with high frequency. Sometimes, even genes determining serious illnesses, such as sickle-cell and mediterranean anemias, that are almost lethal in homozygotes, can become very frequent. The reasons why these two abnormalities are so frequent are now at least partially understood (see Chapter 4). But in general, strong selection for human polymorphisms has not been observed. We may consider for an example, the blood groups. The ABO system is undoubtedly the best-known of the blood groups, mainly in view of its primary importance in blood transfusion. Most populations harbor three alleles of the system *A*, *B*, and *O*, each at a frequency greater than 5 or 10 percent. More than a dozen polymorphic blood-group systems are known for the red blood cells and similar systems are being discovered for the white blood cells. Recently, many other polymorphisms have been detected by analyzing variations in enzymes and other proteins contained in plasma, or in red cells. Undoubtedly, other cells and body tissues will exhibit new polymorphisms, signalling a rapid extension of our knowledge of human polymorphism. (See Table 4.7 and Chapter 11 for summaries of most of the known polymorphisms.)

2.2 Phenotype, Genotype, and Gene Frequencies

The distribution of the frequencies of the various phenotypes constitutes, for our purpose, the essential description of a population.

If the genetic determination of observed phenotypes is known, we can infer from their frequencies those of the corresponding genotypes. Consider, for example, the MN blood-group system. Three phenotypes M, MN, and N can be distinguished. These correspond to the three genotypes *MM*, *MN*, *NN*, namely those of the homozygote for the *M* allele, of the heterozygote, who carries *M* and *N* alleles, and of the homozygote for the *N* allele. Thus, for example, in a sample from a particular population there were 47 M, 52 MN, and 12 N individuals. The frequency distribution of the respective genotypes is the same as that of these phenotypes.

If the frequencies of the genotypes are known, the frequencies of the genes can often be precisely determined.

Thus, in the example just used, we simply count the number of *M* and *N* genes present in the sample of 111 individuals. As every individual is diploid and we are dealing with an autosomal locus, there must be a total of 222 *M* and *N* genes among the 111 individuals. Each M individual carries only *M* genes, so that the 47M individuals contribute $2 \times 47 = 94$ *M* genes. MN individuals each have only one *M* gene and therefore contribute 52 *M* genes. The N individuals have no *M* genes. Our sample thus contains $94 + 52 = 146$ *M* genes, and so, by subtraction, $222 - 146 = 76$ *N* genes, since there are no other alleles. The frequency of the *M* allele in the population of 222 genes derived from these 111 individuals is therefore $146/222 = 0.66$. The frequency of the *N* allele in the population is $1 - 0.66 = 0.34$ as can be verified by a direct count. It is easily seen that the frequency of *M* genes is in fact just the sum of the frequency of the *MM* homozygotes plus one-half the frequency of the *MN* heterozygotes, and similarly for the *N* genes, substituting *NN* for *MM*. Quite generally, the frequency of an allele is equal to the frequency of the homozygotes for that allele plus one-half the frequency of heterozygotes carrying the allele.

Evolution in its most basic form is described by the changes in the frequencies of the genes.

What we observe are phenotypes. However, from the evolutionary point of view we are interested only in the heritable components of the phenotypes, that is, in the genotypes. The frequencies of genotypes are determined to a large extent by the frequencies of the genes in the population. It is thus convenient for the formal description of evolution in a population to consider the *changes in the frequencies of genes* in the population.

The study of gene frequencies is therefore an essential part of the formal analysis of evolution, but it is only a part and not the whole. It is important to know at what rate new alleles arise by mutation, increase in frequency, and invade a population. The phenotypes and not the genes are, however, observed, and it is therefore the frequencies of the phenotypes that should be predicted. Evolutionary prediction, which is an important goal of population genetics, is the prediction of the frequencies of phenotypes in future generations, based on knowledge of their frequencies in present and past generations.

Three basic types of information on the characteristics of a population are needed for the construction of models of the evolutionary process.

1. We must know the **pattern of inheritance** of the phenotypes whose evolution we are studying. This includes knowledge of the rules of the correspondence between genotypes and phenotypes, which provides the necessary basis for making inferences from frequencies of phenotypes to frequencies of genotypes and vice versa. It also includes knowledge of mutation rates and of possible distortions of segregation ratios from those expected on the basis of Mendel's laws. These forces are important in determining changes in frequencies of genes from generation to generation. For characters controlled by more than one locus, the pattern of inheritance must include a description of recombination patterns.

2. We need knowledge concerning the **selective forces** acting on the phenotypes that are being studied. By this is meant knowledge of all the causes that differentially affect the probability of survival or fertility of gametes or zygotes. Selection is probably the most important factor in evolution. Unfortunately it may be very difficult to describe, especially in human populations. A complete description of selection would involve a very large number of parameters, many of which are difficult to measure. Moreover, its validity would be limited to the given environment in which the measurements were made. Changes in the environment take place continuously and they alter radically the set of parameters that must be used to describe selection.

3. Finally, we need to know the **population structure**. By this is meant the probabilities with which matings will occur between the various phenotypes. These frequencies will, of course, affect the frequencies of phenotypes. Moreover, changes in frequencies of genes may also occur as a result of the finite size of populations. In any real population, the transition from one generation to another is accomplished by means of a finite number of gametes. This can be taken, at its simplest, to be a random sample of the genes present in the parental population. Because of the *finite* size of this sample, there will usually be a difference due to random sampling between the gene frequency of the sample of gametes forming the next generation and that of the parental generation. This difference will, on the average, be larger, the smaller the sample. This is the origin of **random genetic drift** (see Chapter 8).

The description of evolutionary changes is, indeed, a formidable task. It might seem that no mathematical theory could be developed that would adequately take account of the many forces since it would require such a large number of parameters, all of which must be estimated. However, we can simplify the situation by taking account of one, or at most of two at a time, of the various forces that are operating and thus make some interpretation of observed data possible. Such a simplification is often adequate for the construction of models appropriate for the level of resolution of our observations. Experience has shown that these models are useful, and thus that evolution is, at least to some extent, amenable to quantitative study. Our success in predicting evolutionary changes will depend, of course, upon our striking the right balance between simplification and reality.

2.3 Prediction of Frequencies of Genotypes under Random Mating: The Hardy-Weinberg Equilibrium

We are interested in predicting frequencies of genotypes and phenotypes on the basis of gene frequencies.

We shall first consider the simplest possible model for one locus with two alleles segregating in a population of diploid organisms, such as man. Let the relative frequencies of the two alleles A_1 and A_2 (calculated as described in Section 2.2) be p_1 and p_2. As there are only two alleles, $p_1 + p_2 = 1$.

We now make the following assumptions.

1. For the *pattern of inheritance* we assume that the genotypes can be distinguished unequivocally, meaning that the frequencies of the phenotypes are the same as those of the genotypes. We also assume that mutation rates are negligible and that segregation occurs according to Mendelian rules, with no segregation distortion.

2. We assume that there is *no selection*, namely that the expected number of fertile progeny from a mating that reaches maturity does not depend on the genotypes of the mates.

3. Finally, the *population structure* assume is, simply, that all matings take place at random with respect to the genetic differences being considered and in a population of infinite size. This means that the probability of mating between individuals is in no way influenced by their genotype at this locus. In addition, we assume that all individuals mate at the same given time and then are completely replaced by their offspring. This, of course, ignores the "age structure" of the population, since individuals do not all reproduce at the same age. This assumption that generations are discrete greatly simplifies population-genetics models. The corresponding frequencies of genes can be thought of as referring to some particular point of time in the life cycle, say the onset of sexual maturity. We will consider in Chapters 6 and

8 some of the complications in taking age structure into account in population-genetics models.

This list of assumptions is such that it is certainly hard to believe that all are ever realized in any "real" population. However, the optimism underlying the use of such a set of simplifying assumptions turns out to be justified, as the theorem that can be derived on the basis of these assumptions fits the facts well in a great number of cases. The general theorem was formulated in 1908 independently by the English mathematician G. H. Hardy and the German physician W. Weinberg.

The assumptions that we have been discussing are often summarized, somewhat loosely, by the words *random mating*. In its simplest form, the Hardy-Weinberg theorem states that:

A population undergoing random mating reaches, in one generation, a distribution of genotype frequencies given by the expansion of $(p_1 + p_2)^2$, *generating the three terms,* p_1^2, $2p_1p_2$, p_2^2, *which give the relative frequencies of the genotypes* A_1A_1, A_1A_2, A_2A_2, *respectively. The quantities* p_1 *and* p_2 *are the respective frequencies of the alleles* A_1 *and* A_2.

We can also expand $(p_1A_1 + p_2A_2)^2$ to obtain

$$p_1^2A_1A_1 + 2p_1p_2A_1A_2 + p_2^2A_2A_2,$$

a notational device showing immediately which genotype is associated with a given frequency.

We shall prove the theorem for a population in which the frequencies of the three genotypes are initially u for A_1A_1, v for A_1A_2 and w for A_2A_2 in both sexes ($u + v + w = 1$). There are nine possible matings between three male genotypes and three female genotypes. Only six of these need to be distinguished for our purpose. A 3×3 table (Table 2.1) with male and female genotypes at the heads of the rows

TABLE 2.1
Mating-type Frequencies in a Population Mating at Random

			Frequency of female genotypes		
			A_1A_1	A_1A_2	A_2A_2
			u	v	w
Frequency of male genotypes	A_1A_1	u	u^2	uv	uw
	A_1A_2	v	uv	v^2	vw
	A_2A_2	w	uw	vw	w^2

and columns respectively, illustrates, simply, the derivation of the frequencies of the various mating types. The products of the frequencies of the corresponding genotypes in males and females are given in the body of the table. These products give the expected frequencies of each mating on the assumption that mating takes place at random.

A set of values like that in Table 2.1, which can be given in a regular arrangement of rows and columns, is called a **matrix**. (A somewhat more complete account of the use of matrices is given in the examples at the end of Chapter 6.) The matrix of frequencies given as Table 2.1 is symmetrical along the main diagonal (from top left to bottom right) because the frequencies of the genotypes are assumed to be the same in males and in females. The symmetry of the matrix shows that it is not necessary, at this stage, to distinguish between matings in which, say, the husband is A_1A_1 and the wife is A_1A_2 and those in which the husband is A_1A_2 and the wife A_1A_1. Such matings, called **reciprocal**, can therefore be pooled.

We may thus obtain the list of the six matings types given in Table 2.2, in which reciprocal matings are grouped together and their respective frequencies summed. The last three columns of Table 2.2 give the expected proportions of the three

TABLE 2.2
Mating Types, Their Frequencies and Progeny Output for an Autosomal Locus with Two Alleles

		Genotypes of progeny		
Mating types	*Frequency*	A_1A_1	A_1A_2	A_2A_2
$A_1A_1 \times A_1A_1$	u^2	all	—	—
$A_1A_1 \times A_1A_2$	$2uv$	1/2	1/2	—
$A_1A_2 \times A_1A_2$	v^2	1/4	1/2	1/4
$A_1A_1 \times A_2A_2$	$2uw$	—	all	—
$A_1A_2 \times A_2A_2$	$2vw$	—	1/2	1/2
$A_2A_2 \times A_2A_2$	w^2	—	—	all

genotypes among the progeny of each mating. From this information, we can compute the frequency distribution of the genotypes among the total progeny of all matings. A convenient graphical representation of this information is given in Figure 2.1.

Continuing to use the same example, we may see that A_1A_1 progeny is obtained from the matings

$$A_1A_1 \times A_1A_1, \quad \text{whose frequency is } u^2;$$

$$A_1A_1 \times A_1A_2, \quad \text{whose frequency is } 2uv;$$

$$A_1A_2 \times A_1A_2, \quad \text{whose frequency is } v^2.$$

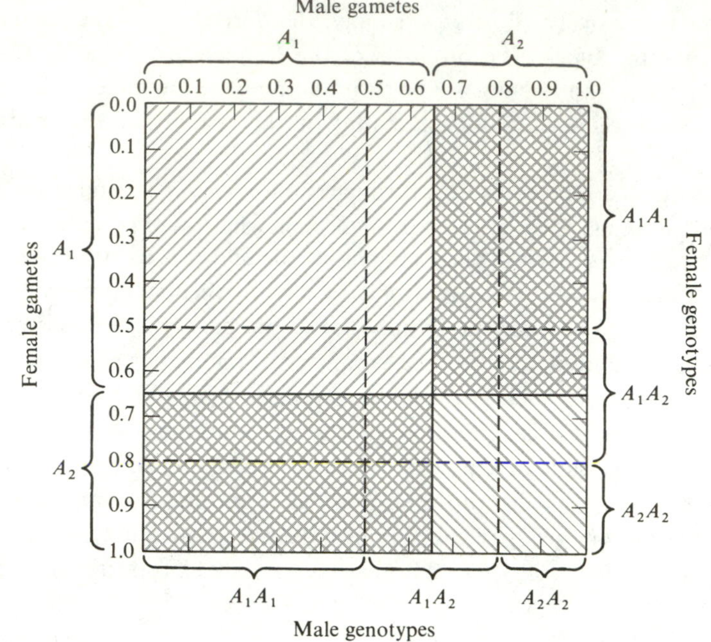

FIGURE 2.1
Hardy-Weinberg theorem in a graphical form. Random mating between individuals of three genotypes leads to expected mating frequencies given by the area of a rectangle whose sides are the gamete frequencies of males and females. For example, starting with genotype frequencies $u = 0.5$ for A_1A_1, $v = 0.3$ for A_1A_2, $w = 0.2$ for A_2A_2 in both of the sexes, gene frequencies and therefore gamete frequencies are, in both sexes: $0.5 + 0.3/2 = 0.65$ for A_1 and $0.3/2 + 0.2 = 0.35$ for A_2. In one generation the following equilibrium proportions are expected among the genotypes:

$$u = 0.65^2 = 0.4225 \text{ for } A_1A_1$$
$$v = 2 \times 0.65 \times 0.35 = 0.4550 \text{ for } A_1A_2$$
$$w = 0.35^2 = 0.1225 \text{ for } A_2A_2$$

These are indicated by the hatched areas.

If we assume, for simplicity that there is only one offspring from each mating, there will be, on the average

u^2	A_1A_1 progeny from $A_1A_1 \times A_1A_1$ (because all the progeny is A_1A_1);
$\frac{1}{2} \times 2uv = uv$	A_1A_1 progeny from $A_1A_1 \times A_1A_2$ (because $\frac{1}{2}$ of the progeny is A_1A_1);
$\frac{1}{4}v^2$	A_1A_1 progeny from $A_1A_2 \times A_1A_2$ (because $\frac{1}{4}$ of the progeny is A_1A_1).

Altogether there thus will be

$$u^2 + uv + \tfrac{1}{4}v^2 = (u + \tfrac{1}{2}v)^2 \ A_1A_1 \text{ progeny};$$

and similarly,

$$uv + \tfrac{1}{2}v^2 + 2uw + vw = 2(u + \tfrac{1}{2}v)(\tfrac{1}{2}v + w) \ A_1A_2 \text{ progeny};$$

and

$$\tfrac{1}{4}v^2 + vw + w^2 = (\tfrac{1}{2}v + w)^2 \ A_2A_2 \text{ progeny}.$$

In the previous section, it was shown that, in general, the frequency of an allele is equal to the frequency of the homozygotes for the allele plus $\frac{1}{2}$ the frequency of the heterozygotes. In this example, the frequencies of the two alleles are therefore $p_1 = u + \frac{1}{2}v$ and $p_2 = \frac{1}{2}v + w$. We can thus write the distribution of genotypes among the progeny as follows

A_1A_1 frequency $p_1^2 = (u + \frac{1}{2}v)^2$;

A_1A_2 frequency $2p_1p_2 = 2(u + \frac{1}{2}v)(\frac{1}{2}v + w)$;

A_2A_2 frequency $p_2^2 = (\frac{1}{2}v + w)^2$.

The gene frequencies in the progeny are the same as those in the parental population.

Thus, for example, the frequency of gene A_1 in the progeny is

$$p_1^2 + \tfrac{1}{2}(2p_1p_2) = p_1^2 + p_1p_2 = p_1(p_1 + p_2) = p_1,$$

since $p_1 + p_2 = 1$. It is clear, then, that if we use these progeny as a new set of parents, the genotype frequencies among their progeny would be exactly the same. The Hardy-Weinberg distribution $p_1^2A_1A_1 + 2p_1p_2A_1A_2 + p_2^2A_2A_2$ is thus achieved in one generation (assuming $u \neq p_1^2$, $v \neq 2p_1p_2$, and $w \neq p_2^2$), and then remains the same in all subsequent generations. This completes the proof of the theorem.

Random mating is equivalent to random union of gametes.

The Hardy-Weinberg equilibrium can also be obtained by picking pairs of genes at random out of a pool containing a proportion p_1 of A_1 genes and p_2 of A_2 genes. Thus the probability that each of the two genes picked out is A_1 is $p_1 \times p_1 = p_1^2$. Similarly, the probability that each is A_2 is $p_2 \times p_2 = p_2$. The probability that we first pick A_1 and then A_2 is p_1p_2, which is the same as the probability of picking

A_2 first and then A_1, giving a total of $2p_1p_2$ for A_1A_2. This process of **random union of gametes** can also be represented as follows:

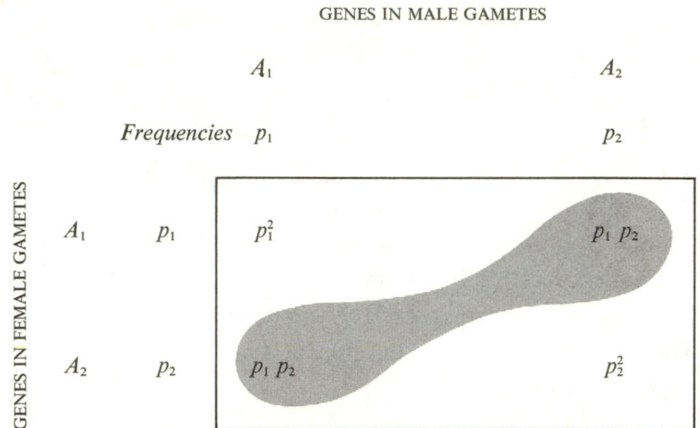

Circled genotypes are indistinguishable and can be pooled. We thus obtain again the standard formula. The Hardy-Weinberg theorem can therefore be interpreted as saying that random mating is equivalent to random union of gametes.

One of the most important features of the Hardy-Weinberg theorem is that it enables us to express the distributions of genotypes in a population entirely in terms of the gene frequencies.

The way in which gene frequencies change when subjected to the forces of mutation, selection, migration, and random drift will be discussed in the following chapters. As we shall see later, the effects of these forces are very rarely large enough to be detectable among genotype distributions. When a Mendelian population is examined it is almost always found to obey Hardy-Weinberg equilibrium. The sample must be very large, or conditions highly unusual, for a deviation from equilibrium to be detectable. The Hardy-Weinberg theorem verifies that, in the absence of mutation, selection, or random-sampling variations due to finite population size, genetic variation is maintained in a population at its prevailing level, and is not eroded, as it would be if inheritance were correctly described by a blending theory.

2.4 Composition of an Equilibrium Population

In this section we shall discuss some of the properties of a population in Hardy-Weinberg equilibrium. The relative porportions of the two homozygotes and of the heterozygote vary continuously with the gene frequencies, with the heterozygote frequency having a maximum of $\frac{1}{2}$ at $p_1 = p_2 = \frac{1}{2}$. The variation of the genotype frequencies as a function of the gene frequencies is illustrated in Figure 2.2. The same data are plotted on a logarithmic scale in Figure 2.3, which illustrates more clearly what happens when gene frequencies are very small. Variation in this range is important because of the existence of large numbers of rare deleterious mutant genes. If the human species actually bred at random, a gene whose frequency was 10^{-5} would probably not occur at all in the homozygous state, since the expected frequency of a homozygote for this gene would be only 10^{-10} and the present size of the human species is between 3 and 4×10^9. Heterozygotes, however, would occur with a frequency of about 1 in every 50,000 individuals. We shall see later that in actual populations there are deviations from random mating that do give rise occasionally to homozygotes for very rare genes.

The frequency of a rare recessive gene is about $\frac{1}{2}$ the frequency of the heterozygotes who carry it.

There are many examples of rare mutant genes that have frequencies in the range 10^{-2} to 10^{-4}. When p_1 (the frequency of allele A_1, as before) is very small the frequency of heterozygotes $2p_1p_2 = 2p_1 - 2p_1^2$ is approximately $2p_1$, since p_1^2 is very much smaller than p_1 and is thus approximately twice the gene frequency. Many more mutant genes are therefore present in heterozygotes ($\sim 2p_1$) than in homozygotes (p_1^2). (Note that "$\sim$" means "approximately equal to.")

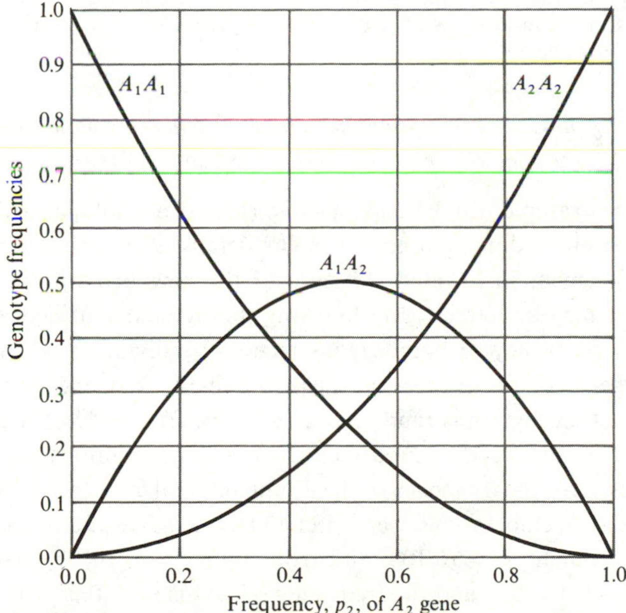

FIGURE 2.2
Plot of the three genotype frequencies for A_1A_1, A_1A_2, and A_2A_2, against the gene frequency of A_2, based on the Hardy-Weinberg law. The ordinate shows the relative proportions of heterozygotes and homozygotes as a function of the gene frequency of A_2.

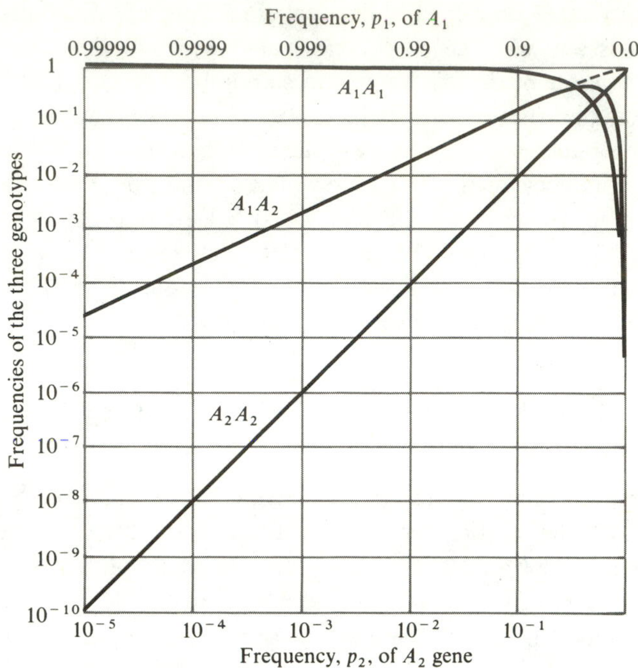

FIGURE 2.3
These are the same data as in Figure 2.2, but here the logarithms of the frequencies are plotted to magnify the results for low gene frequencies.

For dominant genes, the assumption of a Hardy-Weinberg equilibrium provides the simplest way of estimating the frequency of heterozygotes in the population.

Consider, for example, the Rhesus positive (Rh+) phenotype, which is known to be determined by a dominant gene *D*. (The details of the Rhesus blood groups and their involvement in hemolytic disease of the newborn will be discussed in Chapter 5.) We may be interested in knowing the frequency of heterozygotes *Dd*. Given that the frequency of homozygous recessive individuals *dd*, who have the Rh− phenotype, is 16 percent, we can easily obtain the frequency of the gene *d* on the assumption that mating is random. Let p be the frequency of the gene *D* and $q(=1-p)$ the frequency of the gene *d*. (It is a common convention to use, in working with two alleles, p for the frequency of the dominant, and q for that of the recessive.) Using the Hardy-Weinberg law, we expect p^2DD, $2pqDd$, and q^2 *dd* individuals. Thus we can estimate $q^2 = 16/100$, and from this, $q = (0.16)^{1/2} = 0.4$. Therefore $p = 1 - q = 1 - 0.4 = 0.6$, and the frequency of *DD* is $p^2 = 0.6^2 = 0.36$, while the frequency of heterozygotes *Dd*, $2pq = 2 \times 0.6 \times 0.4 = 0.48$. Individuals heterozygous for the Rhesus genes, and thus having the Rh+ phenotype, therefore constitute almost half of the population. The total frequency of individuals having the Rh+ phenotype is, of course, $1 - 0.16 = 0.84$. The frequency of heterozygotes among all individuals having the Rh+ phenotype is thus 0.48/0.84 = 0.57 or 57 percent.

Another interesting example is provided by albinism, the lack of pigmentation in the skin, hair, and eyes, which occurs in certain populations with a frequency of about 1/20,000. Assuming it is determined by a recessive gene a, we can estimate the frequency q of this gene by

$$q = \sqrt{\frac{1}{20{,}000}} = \frac{1}{141.42}.$$

If a mating between an albino and a normal individual gives rise to albino offspring the normal mate must be a heterozygote Aa. The proportion of heterozygotes Aa among normal people (either Aa or AA) is

$$\frac{2pq}{p^2 + 2pq} = \frac{2q}{p + 2q} = \frac{2q}{1 + q} = 0.014027,$$

which is actually approximately

$$2q = \frac{2}{141.42} = 0.014142, \text{ since } q \text{ is small.}$$

Hence, the probability that an offspring from a mating between an albino and a normal individual will be an albino is $\frac{1}{2} \times 0.014$, or 0.7 percent. Most of the albinos in the population have normal parents, who must, of course, be heterozygotes Aa. The mating $Aa \times Aa$ occurs with frequency

$$(2pq)^2 = (0.014042)^2 = 0.00019718,$$

and contributes a proportion $\frac{1}{4} \times 0.00019718 = 0.000049295$ of albinos to the next generation. Thus, the proportion of all albinos that have normal parents is $0.000049295/(1/20{,}000) = 0.9859$. Actually, there is more than one gene for albinism and so precise computations would be more complicated than this.

2.5 Extension of the Hardy-Weinberg Theorem to Multiple Alleles and to Polyploids

When there are more than two alleles at a given locus, say $A_1, A_2, \ldots, A_m$, the number of possible genotypes is $m(m + 1)/2$, of which m are homozygotes (A_1A_1, $A_2A_2, \ldots, A_mA_m$) and $m(m - 1)/2$ are heterozygotes (A_1A_2, $A_1A_3, \ldots, A_{m-1}A_m$).

The Hardy-Weinberg theorem (as shown by Weinberg, 1909) *can be extended quite simply to cover multiple alleles.*

Thus, if we assume that random mating is equivalent to random union of gametes, we may compute the frequencies of the various genotypes by the expansion of

$$(p_1 + p_2 + \cdots + p_m)^2$$

where the p_i's are the frequencies of the genes A_i. To allocate frequencies to genotypes it is useful to consider, by analogy with the way we handle frequencies of two alleles, the expansion of the expression

$$(p_1A_1 + p_2A_2 + \cdots + p_mA_m)^2.$$

Homozygotes A_iA_i occur with a frequency p_i^2; heterozygotes A_iA_j $(i \neq j)$ occur with a frequency $2p_ip_j$. Gene frequencies are determined by summing the frequency of the respective homozygote and half the frequencies of all the heterozygotes containing the allele in question, of which there are $m - 1$.

Complications may arise when, because of dominance, some genotypes are pooled together in one phenotype. This is the case, for example, for the well-known ABO blood-group system. Here there are three alleles, *A*, *B*, *O* that have frequencies of p, q, r, say. Table 2.3 gives the four phenotypes that can be distinguished and their

TABLE 2.3
ABO Blood-group System: Phenotypes and Genotypes

Phenotype	*Genotype*	*Frequencies*
O	*OO*	r^2
A	*AA* + *AO*	$p^2 + 2pr$
B	*BB* + *BO*	$q^2 + 2qr$
AB	*AB*	$2pq$

expected frequencies. When a pair of alleles such as *A* and *B* both show their presence in the heterozygote, they are sometimes referred to as being **codominant**. The estimation of gene frequencies in cases such as this often requires a considerable amount of numerical work and even the use of a computer. The method of estimation generally used is that of maximum likelihood, whose logical basis and practical application are briefly summarized in the appendices of this book.

The following method gives a simple, rough (that is with relatively low statistical efficiency—see Appendix II) estimate of the *ABO* gene frequencies. Let A, B, and O refer to the numbers of individuals with the corresponding blood type and N to the population size. Then the *O* gene frequency is estimated by

$$\hat{r} = \sqrt{\text{O}/\text{N}}.$$

(Note that a caret over the symbol for frequency means that it is an estimate based on observed frequencies—see Appendix I.)
Since the combined frequencies of O and A individuals are

$$r^2 + p^2 + 2pr = (p + r)^2,$$

the *A* gene frequency is given by

$$\hat{p} = \sqrt{\frac{(\text{A} + \text{O})}{\text{N}}} - \hat{r},$$

and thus, by subtraction,

$$\hat{q} = 1 - \hat{p} - \hat{r}.$$

Consider, for example, the following tabulation of numbers of persons having the various ABO blood types among a total of 163 persons; these data were obtained from a Pygmy group in the Central African Republic:

Blood type			
O	A	B	AB
88	44	27	4

Using the simple method of estimation just given, we may determine the O gene frequency,

$$\hat{r} = \sqrt{\frac{88}{163}} = 0.735;$$

while

$$\hat{p} + \hat{r} = \sqrt{\frac{88 + 44}{163}} = 0.900.$$

This gives the A gene frequency, $\hat{p} = 0.9 - 0.735 = 0.165$; and hence, the B gene frequency, $\hat{q} = 1 - \hat{p} - \hat{r} = 0.1$.

A statistically more efficient estimation method, developed by Bernstein (1930), uses preliminary estimates such as those just obtained (also see Appendix II), for $\hat{r}$ and $\hat{q} = 1 - (\hat{p} + \hat{r})$, and, in addition, by analogy with the estimate for $\hat{q}$,

$$\hat{p} = 1 - \sqrt{\frac{\mathrm{B + O}}{\mathrm{N}}} = 1 - \sqrt{\frac{88 + 27}{163}} = 0.16.$$

Then, if we let $D = 1 - \hat{p} - \hat{q} - \hat{r} = 0.005$, corrected estimates p^*, q^*, r^* are given by

$$p^* = \hat{p}\left(1 + \frac{D}{2}\right) = 0.1604,$$

$$q^* = \hat{q}\left(1 + \frac{D}{2}\right) = 0.1003,$$

and

$$r^* = \left(\hat{r} + \frac{D}{2}\right)\left(1 + \frac{D}{2}\right) = 0.7393,$$

which are here found to be very close to the preliminary estimates.

The ABO blood-group polymorphism is used in a number of practical applications, other than the problem of blood transfusion (see Chapter 5), such as paternity testing and determining possible parentage where there is some confusion. There is a very slight chance, for example, that babies in a nursery for the newborn may be interchanged by mistake, and that parents may thus be given the wrong child.

Determination of genetically inherited characteristics, can of course, in some cases, help in correcting such an unfortunate mistake. What can be said about the probability of determining that the wrong child has been given, for example, to parents whose blood types are O and A? Only types B and AB among possible progeny are excluded by the O × A mating, since if the A parent is heterozygous *AO*, O offspring may be produced. The probability of determining that these parents have the wrong baby is thus the combined frequencies of phenotypes B and AB. For Caucasian populations this would be about 0.19. This probability is, of course, a function of the mating types; the chance of making a proper determination could be increased by the use of more genetic differences.

The Hardy-Weinberg law can be extended to cover trisomy.

Down's syndrome, as already mentioned is often the result of trisomy for chromosome 21—that is, chromosome 21 is present in triplicate. This happens when one or other parent transmits a gamete that, as a result of nondisjunction, carries two copies of chromosome 21 instead of one. If the nondisjunction occurs in the first meiotic division, a parent heterozygous for a locus near the centromere will nearly always transmit both allelles to the nondisjunctional gamete (see also Chapter 10). In a population in which mating is random, individuals with Down's syndrome resulting from first-meiotic-division nondisjunction will thus carry a random sample of three genes (instead of two) from the population. This fact can, in principle, be used to see whether a gene is on chromosome 21, because if it is, it should lead to a disturbance in the expected proportion of apparent heterozygotes and homozygotes with Down's syndrome (Bateman, 1960). Thus, let p_1 and p_2 be the gene frequencies of a pair of alleles A_1 and A_2. The frequencies of the various genotypes among individuals with Down's syndrome will be given by

$$(p_1A_1 + p_2A_2)^3 = p_1^3A_1A_1A_1 + 3p_1^2p_2A_1A_1A_2 + 3p_1p_2^2A_1A_2A_2 + p_2^3A_2A_2A_2,$$

since individuals having the genotypes represented carry a random sample of three alleles from the population. The apparent frequencies of the two types of homozygotes are therefore p_1^3 and p_2^3, and those of heterozygotes $3p_1^2p_2 + 3p_1p_2^2 = 3p_1p_2$. The expected ratios of these apparent frequencies in Down's syndrome compared with those in the normal population will therefore be $p_1^3/p_1^2 = p_1$ for A_1A_1, $3p_1p_2/2p_1p_2 = 3/2$ for A_1A_2, and $p_2^3/p_2^2 = p_2$ for A_2A_2. To the extent that nondisjunction does not occur at the first meiotic division, or that the locus in question is not very closely linked to the centromere, the difference between these frequencies will be reduced. A search for disturbances in blood-group frequencies among individuals with Down's syndrome has so far given negative results, indicating that none of the common blood-group loci are on chromosome 21. (See, for example, Goodman and Thomas, 1968; and Shaw and Gershowitz, 1962.)

2.6 Testing Equilibrium and the Meaning of Departures from It

When appropriate data for calculating gene frequencies are available, it is of some interest to test the validity of the Hardy-Weinberg equilibrium. This is done by computing the gene frequencies from the data, and then from these frequencies the expected frequencies of the various phenotypes. These expected frequencies are then compared with the observed frequencies using the chi-square test. As an example, consider the data on ABO blood groups in Pygmies. From the estimated *A*, *B*, and *O* gene frequencies of 0.1604, 0.1003, and 0.7393, we can calculate the expected frequencies of phenotypes:

Phenotype	*Expected frequency*	*Observed frequency*
O	$N(r^*)^2 = 89.1$	88
A	$N(2p^*r^* + (p^*)^2) = 42.9$	44
B	$N(2q^*r^* + (p^*)^2) = 25.8$	27
AB	$N2p^*q^* = 5.2$	4
		N = 163

The chi-square for testing the goodness of fit (with one degree of freedom) is given by

$$\chi^2_{[1]} = \frac{(88 - 89.1)^2}{89.1} + \frac{(44 - 42.0)^2}{42.9} + \frac{(27 - 25.8)^2}{25.8} + \frac{(4 - 5.2)^2}{5.2} = 0.375,$$

which corresponds to a probability between 0.7 and 0.5, and so shows a very good fit of observed with expected. For those unfamiliar with the basic concepts underlying the chi-square test, a brief discussion is given in Appendix I.

The general problem of the evaluation of the number of degrees of freedom in using the chi-square test for the goodness of fit deserves a few comments. There are for example, three phenotypic classes with two alleles at a single locus. To compute the expected frequencies, we use the total number of observations and also another linear function of the observations, namely, the gene frequency of one or other of the two alleles. There is thus only one independent class, and one degree of freedom for the chi-square test.

In the ABO system we were just discussing there are four phenotypic classes. In addition to the total number of observations, we use two other functions of the observed frequencies, namely, the gene frequencies of two of the alleles. The frequency of the third allele follows from the fact that gene frequencies must add up to one. There are thus three independent constraints on the four phenotypic classes, only one of which can therefore vary independently. This leaves once again, one degree of freedom for testing the agreement with a Hardy-Weinberg equilibrium. In general, the degrees of freedom are the number of phenotypic classes minus the number of alleles.

Although observed departures from the Hardy-Weinberg equilibrium are, in fact, rare, there are many possible causes of departures in either direction, toward deficiency or excess of heterozygotes.

We may at this point list possible causes of a deficiency in the observed proportion of heterozygotes, although these causes will not be considered in detail until a later chapter: (1) heterogeneity of the population, which may consist of a number of independent subpopulations (see Example 2.2 at end of this chapter); (2) inbreeding (see Section 7.3); (3) selection against the heterozygotes; (4) presence of a "silent" allele that masks heterozygotes, making them indistinguishable from one of the types of homozygotes (see Example 2.4); (5) errors in the classification of phenotypes; (6) positive assortative mating—that is, a tendency for individuals of like genotype to mate with each other (see Chapter 9).

An excess of observed heterozygotes is a rarer event than a deficiency and could be caused by: selection in favor of heterozygotes (see Chapter 4); errors of classification of genotypes; or negative assortative mating—that is, a tendency of unlike genotypes to mate with each other. Most of the rare cases in which the Hardy-Weinberg equilibrium does not fit observed data may be the consequence of a recent change in the constitution of a population, such as the recent admixture of two different populations. Equilibrium is reestablished in one generation when generations are discrete in time, as in annual plants (see Example 2.2). When generations overlap, as in man, equilibrium is reached gradually. The rate of approach to equilibrium, however, is usually quite rapid (see Example 2.8).

Clearly, the fact that a population does not deviate significantly from a Hardy-Weinberg equilibrium does not mean that all the possible causes of divergence are absent. It simply indicates that they are not present at a level that can be detected with the number of individuals tested. It should be emphasized that the test for the fit of the Hardy-Weinberg equilibrium is not a sensitive test for the presence of forces, such as selection or inbreeding, that may cause departure from it. This is undoubtedly the major reason why it is rare in practice to find deviations from the Hardy-Weinberg equilibrium. It is also possible that disturbing forces have opposite effects, which cancel each other.

Important exceptions to the simplest form of the Hardy-Weinberg equilibrium include distributions of (1) sex-linked genes, which will be discussed in the next section; (2) two or more linked or unlinked loci—independent distribution of the two loci is only reached gradually (see Examples 2.6 and 2.7); (3) Polyploid organisms, for which the equilibrium conditions can be obtained by using as an exponent of the Hardy-Weinberg expression the degree of ploidy, as in the example of Down's syndrome already discussed.

2.7 Sex-linked Genes

A special treatment is needed for genes located on the sex chromosomes, since, in a bisexual organism, *mating is not random with respect to sex*. The rules for sex-linked inheritance can be summarized as follows. Any gene on the Y chromosome is present only in males and is therefore transmitted only from father to son.

Males have only one copy of X-linked genes, which, of course, always comes from their mother. Females have two copies, one of which is of paternal and the other of maternal origin.

Intuitively, we immediately recognize that in females the inheritance of sex-linked genes is the same as that of autosomal loci, and this can be proved rigorously. The frequencies of genotypes are therefore given by the standard Hardy-Weinberg equilibrium. Males, however, each having only a single dose of X-linked genes, will have one or the other of two distinguishable phenotypes corresponding to the two alleles. The frequencies of male phenotypes for linked genes are therefore identical to the frequencies of the respective alleles in the mothers. We thus expect, at equilibrium, the distribution given in Table 2.4.

TABLE 2.4
Equilibrium Genotype Frequencies for X-linked Genes: two alleles A_1 and A_2 with frequencies p_1 and p_2 in both sexes

Sex	*Genotype*	*Frequency*
Male	A_1	p_1
	A_2	p_2
Female	A_1A_1	p_1^2
	A_1A_2	$2p_1p_2$
	A_2A_2	p_2^2

Equilibrium frequencies for the sex-linked genes are reached in an oscillatory fashion.

Suppose that we start with different frequencies for an X-linked pair of alleles in the two sexes, say, p_m and q_m for genes A_1 and A_2 in the males and p_f and q_f in the females. The approach to equilibrium is then somewhat anomalous. We have already seen that in the first generation the male frequencies are equal to the maternal frequencies. Thus if p'_m, p'_f represent first-generation male and female frequencies,

$$p'_m = p_f. \tag{2.1}$$

Females obtain one gene from their fathers and one from their mothers. Their gene frequency will thus be the average of the parental gene frequencies;

$$p'_f = \frac{p_m + p_f}{2}. \tag{2.2}$$

Combining Equations 2.1 and 2.2, we see that

$$p'_m - p'_f = p_f - \tfrac{1}{2}(p_m + p_f) = -\tfrac{1}{2}(p_m - p_f). \tag{2.3}$$

Thus the difference between the male and female gene frequencies decreases by a factor $-\frac{1}{2}$ in each generation. This difference, therefore, tends to become zero with increasing numbers of generations, but as it does it alternates on either side of zero in successive generations. A typical example of an approach to equilibrium for an X-linked gene is given in Figure 2.4. It can easily be shown that the equilibrium gene frequency, which must be the same in both sexes, is given by

$$p_e = \frac{p_m + 2p_f}{3} \tag{2.4}$$

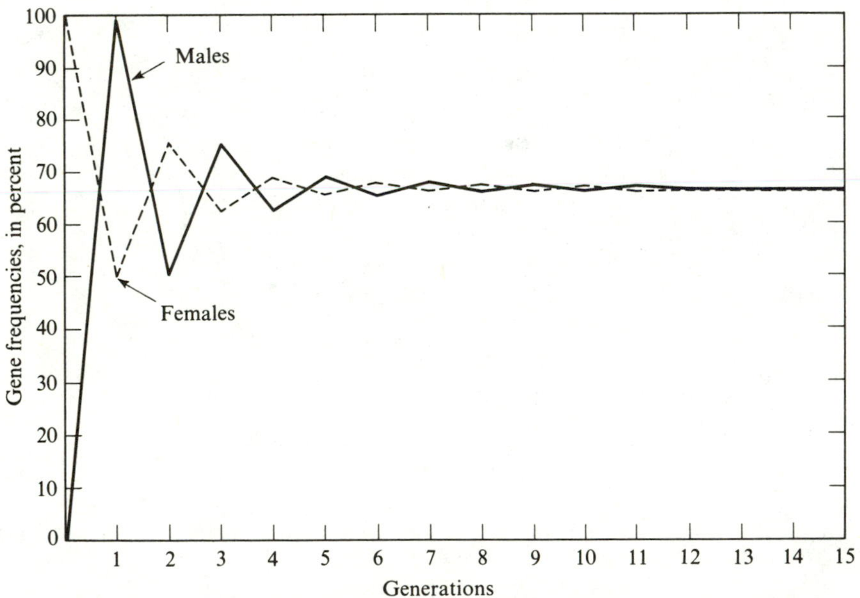

FIGURE 2.4
The approach toward equilibrium of frequencies of a sex-linked gene. It is assumed that, at the beginning, the frequency of the gene is 0 percent in the males and 100 percent in the females. The rule is that, in every generation, the frequency of the gene in the males is equal to that of the females in the previous generation, and that of the females is the mean of the male and female frequencies of the previous generation.

(See Example 2.5. This problem was first studied by Jennings, 1916; see also Haldane, 1924.)

The equilibrium conditions show that a recessive X-linked character is, in general, much more frequent in males than in females.

The frequency in females is, in fact, the square of that in males. Thus, color blindness in women is much less frequent than it is in males, some 0.5 percent versus 8 percent in the males—in fairly good agreement with this expectation. Hemophilia, a rare X-linked recessive disease, has a frequency of less than 1 in 10,000 in males, and is practically unknown in females.

The higher frequency of a trait in males than in females should be considered to indicate only as one possibility that the trait is sex-linked. Quite a number of conditions are sex-limited, or sex-controlled; that is, controlled by autosomal genes that are expressed differently in the two sexes. Such characters may exhibit different frequencies in males and females although they are not determined by X-linked genes. Any suspicion of X linkage must be confirmed by pedigree analysis. The expected pattern of inheritance is so clear-cut that we can usually establish or exclude genuine X linkage. There are now some sixty conditions for which X linkage is considered to be proved or very likely, and another twenty or so for which the evidence is suggestive but not conclusive (see Chapter 10; and McKusick, 1968).

The Xg blood group is a useful example of an X-linked polymorphism. The Xg(a+) phenotype is determined by a dominant X-linked gene (Xg^a).

The following data on the frequency of Xg(a+) individuals were obtained from a sample of Caucasian males and females (Mann, et al., 1962):

Phenotype	♂♂	♀♀
Xg(a+)	95	167
Xg(a−)	59	21
	154	188

There is a highly significant difference in the frequencies of Xg(a−) males and Xg(a−) females in the population ($\chi^2_{(1)} = 34.8$), as would be expected for an X-linked character. Let *Xg* be the recessive allele. The frequencies of Xg^a and Xg, p and q, say, can be estimated from the frequencies among the males of the phenotypes, Xg(a+) and Xg(a−) giving

$$p = \frac{95}{154} = 0.615, q = 1 - p = 0.385.$$

Using these estimates, the expected number of Xg(a−) females is

$$188 \times (0.385)^2 = 27.6,$$

which agrees reasonably well with the observed number, 21. The methods given in the appendices indicate how it is possible to improve the accuracy of such an estimate by also taking into account during the calculation the data on females.

The information contained in this chapter, which is very basic to population genetics, may be summarized as follows: Random-mating populations, in the absence of mutation and selection, rapidly reach an equilibrium distribution of genotypes that is given by a simple theorem named after Hardy and Weinberg. In a diploid organism the distribution of the genotypes for one autosomal locus with two alleles is given by the expansion of $(p_1 + p_2)^2$, where p_1 and p_2 are the frequencies of the two alleles. The equilibrium is reached in one generation if the age structure of the population is neglected. It is, in practice, quite insensitive to disturbing factors and therefore constitutes a basic foundation for many further theoretical developments. Representing a population in terms of the Hardy-Weinberg equilibrium greatly simplifies the analysis of more complex genetic problems, since it allows the frequencies of phenotypes and genotypes to be expressed in terms of the gene frequencies alone. Important special cases of the Hardy-Weinberg equilibrium include those for polyploids, X-linked genes, and the joint distribution for more than one locus. The main factors causing departures from the Hardy-Weinberg equilibrium will be considered in subsequent chapters.

General References

See the references listed at the end of Chapter 1 under the heading Population Genetics.

Worked Examples

2.1 *Segregation in the matings dominant × dominant and dominant × recessive.*

A given trait is determined by a single dominant gene G, which has the population frequency $p = 1 - q$. Assuming that mating is random, what are the expected proportions of the recessive phenotype among the offspring of the matings dominant × dominant and dominant × recessive?

There are three types of dominant × dominant matings:

Mating genotypes	*Mating frequencies*	*Offspring proportions*	
		Dominant	*Recessive*
$GG \times GG$	$p^2 \times p^2 = p^4$	1	0
$GG \times Gg$	$2 \times p^2 \times 2pq = 4p^3q$	1	0
$Gg \times Gg$	$2pq \times 2pq = 4p^2q^2$	$\frac{3}{4}$	$\frac{1}{4}$.

Since $p + q = 1$, the total mating frequency is

$$p^4 + 4p^3q + 4p^2q^2 = p^2(p^2 + 4pq + 4q^2) = p^2(p + 2q)^2 = p^2(1 + q)^2.$$

And, thus, the overall proportion of recessives is

$$\frac{1}{4} \times 4p^2q^2 \times \frac{1}{p^2(1 + q)^2} = \left(\frac{q}{1 + q}\right)^2.$$

There are two types of dominant × recessive matings:

Mating genotypes	*Mating frequencies*	*Offspring proportions*	
		Dominant	*Recessive*
$Gg \times gg$	$2 \times 2pq \times q^2 = 4pq^3$	$\frac{1}{2}$	$\frac{1}{2}$
$GG \times gg$	$2 \times p^2 \times q^2 = 2p^2q^2$	1	0

The total mating frequency is

$$4pq^3 + 2p^2q^2 = 2pq^2(2q + p) = 2pq^2(1 + q).$$

And, thus, the overall proportion of recessives is

$$\frac{1}{2} \times \frac{4pq^3}{2pq^3 \times (1 + q)} = \frac{q}{1 + q}.$$

These are known as "Snyder's ratios" (Snyder, 1932). The first increases from 0 to $\frac{1}{4}$ as q, the frequency of the recessive gene, goes from 0 to 1, while the second ranges from 0 to $\frac{1}{2}$. For the Rh+ phenotype, for example, with $q = 0.4$ we should expect 8.3 percent Rh− individuals among the offspring of Rh+ × Rh+ matings and 28.6 percent Rh− among those of Rh+ × Rh− matings.

2.2. *Outcome of mixing two different populations.*

Two populations, both in Hardy-Weinberg equilibrium for a pair of alleles A_1 and A_2 at a single locus, but having different gene frequencies, $p = (1 - q)$ and $P = (1 - Q)$ where $p \neq P$, are mixed in the respective proportions m and $1 - m$.

Show that the mixed population has a deficiency of heterozygotes, but on further random mating achieves, in one generation, a new Hardy-Weinberg equilibrium with gene frequencies $mp + (1 - m)P$, $mq + (1 - m)Q$.

These gene frequencies with $m = \frac{1}{2}$ hold for a single population if males and females start with frequencies of A_1 genes that are p and P respectively, where again $p \neq P$.

The frequencies of the genotypes A_1A_1 and A_1A_2 in the mixed, parental population are $mp^2 + (1 - m)P^2$ and $2mpq + 2(1 - m)PQ$, respectively. The frequency of the A_1 gene in this population, $\bar{p}$, is thus

$$\bar{p} = mp^2 + (1 - m)P^2 + mpq + (1 - m)PQ = mp + (1 - m)P.$$

Similarly, $\bar{q} = 1 - \bar{p} = mq + (1 - m)Q$. After one generation of random mating, the heterozygotes will, because of Hardy-Weinberg equilibrium, have the frequency

$$2\bar{p}\bar{q} = 2[mp + (1 - m)P][mq + (1 - m)Q].$$

This proportion is larger than that of heterozygotes in the mixed, parental population. In fact the difference in the proportions of heterozygotes between the two generations,

$$2\bar{p}\bar{q} - [2mpq + 2(1 - m)PQ] = 2m(1 - m)(P - p)^2,$$

is always positive if m is different from 0 and 1. A relative deficiency of heterozygotes may, therefore, indicate that there has been a recent admixture of two populations.

This is really a simple example of migration, corresponding to a single event of admixture, as opposed to continuing admixture of two populations generation after generation (see Chapter 8). The result shows that a single event of admixture causes a departure from the Hardy-Weinberg equilibrium for one locus only in the generation in which it occurs.

When the initial gene frequencies in the two sexes are different, the first set of offspring have A_1A_1, A_1A_2, and A_2A_2 genotype frequencies of pP, $pQ + qP$ and qQ respectively (by random union of male with female gametes). The new A_1 gene frequency is therefore

$$p = pP + \tfrac{1}{2}(pQ + qP) = \tfrac{1}{2}(pP + pQ) = \tfrac{1}{2}(pP + qP) = \tfrac{1}{2}(p + P).$$

2.3. *Proof that the Hardy-Weinberg theorem may be extended to account for the frequencies in a population of three alleles at a single locus.*

We note that since the gametic output of an individual of genotype A_1A_2 can be written $\frac{1}{2}A_1 + \frac{1}{2}A_2$, which is sometimes called a **gametic array**, the output of a mating $A_1A_2 \times A_1A_3$ can, for example, be written

$$(\tfrac{1}{2}A_1 + \tfrac{1}{2}A_2) \times (\tfrac{1}{2}A_1 + \tfrac{1}{2}A_3) = \tfrac{1}{4}A_1A_1 + \tfrac{1}{4}A_1A_2 + \tfrac{1}{4}A_1A_3 + \tfrac{1}{4}A_2A_3$$

to give in a simple algebraic expression, which is called the **genotypic array**, the frequencies of the various genotypes. With three alleles A_1, A_2, and A_3 at a single locus there are six genotypes A_1A_1, A_2A_2, A_3A_3, A_1A_2, A_1A_3, and A_2A_3. Suppose that these occur in the population with frequencies u, v, w, x, y, and z; we can then write, formally, the outcome of random mating between these genotypes as

$$u^2A_1^2 + v^2A_2^2 + w^2A_3^2 + x^2(\tfrac{1}{2}A_1 + \tfrac{1}{2}A_2)^2 + y^2(\tfrac{1}{2}A_1 + \tfrac{1}{2}A_3)^2 + z^2(\tfrac{1}{2}A_2 - \tfrac{1}{2}A_3)^2$$
$$+\ 2uvA_1A_2 + 2uwA_1A_3 + 2uxA_1(\tfrac{1}{2}A_1 + \tfrac{1}{2}A_2) + 12 \text{ more similar terms.}$$

It can be shown that this expression is the same as

$$[uA_1 + vA_2 + wA_3 + x(\tfrac{1}{2}A_1 + \tfrac{1}{2}A_2) + y(\tfrac{1}{2}A_1 + \tfrac{1}{2}A_3) + z(\tfrac{1}{2}A_2 + \tfrac{1}{2}A_3)]^2.$$

Rearranging the terms in the square brackets gives

$$[(A_1(u + \tfrac{1}{2}x + \tfrac{1}{2}y) + A_2(v + \tfrac{1}{2}x + \tfrac{1}{2}z) + A_3(w + \tfrac{1}{2}y + \tfrac{1}{2}z)]^2,$$

or

$$(A_1p_1 + A_2p_2 + A_3p_3)^2,$$

where $p_1 = u + \frac{1}{2}x + \frac{1}{2}y$, $p_2 = v + \frac{1}{2}x + \frac{1}{2}z$, and $p_3 = w + \frac{1}{2}y + \frac{1}{2}z$ are the respective frequencies of genes A_1, A_2, and A_3. This way of proving the Hardy-Weinberg equilibrium for three alleles at a locus can easily be extended to take into account more alleles at one or more loci.

2.4. *Deviation from Hardy-Weinberg equilibrium due to silent alleles.*

Two variants of a protein S and T, can be detected in a certain group of individuals. They are such that three recognizable phenotypes exist: S, T, and ST. These occur with frequencies 45S, 21T, and 30ST in a total of 96 individuals. Assuming that mating is random, test the hypothesis that the proteins S and T are determined by a pair of alleles at a single locus. What explanations can you suggest if the hypothesis does not fit the facts?

If we assume that alleles *S* and *T* determine the variant proteins S and T, we can immediately see that genotypes *SS*, *ST*, and *TT* are present with the observed frequencies 45 : 30 : 21. We can compute the frequency of gene *S*,

$$p = \frac{45 + \frac{1}{2}(30)}{96} = \frac{60}{96} = 0.625.$$

The frequency of gene T is $q = 1 - p = 0.375$.

Assuming that mating is random, we can calculate that the expected frequencies of genotypes are

	SS	*ST*	*TT*
	$(0.625)^2 \times 96$	$2(0.625)(0.375) \times 96$	$(0.375)^2 \times 96$
or			
	37.5	45	13.5.

Note the excess of heterozygotes expected. The chi-square test for goodness of fit is, for one degree of freedom,

$$\chi^2_{[1]} = \frac{(45 - 37.5)^2}{37.5} + \frac{(30 - 45)^2}{45} + \frac{(21 - 13.5)^2}{13.5} = 10.7,$$

giving a significance level of about 0.2 percent, and thus making the hypothesis untenable.

Possible explanations include departure from random mating and large selective differences between the genotypes. However, the most likely explanation is the existence of a third "silent" allele *O* such that genotype *SO* makes protein S only, *TO* makes protein T only, and *OO* makes none. (Compare ABO blood groups.) For example, if we assume that the alleles, *S*, *T*, and *O* have frequencies of 0.5, 0.3, and 0.2, respectively, then

phenotypes:	S	T	ST	O
occur with relative frequencies:	$(0.5)^2 + 2(0.5)(0.2)$	$(0.3)^2 + 2(0.3)(0.2)$	$2(0.5)(0.3)$	$(0.2)^2$
or in the percentages:	45	21	30	4

which agree with the observed figures except that the 4 percent of "O" types were not observed. These could have been overlooked or simply not observed because of the comparatively small size of the sample.

2.5. *Approach to equilibrium for X-linked genes.*

Prove that for an X-linked locus with two alleles A_1 and A_2 the equilibrium frequency of gene A_1 is given by $\frac{1}{3}p_m^{(0)} + \frac{2}{3}p_f^{(0)}$ where $p_m^{(0)}$ and $p_f^{(0)}$ are the initial frequencies of the A_1 gene in males and females respectively.

Derive a general expression for the frequencies in males and in females of the A_1 gene in the nth generation in terms of their values when $n = 0$.

From Equations 2.1 and 2.2,

$$p_m^{(n)} = p_f^{(n-1)};\ p_f^{(n)} = \tfrac{1}{2}(p_m^{(n-1)} + p_f^{(n-1)}),$$

where the subscripts n and $n - 1$ refer to the nth and $n - 1^{\text{th}}$ generations respectively. Adding the members of the second identity, after multiplication by $\frac{2}{3}$, to those of the first, after multiplication by $\frac{1}{3}$, and noting that the identity thus obtained is general, we have

$$\tfrac{1}{3}(p_m^{(n)} + 2p_f^{(n)}) = \tfrac{1}{3}(p_m^{(n-1)} + 2p_f^{(n-1)}) = \tfrac{1}{3}(p_m^{(0)} + 2p_f^{(0)}) = p_e,$$

where p_e is the equilibrium frequency. Since

$$p_m^{(n)} - p_f^{(n)} = (-\tfrac{1}{2})^n(p_m^{(0)} - p_f^{(0)}),$$
$$(p_m^{(n)} - p_f^{(n)}) \to 0 \text{ as } n \to \infty$$

with the result that $p_m = p_f$ at equilibrium. Further,

$$p_m^{(n)} = \tfrac{1}{3}(p_m^{(n)} + 2p_f^{(n)}) + \tfrac{2}{3}(p_m^{(n)} - p_f^{(n)}) = p_e + \tfrac{2}{3}(-\tfrac{1}{2})^n(p_m^{(0)} - p_f^{(0)})$$

and

$$p_f^{(n)} = \tfrac{1}{3}(p_m^{(n)} + 2p_f^{(n)}) - \tfrac{1}{3}(p_m^{(n)} - p_f^{(n)}) = p_e - \tfrac{1}{3}(-\tfrac{1}{2})^n(p_m^{(0)} - p_f^{(0)}).$$

If we are given the initial frequencies, we can use these equations to determine the gene frequencies in males and females after n generations of random mating. (See Figure 2.4.)

2.6. *Hardy-Weinberg for two loci.*

Given two linked loci, each with two alleles A_1, A_2 and B_1, B_2, there are four types of gametes A_1B_1, A_1B_2, A_2B_1 and A_2B_2 that can be formed. Let their respective relative population frequencies be x_1, x_2, x_3, and x_4, where $x_1 + x_2 + x_3 + x_4 = 1$.

The gene frequency of A_1 in terms of the gamete frequencies is $p_1 = x_1 + x_2$.

Assuming random mating is equivalent to random union of gametes (could you prove this?) the population frequencies of the genotypes

$$A_1B_1/A_1B_1,\ A_1B_1/A_2B_1, \text{ and } A_1B_2/A_2B_1,$$

for example, are

$$x_1^2,\ 2x_1x_3, \text{ and } 2x_2x_3, \text{ respectively.}$$

Given that the recombination fraction between the two loci is r, the gametic outputs of the genotypes A_1B_1/A_1B_1, A_2B_1/A_2B_2 and A_1B_1/A_2B_2 for example, are, respectively,

$$A_1B_1 \text{ only},\ \tfrac{1}{2}A_2B_1 + \tfrac{1}{2}A_2B_2,\ \tfrac{1}{2}(1-r)A_1B_1 + \tfrac{1}{2}(1-r)A_2B_2 + \tfrac{1}{2}rA_1B_2 + \tfrac{1}{2}rA_2B_1.$$

If generations are discrete and mating is random the contribution of A_1B_1 gametes by an individual of genotype A_1B_2/A_2B_1 to the next generation is, for example,

$$2x_2x_3 \times \tfrac{1}{2}r = rx_2x_3.$$

If mating is random and there is no selection, the gametic contributions of all the ten possible genotypes to the progeny generation can be thus written in the form

$$x_1^2 A_1 B_1 + x_2^2 A_1 B_2 + x_3^3 A_2 B_1 + x_4^2 A_2 B_2 + 2x_1 x_2(\tfrac{1}{2}A_1 B_1 + \tfrac{1}{2}A_1 B_2)$$
$$+ 2x_1 x_3(\tfrac{1}{2}A_1 B_1 + \tfrac{1}{2}A_2 B_1) + 2x_2 x_4(\tfrac{1}{2}A_1 B_2 + \tfrac{1}{2}A_2 B_2)$$
$$+ 2x_3 x_4(\tfrac{1}{2}A_2 B_1 + \tfrac{1}{2}A_2 B_2) + 2x_1 x_4[\tfrac{1}{2}(1-r)A_1 B_1 + \tfrac{1}{2}(1-r)A_2 B_2 + \tfrac{1}{2}rA_1 B_2$$
$$+ \tfrac{1}{2}rA_2 B_1] + 2x_2 x_3[\tfrac{1}{2}rA_1 B_1 + \tfrac{1}{2}rA_2 B_2 + \tfrac{1}{2}(1-r)A_1 B_2 + \tfrac{1}{2}(1-r)A_2 B_1],$$

and hence the equations that show the relationships between gametic frequencies in two successive generations can be written as

$$x_1' = x_1^2 + x_1 x_2 + x_1 x_3 + x_1 x_4(1-r) + r x_2 x_3$$
$$= x_1(x_1 + x_2 + x_3 + x_4) + r(x_2 x_3 - x_1 x_4)$$

or

$$x_1' = x_1 + r(x_2 x_3 - x_1 x_4), \text{ since } x_1 + x_2 + x_3 + x_4 = 1,$$

and

$$x_2' = x_2 - r(x_2 x_3 - x_1 x_4),$$
$$x_3' = x_3 - r(x_2 x_3 - x_1 x_4),$$
$$x_4' = x_4 + r(x_2 x_3 - x_1 x_4).$$

Show that the expression $D = x_2 x_3 - x_1 x_4$ decreases by a factor $(1 - r)$ in each generation.

$$D' = x_2' x_3' - x_1' x_4' = (x_2 - rD)(x_3 - rD) - (x_1 + rD)(x_4 + rD)$$
$$= D - rD(x_2 + x_3 + x_1 + x_4) = (1-r)D.$$

Show that $x_1 = p_1 p_2 - D$, and hence that at equilibrium the gametic frequencies of A_1B_1, A_1B_2, A_2B_1, and A_2B_2 take the form p_1p_2, p_1q_2, q_1p_2, and q_1q_2, respectively, where p_1, p_2, q_1, q_2 are the frequencies of genes A_1, B_1, A_2 and B_2, respectively.

We have

$$p_1 p_2 - x_1 = (x_1 + x_2)(x_1 + x_3) - x_1(x_1 + x_2 + x_3 + x_4) = x_2 x_3 - x_1 x_4 = D,$$

so that

$$x_1 = p_1 p_2 - D,$$

and, similarly,

$$x_2 = p_1 q_2 + D,$$
$$x_3 = q_1 p_2 + D,$$
$$x_4 = q_1 q_2 - D.$$

We have

$$D^{(n)} = (1-r)D^{(n-1)} = (1-r)^n D^{(0)} \to 0 \text{ as } n \to \infty,$$

so that at equilibrium $x_1 = p_1 p_2$, etc. (These results were first obtained by Jennings, 1917; see also Haldane, 1926.)

2.7. *Association between two loci due to recent admixture.*

Two random-mating populations are in equilibrium for two alleles at each of two linked loci. The frequencies of the genes under consideration are different from one population to the other. Show that, if the populations are mixed in the proportions m to $1 - m$, in the mixed population the two loci are no longer independent.

Let the frequencies of alleles A_1 and A_2 at the first locus be p_1 and q_1 in the first population and P_1 and Q_1 in the second. Similarly let p_2, q_2 and P_2, Q_2 be the corresponding frequencies of alleles B_1 and B_2 at the second locus. The frequencies of the four gametes A_1B_1, A_1B_2, A_2B_1, and A_2B_2 in the mixed population are then $mp_1p_2 + (1 - m)P_1P_2$, $mp_1q_2 + (1 - m)P_1Q_2$, $mq_1p_2 + (1 - m)Q_1P_2$, and $mq_1q_2 + (1 - m)Q_1Q_2$, respectively (see previous example). The measure of association between the loci is thus

$$D = [mp_1p_2 + (1 - m)P_1P_2][mq_1q_2 + (1 - m)Q_1Q_2] \\ - [mp_1q_2 + (1 - m)P_1Q_2][mq_1p_2 + (1 - m)Q_1P_2],$$

or

$$D = m(1 - m)(p_1p_2Q_1Q_2 + P_1P_2q_1q_2 - p_1q_2Q_1P_2 - P_1Q_2p_2q_1) \\ = m(1 - m)(p_1Q_1 - P_1q_1)(p_2Q_2 - P_2q_2).$$

Now

$$p_1Q_1 - P_1q_1 = p_1(1 - P_1) - P_1(1 - p_1) = p_1 - P_1,$$

therefore

$$D = m(1 - m)(p_1 - P_1)(p_2 - P_2),$$

which is zero only if $p_1 = P_1$ and $p_2 = P_2$, that is, if either or both of the gene frequencies are the same in the two populations. The two genes are therefore associated in the mixed population and only approach the independent equilibrium asymptotically at a rate $(1 - r)$, where r is the recombination fraction between the loci, as shown in the previous example. This result emphasizes the distinction between linkage (which is association of genes in families), and association of genes in populations. Even unlinked genes may show a population association due to recent admixture because for unlinked genes $r = \frac{1}{2}$. Linked genes are not associated in an equilibrium population.

2.8. *Approach to equilibrium with overlapping generations.*

We may now consider nonoverlapping generations and gene frequencies varying continuously with time. (Moran, 1962). Let $u(t)$, $v(t)$, $w(t)$ be the relative proportions of the genotypes A_1A_1, A_1A_2, and A_2A_2 at time t. Assume that, in the small interval of time dt, a proportion dt of the population dies and is replaced by the offspring of random matings within the remaining part of the population; assume that there is no selection. Prove that the frequencies of genotypes approach the Hardy-Weinberg equilibrium at a rate e^{-t}.

We can write

$$u(t + dt) = u(t)(1 - dt) + dt[u(t) + \tfrac{1}{2}v(t)]^2 + 0(dt)^2,$$

where the first term on the right hand side represents those individuals who have not mated during the time interval dt, the second term represents the contribution to $u(t)$ of those who have mated (at random) during this time interval, and the third term represents the fact that we ignore the possibility of mating twice in the interval.

Rearranging, dividing by dt, and taking the limit as $dt \to 0$, gives

$$\frac{du}{dt} = -u + (u + \tfrac{1}{2}v)^2;$$

similarly,

$$\frac{dv}{dt} = -v + 2(u + \tfrac{1}{2}v)(\tfrac{1}{2}v + w)$$

and

$$\frac{dw}{dt} = -w + (w + \tfrac{1}{2}v)^2.$$

If we write $p(t) = u(t) + \tfrac{1}{2}v(t)$ = the gene frequency of A, at time t, then

$$\frac{dp}{dt} = \frac{du}{dt} + \tfrac{1}{2}\frac{dv}{dt} = -(u + \tfrac{1}{2}v) + (u + \tfrac{1}{2}v)(u + \tfrac{1}{2}v + \tfrac{1}{2}v + w) = 0$$

since $u + v + w = 1$, so that $p(t)$ = constant = p_0, say, the gene frequency at time $t = 0$.

Therefore,

$$\frac{du}{dt} = -u + p_0^2 \text{ or } \frac{d}{dt}(u - p_0^2) = -(u - p_0^2)$$

so that $u(t) - p_0^2 = [u(0) - p_0^2]e^{-t}$.

Similarly,

$$v(t) - 2p_0q_0 = (v(0) - 2p_0q_0)e^{-t}$$

and

$$w(t) - q_0^2 = (w(0) - q_0^2)e^{-t},$$

where $q_0 = 1 - p_0$.

Note that, by definition, t is measured in generations, since we assume an exponential life time distribution with mean 1. Thus the departure from Hardy-Weinberg equilibrium will decrease by a factor $1/e$ in one generation and $(1/e)^2 \sim 0.1$ in two generations.

3

Deleterious Mutations and the Estimation of Mutation Rates

3.1 General Characteristics of Deleterious Mutations

Most new mutations are deleterious, and therefore selected against.

There is a small probability that an allele, newly arisen at any given locus confers a reproductive advantage over previously existing alleles, as was pointed out before. In an environment that has been stable for a long time, many " new " advantageous alleles will already have been selected for. However, in any finite population it will be a very long time before a significant proportion of all possible mutations at any given locus have arisen often enough to be given the chance to increase, if advantageous. There is always a definite probability that a new advantageous allele will be lost, due to chance fluctuations in its frequency (see Chapter 8 for further discussion). On the average, however, the increase in frequency of a new favorable mutation follows inevitably from the definition of favorable or advantageous, namely, the greater reproductivity of individuals carrying the mutation. In this chapter we discuss the nature and fate of the majority of new mutations, which are deleterious, and which are therefore subject to the opposing pressures of recurrent mutation, which produces them, and selection, which eliminates them.

"Fitness" is the overall measure of natural selection for the mutant, and therefore includes both the effects of viability and fertility.

The biological **fitness**, or overall expectation of progeny of the normal type, w_N, can be taken as a standard with which the fitness w_M of a given mutant phenotype can be compared. The concept of fitness and the factors that determine it will be discussed in much more detail in Chapter 6. For most purposes, when discussing the simpler models of population genetics, we may think of the fitness of an individual as the average number of his progeny that reach reproductive age. Fitness has two main components, namely viability, or the probability of the progeny's survival to maturity, and fertility, or the rate at which progeny are produced, which depends, in part, on the probability of marriage. It is often convenient to think of fitness just in terms of viability, as this greatly simplifies the construction of population-genetics models. Fortunately, it seems that, in general, theoretical results formulated only in terms of a viability differential can often safely be applied.

For a deleterious mutant, w_M is, by definition less then w_N and therefore takes on values in the range 0 to w_M. When $w_M = 0$, the individual of mutant phenotype is incapable of leaving progeny. This may be due to an incapacity to survive, or to reproduce. A mutant gene that is incompatible with survival to reproductive age is called a **lethal**. Individuals carrying early lethals are lost before birth or shortly after it. If the loss is at a very early stage of embryonic development, it might even pass unnoticed (that is, it might occur even before the pregnancy was diagnosed).

If for any reason the frequency of a deleterious mutant gene increases slightly, the mutant gene will eventually be eliminated from the population at a rate that increases, the larger the disadvantage of the mutant—that is, the lower the value of w_M. No data are yet available on the frequency distribution of w_M values of newly produced mutations.

It is perhaps surprising to find that in spite of their being deleterious the total number of such mutant genes in a population is considerable. Although at any given locus deleterious alleles occur at very low frequencies there are so many loci that when the frequencies are summed over all loci, it is found that each individual on the average carries more than one such deleterious recessive allele in a heterozygous state (see Chapter 7).

Many deleterious mutations affect the synthesis or function of a protein, usually an enzyme, either preventing its synthesis or rendering it nonfunctional, or at least only partly functional.

The amount of protein synthesized by *one* normal allele is usually ample for the normal functioning of a diploid individual. With deleterious mutations affecting loci for which this is true, the defect is not generally expressed in the heterozygote, whose fitness thus remains equal, or almost equal, to that of the normal homozygote. If we let the fitness of the latter be one, that of the heterozygote will also be one, or very nearly so, while that of the mutant homozygote will be less than one. Such a mutation is called a **deleterious recessive**.

It should be emphasized, however, that recessivity for fitness is not necessarily the same as recessivity for an overt phenotypic effect of a gene. A mutation to an allele that in the homozygous state causes phenylketonuria, occurs at the locus controlling the synthesis of an enzyme that, apparently transforms phenylalanine into tyrosine (phenylalanine hydroxylase). Homozygosity for the deleterious mutant gene renders the enzyme inactive, resulting in an excess of phenylalanine,

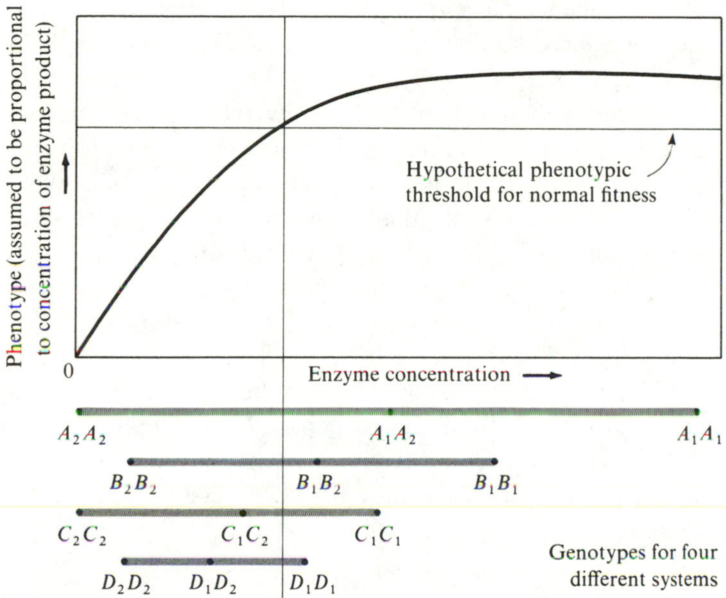

FIGURE 3.1
Effect of gene dosage on enzyme and on enzyme-product concentration. It is assumed that the concentration of enzyme is the sum of the effects of two alleles, and is thus linear with gene dosage. The concentration of the enzyme product, on which the phenotype depends, follows a law of diminishing returns with respect to the enzyme concentration in the range of higher enzyme concentrations. Phenotypic and fitness effects are described for some types of genes. The four segments indicate enzyme concentrations in the three genotypes for four different hypothetical two-allelic loci, assuming that the enzyme production is proportional to gene dosage. According to the figure, A_2 and B_2 will be recessive alleles, and C_2 and D_2 dominant, because for the first two genotypes the enzyme concentrations of the heterozygotes are above the threshold for normal fitness and for the other two, are below it. The A and B genes represent common types of situations, as indicated by the data given in Table 3.1. There is no need to have complete disappearance of function in the mutant (see, for example, the effects with the B and D genes).

which is in part transformed to phenylpyruvic acid and excreted in the urine, but which also accumulates in the cells of the brain and, if enough accumulates, seriously impairs their metabolism. Permanent brain damage ensues if the accumulation of phenylalanine is not prevented, for example by a phenylalanine-deficient diet, at the appropriate stage of life. In the heterozygote, however, the single normal allele manufactures enough normal enzyme to cope with the required rate of conversion of phenylalanine to tyrosine. If an excess of phenylalanine is given to heterozygotes (a loading test), they may be unable to dispose of it all quickly enough, resulting in a high level of phenylalanine in the blood, which can easily be detected. Heterozygotes can thus be specifically identified, under appropriate conditions, although they do not necessarily have impaired fitness.

The rationale of the relationship of gene dosage to effect was clearly understood by Haldane some forty years ago (see Haldane, 1954; and Goldschmidt 1927, 1958). It is not the amount of an enzyme present in an individual that is crucial, but that of the product of the reaction catalyzed by the enzyme. The concentration of the product is proportional to the concentration of the enzyme over a considerable range of enzyme concentrations. But when the enzyme is in excess, then the enzyme product follows a law of diminishing returns. In simple cases, the mathematical form of their relationship is accurately predictable, and is roughly described in Figure 3.1. The concentration of the enzyme in the body may well be directly proportional to the concentration of genes, as has been shown to be true in some instances. Thus, the enzyme concentration in a heterozygote for a normal and a nonfunctional allele, will be about half that in the normal homozygote (for example, in phenylketonuria—see also Chapter 9). In Figure 3.1 the postulated gene and enzyme concentrations in the three genotypes, and the phenotypic effects in terms of enzyme-product concentration are described for a number of different hypothetical mutations.

Biochemical knowledge of genetic differences is increasing rapidly.

The prediction of the relation between gene dosage and amount of product by Goldschmidt and Haldane was little more than an informed guess. Even earlier, at the beginning of this century, Garrod (1902) had correctly interpreted the biochemical nature of human "inborn errors of metabolism," on the basis of a selected number of examples. The detailed clarification of the abnormalities involved was to come much later and the availability of knowledge about them is now in an explosive phase, as is apparent from the list of genetically determined enzymatic deficiencies given in Table 3.1, where traits and diseases caused by the deficiencies are listed, along with the year of their discovery, up to 1967. It is clear, from the rapid increase of knowledge in this area, that a revolution in medicine is taking place. A large number of abnormalities that were previously described merely as rare, probably inherited defects, are now being precisely described, each in terms of the particular biochemical step affected by the given mutational change. Such descriptions assist both in the development of therapy, and in mapping the human chromosomes (see Appendix II). The number of diseases and conditions still to be described in biochemical terms remains, however, extremely large.

Almost all of the characters listed in Table 3.1 are recessive and very rare, in agreement with expectation. Even though the performance of heterozygotes may be normal or almost normal from a clinical and physiological point of view, loading tests such as those described for phenylketonuria or direct, accurate estimations of enzyme concentrations can often distinguish heterozygotes from normal homozygotes. The table indicates for which of these "recessive" conditions heterozygotes are detectable by means of appropriate laboratory tests.

Partially dominant mutant alleles may appear at loci where the product of a single gene is not enough for normal functioning.

For a number of loci, the amount or quality of protein manufactured by a single gene dose or, more generally, the activity of a single normal gene, is not adequate for the normal functioning of a diploid individual. The heterozygote's fitness is then usually lower than that of the normal homozygote and the mutant homozygote may be very seriously affected. A deleterious mutant gene at such a locus is called a **deleterious dominant**. The mutant gene is exposed to natural selection in the heterozygote immediately after it has arisen. As we shall see later, it will therefore be more easily prevented from reaching frequencies as high as those accessible to deleterious recessives.

McKusick in cataloguing the known human genetic variations at defined loci listed 793 autosomal dominants, 629 autosomal recessives, and 123 sex-linked genes (McKusick, 1968). The frequency of sex-linked genes is perhaps a little higher than the expectation, which is 5 percent on the basis of the length of the X chromosome relative to the lengths of the others. The dominant or recessive modes of inheritance are not clearly determined for all abnormalities known to be genetic in origin, and for some the classification is arbitrary or uncertain. By far the majority of these inherited conditions are diseases, most of them rare and poorly understood. Only for about one out of five of these is some physiological or biochemical knowledge available.

Chromosomal aberrations are effectively dominant mutations.

Many chromosomal aberrations also belong to the class of deleterious dominants. Though they constitute a different category of mutations from the usual gene mutations, they behave similarly with respect to their formal genetics. As discussed in Chapter 1, the presence in a single or in a triple dose of a large block of genetic

TABLE 3.1

Genetic Conditions in Which the Enzyme Defect Was Known by 1967. AR = autosomal recessive; AD = autosomal dominant; SLR = sex-linked recessive. Catalogue numbers refer to McKusick, 1968.

Condition	Seriousness of disease	Enzyme	No. in catalogue	Date of discovery	Trans-mission	Heterozygote distinguishable[a]
Hypophosphatasia	early lethal	Alkaline phosphatase	2331	1948	AR	yes
Glycogen-storage disease I	usually lethal	Glucose-6-phosphatase	2246	1952	AR	
Acatalasemia	asymptomatic	Catalase	2003	1952	AR	yes
Phenylketonuria	early mental retard.	Phenylalanine hydroxylase	2493	1953	AR	yes
Galactosemia	cataracts, mental retardation	Galactose-1-phosphate uridyl transferase	2230	1956	AR	yes
Hemolytic anemia	usually mild	Glucose-6-phosphate dehydrogenase	3053	1956	SLR	yes
Goitrous cretinism	retardation	Deiodinase	2597	1956	AR	
Glycogen-storage disease III	usually lethal	Amylo-1,6-glucosidase(debrancher)	2248	1957	AR	
Familial nonhemolytic jaundice	not always lethal	Glucuronyl transferase	2142	1957	AR	yes
Adrenogenital syndrome I	mild and severe forms	21-hydroxylase	2014	1958	AR	
Alkaptonuria	usually none	Homogentisic acid oxidase	2029	1958	AR	
Glycogen-storage disease VI	mild disease	Phosphorylase (liver)	3054, 2051	1959	SLR, AR	
Glycogen-storage disease V	mild disease	Phosphorylase (muscle)	2250	1959	AR	
Methemoglobinemia	mental def.	NAD diaphorase	2405	1959	AR	
Hemolytic anemia	mild and severe	Pyruvate kinase	2528	1961	AR	
Orotic aciduria	anemia	Oritodine 5′-phosphate pyrophosphorylase Oritodine 5′-phosphate decarboxylase	2470	1961	AR	
Fructose intolerance	not serious	Fructose-1-phosphate aldolase	2226	1961	AR	
Histidinemia	mental retard.	Histidase	2282	1962	AR	
Fructosuria	asymptomatic	Fructokinase	2227	1962	AR	
Hyperammonemia I	mental deterioration	Ornithine transcarbamylase	2293	1962	AR	
Suxamethonium sensitivity	apnea in anesthesia	Pseudocholinesterase	2578	1962	AR	yes

Hemolytic anemia	mild	Glutathione reductase	2243	1963	AR	
Glycogen-storage disease II	early lethal	Alpha-1,4-glucosidase	2247	1963	AR	
Maple-syrup-urine disease	retardation	Branched-chain ketoacid decarboxylase	2386	1963	AR	
Xanthinuria	mild	Xanthine oxidase	2626	1964	AR	
Homocystinuria	mild mental retard.	Cystathionine synthetase	2285	1964	AR	
Argininosuccinicaciduria	mental, phys. retard.	Argininosuccinase	2065	1964	AR	
Citrullinuria	mental retard.	Argininosuccinic acid synthetase	2118	1964	AR	
Hemolytic anemia	mild disease	Triose-phosphate-isomerase	2604	1964	AR	yes
Hemolytic anemia	severe	2,3-diphosphoglycerate mutase	2176	1965	AR	yes
Hemolytic anemia	mild	Glutathione synthetase	2242	1965	AR	
Glycogen-storage disease XI	mild	6-phosphofructokinase	2256	1965	AR	yes
Hydroxyprolinemia	mental retard.	Hydroxyproline oxidase	2291	1965	AR	
Hyperprolinemia	mental retard.	Proline oxidase	2316	1965	AR	
Metachromatic leukodystrophy	early lethal	Aryl sulfatase	2400	1965	AR	
Cystathioninuria	mental retard.	Cystathioninase	2149	1965	AR	
Tyrosinemia	early lethal	p-hydroxyphenylpyruvic acid oxidase	2149	1965	AR	
Niemann-Pick disease	mental retard.	Sphingomyelin cleaving enzyme	2455	1966	AR	
Isovaleric acidemia	early lethal	Isovaleric acid CoA dehydrogenase	2343	1966	AR	
Gaucher's disease	complex syndrome	Glucocerebrosidase	2232	1966	AR	
Glycogen-storage disease IV	cirrhosis	Amylo 1,4→1,6 transglucosidase (brancher)	2249	1966	AR	
Hypervalinemia	growth retard.	Valine transaminase	2615	1967	AR	yes
Hemolytic anemia	mild	Hexokinase	2281	1967	AR	
Hyperuricemia	Lesh-Nyhan syndrome, early lethal	Hypoxanthine-guanine phosphoribosyl transferase	3077	1967	SLR	yes
Cataract	juvenile	Galactokinase	2281	1967	AR	yes
Hyperoxaluria	kidney stones	2-oxo-glutarate glyoxalate carboligase	2313	1967	AR	
Central-nervous-system disease with cataracts	mental retard.	Sulfite oxidase	2576	1967	AR	
Methemoglobinemia	no disease	NADP methemoglobin reductase	2404	1967	AR	
Refsum's disease	retinitis and polyneur.	Phytanic acid hydroxylating enzyme	2531	1967	AR	

Source: Based on Childs and de Kaloustian (1968) and McKusick (1968).

[a] Distinguishable by laboratory tests.

material usually causes a gross alteration in the functional balance of many genes, often with striking phenotypic effects. Other chromosomal aberrations such as inversions, which are alterations in the gene order, may have, as shown in experimental animals, phenotypic effects in the heterozygous state and thus behave as dominants with respect to phenotype and fitness.

The persistence of a deleterious mutant phenotype is the inverse of the selection coefficient.

All deleterious mutant genes are lost more-or-less rapidly depending on the severity of the disadvantage they confer upon their carriers. All such mutant genes that are found in a population are therefore likely to be the result of relatively recent mutational events. The average lifetime of a mutation is a simple function of its fitness and can easily be predicted. We define the **selection coefficient** or selective disadvantage of a mutant phenotype to be the loss in fitness relative to that of the standard normal phenotype. If w_N is the fitness of the normal phenotype, and w_M the fitness of the mutant (the heterozygote if the mutant gene is dominant, the homozygote if it is recessive) the **relative fitness** of the mutant is $w_M/w_N = 1 - s$, where s is the selection coefficient, or the relative disadvantage, of the mutant phenotype. When $s = 0$ the mutant phenotype is, of course, equivalent to the normal with respect to fitness. The expected persistence of the phenotype—that is, the average number of generations during which it is present in the population before becoming eliminated—is $1/s$ (see Muller, 1950). This can be seen intuitively through the fact that the probability of elimination in each generation is s and of survival is $1 - s$ (see also Example 3.2 at the end of this chapter).

When several mutants of the same phenotype are found in a population they could all be the result of a single mutation, which occurred at some earlier time. A complete genealogical analysis would be required to establish this and, in addition, to give an estimate of the average persistence $1/s$—doing such an analysis is rarely feasible. Often, however, the same or similar mutations will recur in a population. Recurrent mutation produces a fresh supply of mutant genes, so that the number eventually found in any population will be higher, the higher the mutation rate and the lower the selection against them.

3.2 Models Describing the Balancing Forces of Mutation and Selection

It is possible to predict the level at which a balance will be established between the two opposing forces: mutation creating new types and selection weeding them out. In this and following sections we shall consider the statics and the kinetics of such opposing systems of forces and in addition how the results obtained can be used to estimate mutation rates. For the construction of models, we assume a Hardy-Weinberg equilibrium. This is not unreasonable because random mating is in general a good approximation to reality and selection usually takes place after zygotes have been formed and before the new individuals that develop from them can participate in mating. Most deleterious mutations in fact increase the likelihood that the new individual who receives them at conception will die during the prenatal or the postnatal stage, and in any case before the onset of reproductive maturity. Other deleterious mutations presumably lower fitness by decreasing the probability of marriage. The simplified sequence of events is therefore

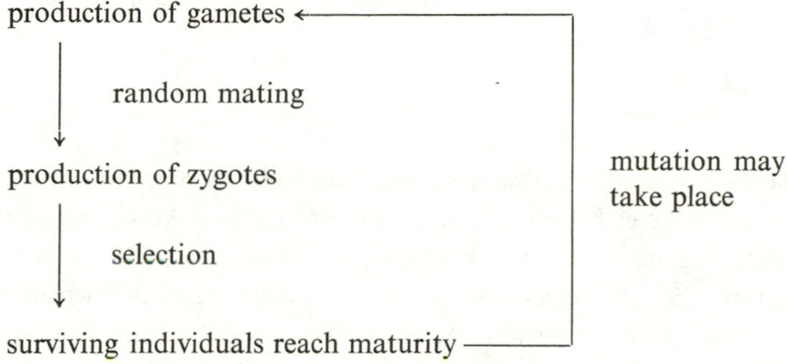

BALANCE BETWEEN MUTATION AND SELECTION FOR AUTOSOMAL GENES

We shall describe models for three specific sorts of mutation, according to the kinds of alleles they produce: (1) deleterious recessives, (2) deleterious dominants, and (3) partially dominant deleterious recessives. For each model we assume a Hardy-Weinberg equilibrium and let μ be the rate of mutation of the normal allele A or a to the mutant form a or A, p the frequency of allele A, and $q = 1 - p$, the frequency of allele a, and s a selection coefficient to be defined. When the mutant allele is recessive, we refer to it as a, and when dominant as A.

At equilibrium, when mutation balances selection, the gene frequency of a deleterious recessive is $\sqrt{\mu/s}$.

The loss per generation is calculated by comparing the gene frequencies in successive generations. For a deleterious recessive we assume the set of fitnesses given in Table 3.2. (Note the distinction between absolute and relative fitnesses.) On the basis of the Hardy-Weinberg equilibrium, we can write the frequencies of genotypes before and after selection, as given in Table 3.2. The fitness of the heterozygote is, by definition, equal to that of the normal homozygote, so that only the homozygous recessive has a reduced fitness.

TABLE 3.2
Selection Against Deleterious Recessives

	Fitness		Frequencies	
Genotype	Absolute	Relative	Before selection	After selection
AA	w_{AA}	1	p^2	p^2
Aa	w_{Aa}	$\frac{w_{Aa}}{w_{AA}}=1$	$2pq$	$2pq$
aa	w_{aa}	$\frac{w_{aa}}{w_{AA}}=1-s$	q^2	$q^2(1-s)$
Total			1	$1-sq^2$

The relative frequencies of the genotypes after selection are obtained simply by multiplying frequencies before selection by their fitnesses. According to the model, only the genotype *aa* is reduced in fitness, by an amount sq^2. The affected individuals can be thought of as falling into two categories: a fraction $1 - s$ that contributes genes to the next generation, and a fraction s that does not. Thus a fraction sq^2 of *aa* individuals, and thus the same fraction of *a* genes, is lost every generation. This follows from the fact that the gene frequency of *a* is the frequency of homozygotes $aa + \frac{1}{2}$ the frequency of heterozygotes. (Here we ignore the factor $1/(1 - sq^2)$ needed to make the frequencies after selection add up to 1, since q^2 is very small. The exact formula is given in the footnote to Table 3.4 and derived in Example 3.2.)

For an equilibrium in which the gene frequency remains constant, the loss of sq^2 per generation must be exactly compensated by the frequency of fresh mutations arising every generation. The frequency of new *a* mutant alleles is $p\mu$ since by the definition of the mutation rate μ, of the proportion p of A genes in each generation a fraction μ mutate to *a*. However, since p is very near 1, and hence q is very small,

$$\mu p = \mu(1 - q) \sim \mu,$$

neglecting the product μq; that is, neglecting the fact that not quite all the genes are A. Mutation from *a* to A is also not considered. Thus, if the gene frequency q is approximately such that

$$sq^2 = \mu, \tag{3.1}$$

the *a* genes lost per generation by selection will equal those newly arisen by mutation, and the population will be at an equilibrium. The gene frequency q_e at equilibrium is therefore given from Equation 3.1, as

$$q_e = \sqrt{\frac{\mu}{s}}. \tag{3.2}$$

(This result was, as far as we know, first derived by Haldane in 1927.) For a recessive lethal, namely a gene for which $s = 1$ and $w_{aa} = 0$, $q_e = \mu^{1/2}$. Thus with mutation rates of the order of 10^{-6}, the frequency of a lethal gene can increase only to 0.1 percent, while a less deleterious mutant gene may become more common.

Since albinism, as noted before, occurs with a frequency of about 1/20,000, if q is the frequency of the mutant albino gene, $q^2 = 1/20{,}000$. If we assume that carriers of the albino gene have normal fitness, while albinos have a relative fitness of 0.9, then from Equation 3.1 we can estimate the mutation rate to the albino gene, on the assumption of equilibrium, by

$$\mu = sq^2 = 0.1 \times \frac{1}{20{,}000} = 5 \times 10^{-6}.$$

Alternatively, if we assume a mutation rate of, for example, 10^{-5}, we can estimate the selective disadvantage of the albinos by

$$s = \frac{\mu}{q^2} = 10^{-5} \times 20{,}000 = 0.2.$$

A dominant deleterious mutant gene has an equilibrium frequency μ/s.

The dominant deleterious mutant gene is exposed to selection in the heterozygote, thus preventing any appreciable increase in numbers, so that the formation of mutant *AA* homozygotes is very rate. (Here the normal gene is *a*.) Their fitness is therefore largely immaterial and they can be safely neglected. Even if they are fitter than the normal homozygote, there is still little or no chance that the *A* gene will increase in frequency, so long as the heterozygote is less fit than the normal.

The fitnesses and frequencies before and after selection are given in Table 3.3 using the approximation that when q is nearly equal to 1, and hence p is very small, p^2 can be neglected in comparison with p. The loss of *Aa* individuals is approximately $2ps$. Since only half of their genes are *A*, the loss of *A* genes is $\frac{1}{2}(2ps) = ps$. This is now the quantity to be equated to the frequency of new *a* genes arising by

TABLE 3.3
Selection Against Deleterious Dominants

Genotype	*Fitness*	*Frequencies* — *Before selection*	*Frequencies* — *After selection*
aa	1	$q^2 \sim 1 - 2p$	$1 - 2p$
Aa	$1 - s$	$2pq \sim 2p$	$(1 - s)\ 2p$
Total		1	$1 - 2ps$

mutation, which, as before, is approximately μ. We therefore obtain the equilibrium condition

$$ps = \mu, \tag{3.3}$$

giving the equilibrium frequency of A

$$p_e = \frac{\mu}{s}. \tag{3.4}$$

This is of the order of the square of the equilibrium frequency for deleterious recessives. For a dominant lethal gene for which $s = 1$, $p_e = \mu$, the mutation rate. This is as expected, since no new gene survives to the next generation.

The equilibrium frequency of a partially dominant deleterious recessive gene is dominated by the heterozygote fitness.

Finally, we consider the model in which the heterozygote has only a small disadvantage compared with the fitness of the normal homozygote, although the mutant homozygote may be at a severe disadvantage. We assume that the mutant homozygote has a fitness $1 - t$ and the mutant heterozygote a fitness $1 - ht$. When $h = 0$ this reduces to the model for a deleterious recessive and when $h = 1$ it is the same as a deleterious dominant (with, however, A and a interchanged). If h is very small, then there is only a slight decrease in the fitness of the heterozygote. The consequence of one generation of selection is summarized in Table 3.4.

Consider, again, as an example, albinism. Suppose now that the heterozygote

TABLE 3.4
Selection Against a Partially Dominant Deleterious Recessive Gene

Genotype	Fitness	Frequencies: Before selection	Frequencies: After selection	Approximate relative loss of *a* genes due to selection
AA	1	p^2	p^2	
Aa	$1 - ht$	$2pq$	$2pq(1 - ht)$	$(\frac{1}{2})2pqht \sim qht$
aa	$1 - t$	q^2	$q^2(1 - t)$	tq^2
Total		1	$1 - 2pqht - q^2t$	$pqht + tq^2$

Note: The loss of heterozygotes as given neglects a term in q^2 which arises from the fact that the heterozygote frequency after selection is actually $[2pq(1 - ht)]/(1 - 2pqht - q^2t)$, so that the loss of a genes from heterozygotes is

$$pq - \frac{pq(1 - ht)}{1 - 2pqht - q^2t} = \frac{qht - 3q^2ht + [\text{terms in } q^3 \text{ and } q^4]}{1 - 2qht - q^2t(1 - 2h)}$$

To this same order of approximation, mutation and selection balance each other when $\mu = tq^2 + qht - q^2ht(2 - ht)$ (see Example 3.3 and Kimura, 1961).

carriers have a fitness of 0.99. The loss of albino genes due to the heterozygote disadvantage is

$$qht = \sqrt{\frac{1}{20{,}000}} \times 0.01 = 7.1 \times 10^{-5}.$$

This may be compared with a loss of 5×10^{-6} from the homozygous albinos, assuming that their fitness is 0.9. The loss from the heterozygotes is thus more than ten times higher than that from the homozygotes.

It is possible to derive exact equilibrium gene frequencies for partially dominant deleterious genes.

The treatment given in Table 3.4 is approximate. The exact equations of change in gene frequencies for one generation are given in Example 3.3 (see also Kimura, 1961). From these equations, the equilibrium gene frequencies have been obtained and are tabulated in Table 3.5 for a set of μ, h, and t values.

TABLE 3.5

Equilibrium Gene Frequencies (in Percentages) for Mutation-Selection Balance as a Function of the Mutation Rate μ, with Selection t Against the Homozygote and ht Against the Heterozygote

		h									
	t	0	0.001	0.003	0.01	0.03	0.1	0.2	0.3	0.5	1.0
$\mu = 10^{-4}$											
	1.0	1.00	0.952	0.863	0.621	0.304	0.099	0.050	0.033	0.020	0.010
	0.5	1.41	1.366	1.276	1.007	0.566	0.197	0.100	0.067	0.040	0.020
	0.2	2.24	2.189	2.097	1.805	1.209	0.481	0.248	0.166	0.100	0.050
	0.1	3.16	3.116	3.024	2.725	2.035	0.931	0.493	0.332	0.200	0.100
	0.05	4.47	4.427	4.337	4.036	3.285	1.756	0.972	0.661	0.400	0.200
$\mu = 10^{-5}$											
	1.0	0.316	0.270	0.200	0.092	0.033	0.010	0.005	0.003	0.002	0.001
	0.5	0.447	0.400	0.322	0.171	0.065	0.020	0.010	0.007	0.004	0.002
	0.2	0.707	0.659	0.574	0.368	0.159	0.050	0.025	0.017	0.010	0.005
	0.1	1.000	0.952	0.863	0.621	0.304	0.099	0.050	0.033	0.020	0.010
	0.05	1.414	1.366	1.275	1.007	0.566	0.197	0.100	0.067	0.040	0.020
$\mu = 10^{-6}$											
	1.0	0.100	0.062	0.030	0.010	0.0033	0.0010	0.0005	0.0003	0.0002	0.0001
	0.5	0.141	0.100	0.056	0.020	0.0067	0.0020	0.0010	0.0007	0.0004	0.0002
	0.2	0.224	0.179	0.119	0.048	0.0166	0.0050	0.0025	0.0017	0.0010	0.0005
	0.1	0.316	0.270	0.200	0.092	0.0330	0.0100	0.0050	0.0033	0.0020	0.0010
	0.05	0.446	0.400	0.322	0.171	0.0653	0.0200	0.0100	0.0067	0.0040	0.0020

Note: Computations are based on the equations given in Example 3.3. The dotted lines indicate the position of the critical values discussed in the text, which can be used to separate "recessive" (at the left) from "dominant" (at the right) deleterious mutants.

It can be seen that even relatively small h values, for example of the order of 0.01 can have a marked effect on the equilibrium value, the more so, the larger the t value and the smaller the mutation rate. If the mutation rate is inferred from the equilibrium gene frequency for a lethal recessive gene ($t = 1$), the error can be very serious. Thus, an equilibrium gene frequency of 0.3 percent (about 1/100,000 defectives) can correspond (see Table 3.5) to a mutation rate of 10^{-5} if the heterozygote has a fitness identical to that of the normal, but to a mutation rate ten times higher if the heterozygote has a fitness only 3 percent below that of the normal homozygote.

For a better understanding of the relative importance of the homozygote and heterozygote disadvantages, part of the data of Table 3.5 are plotted in Figure 3.2. Three curves are shown, for a chosen set of μ and t values, giving as abscissa the value of h and as ordinate the equilibrium values that would obtain if only homozygote disadvantage mattered (descending straight lines at right). The actual values

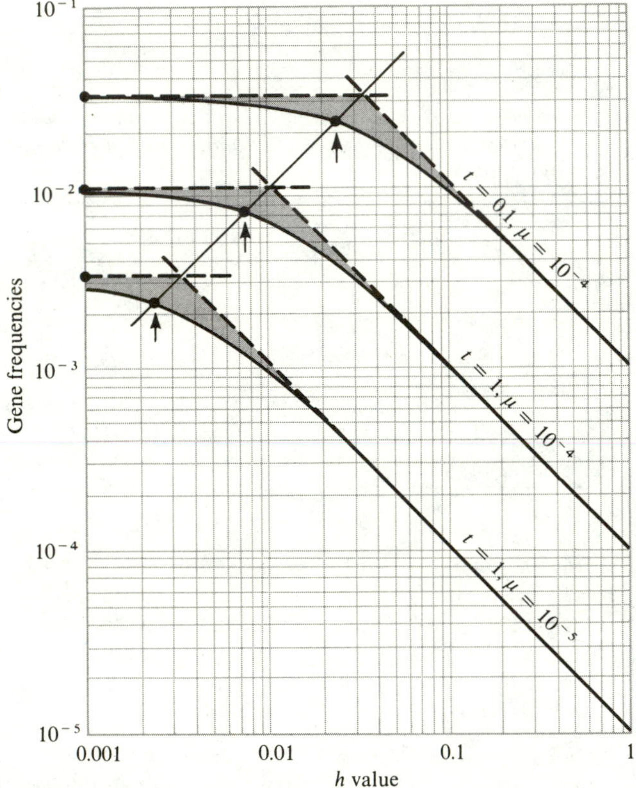

FIGURE 3.2
Equilibrium in gene frequencies (ordinates) for a partially deleterious dominant mutation as a function of the h value (abscissa). The continuous curves show the actual values. The arrows indicate critical values of h, such that it is advisable to consider the deleterious mutant genes to the left of them as recessives and those to the right as dominants.

shown by the continuous curves are the combination of the two effects. The points at which the two dotted lines meet indicate where the joint effect is largest. The shaded area indicates the deviation from the two hypotheses given by the two dotted lines.

It is useful to consider a hypothetical "critical value" of h, such that lower actual values may be taken to indicate that the deleterious mutation is recessive, and higher values, that it is dominant.

One meaningful way of choosing this value of h is that of taking it as the h value for which the loss of mutant genes from homozygotes equals that from heterozygotes. The former is q^2t, and the latter qht (more exactly, $pqht$; see Table 3.4). The two will be approximately equal at equilibrium when

$$q_e^2 t = q_e ht \text{ or } q_e = h, \tag{3.5}$$

that is, when h is equal to the equilibrium frequency. This value of h can be obtained in Figure 3.2 by drawing a straight line at a 45° angle from the point at which the two dotted lines meet and taking q and h where this straight line meets the continuous curve. It is indicated by arrows in the figure. The exact value of h would be very slightly greater than this.

The positions of these critical values are also indicated by the dotted segments in Table 3.4. Unless h is very small (of the order of 0.01 or less) the deleterious mutant genes can be considered as "dominants" and their equilibrium will be dictated by the heterozygote disadvantage. Unfortunately, small values of h are very difficult to measure. Thus, the selection patterns of "recessives," or near-recessives are very difficult to study from the point of view of their equilibria.

The discussion given so far for recessives does not consider the effect of inbreeding. Further consideration of this subject will therefore be made when inbreeding effects are discussed (Chapter 7).

Danforth in 1921 was the first to consider the case of mutation-selection balance for a dominant deleterious mutant gene. Haldane in 1927, in one of the famous papers he published in *The Proceedings of the Cambridge Philosophical Society*, first, to our knowledge, gave a fairly complete treatment of this problem.

3.3 Kinetics of the Approach to Equilibrium

In this section, we shall investigate the speed with which gene frequencies approach the equilibrium values determined by the balance between mutation and selection.

Mutation pressure changes gene frequencies very slowly.

We may consider first the change in gene frequency due to mutation alone. Assuming only mutation from A to a at a rate μ per generation, the A gene frequency in generation $n + 1$, p_{n+1}, is given in terms of the frequency in the previous generation n, p_n, by

$$p_{n+1} = p_n - \mu p_n = p_n(1 - \mu). \tag{3.6}$$

since, as before, there are p_n genes that can mutate from A to a and of these a proportion μ mutate per generation, so that p_n decreases by the amount μp_n per generation. The ratio of the frequencies in two successive generations is given by

$$\frac{p_{n+1}}{p_n} = 1 - \mu,$$

so that

$$\frac{p_{n+1}}{p_{n-1}} = \frac{p_{n+1}}{p_n}\frac{p_n}{p_{n-1}} = (1 - \mu)^2, \tag{3.7}$$

and finally, after n generations

$$\frac{p_n}{p_0} = (1 - \mu)^n, \tag{3.8}$$

where p_0 is the initial gene frequency, when $n = 0$. If mutation from A to a is unchecked by selection against a or by back mutation from a to A, it will eventually lead to the complete substitution of gene A by a, although at a very slow rate. When there is selection against a the process will, of course, be stopped long before that end point.

It is convenient to assume that at the beginning there are no mutant genes in the population, and that the gene frequency in the first generation we consider is therefore equal to the mutation rate. Such a situation would not be applicable to a population whose size N is smaller than the inverse of the mutation rate, for then several generations might elapse before the first mutation appeared (approximately $1/N\mu$ generations, see Chapter 11).

It is characteristic of all processes in which the force changing the gene frequencies in one direction is mutation pressure, that the rate of change is very slow. The magnitude of the change per generation is, as we have seen, of the order of μ, so that the number of generations required to go a reasonable part of the way to equilibrium may be of the order of $1/\mu$, for example one hundred thousand to one million generations for usual mutation rates. (When μ is small, $(1 - \mu)^n \sim e^{-n\mu}$ so that the gene frequency p_n changes by a factor $1/e \sim 37$ percent in $n = 1/\mu$ generations.) The slow rate of change due to mutation is compensated by the fact that mutation rates probably rarely change, so that the process continues more-or-less indefinitely and the time available for approaching an equilibrium may therefore be very long.

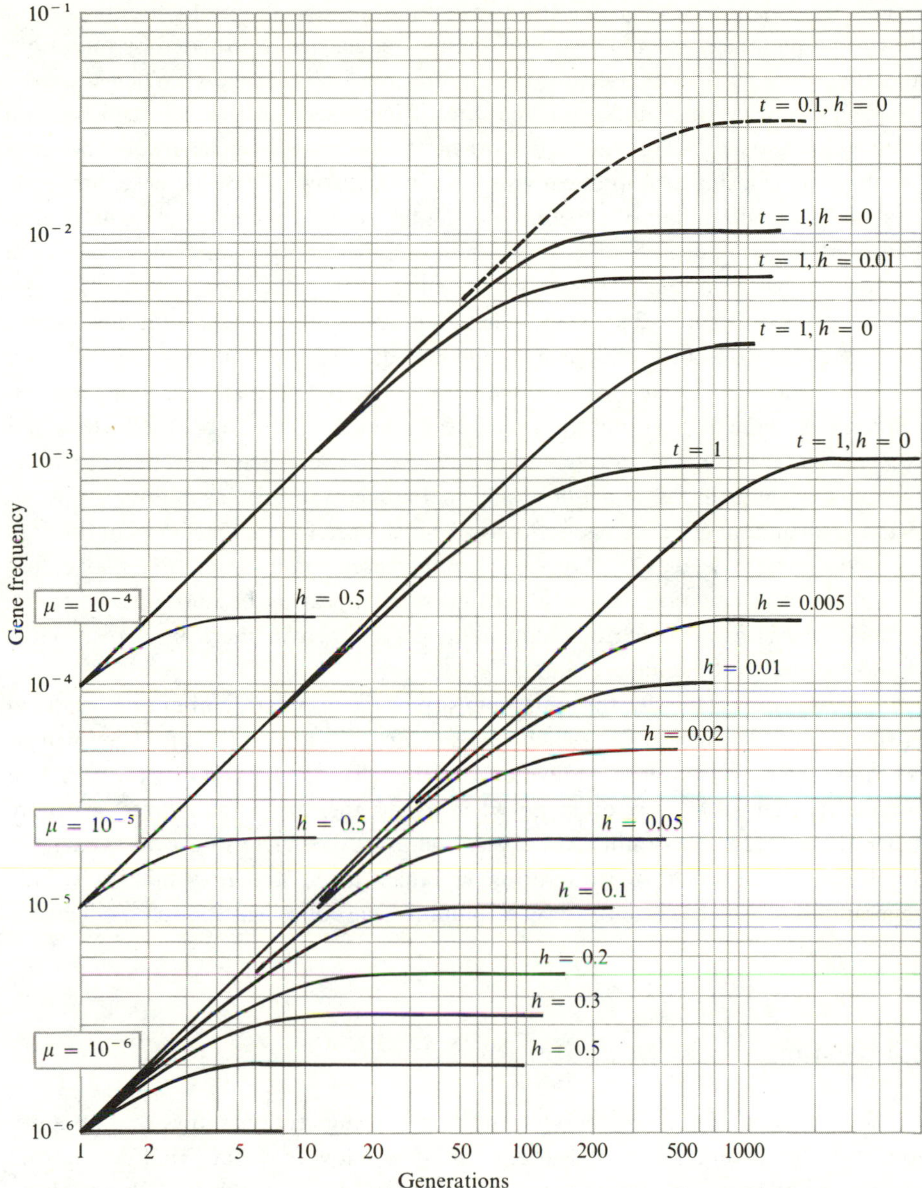

FIGURE 3.3
Kinetics of accumulation of deleterious mutant genes starting from an initial gene frequency equal to the mutation rate. Curves are shown for three values of μ, the mutation rate, and various values of t, the selective disadvantage of the deleterious homozygote, and h, the dominant effect of the deleterious gene (heterozygote viability $1\text{-}ht$). Note the large effect of small differences in the value of h. The larger h the sooner equilibrium is reached.

Some numerical examples illustrating the approach to equilibrium for genes subject to the balance between selection and mutation are shown in Figure 3.3. In these examples the pressure toward the higher frequencies is provided by mutation, and thus the rate of increase towards equilibrium is rather slow. When the gene frequency is near the equilibrium, the rate of increase is further reduced because selection acts as a brake and slows down the increase in gene frequency until eventually it vanishes asymptotically. In spite of the slow rate of increase, however, a relatively small number of generations may be required to approach equilibrium. This is true when the selection is strong because, then, only a small change in gene frequency from an initial frequency of μ is necessary.

The curves of Figure 3.3 show the kinetics of accumulation of deleterious mutant genes starting from an initial gene frequency equal to the mutation rate. A fully recessive lethal ($t = 1$) always takes a long time to approach equilibrium: 200 generations at the highest mutation rate, 10^{-4}, which is, of course, the fastest, and 2000 at the lowest mutation rate considered, 10^{-6}. The process described is, however, somewhat artificial. It might occur if some special event has deprived a population of all mutant genes at a given locus: for instance, if a population went through a sudden extreme decrease in size, and then increased again. Some time may elapse however, in this situation, before mutations start appearing because only when the population has again reached a sufficiently large size will there be a reasonable chance of again producing the mutant gene.

Problems connected with the real, finite size of the population are not considered in the treatment given so far, which refers to an infinite population. They will be taken up later, in Chapter 8. Kinetic problems also come up when the selection conditions change, as they often do when a new therapy is introduced, or more generally, when the environment changes. Finally, if the deleterious gene is partially dominant, the approach to equilibrium is much faster, and may take only 5–50 generations. This is because the level of mutant-gene concentration to be reached is much lower.

3.4 Estimation of Mutation Rates

The discussion of the three models in the preceding section shows that an autosomal deleterious mutant gene will equilibrate at a low frequency that depends on the pattern of fitness values and the mutation rate. For a condition whose inheritance is known, the mutation rate can sometimes be estimated from the frequency of the condition and its fitness. This approach to the estimation of mutation rates was pioneered by Haldane (1935) and Gunther and Penrose (1935). The frequency of a mutant condition is defined, in what follows, as the frequency after fertilization and before selection. The situation is especially simple if the condition can be recognized at birth and can be assumed to cause no mortality *in utero*, for then the fitness of the phenotype showing the condition may be comparatively easy to determine. Data collected over a relatively long period of time are generally needed, however, as the selective disadvantage of the mutant phenotype may not be due, or not entirely, to early mortality, but also to lowered fertility. For this, the examination of the whole life history of the affected individual from birth until the end of the fertile period is required.

For dominant or fully recessive conditions, the mutation rate can sometimes be estimated from the product of the frequency of the condition at birth m and its selection coefficient s: (ms)/2 for dominants, ms for recessives.

For a dominant condition, assuming equilibrium, full penetrance, and loss of fitness only after birth, we know, from Equation 3.4 and Table 3.2, that the relative frequency of affected individuals m, among all newborns will be given by

$$m = \frac{2\mu}{s}, \tag{3.9}$$

where s is the selective disadvantage of the affected individuals. Consider, for example, the dominant dwarfing trait, chondrodystrophy. This occurs with a frequency of 1/10,000 at birth and affected individuals have an estimated fitness that is 0.2, or 20 percent of the normal. Thus, $m = 1/10{,}000$ with $s = 1 - 0.2 = 0.8$, and we estimate

$$\mu = \frac{1}{2}ms = \frac{1}{2} \times \frac{1}{10{,}000} \times 0.8 = 4 \times 10^{-5}$$

which is a rather high mutation rate.

For a fully recessive condition, namely when the fitness of the heterozygote is equal to that of the normal homozygote, we have from Equation 3.1 and Table 3.2.

$$m = \frac{\mu}{s}, \tag{3.10}$$

where s is the selective disadvantage of the mutant homozygote. For fixed values of μ and s, this gives just half as many affected individuals as there would be for a dominant condition. For phenylketonuria, for example, $m = 1/10{,}000$ and, until therapy became available lately, $s = 1$, giving $\mu = ms = 10^{-4}$, which is an almost unbelievably high mutation rate.

Estimates of recessive mutation rates based on there being a balance between mutation and selection are subject to serious sources of error.

Mutation-rate estimates for recessive traits are subject to some major complications. In the first place, in our discussion thus far the contribution due to inbreeding has been ignored—this will be dealt with in Chapter 7. In the second place, we are seriously hampered by the usual lack of knowledge concerning the fitness of the heterozygote. We already know that if the fitness of the heterozygote is lower than that of the normal homozygote, the gene frequency at equilibrium will be largely determined by this fitness rather than that of the mutant recessive homozygote. If h, the measure of dominance for fitness, is much larger than $(\mu/2t)^{1/2}$ (see Equation 3.5) the equilibrium frequency q_e will depend on ht rather than on the disadvantage, t, of the affected homozygote and will thus be approximately $q_e = \mu/ht$. It should be realized that h, even if it is larger than $(\mu/2t)^{1/2}$, may still be a very small quantity. For example, for a lethal gene ($t = 1$), when $\mu = 2 \times 10^{-6}$ the critical value of h is $(\mu/2t)^{1/2} = 0.001$, so that even when h is 10 times this value, it is still quite small.

Unfortunately, information on the fitness of a heterozygote relative to that of the normal homozygote is not easy to obtain. The phenotypic manifestation of the condition being studied may be completely absent in the heterozygote as far as can be determined, but nevertheless the heterozygote's fitness may be different from that of the normal homozygote. As we have already emphasized, very small fitness differences can have quite substantial effects. On the other hand, even if there is partial manifestation of the mutant phenotype in the heterozygotes, making them readily distinguishable, their fitness may not necessarily be affected. The fitness of the individual who is heterozygous for a deleterious recessive gene should be estimated quite independently from any consideration of its phenotype. The phenotype we see is not, by any means, necessarily positively correlated with the selective response of the mutant gene. Unfortunately, even if the heterozygote can be distinguished from either homozygote, the estimation of its fitness remains a major problem. It may be shown (see Chapter 6) that it requires the study of more than ten thousand matings to be reasonably sure of detecting a selective disadvantage, ht, of 0.01, which is already quite large.

A low fitness of the heterozygote will so decrease the frequency of mutant genes that the mutation rate to a recessive allele will be underestimated using Equation 3.10. For then, instead of $\mu = mt$, we should have $\mu = q_e(ht)$, giving

$$\mu = ht\sqrt{m}, \tag{3.11}$$

where $1 - ht$ is the fitness of the heterozygote, $1 - t$ that of the homozygote and q_e the equilibrium gene frequency, so that $q_e^2 = m$, the frequency of the recessive phenotype. The ratio of the proper estimate of the mutation rate as given by Equation 3.11 to the spurious estimate given by Equation 3.10 is $ht(m)^{1/2}/mt = h/(m)^{1/2}$, which may be quite small depending on the relative magnitude of h and of m. It can, perhaps, be intuitively accepted at this stage, that if the reverse is true, namely if the heterozygote has a higher fitness than the normal homozygote, mutation rates will be overestimated using Equation 3.10. A more rigorous and complete treatment of this possibility will be given later (see Chapter 4). These and other considerations lead us to doubt the validity of all estimates of mutation rates for autosomal recessive alleles based on the assumption that there is an equilibrium between mutation and selection.

Mutation-rate estimates for dominant genes are more reliable, but still subject to many uncertainties.

Estimation of rates of mutation to dominant alleles is clearly more satisfactory than that for recessives. In addition to the indirect method of estimation, based on Equation 3.9, we can estimate μ *directly* for dominant alleles whenever the genotypes of the parents are known. Both parents of the carrier of a new mutant gene, as pointed out earlier, are normal. If there is no mortality due to the condition before birth (or, more generally, before observation) and no lack of penetrance, the observed number of affected individuals, m, will fall into two moieties, a proportion n having normal parents and having the mutant allele because of new mutations and a proportion $1 - n$ having at least one affected parent. As $nm = \mu$, from Equation 3.9, we should have $2n = s$, the selective disadvantage of the affected, which can be estimated quite independently. This internal check provides a test of the validity of the hypotheses used to estimate mutation rates for dominants including, in particular, whether the equilibrium condition has been reached.

A list of traits specified by dominant mutant alleles whose mutation rates have been estimated is given in Table 3.6. They are all probably determined by single

TABLE 3.6
Mutation Rates for Autosomal Dominants in Man

Trait	*Mutation rate per gamete*	*Remarks*
Epiloia	0.8×10^{-5}	One of the earliest measurements
Aniridia	0.5×10^{-5}	Corrected value from Penrose
Microphthalmus	0.5×10^{-5}	
Waardenburg's syndrome	0.4×10^{-5}	
Facioscapular muscular dystrophy	0.5×10^{-5}	Direct
	$<0.5 \times 10^{-6}$	Indirect
Pelger anomaly	0.9×10^{-5}	Indirect
Amyotrophic lateral sclerosis	3×10^{-5}	
Myotonia dystrophica	1.6×10^{-5}	
Myotonia congenita	0.4×10^{-5}	
Huntington's chorea	0.5×10^{-5}	Upper limit
	0.2×10^{-5}	More probable estimate
	1.0×10^{-5}	Indirect, large error
Chondrodystrophy	4×10^{-5}	Probably more than one locus, and overestimates for other reasons also
	6×10^{-5}	
	6×10^{-5}	
Retinoblastoma	1.5×10^{-5}	Probably include phenocopies
	2.3×10^{-5}	
	0.6×10^{-5}	Bilateral cases only
	0.4×10^{-5}	Corrected for phenocopies
Neurofibromatosis	13.25×10^{-5}	Direct
	8.10×10^{-5}	Indirect
Deafness	4.7×10^{-5}	Semidirect. Estimate of all loci causing dominant deafness

Source: From Crow (1961b).

gene mutations, because they are unaccompanied by visible "gross" chromosomal aberrations, and because they segregate regularly in families. The phenotype in the homozygous condition of many of these mutants is not known, because of its rarity, and homozygosity may well be lethal. The estimates are subject to various errors:

1. They may include some cases that are the consequences of somatic, rather than gametic, mutation. (Retinoblastoma, a special form of eye tumor, can be determined by a dominant gene, but some cases are believed to be due to somatic mutation.)
2. Phenocopies—that is, individuals simulating the mutant phenotype as a consequence of environmental disturbances of development—will also tend to inflate the estimate of mutation rate. A broad definition of phenocopy can include changes due to somatic mutation.
3. They may include mutations at several different loci, all of which determine a similar condition. The mutation rate, if taken to be estimated *per locus*, will thus be artificially inflated.
4. They may be disturbed by incomplete penetrance.

A further discussion of the estimates obtained will be given in the final section of this chapter.

3.5 Chromosomal Aberrations

The discovery that many pathological conditions are due to chromosomal abnormalities has stimulated a great deal of work in the field of human cytogenetics. The overall frequency of chromosomal anomalies in unselected adult populations is of the order of 1 percent, about a quarter to a fifth of which involve the X and Y chromosomes (see Chapter 10). It is probably not much higher than this at birth. The frequencies of the most important autosomal anomalies among live births are given in Table 3.7.

TABLE 3.7
The Frequency (in Percentages) of Newborn Subjects with Various Chromosome Abnormalities

Abnormality	*Males*	*Females*
Sex-chromosome aneuploidy	0.20	0.16
Trisomy 21	0.15	0.15
Trisomy 13/15, 17/18	0.07	0.07
Autosomal rearrangements	0.50	0.50
Totals	0.92	0.88

Source: From Court-Brown (1967).

Since chromosomal anomalies behave formally as dominants, their mutation rate can be estimated but may be subject to considerable uncertainty because of in utero *mortality and parental age and sex biases.*

Chromosomal anomalies behave formally as dominants when they have a phenotypic effect. Most of these anomalies have a fitness that is 0 or nearly 0, and therefore $s \sim 1$. Thus, the frequency of affected individuals should be twice the mutation rate estimated from Equation 3.9 unless, as seems likely, there is appreciable *in utero* mortality. Trisomy 21, for example, occurs with an overall frequency of about 1.5/1,000, suggesting an effective mutation rate, namely rate of production of nondisjunctional gametes, of 7.5×10^{-4} per gamete per generation. Estimates based on Equation 3.9 assume an equal mutation rate from both sexes The frequency of trisomy 21 increases, however, some 40-fold as the mother's age increases to the onset of menopause and thus, a large proportion of trisomy 21 originates from the relatively small proportion of zygotes contributed by older mothers (see Section 3.7 and Fig 3.6). The mutation-rate estimate just given must, therefore, refer predominantly to female gametes. Thus, the mutation rate in female gametes must be almost twice the estimate of 7.5×10^{-4} and that in male gametes correspondingly less. The mutation rate will also be materially affected by the distribution of maternal ages in a population. A mutation rate (average for the two sexes) of 1.45×10^{-4} would be obtained for Turner's syndrome (XO—see Chapter 10).

However, it has been found that about 22 percent of the embryos whose development is terminated by early spontaneous abortions contain chromosomal anomalies and that approximately 25 percent of these are Turner's or related syndromes (Carr, 1967; see also Chapter 10). Assuming an overall frequency of spontaneous abortion of 0.25 per fertilized zygote (which may be too low) these data imply that the overall frequency of Turner's syndrome among fertilized zygotes is about $0.25 \times 0.25 \times 0.22 = 1.6$ percent, which is 47 times the observed frequency at birth. This suggests an *in utero* (either early embryonic or fetal) mortality of at least 97–98 percent and leads to an estimate of about 7×10^{-3} for the frequency of nondisjunction leading to Turner's syndrome. This is indeed remarkably high. It emphasizes the pitfalls involved in using data on frequency at birth for mutation-rate estimates and at the same time the risks in considering the observed living Turner's syndrome cases as typical products of an XO chromosomal constitution. This point is further brought out by the fact that no monosomics other than XO and no trisomics for any of the larger chromosomes have been observed. These grosser anomalies are presumably sometimes produced by errors in the meiotic process, but it is likely that they are lethal so early in development, perhaps even before implantation, that there is little, if any, hope of ever obtaining satisfactory estimates of their rate of production.

A further complication in estimating the rate of chromosomal aberrations arises from the fact that in some cases there is reason to believe that an abnormal gamete is contributed by one sex only. This is certainly true, for example, in the production of XYY individuals, which must be formed by the combination of abnormal YY gametes of male origin and a normal X gamete from the female. It is also true, to a large extent, of 21 trisomy, which, as was already mentioned, is known to be largely derived from eggs produced by women near the end of their reproductive period.

When mutation rates are significantly affected by environmental factors, it is clear that they may vary greatly from one sample of the population to another according to variations in the relevant factors. It is worth noting that the frequencies of production of the most-frequent chromosomal aberrations, which reflect the accuracy of the meiotic process, are apparently much higher than those of gene mutations.

3.6 Sex-linked Genes

The mutation rate of X-linked recessives can also be estimated by equating the balancing forces of mutation and selection.

A recessive sex-linked deleterious gene is in the special situation of being protected by recessiveness only in females, and being exposed to selection as a dominant in males. It does not, therefore, increase to relatively high frequencies as a deleterious autosomal recessive gene may. An X chromosome carrying a mutant gene that is completely recessive with respect to fitness in the female X', will have genotype frequencies and selection coefficients as given in Table 3.8 if selection is equal in the two sexes. (X represents the normal, non-mutant-carrying, X chromosome.)

TABLE 3.8
Selection Against an X-linked Recessive

	Females			*Males*	
Genotypes	XX	XX′	X′X′	XY	X′Y
Relative fitnesses	1	1	$1-s$	1	$1-s$
Frequencies	p^2	$2pq$	q^2	p	q

The loss of X′ genes per generation is sq in males and sq^2 in females. Since q is so small that terms involving q^2 can be neglected, only the loss from males (sq) need be considered for equating the selection and mutation pressures in order to calculate the equilibrium frequency.

For every male sex-linked gene there are two female sex-linked genes, so that with the same mutation rate in the two sexes, μ, there will be μ mutant genes per generation in the X chromosomes in males and 2μ in X chromosomes in females. Equating the overall mutation rate 3μ to the loss of X′ genes, sq, gives the equation

$$sq = 3\mu \tag{3.12}$$

for the equilibrium condition. The gene frequency at equilibrium is therefore

$$q_e = \frac{3\mu}{s}, \tag{3.13}$$

which is of the same order of magnitude as the frequency of autosomal dominant deleterious mutant genes. This treatment is originally due to Haldane (1935) who gave one of the first estimates of mutation rates in man—that for hemophilia.

A more general treatment would take account of possible differences in the mutation rates in males and females (which should be more easily detectable for sex-linked genes, if differences exist) and of possible fitness differences between the sexes. A discussion of a more general scheme of selection will be given when we discuss polymorphism for X-linked genes in Chapter 4.

Direct estimates of mutation rates for sex-linked recessives can be obtained.

Most carriers of rare sex-linked characters are males. A male carrier of a rare sex-linked recessive gene is either an isolated case in a family or not. If he is, the mutation may be new in the mother. He could also arise from a heterozygous mother whose family has no record of the mutant gene and who has had no other mutant child. To determine the expected frequency of these two situations, models for the expected distributions of normal and affected sibs per family must be analyzed. A complete analysis requires the building and testing of distributions of progeny size, the estimation of the segregation frequency, which may differ from the expected Mendelian ratio because of viability disturbances, and the estimation of the "ascertainment" probability (the probability of finding existing cases) and its possible heterogeneity. Only then can the frequency of *sporadic cases*, namely those mutants that are not the progeny of a heterozygous mother, be estimated. Sporadic cases include true new mutants, and phenocopies. Further analysis, and the availability of data over several generations may help distinguish these situations. The methods used are given in Appendix II under the heading of "segregation analysis."

Perhaps the best-known deleterious sex-linked recessive allele is the one that causes hemophilia. There are two clinically distinguishable varieties of hemophilia, called A and B, which lack different components of the blood-coagulation system, both of which are sex-linked but which are not allelic, as shown by pedigree linkage studies. Hemophilia A is more serious and more frequent. It has, apparently, a higher mutation rate, $1.3 \pm 0.1 \times 10^{-5}$, while the estimate for hemophilia B is $0.55 \pm 0.04 \times 10^{-6}$ or 22 times smaller (Barrai et al., 1968). Actually hemophilia B is only about four times rarer than hemophilia A, but its selective disadvantage is much smaller, which explains the discrepancy.

A survey of data on sex-linked recessives with approximate mutation rates is given in Table 3.9.

TABLE 3.9

X-Linked Traits: Estimates of Varying Validity of Birth Frequencies and Mutation Rates

No. in catalogue of X-*linked genes* (*McKusick,* 1968)	*Trait*	*Approx. relative reproductive fitness of trait bearers* (f)	*No. of families reported in literature*	*Orders of size of*	
				Birth frequency ($\times 10^{-6}$) (x)	*Mutation rate* ($\times 10^{-6}$) $\mu = \frac{1}{3}(1-f)x$
6	Duchenne type muscular dystrophy	0.0	Very many	200–220	70–90
7	Becker type muscular dystrophy	0.5	Many	9–12	1–2
8	Factor VIII Deficiency (Haemophilia A)[a]	0.6–0.9	Very many	100–120	20–40
9	Factor IX Deficiency (Christmas Disease—Haemophilia B)[a]	0.6–0.9	Very many	20–30	5–10
10	Hypogammaglobulinaemia	0.0	Many	10–15	3–5
11	Hurler's syndrome	0.0	Few	<1.0	<0.1
12	Late spondylo-epiphyseal dysplasia	High	~5	<1.0	<0.1
13	Wiskott-Aldrich syndrome (thrombocytopenia)	0.6–0.8	~10	<1.0	<0.1
14	Hypophosphataemia and Vitamin D-resistant rickets	0.3–0.5	Many	5–15	1–4
15	Hypoparathyroidism (Neonatal)	0.2–0.4	Few	<1.0	<0.1
16	Diabetes insipidus (Nephrogenic)	0.3–0.5	Few	<1.0	<0.1
17	Diabetes insipidus (Neuro-hypophyseal)	0.8–0.9	Few	1–5	~1.0
18	Oculo-cerebro-renal syndrome of Lowe	0	~10	<1.0	<0.1
19	Hypochromic anaemia (Pyridoxine responsive type)	High	Few	<1.0	<0.1
20	Fabry's syndrome (angiokeratosis corporis diffusum)	0.7–0.9	Few	2–5	<1.0
22	Macular bullous dystrophy	Low	1	<0.1	<0.1
23	Keratosis follicularis spinulosa decalvans	High	10–15	<1.0	<0.1
24	Ichthyosis	No reduction detected	Very many	200	~1.0
25	Anhidrotic ectodermal dysplasia	0.6	Many	10	1–5
26 and 27	Amyelogenesis imperfecta	No reduction detected	Many	10	<1.0
29 and 30	Deafness	0.8	Many	15–25	3–6
31	Mental defect (severe not specific)	0.0	Many	40–80	10–20
32	Borjeson's syndrome	?	2	<0.1	<0.1

[a] The figures here given are those of the survey by Stevenson and Kerr. Those cited in the text are from a later survey and analysis.

TABLE 3.9 (*continued*)

No. in catalogue of X-linked genes (McKusick, 1968)	Trait	Approx. relative reproductive fitness of trait bearers (f)	No. of families reported in literature	Orders of size of: Birth frequency ($\times 10^{-6}$) (x)	Orders of size of: Mutation rate $\mu =$ ($\times 10^{-6}$) $\frac{1}{3}(1-f)x$
34	Cerebellar ataxia with extra pyramidal signs	Low	1	< 0.1	< 0.1
35	Spastic paraplegia	?	1	< 0.1	< 0.1
36	Progressive bulbar palsy	Low	3	< 0.1	< 0.1
37	Progressive muscular atrophy	?	3	< 0.1	< 0.1
38 and 39	Pelizaeus-Merzbacher leucodystrophy		Few	< 0.1	< 0.1
40	Hydrocephalus (stenosis aqueduct of Sylvius)	0 +	Few	5–15	1–5
41	Parkinsonism	1	2	< 0.1	< 0.1
42	Ocular albinism	High	Many	10–50	~ 1.0
43	External ophthalmoplegia and myopia	?	1	< 0.1	< 0.1
44	Microphthalmia/anophthalmia	?	Few	< 1.0	< 1.0
45	Microphthalmia/anophthalmia and digital anomalies	?	1	< 0.1	< 0.1
46	Nystagmus	High	Many	10–50	5
47	Megalocornea	?	Few	< 1.0.	< 1.0
48	Hypoplasia of iris with glaucoma	?	1	< 0.1	< 0.1
49	Total cataract	?	Few	< 0.1	< 0.1
50	Cataract with microcornea or slight microphthalmia	?	1	< 0.1	< 0.1
51	Congenital night blindness with myopia	?	Few	< 1.0	< 1.0
52	Choroideremia	High	Many	1–4	< 1.0
53	Retinitis pigmentosa	High	Many	1–5	< 1.0
54	Macular dystrophy (juvenile) and dyschromatopsia	?	Few	<0.1	< 1.0
55	Retinoschisis	?	Few	< 1.0	< 1.0
56	Pseudoglioma	?	Few	< 1.0	< 1.0
57	Van den Bosch syndrome	?	1	< 0.1	< 0.1
58	Menkes' syndrome	?	1	<0.1	< 0.1
Addendum by McKusick	Deafness and albinism	?	1	<0.1	< 0.1
Addendum to McKusick	Imperforate anus	High	Few	~ 10	< 1.0

Source: Adapted from Stevenson and Kerr (1967).

3.7 Parental Age and Mutation

A significant fraction of mutations may be expected to occur at rates that are dependent on parental age.

Mutations in gamete-producing cells give rise to clones that may persist and increase in size during the lifetime of an individual. In males, spermatogenesis, or the formation of male gametes, continues from puberty to old age, though at a declining rate. If there is a constant mutation rate, α, per unit time in the spermatogonia (the gamete-producing stem cells) in the testes of an adult male, the relative frequency, μ, of mutant genes among gametes increases linearly with time according to the equation

$$\mu = \alpha x + \mu_0, \tag{3.14}$$

where μ_0 is the frequency of mutant genes at the beginning of the reproductive period (puberty) and x is the time elapsed since puberty. This assumes that mutated spermatogonia have the same reproduction rate as nonmutated ones (see Example 3.4). Even at times when spermatogonia do not produce gametes, or produce only a few, mutations may still arise and accumulate linearly with time, although they will probably do so at a slower rate.

Oogenesis, the production of female gametes, may show a different picture. New egg cells are not formed after birth. They are almost mature in the ovaries of a newborn female. They become fully mature much later, one by one, when during each menstrual cycle one (rarely more than one) egg is made available for fertilization.

There is thus less opportunity for the accumulation of mutant genes in female than in male gametogenesis. If, however, a mutagenic agent acts on resting gametes and its intensity is constant in time, even in the absence of division of gamete-forming cells, a linear accumulation of mutant genes with age may also be expected in females. Thus, mutation rate may also increase with maternal age, although at a presumably lower rate than for paternal age, because in females there is no possibility of accumulation of mutant genes during the reproductive years as there is in males. This may be a cause of higher mutation rates in males than in females.

The effect of parental age on mutation can be analyzed by comparing the mean age of parents of mutants with those of controls.

The correlation between parental age and mutation can be analyzed in several ways. When the data are scanty, as is often the case with mutants, a direct comparison of the mean age of parents of mutants with that of controls can be made (see Example 3.5). The effect of birth order may be considered in a similar way. Birth order is very highly correlated with parental age and abundant material is required to distinguish the two effects.

The analysis of the effects of parental age on mutation has been carried out essentially only on dominant and recessive sex-linked mutations, as their detection is so much easier. Some work is needed to disentangle paternal from maternal effects, because of the typical close correlation between the ages of husband and wife. The analysis is usually carried out by regression and correlation (see Appendix I). Of the dominant mutations investigated, the most striking effect is found in chondrodystrophy, which shows a high, practically linear increase of mutation rate with age of fathers (Figure 3.4). Age of mother is not involved, as shown by partial correlation studies. Other dominant mutations tested (with the exception of chromosomal aberrations) usually do not show the same pattern, or only to a much smaller degree (see, again, Figure 3.4). There is, however, at least one other clear-cut example of paternal age effect for a dominant mutation, namely acrocephalosyndactyly, or Apert's syndrome, a disease with skull and limb deformities, which,

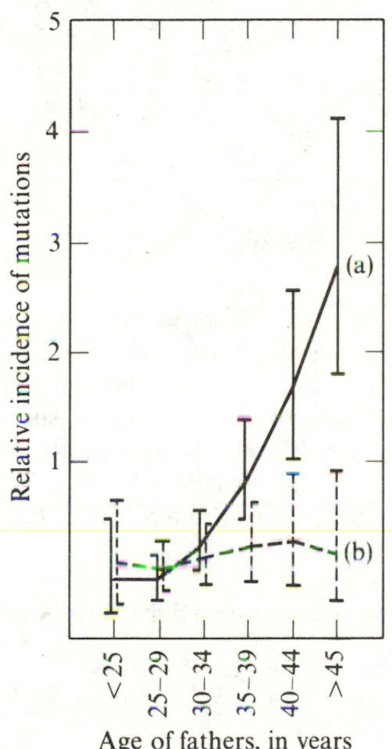

FIGURE 3.4
Relative incidence of dominant mutations among individuals grouped according to their fathers' ages (at birth of offspring) in (a) chondrodystrophy (from a study of 175 cases done by Mørch, Greve, and Stevenson), and in (b) neurofibromatosis, tuberous sclerosis, and osteogenesis imperfect (from a study of 108 cases done by Borberg and Seedorff). Ninety-five percent confidence limits are indicated in the figure by the vertical bars. Relative incidences are based on the Danish age distribution of fathers in 1930 (see Mørch, 1941). These data illustrate a striking effect of father's age on the rate of occurrence of the dominant mutation for chondrodystrophy. The other data are given for comparison. (From Vogel, 1963.)

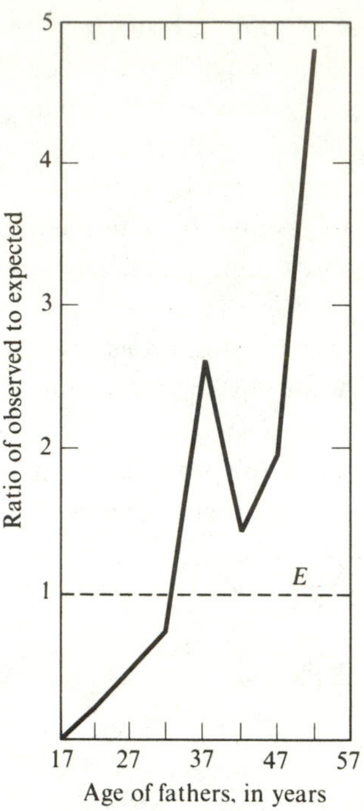

FIGURE 3.5
Relative incidence of Apert's syndrome (acrocephalosyndactyly) among individuals grouped according to their fathers' ages. This graph is based on a study of 37 cases; data on patients and normal controls are quoted from Blank, 1960. The dashed line, marked *E*, indicates the value that would be expected if there were no age dependence. This is another striking example of a dependence of mutation rate on father's age. (From Vogel, 1963.)

like chondrodystrophy, is not accompanied by any visible cytological aberrations (Figure 3.5).

Chromosomal aberrations have anomalous correlations with parental age.

There are, as already mentioned, special effects of parental age for some chromosomal aberrations. We have already noted that in Down's syndrome, mother's age has a marked effect. Women who are approaching menopause have an increasingly higher chance than younger women of having children affected by Down's syndrome. The change in the frequency of such offspring with maternal age is shown in Figure 3.6. Paternal age has been shown to have no effect. The effect of increasing maternal age is far from linear. The hormonal imbalance resulting from approaching menopause may be a factor contributing to the striking increase in incidence of Down's syndrome. Maternal age effects are less striking for other chromosomal

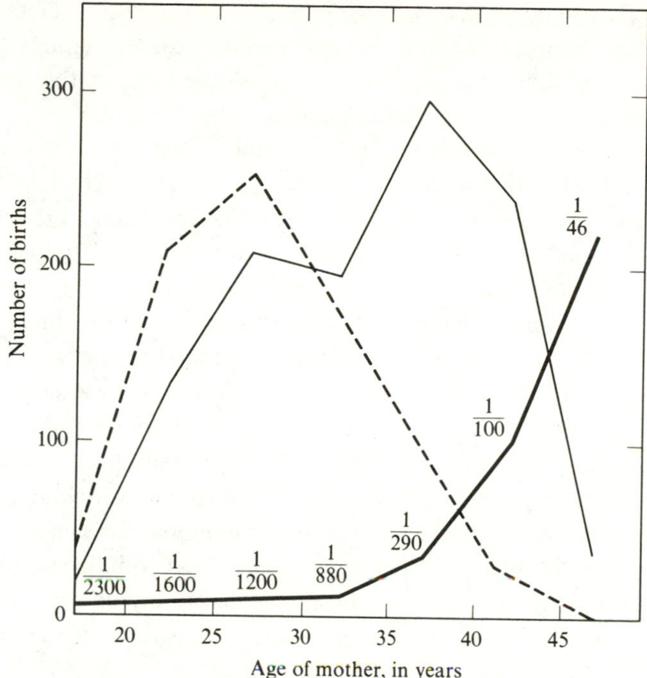

FIGURE 3.6
Incidence of Down's syndrome at birth as a function of mother's age. The dashed line plots all births, in thousands. The lighter solid line plots the number of babies born with Down's syndrome. The darker solid line shows the incidence of Down's syndrome relative to maternal age. (From Penrose and Smith, *Down's Syndrome*, 1966.)

aberrations, but have been noted for trisomy 17 and for XXY (Klinefelter's syndrome).

Special techniques may detect parental age effects for mutations to sex-linked recessive alleles.

Sex-linked recessive mutations pose special problems. Almost all mutant genes first appear in males, though a substantial fraction of these must have originated in the mother's gametes. If there is an accumulation of mutant genes with time, there should be an increase in mutation frequency with mother's age. This might be detected, for example, by an increase in the mean age of mothers of female carriers of sex-linked recessives. No such effect is, however, detectable for two mutations so far tested, namely Duchenne-type muscular dystrophy and hemophilia.

A correlation has been found in some bodies of data between the age of the mother and the sex-ratio in her progeny, as we shall discuss more completely in

Chapter 10. This could, however, largely be due to nongenetic factors. It is also conceivable that nongenetic effects may be involved in correlations between maternal age and mutation rates. An influence of mother's age on the penetrance of a gene in the progeny is not inconceivable, especially in view of the metabolic exchanges between mother and fetus. Thus, mother's age does not lend itself to a satisfactory analysis of this kind.

Using the ages of maternal grandparents (that is, their ages at the birth of their children), is less likely to produce side-effects. Grandparental age is as important for sex-linked mutations as maternal age. In fact, the mutation to a sex-linked recessive gene at equilibrium may have originated with a probability of 1/3 in the mutant's mother, 1/3 in one of the two maternal grandparents and 1/3 in one of the earlier ancestors in the female line. This may, however, be altered if the overall mutation rates are not the same in the two sexes. It should be noted that the analysis of how old the grandparents were at the birth of their children has the advantage of making it possible to analyze the dependence of mutation rate on age in both males and females. Such an analysis has been attempted for hemophilia, and for the sex ratio (see Chapter 10). Negative results were obtained for hemophilia (Barrai et al., 1968) though the available data could not exclude an age dependence (with respect to the maternal grandfather) which was such that the mutation rate doubled over a period of 20 years or more.

In conclusion, effects of parental age on mutation rates have been observed, but with rather puzzling results. Most dominant mutations and the few sex-linked ones examined show very little or no effect. A few dominants show a strong linear effect of paternal age. It is not inconceivable but it is certainly difficult to prove, that some mutations—for example that to the gene that produces chondrodystrophy—confer a slight advantage on the spermatogonia in which they occured, so that selection at the intercellular level occurs among spermatogonia, a phenomenon known as **gametic selection.** Another possible explanation of this very high rate is given in the next section.

Effects of parental age noted in some chromosome aberrations are likely to be the consequence of changes, with age, in physiological conditions of fathers and especially of mothers capable of affecting the frequency of nondisjunction in the gamete-forming cells. We have seen that parental-age effects might bias the mutation rate in males and make it higher than that in females. However, there is at the moment no clear hint of a sex difference in mutation rates in man.

3.8 Average Mutation Rate per Locus per Generation

The estimates of mutation rates that we discussed in Sections 3.4 and 3.6 are meant to apply to a single locus. The rates are measured using a generation as the time unit because mutations are detected in the progeny of the individuals whose mutant gametes are being counted. The validity of the definition *per locus* is somewhat doubtful. Practically all estimates are based on strictly phenotypic consideration. It is not impossible that many mutant phenotypes are composite, representing mutations at more than one locus. This has been shown to be true for the rare phenotype *elliptocytosis*, which is observed as oval-shaped red blood cells. This phenotype is dominant and is not associated with obvious pathological signs. Pedigrees of carriers of this trait show a heterogeneous behavior; in some families the phenotype is definitely linked with Rh blood type, in others it definitely is not (see Appendix II). Thus at least two loci, either on different chromosomes or located fairly far apart on the same chromosome, can determine the same trait. A similar conclusion holds for hemophilia A and B, between which, however, there are various differences that allow them to be distinguished at the clinical level. They are both sex linked but show different linkage with other X chromosome markers (see Chapter 10). There are not enough genes that have been mapped in man to test if this is true for many of the other mutations that have been investigated.

Further uncertainties are to be kept in mind with mutant genes such as the one that produces chondrodystrophy. The mutation to this dominant has a suspiciously high rate and is also highly affected by increasing paternal age; these features are common to some chromosome aberrations. It may be that chondrodystrophy is a chromosome aberration, perhaps a deficiency, too small to be detectable with present cytological techniques.

Early estimates of average mutation rates in man are too high.

In the genetic literature are found lists of mutation rates compiled from the available data. Averages taken from such data, which include rates for autosomal dominants and recessives, and sex-linked recessives, converge to an average estimate of mutation rate per generation per locus of approximately 3×10^{-5}. Some authors, however, have warned that this estimate could be too high. Almost all sources of error that we have discussed before tend to inflate mutation-rate estimates per single locus. Thus the fact that phenocopies, or somatic mutations may be mistaken as true mutations, or the possibility that many similar phenotypes may be due to mutations at more than one locus would account for contributions to an estimate's being too high. Also, incomplete penetrance may have the same consequence. There is, however, a more serious source of error: namely, the fact that the sample of mutation rates considered is very probably biased.

If we compute the average of the most-reliable mutation rate estimates, namely those for the autosomal dominants listed in Table 3.6, we obtain 2.6×10^{-5}. These however, are all estimates that come from nonsystematic surveys in which one mutation at a time was considered.

There is only one systematic survey of mutation rates that has been done so far.

The results of this survey, which was for *X*-linked mutations, were given in Table 3.9. These results should be far less biased by the choice of frequent mutations than any other existing data as they are based, in part, on a survey carried out by screening a well-defined population for known defects chosen in advance. Naturally, mutations too rare to be detected in the sample analyzed, which included approximately one million persons can only be given as less than a certain value ($<10^{-6}$). The survey also includes data from the literature, in which the estimates of very rare mutation rates are given somewhat arbitrarily as $<10^{-7}$. A summary of the data from Table 3.9 is given in Table 3.10.

TABLE 3.10
Distribution of 49 Traits by Estimated Frequency at Birth and by Mutation Rate

Approximate frequencies (all $\times 10^6$*)*	*Numbers in each group*		*Cumulative mutation frequency (percent)*
	Birth frequency	*Mutation frequency per gamete per generation*	
> 100 +	3	0	100
50–100	1	1	100
20–50	4	1	98.0
10–20	5	1	95.9
5–10	2	2	93.9
1–5	4	9	89.8
<1.0	13	11	71.4
< 0.1	17	24	49.0
Total	49	49	

Source: Based on data from Stevenson and Kerr (1967).

It is impossible to compute exactly the arithmetic mean of these mutation rates because many of them are simply given as being below threshold. The median can however, be computed, and this is done in Figure 3.7 on the assumption that the distribution of the mutation rates is approximately "lognormal," a word which indicates that the logarithms of the rates are normally distributed. (The median of a distribution is the value that so divides the distribution that half lies above it, and half below. See also Appendix II.)

This assumption seems satisfactory, as far as this limited sample of mutation rate goes. Assumptions on the distribution of mutation rates allow us to make some further predictions, which may be of interest.

A simple prediction can be made concerning the difference between the median and the average (arithmetic-mean) rates. These two values coincide if the distribution is symmetrical, but they may diverge widely if the distribution, as is likely to be true of mutation rates, is skew. Only the median can be estimated from data, such as those in Table 3.10, but we require the arithmetic mean. In a distribution having the skewness that seems characteristic of mutation rates, the mean may be much higher than the median if the variance of the distribution is high. Thus, the median mutation rate from Figure 3.7 is about 1.6×10^{-7}, but the mean, computed approximately from the observed distribution (Table 3.10), which is truncated at the lower end, is about twenty-two times higher (3.6×10^{-6}). Actually, in a log-

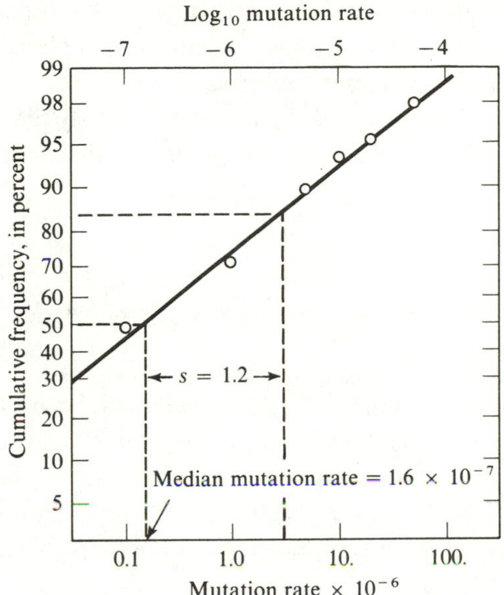

FIGURE 3.7
Cumulative distribution of mutation rates for sex-linked genes in man (data are from tables 3.9 and 3.10), plotted on probability paper with logarithmic abscissa. Probability paper has ordinates graduated on the basis of the probability integral, in such a way that, if the cumulative frequencies of an observed distribution are plotted on it, they will be distributed around a straight line if the distribution is normal. Here the variate, mutation rate, has been plotted on a logarithmic abscissa and the graph shows that the logarithms of the mutation rates are normally distributed; we may say that the mutation rates have a "lognormal" distribution. The median, corresponding to the arithmetic mean of the logarithms, is given by the abscissa value corresponding to the point at which the straight line cuts the 50 percent frequency value, and is here 1.6×10^{-7}. The distance on the abscissa between this and the point of the 84 percent frequency corresponds to the standard deviation s. The standard deviation $s = 1.2$ (in $\log_{10}$ units).

normal distribution, the arithmetic mean is expected to be $10^{1.15s^2} = K$ times higher than the median, where s is the standard deviation of the logarithmic values (to base 10) of the variate (see Example 3.6). The standard deviation of the logarithmic variate is, from Figure 3.7, $s = 1.2$, which gives $K = 44$. Thus, the median mutation rate is much smaller than the arithmetic mean of the mutation rates. This is true apart from any consideration of bias in selecting mutants. It is worth noting, however, that the estimate of the *mean* mutation rate from Stevenson and Kerr's collection of X-linked mutation data, is about 4×10^{-6}, and is at least one order of magnitude larger than the median from the same data, but also about one order of magnitude less than previous estimates of mean mutation rates.

So far, we have discussed the bias that is common to almost all studies of mutation rates and that has to do with the automatic choice for specific studies of some more-frequent mutations because they supply enough mutants to be counted. This automatic choice has necessarily concentrated the interests of research workers on the most-frequent mutations, and thus most studies in the literature refer to them. The resulting bias can be avoided when a sample of people is investigated for all mutations known to occur, common and rare, as was done by Stevenson and Kerr (1967). Nevertheless possible sources of bias are not entirely removed, as recognized by these authors, and as we discuss in the paragraphs that follow.

If we want to study mutation rates and if we choose for this purpose to observe loci at which we already know mutations have occurred, we will be working with a biased sample.

The probability that a mutation at a locus will have been observed earlier and thus have been included in the sample is roughly proportional to the mutation rate of the locus. This simple relationship allows us to predict the increase in the mean mutation rate in the sample that has been subjected to selection as compared with an unselected one. As before, assumptions about the shape of the distribution have to be made. Given that the distribution is lognormal, we can show (see Example 3.6) that if M_e is the median mutation rate in the unselected sample; M is the arithmetic mean of the mutation rates in the unselected sample; M_s is the arithmetic mean of the mutation rates in the sample of loci at which mutations are known to have occurred before the study was begun; and K is the ratio between M and M_e, which we have used before; then the ratio between M_s and M is equal to K^2, and depends therefore very much on the variance of mutation rates.

Practically all mutation-rate estimates in Drosophila and in mice (to be discussed in the paragraphs that follow), as well as in man, are based on loci at which mutations were known to have occurred before the estimates were made. This means that these estimates contain the bias we have been discussing. The only way to avoid it would be to take loci known from sources other than earlier mutation studies, known, for example, because the protein they produce has been studied, and analyze these for mutation rates. This, however, has not yet been done systematically for any locus.

From the mean and variance of the mutation rates observed for loci chosen because mutations had been observed previously in them, it is theoretically possible to compute the mean and variance of the mutation rates of unselected loci: that is to compute M and V given M_s and V_s. This estimation would depend, however, on assumptions about the distribution, and on the precision of the estimated variance of observed mutation rates. This precision would usually be small, because the sample of loci studied would be small. It is in any case worth remembering that all present estimates of mean mutation rates are subject to this bias. Although several authors have discussed sources of bias in estimating mutation rates it does not seem that this particular source has been clearly understood, and its magnitude assessed until now.

If we know the real distribution, we may compute the unselected median, mean, and variance from the selected data. Using the data of Table 3.10 as a selected sample, we compute the mean and variance of the *selected (observed) mutation rates*:

$$M_s = 3.6 \times 10^{-6};$$

$$V_s = 138 \times 10^{-12}.$$

From these, and from the formulas of Example 3.6, we obtain, for the *unselected* mutation rates:

$$Me = \text{median} = 9.1 \times 10^{-8};$$

$$M = \text{mean} = 3.1 \times 10^{-7}.$$

The variance of the logarithms to base 10 of the unselected mutation rates is 0.46. It is interesting to note that the mean of the unselected sample is about ten times lower than the mean of the selected sample. Moreover, in the unselected sample only 6 percent of the loci have mutation rates above 10^{-6}, while in the selected sample this fraction is 25 percent. Also, less than 0.2 percent of the unselected mutation rates are above 10^{-5}, while some 6 percent of them are greater than this value in the selected sample.

The comparison with estimates from other animals indicates 10^{-6} *(or less) as a common reasonable order of magnitude for mean mutation rates per locus per generation.*

The animal most thoroughly investigated for mutation rates is *Drosophila melanogaster*. Drosophila rates comparable to those for the human deleterious mutant genes are those for the so-called "visible" mutant genes—that is, mutant genes that are not lethal but that give visible phenotypic effects either in the homozygote (which is more common with Drosophila mutations) or in the heterozygote. There is still some doubt whether there is a difference in mutation rates between the sexes in Drosophila. The average mutation rate per locus per generation is approximately 10^{-6}. It is possible in Drosophila to estimate the rate of mutation to lethal genes. Lethal mutation rates give figures fairly close to those for "visibles." The total observed mutation rate is the sum of the two, the rate for "visible" plus that for lethals, and is thus roughly twice that for "visibles."

The only mammal other than man for which mutation rates have been investigated is the mouse. A group of seven autosomal loci has been examined for recessive mutations, which were preselected and incorporated into a multiple recessive strain, with which normal mice whose gamates were being tested for mutations from normal to recessive were mated. The average spontaneous mutations rate obtained for the seven loci was 8.4×10^{-6} in males and 1.4×10^{-6} in females (W. L. Russell, 1963). It is interesting to note that this difference between the sexes is significant and in the expected direction (see Section 3.7). The average for the two sexes is 4.9×10^{-6}, which must be biased upwards, for the reasons already discussed and explained further in Example 3.6.

It would seem, from these studies, that 10^{-6} (or less) is a reasonable order of magnitude for average spontaneous mutation rates per locus per generation for a wide range of animals.

There exist more, and less, mutable loci. Experiments with organisms other than man (Drosophila, bacteria, maize) also indicate the existence of *genotypes* that may be appreciably more mutable than the average.

Mutation rates per nucleotide may be of the order of 10^{-8} *or less.*

For a locus that specifies production of a protein that consists of, say, 100 amino acids, there are 300 nucleotides, and the substitution of any one of them is a mutation although it is not necessarily detectable, even as an amino acid substitution. It would seem, therefore, that the mutation rate per nucleotide may be between 10^{-8} and 10^{-9} if the rate per locus is 10^{-6}. The rate per nucleotide would be lower if the average protein specified by a single locus comprises more than 100 amino acids. One factor that might, on the other hand, possibly increase this rate (at most threefold) is the fact that many nucleotide substitutions may be silent. Thus, because of duplicate specifications in the genetic code (see Chapter 1) it is known that as many as thirty percent of all nucleotide pair substitutions may not change the acid coded for; we refer to these as **isonymous substitutions**. It is, however, possible, that the transition from a given triplet to an isonymous one might sometimes be accompanied by some phenotypic change. Different isonymous triplets use different soluble RNA's and the relative concentrations of these may influence the rate of protein synthesis, and thus the final concentrations of proteins.

The number of different mutations possible at each locus is certainly large as is obvious when the structure of DNA and the variety of known changes is considered. In a sense, it is likely that every single mutation is unique, even though many—in particular all of the mutations at one locus that give the same phenotype and selective disadvantage, as, for instance, lethals—may be indistinguishable in terms of evolutionary effects. Molecular genetics has provided the means for considerably refining the analysis of the mutational event.

There are other possible ways of estimating mutation rates in man.

Well-known molecules such as hemoglobin would be fairly satisfactory for a quantitative molecular study of mutation rates in man. Electrophoresis, a quick and cheap method of detecting amino acid changes in a protein molecule, probably detects about $\frac{1}{4}$ to $\frac{1}{3}$ of the changes. This is based on the probability that one amino acid is replaced by another one that has a different electric charge, which is the main requirement for an electrophoretically detectable change in the protein. Electrophoresis can be applied directly to the content of red blood cells because hemoglobin is by far the most abundant protein there, and thus no purification is necessary. Further chemical methods of analysis are well standardized, and it is relatively easy to show which amino acid, if any, is changed in the molecule.

At present, the number of known different mutations affecting the different polypeptide chains of human hemoglobin (mainly alpha and beta chains) is 62 (see Chapter 4). Almost all of these have been detected by electrophoresis, meaning that they are likely only $\frac{1}{4}$ to $\frac{1}{3}$ of the mutations actually in existence. There is good evidence that many of these mutations truly recur independently. Almost all changes known are single amino acid substitutions, and can be explained by the substitution of one nucleotide. In addition, there is one amino acid deletion and two or more mutations that can be explained by the occurrence of anomalous crossovers between the beta and delta chain genes, as in hemoglobin-Lepore (see also the discussion on haptoglobins in Chapter 4). More destructive changes are probably not easily observed.

It is unfortunately impossible to give the denominator for the fraction that would provide an estimate of mutation rates to hemoglobin changes using the observed number of mutations as the numerator. This denominator should be the number of individuals screened for hemoglobin changes, but it is not usually known whether the changes studied were the result of fresh mutations or whether they were carried for a time in the population before being observed, in an earlier generation. Only in very few families (notably those with hemoglobin M mutations) have there been reports of a possible mutational origin of the new hemoglobin. Surveys expressly designed to estimate mutation rates for a molecule such as hemoglobin should screen many hundreds of thousands of individuals systematically. If their hemoglobin turns out to be abnormal, their parents and other relatives should also be tested. This is not an impossible proposition although it is certainly, using present techniques, a fairly expensive one. It is not inconceivable that a study based on it will be carried out in the future for this or some other well-known molecule, which is easy to analyze from both biochemical, and genetic points of view.

Other methods of estimating mutation rates that have been employed rely on the *global* estimation of rates for a relatively large number of genes: for example, all lethal autosomal recessives (from consanguinity studies), all sex-linked lethal recessives (from sex-ratio studies), all detrimental recessives determining a common genetic defect such as deaf mutism, or mental deficiency (from consanguinity studies). They depend on a special theoretical treatment, discussion of which is deferred to Chapter 7, and which relies heavily on assumptions that are difficult to prove. Comparison of estimates made using these methods with the per locus estimates requires knowledge of the number of loci involved, which again is not easy to obtain.

General References

Haldane, J. B. S., "A mathematical theory of natural and artificial selection." *Transactions of the Cambridge Philosophical Society*, 23, No. 11, 19–41, 1924. (This paper and the two listed immediately below laid the basis for Haldane's major contributions to population genetics.)

Haldane, J. B. S., "A mathematical theory of natural and artificial selection. Part III." *Proceedings of the Cambridge Philosophical Society*, **23**: 363–372, 1926.

Haldane, J. B. S., "A mathematical theory of natural and artificial selection. Part V." *Proceedings of the Cambridge Philosophical Society*, **23**: 838–844, 1927.

Haldane, J. B. S., "The rate of spontaneous mutation of a human gene." *Journal of Genetics* **31**: 317–326, 1935. (The first example of the estimation of a mutation rate in man for hemophilia.)

Worked Examples

3.1 *Time necessary for change under mutation pressure alone.*

Suppose that a gene A mutates to a with a frequency of 2×10^{-6} per generation. Mating is at random and there are no other forces acting on the two alleles. How many generations are needed to increase the frequency of gene a from 1 percent to 2 percent?

From equation 3.8

$$p_n = p_0(1 - \mu)^n, \tag{1}$$

where μ is the rate of mutation from A to a, and p_n is the frequency of A, and $q_n = 1 - p_n$ the frequency of a in the nth generation. We want n such that

$$0.98 = [1 - (2 \times 10^{-6})]^n \times 0.99, \tag{2}$$

which is approximately

$$[1 - (n \times 2 \times 10^{-6})] \times 0.99$$

which may be solved by the binomial expansion, giving

$$n = 1 - \frac{0.98}{0.99} / 2 \times 10^{-6} = 5050 \text{ generations.}$$

The binomial expansion is, of course, only valid as long as $n\mu$ is small, and thus, as long as $q_n(=1 - p_n)$ and $q_0(=1 - p_0)$, the initial and final frequencies of a are also small. Equation 1 gives, approximately,

$$(1 - q_n) = (1 - q_0)(1 - n\mu) \tag{3}$$

so that

$$n = \frac{1}{\mu}\left[1 - \left(\frac{1 - q_n}{1 - q_0}\right)\right] \sim \frac{1}{\mu}[1 - (1 - q_n)(1 + q_0)] \sim \frac{1}{\mu}[1 - 1 + q_n - q_0].$$

since when q_0 is small

$$\frac{1}{1 - q_0} \sim 1 + q_0,$$

and quadratic terms in q_n and q_0 can be neglected. Equation 3 therefore reduces to

$$n = \frac{q_n - q_0}{\mu}. \tag{4}$$

3.2 *Persistence of a mutant gene as a function of its selection coefficient.*

If in each generation the individual carrying a particular mutant gene has a probability $1 - s$ of surviving to reproduce, then the average number of generations during which the phenotype representing the mutation remains in the population, the *persistence*, is $1/s$.

The probability that the number of generations is exactly n is $(1-s)^{n-1}s$, this being the probability of the phenotype's surviving for $n - 1$ generations but then succumbing at the nth generation. The persistence is therefore given by

$$\sum_{n=1}^{\infty} n(1-s)^{n-1}s$$

Now

$$\sum_{n=1}^{\infty} n(1-s)^{n-1} = -\frac{d}{ds}\sum_{n=1}^{\infty}(1-s)^n = -\frac{d}{ds}\left(\frac{1}{1-(1-s)}\right) = \frac{1}{s^2} \tag{1}$$

since $\sum_{n=1}^{\infty}(1-s)^n$ is a geometric progression with an index of $1 - s$. (An alternative solution would expand

$$\frac{1}{s^2} = \frac{1}{[1-(1-s)]^2}$$

by the binomial expansion in terms of $1 - s$.) The persistence is, therefore,

$$s\sum_{n=1}^{\infty} n(1-s)^{n-1} = s \times \frac{1}{s^2} = \frac{1}{s}. \tag{2}$$

3.3 *Exact treatment of the balance between mutation and selection.*

Assuming a mutation rate $A \underset{\mu}{\rightarrow} a$ and relative fitnesses of 1, $1 - ht$, and $1 - t$ for the genotypes AA, Aa, and aa, respectively, write down the general equation relating the gene frequencies of a in successive generations on the basis of the scheme given at the end of the first paragraph in Section 3.2.

	Genotype			
	AA	*Aa*	*aa*	*Total*
Fitness	1	$1 - ht$	$1 - t$	
Frequency before selection	p^2	$2pq$	q^2	1
Relative frequency after selection	p^2	$(1-ht) \times 2pq$	$(1-t) \times q^2$	$p^2 + 2(1-ht)pq + (1-t)q^2 = R$

The gene frequencies of a and A after selection, but before mutation, are given by

$$\frac{\frac{1}{2}(1-ht)\times 2pq}{R}+\frac{q^2(1-t)}{R}$$

and

$$\frac{p^2}{R}+\frac{1}{2}\frac{(1-ht)\times 2pq}{R},$$

respectively. The gene frequency of a in the next generation, which is the frequency after selection *and then mutation*, is thus given by

$$q' = \frac{pq(1-ht)+q^2(1-t)}{R}+\mu\frac{p^2+(1-ht)pq}{R},$$

or, since $p+q=p^2+2pq+q^2=1$,

$$q' = \frac{q(1-tq-pht)+\mu p(1-qht)}{1-2htpq-tq^2}. \tag{1}$$

This is the equation that forms the basis for the calculation of the curves given in Figure 3.1. Equilibrium is given by putting $q'=q$. Similar equations can be derived allowing a mutation rate for a to A, and allowing for sex-linked genes (see Kimura, 1961).

3.4. *Accumulation of mutant cells with time in a population of gamete-forming cells.*

The argument is taken from microbial genetics, as the gamete-forming cells, spermatogonia or oogonia, multiply asexually by mitosis. The process is thus quite comparable to that taking place in a bacterial population. Let the rate of mutation per cell division be α, and m the frequency of mutant cells at time t measured in cell generations. Mutation is considered to be irreversible and the population to be made up of one nonmutant cell at time zero. It will be noted that what is called "mutation rate" in man is the frequency of mutant cells among gametes, and thus is the same as m and not α.

Suppose that there are, at the beginning ($t=0$), $N=N_0$ total cells, of which m_0 are mutant. During one generation, αN new mutant cells form since the probability of mutation is α per cell per generation. We neglect the fact that some of the N cells are already mutant because this fraction is small. But, in addition to the new mutant cells, those already in existence, m_0, will multiply, and, assuming that they have the same growth rate as normal cells, will form $2m_0$ cells. The total number of mutant cells is now, therefore, $m_1=\alpha N_1+2m_0$. The ratio of the number of mutant cells to the total number of cells, at the end of one generation ($t=1$) when cells have doubled and are $N_1=2N_0$, is:

$$\mu=\frac{m_1}{N_1}=\frac{\alpha N_1+2m_0}{2N_0}=\alpha+\frac{m_0}{N_0}=\alpha+\mu_0\text{, if } \mu_0=\frac{m_0}{N_0}.$$

In the next generation, again m_1 will become $2m_1$ but in addition, αN_2 new mutant cells will form, giving a total of $\alpha N_2 + 2m_1$, out of a total of $N_2 = 2^2 N_0$ cells. The ratio $\mu_2 = m_2/N_2$ will be

$$\frac{\alpha N_2 + 2m_1}{N_2} = 2\alpha + \mu_0 .$$

The successive increase of mutant cells is given in the table that follows and corresponds to

$$\mu_t = \alpha t + \mu_0 .$$

Generation	*Number of cells at beginning of generation*	*New mutant cells formed*	*Mutant cells already in existence*	*Total number of mutant cells*	*Ratio of mutant cells to total number of cells*
0	N_0	—	m_0	m_0	μ_0
1	$N_1 = 2N_0$	$\alpha N_1 = 2\alpha N_0$	$2m_0$	$m_0 + 2\alpha N_0 + 2m_0$	$\alpha + \mu_0$
2	$N_2 = 4N_0$	$\alpha N_2 = 4\alpha N_0$	$4\alpha N_0 + 4m_0$	$8\alpha N_0 + 4m_0$	$2\alpha + \mu_0$
3	$N_3 = 8N_0$	$\alpha N_3 = 8\alpha N_0$	$16\alpha N_0 + 8m_0$	$24\alpha N_0 + 8m_0$	$3\alpha + \mu_0$
4	$N_4 = 16N_0$	$\alpha N_4 = 16\alpha N_0$	$48\alpha N_0 + 16m_0$	$64\alpha N_0 + 16m_0$	$4\alpha + \mu_0$

3.5. *Age effect on mutation rates.*

Given that $\phi(x)$ is the distribution of father's age x at the birth of a child and assuming that mutation rates increase linearly with paternal age according to the formula (Equation 3.14 and Example 3.4.)

$$\mu = \alpha x + \mu_0 ,$$

where μ, μ_0 are the mutation rates at age x and 0 respectively, calculate the mean age of fathers at the birth of mutant offspring.

The mean paternal age $\bar{x}$ of controls is

$$\bar{x} = \sum_x x\phi(x),$$

where the sum is extended to all fertile ages. The relative frequency of individuals of age x who are parents of mutants is proportional to

$$\mu\phi(x).$$

In order to make these frequencies sum to 1 for all ages, the actual frequency must be divided by the sum of frequencies at all ages:

$$\Psi(x) = \frac{\mu\phi(x)}{\sum_x \mu\phi(x)}. \tag{1}$$

Thus, the mean age among the parents of mutants is

$$\bar{x}_m = \sum_x x\psi(x).$$

On the assumption that the relationship between μ and x is linear as given above, substituting from Equation 3.14 to Equation 1 for $\Psi(x)$, and remembering that

$$\sum_x x^2\phi(x) = \text{var}(x) + \bar{x}^2 \quad \text{and} \quad \sum_x \phi(x) = 1,$$

we obtain

$$\bar{x}_m = \frac{\alpha\ \text{var}(x) + \alpha\bar{x}^2 + \bar{x}_0}{\alpha\bar{x} + \mu_0}, \tag{2}$$

where var(x) is the variance of paternal age at birth of a child.
When $\mu_0 = 0$, the frequency of mutants at age $x = 0(\bar{x}_0)$ is zero, and so

$$\bar{x}_m = \frac{\text{var}(x)}{\bar{x}} + \bar{x}, \tag{3}$$

a formula that was given by Penrose (1955).

The effect of parental age on mutation thus depends on the variance of parental age in the general population. This is not unexpected, since if there were no variation in parental age, no difference in effect on mutation would be observable.

3.6. *Bias due to choice of loci at which mutations were already known to exist for estimating average mutation rates.*

Loci at which mutations are already known are more likely to be included in a sample of loci to be studied for rate of mutation. The probability of such a locus being included is simply proportional to its mutation rate. If the mutation rate of a locus is m, with frequency distribution $P(m)$, and arithmetic mean M, then the probability of its being included in a sample is

$$P_s(m) = \frac{mP(m)}{\sum mP(m)} = \frac{mP(m)}{M},$$

where the sum $\sum$ is extended over the whole distribution and the subscript s refers to the selected sample. The argument is identical to that previously used in another context (see Example 3.5). We can thus conclude that the average mutation rate in the selected sample of loci, M_s, is

$$M_s = \sum mP_s(m) = \frac{\sum m^2P(m)}{M} = \frac{V + M^2}{M} = M + \frac{V}{M},$$

where V is the variance of the original distribution of m values.

The variance of the selected sample, will be similarly,

$$V_s = \sum m^2 P_s(m) - M_s^2 = \frac{\sum m^3 P(m)}{M} - M_s^2 = \frac{\mu_3}{M} - M_s^2,$$

where μ_3 is the third moment about zero of the original m distribution.

For predicting the value of the third moment, knowledge of the original m distribution must be available. We assume that the distribution is lognormal; that is $\log m$ is normally distributed with mean $\overline{\log m}$ (equal to the logarithm of the median) and variance $V_{\log m}$. If logarithms are given to base 10, $\overline{\log_{10} m} = 0.434\ \overline{\log\ m}$ and $V_{\log_{10} m} = 0.434^2 V_{\log m}$.

The rth moment about zero of the original variate, m, in a lognormal distribution is given by (see Kendall and Stuart, 1958, Vol. 1):

$$\mu_r = \exp(\tfrac{1}{2} r^2 V_{\log m} + r\ \overline{\log m}).$$

The arithmetic mean of m, the first moment, is

$$\mu = M = \exp(\tfrac{1}{2} V_{\log m} + \overline{\log m})$$

and is greater than the median value of m, Me, which is equal to $\exp(\overline{\log m})$, by a quantity $K = M/Me = \exp\,(V_{\log m}/2) = 10^{1.5s^2}$ if s is the standard deviation in $\log_{10}$ units. The variance V is equal to the second moment about zero minus the mean squared, and hence

$$V = \mu_2 - M^2 = M^2(e^{V \log m} - 1).$$

It will be seen that the mean of the selected sample, M_s, is larger than the unselected mean M, by a factor that is

$$\frac{M_s}{M} = 1 + \frac{V}{M^2} = e^{V\ \log m},$$

corresponding to the square of K given before.

Similarly V_s can be computed from the third moment value:

$$V_s = \exp(4\ V_{\log m} + 2\ \overline{\log m})\, M -{}_s^2 = M_s^2\ \frac{M_s}{M - 1}.$$

The original, unselected, mean and variance of the logarithmic values can be computed from the system of equations:

$$\log(V_s + M_s^2) = 4\ V_{\log m} + 2\ \overline{\log m};$$

$$\log M_s = \tfrac{3}{2} V_{\log m} + \overline{\log m}.$$

Numerical values are given in the text.

This approach depends on the assumption of a particular distribution for the original values, $P(m)$. An alternative approach makes, in principle, no such assumption. For this it is enough to note that $P(m)$ must be proportional to $P_s(m)/m$, and the proportionality constant must be the reciprocal of $\sum(P_s(m)/m)$. This constant is simply the harmonic mean M_h of the selected, observed values, and thus $P(m) = M_h P_s(m)/m$. The mean of the original, unselected distribution is $M = \sum mP(m)$, and, since $\sum P_s(m) = 1$, M is simply equal to the harmonic mean of the observed values. In practice, when observations grouped in class intervals as rough as those given, by necessity, in Table 3.10, are used, assumptions about the distributions are still implicit in the computation of the harmonic mean, which is very sensitive to the choice of midclass values (especially at the lowest end). A conservative estimate of the harmonic mean from the data of Table 3.10, taking the class mid-values on an arithmetic scale, gives for $M = M_h$, the unselected mean, the value 1×10^{-7}, an even lower value than that obtained by the approach described earlier (3.1×10^{-7}).

4

Transient and Balanced Polymorphisms

4.1 Definition of Polymorphism, Balanced (Stable) and Transient

Genetic polymorphism is the occurrence in the same population of two or more alleles at one locus, each with appreciable frequency.

It is not easy to give a universally acceptable definition of polymorphism. A formal definition such as the one above, based on the frequency of the genes that are found in a population is likely to be most satisfactory. Thus, if in any population we find, in addition to the normal gene, another allele that has an appreciable frequency, we shall call this a polymorphism. The definition of "appreciable frequency" is arbitrary, but for practical purposes mainly relating to the size of readily available samples of human populations, it can be taken to be of the order of one percent. A definition based entirely on allelic frequency has the advantage of not requiring any knowledge of the forces that originally determined, or are now maintaining, the polymorphism. The nature of these forces is very rarely known, and therefore can hardly be useful in a definition. This is the reason why it is difficult to accept unequivocally Ford's definition: genetic polymorphism is the occurrence together in the same locality of two or more discontinuous forms of a species in such proportions that the rarest of them cannot be maintained merely by recurrent mutation. However, it is likely that Ford's definition applies to many, if not most, instances of polymorphism (see Ford, 1964).

Most of the polymorphisms encountered in human populations so far fall into two main categories: blood-cell antigens and blood proteins.

Genetic polymorphism usually refers to alternative hereditary forms that can easily be distinguished from each other and whose inheritance is clearly understood. Most of the polymorphisms observed in man fall into two general categories; an easy way of differentiating between the two is to note that two different kinds of techniques are used for the detection of the two kinds of polymorphisms. Both categories refer to blood, the tissue that is easiest to obtain and analyze for genetic purposes.

The first class of polymorphisms, the kind detected by immunological techniques, is that of blood groups, or blood-cell antigens of which, the ABO blood groups are an outstanding example. The ABO blood groups are so important for medical practice that they are widely known even to the layman. The Rh (rhesus) system has also become well known—because of its association with hemolytic disease of the newborn (see Chapter 5). A dozen or so other red-cell systems, most of which have been identified only in the last thirty years, are much less well known, among other reasons because they are generally less important for the clinician. Blood groups on the white cells have been discovered quite recently. These are turning out to be of considerable significance for organ transplantation (see Chapter 5).

Some blood-group and other polymorphisms detected by immunological techniques are potentially subject to a special type of selection called incompatibility. This is the consequence of immunological reactions in the mother against a fetus having an immunological difference. The fetus may thus be selected against. For this reason, blood groups and potentially incompatible systems are treated together, in the next chapter.

In this chapter we will concentrate on polymorphisms for which complications due to incompatibility are not known, or are not likely to occur. Most of these polymorphisms belong in the second category, being proteins found in the blood, either in the free, liquid portion (serum or plasma) or in its cells (red or white). Most of them have been detected by electrophoretic techniques that depend on the separation of proteins according to electric-charge differences, which make them travel at different rates in an electric field. An example of an electrophoretic pattern of hemoglobins is shown in Table 4.1.

Table 4.2 contains a list of blood proteins, from both plasma and cells, with the number of variants known by early 1968. Many of these proteins are polymorphic for one or a few variants, all the other variants being rare. For some, such as phosphoglucomutase, variants at more than one locus have been found.

TABLE 4.1
Electrophoretic Patterns of Human Hemoglobins A and S.

Phenotype	Genotype		Fitness	Hemoglobin Electrophoretic Pattern (origin →+)	Hemoglobin Types Present
Normal	Hb^A	Hb^A	0.98		A
Sickle-cell trait	Hb^A	Hb^S	1.24		S and A
Sickle-cell disease	Hb^S	Hb^S	0.19		S

Source: From Lerner (1968), by Anthony C. Allison.

TABLE 4.2
Blood Proteins for Which More than One Variety Exists in the Cells or Plasma of Human Beings

Protein	No. of Variants	Has >1 Common Variety	Usual Distinguishing Method	Yr. of Report of First Variant
Hemoglobin:				
Alpha chain	26		Amino acid sequence	1949
Beta chain	46	+		
Gamma chain	5			
Delta chain	4			
Deletions	3			
Haptoglobin	13		Electrophoresis	1955
Alpha chain	10	+		
Beta chain	3			
Gm groups	26	+	Immunologic	1956
Transferrin	18	+	Electrophoresis	1957
Pseudocholinesterase			Inhibition	1957
E_1 locus	3	+	Electrophoresis	
E_2 locus	2	+		
G-C protein	6	+	Electrophoresis	1959
Albumin	7	+	Electrophoresis	1960

TABLE 4.2 (*continued*)

Protein	*No. of Variants*	*Has >1 Common Variety*	*Usual Distinguishing Method*	*Yr. of Report of First Variant*
Inv	3	+	Immunologic	1960
Placental alkaline phosphatase	9	+	Electrophoresis	1961
AG lipoproteins	5		Immunologic	1961
LP lipoproteins	2		Immunologic	1961
Glucose-6-phosphate dehydrogenase	25	+	Electrophoresis	1962
Carbonic anhydrase	5		Electrophoresis	1962
Erythrocyte esterase	2		Electrophoresis	1962
Acid phosphatase	5	+	Electrophoresis	1963
Catalase	3		Electrophoresis	1963
6-phosphate gluconate dehydrogenase	7	+	Electrophoresis	1963
Myoglobin	2		Electrophoresis	1963
Phosphoglucomutase			Electrophoresis	
PGM I	8	+		1964
PGM II	5	+		1965
PGM III	2	+		1968
Fibrinogen	3		Electrophoresis	1965
Adenylate kinase	4	+	Electrophoresis	1965
Lactate dehydrogenase			Electrophoresis	
A subunit	5			1965
B subunit	1			1965
Amylase	4	+	Electrophoresis	1965
Alpha, acid glycoprotein	2	+	Electrophoresis	1965
Beta lipoprotein	2		Electrophoresis	1965
Galactose-1-phosphate uridyl transferase	2	+	Electrophoresis	1966
Malate dehydrogenase (soluble)	2		Electrophoresis	1966
Xm protein	2	+	Immunologic	1966
Alpha, antitrypsin	3		Electrophoresis	1966
Malate dehydrogenase (mitochondrial)	2	+	Electrophoresis	1967
Glutathione reductase	2	+	Electrophoresis	1967
Ceruloplasmin	3	+	Electrophoresis	1967
Peptidase A	4	+	Electrophoresis	1967
Peptidase B	4		Electrophoresis	1967
Hypoxanthine-guanine PR transferase	3		Heat stability	1967
Glutamic oxalacetic transaminase	2		Electrophoresis	1967
NADH diaphorase	2		Electrophoresis	1968
Phosphohexose isomerase	10	+	Electrophoresis	1968

Note: Each variant has been shown to be distributed in families in such a way as to suggest its genetic origin. Common variants exist in frequencies of 1 percent or more in at least 1 population.
Source: From Childs and Kaloustian (1968).

There are, in addition to the polymorphisms that fit into these two major categories, several other polymorphisms for traits that were investigated using other techniques. One of the oldest known polymorphisms is that for capacity to taste phenylthiocarbamide (PTC). For some people PTC has only a faint taste or no taste at all; for others it has a very bitter taste. More specifically there is a single dominant gene T (with incomplete penetrance) that determines a high sensitivity for the taste of PTC. Nontasters are homozygous for the recessive allele t.

What proportion of genes is polymorphic?

Until a few years ago the answer would have probably been—very few. Recently, however, electrophoretic analysis of proteins, mainly by Harris (1966, 1969) in man, and by Lewontin and Hubby (1966) in *Drosophila*, has made it necessary to reconsider this question, which will be examined more fully in Chapter 11. Here it will be enough to glance at the third column of Table 4.2 which shows that a large proportion of known variants among blood proteins is polymorphic.

It is worth noting that most human polymorphisms are present in almost all racial groups, although the frequencies of the various types may differ considerably. They thus provide a very interesting tool for anthropological research, which has already been widely applied and will undoubtedly be used increasingly in the future. This use of polymorphism will be discussed further in Chapter 11.

It is difficult to establish whether a polymorphism is stable or transient.

A major problem posed by the presence of several alleles at one locus in any population is the following: is the polymorphism a stable one, or are we observing a transient stage in an evolutionary process, destined to end relatively soon, in which one allele is replacing another?

This question is unfortunately very difficult to answer. The obvious way is to look for changes in the gene frequencies with time, which would require data from the past and, possibly, the very remote past. The only examples in human genetics of such a study are a few observations on ABO blood groups from Egyptian and American Indian mummies (Boyd and Boyd, 1937). They refer to individuals who lived only a few thousand years ago. Also, the test for a mummy's blood type may be invalidated by the fact that bacteria often have antigens resembling those of ABO blood groups, as well as enzymes that may destroy ABO antigens. In the absence of direct evidence, the assessment regarding stability of a polymorphism can only be tentative, and we are in fact still unable to say which, if any, of the polymorphisms that we now know are really stable.

Direct evidence for at least one balanced polymorphism is available: the polymorphism for the group of hemoglobins, including hemoglobin S, in the presence of malaria.

There are some instances of genetic variation sufficiently well investigated to show conclusively that a balanced polymorphism can be maintained by **heterosis** (heterozygote advantage) in the presence of malaria. This disease is widespread in the tropics and was common also in temperate areas until a short time ago. It is probably, where uncontrolled, the greatest single cause of human mortality. It is not surprising, therefore that malaria is a very powerful selective agent, favoring genetic variants endowed with an increased resistance to it. Some of the most dramatic selective situations known are connected with malaria susceptibility. The best studied is hemoglobin variation, but several others are known and will be discussed in the second part of this chapter.

4.2 Opposing Forces That Produce and Maintain Polymorphism

For a polymorphism to be stable there must be at least two opposing forces acting upon it, and they must be balanced in their effects on it.

It was shown in the previous chapter that when mutation and adverse selection oppose each other an equilibrium results. The gene frequencies at equilibrium are usually too low to be considered polymorphic by our definition, which is the basis of Ford's definition of polymorphism. It must, however, be remembered that since we do not know how most of the presently known polymorphisms are maintained, recurrent mutation balanced by very weak selection cannot really be excluded as the mechanism that might maintain many, if not most, of them. Present knowledge may make this mechanism unlikely, but can almost never exclude it. Equilibrium can be the result of many other combinations of balancing forces, which will be the subject of the discussion in the rest of this chapter.

The balance between opposing mutation pressures may give rise to a polymorphism.

Consider a population with two alleles: A with a frequency p, and a with a frequency q, or $1 - p$. Let μ and ν be the mutation rates:

$$A \underset{\nu}{\overset{\mu}{\rightleftarrows}} a$$

Each generation a fraction μ of the pA genes will mutate to give $p\mu$ a genes, and a fraction ν of the a genes will mutate to give $q\nu$ A genes. Thus, when

$$p\mu = q\nu, \tag{4.1}$$

the number of A alleles mutating to a is equal to the number a alleles mutating to A and no further change in gene frequencies will take place. The frequency of a at equilibrium, q_e, is thus given by

$$\frac{p_e}{q_e} = \frac{v}{\mu} \quad \text{and} \quad q_e = 1 - p_e = \frac{\mu}{\mu + v}. \tag{4.2}$$

The ratio of the equilibrium frequencies therefore depends on the ratio of the mutation rates. If forward ($A \rightarrow a$) and backward ($a \rightarrow A$) mutation rates are equal,

$$p_e = q_e = 50 \text{ percent.}$$

A polymorphism by our definition, will result whenever μ/v lies approximately between 1/100 and 99/100.

A balance between opposing mutation pressures seems somewhat artificial from a molecular point of view, because mutations at a given locus can take place at any of its nucleotide sites and the mutation rate for the average nucleotide is in general very small (Chapter 3). However, the existence of mutational "hot spots" (namely, highly mutable sites) has been suggested for bacteriophage by Benzer (1961). If such hot spots exist in man, then forward and backward mutation rates at what is really the same site may be high enough to lead to a mutation-balance polymorphism. In addition, it may be possible that forward and backward mutations for many nucleotide sites in the same locus may give rise to similar phenotypes or, more precisely, phenotypes with similar selective values. However, it seems rather unlikely that many polymorphisms are the result of a balance between mutations because there are frequently other factors that override the effect of mutation rates in changing gene frequencies.

Heterozygote advantage is the classical explanation for a balanced polymorphism.

Given two alleles at one locus, selection can favor either or both of them at different times in various ways, making possible a number of different types of polymorphisms. In the simplest situation, the heterozygote is, on the average, at an advantage over both homozygotes. This is sometimes referred to as "overdominance" (with respect to fitness).

It is fairly easy to see that selection for the heterozygote, or, equivalently, selection against both homozygotes, may give rise to a polymorphic equilibrium. This was first pointed out by Fisher in 1922. Consider first of all, for simplicity, the extreme case in which both homozygotes have zero fitness (**balanced lethals**). Since only heterozygotes survive, $p = q = \frac{1}{2}$ in all generations.

For a more general demonstration that heterozygote advantage leads to a polymorphism, we shall follow essentially the treatment originally given by Fisher in 1922. We assume that there is a Hardy-Weinberg equilibrium with the constant genotype fitnesses and the genotype frequencies indicated in Table 4.3.

TABLE 4.3
Genotype Fitnesses and Frequencies Before and After Selection for One Locus with Two Alleles

	Genotypes			
	AA	*Aa*	*aa*	*Totals*
Fitnesses	w_1	w_2	w_3	
Frequencies				
Before selection	p^2	$2pq$	q^2	1
After selection	$w_1 p^2$	$w_2 2pq$	$w_3 q^2$	$w_1 p^2 + 2w_2 pq + w_3 q^2$

The frequency, p', of gene A after selection will be proportional to

$$w_1 p^2 + \tfrac{1}{2} w_2 \, 2pq = p(w_1 p + w_2 q);$$

while that of gene a ($q' = 1 - p'$) will be proportional to

$$\tfrac{1}{2} w_2' \, 2pq + w_3 q^2 = q(w_2 p + w_3 q).$$

Thus the ratio of the two gene frequencies after selection will be

$$\frac{p'}{q'} = \frac{p(w_1 p + w_2 q)}{q(w_2 p + w_3 q)}. \tag{4.3}$$

For equilibrium we must have $p' = p$, and hence also $q' = q$, so that there is no change in gene frequencies. This can happen when $p = 0$ or $q = 0$, either of which would indicate, of course, that the population is not polymorphic. When neither p nor q is 0, the equilibrium condition $p' = p$ and $q' = q$ gives

$$\frac{w_1 p + w_2 q}{w_2 p + w_3 q} = 1.$$

Solving for p gives p_e, the equilibrium frequency of A, which is

$$p_e = \frac{w_2 - w_3}{(w_2 - w_1) + (w_2 - w_3)}. \tag{4.4}$$

If we write the relative fitnesses of the genotypes AA, Aa, aa in the simpler form $1 - s$, 1, and $1 - t$, respectively, then we must have

$$1 - s = \frac{w_1}{w_2}; \; s = \frac{w_2 - w_1}{w_2};$$

$$1 - t = \frac{w_3}{w_2}; \; t = \frac{w_2 - w_3}{w_2}.$$

In this form, the equilibrium frequencies given by Equation 4.4 become

$$p_e = \frac{t}{s+t} \quad \text{and} \quad q_e = \frac{s}{s+t}, \tag{4.5}$$

and the ratio of the two equilibrium gene frequencies is the reciprocal of the ratio between the selective disadvantages of the opposite homozygotes,

$$\frac{p_e}{q_e} = \frac{t}{s}.$$

This equilibrium formula is still valid if both s and t are negative, that is, if the heterozygote is at a disadvantage relative to both homozygotes. If, however, s and t are negative, the equilibrium is unstable, as we shall see below. When s and t have opposite signs, p_e/q_e becomes negative and so either p_e or q_e is negative. Thus in such cases Equation 4.4 does not give valid equilibrium frequencies (which must always be between 0 and 1). It can be shown that when s is negative and t positive $p_e = 1$ and $q_e = 0$. These cases will be discussed more fully in Section 4.4.

4.3 Stability of a Polymorphism

A valid equilibrium gives rise to a balanced polymorphism only if it is stable: that is, if gene frequencies, when disturbed, tend to return to their equilibrium. By analogy with simple principles of mechanics, we recognize the existence of stable, indifferent, and unstable equilibria. The first is exemplified by a cone standing on its base, the second by one lying on a side and the third by a cone standing on its apex. We shall show that for the simple genetic model we are considering the three corresponding effects are heterozygote advantage, lack of fitness differences between the genotypes, and heterozygote disadvantage (see Table 4.4).

To examine the stability of an equilibrium, we determine whether, after a change in gene frequency near the equilibrium value, the gene frequency always moves back toward the equilibrium, which is the essential characteristic of stability. Let

$$\Delta p = p' - p,$$

where p' and p are the gene frequencies in two successive generations.

Clearly if Δp is positive, p' is greater than p, and the gene frequency increases, while if Δp is negative, p' is less than p, and the gene frequency decreases.

Stability can be tested by examining the change in gene frequency per generation.

Given the selection coefficients indicated in Table 4.3 (w_1 for AA, w_2 for Aa, and w_3 for aa) Equation 4.3 gives the ratio p'/q' after one generation of selection. As $p' + q' = 1$, if $p'/q' = a/b$, $p' = a/(a+b)$ and thus one can write, from Equation 4.3

$$p' = \frac{p(w_1 p + w_2 q)}{p(w_1 p + w_2 q) + q(w_2 p + w_3 q)}. \tag{4.6}$$

To simplify, without loss of generality, we may rewrite Equation 4.6 using the alternative notation for fitnesses shown in Table 4.4, that is, $w_1 = 1 - s$, $w_2 = 1$, $w_3 = 1 - t$.

Equation 4.6 then becomes

$$p' = \frac{p[(1-s)p + q]}{p[(1-s)p + q] + q[p + (1-t)q]} = \frac{p(1-sp)}{1 - sp^2 - tq^2}, \tag{4.7}$$

and the change in gene frequency is

$$\Delta p = p' - p = \frac{p(1-sp) - p(1 - sp^2 - tq^2)}{1 - sp^2 - tq^2} = \frac{pq(-sp + tq)}{1 - sp^2 - tq^2} = A(-sp + tq). \tag{4.8}$$

The quantity $A = pq/(1 - sp^2 - tq^2)$ is always positive and so the sign of Δp depends only on the sign of $(-sp + tq)$, which can be rewritten, using Equation 4.5, as

$$\begin{aligned}(-sp + tq) &= (s + t)\left[-\frac{s}{s+t}p + \frac{t}{s+t}(1-p)\right] \\ &= (s+t)(qp_e - pq_e) = (s+t)(p_e - p)\end{aligned}$$

Thus,

$$\Delta p = p' - p = A(s+t)(p_e - p) = \frac{pq}{1 - sp^2 - tq^2}(s+t)(p_e - p). \tag{4.9}$$

We can now consider the following three cases.

1. If s and $t > 0$ or w_1 and $w_3 < w_2$ there is *heterozygote advantage*. The equilibrium is stable because when the *a* gene frequency is moved away from equilibrium it always tends to return towards it. Suppose (see Figure 4.1) that the gene frequency is p_1, which is smaller than p_e. Then $p_1 < p_e$, $p_e - p_1 > 0$, and

$$\Delta p_1 = p' - p_1 = A(s+t)(p_e - p_1) \tag{4.10}$$

is a positive quantity, because both $s + t$ and $p_e - p_1$ are positive. Therefore $p' > p_1$, and after one generation the gene frequency will have moved towards equilibrium. We can also satisfy ourselves that the new gene frequency will not "overshoot" the equilibrium value. It could do so, only if $\Delta p_1 > p_e - p_1$, but this would obtain only if $A(s + t)$ were greater than 1. Now A is $pq/(1 - sp^2 - tq^2)$/and s and t can be at most equal to 1, because fitnesses cannot be less than zero. In this extreme case $A(s + t)$ would be $2pq/(1 - p^2 - q^2) = 1$, and the equilibrium would be reached in one generation (this is the case of balanced lethals, mentioned before). In all other cases, $A(s + t)$ is less than 1 and therefore p' will lie between p_1 and p_e.

The same considerations are valid if we start our approach to the equilibrium from the other side (Figure 4.1). Let $p_2 > p_e$. Then

$$\Delta p_2 = p' - p_2 = A(s+t)(p_e - p_2). \tag{4.11}$$

The quantity $p_e - p_2$ is now negative and so is Δp_2. Therefore $p' < p_2$ and the gene frequency will move towards the left; that is, towards the equilibrium (again without danger of overshooting).

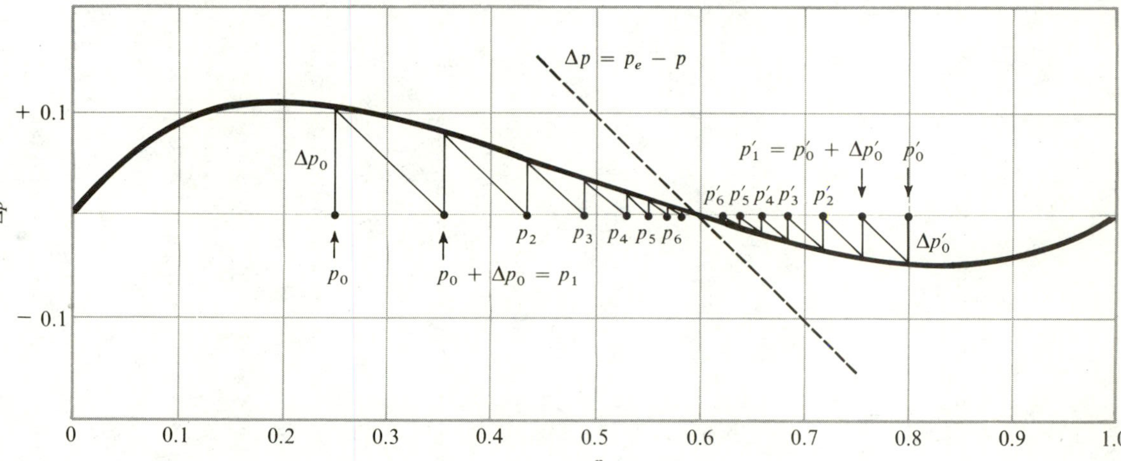

FIGURE 4.1

The graph of the change in gene frequency, Δp, as a function of p, for heterozygote advantage. This shows that if the slope of Δp at or near equilibrium is negative, then the equilibrium is stable (see also Figure 4.2). The change Δp is given by $pq(-sp + tq)/(1 - sp^2 - tq^2)$, and the example illustrated has $s = 0.4$, and $t = 0.6$, making the equilibrium gene frequency $p_e = t/(s + t) = 0.6$. A population with gene frequency $p_0 = 0.25$ will have $\Delta p = 0.103$; and $p_0 + \Delta p_0 = 0.353$ which is nearer to the equilibrium p_e, will be the gene frequency in the next generation. With gene frequency above p_e, such as $p_0 = 0.8$, Δp is negative ($\Delta p_2 = -0.044$) and thus p will, again, move towards p_e, becoming $p_0 + \Delta p_0 = 0.756$. There is no danger of overshooting because Δp is always smaller than the distance of p from equilibrium, $p_e - p$. The case when $\Delta p = p_e - p$ is shown by the dotted oblique straight line. Only if Δp were above this line on the left or below on the right could overshooting occur. If Δp were at any time equal to $p_e - p$, equilibrium would be reached in one generation. It is actually reached by successive steps as shown in this figure, where $p_0, p_1, p_2, \ldots p_e$, represent gene frequencies at times $t = 0$, $t = 1, \ldots$ t $= \infty$, starting with $p_0 = 0.25$ at time $t = 0$, and similarly for $p'_0 = 0.8$.

2. If $s = t = 0$, there is *no selection*. The gene frequencies will not change. In this case, $s + t = 0$ and thus $\Delta p = 0$. As we have seen for the Hardy-Weinberg equilibrium, there is no change in gene frequencies. If, for some reason, the gene frequencies were disturbed, they would *not* return to their original values but would remain at their new values.

3. If s and $t < 0$ or w_1 and $w_3 > w_2$, there is *heterozygote disadvantage*. The equilibrium is unstable because gene frequencies tend to move away from it rather than towards it.

If the gene frequency of a population happens to be exactly the equilibrium gene frequency, p_e, it will not change, because then $p' = p$, in Equation 4.9, and $\Delta p = 0$. But if it has a different value, say $p_1 < p_e$, then in Equation 4.10, $p_e - p_1$ is positive, $s + t$ is negative, and Δp_1 is negative; that is p' is less than p_1. If $p_2 > p_e$, Δp_2 will be greater than zero and thus $p' > p_2$. The new gene frequency will always move away from the equilibrium, towards 0 if $p_1 < p_e$ and towards 1 if $p_2 > p_e$ (see Figure 4.2).

A summary of these equilibria is given in Table 4.4.

TABLE 4.4
Intermediate Gene Frequency (That Is, $p \neq 0$) or 1 Equilibria

Nature of Equilibrium	*AA* w_1 1–s	*Aa* w_2 1	*aa* w_3 1–t	
	Genotypes and Fitnesses			
Stable s, t > 0 (heterozygote advantage)	< 1	1	< 1	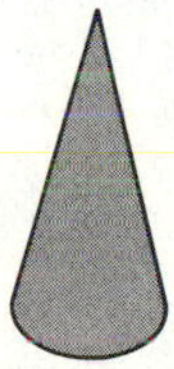
Indifferent s, t = 0 (no selection)	1	1	1	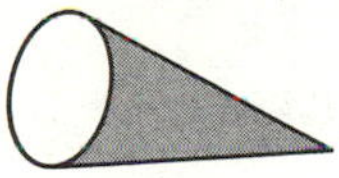
Unstable s, t < 0 (heterozygote disadvantage)	> 1	1	> 1	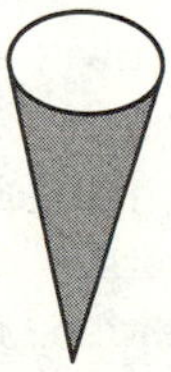

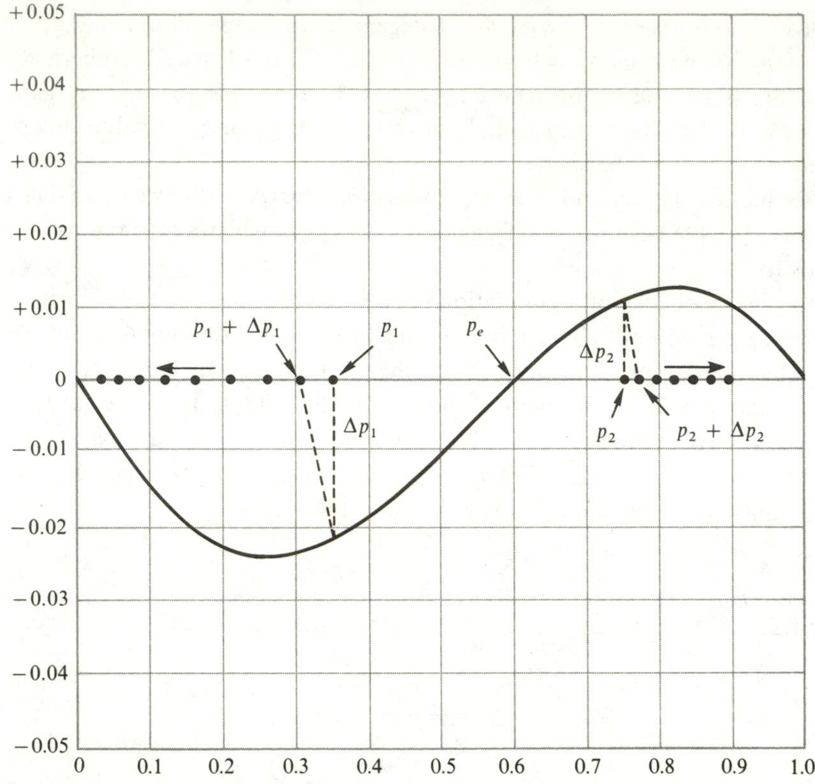

FIGURE 4.2
The graph of the change in gene frequency for heterozygote disadvantage (see Figure 4.1). The slope at equilibrium is now positive and the equilibrium is unstable. The actual s and t values used to compute the graph are now $s = -0.4$, $t = -0.6$, so that $p_e = t/(s + t) = 0.6$. The sign of Δp is here such that, throughout the range of p values from 0 to 1 except for $p = p_e$, the gene frequency in the next generation will always be farther removed from the equilibrium than in the previous one. The evolutionary process will end up at $p = 0$ starting for values of $p < p_e$ and at $p = 1$ starting for values $p > p_e$. The points on either side of p_e indicate successive steps of the process starting from $p_1 = 0.35$ and from $p_2 = 0.75$.

The graph of Δp versus p indicates the stability of equilibrium, and the slope of Δp at equilibrium measures the rate of return to it.

The graph of Δp against p provides a simple overall picture of the behavior of any genetic system with respect to changes in gene frequency. The sign of the slope of Δp at the equilibrium tells us if the equilibrium is stable or unstable. When it is negative the equilibrium is stable and when positive unstable. The graph is especially useful in complex situations, in which there may be more than one equilibrium point. In addition, the magnitude of the slope at equilibrium is a measure of the "resilience" of a stable equilibrium—that is, of the rate at which the gene frequency returns to the equilibrium after being displaced from it. This rate has been called by Malécot the "coefficient of recall" to equilibrium. The larger this coefficient, the more rapid is the return to equilibrium, and therefore the greater the resilience.

If Δp were linear with respect to p, that is, if it could be given by the equation

$$\Delta p = -b(p - p_e),$$

then the resilience would be the coefficient b for the whole range of values of p. The curve of Δp is more complicated than a straight line except in some special instances, but in Figure 4.1 and 4.2 it is almost linear for a wide range around p_e.

A purely mutational balance is stable but has low resilience.

Consider now the application of these criteria to mutational balance. The change in gene frequency is given by

$$\Delta p = -\mu p + vq = -(\mu + v)p + v, \qquad (4.12)$$

where μp is the loss of A genes due to mutation to a and vq is the gain due to mutation from a to A. The change in gene frequency is plotted by the line A in Figure 4.3. With mutation pressure, the graph of Δp is linear for the whole range. Since Δp has a negative slope, the equilibrium for a mutational balance is always stable. Equation 4.12 can be put in the form

$$\Delta p = -(\mu + v)(p - p_e), \qquad (4.13)$$

where $p_e = v/(\mu + v)$. The magnitude of the recall coefficient, $b = (\mu + v)$ is exceedingly small because of the smallness of mutation rates, indicating the low resilience of a mutation-balanced polymorphism.

Mutation, in general, hardly affects selection balance.

The joint effects of mutation and selection are illustrated by the curves B in Figure 4.3, which show Δp versus p for a case of selection balance and the same case with, in addition, mutation. The selection balance is chosen so as to give an equilibrium value $p = \frac{1}{3}$ with relative selective values of 0.998, 1, and 0.999 for the three genotypes AA, Aa, and aa, while the mutation balance has an equilibrium at $p = \frac{1}{2}$ when the mutation rates are $\mu = v = 10^{-5}$. It can be seen that mutation hardly changes either the equilibrium value or the shape of the curve of Δp versus p for selection alone, in spite of the low selection coefficients chosen (of order 0.001) and the relatively high mutation rates (10^{-5}).

It can be shown that for a balanced polymorphism due to heterozygote advantage with respective relative fitnesses of $1 - s$, 1, and $1 - t$ for the three genotypes AA,

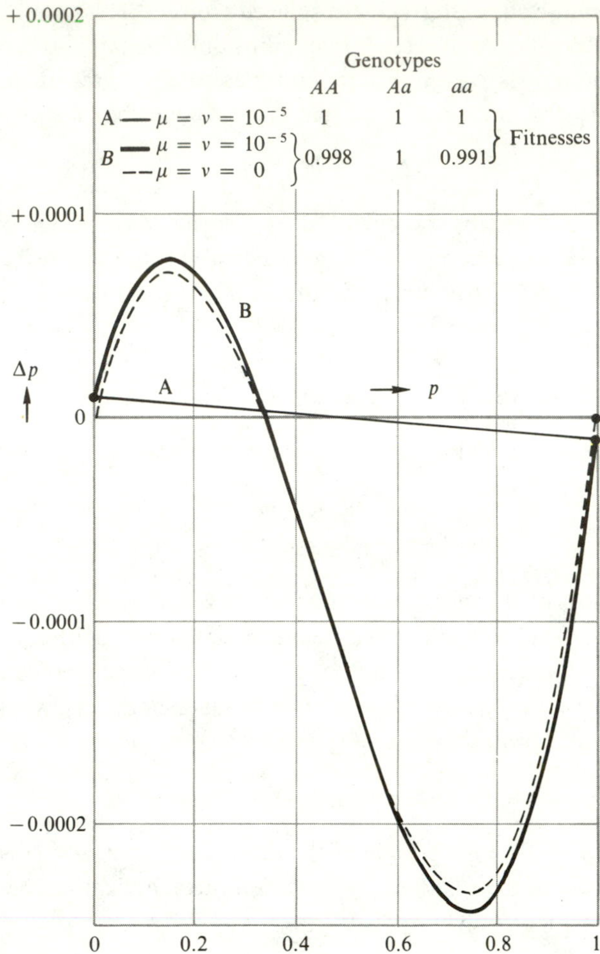

FIGURE 4.3
Graph of Δp versus p in mutation and selection balance. Line A gives mutation balance, assuming, $\mu = \nu = 10^{-5}$ and $p_e = \frac{1}{2}$. Curves B give selection balance with weak selection ($s = .002$, $t = .001$) without mutation pressure (dashed curve) and with mutation pressure (solid curve). The effect of mutation pressure on the equilibrium of the selection balance is negligible. The equation used is $\Delta p = [pq(tq - sp) - \mu p(1 - sp) + \nu q(1 - tq)]/(1 - sp^2 - tq^2)$. (See Kimura, 1961.)

Aa, and *aa*, the magnitude of the slope of Δp at equilibrium is $st/(s + t - st)$, which is approximately $s/2$ when $s = t$ and s is small. This is, in general, of the same order of magnitude as the selection coefficients s and t. In so far as selection coefficients are generally much higher than mutation rates, this demonstrates that selection-balanced polymorphisms have much greater resilience than mutation-balanced polymorphisms. Mutation has only a second order effect, unless the magnitudes of the selection coefficients are of the same order as the mutation rates.

4.4 Kinetics of the Selection Process for Transient Polymorphisms

When the fitness of AA is greater than that of Aa and aa and in the absence of mutation, the frequency of A will increase steadily to one with the result that gene A will eventually be fixed and gene a will be lost.

Conversely, when the fitness of *aa* is greater than that of *Aa* and *AA*, *a* will eventually be fixed and *A* lost. These effects can be summarized as follows:

	Genotypes			*Genes at equlibrium*	
	AA	*Aa*	*aa*	*A*	*a*
Fitnesses	>1	1	$\leqslant 1$	fixed	lost
	$\leqslant 1$	1	>1	lost	fixed

Consider now the example in which A is destined to become fixed. The selection coefficient s is negative and t is positive or zero. The equation for the change in gene frequency (4.8) takes the form

$$\Delta p = p' - p = \frac{pq(|s|p + tq)}{1 + |s|p^2 - tq^2}, \tag{4.14}$$

where $|s|$ is the positive value of s. In this example the quantity Δp given by Equation 4.14 is always positive, showing that the frequency p, of A, will increase steadily until it reaches the value 1.

When $t = |s|$ Equation 4.14 reduces to

$$\Delta p = \frac{|s|pq}{1 + |s|(p^2 - q^2)}. \tag{4.15}$$

If s is not large, the denominator can be neglected, and the change per generation will be, approximately,

$$\Delta p = |s|pq. \tag{4.16}$$

These equations, which were first studied by Haldane (1924, 1926, and later), however simple, cannot be solved for the general case, to give the expected gene frequency p_n after n generations starting from p_0. There are two possible ways of obtaining the desired result. One is to use a computer, and, given the initial gene frequency and the chosen selective values, calculate the gene frequency generation after generation. Several curves (see Figures 4.6 and 4.7) are obtained by this approach. The other is to approximate the equation of change per generation by a differential equation, which is soluble (see Example 4.3 and figures 4.4 and 4.5). In this way, time is turned from a discrete variable (number of generations) into a continuous one. The approximation is usually satisfactory, and sometimes even more satisfactory than the exact one with discrete time, considering that in man and many other organisms, generations overlap.

The simplest example of selective advantage of a newly arisen allele is that in which the new allele has intermediate dominance in fitness, as in Equation 4.16.

Using the corresponding differential equation (see Example 4.3), the gene frequency p of the new allele A at time T (in generations) can be shown to be approximately given by

$$p = \frac{p_0 e^{sT}}{q_0 + p_0 e^{sT}}, \tag{4.17}$$

where now for notational simplicity we consider the fitness of AA as $1 + s$, where s is positive; thus we no longer need $|s|$. This curve is plotted for various values of s in Figure 4.4: for most of the range it is a simple exponential, because up to large values of p the denominator is very nearly equal to one.

This type of selection is a great interest to us because it represents the simplest case of a *transient* polymorphism. From the appearance of the new mutant gene, and as long as it is rare in the population, its frequency remains too low to define that there is a polymorphism at the locus. For $s = 0.1$, about 70 generations will be required for the increase from the initial gene frequency of 10^{-5} to the threshold for polymorphic range, that is, $p = 0.01$. After p reaches the value 0.99, the now rare allele a will continue to decrease. Thus the transient polymorphism will be observable only for about one-third of the time necessary for the frequency of A to go from $p_0 = 0.00001$ to $p = 0.99999$. This fraction of the time is, incidentally, approximately independent of the selection coefficient. Thus, ignoring the fact that the choice of the initial and final gene frequencies is somewhat arbitrary, and that we do not consider random fluctuations, a transient polymorphism will be observable only for about one-third of the time during which both alleles are present in the population.

The time for selection in this process is easy to compute. The change is symmetrical with respect to $p = 0.5$. The time needed to reach $p = 0.5$ starting from a gene frequency p_0 is approximately $T = -\log_e(p_0/q_0)/s$ (see Example 4.3), and is thus

$$\frac{2.3}{s} \text{ for } p_0 = 0.1, \frac{4.6}{s} \text{ for } p_0 = 0.01, \frac{6.9}{s} \text{ for } p_0 = 0.001,$$

$$\frac{9.2}{s} \text{ for } p_0 = 0.0001, \text{ and } \frac{11.5}{s} \text{ for } p_0 = 0.00001.$$

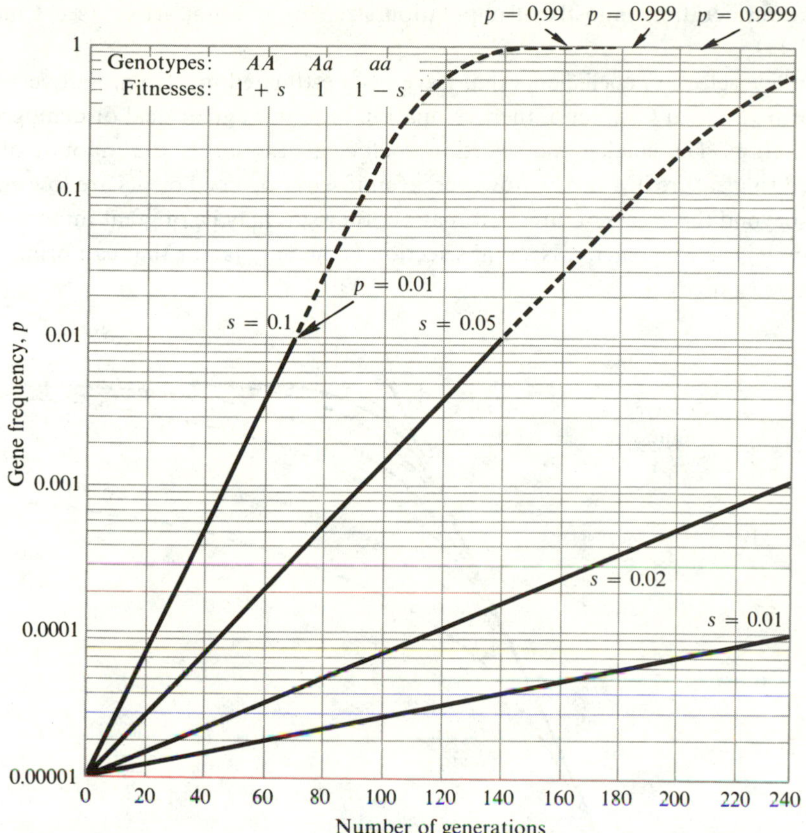

FIGURE 4.4
Selection with intermediate dominance in favor of the new A allele, starting with p_0 equal to 10^{-5}, with various selection coefficients. For lower selection coefficients, the number of generations may be multiplied by the same quantity used to divide the selection coefficient. For instance, for $s = 0.001$ we may use the curve of $s = 0.1$, multiplying the generations by 100. The dashed portions of the lines indicate that during this period the gene shows a transient polymorphism.

To compute the time, say, for reaching $p = 90$ percent from $p = 0.0001$, for $s = 0.001$, we calculate $9.2/0.001 = 9200$ generations to reach $p = 0.5$, and then because of the symmetry, $2.3/0.001 = 2300$ generations to reach 0.9 from $p = 0.5$. Thus, in total, the time needed is 11500 generations.

The magnitude of selection coefficients is known in a very few instances. Naturally, the rate of change is a function of them, and with selection coefficients of order 0.001 the change is slow. It is, however, always much faster than the change that could take place under mutation pressure alone. The ratio of change by selection versus that by mutation is approximately the ratio of the selection coefficients to the mutation rates per generation. Thus, even selection coefficients as low as 10^{-5} are still significantly more important than mutation rates that are, on an average, one order of magnitude lower. With such very low selection coefficients, however, other considerations, mostly of population size, become important (see Chapters 8 and 11).

The few selection coefficients that have been estimated in man are all very high. As we shall see in Chapter 6, their estimation requires a great deal of demographic information. The smaller the selection coefficient, the larger the amount of data needed to measure it. Clearly, however, even if selection coefficients are low enough to be beyond the access of direct estimation, they still may be of great importance in changing gene frequencies. Natural selection is the only factor that can bring about adaptive changes.

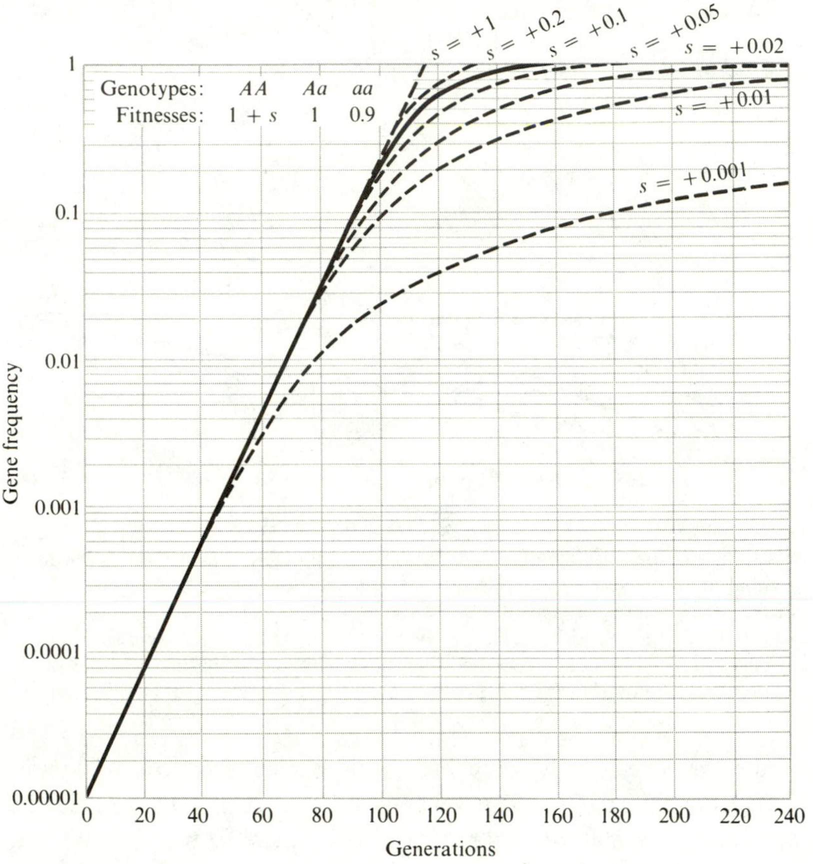

FIGURE 4.5
Effects of dominance for fitness on rate of selection. A new allele A has initial gene frequency $p_0 = 10^{-5}$ at time zero. The fitnesses are indicated on the figure. The solid line corresponds to a 10 percent difference in fitness between AA and Aa, and also between Aa and aa; there is thus no dominance of fitness ($s = 0.1$). If, however, AA is fitter than 1.1, the rate of disappearance of aa remains high until the end (dotted curves at the left of the solid line). If AA is less fit than 1.1 the rate is lower (curves to the right). Dominance in fitness, affecting the s value, modifies only the second part of the curve. Curves were calculated using the differential equations (see Example 4.3). Results differ only slightly from those obtained with the exact equations, as used in Figure 4.6.

Dominance hardly affects the initial rates of selection for a newly arisen allele.

The curves given in Figure 4.4 refer to a selective process uncomplicated by dominance in fitness, because the heterozygote was considered to be intermediate between the two homozygotes. For a large part of the process, however, as long as the newly arisen allele is rare, the rate of selection is dominated by the magnitude of the selective advantage of the heterozygote, relative to the original homozygote. This is because no or few homozygotes for the new allele have yet appeared. They will, however, appear, sooner or later, when p becomes large enough. The selection process will then be affected by the fitness of the new homozygote. It will become faster if the homozygote for the new allele is fitter, slower if it is less fit than the heterozygote. If the new homozygote is less fit than the heterozygote, an equilibrium gene frequency will be reached and the process of gene substitution will not go to completion; a balanced polymorphism will thus result. Figure 4.5 illustrates the effects of the fitness of the homozygote for a new allele for $s = 0.1$ and various values of t.

4.5 Kinetics of the Selection Process for Balanced Polymorphisms

Sometimes, even when a heterozygote for a new mutant gene is fitter than the original homozygote, the new homozygote has a very low fitness or the homozygous state may even be lethal. This is the case, for instance, for the best known polymorphisms in man, which are connected with resistance to malaria. An example of a lethal homozygous condition is thalassemia, a hereditary anemia with abnormalities in the synthesis of hemoglobin, which is found mostly among populations that live around the Mediterranean. In some areas in which malaria is prevalent, the frequency of thalassemics is very high, and the gene determining thalassemia may reach frequencies of 10 percent or even higher. Thus, 1 percent or more of all babies born in such an area carry the disease and die early, while the heterozygotes, which are 20 percent or more of the population, show no pathological signs, and are more resistant to malaria than are persons who do not carry the gene. The homozygote, being inviable, has zero fitness. The equilibrium gene frequency of the lethal gene will then be, from Equation 4.5, for $t = 1$

$$q_e = \frac{s}{(1 + s)}.$$

For a lethal gene to reach polymorphism, that is $q_e = 0.01$, s must be higher than about 0.01. For example for $q_e = 10$ percent, assuming that populations showing this high gene frequency are at equilibrium, s must be $q_e/(1 - q_e) = 0.1/0.9 = 0.11$. A reduction in fitness of the normal homozygote of 11 percent is fairly high and should be detectable (see Chapter 6).

Let us assume that a lethal gene such as the one for thalassemia is introduced at a very low frequency into a population in which conditions for the balanced polymorphism exist—that is, there is malaria. The fitness of the normal homozygote will then become $1 - s$ with respect to thalassemic heterozygotes. Lethality of the thalassemic homozygote will prevent the process from going to completion, and the thalassemia gene will stabilize at a frequency of $s/(1 + s)$. Only relatively few generations will be necessary for this process, because, for a lethal gene to be able to form a balanced polymorphism, the difference in fitness between the heterozygote and normal homozygote must be high. Actual curves of the approach to equilibrium are shown in Figure 4.6. The equilibrium is, in theory reached only asymptotically (at infinite time), as usual, but gene frequencies will approach it in fewer than 100 generations for s values, and hence also q_e values, above 0.1. The lowest curve in Figure 4.6 (for $s = 0.05$) takes 250 generations to reach $q = 0.047$; the actual equilibrium value is $q_e = 0.04762$.

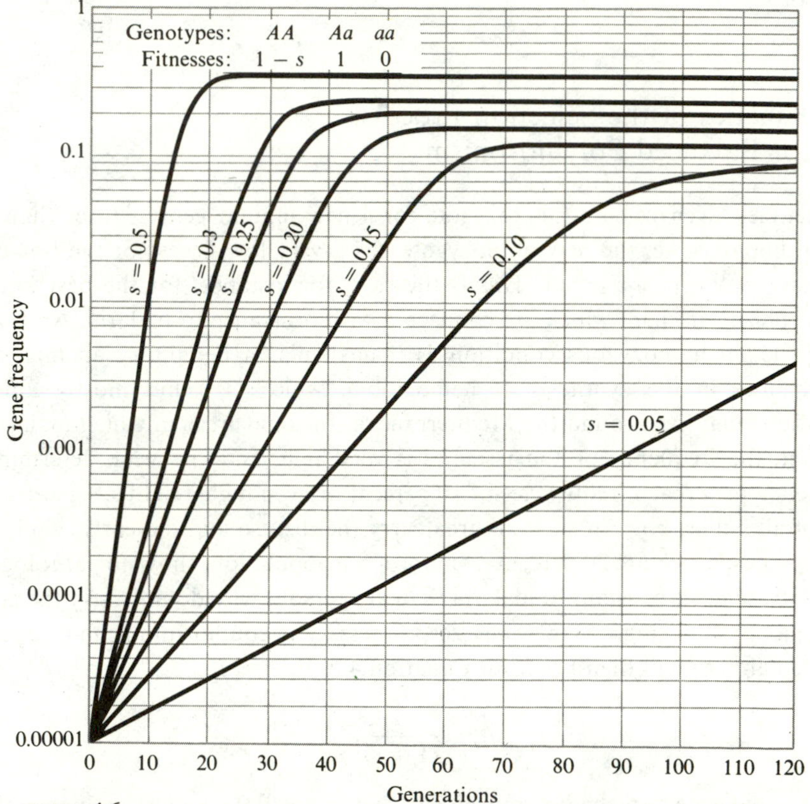

FIGURE 4.6
Approach to equilibrium for a lethal allele capable of establishing a balanced polymorphism ($t = 1$). The situation corresponds to that of selection for the thalassemic gene in a malarial environment; s is the selection coefficient against the normal homozygote.

It is worth mentioning, incidentally, that the curve of approach to equilibrium for a lethal gene is the only one that has a complete analytical solution in the discrete case. The mathematical treatment is given in Example 4.1. The curves of Figure 4.6 were computed using the numerical iteration procedure, which, apart from rounding off errors, gives the same results as the exact discrete solution.

A related example is that of a quasi-lethal condition, sickle-cell anemia, to be discussed in the next section. The fitness of the homozygote for the sickle-cell gene ranges between being slightly higher than zero and at most $\frac{1}{3}$ that of the normal homozygote (which would give $t = 0.67$). The gene frequency at equilibrium is often as high as 0.15, that is, heterozygotes constitute about 30 percent of the population. An equilibrium at $q = 0.15$ requires an s value such that $q_e = 0.15 = s/(s + t)$: with $t = 0.67$ this gives $s = tq_e/(1 - q_e) = 0.117$. Curves for various values of s are given in Figure 4.7 and are very similar to those of

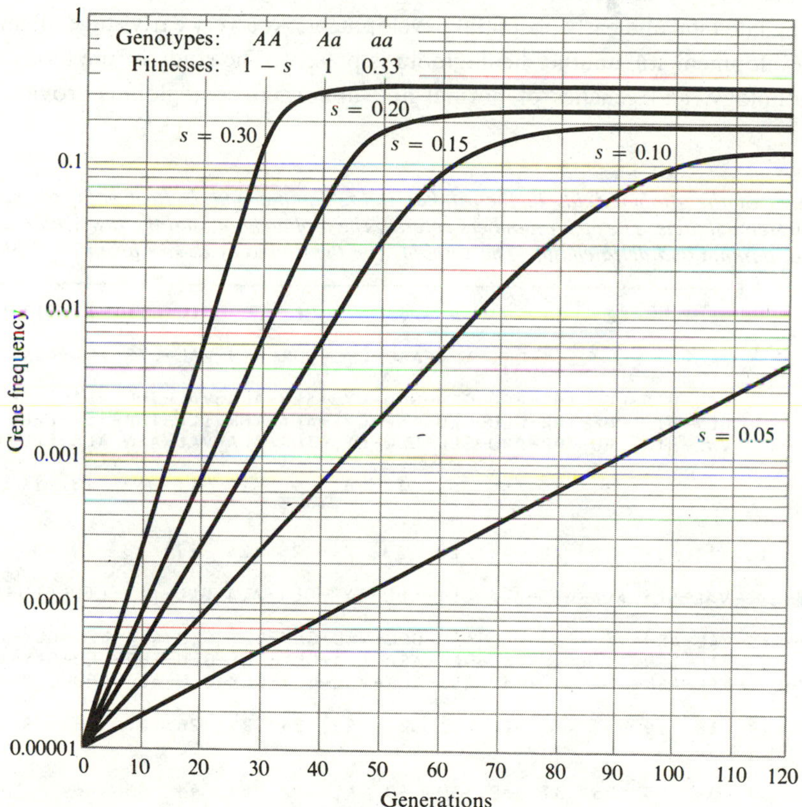

FIGURE 4.7
Approach to equilibrium for a polymorphic semilethal, whose homozygote has a fitness one-third that of the heterozygote ($t = 0.67$) where $1 - s$ is the fitness of the normal homozygote with respect to the heterozygote. This is similar to the case of selection for sickle-cell anemia in a malarial environment.

Figure 4.6. The process is only slightly slower. The curve for $s = 0.05$ reaches a value of 0.069 after 270 generations; the equilibrium is at 0.069440.

4.6 The Sickle-cell Polymorphism

One of the best-known proteins, hemoglobin, gives rise to the most thoroughly studied polymorphism in man, the polymorphism that includes the gene for sickle-cell anemia.

One-hundred milliliters of human blood contain about 15 grams of the protein hemoglobin, which is the major component of red cells and is the major vehicle for oxygen transport in the blood. Normal adult red cells contain for the most part a type of hemoglobin called A, and in addition a small amount of a minor component called A_2. Fetal red cells have hemoglobin F, which is gradually replaced by A during late fetal and early postnatal development. At a very early stage of embryonic development still another hemoglobin is present. The study of the biochemistry, pathology, and genetics of normal and abnormal hemoglobins provides the

TABLE 4.5
Sequences of the Amino Acids in the Alpha, Beta, Gamma and Delta Chains of Normal Human Hemoglobins. The upper numbers refer to the alpha chain, and the lower numbers to the beta, gamma and delta chains. The symbols for the amino acids are given in Table 1.1

	1		2	3	4	5	6	7	8	9	10	11	12	13	14
ALPHA	VAL-		-LEU-	SER-	PRO-	ALA-	ASP-	LYS-	THR-	ASN-	VAL-	LYS-	ALA-	ALA-	TRY-
BETA	VAL-	HIS-	LEU-	THR-	PRO-	GLU-	GLU-	LYS-	SER-	ALA-	VAL-	THR-	ALA-	LEU-	TRY-
GAMMA	GLY-	HIS-	PHE-	THR-	GLU-	GLU-	ASP-	LYS-	ALA-	THR-	ILE-	THR-	SER-	LEU-	TRY-
DELTA	VAL-	HIS-	LEU-	THR-	PRO-	GLU-	GLU-	LYS-	THR-	ALA-	VAL-	ASN-	ALA-	LEU-	TRY-
	1	2	3	4	5	6	7	8	9	10	11	12	13	14	15

	15	16	17	18	19	20	21	22	23	24	25	26	27	28	29	30	31
α	GLY-	LYS-	VAL-	GLY-	ALA-	HIS-	ALA-	GLY-	GLU-	TYR-	GLY-	ALA-	GLU-	ALA-	LEU-	GLU-	ARG-
β	GLY-	LYS-	VAL-	ASN-	-		-VAL-	ASP-	GLU-	VAL-	GLY-	GLY-	GLU-	ALA-	LEU-	GLY-	ARG-
γ	GLY-	LYS-	VAL-	ASN-	-		-VAL-	GLU-	ASP-	ALA-	GLY-	GLY-	GLU-	THR-	LEU-	GLY-	ARG-
δ	GLY-	LYS-	VAL-	ASN-	-		-VAL-	ASP-	ALA-	VAL-	GLY-	GLY-	GLU-	ALA-	LEU-	GLY-	ARG-
	16	17	18	19			20	21	22	23	24	25	26	27	28	29	30

	32	33	34	35	36	37	38	39	40	41	42	43	44	45	46		47
α	MET-	PHE-	LEU-	SER-	PHE-	PRO-	THR-	THR-	LYS-	THR-	TYR-	PHE-	PRO-	HIS-	PHE-		-ASP-
β	LEU-	LEU-	VAL-	VAL-	TYR-	PRO-	TRY-	THR-	GLN-	ARG-	PHE-	PHE-	GLU-	SER-	PHE-	GLY-	ASP-
γ	LEU-	LEU-	VAL-	VAL-	TYR-	PRO-	TRY-	THR-	GLN-	ARG-	PHE-	PHE-	ASP-	SER-	PHE-	GLY-	ASN-
δ	LEU-	LEU-	VAL-	VAL-	TYR-	PRO-	TRY-	THR-	GLN-	ARG-	PHE-	PHE-	GLU-	SER-	PHE-	GLY-	ASP-
	31	32	33	34	35	36	37	38	39	40	41	42	43	44	45	46	47

TABLE 4.5 (*continued*)

```
    48  49  50  51  52  53                      54  55  56  57  58  59

α  LEU-SER-HIS-GLY-SER-ALA-  -   -   -   -   -GLN-VAL-LYS-GLY-HIS-GLY-

β  LEU-SER-THR-PRO-ASP-ALA-VAL-MET-GLY-ASN-PRO-LYS-VAL-LYS-ALA-HIS-GLY-
γ  LEU-SER-SER-ALA-SER-ALA-ILE-MET-GLY-ASN-PRO-LYS-VAL-LYS-ALA-HIS-GLY-
δ  LEU-SER-SER-PRO-ASP-ALA-VAL-MET-GLY-ASN-PRO-LYS-VAL-LYS-ALA-HIS-GLY-

    48  49  50  51  52  53  54  55  56  57  58  59  60  61  62  63  64


    60  61  62  63  64  65  66  67  68  69  70  71  72  73  74  75  76

α  LYS-LYS-VAL-ALA-ASP-ALA-LEU-THR-ASN-ALA-VAL-ALA-HIS-VAL-ASP-ASP-MET-

β  LYS-LYS-VAL-LEU-GLY-ALA-PHE-SER-ASP-GLY-LEU-ALA-HIS-LEU-ASP-ASN-LEU-
γ  LYS-LYS-VAL-LEU-THR-SER-LEU-GLY-ASP-ALA-LLE-LYS-HIS-LEU-ASP-ASP-LEU-
δ  LYS-LYS-VAL-LEU-GLY-ALA-PHE-SER-ASP-GLY-LEU-ALA-HIS-LEU-ASP-ASN-LEU-

    65  66  67  68  69  70  71  72  73  74  75  76  77  78  79  80  81


    77  78  79  80  81  82  83  84  85  86  87  88  89  90  91  92  93

α  PRO-ASN-ALA-LEU-SER-ALA-LEU-SER-ASP-LEU-HIS-ALA-HIS-LYS-LEU-ARG-VAL-

β  LYS-GLY-THR-PHE-ALA-THR-LEU-SER-GLU-LEU-HIS-CYS-ASP-LYS-LEU-HIS-VAL-
γ  LYS-GLY-THR-PHE-ALA-GLN-LEU-SER-GLU-LEU-HIS-CYS-ASP-LYS-LEU-HIS-VAL-
δ  LYS-GLY-THR-PHE-ALA-THR-LEU-SER-GLU-LEU-HIS-CYS-ASP-LYS-LEU-HIS-VAL-

    82  83  84  85  86  87  88  89  90  91  92  93  94  95  96  97  98


    94  95  96  97  98  99 100 101 102 103 104 105 106 107 108 109 110

α  ASP-PRO-VAL-ASN-PHE-LYS-LEU-LEU-SER-HIS-CYS-LEU-LEU-VAL-THR-LEU-ALA-

β  ASP-PRO-GLU-ASN-PHE-ARG-LEU-LEU-GLY-ASN-VAL-LEU-VAL-CYS-VAL-LEU-ALA-
γ  ASP-PRO-GLU-ASN-PHE-LYS-LEU-LEU-GLY-ASN-VAL-LEU-VAL-THR-VAL-LEU-ALA-
δ  ASP-PRO-GLU-ASN-PHE-ARO-LEU-LEU-GLY-ASN-VAL-LEU-VAL-CYS-VAL-LEU-ALA-

    99 100 101 102 103 104 105 106 107 108 109 110 111 112 113 114 115


   111 112 113 114 115 116 117 118 119 120 121 122 123 124 125 126 127

α  ALA-HIS-LEU-PRO-ALA-GLU-PHE-THR-PRO-ALA-VAL-HIS-ALA-SER-LEU-ASP-LYS-

β  HIS-HIS-PHE-GLY-LYS-GLU-PHE-THR-PRO-PRO-VAL-GLN-ALA-ALA-TYR-GLN-LYS-
γ  ILE-HIS-PHE-GLY-LYS-GLU-PHE-THR-PRO-GLU-VAL-GLN-ALA-SER-TRY-GLN-LYS-
δ  A9G-ASN-PHE-GLY-LYS-GLU-PHE-THR-PRO-GLN-MET-GLN-ALA-ALA-TYR-GLN-LYS-

   116 117 118 119 120 121 122 123 124 125 126 127 128 129 130 131 132


   128 129 130 131 132 133 134 135 136 137 138 139 140 141

α  PHE-LEU-ALA-SER-VAL-SER-THR-VAL-LEU-THR-SER-LYS-TYR-ARG

β  VAL-VAL-ALA-GLY-VAL-ALA-ASN-ALA-LEU-ALA-HIS-LYS-TYR-HIS
γ  MET-VAL-THR-GLY-VAL-ALA-SER-ALA-LEU-SER-SER-ARG-TYR-HIS
δ  VAL-VAL-ALA-GLY-VAL-ALA-ASN-ALA-LEU-ALA-HIS-LYS-TRY-HIS

   133 134 135 136 137 138 139 140 141 142 143 144 145 146
```

Source: From McKusick (1968).

most complete information, in any organism, linking molecular understanding of the genetic control of a protein to the evolutionary fate of new mutations. (See, for example, Ingram, 1963; Huehns and Shooter, 1965.)

Each of the three hemoglobins A, A_2, and F has a molecular weight of about 64,000 and is made up of two pairs of identical subunit polypetide chains each with a molecular weight of about 16,000. The two polypeptide chains in hemoglobin A are called alpha (α) and beta (β), those in F are alpha and gamma (γ), and in A_2 alpha and delta (δ). Thus, all the hemoglobins have one common chain, alpha, and one unique chain, beta, gamma, or delta. The molecular formulae of the three hemoglobins can be written in the form $\alpha_2 \beta_2$, $\alpha_2 \gamma_2$, and $\alpha_2 \delta_2$. The three hemoglobins can be distinguished by electrophoresis and other methods of protein separation. The amino acid sequence of the four chains is fully known and is given in Table 4.5. The alpha chain is formed by 141 amino acids, and the beta, gamma, and delta chains are each formed by 146 amino acids. There is considerable similarity between the four chains, showing their common evolutionary origin, which will be further discussed in Chapter 11.

Sickle-cell anemia was the first "molecular disease" to be described.

Using the electrophoretic technique of protein analysis, Pauling, Itano, Singer, and Wells were able to show in 1949 that a hereditary anemia, especially common among Black Americans, and first described some 50 years ago in Chicago by an American physician, Herrick, was characterized by the presence of an abnormal hemoglobin that has a slower electophoretic mobility than A (see Table 4.1). Also in 1949, Neel demonstrated the regular Mendelian inheritance of this hemoglobin difference. Sickle-cell anemia was the first example of a "molecular disease," a term coined by Pauling and his colleagues.

The condition was originally called *sickle-cell* anemia because red cells of the affected carriers are easily deformed to a sickle shape under reduced oxygen tension. Carriers of the abnormal hemoglobin, called hemoglobin S, fall into two disease categories.

1. A mild form of anemia, hardly a pathological condition, but recognizable by the capacity of cells to sickle under reducing treatment occurs in individuals whose red cells contain both normal A and abnormal S hemoglobins. Genetically, they are found to be heterozygous for the gene *S* that determines the synthesis of the abnormal hemoglobin S.

2. A severe form of anemia occurs in individuals who are homozygous for the *S* gene. These individuals, when adults, in general, only have hemoglobin S and no A. They have a relatively low fitness. Clearly in the heterozygotes, both the allelic genes *A* and *S* are active, with each producing, at a somewhat different rate, its own hemoglobin.

Hemoglobin S differs by one amino acid from A: this simple change accounts for the phenotypes of AS and SS genotypes.

Hemoglobin S was the first mutant protein that was subjected to a complete chemical analysis. Ingram (1957) showed that it differed from its normal counterpart, A by a single amino-acid substitution. The beta polypeptide chain of hemoglobin A is a sequence of 146 amino acids with glutamic acid in position 6. In hemoglobin S, this amino acid is replaced by valine. As valine, in contrast with glutamic acid, has no free acidic group, this change causes a difference in the electrical charge, which is the basis for the separation and characterization of hemoglobins S and A by electrophoresis. It is interesting to note that hemoglobin S was, in fact, the first clear-cut example of the relationship between a gene mutation and an amino-acid substitution in a protein.

The molecular basis for the sickling is now, at least partially, understood. Hemoglobin S forms large crystal aggregates under conditions of low oxygen tension. The aggregation is, apparently, the result of a tendency of that end of the beta chain near position 6 to form a ring of amino acids when valine replaces glutamic acid in this position. These rings result in an interlocking of adjacent hemoglobin molecules, which leads to the formation of the crystalline aggregates.

As a consequence of the sickling, the red cells often lyse. *In vivo* this tends to occur especially in capillaries. Many symptoms of the disease are explained by this behavior, especially the anemia, which is hemolytic, and the thrombotic symptoms, which are started by the accumulation of red cell ghosts in smaller blood vessels.

Many other mutant hemoglobins are known.

Application of the electrophoretic technique has permitted the detection of a large number of other mutant hemoglobins, most of which, like hemoglobin S, were later shown to differ from normal hemoglobins in having an amino acid substitution presumably due to a nucleotide substitution in the DNA. Present knowledge of the amino acid changes accompanying known mutations is summarized in Table 4.6. Of these changes, only three are clearly polymorphic in some well-specified human populations: hemoglobin S, hemoglobin C, which is specified when lysine replaces glutamic acid at position 6 in the beta chain (the same position at which the hemoglobin S substitution occurs), and hemoglobin E, which is specified when lysine replaces glutamic acid at position 26 of the beta chain. Pedigree studies have shown that the alpha and beta chains are synthesized by unlinked genes. The delta chain gene is closely linked to the beta chain gene, as shown by pedigree and other studies. The gamma chain gene is also probably linked to the beta and delta chain genes. The three polymorphisms known for hemoglobin, S, C, and E thus all arise

TABLE 4.6
Mutants Carrying Single Amino Acid Substitutions in Hemoglobin

HEMOGLOBIN ALPHA CHAIN

POSITION	FROM	TO	HEMOGLOBIN
5	ALA	ASP	J(TORONTO)
12	ALA	ASP	J(PARIS-1)
15	GLY	ASP	J(OXFORD)
15	GLY	ASP	I(INTERLAKEN)
16	LYS	ASP	I
22	GLY	ASP	J(MEDELLIN)
23	GLU	GLN	MEMPHIS
23	GLU	GLN	CHAD
30	GLU	GLN	G(HONOLULU)
30	GLU	GLN	G(SINGAPORE)
30	GLU	GLN	G(HONGKONG)
30	GLU	GLN	G(CHINESE)
47	ASP	HIS	SEALY
47	ASP	HIS	SINAI
47	ASP	HIS	HASHARON
47	ASP	GLY	UMI
47	ASP	GLY	L(FERRARA)
47	ASP	GLY	KOKURA
47	ASP	GLY	TAGAWA II
47	ASP	GLY	BEILINSON
51	GLY	ARG	RUSS
54	GLN	ARG	SHIMONOSEKI
54	GLN	ARG	HIROSHIMA
54	GLN	GLU	MEXICO
54	GLN	GLU	J(PARIS-2)
57	GLY	ASP	NORFOLK
57	GLY	ASP	G(IBADAN)
57	GLY	ASP	NISHIKI
57	GLY	ASP	KAGOSHIMA
58	HIS	TYR	M(BOSTON)
58	HIS	TYR	M(OSAKA)
58	HIS	TYR	M(GOTTENBERG)
58	HIS	TYR	M(LEIPZIG-2)
68	ASN	LYS	G(BRISTOL)
68	ASN	LYS	G(PHILADELPHIA)
68	ASN	LYS	D(ST. LOUIS)
68	ASN	LYS	KNOXVILLE-1
68	ASN	LYS	STANLEYVILLE-1
68	ASN	LYS	
87	HIS	TYR	M(KANKAKEE)
87	HIS	TYR	M(IWATE)
87	HIS	TYR	M(SHIBATA)

TABLE 4.6 (*continued*)

HEMOGLOBIN ALPHA CHAIN

POSITION	FROM	TO	HEMOGLOBIN
92	ARG	LEU	CHESAPEAKE
92	ARG	GLN	J(CAPE TOWN)
112	HIS	ASP	HOPKINS-2
115	ALA	ASP	J(TONGARIKI)
116	GLU	LYS	O(INDONESIA)

HEMOGLOBIN BETA CHAIN

POSITION	FROM	TO	HEMOGLOBIN
2	HIS	TYR	TOKUCHI
	GLU	VAL O	S
6	GLU	LYS	C
6	GLU	LYS	X
6	GLU	VAL	C(HARLEM)
(THE ABOVE HAS A SECOND CHANGE AT BETA 73, Q.V.)			
6	GLU	VAL	C(GEORGETOWN)
(THE ABOVE HAS A SECOND CHANGE IN THE BETA CHAIN.)			
7	GLU	GLY	G(SAN JOSE)
7	GLU	LYS	SIRARAJ
16	GLY	ASP	J(BALTIMORE)
16	GLY	ASP	N(NEW HAVEN-2)
16	GLY	ASP	J(TRINIDAD)
16	GLY	ASP	J(IRELAND)
16	GLY	ARG	D(BETA-BUSHMAN)
22	GLU	ALA	G(COUSHATTA)
26	GLU	LYS	E
30	ARG	SER	TACOMA
42	PHE	SER	HAMMERSMITH
43	GLU	ALA	G(GALVESTON)
43	GLU	ALA	G(TEXAS)
43	GLU	ALA	G(PORT ARTHUR)
46	GLY	GLU	K(IBADAN)
47	ASP	ASN	G(COPENHAGEN)
61	LYS	ASN	HIKARI
61	LYS	ASN	N(SEATTLE)

TABLE 4.6 (*continued*)

HEMOGLOBIN BETA CHAIN

POSITION	FROM	TO	HEMOGLOBIN
63	HIS	TYR	M(SASKATOON)
63	HIS	TYR	M(EMORY)
63	HIS	TYR	M(KURUME)
63	HIS	TYR	M(CHICAGO)
63	HIS	TYR	M(HAMBURG)
63	HIS	ARG	ZURICH
67	VAL	GLU	M(MILWAUKEE-1)
67	VAL	ALA	SYDNEY
69	GLY	ASP	J(RAMBAM)
69	GLY	ASP	J(CAMBRIDGE)
70 OR 76	ALA	GLU	SEATTLE
73	ASP	ASN	C(HARLEM)
77	HIS	ASP	J(IRAN)
79	ASP	ASN	G(ACCRA)
87	THR	LYS	D(IBADAN)
90	GLU	LYS	AGENOGI
92	HIS	TYR	M(HYDE PARK)
94	ASP	ASN	OAK RIDGE
95	LYS	GLU	N(BALTIMORE)
95	LYS	GLU	N(JENKINS)
95	LYS	GLU	HOPKINS-1
98	VAL	MET	KOLN
99	ASP	HIS	YAKIMA
102	ASP	THR	KANSAS
113	VAL	GLU	NEW YORK
120	LYS	GLU	HIJIYAM
121	GLU	LYS	O(ARABIC)
121	GLU	GLN	D(PUNJAB)
121	GLU	GLN	D(LOS ANGELES)
121	GLU	GLN	D(CYPRUS)
121	GLU	GLN	D(PORTUGAL)

TABLE 4.6 (*continued*)

HEMOGLOBIN BETA CHAIN

POSITION	FROM	TO	HEMOGLOBIN
130	ALA	GLU OR ASP	K(CAMEROON)
132	LYS	GLN	K(WOOLWICH)
136	GLU	ASP	HOPE
143	HIS	ASP	KENWOOD
145	TYR	HIS	RAINIER

OF THE ABOVE, HEMOGLOBINS C(HARLEM) AND C(GEORGETOWN) HAVE TWO SUBSTITUTIONS IN THE BETA CHAIN, AND HEMOGLOBIN X HAS A SUBSTITUTION IN BOTH THE ALPHA AND THE BETA CHAIN.

HEMOGLOBIN GAMMA CHAIN

POSITION	FROM	TO	HEMOGLOBIN
UNCERTAIN	GLN	ALA	F(HOUSTON)
5 OR 6	GLN	LYS	F(TEXAS)
121	GLU	LYS	F(HULL)

HEMOGLOBIN DELTA CHAIN

POSITION	FROM	TO	HEMOGLOBIN
2	HIS	ARG	SPAKIA
16	GLY	ARG	A(2)PRIME OR B(2)
22	ALA	GLU	FLATBUSH

Source: From McKusick (1968).

at the same locus and the first two of them differ at the same amino acid site. The C hemoglobin involves a change in the first nucleotide of the relevant triplet, and the S hemoglobin in the second, as can be checked by reference to the table of the genetic code (Table 1.1).

In addition to amino acid substitutions, other types of mutations affecting the hemoglobins are known in which a chain is not synthesized at all, exceptional hemoglobins are found, or both. Some changes are thought to be due to mutations in "regulatory" genes—that is, genes that are not connected with the structure of the protein but with the regulation of the production of the chains that form it.

4.7 Relationship Between the Sickle-cell Polymorphism and Malaria

Heterozygous sickle-cell trait carriers are probably more resistant to malaria than normal homozygotes.

The sickle-cell polymorphism is mostly limited to colored people of African origin, although it is found in populations in many geographical areas. More specifically, the gene for hemoglobin S is especially common in West and Central Africa, where it reaches frequencies as high as 16 percent. Heterozygotes may thus constitute almost 30 percent of the population before selection and up to 35 percent after selection, while the frequency of homozygotes at birth can be as high as 2.5 percent. In most of West Africa the gene frequency is around 10 percent and the frequency of homozygotes thus about 1 percent. Among black Americans the frequency of the trait is about 9 percent, corresponding to a gene frequency of approximately 0.05. A lower incidence of the trait is found in some Mediterranean areas (Greece and Sicily) and among some aboriginal Indian tribes. (See Figure 4.8.)

The areas in which the polymorphism occurs all have a high endemic incidence of malaria, especially the more malignant types caused by *Plasmodium falciparum*. (See Figure 4.9.) On the basis of this, Haldane (1949a) suggested that the maintenance of this polymorphism (or more precisely of a similar polymorphism, that in which the gene for thalassemia occurs) was due to the heterozygote *AS* being at an advantage in malarial areas in having an increased resistance to malarial infection. The homozygote *SS*, having sickle-cell anemia, is at a very severe selective disadvantage. Estimates of the fitnesses of the different genotypes will be given later.

Following the analysis of the simplest conditions for the existence of a balanced polymorphism, it is difficult to understand how the sickle-cell polymorphism could be maintained at such a high frequency in the face of very strong selection against the homozygote, unless there were selection in favor of the heterozygote.

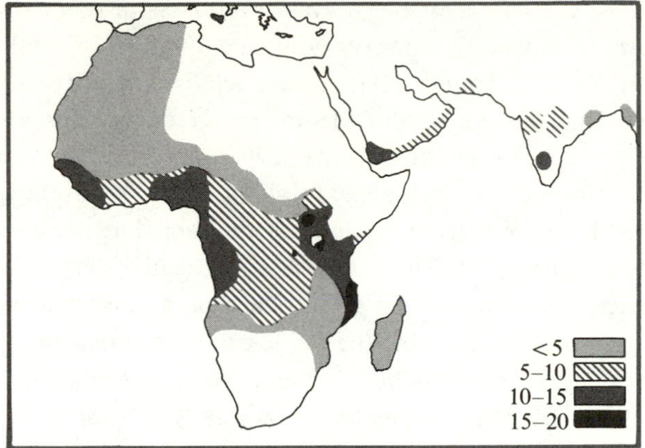

FIGURE 4.8
Frequency of the sickle-cell gene Hb_{β}^{s} in various parts of the Old World. The key in the right-hand corner shows the percentages of the populations that bear the gene. (After Allison, 1961.)

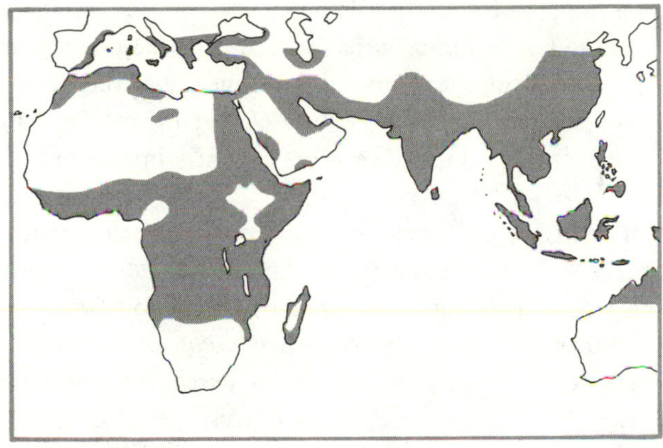

FIGURE 4.9
Distribution of falciparum malaria before 1930. (From M. F. Boyd's *Malariology*.)

Since Haldane's original suggestion, much evidence has accumulated in support of the idea that the heterozygote for the sickle-cell gene is at an advantage over both homozygotes and that this advantage is due to an increased resistance to malaria (following, especially, the studies of Allison 1954a, b, and later). The evidence in favor of the heterozygote's being at an advantage comes from two main sources.

1. In areas where malaria is prevalent, the frequency of the sickle-cell trait increases with increasing age. Differential mortality favoring the heterozygote would be expected to have just this effect. The ratio of the frequency of the sickle-cell trait carriers among newborns to that among reproducing adults should, in fact, supply

a direct estimate of the fitness of the normal homozygote relative to that of the heterozygote. This estimate is, however, only complete if mortality is the sole contributor to the fitness differences. An average relative viability for *AA* relative to *AS* of approximately 0.85 was obtained from seven studies summarized by Rucknagel and Neel (1961). Although there are wide fluctuations in the viability estimate from one study to another, they all agree in giving a lower viability to the normal homozygote than the heterozygote. In spite of the fact that these studies were all done in areas where malaria is an important cause of death, it is hard to measure the relative importance of the disease. In the absence of a direct correlation between malarial deaths and the frequency of the sickle-cell trait, this cannot even be considered sufficient evidence that malaria is the cause of the differential survival.

If we assume the fitness of the normal homozygote is 0.85 relative to a fitness of 1 for the heterozygote, $s = 1 - 0.85 = 0.15$, and the expected equilibrium gene frequency, as given by Equation 4.5 with $t = 1$, is $q_e = 0.15/(1 + 0.15) = 0.13$. This is certainly of the right order of magnitude, suggesting that the observed differential survival may be a major part of the explanation for the polymorphism. We have already seen that an *SS* homozygote fitness higher than zero, and therefore with $t < 1$, does not seriously affect this semiquantitative consideration. This rough estimation will be improved upon in a later section.

Black Americans now live in areas that are largely free of tropical diseases. As might be expected, therefore, the trend in the frequency of the sickle-cell trait with age is reversed: fewer older people carry the sickle-cell trait. This suggests that, in the absence of malaria, the sickle-cell trait may slightly impair fitness. This suggestion is further discussed in Chapter 8.

2. At least a part of the difference in fitness between heterozygotes and homozygotes may be due to differential fertility. Some studies do suggest that heterozygous women have a higher fertility than normal homozygous women. The observed increase in the number of children born to heterozygous mothers varies between studies from nearly 0 to 45 percent. The latter percentage would be more than enough to guarantee that there were a polymorphism even if there were no differential survival. In fact, it leads to an estimate of the equilibrium frequency that is higher than that found in the population of Black Caribs from which the estimate of greater fertility was obtained (Firschein, 1961). The implications of this will be discussed further in the paragraphs that follow. A possible explanation for the higher fertility of heterozygous women is that they have a lower abortion rate. Malarial infection of the placenta, which often causes fetal suffering and death, is sensibly believed to be less severe in sicklers than in normals, thus possibly accounting for the increased fertility.

It must be remembered, when evaluating these studies, that demographic data are less readily available, more difficult to collect, and, at the same time, less reliable in the areas where malaria is hyperendemic.

There are four lines of evidence that it is actually malaria which is responsible for the heterozygote advantage.

1. The geographic distribution of malaria is positively correlated with that of the sickle-cell gene, as has already been discussed. This correlation cannot be expected to be too close for the following reasons: There may be fluctuations in the intensities of local infections. People may move away from highly infested regions to healthier surroundings. Some time, perhaps even centuries or millenia, is necessary for genetic adaptation to new local conditions. Finally, there may be genetic adaptations other than S hemoglobin contributing to decrease the correlation.

2. Direct experimentation has provided preliminary evidence that *AS* sicklers have increased resistance to malaria. Allison (1954b) infected 15 sickler volunteers and 15 normal volunteers from the Luo population in Africa with *Plasmodium falciparum*. Malarial parasites were subsequently found in the blood of 14 of the normals but in only 2 of the sicklers, a highly significant difference. This evidence was apparently contradicted by later tests on a group of black Americans reported by Beutler and others (1955) who were not able to find any significant difference in the numbers of sicklers and normals whose blood contained the parasites after experimental infection. It must be emphasized, however, that the black Americans were not pre-immunized by previous contacts with malaria, in marked contrast to the African group treated by Allison. It is also noteworthy that the American normals had to be treated for malaria much sooner than the Africans.

3. There is a small but significant difference between the parasite counts per unit blood volume following spontaneous infection in sicklers and normals, the latter giving higher counts than the former (Allison, 1954b).

4. Perhaps the most striking and direct evidence for an association between resistance to malaria and sickling comes from hospital investigations of malarial mortality. Though *mild* malarial morbitidy is not, in general, significantly lower in sicklers than in normals, the relative incidence of *severe or fatal* infection is strikingly lower among sicklers. Thus, in a body of data summarized by Allison in 1964, the incidence of severe infections in children was about twice as high among those of *AA* genotype (about 20 percent) as among those of *AS* genotype (about 10 percent). Similarly Motulsky (1964) reported (summarizing five independent studies) that of 100 children who died of malaria only one was *AS*; from the observed population frequencies of the *S* gene it might have been expected that 22 or 23 of the 100 children would be *AS*—were there no relation between the genotype and the disease. There can, thus, be little doubt that sickling confers a very significantly increased resistance to at least the severer forms of malaria such as the cerebral and renal forms of the disease. Allison has emphasized that the first two years of life—the period before the onset of acquired immunity—are the stage of greatest susceptibility, in spite of the fact that the infant derives some immunity from the mother. Studies aimed at detecting changes in the proportion of the *AS* genotypes

with age must, therefore, to be satisfactory, start with newborn infants. The rate of malarial infection in hyperendemic areas is very high—in Allison's words, "one or more infected bites per night."

There is evidence for some differential fertility of AS heterozygotes, and combined mortality and fertility estimates adequately account for observed frequencies of the sickle-cell gene, assuming a stable balanced polymorphism.

Selection may, of course, take place both through differential mortality and differential fertility. It can be shown that the equilibrium values calculated before, assuming constant fitness differences due to mortality, also apply if fitness differences are due solely to differential fertility in one sex. Although there are not many good estimates of fertility differences, as already pointed out, the general indication is that observed differences are not adequate, on their own, to maintain the polymorphism. Assuming that the fertility of a mating is proportional to the product of the fertilities of the two mates, plausible models can be constructed that combine the effects of fertility and mortality. This assumption is reasonable, since fertility effects, in this case, probably act mostly through the female. If we take 0.85 as the overall relative viability of genotype AS as compared to AA, then it can be shown that to reach an equilibrium frequency of 25 percent for S requires that $AS \times AA$ matings be of 4 percent higher fertility than $AA \times AA$ matings (Bodmer, 1965). Such a small fertility difference would be hard to exclude and is certainly not contradicted by available data. The high relative fertility of 1.45 for $AS \times AA$ as compared with $AA \times AA$ matings among the Black Caribs of British Honduras is, in fact, incompatible with the frequency of sicklers in that population, namely 24 percent (Firschein, 1961). Taken at their face value the data imply a 16 percent viability *deficiency* of AS as compared with AA that is overcompensated by the excess fertility of $AS \times AA$ matings. The possibility that fertility and viability act in opposite directions must not be overlooked. As will be discussed later in this chapter, such opposition can lead to polymorphisms when the heterozygote is not obviously at an overall advantage.

Though the available estimates of fitness are not very satisfactory, they are nevertheless remarkably close to the values required to explain this polymorphism even on the basis of the simplest models. The implied selection coefficients are rather high, so that there should be a relatively rapid approach to equilibrium. It may therefore be reasonably safe to assume that the populations investigated are in equilibrium. Clearly, however, if there have been large fluctuations in the prevalence of malaria from time to time, this assumption will be wrong. Recent displacements in populations, which are especially frequent in Africa, may also jeopardize the assumption of equilibrium. On the other hand, the fact that data and theory fit reasonably well certainly supports the contention that it is reasonable to assume that certain African populations are near equilibrium for these genes.

Some numerical examples of the approach to equilibrium for selective values appropriate to the sickle-cell polymorphisms were shown in Figure 4.7. The time necessary for reaching equilibrium gene frequencies of the order of 10–15 percent, as observed in most of Central and West Africa (Figure 4.8) is of the order of 70–100 generations, as can be obtained from the curves given in Figure 4.7. This is not in disagreement with present guesses on the history of malaria in Africa. It is believed that the diffusion of the disease was greatly enhanced by the development of agriculture (Wiesenfeld, 1967). By clearing the forest and creating greater concentrations of people in villages, this development favored both the spread of the vector and of the parasite. Agriculture in West Africa started ca. 2000 B.C., thus giving more time than would be required for reaching the polymorphic equilibrium. If *S* mutants were there at the beginning, equilibrium may have been reached a thousand or more years ago. The Bantu speakers who spread to central and southern Africa from a western habitat some centuries ago probably brought the gene with them already at an equilibrium concentration. In agreement with these ideas, in areas of Africa where the forest is still relatively undisturbed, forest dwellers (Pygmies) tend to have lower frequencies of sickling.

We will consider in Chapter 12 the consequences of relaxed selection due to disappearance of malaria, as for example, is probably the situation with black Americans.

There are practically no data on the mutation frequencies of the sickle-cell hemoglobin gene. The geographic distribution of the *S* gene suggests that the mutation may have arisen independently three times at most, in western or central Africa, in the Mediterranean area, and in Central Asia—in other words, no more than three mutations have occurred, which were successful in spreading throughout the population. It is possible, of course, that populations in which the gene is not found, but which were nevertheless exposed to malaria, did not provide suitable conditions for the increase of the gene, even though the mutation did occur or was introduced by migration. Most populations exposed to malaria in which the *S* gene is not present seem to possess other genetic adaptations to malaria.

The physiological basis of the resistance to malaria can be explained in terms of the structural difference between hemoglobins A and S.

The basic cause of the resistance to malaria is presumably the reduced vigor of the malarial parasite in the *AS* sickler environment. Though the direct evidence is not absolutely conclusive, it is indeed hard to think of any reasonable alternative hypothesis. *Plasmodium falciparum* is known to be very sensitive to its environment and quite host-specific. Two plausible explanations for the reduced vigor of the parasite in sicklers have been suggested.

1. The parasite probably derives most of its nutrition from the hemoglobin in the red blood cells. It seems quite likely that the plasmodium is not able to digest the sickle-cell hemoglobin when it has formed crystalline aggregates under conditions of low oxygen tension.

2. The presence of the parasites in the red blood cells may cause the cells to stick to the walls of the capillaries long enough for there to be a significant reduction in oxygen tension, which leads to sickling. The sickled cells are more likely to be phagocytized, which brings about the preferential removal of the parasite (Motulsky, 1964).

Perhaps both of these mechanisms and others that have been suggested play a role in the increased resistance of the *AS* genotype to malaria.

We have now completed the chain of inference from the change in the nucleotide sequence in the DNA, to the change in amino acid sequence, the consequent change in hemoglobin structure leading to sickling of the red blood cells, which confers an increased resistance to malaria in sickle-cell trait carriers but also pathological consequences in the homozygote. The resulting selective advantage to the *S* gene leads to the observed balanced polymorphism. This situation has been discussed in great detail because it is the unique prototype for studies linking molecular mechanisms of genetic defects to their evolutionary fate. In spite of its superficial simplicity, the story is complicated and involved in detail and still not fully understood. Yet for no other polymorphism at the present time do we have the depth of knowledge that can rival our present understanding of the sickle-cell polymorphism.

4.8 Other Polymorphisms That May Be Adaptations to Malaria

In this section, we shall outline briefly the details of other polymorphisms whose selective bases are believed to be resistance to malarial infection.

1. In Hemoglobin C, as we have seen, lysine replaces the glutamic acid in position 6 of the beta chain—the same position at which the change to valine occurs in the mutation that specifies hemoglobin S. The *C* gene occurs in west Africa, though generally at lower frequencies than *S*. Homozygotes for the *C* gene have a much less serious anemia than the sickle-cell homozygotes. Hemoglobins C and S will be discussed further in the next section in regard to the problem of determining the conditions for the polymorphism of three alleles at a single locus.

2. Hemoglobin E, already referred to, is distributed throughout southeast Asia, through the eastern part of India to Borneo, Sumatra, and the neighboring islands (see Figure 4.10). This hemoglobin is the result of a mutation causing a substitution of another amino acid of the beta chain of hemoglobins. Homozygosity for

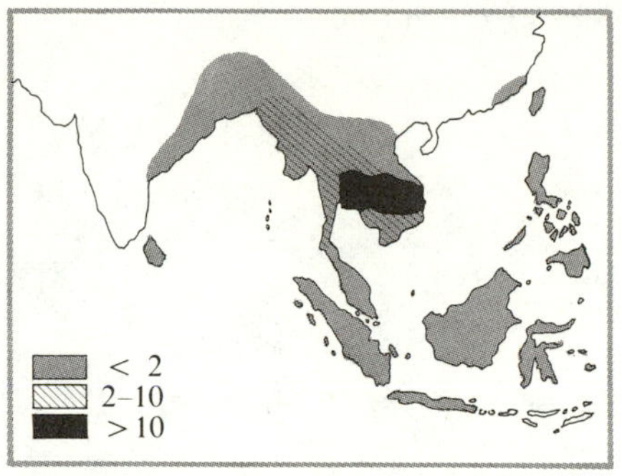

FIGURE 4.10
Frequency in Southeast Asia of the gene that specifies hemoglobin E. The key in the left-hand corner shows the percentages of the populations that bear the genes. (After Allison, 1961.)

this gene causes a mild anemia analogous to that found in homozygotes for the *C* gene. No direct evidence is available to show that carriers of the gene *E* have an increased resistance to malaria.

3. Thalassemia, already mentioned, is a condition that is recognized by a severe anemia (Cooley's anemia) which is almost invariably fatal at an early age in homozygotes. The thalassemia trait, appearing in the heterozygote, is usually nonpathological. There is a lowered cell volume mainly as a result of the smaller diameter of the red blood cells. An increase in the number of red cells compensates for the decreased cell volume. There is also an increased osmotic fragility of red cells.

There are at least two common types of thalassemia, alpha and beta, which are inherited as allelic to alpha and beta chain genes, respectively. The heterozygotes of the beta type have double the amount of hemoglobin A_2 that normals have. The alpha-beta thalassemia double heterozygotes show very mild signs of anemia, confirming that the alpha and beta genes are not allelic.

There are, perhaps, several alleles at each locus that determine different degrees of disease. The typical thalassemia alleles, however, cause a very severe reduction in synthesis of the respective chain, alpha or beta. Several hypotheses have been put forward to explain thalassemia: Some are based on a structural change; more have assumed a change at the regulatory level, which decreases or blocks completely the function of the affected allele.

The connection of thalassemia with malaria is shown by the correlation of the distributions of the two diseases. Studies on the Italian peninsula, and, in greater detail in Sardinia, where a fine mapping of the endemicity of the malarial infection was available (Figure 4.11), have shown that the correlation between the incidence of malaria and that of thalassemia is fairly close. This implies a low mobility of the population, which is in agreement with what is known of the local demography, and

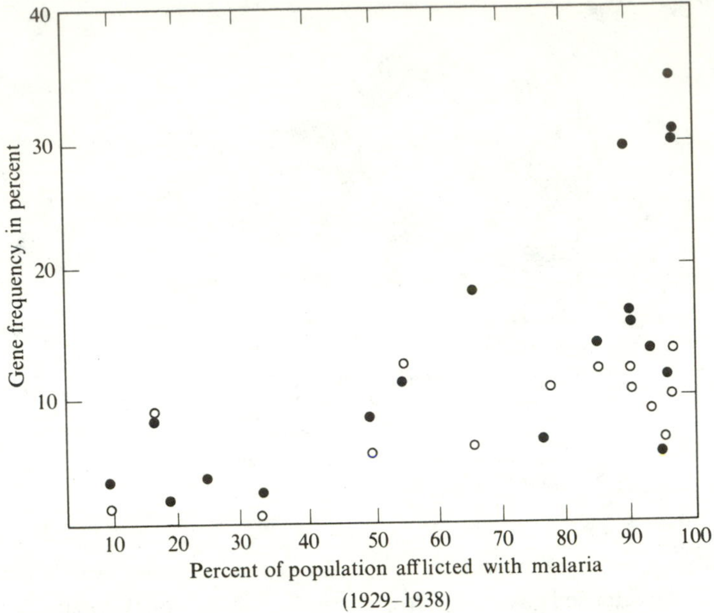

FIGURE 4.11
Correlation between malarial incidence (abscissa) and frequencies of the genes for thalassemia (open circles) and for GP6D deficiency (solid circles) in Sardinian villages. (From Siniscalco and others, 1961).

a stability of malaria with time. The latter consideration is in agreement with the high resistance of the vector to experimental eradication, which failed in spite of concentrated efforts. The malarial parasite was probably imported to Sardinia in Roman times, and may have spread very rapidly. This gives just about enough time (about 70 generations) for establishing equilibrium, especially in the more heavily infested areas with higher s values, as shown by the kinetics of increase in Figure 4.6.

Selection relaxation due to eradication of the malarial parasite has been accompanied by the expected decrease in thalassemia incidence in the area of Ferrara in northern Italy.

4. G6PD deficiency disease is a sex-linked condition resulting in the loss of activity of the enzyme glucose-6-phosphate dehydrogenase. This disease is found mainly in the Mediterranean area (Sardinia, Greece, Israel), in Africa (Congo, West and East Africa), and in India. In some places, it reaches very high frequencies. Geographic correlation with malarial endemicity is well documented (see Figure 4.11). A recent, very elegant observation by Luzzatto, Usang, and Reddy (1969) has given a direct proof that G6PD-deficient cells are more resistant to the malarial parasite than normal cells. As will be discussed in Chapter 10, at least a major section of one of the two X chromosomes is inactivated in females. Red cells

of a female heterozygous for a G6PD-deficient allele are thus of two types: those in which the X chromosome carrying the normal allele has been inactivated, which do not have the enzyme, and those in which the deficient allele is inactivated, which have a normal amount of the enzyme. The two types of cells are in approximately equal proportions. In a malarial attack the cells carrying the normal allele are more prone to rapid destruction by the parasite.

As to the biochemical mechanism of the increased malarial resistance of G6PD-deficient cells, it is known that the deficiency of the enzyme G6PD is responsible for a lower concentration in the red cells of a substance, reduced glutathione (GSH), which is necessary for the growth of malarial parasites. The GSH and G6PD deficiency is especially noticeable in old cells, which are attacked preferentially by *falciparum* parasites. They are not preferentially attacked by *Plasmodium vivax* which is responsible for tertian malaria, to which G6PD-deficient individuals are probably no more resistant than normals.

5. A polymorphism for the serum protein haptoglobin, which is capable of binding hemoglobin, has been detected in practically all human populations. This protein is made of two polypeptide chains, alpha and beta, of which only the gene for the first shows polymorphic alleles. Two of these Hp_1^F and Hp_1^S, differ by a single amino acid. A third common allele at the same locus, Hp_2 shows a more conspicuous difference, the corresponding alpha chain being almost twice as heavy as that produced by Hp_1 and being in all probability the product of a duplication of the alpha chain gene, that connects (apart from a minor alteration at the junction) an Hp_1^F and an Hp_1^S chain. The duplicated structure of this chain confers on it special polymerizing properties, which explain its peculiar appearance on electrophoresis (see Smithies, 1964). It also generates the possibility of unequal crossing-over, as in the hypothetical scheme below:

Hp_2 homozygote: the allele Hp_2 is represented as the combination, in tandem, of Hp_1^F and Hp_1^S.

Unequal crossing-over can produce one triplicated gene plus one normal Hp_1 gene, Hp_1^F in the example shown.

The result of an unequal crossover in an Hp_2 homozygote can be a triplicated new type and an original Hp_1 variant. The triplicate type has probably been found in the form of rare alleles called the Johnson phenotypes.

Some studies have shown an apparent excess of heterozygotes, but no clear-cut data exist. A rare presumed genotype, Hp_0, which is unable to form the protein, has also been postulated to exist in some families of African origin (see Giblett,

1969). In Central Africa however, a few populations are found in which very large proportions of individuals have no haptoglobins, very probably as a result of environmental factors, rather than because of a " null " allele (Cavalli-Sforza et al., 1969).

The connection of haptoglobins with malaria, if any, is only indirect. These proteins bind to hemoglobin, and bound hemoglobin is degraded by a specific enzyme. They probably serve an important function in the disposal of hemoglobin freed during the destruction of red cells, and thus may disappear during hemolytic crises. The slight difference in capacity for hemoglobin removal between Hp_1 and Hp_2 may be a cause of selection in hemolytic anemias. It has been suggested that there is an equilibrium between such selection for Hp_2 (in areas where diseases determining hemolysis are frequent) and the production of the Hp_1 type haptoglobin from crossing-over in Hp_2/Hp_2 individuals. This seems unlikely, since the frequency of this particular crossover is totally unknown and must be low, perhaps comparable to a mutation rate.

4.9 The Conditions for Simultaneous Polymorphism of Three Alleles at One Locus

For a stable equilibrium with three alleles at one locus and constant fitness values, heterozygotes in general must be fitter than homozygotes but it is not necessary that all heterozygotes be fitter than all homozygotes.

A number of the known human polymorphisms comprise more than two alleles at a locus. Notable examples may be drawn from the ABO, Rh, and some other blood groups, and the ASC hemoglobins. Assuming all the genotypes have constant fitness values, and mating is at random, it is possible to obtain the general conditions on the fitness values for the coexistence of all alleles in a stable polymorphism (Owen, 1954; Kimura, 1956; Mandel, 1959). These conditions are closely related to the conditions for the increase of a new allele at a locus that is already polymorphic (Bodmer and Parsons, 1960; Haldane, 1957). They imply a certain average advantage of heterozygotes, though it is not necessary for every heterozygote to be fitter than every homozygote. We shall illustrate the nature of these conditions by an analysis of the three-allele case and its application to the ASC hemoglobin polymorphism.

Six genotypes can be formed with three alleles at one locus: three homozygotes and three heterozygotes. The conditions for polymorphism thus depend on the six relative fitness values of these genotypes. Let $\overline{\mathrm{AA}}$, $\overline{\mathrm{SS}}$, $\overline{\mathrm{CC}}$ be respective fitnesses of the homozygotes *AA*, *SS*, and *CC*, and $\overline{\mathrm{SC}}$, $\overline{\mathrm{AC}}$, $\overline{\mathrm{AS}}$ of the heterozygotes *SC*, *AC*, and *AS*. There are a number of forms in which the equilibrium conditions can be given. The one most symmetrical and convenient for numerical calculations is

TABLE 4.7
Fitnesses and Equilibrium and Initial Increase Conditions for Three Alleles at One Locus

Genotypes	*Corresponding Fitnesses*
AA	$\overline{\mathrm{AA}}$
SS	$\overline{\mathrm{SS}}$
CC	$\overline{\mathrm{CC}}$
AS	$\overline{\mathrm{AS}}$
AC	$\overline{\mathrm{AC}}$
SC	$\overline{\mathrm{SC}}$

The quantities A, B, C, F, G, H are defined in terms of the fitnesses as follows:

$$\mathrm{A} = \overline{\mathrm{SC}}^2 - (\overline{\mathrm{SS}} \times \overline{\mathrm{CC}}) \qquad \mathrm{F} = (\overline{\mathrm{AS}} \times \overline{\mathrm{AC}}) - (\overline{\mathrm{AA}} \times \overline{\mathrm{SC}})$$
$$\mathrm{B} = \overline{\mathrm{AC}}^2 - (\overline{\mathrm{AA}} \times \overline{\mathrm{CC}}) \qquad \mathrm{G} = (\overline{\mathrm{SC}} \times \overline{\mathrm{AS}}) - (\overline{\mathrm{SS}} \times \overline{\mathrm{AC}})$$
$$\mathrm{C} = \overline{\mathrm{AS}}^2 - (\overline{\mathrm{AA}} \times \overline{\mathrm{SS}}) \qquad \mathrm{H} = (\overline{\mathrm{SC}} \times \overline{\mathrm{AC}}) - (\overline{\mathrm{CC}} \times \overline{\mathrm{AS}})$$

The quantities K, L, M are now defined as follows:

$$\mathrm{K} = \mathrm{G} + \mathrm{H} - \mathrm{A}, \quad \mathrm{L} = \mathrm{F} + \mathrm{H} - \mathrm{B}, \quad \mathrm{M} = \mathrm{F} + \mathrm{G} - \mathrm{C}.$$

The increase and equilibrium conditions can then be given in terms of A, B, C, K, L, and M as follows:

I. A three-allele balanced polymorphism exists if A, B, C, K, L, and M are all > 0.

II. When newly introduced in a population:
allele *C* will increase in the presence of *A* and *S* if $\mathrm{M} > 0$;
allele *S* will increase in the presence of *A* and *C* if $\mathrm{L} > 0$;
allele *A* will increase in the presence of *S* and *C* if $\mathrm{K} > 0$.

III. At a three-allele equilibrium, when it exists, $p_1 = \mathrm{K/D}$, $p_2 = \mathrm{L/D}$, $p_3 = \mathrm{M/D}$, where $\mathrm{D} = \mathrm{K} + \mathrm{L} + \mathrm{M}$, and p_1, p_2, p_3 are the frequencies of *A*, *S*, and *C*, respectively.

given in Table 4.7. The equilibrium conditions and frequencies are expressed in terms of the six quantities A, B, C, F, G, H given in Table 4.7. It can be shown that these conditions imply that the viability of all homozygotes must be less than the mean population fitness at equilibrium. This represents a kind of average advantage of heterozygotes. There are only four different systems of viabilities that can give rise to a stable equilibrium for all three alleles. They are typified by the following conditions, none of which is, however, by itself sufficient for stable polymorphism.

1. $\overline{\mathrm{AA}}, \overline{\mathrm{SS}}, \overline{\mathrm{CC}} < \overline{\mathrm{AS}}, \overline{\mathrm{AC}}, \overline{\mathrm{SC}}$: Each of the heterozygotes is fitter than each of the homozygotes.

2. $\overline{\mathrm{AA}}, \overline{\mathrm{SS}} < \overline{\mathrm{SC}} < \overline{\mathrm{CC}} < \overline{\mathrm{AC}}, \overline{\mathrm{AS}}$: One heterozygote (*SC*) is less fit than one of its associated homozygotes (*CC*).

3. $\overline{\text{SS}}, \overline{\text{CC}} < \overline{\text{SC}} < \overline{\text{AA}} < \overline{\text{AC}}, \overline{\text{AS}}$: One heterozygote (*SC*) is less fit than the nonassociated homozygote (*AA*).

4. $\overline{\text{SS}} < \overline{\text{SC}} < \overline{\text{AA}}, \overline{\text{CC}} < \overline{\text{AC}}, \overline{\text{AS}}$: One heterozygote (*SC*) is less fit than two homozygotes, the nonassociated homozygote (*AA*) and one of the associated homozygotes (*CC*).

In each case, at least two heterozygotes must be fitter than all of the homozygotes, and no heterozygote can be less fit than all the homozygotes, or even than each of its two associated homozygotes.

Many three-allele systems presumably have evolved from two-allele systems. It is therefore of some interest to consider the conditions under which a third allele, introduced into a population already at a balanced polymorphism for two alleles, increases in frequency. These conditions, which are given in Table 4.7, are closely related to the equilibrium conditions. Thus, a necessary condition for equilibrium is that any one of the three alleles can increase in the presence of the other two.

The combined fate of the S and C genes cannot be predicted with certainty in the absence of satisfactory data on the fitnesses of the genotypes AC, CC, and SC.

It is clear that full knowledge of the fitnesses of the six genotypes formed by the *A*, *C*, *S* alleles is the principal basis for an understanding of evolution of the ACS hemoglobin system. This knowledge can be acquired in several ways. The most direct is the complete demographic investigation of viabilities and fertilities of the genotypes, and even better, of the various mating types. Such data are very scant and insufficient, and difficult (but not impossible) to obtain in developing countries such as those where malaria is still prevalent. Another possibility is to compare gene frequencies before and after selection, on the assumption that selection occurs mainly between birth and reproductive age. This method was used in earlier investigations, but it was later recognized that some selection may occur before birth and much selection in the very first weeks after it, at a time for which data are difficult to obtain and evaluate (the adult phenotypes are not yet fully developed, among other things). Moreover, differential fertility is not taken into account. A third possibility is that of computing fitnesses of the genotypes assuming equilibrium, and using the gene frequencies as indicators of the fitnesses. Such an approach can be used if some homozygous genotypes have known fitnesses—for example, are lethal. Otherwise, the number of unknowns is greater than the number of equations available for estimation. Thus, in the polymorphism that includes the gene for thalassemia, which is known be be lethal in the homozygous state, the frequency of the normal allele at equilibrium is $p_e = 1/(1 + s)$. The selective disadvantage of the normal homozygote can be estimated from the gene frequency assuming equilibrium. If the gene frequency is not at equilibrium, however, the estimates thus obtained will differ widely from the true values. Finally there is a fourth possibility based on the examination of deviations from Hardy-Weinberg equilibria. In general we have seen that these deviations are not very large, but, when selection is very strong, they may be appreciable if sufficiently large samples are investigated. These deviations will, however, measure, predominantly, viability rather than fertility effects.

It is clear that, in the ACS hemoglobin system, strong selection is involved, because frequencies of near-lethals like the sickle-cell anemia gene can attain rather high values. The approach therefore deserves consideration, provided that three important limitations are remembered. The first of these is the fact that populations may be inbred, generating a shortage of heterozygotes (see Chapter 7). However, it is known that the social customs of Africans usually prevent inbreeding, so that it is doubtful that recent inbreeding will determine appreciable deviations from a Hardy-Weinberg equilibrium. The second limitation is that populations may be heterogeneous. We know that, if we pool data from populations having different gene frequencies even if each of them is in Hardy-Weinberg equilibrium, a shortage of heterozygotes will result (Chapter 2 and Example 4.2). This is a more serious limitation than the first one, because it is difficult to collect a large body of data without making recourse to widely different populations. The objection can be met, however, if observations of different origin are kept distinct and Hardy Weinberg equilibria are computed for each population separately. Results of the analysis can then be pooled. They will, of course, present an average picture, and this is to be kept in mind as there could be heterogeneity of fitness values in the geographical area considered. The third limitation of this approach is that fertility effects may not be adequately accounted for.

There is a deviation from Hardy-Weinberg equilibrium for ACS hemoglobin alleles.

Altogether samples from 72 populations of varying size are available from a recent world survey of hemoglobins (Livingstone, 1967). We confine our attention to West Africa, where both *S* and *C* occur with detectable frequencies. The total number of individuals in the sample is in the neighborhood of 33,000. Gene frequencies vary considerably throughout the area, and a correlation diagram for *S* and *C* gene frequencies is given in Figure 4.12. It will be seen that some populations have very few or no *C* genes. There is almost no population that has *C* genes but no *S* genes, and there is a negative correlation between *S* and *C* frequencies when both genes are present. This is not entirely surprising, since they are proportions from the same total. The correlation observed (-0.20) is, however, significantly different from that expected because of the allelic relationship between *S* and *C*.

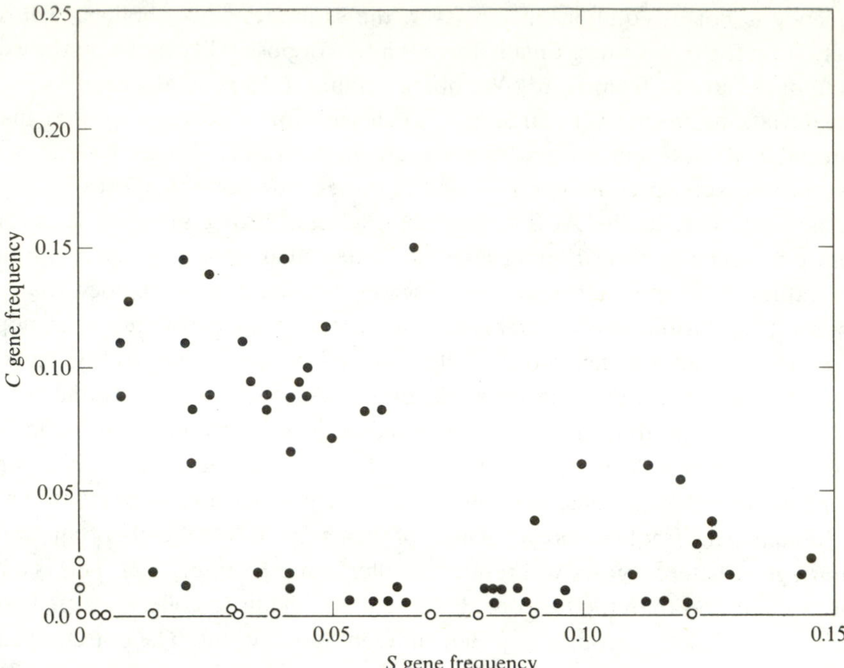

FIGURE 4.12
Frequency of C and S genes in 72 West African populations. (Based on a literature survey by Livingstone, 1967.)

By computing Hardy-Weinberg expectations for each population and summing over all populations, the figures given in Table 4.8 are obtained. The deviation from Hardy-Weinberg equilibrium is highly significant ($\chi^2 = 265.97$ with 5 degrees of freedom). The main features indicate a shortage of AA and SS homozygotes, and an excess of AS heterozygotes, as expected because of the advantage of AS individuals in malarial areas. Thus the genotype distribution confirms the predictions made through other approaches. The situation with respect to C is less clear and must be examined further. (The chi square just given has 5, not 3, degrees of freedom because the gene frequencies were not computed from the six totals given in Table 4.8, but from the totals of each of the populations that were pooled to obtain both the observed and expected totals given in the table.)

It is possible to obtain estimates of fitness (on an arbitrary scale) from these data simply by dividing the observed frequencies by the expected on the assumption that selection is over by the age at which genotypes are counted (which is usually reproductive age). (See also Roberts and Boyo, 1960.) It can be shown that, even if the gene frequencies are far from equilibrium, the error involved is hardly detectable. Unfortunately very large numbers are necessary and this usually introduces heterogeneity. Even the large number of observations on which the data given in Table 4.8

TABLE 4.8

Observations of Hemoglobin A, S, C genotypes and Hardy-Weinberg Expectations, Summed over 72 West African Populations: Computation of Fitnesses (Observed/Expected Frequencies) and Their Confidence Intervals and Standard Errors

Genotype	*AA*	*SS*	*CC*	*AS*	*AC*	*SC*	*Total*
Observed	25,374	67	108	5482	1737	130	32898
Expected	25,615.5	306.87	74.69	4967.2	1768.6	165.01	~32898
Fitness	0.991	0.218	1.446	1.104	0.982	0.788	
99% CONFIDENCE LIMITS							
Lower	1.001	0.432	2.203	1.197	1.021	0.932	
Upper	0.977	−0.118	0.724	1.040	0.938	0.638	
Standardized fitnesses[a] and their standard errors	0.89 ± 0.03	0.20 ± 0.11	1.31 ± 0.29	1	0.89 ± 0.035	0.70 ± 0.07	

[a] Standardized by taking $\overline{AS} = 1$.

are based are not sufficient to make errors adequately small for a complete evaluation of the stability of the system. There is no *direct* evidence that *AC* heterozygotes are more resistant to malaria. The relatively low fitness of *AC* found in Table 4.8, which is comparable to the fitness of *AA*, is not unexpected, but the margin for error is still considerable. Homozygotes for the *C* gene have a much milder anemia than *SS* homozygotes. According to at least one author (Edington, 1959) "patients suffering from pure hemoglobin C disease are at a very slight disadvantage." They do not appear to have "unduly shortened" life expectancy and "women suffering from the condition can successfully bear children." Early estimates or assumptions of low fitness for *CC* were probably incorrect. The summary of available data in Table 4.8 assigns a surprisingly high fitness to *CC*, but the standard error is still large. The *SC* heterozygotes appear to have a mild form of sickle-cell anemia, whose severity is intermediate between *SS* and *CC*. Clinical evidence is not clear-cut. Apparently some *SC* individuals have no symptoms, live a long life, and bear children. For other *SC* individuals, however, the condition is, in general "especially hazardous" for pregnancy (Edington, 1959). The observed fitness probably reflects this variable clinical picture. Homozygotes *SS* fare badly, and the fitness given here, 0.2, is in agreement with the clinical impression of a severe and almost completely lethal condition, if untreated.

In summary, present evidence suggests the following order of fitness values:

$$\overline{\text{AS}} > \overline{\text{AA}}, \overline{\text{AC}} > \overline{\text{SC}} > \overline{\text{SS}}.$$

The fitness of *CC* can be anywhere in this range above that of *SS*, and is probably above that of *SC*. Especially because of the absence of a clear-cut difference between $\overline{\text{AA}}$ and $\overline{\text{AC}}$ (which should be in the order $\overline{\text{AC}} > \overline{\text{AA}}$) none of the conditions for stable equilibrium is clearly met. A more complete analysis, given below, shows how far our present knowledge still is from answering the problem.

The intensity of malaria can hardly be the same over such a vast area and this is likely to lead to heterogeneity of the fitnesses, which should therefore be taken as average values. The presence of some heterogeneity is disclosed by the fact that the chi square, obtained by subtracting from the sum of the Hardy-Weinberg chi squares for each population ($\chi^2 = 622.94$ with 93 degrees of freedom) the one computed on the sums given in Table 4.8 ($\chi^2 = 265.97$ with 5 degrees of freedom), is 356.97 with 88 degrees of freedom, which is very significant. It remains high even if very small populations are excluded in order to increase the accuracy of the χ^2 test.

We shall now test the stability of the three-allele equilibrium. The conditions for equilibrium given in Table 4.7 can be tested on the basis of the computed fitnesses. The six quantities, A, B, C, K, L, M, computed from the data of Table 5.8, are given, with their approximate standard errors, in Table 4.9. Taken at face value, the three-allele equilibrium would not be stable. But, except for C, none of these quantities is significantly different from the threshold for stability (which is zero), and hence the material so far available is insufficient for answering the problem. Perhaps 10 times as many observations might be adequate. Analysis of the distributions of the above values reveals heterogeneity, especially for the equilibrium of *A* and *C*. The values of B and to a minor extent those of C and A tend to be negative in Upper Volta, the region with highest hemoglobin C frequency, and positive elsewhere.

TABLE 4.9
Mean Values with Standard Errors of the Quantities Necessary for Testing the Stability of the Equilibrium of the A, S, C Gene Frequencies

	Mean	*Standard Error*
A	+0.251	±0.184
B	−0.384	±0.268
C	+0.839	±0.098
K	−0.388	±0.365
L	−0.042	±0.408
M	−0.036	±0.172

Note: See Table 4.7 for the definition of A, B, etc. The estimates are based on data given in Figure 4.12 and Table 4.8).

It seems, therefore, that present data give no clear indication of whether the *A*, *C*, and *S* alleles are at a stable equilibrium or not, though at face value an equilibrium situation appears unlikely, contrary to the opinion previously expressed by others on the basis of tenuous evidence. This opposite conclusion was, in fact, in some studies based on a small fraction of the data considered here.

Considerations based on its geographic distribution favor the idea that the C gene is increasing.

The *S* gene is spread relatively homogeneously throughout black populations over the whole of Africa, but the distribution of the *C* gene is quite different. It shows a peak in the Upper Volta region, and from there decreases fairly regularly in all directions. The map of the *C* gene in Africa is far from complete and surprises are possible, but what is known at present suggests that the *C* gene is spreading from a

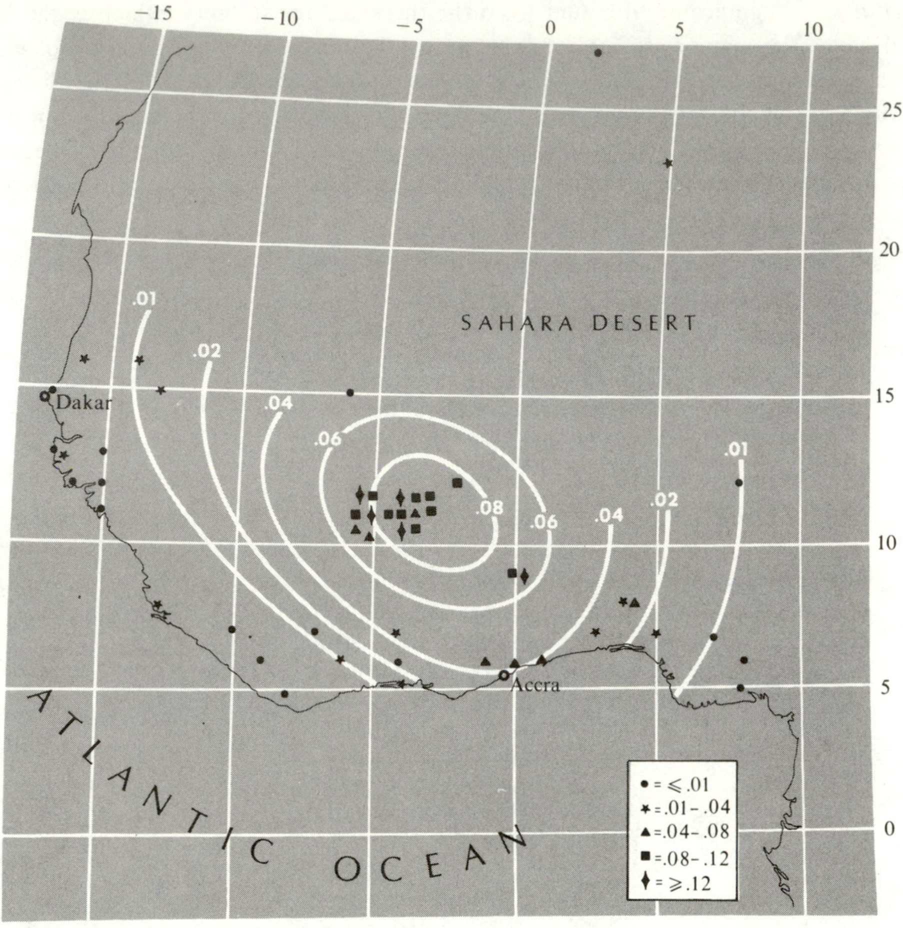

FIGURE 4.13
The spread of the *C* gene in West Africa. Ellipses indicate expected *C* gene frequencies on the basis of a simple diffusion model, described in the text. Symbols are used to indicate *C* gene frequencies of the population samples: a dot stands for a frequency of less than 0.01; a star, 0.01–0.04; a triangle, 0.04–0.08; a square, 0.08–0.12; and a diamond, more than 0.12.

center of origin, probably located near its highest frequency. Where *C* is highest, *S* is lowest, as if *C* were gradually supplanting *S* in a process that is still continuing.

To emphasize this suggestion, we have depicted the distribution of *C* in West Africa (Figure 4.13) having fitted, to the data, a diffusion-by-migration hypothesis represented by a bidimensional gaussian surface (see Appendix I). This is the shape that a migration process would give to the gene frequencies in continuous space (see Chapter 8) if a mutation which had arisen at the center spread gradually under positive selection pressure. The diffusion model used allowed migration to proceed at different rates in the two directions, in order to allow for geographic irregularities.

The ellipses are drawn through points of equal gene frequencies according to the diffusion model. The simple model used fits the data well, but it should be emphasized that a fairly good fit can only be taken as suggestive of a diffusion process, and is not sufficient to demonstrate it.

It is interesting to test whether the necessary conditions for a diffusion process of the *C* gene, which is presumably still going on, are compatible with present knowledge of the fitness values. We have already seen that the value *M*, expressing (Table 4.7) the condition for a newly introduced *C* gene to increase in the presence of *A* and *S*, is not significantly different from zero (Table 4.9) and thus our hypothesis cannot be considered as proved. It is interesting to check if the confidence limits of present estimates are at least compatible with the idea that the *C* gene is gradually increasing, probably at the expense of *S*.

The minimum fitness of AC necessary for C to increase is surprisingly low.

For this analysis, it is convenient to express the value M in a slightly different form, so as to obtain the minimum values of the fitness of *AC* necessary for an increase of *C*. As we are considering the conditions for initial increase, the fitness of *CC* is immaterial, but those of competing genotypes are important, and we will take for them the observed values. We will, however examine a whole range of ratios of $\overline{\text{AS}}$ to $\overline{\text{AA}}$ fitnesses, because the relative fitnesses of these genotypes depend, probably more than those of others, on the intensity of malarial infection, and the conditions at the time when *C* may have arisen are not known. Also, it will thus be possible to predict the behavior of *C* in different ecological niches that have various intensities of malarial infestations.

The formula for M (table 4.7) can be turned into the following condition for *C* to increase in the presence of *A* and *S*:

$$\overline{\text{AC}} > \frac{\overline{\text{AA}}(\overline{\text{SC}} - \overline{\text{SS}}) + \overline{\text{AS}}(\overline{\text{AS}} - \overline{\text{SC}})}{\overline{\text{AS}} - \overline{\text{SS}}}.$$

We will take the fitness of $\overline{\text{AS}}$ as equal to 1, and determine first the fitness of $\overline{\text{AA}}$ corresponding to various equilibrium gene frequencies of the *S* gene, in the presence of *A* only. We assume $\overline{\text{SS}} = 0.2$ on the basis of Table 4.8, using the value $\overline{\text{SS}}/\overline{\text{AS}}$. From the formula for equilibrium, $\overline{\text{AA}}$ must be equal to $1 - (1 - 0.2)q_e/(1 - q_e)$, where q_e is the gene frequency of *S* at equilibrium (see Table 4.10). Given the value of the *AA* fitness corresponding to a certain equilibrium and assuming $\overline{\text{SC}} = 0.7$ (from Table 4.8, using the value of $\overline{\text{SC}}/\overline{\text{AS}}$), we can obtain, from the formula just given, the minimum value of the *AC* fitness for which *C* can increase in the presence of *A* and *S*. With the numerical values given, the formula becomes

$$\overline{\text{AC}} > (0.5\overline{\text{AA}} + 0.3)/0.8,$$

and actual estimates are given in Table 4.10.

TABLE 4.10

Conditions for the Increase of Gene C in the Presence of A and S, on the Assumption that the Fitnesses of AS, SS, and SC are Respectively 1, 0.2, and 0.7, for Various Equilibrium Gene Frequencies of S in the Presence of A

Frequency of S Gene at Equilibrium with the A Gene	*Fitness of AA is Then*	*Minimum Fitness of AC Permitting Increase of C*
0.010	0.992	0.995
0.020	0.984	0.990
0.030	0.975	0.985
0.040	0.967	0.979
0.050	0.958	0.974
0.060	0.949	0.968
0.070	0.940	0.962
0.080	0.930	0.957
0.090	0.921	0.951
0.100	0.911	0.944
0.110	0.901	0.938
0.120	0.891	0.932
0.130	0.880	0.925
0.140	0.870	0.919
0.150	0.859	0.912
0.160	0.848	0.905
0.170	0.836	0.898
0.180	0.824	0.890
0.190	0.812	0.883
0.200	0.800	0.875

It will be noted that $\overline{\mathrm{AC}}$ need be only slightly larger than $\overline{\mathrm{AA}}$ for *C* to increase, and the difference required is well within the margin of error of present estimates.

Thus, in the average conditions of West Africa, the fitness of *AA* is 0.89. Using the observed fitnesses of *SC* and *SS*, the minimum fitness of *AC* that will permit *C* to increase in the presence of only *A* and *S* is 0.932, which is one standard deviation above the observed fitness and therefore compatible with it.

It should be added that, if the fitness of *AC* were just slightly above this threshold value for the increase of *C*, the process of increase would be very slow. Thus, with a fitness of *AC* higher than the threshold by only one percent it would take 500 generations for the frequency of the *C* gene to increase from an initial value of, say, 10^{-4} to 10^{-2} (see Figure 4.14). A fitness of *AC* 5 percent higher than the threshold ($\overline{\mathrm{AC}} = 0.978$) as shown in the figure, would give a kinetics compatible with the time available for selection, as it would bring the gene frequency from 10^{-4} to the present values in 160 generations. The choice of the initial value for the

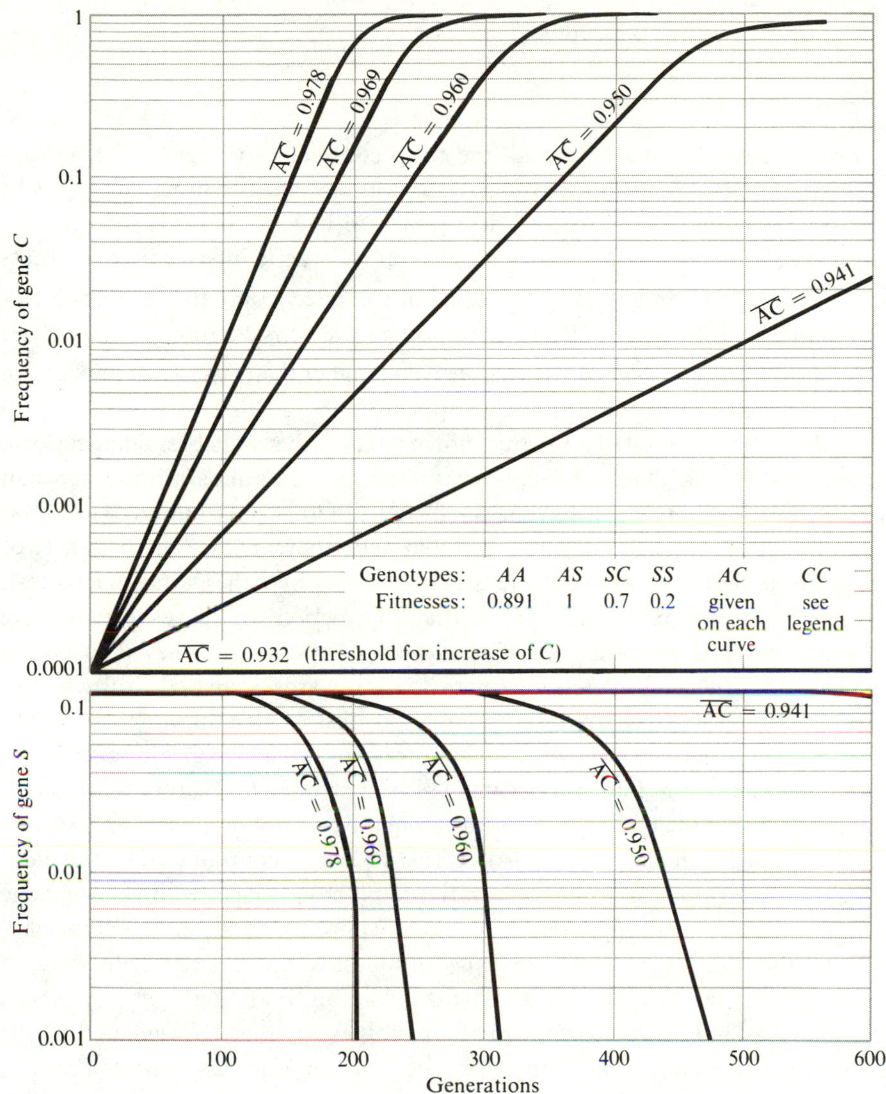

FIGURE 4.14
Kinetics of increase of the *C* gene (upper half) and decrease of the *S* gene (lower half), starting at a *C* gene frequency of 10^{-4} in a population of which *A* and *S* are at equilibrium, with an *S* gene initial frequency of 0.12. Fitnesses of *AC* used in the figure are based on the AC threshold for increase of *C*, which is 0.932, augmented in 1–5 percent. The fitnesses of *CC* are computed as $\overline{AA} + 2(\overline{AC} - \overline{AA})$.

gene frequency (10^{-4}) is based on there being one initial mutant in a group (tribe) of approximate size 10^4.

The polymorphism for C may be transient.

It should be emphasized that in all the cases considered in Figure 4.14, *C* supplants both *S* and *A*, and the polymorphism is transient. The fitness of *CC* with respect to *AC* is not sufficiently well determined to be sure that this is so. It was assumed, in all the curves of Figure 4.14, that *AC* is exactly intermediate in fitness between *AA* and *CC*. In any case, the value of $\overline{\text{CC}}$ affects only the later course of selection and is of no importance with respect to the considerations made so far, except that if it is lower than $\overline{\text{AC}}$ (the latter being higher than $\overline{\text{AA}}$) a polymorphism for *AC* will result, from which *AS* has disappeared.

It should be remembered that we are ignoring many other possible complications. It is possible that hemoglobin *C* supplies resistance to malarial parasites other than *Plasmodium falciparum*, which have a somewhat different distribution. The vagaries of infestation of malarial parasites from year to year, and the absence of satisfactory surveys over such a large area make this hypothesis difficult to test. Moreover, it is not inconceivable that in this long period, the parasite may have undergone an evolutionary change. *Plasmodium falciparum* is believed to be the lastcomer, and, in general, evolution of parasites is towards decreased virulence. But evolution for increased activity—for example, against carriers of *C* genes—might also be possible. Ignorance of other polymorphisms, which might interact in an important way with the known ones, may also contribute to obscure the picture. The hypothesis that the *C* gene is spreading from the point of origin and is, partially at least, supplanting the *S* gene seems attractive and is not contradicted by the data, although it is far from proved. The alternative hypothesis, suggested by Livingstone (1967), that the *S* gene is supplanting *C*, is equally unproved and perhaps somewhat less compatible with the present knowledge of fitnesses. But it is especially difficult to reconcile with the observed geographic distribution.

Perhaps the simplest way to present the hypothesis is that *C* originated at the point where it now has its maximum frequency (Upper Volta or nearby) and *S* at some other unknown location probably at a different time (but this is immaterial, as long as they both appeared more than 3000 years ago). Gene *S*, having a greater heterozygous fitness, has spread much more rapidly and is at equilibrium, or near it, over the whole area, except where it has met the more slowly increasing *C* gene. This allele has a slower kinetics because of the smaller heterozygote advantage, which is nevertheless sufficient to give it a chance to increase even in the presence of *S* and, though slowly, to supplant it. Clearly, only some of the conditions for this theory have been proved; one, namely that $\overline{\text{AC}}$ is larger than the minimum threshold for increase, awaits further data. It is also clear that a full explanation requires the analysis of the interaction between migration and selection, which will be examined in Chapter 8.

4.10 X-linked Polymorphisms

A selectively balanced polymorphism at an X-linked locus can exist either if the female heterozygote is at an advantage or if selection acts in opposite directions in the two sexes.

The theoretical treatment of the conditions for polymorphism at an X-linked locus is complicated by the fact that the two sexes must be treated separately and different fitness values in the sexes must be specified. The conditions were first given in general form by Bennett (1958; see also Edwards, 1961; and Haldane and Jayakar, 1964). Since all matings are necessarily between a male and a female, only relative fitnesses *within* each sex need to be specified. The conditions can thus be given in terms of three parameters, two fitnesses in females and one in males. A fitness scheme for two alleles at an X-linked locus (X and X') can be given in the following form:

	Females			*Males*	
Genotypes:	XX	XX'	$X'X'$	XY	$X'Y$
Fitnesses:	$1-s$	1	$1-t$	$1+m$	$1-m$

The intermediate equilibrium gene frequencies in males and females can be obtained in terms of s, t and m from the equations derived in Example 4.5. The intermediate ($p \neq 0$ or 1) equilibrium frequency of X in males, p, is given by

$$p = \frac{t + m + m^2(1-t)}{t + s + m^2(2 - t - s)},$$

and the frequency P in females is

$$P = \frac{t + m - tm}{s + t + m(s - t)}.$$

When s, t, and m are small,

$$P \sim p \sim \frac{t + m}{t + s}.$$

It can be shown (see Example 4.5) that the intermediate equilibrium exists and is stable only if $(1-t)(1-m) < 1$ and $(1-s)(1+m) < 1$. If $(1-t)(1-m) > 1$ then X cannot increase in a population that is predominantly X' and vice versa if $(1-s)(1+m) > 1$.

Once again, the equilibrium conditions are related to the conditions for increase of a newly arisen allele. Thus only if X can increase when introduced into a predominantly X' population and also if X' can increase when introduced into an X population, will there exist a selectively balanced polymorphism. The equilibrium gene frequencies in males and females are in general different, though the difference is small if selective differences between the genotypes are small. There are four sets of values of the selection coefficients s, t, and m determining different combinations of existence and stability of the three possible equilibria "X-fixed" ($P = p = 1$), "X'-fixed" ($P = p = 0$), and the intermediate equilibrium. These four sets are distinguished according to the signs of s and t—that is, according to the relative fitnesses of the female genotypes. We assume without loss of generality *that* XY *is fitter than* $X'\ Y$, that is $m > 0$.

The four cases, and the corresponding conditions for stability of the various equilibria are as follows:

			X' fixed ($P = p = 0$) is stable	*X fixed ($P = p = 1$) is stable*	*X and X' both present—the intermediate equilibrium is stable*
I	$s<0$	$t<0$	if $m<\dfrac{\lvert t\rvert}{1+\lvert t\rvert}$	always	never
II	$s<0$	$t>0$	never	always	never
III	$s>0$	$t>0$	never	if $m>\dfrac{s}{1-s}$	if $m<\dfrac{s}{1-s}$
IV	$s>0$	$t<0$	if m $<\dfrac{\lvert t\rvert}{1+\lvert t\rvert}$	if $m>\dfrac{s}{1-s}$	if $\dfrac{\lvert t\rvert}{1+\lvert t\rvert}<m<\dfrac{s}{1-s}$

The notation is as given on page 171. The expressions given under the three equilibria are the conditions on m for the existence and stability of the corresponding equilibrium. Note that in Case IV the balanced polymorphism (intermediate equilibrium) can be stable only if

$$\frac{|t|}{1+|t|} < \frac{s}{1-s}. \quad \text{If} \frac{s}{1-s} < \frac{|t|}{1+|t|}, \text{ then for } \frac{s}{1-s} < m < \frac{|t|}{1+|t|}$$

both trivial equilibria are stable (X fixed and X' fixed), and the intermediate equilibrium is never stable.

In words, these four cases can be summarized as follows:

I. There is heterozygote disadvantage in the female. In this case, "X-fixed" is always a stable equilibrium. "X'-fixed" is stable if

$$m < \frac{|t|}{1+|t|}$$

(where $|t|$ is the positive value of t), which when t is small, is approximately $m < |t|$. The intermediate equilibrium is never stable.

II. When X is at an absolute advantage in both males and females only the equilibrium with X fixed is stable.

III. When there is heterozygote advantage in the female, a balanced polymorphism exists if $m < s/(1 - s)$—that is, provided the difference between the fitness of the male genotypes XY and $X'Y$ is not too large. If $m > s/(1 - s)$, then only the equilibrium X is stable.

IV. When X' is at an absolute advantage in females while XY is fitter than $X'Y$ in males, a balanced polymorphism can exist provided

$$\frac{|t|}{1 + |t|} < m < \frac{s}{1 - s},$$

or, for small $|t|$ and s, approximately, $|t| < m < s$. The equilibrium at X is stable if $m > s/1 - s$ and that at X' if

$$m < \frac{|t|}{1 + |t|}$$

so that when

$$\frac{|t|}{1 + |t|} > \frac{s}{1 - s},$$

both X and X' fixed are stable equilibria if

$$\frac{|t|}{1 + |t|} > m > \frac{s}{1 - s},$$

while then the intermediate equilibrium cannot be stable. This case is of interest as an example of how selection acting in opposite directions in the two sexes can, under certain conditions, lead to a balanced polymorphism.

There are, thus, two ways in which a balanced polymorphism can occur at an X-linked locus, assuming that fitness values are constant. Either the female heterozygote is at an advantage or selection acts in opposite directions in the two sexes. In either case, a polymorphism only exists if the difference between the male fitnesses lies within a given range.

There are four important human polymorphisms that are X-linked: color blindness, G6PD deficiency, the Xg blood group, and *Xm* (a serum protein detected immunologically). The last two are not known to be associated with any selective effects so that little or nothing can be said about the forces that maintain them. In the following paragraphs, we will discuss briefly current ideas about the maintenance of the first two of these polymorphisms.

G6PD *deficiency may be a transient polymorphism in malarial areas, and provisional estimates of the selective coefficients associated with it can be obtained from fitting the probable rate of change of* G6PD *frequencies in certain migrant Jewish populations.*

As has already been mentioned, G6PD deficiency occurs mainly in populations that live in those Mediterranean areas in which malaria is prevalent and is probably associated with an increased resistance to malarial infection. No good direct estimates of the genotype fitnesses are, however, available. Published pedigree data show no significant departures from Mendelian expectations, which would be expected only if there were large viability effects. On the other hand, high frequencies of the gene for G6PD deficiency are known in some Jewish populations, which must have originated from populations with very low frequencies for this gene in a comparatively short time, and suggest appreciable selection coefficients associated with G6PD deficiency. The incidence among some Jewish populations and those of some other groups are shown in Table 4.11. The highest frequency, 58.2 percent, is found in Kurdistan. All the populations referred to in Table 4.11 probably originated from emigrants who left Palestine 2000–2500 years ago. Since Palestine is not an area with a high incidence of malaria, it seems most likely that the G6PD deficiency frequencies among the original migrants were all quite low, probably at the level expected for a mutation-selection balance. The high frequencies now found in some areas would then have to be explained by subsequent selection for G6PD deficiency in malarial areas, implying that the selection coefficients were strong enough to change the frequency from about 10^{-3} or 10^{-4} to 50–60 percent in a period of 2,000 years. Using Equation 3.12 for an X-linked trait maintained by mutation-selection balance, namely,

$$qs = 3\mu,$$

where q is the gene frequency, s the selective disadvantage, and μ the mutation rate, gives, for $\mu = 10^{-5}$, $q = 3 \times 10^{-3}$ if $s = 0.01$, and $q = 3 \times 10^{-4}$ if $s = 0.1$. We can form an approximate idea of the value of s needed to change a gene frequency from 10^{-4} to 0.6 in 60 generations (which for humans is about 2,000 years), considering that the process requires a geometric rate of increase of about 1.15 per generation. This corresponds to a selective advantage of about 15 percent. The exact kinetics of the change are shown in Figure 4.15, in which selective advantages are $s = -\alpha$, $t = m = \alpha$, where α is positive and equal to 0.1 and to 0.05, respectively, in the two curves. These were constructed using the equations given in Example 4.5. This corresponds to a situation in which G6PD deficiency has an absolute advantage and the presently observed polymorphism is transient. It will be noted that the selection coefficient postulated is large, but of a similar magnitude to that estimated for the advantage of *AS* over *AA*. Direct estimates will be of great interest.

TABLE 4.11
Incidence of G6PD Deficiency among Presumably Random Samples of Male Jews from Various Communities

Community	*Sample Size*	*% Deficient*
ASHKENAZIM	819	0.4
NON-ASHKENAZIM		
Europe (Sephardim)		
Turkey	256	1.9
Greece and Bulgaria	152	0.7
Others	93	2.2
Asia		
Kurdistan	196	58.2
Iraq	902	24.8
Iran	557	15.1
Caucasus	25	28.0
Afghanistan	29	10.3
Yemen and Aden	415	5.3
Bukhara	46	—
Syria and Lebanon	80	6.3
India (Cochin)	58	10.3
India (Bnei Israel)	102	2.0
North Africa		
Egypt	112	3.8
Morocco	219	0.5
Atlas Mountains	23	1 case
Algiers and Tunisia	112	0.9
Libya	219	0.9
Gerba	52	—
Ethiopia		
Fallasha	208	—
OTHER GROUPS		
Samaritans	69	—
Karaites	18	—
Arabs	264	4.4
Druzes	92	4.4
Circassians	57	—

Source: From Szeinberg (1963).

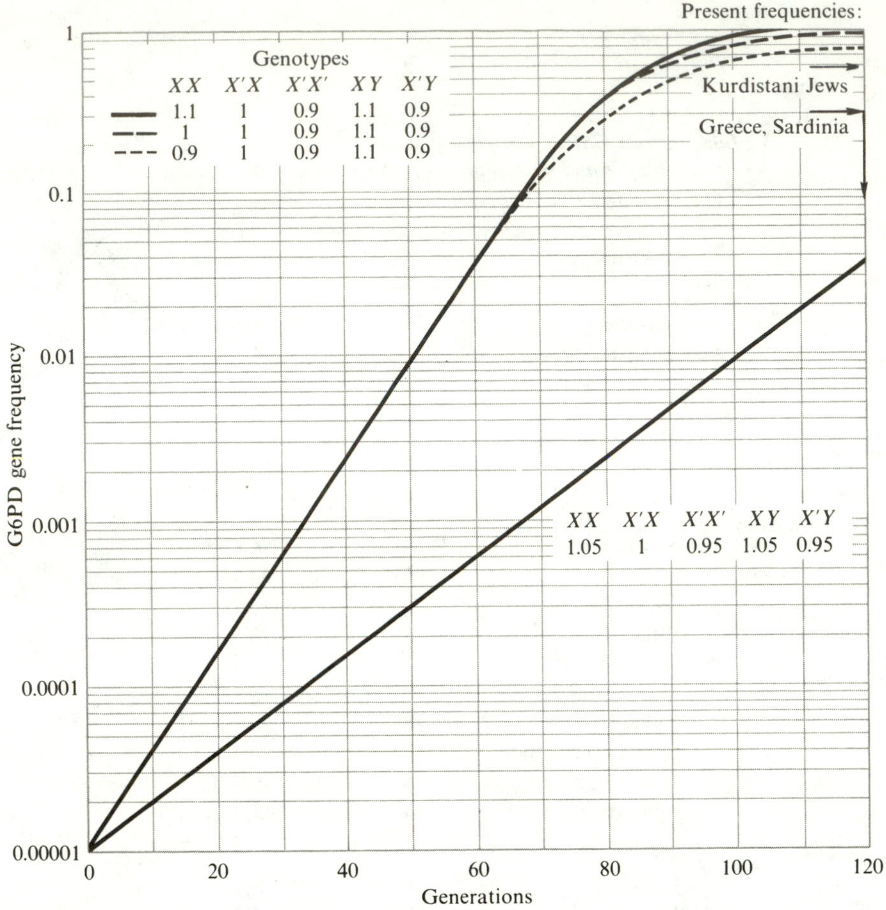

FIGURE 4.15
Kinetics of selection for a sex-linked gene, such as the G6PD-deficiency allele, in malarial areas. Values of s, t, and m as indicated on page 171 can be easily computed from the fitness values given in the figure (see text). Sets of t, m, and s values different from these, but proportional to them, will give rise to curves indentical in shape but with a time scale divided by the coefficient of proportionality.

The kinetics of the change for a set of selection coefficients leading to a balanced polymorphism ($t = s = m = 0.1$) are also shown (as a dotted curve) in Figure 4.15. The selective advantage of the female homozygote affects the kinetics of selection only in the last stages, near equilibrium. For the set of values given, the gene frequency eventually attained will be 0.953. For equilibrium other than fixation, if $t = m$, s must be positive and larger than $m/(1 + m)$. The equilibrium frequency expected is given on page 171.

The data on G6PD deficiency are not inconsistent with a transient polymorphism (in malarial areas). The wide variation in observed frequencies of the gene for G6PD deficiency would be expected for the intermediate states of a transient, polymorphism. In other groups and areas (for example, Sardinia), G6PD-deficiency gene frequencies have lower values than among Kurdistani Jews, although they are sometimes fairly high (see Table 4.11). What are the possible causes of the discrepancies? There are several possible answers. Many different G6PD alleles have been described, and each may have different selective coefficients. The intensity of selection due to malaria and the time available for selection since the first mutant gene came into the population (by mutation or by immigration), may vary from place to place and from time to time. Moreover, in areas where ecological conditions vary considerably even at short distances, as is true for instance of Sardinia and Greece, there may be high and low malarial zones located near each other. This will create ecological niches with different selection intensities, and complex relations will therefore be expected because of the interaction of selection and migration. Finally, the possible coexistence of several different genetic adaptations to malaria, in part allelic (hemoglobins S and C, β-thalassemia), in part nonallelic but possibly interacting with them (α-thalassemia, G6PD-deficiency, and others) will inevitably complicate the picture.

Color blindness may be selected against more strongly in primitive hunting and gathering societies than in more advanced societies; its present relatively high incidence in most parts of the world might be a product of relaxed or reversed selection.

There are two kinds of color blindness, protan and deutan, both of which are X-linked. In most populations there are about three times as many deutan males as protan males. There are very significant differences in the total frequency of color blindness between populations. In general, the more "primitive" the population, the lower the frequency. Thus, in hunting and gathering communities the frequency of color blindness is about 1–2 percent whereas in an industrial society it is around 7 percent. It has been suggested by Neel and Post (1963; see also Post, 1962) that color blindness may be a disadvantage to primitive people because it impairs their ability to distinguish the flora and fauna that are essential for their livelihood. The negative selection against color blindness has most likely been relaxed in more modern societies, which would explain its higher frequency in such societies.

If the present polymorphism is transient rather than stable, this evolutionary trend may, perhaps, have been reversed by the relaxation of selection against color blindness in modern society. If color blindness were a stable polymorphism in primitive societies and is also now in modern society, it would seem most likely that it is maintained by female heterozygote advantage, since it is hard to believe that selection with respect to color blindness in the two sexes could be in opposite directions. It should be noted, however, that for mutation alone to cause the increase from 1 to 6 percent would require a time that is much greater (by a factor of 10–20, assuming the fairly high mutation rate of 10^{-5}) than the length of time separating primitive and modern societies.

Let us assume that in primitive societies $q = 0.01$ is due to equilibrium between mutation and selection. Then $qs = 3\mu$. For a mutation rate $\mu = 10^{-5}$, s would be 0.003. If selection is completely relaxed, $s = 0$, and the gene frequency increases under mutation pressure alone. The number of generations necessary to increase it from $q = 0.01$ to a present value of $q = 0.06$ is given approximately (see Example 3.1) by $n = (q_n - q_0)/\mu = (0.06 - 0.01)/10^{-5} = 5000$ or 150,000 years. This would imply that selection against color blindness was relaxed at a time much earlier than that of the transition from hunting and gathering to agriculture, which probably took place from 10,000–15,000 years ago, or else it would require a mutation rate much higher than average; for example, such that $\mu = (q_n - q_0)/n = (0.06 - 0.01)/(15{,}000/30) = 10^{-4}$. This is not, however, inconceivable, considering that at least two sex-linked loci, and perhaps more, contribute to the total mutation rate. If that is true, then the selective disadvantage in primitive societies must be $s = 0.03$, a relatively high value. If this is not true, however, we must imagine that today color blindness has a slight selective advantage for males or perhaps also for heterozygous females, for entirely unknown reasons. Another very different possibility is that color blindness is the last vestige of evolution towards the present state of color vision in humans.

Clearly, much more information on relative fitness values is needed for a deeper understanding of these two X-linked polymorphisms.

4.11 Other Mechanisms That Lead to Balanced Polymorphism

Other mechanisms that can give rise to a balanced polymorphism include fitness as a function of gene frequency, differential fertility opposing differential viability in the same or in different sexes, selection varying in direction with time or with place, and selection acting in opposite directions in gametes and zygotes.

Many different combinations of balancing forces can give rise to stable polymorphisms. In this section we shall briefly outline some of the simplest situations, which can, at least to some extent, be treated analytically.

1. When fitnesses are functions of the gene frequency, even in the simple case of two alleles at one locus and random mating, it is no longer necessary for the heterozygote to be fitter than both homozygotes for there to exist a balanced polymorphism. Intuitively, a sufficient (though not a necessary) condition for the existence of a polymorphism is that either allele should be at an advantage when rare, and so able to increase in frequency when introduced into a population carrying primarily the other allele. As a specific example, assume that the relative fitnesses of

genotypes *AA*, *Aa*, and *aa* are $1 + s_1 p$, $1 + \frac{1}{2}(s_1 p + s_2 q)$ and $1 + s_2 q$, respectively, and $p = (1 - q)$ is the frequency of *A*. In this example the fitness of the heterozygote is always exactly halfway between the fitnesses of the two homozygotes. It was shown by Wright (1955) that this model system leads to a stable intermediate equilibrium with $p = s_2/(s_1 - s_2)$ so long as both s_1 and s_2 are negative (see Example 4.2). This selective scheme for a balanced polymorphism is therefore a form of **frequency dependent selection**, giving the advantage always to the rare gene. A possible hypothetical example in human populations could be sexual selection (that is, mating preference) for a rare or unusual phenotype, such as fair hair and blue eyes!

2. When differential fertility and differential viability are distinguishable in population genetic models, selection parameters that are different in the two sexes must be considered. It can then be shown that differential fertility acting in an opposite direction, in the two sexes, to differential viability can lead to a stable balanced polymorphism (Owen, 1953; Bodmer, 1965). This is exactly analogous to selection acting in opposite directions in the two sexes, which as we saw can lead to a balanced polymorphism for an X-linked gene. It can be shown that the same is also true for an autosomal gene. An application of this situation was discussed in relation to the sickle-cell polymorphism, where it was pointed out that the data on the fertility of $AS \times AA$ matings among the Black Caribs of British Honduras were compatible with the observed frequency of the sickle-cell genes—only if the high observed fertility of these matings was offset by a viability deficiency of the *AS* sickle-cell trait carrier heterozygotes. Such overcompensation by fertility for a viability deficiency may not be too uncommon in human populations.

3. Selection of varying intensity in different generations or in different geographical regions can also give rise to a balanced polymorphism (Levine, 1953; Dempster, 1955; Haldane and Jayakar, 1963). For example, it can be shown that if selection varies in direction arbitrarily from one generation to the next in such a way that the heterozygote is not at an advantage in any particular generation but is, on the average over all generations, at an advantage, then a stable polymorphism can result. This may be true even if the heterozygote is mostly at a disadvantage, but rarely, perhaps due to occasional epidemics, assumes a considerable positive advantage. Even when selection of varying intensity in time and space does not actually lead, in the long run, to a stable polymorphism, it may contribute very significantly to the length of time that two alleles remain together in a population. From some points of view, this may perhaps be the most important mechanism for the maintenance of polymorphisms, since on an evolutionary time scale it is hard to consider any polymorphism as absolutely stable.

4. Selection may act in opposite directions in gametes and zygotes. Thus, for example, situations are known in which a gamete carrying a particular allele is always at an advantage over gametes carrying an alternative allele with respect to fertilization in heterozygotes. In some cases such advantages can be explained in

terms of disturbance of segregation at meiosis; this type of selective force has been referred to as meiotic drive. It may also, for example, be due to a selective advantage in sperm attached to a given haploid genotype, though it seems unlikely that many genes express themselves adequately in sperm for the effects of such selection to be important. While no examples of gametic selection have been firmly established in humans, an outstanding example is known in another mammal, the mouse, in which certain alleles at the *T*-locus have an appreciable advantage in fertilization (Lewontin and Dunn, 1959). Gametic selection may act in the opposite direction to the effects of the alleles on the zygote. Thus a gene whose effect is detrimental to zygotes may be selected for in gametes, and a polymorphism may then result when these two opposing forces balance each other. It has been suggested that segregation distortion may be a mechanism that could explain the maintenance of the blood-group polymorphisms in man though evidence for this is very hard to obtain, and is so far not very convincing (see Chapter 5).

Many other complex selection models can, of course, be constructed that may lead to balanced polymorphisms. Most of them will be some combination of the mechanisms that we have discussed in this chapter. In the next chapter we shall consider some special models relating to patterns of selection involving an interaction between mother and child, which correspond to some situations observed for the Rhesus and ABO-blood-group systems in man. The explanation for the maintenance of balanced polymorphisms is really just one facet of the explanation for the maintenance of genetic variability in populations. Evolutionary theory suggests that this genetic variability, while it may have some immediate disadvantages, is essential for the further evolution of a species, which must take place when new environmental situations present themselves.

General References

Giblett, E. R., *Genetic Markers in Human Blood.* Philadelphia: F. A. Davis, 1969. (A recent major survey of genetic polymorphisms for enzymes and other proteins.)

Harris, H., "Enzyme polymorphisms in man." *Proceedings of the Royal Society, Series B* **164**: 298–310, 1966. (The first major assessment of the overall extent of polymorphism in man.)

Huehns, E. R., and E. M. Shooter, "Human haemoglobins." *Journal of Medical Genetics*, **2**: 48–90, 1965. (A comprehensive review of the genetics and biochemistry of the human hemoglobins.)

Ingram, V. M., *The Hemoglobins in Genetics and Evolution.* New York: Columbia University Press, 1963.

Livingstone, F. B., *Abnormal Hemoglobins in Human Populations.* Chicago: Aldine, 1967. (A comprehensive survey of frequencies in different populations, especially African, with a useful review of the data.)

Worked Examples

4.1 *Exact solution for a lethal with heterozygote advantage or disadvantage.*

Genotypes AA, Aa, and aa have relative fitnesses 1, $1 - s$, and 0. Find a general expression for the frequency of the lethal gene a after n generations, in terms of its initial frequency.

The frequencies of the genotypes after selection are

AA	Aa	aa	Total
p_0^2	$2p_0 q_0(1-s)$	0	$p_0^2 + 2p_0 q_0(1-s)$,

where $q_0 = 1 - p_0$ is the initial frequency of a. The frequency of a in the next generation is given by

$$q_1 = \frac{p_0 q_0(1-s)}{p_0^2 + 2p_0 q_0(1-s)} = \frac{q_0(1-s)}{1 + q_0(1-2s)}. \tag{1}$$

Equation 1 can be written in the form

$$\frac{1}{q_1} = \frac{1-2s}{1-s} + \frac{1}{1-s}\frac{1}{q_0}. \tag{2}$$

Thus in the next generation we have

$$\frac{1}{q_2} = \frac{1-2s}{1-s} + \frac{1}{1-s}\frac{1}{q_1} = \frac{1-2s}{1-s}\left[1 + \frac{1}{1-s}\right] + \frac{1}{(1-s)^2}\frac{1}{q_0},$$

and, more generally,

$$\frac{1}{q_n} = \frac{1-2s}{1-s}\left[1 + \frac{1}{1-s} + \cdots + \frac{1}{(1-s)^{n-1}}\right] + \frac{1}{(1-s)^n}\frac{1}{q_0}. \tag{3}$$

The term in brackets is a geometric progression with the sum

$$\frac{\dfrac{1}{(1-s)^n} - 1}{\dfrac{1}{1-s} - 1} = \frac{1-s}{s}\left[\frac{1}{(1-s)^n} - 1\right].$$

Thus Equation 3 gives

$$\frac{1}{q_n} = \frac{1-2s}{s}\left[\frac{1}{(1-s)^n} - 1\right] + \frac{1}{(1-s)^n}\frac{1}{q_0},$$

or

$$q_n = \frac{(1-s)^n q_0}{1 + \dfrac{(1-2s)[1-(1-s)^n]q_0}{s}}. \tag{4}$$

When $s > 0$, and q_0 is small Equation 4 is approximately

$$q_n = (1-s)^n q_0, \tag{5}$$

showing that the gene frequency of a decreases, by a factor of approximately $1 - s$ in each generation. When $s < 0$, then as n becomes large so does

$$(1-s)^n = (1+|s|)^n.$$

It can then be shown from Equation 4 that as n becomes large, q_n approaches the limit $|s|/(1 + 2|s|)$ as expected from the simple theory for a balanced polymorphism due to heterozygote advantage. The ultimate rate at which this equilibrium is approached is $(1 + |s|)^n$. This is the only case of the simple two-allele one-locus model that can be solved completely.

4.2 *Polymorphism when fitnesses are a function of the gene frequencies.*

Suppose that genotypes AA, Aa, and aa have fitnesses $1 + s_2 p$, $1 + \frac{1}{2}(s_1 p + s_2 q)$, and $1 + s_2 q$, respectively, with the gene frequency of A being $p(=1-q)$. Find the gene-frequency equilibria for this system of selection, assuming random mating, and determine their stability (Wright, 1955).

The frequencies of the three genotypes after selection are

$$\begin{array}{ccc} AA & Aa & aa \\ p^2\dfrac{1+s_1p}{R}, & \dfrac{2pq + pq(s_1p + s_2q)}{R}, & q^2\dfrac{1+s_2q}{R}, \end{array}$$

where

$$\begin{aligned} R &= p^2 + s_1p^3 + 2pq + s_1p^2q + s_2pq^2 + q^2 + s_2q^3 \\ &= 1 + s_1p^2(p+q) + s_2q^2(p+q) = 1 + s_1p^2 + s_2q^2. \end{aligned}$$

The new frequency of gene A is given by

$$p' = \frac{p^2(1+s_1p) + pq + \frac{1}{2}pq(s_1p + s_2q)}{R}, \tag{1}$$

so that the change in gene frequency over one generation is

$$\Delta p = p' - p = \frac{1}{R}(p^2 + s_1p^3 + pq + \tfrac{1}{2}p^2qs_1 + \tfrac{1}{2}pq^2s_2 - p - s_1p^3 - s_2pq^2),$$

which reduces to

$$\Delta p = \tfrac{1}{2}pq\,\frac{s_1p - s_2q}{1 + s_1p^2 + s_2q^2}. \tag{2}$$

(Note that $p^2 - p + pq = -pq + pq = 0$.)

Equation 2 is, apart from a factor $\frac{1}{2}$, equivalent to the equation for Δp obtained with constant fitnesses $1 - s$, 1, and $1 - t$ for the genotypes AA, Aa, and aa (see Equation 4.8), replacing s by $-s_1$ and t by $-s_2$. Thus we can say that apart from the trivial equilibria $p = 0$ and $q = 1$ (all a), and $p = 1$ and $q = 0$ (all A), the intermediate equilibrium

$$p = \frac{s_2}{s_1 + s_2} \tag{3}$$

is stable provided both s_1 and s_2 are negative. In this case we have a balanced polymorphism although the heterozygote is *never* fitter than both homozygotes. In fact, the fitness of the heterozygote is always exactly halfway between those of the two homozygotes. When s_1 and $s_2 < 0$, the heterozygote is always fitter than the more common homozygote if either A or a is rare. The rarer gene is then always favored over the common allele.

4.3 *Approximate number of generations needed for a given gene-frequency change.*

Assuming genotypes AA, Aa, and aa have fitnesses $1 - s$, 1, and $1 - t$ find the approximate number of generations, n, needed to change the frequency of gene A from p_0 to p_n.

From Equation 4.8 the change in p in one generation is given by

$$\Delta p = pq\,\frac{tq - sp}{1 - sp^2 - tq^2}. \tag{1}$$

If t and s are small enough that we can ignore terms in s^2 and t^2, then from Equation 1, we can obtain as an approximation,

$$\Delta p = pq(tq - sp). \tag{2}$$

Following an approach originally used by Haldane (1924, and later), we can approximate Equation 2 by the differential equation

$$\frac{dp}{dn} = pq(tq - sp), \tag{3}$$

where n is the time measured in generations. This approximation is valid so long as the change in p per generation is slow. Equation 3 can be solved by separating variables to give

$$dn = \frac{dp}{pq(tq - sp)} = dp\left\{\frac{1}{tp} - \frac{1}{sq} + \frac{(s+t)^2}{st(tq - sp)}\right\}, \tag{4}$$

and so by integration

$$\int_0^n dn = n = \int_{p_0}^{p_n} \left[\frac{1}{tp} - \frac{1}{sq} + \frac{(s+t)^2}{st(tq - sp)}\right] dp. \tag{5}$$

This gives

$$\begin{aligned} n &= \left(\frac{1}{t}\log p + \frac{1}{s}\log q - \left|\frac{s+t}{st}\right| \log\left|tq - sp\right|\right)_{p_0}^{p_n} \\ &= \frac{1}{t}\log\frac{p_n}{p_0} + \frac{1}{s}\log\frac{q_n}{q_0} - \left|\frac{s+t}{st}\right| \log\left|\frac{tq_n - sp_n}{tq_0 - sp_0}\right| \end{aligned} \tag{6}$$

for the approximate number of generations needed for the change from p_0 to p_n, given these values. So long as p_0 and p_n are not near 0, 1 or $t/(s + t)$, n as given by Equation 6 depends mainly on s and t, increasing as they decrease. When $-s = t > 0$, so that *Aa* is exactly intermediate in fitness between *AA* and *aa*, Equation 6 reduces to the simple form

$$n = \frac{1}{t}\log\frac{p_n(1 - p_0)}{p_0(1 - p_n)}. \tag{7}$$

Suppose that $p_0 = 1 - p_n$, as might be the case if gene A is replacing a, with A at a mutation-balance equilibrium at the beginning, and a at a similar equilibrium at the end. Then Equation 7 becomes, approximately, ignoring terms in p_0^2 (for small p_0),

$$n = -\frac{2}{s}\log p_0. \tag{8}$$

This represents, approximately, the number of generations needed for one gene to replace another, which increases as s and p_0 decrease. If, for example, $s = 0.01$ and $p_0 \sim 1/1000$ then n is approximately 1400.

When both s and $t > 0$, an equilibrium $p_e = t/(s + t)$ is reached asymptotically. Equation 6 can then be recast into the form

$$n = \frac{1}{s+t}\left(\frac{1}{p_e}\log\frac{p_n}{p_0} + \frac{1}{q_e}\log\frac{q_n}{q_0} - \frac{1}{p_e q_e}\log\left|\frac{p_n - p_e}{p_0 - p_e}\right|\right).$$

4.4. *The three-allele polymorphism.*

Let the six genotypes formed by three alleles at one locus have fitnesses as follows:

$$\begin{aligned} \text{fitness of } AA &= a \\ SS &= b \\ CC &= c \\ SC &= f \\ AC &= g \\ AS &= h \end{aligned}$$

Assuming random mating find the condition for C to increase in frequency given that A and S are being maintained polymorphic by heterozygote advantage (see Table 4.7).

Assume alleles A, S, and C have frequencies p_1, p_2, and p_3, respectively. The frequency of C after one generation of random mating is

$$p_3' = \frac{p_3 p_1 g + p_3 p_2 f + p_3^2 c}{R}, \tag{1}$$

where

$$R = ap_1^2 + bp_2^2 + cp_3^2 + 2hp_1p_2 + 2gp_1p_3 + 2fp_2p_3. \tag{2}$$

(Note that the frequency of genotype AC, for example, is $2p_1p_3$, its fitness is g, and half the alleles it produces are C, hence the contribution $p_3p_1\ g$ to p_3'; similarly for $p_3p_2 f$. The frequency of CC is p_3^2, its fitness c, and all the alleles it produces are C, hence the contribution $p_3^2 c$.) The frequencies of A and S in the next generation are, similarly, given by

$$p_1' = \frac{p_1^2 a + p_1 p_2 h + p_1 p_3 g}{R} \tag{3}$$

and

$$p_2' = \frac{p_2 p_1 h + p_2^2 b + p_2 p_3 f}{R}, \tag{4}$$

where R is a normalizing factor such that

$$p_1' + p_2' + p_3' = 1.$$

In the absence of C, A and S would be at the polymorphic equilibrium given by

$$p = \frac{h-b}{h-a+h-b} \qquad q = \frac{h-a}{h-a+h-b} \tag{5}$$

provided $h > a, b$ (see Equation 4.4).

When p_3 is very small

$$p_1 = p + 0(p_3),\ p_2 = q + 0(p_3), \tag{6}$$

where $0(p_3)$ means terms of the same order as p_3. If, assuming p_3 is small, we ignore terms in p_3^2 and higher powers of p_3, then Equation 1 for p_3' becomes, approximately,

$$p_3' = \frac{p_3(pg + qf)}{ap^2 + 2pqh + bq^2}, \tag{7}$$

where p and q are as given in Equation 5.
Now

$$\begin{aligned} ap^2 + 2pqh + bq^2 &= p(ap + hq) + q(hp + bq) \\ &= \frac{h^2 - ab}{h - a + h - b} \end{aligned} \tag{8}$$

since $ap + hq = hp + bq = \dfrac{h^2 - ab}{h - a + h - b}$.
Equation 7 can therefore be written in the form

$$\frac{p_3'}{p_3} = \frac{hg - af + fh - bg}{h^2 - ab}. \tag{9}$$

Thus $p_3' > p_3$, and so C increases in frequency provided that

$$\frac{hg - af + fh - bg}{h^2 - ab} > 1$$

or

$$hg - af + fh - bg + ab - h^2 > 0$$

or

$$F + G - C > 0, \tag{10}$$

where $F = hg - af$, $G = fh - bg$, $C = h^2 - ab$ as given in Table 4.7. The conditions for S to increase in the presence of A and C, and for A to increase in the presence of C and S are, similarly,

$$F + H - B > 0 \tag{11}$$

and

$$G + H - A > 0, \tag{12}$$

respectively (Haldane, 1957; Bodmer and Parson, 1960). In each case the condition for either A and C or S and C to be able to coexist in a balanced polymorphism must be satisfied. These are, respectively $B > 0$ and $A > 0$.

If whenever two of the alleles coexist in a population the third will increase in frequency when introduced, then the only possible stable equilibrium state is an "internal" one, namely with none of p_1, p_2, and p_3 equal to zero. It can, in fact, be shown that this is also a necessary condition for a stable internal equilibrium, leading from Equations 10, 11, 12, and the conditions A, B, $C > 0$, to the conditions given in Table 4.7.

The equations for the equilibrium frequencies can be obtained as follows. Equation 1, for p_3', can be rewritten in the form

$$p_3' = p_3 \frac{gp_1 + fp_2 + cp_3}{R}, \tag{13}$$

and R can be expressed in the form

$$R = p_1(ap_1 + hp_2 + gp_3) + p_2(hp_1 + bp_2 + fp_3) + p_3(gp_1 + fp_2 + cp_3). \tag{14}$$

The equilibrium condition $p_3' = p_3$ leads from Equation 13 to the equation

$$R = gp_1 + fp_2 + cp_3 .$$

Extending the argument similarly to p_1 and p_2 leads to the equilibrium equation

$$R = ap_1 + hp_2 + gp_3 = hp_1 + bp_2 + fp_3 = gp_1 + fp_2 + cp_3 . \tag{15}$$

It can readily be shown that the equilibrium values given in Table 4.7 satisfy Equation 15 (Mandel, 1959).

4.5. *Selectively balanced polymorphisms for an X-linked gene.*

Assuming the selective scheme for an X-linked gene given on page 171 show that the condition for gene X to increase in frequency, when it is rare is

$$(1 - t)(1 - m) < 1$$

(Haldane and Jayakar, 1965).

Let the frequencies of X and X' in adult females be P and Q and in males p and q. Assuming random mating, the frequencies of genotypes XX, XX', and $X'X'$ among female offspring before selection are, respectively, Pp, $Pq + Qp$, and Qq. The gene frequency of X in the next generation of females, after selection, is therefore

$$P' = \frac{(1 - s)Pp + \frac{1}{2}(Pq + Qp)}{(1 - s)Pp + (Pq + Qp) + (1 - t)Qq} = 1 - Q'. \tag{1}$$

The frequencies of male offspring X and X' before selection are simply P and Q, so that the new frequency of X in males is

$$p' = \frac{(1 + m)P}{(1 + m)P + (1 - m)Q} = 1 - q'. \tag{2}$$

If we let $u_1 = p/q$ and $u_2 = P/Q$ then Equations 1 and 2 can be rewritten in the form

$$u_2' = \frac{(1-s)u_1u_2 + \frac{1}{2}(u_1+u_2)}{1-t+\frac{1}{2}(u_1+u_2)} \tag{3}$$

$$u_1' = \frac{1+m}{1-m}\,u_2. \tag{4}$$

Note that

$$Q' = \frac{\frac{1}{2}(Pq+Qp)+(1-t)Qq}{(1-s)Pp+Pq+Qp+(1-t)Qq},$$

so that

$$\frac{P'}{Q'} = \frac{(1-s)Pp+\frac{1}{2}(Pq+Qp)}{(1-t)Qq+\frac{1}{2}(Pq+Qp)},$$

which may be reduced to Equation 3 by dividing the top and bottom of the right-hand side by Qq. Similarly,

$$q' = \frac{(1-m)Q}{(1+m)P+(1-m)Q},$$

and so

$$\frac{p'}{q'} = \frac{1+m}{1-m}\frac{P}{Q}.$$

If X is rare, then P and p and thus also u_1 and u_2 are small. Ignoring terms of order u_1^2, u_2^2, and u_1u_2, Equation 3 becomes

$$u_2' = \frac{1}{2(1-t)}(u_1+u_2). \tag{5}$$

Combining Equations 4 and 5 gives

$$u_2'' = \frac{1}{2(1-t)}(u_1'+u_2') = \frac{u_2'}{2(1-t)} + \frac{1}{2(1-t)}\frac{1+m}{1-m}u_2,$$

or

$$2(1-t)(1-m)u_2'' - (1-m)u_2' - (1+m)u_2 = 0. \tag{6}$$

Equation 6 is a linear difference equation of the second order that can be solved by trying to fit solutions of the form

$$u_2 = \text{constant} \times \lambda^n. \tag{7}$$

The exponent λ must satisfy the quadratic equation

$$2(1-t)(1-m)\lambda^2 - (1-m)\lambda - (1+m) = 0. \tag{8}$$

(Substitute $u_2'' = \text{constant } x\lambda^{n+2}$, $u_2' = \text{constant } x\lambda^{n+1}$ and $u_2 = \text{constant} \times \lambda^n$ into Equation 6 and cancel out λ^n.)

The solutions of Equation 8 are

$$\lambda_1 = \frac{(1-m) + \sqrt{(1-m)^2 + 8(1-t)(1-m^2)}}{4(1-t)(1-m)} \tag{9a}$$

$$\lambda_2 = \frac{(1-m) - \sqrt{(1-m)^2 + 8(1-t)(1-m^2)}}{4(1-t)(1-m)}, \tag{9b}$$

where $\lambda_1 > \lambda_2$ (Edwards, 1961). The frequency of X will increase provided u_2 (and thus also u_1) increases, which will be true if $\lambda_1 > 1$. It can be shown with a little algebraic manipulation that this condition is satisfied if

$$(1-t)(1-m) < 1. \tag{10}$$

In fact,

$$\lambda_1 > 1 \text{ if } \sqrt{(1-m)^2 + 8(1-t)(1-m^2)} > 4(1-t)(1-m) - (1-m);$$

that is

$$(1-m)^2 + 8(1-t)(1-m^2) > (1-m)^2(3-4t)^2;$$

that is,

$$8(1-t)(1+m) - (1-m)[8 + 16t^2 - 24t] > 0,$$

which reduces to $(-mt + m + t)(1-t) > 0$, and so $-mt + m + t > 0$, which is equivalent to Equation 10. In a similar way it can be shown that X' increases provided

$$(1-s)(1+m) < 1. \tag{11}$$

Provided both Equations 10 and 11 hold, if a stable equilibrium exists it must be such that neither X nor X' are absent. As indicated on page 171 it can, in fact, be shown that Equations 10 and 11 are both necessary and sufficient conditions for the existence of a nontrivial balanced polymorphism for an X-linked gene.

5

Polymorphisms for Blood Groups, Transplantation Antigens, and Serum Proteins: Incompatibility Selection

5.1 Polymorphisms of Blood Components

In this chapter, we shall discuss in detail four human polymorphic systems that are among those which have been most widely studied and which have great practical importance. They are the ABO and Rh red-blood-cell groups, the HL-A tissue antigens and the Gm groups found on antibody molecules. These are all detected in or on the various components of the blood through the use of serological techniques (antibody-antigen reactions which will be discussed in the next section). Three of them at least (ABO, Rh, HL-A) may be associated with a special form of selection connected with fetal-maternal incompatibility that, for Rh and ABO, is the cause of hemolytic disease of the newborn.

Blood is the most readily available of human tissues and so, not surprisingly, has been studied most intensively from a genetic point of view. As a result many of the best-known polymorphisms pertain to blood components (see Chapter 4). There is, however, no *a priori* reason for expecting that blood components should be more polymorphic than components of other tissues.

More than a dozen red-blood-cell polymorphisms have been described (see Race and Sanger, 1968), starting with the ABO system discovered by Landsteiner in 1900. By far, the most intensively studied of these are the ABO and Rh systems.

Both are of great clinical importance, the ABO system in blood transfusion and the Rh with respect to hemolytic disease of the newborn. The solution to the problem of organ transplantation in man depends to a considerable extent on the understanding of the genetics of the differences between individuals that are responsible for graft rejection. The recently described complex HL-A white-blood-cell polymorphism has been shown to be of major importance in matching for transplantation. (see *Histocompatibility Testing* 1967, 1970). The last of the polymorphisms we shall discuss, the Gm system, brings about differences with respect to the antibody molecules. Though it has not yet been shown to have any major clinical significance, this polymorphism has contributed very substantially to the understanding of the genetic control of antibody structure.

5.2 Antigens and Antibodies

When the body is challenged with a foreign substance, which may be a virus, a bacterium, a purified chemical substance, or someone else's blood cells, it responds by manufacturing **antibodies**. The antibodies are protein molecules that have the capacity to bind specifically to the foreign substance, or **antigen**, that elicited their production. The antibody molecules are mostly present in **serum**, which is the fluid component of the blood. Serum is actually the supernatant fluid obtained by letting fresh blood clot and then centrifuging to remove the clotted red cells and any other remaining unclotted cells. Serum contains many important proteins which are polymorphic, including the antibody molecules themselves. Antibodies are made mainly by the lymphocytes, which are one of the two major types of white cells in the blood.

The combination of antibody with invading organism enhances the body's ability to destroy bacteria, viruses, and other infectious pathogenic agents. The antibody response is thus a most important mechanism for protection against infectious diseases. It is, however, also the response that may eventually give rise to hemolytic disease of the newborn if a mother is "challenged" with the foreign antigen of her fetus. Blood-transfusion reactions and the rejection of organ transplants are also due to the production of antibodies in response to a challenge from foreign cells or tissues.

An essential feature of the immune response, as the production of antibodies is sometimes called, is that it only happens following a challenge with a foreign substance. Foreign must here be defined as something not found within the individual being challenged. An immune response thus implies the ability to recognize *self*, namely not to make antibodies against substances found in oneself. The mechanisms responsible for this are not known. It is however, sometimes possible to "trick" an individual into "thinking" a foreign substance is not foreign by, for example, challenging the individual with it at a very early age, before the immune response is fully developed. The individual will then not subsequently form antibodies to the corresponding substance and is said to have developed tolerance to it.

There is an enormous variety of antigens, essentially unlimited, to which an individual can respond with specific antibodies. The number of different antibodies that can be produced by a single individual must therefore be in the thousands at least, and is probably in the tens or hundreds of thousands. As the primary structure, or amino-acid sequence, of a protein determines its shape, each of these many antibodies having the capacity to bind specifically with some substance must have a different amino-acid sequence. It is not known whether the genetic information for the very large variety of antibody molecules exists as such in the germ line, or whether the information is produced by somatic mutation or recombination during the development and differentiation of the antibody-producing cells. The molecular mechanism of the immune response to foreign substances remains one of the most tantalizing problems of molecular biology.

The specificity of the antibody-antigen reaction has proved to be a very useful technique for the detection of polymorphic differences.

Specificity of the antibody-antigen reaction is, in fact, used for distinguishing the phenotypes in all of the four polymorphisms discussed in this chapter. The reaction of an antibody with its specific antigen can be detected in a variety of ways. Perhaps the simplest reaction is that called **agglutination**, which takes place when the antigen is on the surface of a cell. Agglutination is simply the formation of clumps of cells held together by antibody molecules attached to the antigens on the cell surface. For example, an antibody specific for the A antigen of the ABO red-cell system, often called anti-A, will clump or agglutinate A and AB cells but not B or O cells. Anti-B will agglutinate B and AB cells but not O or A cells (see Figure 5.1). Sometimes the attachment of antibody to cell surfaces is detected by the fact that, in the presence of a series of proteins from the serum called **complement**, the attached antibody causes the cell to be burst open and destroyed. Cells that do not have any antibody attached to their surface will not be affected by complement.

The antibodies used for the detection of human red-blood-cell groups and other antigenic polymorphisms (except for ABO, for which there exist natural antibodies) mostly come from people who have had several blood transfusions or from mothers who have made antibodies to the antigens of their offspring. In some cases antibodies have been made by immunizing volunteers with blood from a specifically chosen donor. The donor is chosen to have the antigen against which antibody is to be made, say Rh(+), while the recipient, of course, must not have this antigen, in this example must be Rh(−).

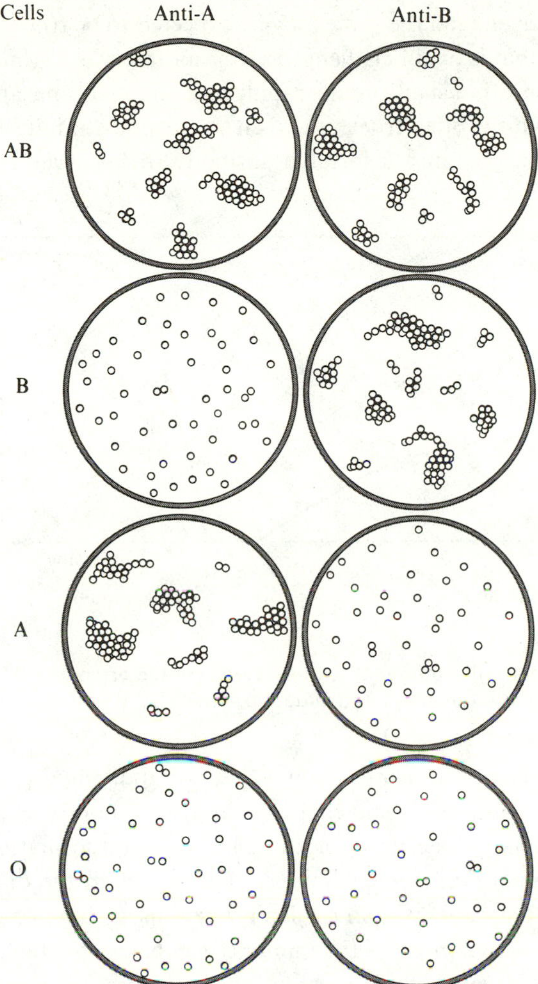

FIGURE 5.1
Reaction pattern of anti-A and anti-B with A, B, AB, and O cells. (From Hardin, 1966.)

The amount of antibody activity in a serum is measured by the highest dilution at which there is still a reaction with the specific antigen. This dilution is called the serum's ***titer***.

Usually the serum is tested systematically at a series of 1 in 2 or 1 in 10 dilutions and the highest dilution still reacting noted. The antibody titer reached in reaction to an antigenic stimulus depends on the antigen, the amount used to stimulate, and also, to some extent, on the capacity of the individual to react. It has been shown, at least in mice, that there may be genetic differences with respect to the ability to react to a given antigen, and the same may be expected to be true in human populations. If an individual is again challenged, after an interval of time, with the same antigen, his response is usually more rapid, and the resulting antibody titer appreciably higher. This is often true even when the second stimulus is given after the individual no longer has any detectable antibody in his serum (see Figure 5.2).

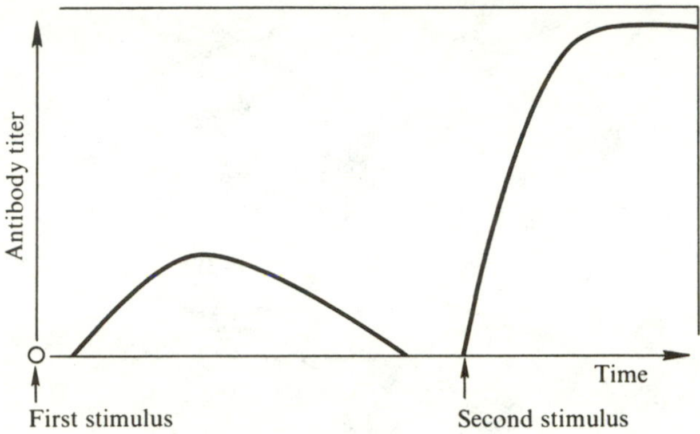

FIGURE 5.2
Schematic illustration of the time course of the primary and secondary responses to an antigenic stimulus.

This so-called **secondary response** shows that an individual's immune system, "remembers" a previous antigenic challenge and so responds much more quickly the second time. This is the basis for the efficiency of vaccination, which is the equivalent of a primary antigenic stimulus, challenge with the disease itself being the secondary stimulus. As we shall see later when discussing hemolytic disease of the newborn, this characteristic of the immune response is the main reason why so few of the babies that suffer from hemolytic disease of the newborn are firstborn offspring.

5.3 Rhesus Blood Groups and Hemolytic Disease of the Newborn

A simple case report by Levine and Stetson in 1939 of an unusual complication during childbirth, paved the way toward an understanding of the basis for hemolytic disease of the newborn, in which the infant's red-blood-cells are destroyed. Their discovery represents one of the most important contributions of genetics to clinical practice.

A woman whose second pregnancy terminated with a stillbirth reacted severely to transfusion with her husband's blood. It was found that she had an antibody in her serum that agglutinated her husband's blood cells and in fact agglutinated red-blood-cells from about 85 percent of a random sample of American Caucasians. The reactions of the antibody showed no correlation with the then-known blood groups, ABO, MN, and P. Levine and Stetson correctly hypothesized that the antibody had been formed, during pregnancy, in reaction to the paternal antigen carried by the fetus, which was, of course, absent in the mother, and that the antibody was responsible for the destruction of fetal tissue, leading to stillbirth. In the following year, 1940, Landsteiner and Wiener reported the discovery of an antibody, produced by immunizing rabbits with blood from the rhesus monkey, that also reacted with the blood of about 85 percent of the human population and was not related to the ABO, MN, or P systems. They called this antibody anti-Rh, Rh standing for rhesus. It was soon discovered that the antibody described by Levine and Stetson had essentially the same specificity as anti-Rh and that many cases of hemolytic disease of the newborn could be attributed to its activity. People who react to this antibody are called Rh-positive and those who do not Rh-negative. The Rh(+), Rh(−) difference is inherited as a simple dominant determined by alleles D and d; genotypes DD and Dd are Rh(+) and dd is Rh(−). Many other specificities have subsequently been found that belong to the rhesus system, but the difference first described remains the one of greatest significance for hemolytic disease of the newborn—accounting for about 90 percent of the observed cases of this disease. The complex genetics of this system will be reviewed later in this chapter. Only the Rh(+) and Rh(−) difference just described really needs to be taken into account in discussing the selective consequences associated with hemolytic disease of the newborn.

Hemolytic disease of the newborn is effectively selection against heterozygotes.

Only matings in which the mother is Rh(−) and the father Rh(+) are at risk with respect to hemolytic disease of the newborn. In terms of genotypes, there are two such matings ♀dd × ♂ Dd and ♀dd × ♂ DD. The first produces $\frac{1}{2}Dd$ and $\frac{1}{2}dd$ offspring, of which the dd, being Rh(−), are not affected and also do not immunize the mother. The second mating produces only Dd offspring, which are all Rh(+) and so potentially at risk. Since affected children from either type of mating are always heterozygotes Dd, the "incompatibility" selection is directed against that fraction of the heterozygotes that comes from dd, Rh(−) mothers. The proportion of heterozygotes at risk clearly depends on the population frequency of the D and d genes. Thus, if, for example, d is very rare, it will occur mostly in heterozygotes Dd. Since there are then almost no dd homozygotes, incompatibility selection against Dd is negligible. If, on the other hand, D were very rare, approximately half of the matings involving D would be dd × Dd, and about half of the individuals carrying the D gene would be at risk with respect to incompatibility selection. Thus, in the absence of other selective forces, the *d* gene would be neutral when introduced into a predominantly *D* population, but *D* would be selected against if introduced into a predominantly *d* population.

Only a small proportion of heterozygotes Dd at risk are affected by hemolytic disease of the newborn.

Assuming the Hardy-Weinberg law applies, the frequency q of gene d is the square root of the genotype frequency of *dd*, so that approximately $q = (0.15)^{1/2} = 0.39$. The expected frequencies of the genotypes *DD* and *Dd* are therefore $p^2 = 0.37$ and $2pq = 0.48$, respectively, as given in Chapter 2. Assuming random mating, the total proportion of *Dd* offspring coming from $dd \times DD$ or $dd \times Dd$ matings is

$$q^2 \times p^2 + q^2 \times 2pq \times \tfrac{1}{2} = q^2p(p+q) = q^2(1-q) \qquad (5.1)$$

which gives 9.15 percent when $q = 0.39$. This would therefore be the expected proportion of conceptions giving rise to hemolytic disease of the newborn if all *Dd* heterozygotes at risk—that is, with Rh(−) mothers—were affected. The observed frequency of hemolytic disease of the newborn in Caucasian populations until recently was about one per 150 births, indicating that only a fraction $1/150 \div 0.0915$, or about 6 percent, of *Dd* heterozygotes at risk actually contract the disease. In the absence of preventive therapy, a large proportion, perhaps as many as 75 percent, of affected infants die (see Ceppellini, 1952; and P. Levine, 1958).

Only about 5 percent of offspring with hemolytic disease are first born and 40–50 percent of first appearances of hemolytic disease in a family occur at the third or later births.

This suggests that, on the average, more stimuli than those from one Rh(+) fetus are needed to produce enough antibody to give rise to clinically observable disease. Some data on the relative probabilities of Rh(−) mothers having a first-affected child as a function of birth order are shown in Figure 5.3. Once an Rh(−) mother has had an affected offspring, the chances of subsequent Rh(+) offspring being affected are very high. They are usually, moreover, affected more severely. This is expected from the principle of the secondary response to an antigenic stimulus. The fact that on average only half the offspring of a $dd \times Dd$ mating are potentially able to sensitize the mother means that incompatibility selection against offspring from these matings is likely to be less severe than that against offspring from the $dd \times DD$ mating. Selection against *Dd* heterozygotes in $dd \times Dd$ matings is lessened still further by the fact that the average interval between births of *Dd* children is approximately twice the normal birth interval, thus presumably

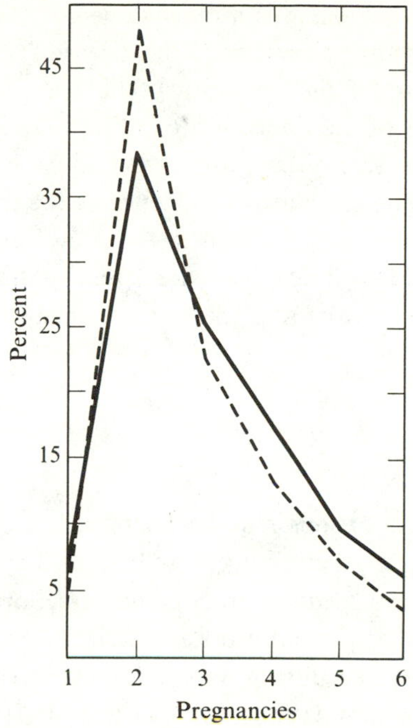

FIGURE 5.3
Relation between birth order (abscissa) and the probability of first observation of Rh hemolytic disease in a family (ordinate). Data of Morganti and Beolchini are plotted by the solid line; those of Van Loghem by the dashed line. (From R. Ceppellini, 1952.)

minimizing the chances of a secondary antibody response. The expected proportion of secondborn offspring that are the second of two Rh(+) offspring of Rh(−) mothers is

$$\underset{(\text{from } \female\, dd \times \male\, DD)}{q^2p^2} + \underset{(\text{from } \female\, dd \times \male\, Dd)}{\tfrac{1}{4}q^2(2pq)} = pq^2(p + \tfrac{1}{2}q)$$

since the probability that the first two offspring of a mating ♀ *dd* × ♂ *Dd* are both *Dd* is $\frac{1}{4}$, while all offspring of the mating ♀ *dd* × ♂ *DD* are *Dd*. When $q = 0.39$ this proportion is 0.074, which is still much higher than the observed proportion, about one percent, of offspring with hemolytic disease among the secondborn. The variability in the birth order of a first-affected child in a family and the small proportion of offspring at risk that are affected indicate that there must be factors other than the rhesus difference between mother and offspring that contribute to determining whether there is a significant antigenic stimulus. This problem will be discussed further when we consider the interaction between the rhesus and the ABO systems. As might be expected, there is a reasonable correlation between the titer of antibody in the mother and the severity of the disease. (See Mollison, 1967, for a review of hemolytic disease of the newborn.) The correlation is, however, not absolute, and the detection of the appropriate antibody may be a matter of technique, since some cases of severe disease are known in which the titer before birth is quite low or negligible. The population incidence of the disease also depends on average family size. The larger average family size prevailing in the less industrial societies that have high infant mortality rates predisposes to a higher rate of hemolytic disease of the newborn than the current small family sizes of western industrialized societies.

Precise estimates of death rates from hemolytic disease of the newborn are hard to obtain for at least three reasons.

First, a number of early abortions may be unnoticed. Second, as most data are obtained from hospital records, they are likely to be biased by an overrepresentation of severe cases. Last, modern medical practice has ameliorated the effects of the disease to such an extent, that present estimates of death rates bear no relation to those which must have operated during most of the time that the Rh genes have been segregating in human populations. This is well illustrated by the fact that Levine and co-workers in 1941 reported 18 out of 24, or 75 percent of cases of diagnosed hemolytic disease as leading to stillbirths, abortions, or neonatal deaths. Bentley Glass in 1950 (see also Glass, 1949) reported 54 percent of 119 cases diagnosed between 1945 and 1947 as fatal but only 33 percent fatality among diagnosed cases at the time of his writing, whereas Walker and Murray in 1956 report only 12—15 percent fatality from hemolytic disease of the newborn. The chance that a *Dd* heterozygote at risk will die is thus reduced to less than 1 percent since, as discussed before, less than 10 percent of heterozygotes at risk are affected. These figures are subject to additional fluctuations due to inadvertent transfusion of Rh(+) blood to Rh(−) mothers in ignorance of the rhesus blood groups. Such transfusions greatly increase the risk of hemolytic disease. Thus, Levine and Waller reported in 1946 that 50 out of 66 mothers whose *firstborn* children had hemolytic disease, had received previous blood transfusions, whereas no more than 1–2 per cent of a random sample of the population, at that time, had received transfusions.

Incompatibility selection against heterozygotes in the absence of other balancing forces gives rise to an unstable equilibrium.

Assuming that about 5 percent of *Dd* heterozygotes at risk actually die of hemolytic disease of the newborn, from Equation 5.1 we see that incompatibility selection causes a loss of about $0.05q^2(1 - q)$ *Dd* heterozygotes, and thus half this many *d* or *D* genes. Ignoring, for simplicity, the difference between ♀ *dd* × ♂ *Dd* and ♀ *dd* × ♂ *DD* matings and assuming that the Hardy-Weinberg law holds, the change in q (the frequency of *d*), is given approximately by

$$q' = \frac{q - \frac{1}{2} \times 0.05q^2(1-q)}{1 - 0.05q^2(1-q)}$$

or

$$\Delta q = q' - q = \frac{0.05q^2(1-q)(q-\frac{1}{2})}{1 - 0.05q^2(1-q)}. \tag{5.2}$$

The normalizing factor needed to make $p' + q' = 1$ is $1 - 0.05q^2(1-q)$ since p and q are each reduced by an amount $\frac{1}{2} \times 0.05q^2(1-q)$. This equation gives $\Delta q = 0$ when $q = 0$, 1, or $\frac{1}{2}$. Since Δq is negative when $q < \frac{1}{2}$ and positive when $q > \frac{1}{2}$, q always moves away from the value $\frac{1}{2}$ and so $q = \frac{1}{2}$ is an unstable equilibrium (see Chapter 4 for a discussion on stability). Thus under these assumptions, when the frequency of d is greater than 0.5 it will tend to increase to 1 and when it is less than 0.5 it will decrease to 0. In the absence of counterbalancing selection, the situation is unstable and the d gene would probably be eliminated from most human populations. It can be shown that this is still true for more realistic and complicated models, which take into account differences in incompatibility selection between the two relevant types of matings (see Example 5.1).

Rh(−) mothers may balance their incompatibility losses by an excess fertility.

Haldane in 1942 suggested that the present D and d frequencies in human populations could be explained by relatively recent admixture of populations that contained only D or only d genes. Others have suggested that mutation-selection balance could be the explanation. This would, however, clearly require unusually high mutation rates for the D and d alleles, especially in view of the considerable differences in the frequency of d in different populations. Fisher and others (1944), on the other hand, have suggested that Rh(−) mothers might compensate for the loss of Rh(+) children through hemolytic disease by an increase in their fertility. Such "compensation" has already been discussed briefly in relation to the population genetics of the sickle-cell trait. Data on fertility of Rh(+) and Rh(−) women obtained by Bentley Glass (1950) in Baltimore, Maryland, are shown in Table 5.1. For both blacks and whites, sensitized Rh(−) women, that is, women who have produced anti-Rh antibody, had significantly higher mean numbers of pregnancies and of living children than their nonsensitized Rh(−) counterparts. The mean number of *live births* per Rh(+) black woman was, however, significantly higher than that of Rh(−) black women, 2.4 compared to 2.2. In the white group, on the other hand, the mean number of living children per Rh(−) woman exceeded that per Rh(+) woman, 1.45 compared to 1.38. Though these data have been criticized as possibly coming from a biased sample, they do suggest the existence of some compensation, possibly overcompensation, in whites for loss of Dd heterozygotes due to hemolytic disease.

TABLE 5.1
Mean Numbers of Pregnancies and Living Children of Rh-positive and Rh-negative Women (with Standard Errors or Means)

Group Observed	*Mean Number of Pregnancies*	*Mean Number of Living Children (Per Woman)*
Whites	$t = 4.65$, $P < 0.001$	$t = 3.05$, $P < 0.01$
Rh+	1.777 ± 0.017	1.378 ± 0.016
Rh−: total	1.908 ± 0.023	1.454 ± 0.020
nonsensitized	1.83	1.43
sensitized	2.46	1.62
Blacks	$t = 1.35$, $0.10 < P < 0.20$	$t = 2.41$, $0.01 < P < 0.02$
Rh+	2.865 ± 0.048	2.412 ± 0.042
Rh−: total	2.725 ± 0.085	2.195 ± 0.074
nonsensitized	2.61	2.11
sensitized	3.6	2.92

Source: From Glass (1950).

In a careful study of child-spacing following stillbirth and infant death, Newcombe and Rhynas (1962) found, in general, a slightly, increased number of pregnancies within the first year following a stillbirth but this tendency was not evident for subsequent years. Their data were, unfortunately, woefully inadequate to detect compensation specifically with respect to Rh hemolytic disease, but their approach is certainly one to be recommended.

Incompatibility selection can, in principle, be balanced by overcompensation by Rh(−) mothers, giving rise to a stable polymorphism.

Simple models for Rh incompatibility selection can be constructed that allow for fertility compensation by Rh(−) mothers in the matings ♀ *dd* × ♂ *Dd* and ♀ *dd* × ♂ *DD*. Parameters k, l and f are used to define the combination of incompatibility selection and fertility compensation as follows (see Example 5.1). It is assumed that the mating ♀ *dd* × ♂ *DD* produces $1 - k$ *Dd* offspring (relative to 1 in the absence of any selective effect) and that ♀ *dd* × ♂ *Dd* produces $\frac{1}{2}(1 - l)f$ *Dd* and $\frac{1}{2}f$ *dd*. If $k > 0$, then incompatibility selection prevails in the mating ♀ *dd* × ♂ *DD*; while if $k = 0$, there is exact compensation for the incompatibility loss; and if $k < 0$, there is overcompensation. For the mating ♀ *dd* × ♂ *Dd*, l measures incompatibility selection against *Dd* and f measures the fertility compensation. When $f(1 - l) < 1$, incompatibility selection prevails in this mating; when $f(1 - l) = 1$, there is exact compensation; while when $f(1 - l) > 1$, there is overcompensation. It can be shown, using this model and ignoring any other potential selective differences, that a stable polymorphism can only exist if there is *over*compensation in at least one of the two relevant matings ($k < 0$ and/or $f(1 - l) > 1$). When there is overcompensation in both matings a stable polymorphism with both alleles present always can exist, while when $k \geqslant 0$ and $f(1-l) \leqslant 1$, a stable polymorphism can never exist.

As discussed earlier, it seems that in general, following sensitization of an Rh(−) mother, each subsequent Rh(+) offspring will be subject to increasingly severe hemolytic disease of the newborn. This seriously limits the opportunities for overcompensation, especially in the mating ♀ *dd* × ♂ *DD*. Numerical calculations indicate, in any case, that no reasonable values of k, l, and f lead to stable equilibria in the observed range of d gene frequencies. Thus, in spite of the inevitable oversimplification, the model clearly does indicate that, if the Rh polymorphism were stable, forces other than incompatibility selection and compensation would have to be operating (Feldman et al., 1969).

In the absence of forces other than the incompatibility selection the frequency of d would decrease.

The rate of decrease would depend, of course, on the amount of incompatibility selection. As already indicated, modern medical practice has changed enormously the magnitude of incompatibility selection against *Dd* Rh(+) heterozygotes in developed industrialized societies. The present distribution of *D* and *d* must, of course, have been molded under very different conditions. Perhaps it is not yet too late to obtain reliable information on the magnitude of incompatibility selection due to Rh hemolytic disease in primitive populations not yet affected by modern hygiene and medical practice. Some curves illustrating the decrease in the frequency of d from an initial value of 0.4 due to incompatibility selection, for reasonable values of k and l, are shown in Figure 5.4. In all cases d disappears and D increases

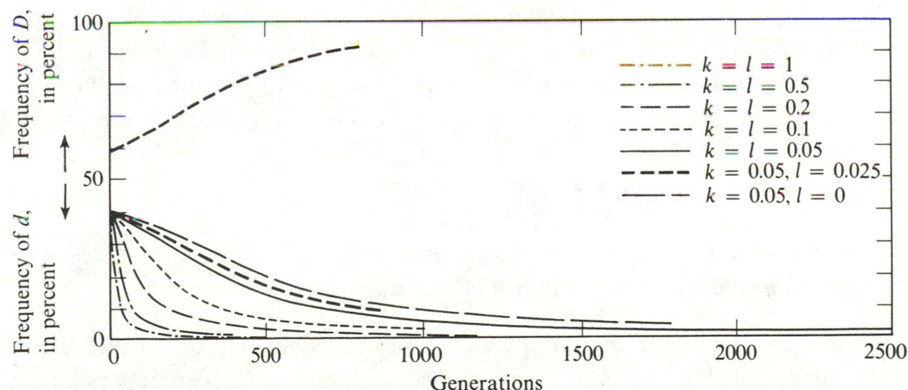

FIGURE 5.4
Change in the frequencies of rhesus alleles d (lower half of the figure) and D (upper half) with various values of the incompatibility selective coefficients k and l. Genotypes *DD* and *Dd* are Rh(+) and *dd* is Rh(−). The coefficient k measures selection against *Dd* in ♀ *dd* × ♂ *DD* matings, l measures selection in ♀ *dd* × ♂ *Dd* matings. Selection is absent when k and l are zero. in all cases an initial frequency of 0.4 for d is assumed.

to 1. The rate of change depends on the magnitude of the incompatibility selection coefficients. Values of $k = 0.05$ and $l = 0.025$, probably present the most realistic situation.

A prophylactic method for preventing Rh immunization has recently been tried out.

Purified anti-Rh(+) antibody is injected, in relatively large amounts, into Rh(−) mothers within 72 hours after delivery of an Rh(+) baby. It appears that the coating of the fetal Rh(+) red cells, which have passed into the mother's blood stream during delivery, by the passively administered antibody will prevent them from actively immunizing the mother. In two different clinical trials none out of 75 "protected" mothers developed an antibody, while 16 out of 88 controls did develop immune Rh(+) antibodies. (See McConnell, 1966, for a review of this method; see also Freda and Gorman, 1966.) If generally applicable, this technique could in principle eliminate almost completely the problem of hemolytic disease of the newborn, leaving the *d* gene to the fate of selective forces other than incompatibility. Its successful application depends, of course, on the availability of adequate amounts of anti-Rh(+) antibody to treat prophylactically large numbers of Rh(−) mothers. One calculation suggests that 50–200 liters of serum would be needed per year for just the population of the United Kingdom. This serum is, of course, very precious. It may be obtained from Rh(−) individuals (or mothers) who have been immunized inadvertently by transfusion of Rh(+) blood (or Rh(+) offspring) or by planned immunization.

These protection studies verify, incidentally, that leakage of fetal blood into the maternal circulation during delivery is the major source of immunization. Small-scale transplacental leakage of fetal blood into the maternal circulation during pregnancy is the second route of immunization; it probably brings about fewer than half of the cases of primary Rh sensitization. Both of these mechanisms are likely to be influenced by chance accidents during delivery or pregnancy, which might account, at least in part, for the variability with which infants at risk are actually affected by hemolytic disease of the newborn.

Heterozygote advantage in some form, independent of incompatibility selection, may be an important factor in maintaining the Rh polymorphism.

It seems unlikely that mutation pressure could be as high as 10^{-3}, which would be the value needed to account for the high frequency of *D* in Caucasian populations on the assumption that *D* is maintained at equilibrium by a balance of mutation and selection. This and other similar considerations indicate that selective factors other than those directly connected with incompatibility selection were most probably the cause for the increase of the *D* or *d* alleles in different human populations. We have already indicated that when D is common, there is negligible incompatibility selection against d and so under these circumstances any intrinsic advantage of the heterozygote Dd will usually be enough to insure the increase of d. On the other hand, when d is common and D rare, incompatibility selection against D must be overcome, either by compensation or by heterozygote advantage or by both in order for D to increase in frequency. These considerations alone suggest that D may be the original allele and d a new mutation that has increased in frequency because of an intrinsic advantage of the heterozygote Dd. We will come back to a consideration of the evolution of the Rh polymorphism after we have discussed the complex genetics of the antigens other than Rh(+) that are a part of this polymorphism.

5.4 ABO Blood Groups and Blood Transfusion

There are four main types of human individuals A, B, AB, and O, distinguished by the antigens they have on their red blood cells and also by the antibodies in their sera. The relationship between antigens, antibodies, and genotypes is indicated in Table 5.2 (see also Table 2.3). A peculiarity of the ABO system is that an individual

TABLE 5.2

ABO Blood Types, Antigens, Antibodies, and Genotypes

Blood Type Corresponding to Antigens on Red Cells	*Antibodies in Serum*	*Genotype*	*Reaction to*		
			Anti-A	*Anti*-B	*Anti*-A *and Anti*-B
O	anti-A and anti-B	*OO*	−	−	−
A	anti-B	*AO* or *AA*	+	−	+
B	anti-A	*BO* or *BB*	−	+	+
AB	—	*AB*	+	+	+

has "naturally occurring" in his sera the antibodies corresponding to the antigens other than those on his own red cells. The basis for Landsteiner's initial recognition of the ABO system was in fact the regular pattern of agglutination observed when cells and serum from different normal individuals were mixed. Initially, owing to the fact that AB is generally rare, he observed just three patterns of reaction, which correspond to blood types O, A, and B. It is interesting to note that the recognition by Bernstein (1924), of the genetic control of the ABO system by the three alleles, rather than by two alleles at each of two loci, depended on an analysis of the **population association** between A and B antigens. Family data had yielded equivocal results because of appreciable misclassification of the ABO types. Bernstein, however, recognized that the A and B antigens were significantly negatively correlated in population data and that this could only be explained reasonably if A, B, and O were controlled by three alleles at a single locus, a hypothesis amply confirmed by subsequent family studies (see Example 5.4). We shall see later how the analysis of population associations has greatly helped the initial determination of the genetic control of antigens on human white blood cells.

Blood-transfusion reactions are generally caused by the agglutination and destruction of donor red cells by antibodies in the recipient. Antibodies in the donor do not, in general, seem to be of much significance in transfusion probably because the amount of blood transfused is small relative to the total blood volume of the recipient and thus the donor antibodies are diluted out. Following this rule, O type blood can usually be transfused without harmful consequences into all ABO types and AB individuals can generally receive blood of any type. For this reason, O is sometimes referred to as a " universal donor " and AB as the " universal recipient." It is, however, always best whenever possible to use a donor whose ABO blood type is the same as that of the recipient because of the occasional effects of the donor antibodies.

Nowadays all transfusions are preceded by cross matching, namely the *in vitro* testing of the compatibility of the bloods of donor and recipient by mixing the serum of each with the red cells of the other to determine the absence of agglutinating reactions. Though ABO typing is adequate in most cases, occasionally a donor has blood-group antibodies for other systems (for example, the rhesus or Kell systems) resulting either from immunization by a fetus or previous blood transfusions, which could give rise to serious reactions. Any such antibodies directed against an antigen on the recipient's cells would be detected by the cross matching test. They are, as indicated before, a major source of reagents for blood typing for presently known blood groups and for the discovery of new antigens.

Blood type A is now known to be subdivided into at least two subgroups called A_1 and A_2, so that there are six types O, $\mathrm{A_1}$, $\mathrm{A_2}$, B, $\mathrm{A_1B}$, and $\mathrm{A_2B}$. $\mathrm{A_1}$ is determined by an allele A_1 dominant to A_2 (which determines antigen $\mathrm{A_2}$) and, of course, to O. Thus genotypes A_1A_1, A_1A_2, A_1O are all of type A_1, and only genotypes A_2A_2 and A_2O are of type $\mathrm{A_2}$. $\mathrm{A_2}$ is in general a weak antigen and is recognized by the use of an antibody that reacts with A_1 and $\mathrm{A_1B}$ cells only. The frequency of type $\mathrm{A_2}$ is generally about one-third the frequency of $\mathrm{A_1}$ in Caucasian populations. A number of other blood groups are now also known to be related to the ABO system. The chemistry and genetics of these systems will be reviewed in a later section.

5.5 ABO Blood Groups and Incompatibility Selection

The possible association between ABO blood groups and disease, especially disease due to ABO incompatibility, has been a subject for some discussion since the discovery of the ABO system. Thus Hirszfeld and Zborowsky in 1925 claimed to find a deficiency of A offspring in ♀O × ♂A as compared with ♂O × ♀A matings. But their analysis was clouded by the then-prevailing incomplete understanding of the genetic determination of the ABO blood types, as is evidenced by the fact that they also claimed an immune mechanism was responsible for the deficiency of AB offspring in AB × O matings.

Following his discovery of the relation between hemolytic disease of the newborn and the rhesus factor, P. Levine, in 1943, showed that parents of diseased offspring were more often ABO compatible than incompatible. A mating is said to be compatible when the father does not have antigens that the mother lacks and to be incompatible if the father does have antigens not present in the mother. A compatible mating will not produce offspring that provide the opportunity for the production of maternal antibodies by fetal-maternal stimulation. Subsequent workers have confirmed the association between ABO and Rh, discovered by Levine in 1943, but have found rare cases of hemolytic disease due only to ABO incompatibility (see, for example, Levene and Rosenfield, 1961). Many claims also exist substantiating Hirszfeld's original observation of a deficiency of A types in ♀O × ♂A matings, which is not specifically associated with clinically observable disease. In addition, a number of significant associations between the ABO blood groups and specific diseases have been shown in recent years.

ABO incompatibility is sometimes a cause of hemolytic disease of the newborn.

Though long suspected, hemolytic disease of the newborn due to ABO incompatibility has only recently been firmly established. It is a rare disease, occurring in about 0.1 percent of the newborn and seems to be largely restricted to the combination of O mothers and A_1 or B fathers. The estimation of A, B, and O gene frequencies, which was discussed in Chapter 2, can easily be extended to include the subdivision of A into A_1 and A_2. Given A_1, B, and O gene frequencies p_1, q, and r, respectively, the expected frequency of O mothers with A_1O or BO offspring, assuming random mating, is $r^2(p_1 + q)$ since r^2 is the frequency of type O and p_1 and q are the probabilities a gamete will be fertilized by gametes carrying A_1 or B genes, respectively. As approximations, in Caucasian populations $p_1 = 0.21$, $q = 0.07$, and $r = 0.66$, giving $r^2(p_1 + q) = 0.12$. Thus, about 0.001/0.12 or only 0.8 percent of offspring at risk with respect to ABO incompatibility develop hemolytic disease. Hemolytic disease due to ABO incompatibility tends to be more self-limiting and, in marked contrast to the rhesus disease, frequently occurs among firstborn. It also has an incidence and a severity that are only slightly correlated with parity. The antibodies responsible for the ABO hemolytic disease are not the same as those in normal, unimmunized individuals. They are the so-called "immune," low molecular-weight antibodies, which apparently cross the placental barrier quite easily and which are formed in response to a specific immune A or B stimulus, amounting effectively to a secondary antibody response. This, presumably explains why the disease may often occur in the firstborn. Normal naturally occurring anti-A and anti-B are thought to be the result of an early immune stimulus by food substances or bacteria, many of which are known to have antigens that cross-react with the human A and B specificities.

ABO incompatibility strongly protects against Rh incompatibility.

Following Levine's original observation in 1943, many workers have confirmed the strongly protective effect of ABO incompatibility against Rh incompatibility. From the definition of compatibility, the nine matings (female first) O × O, A × O, B × O, AB × O, A × A, B × B, AB × A, AB × B, and AB × AB are compatible while the seven matings O × A, O × B, O × AB, A × B, B × A, A × AB and B × AB are incompatible. Thus in, for example, a typical Caucasian population assuming random mating, the expected frequency of incompatible matings can be calculated to be about 34 percent. Frequencies of ABO incompatible matings among those giving rise to Rh hemolytic disease, have been reported to be as low as 15–20 percent, which is appreciably lower than expected for the general population. Not all children of incompatible matings will be incompatible with their mother. Thus an even larger discrepancy in the ABO incompatible mother-child combinations giving rise to Rh hemolytic disease would be expected. The mean expected total frequency of ABO incompatible mother-child combinations in Caucasian populations is about 20 percent, which corresponds well with observations on random samples of the population. The reported frequencies of ABO incompatible combinations among Rh(−) mothers of infants with hemolytic disease, range from 4–13 percent, the variation presumably reflecting differences in the details of pregnancy histories. The interaction of ABO incompatibility and Rh incompatibility is also reflected in complementary differences between the ABO frequencies of mothers, fathers and offspring of matings in which Rh(−) mothers have been sensitized (that is, have produced antibody) as compared to matings in which the Rh(−) mothers are not sensitized. Further direct confirmation of this interaction comes from a study by Vos (1965), showing that the frequency of ABO incompatible matings is correlated with the anti-Rh titer. When this was 1 in 64 or less, the frequency of ABO incompatible mother-child combinations was 26.5 percent, whereas when the titer was greater than 1 in 1024 the frequency was only 4.27 percent. This study also showed a parallel increase in the frequency of stillbirths and neonatal deaths, from one percent for a titer of 1 in 64 and less, to 52 percent when the titer was greater than 1 in 1024.

A direct demonstration of the protective effect of ABO incompatibility against Rh immunization comes from the work of K. Stern and co-workers (1956), who injected ABO incompatible and ABO compatible Rh(+) blood into volunteer

Rh(−) recipients. Of those receiving ABO compatible blood, 10 out of 17 formed antibodies with a median titer of 1 in 128–256, whereas only 2 out of 22 of the group receiving ABO incompatible blood formed antibodies with titers of 1 in 2 and 1 in 8. The most likely mechanism for the interaction between ABO incompatibility and Rh incompatibility is the destruction and elimination of fetal red cells that have passed through the placental barrier into the circulation of ABO incompatible mothers before they have a chance to immunize. The passage of fetal red cells through the placenta during pregnancy may to a large extent be determined by minor accidental ruptures of the placental membrane. The presence of fetal red cells in the maternal circulation can be readily demonstrated. Considerable variations in the incidence and severity of fetal-maternal antibody stimulation have been found in the mouse using inbred lines with little or no residual genetic variation. In this case the variable immunological stimulus must be attributable to nongenetic factors.

Since the ABO and Rh loci are known not to be closely linked, it is reasonable to assume that each Rh mating type defines a random sample of ABO mating types. (Genetic associations due to close linkage will be discussed later in this chapter.) The interaction between Rh and ABO incompatibilities should not, therefore, significantly affect the evolution of the Rh system, when it is considered alone.

There is some evidence for significant fetal wastage due to ABO incompatibility that is not related to hemolytic disease of the newborn.

Though hemolytic disease of the newborn due to either Rh or ABO is an important medical problem, its total incidence is relatively small, probably less than one percent of all births. Evidence for much more significant fetal wastage resulting from ABO incompatibility has been presented by a number of workers. Following Hirzsfeld's original suggestion, Waterhouse and Hogben (1947) found a 20.6 percent deficiency of A children in ♀ O × ♂ A as compared with ♀ A × ♂ O matings, using data collected from various published papers. While their analysis was of heterogeneous data and subject to some questions of interpretation, later studies have, in part, confirmed that there is such a deficiency, though there is still a great deal of heterogeneity between different reports. An example of some conflicting Japanese data is shown in Table 5.3. The first report quoted in this table was based on a study of the populations of two mining towns in Japan and showed a striking effect of ABO incompatibility on overall fertility. The differential loss of incompatible genotypes amounted to more than 20 per cent and was quite significant. In a later and more comprehensive study, also in Japan, no evidence was found for any incompatibility selection. It seems possible (Matsunaga, personal communication) that the earlier studies were subjected to biases avoided in the later study. Another Japanese study (Haga, 1959) also showed an incompatibility effect that was much

TABLE 5.3

Contrasting Data on the Fertility of ABO Compatible and Incompatible Matings

	ABO Compatible	*ABO Incompatible*
MATSUNAGA AND ITOH (1958)		
Number of matings	812	617
Mean number of pregnancies per mating	3.25	3.12
Proportion of childless marriages	0.10	0.18
Mean number of living children per mating	2.6	2.2
MATSUNAGA ET AL. (1962)		
Number of matings	1389	1056
Mean number of pregnancies per mating	3.19	3.14
Proportion of childless marriages	0.030	0.032
Mean number of living children per mating	2.59	2.57

smaller than that of the first Japanese study, only 3 percent loss due to incompatibility selection. Thus, although earlier data showed very significant ABO incompatibility effects, later data from Japan and also from the United States have not always shown such significant effects.

One recent American study from New York is based on a large sample of mothers who were typed for ABO and Rh and for whom fetal deaths had been recorded during a six-year period. The data were obtained by combining information from birth certificates with blood-group data obtained by the New York City Health Department (see Cohen and Sayre, 1968). The fetal death index (ratio of fetal deaths after less than 20 weeks gestation to live births, expressed in percent) for white mothers of all ages was significantly higher if they were O than if they were not (ratio of O to non-O index, 1.14, $P < 0.0001$). This is expected if there is incompatibility selection, since O mothers will much more often carry incompatible fetuses than mothers of other blood types. Morton, Krieger, and Mi (1966), in a recent study of a mixed Brazilian population with high fecundity and high infant mortality, found significant deficiencies associated with A_1 incompatibility but not with A_2 or B. Thus while there is no doubt of the validity of some of the reported significant effects of ABO incompatibility selection, the overall picture is not clear-cut and suggests either heterogeneity in the effects or biases in the sources and methods of analysis of the data. The need for *large* bodies of *homogeneous* data collected in great detail is self-evident, but whether this need can be adequately satisfied is open to some question. It should be noted that the incompatibility effects, when claimed, are much larger than the losses due to established hemolytic disease of the newborn. This suggests that these losses occur during a relatively early stage of development, perhaps by an immune mechanism other than that which causes hemolytic disease of the newborn. These studies have provided no further support for the mechanism of compensation suggested by Fisher and claimed by Bentley Glass in his Baltimore data.

Prezygotic selection could explain observed departures from Mendelian segregations—but it is almost impossible to distinguish from a combination of incompatibility selection and an intrinsic advantage of the heterozygote.

It has been suggested (Matsunaga and Hiraizumi, 1962; see also Hiraizumi, 1964) that some of the observed disturbances in familial ABO blood-group segregations may be due to prezygotic selection, namely preferential transmission of certain gametes according to the ABO genes they carry. As mentioned in the previous chapter, a combination of prezygotic and postzygotic selection acting in opposite directions can lead to a balanced polymorphism. Given certain simplifying assumptions, we can try to estimate the preferential transmission of *A* (or *B*) gametes by *AO* (or *BO*) heterozygotes, from the ratios of A and O phenotypes in appropriate segregating matings. Distortions that are probably more than adequate to counteract incompatibility selection have been estimated in this way. It is, however, almost impossible to distinguish prezygotic selection from a combination of incompatibility selection and an excess viability of heterozygotes, so that the possibility of establishing prezygotic selection remains an unsolved, perhaps even unsolvable, problem. It is known that A and B antigens may be found on the sperm and that anti-A and anti-B antibodies are found in the cervix, but no evidence has yet been obtained for the *segregation* of A and B (or O) antigens among the sperm of a heterozygote (see Race and Sanger, 1968).

5.6 Associations Between ABO Blood Groups and Disease

A number of chronic diseases show significant associations with ABO blood types, but these associations are unlikely to be of much selective importance.

Prompted by the apparent need for some selection associated with the ABO blood groups in order to explain their apparent maintenance as a balanced polymorphism, Ford in 1945 urged a search for associations between blood groups and specific diseases. The idea that there might be such associations must have been present in

Landsteiner's thinking at the time he discovered the ABO blood groups but was discarded (or discredited) in the intervening years. Aird and co-workers discovered in 1953 a significant association between blood type A and cancer of the stomach. Following this a number of other significant associations have been reported. A summary of the most firmly established associations is given in Table 5.4. All of

TABLE 5.4

The Associations Between ABO Blood Groups and Some Diseases

Disease	*Mean Relative Incidence*[a]		*Chi Square*[b]
Duodenal ulcer	1.4	O : A, B and AB	200
Cancer of the stomach	1.25	A : O	49
Pernicious anemia	1.5	A : O	17
Stomach ulcer	1.82	O : A, B and AB	37
Cancer of the pancreas	1.27	A : O and B	8

Source: Based on Clarke (1961).

[a] The mean relative incidence is the ratio of, for example, O : A in diseased patients divided by O : A in a control series.

[b] The χ^2 for one degree of freedom tests the overall significance of the association, pooling all available data.

them have been amply confirmed and are highly significant (see Clarke, 1961 for a review). Stratification, namely the existence of a part of the population, defined by some criteria other than the disease incidence or ABO blood-group frequencies, that is more liable to get the disease and at the same time has a higher frequency of say group A, could cause such associations. The magnitude of the observed associations, however, makes this very unlikely. For if this were the case, either the difference between the specific portion and the rest of the population would have to be almost complete, which would be as interesting as the blood-group association itself, or the stratification would have to be very marked. There is no indication of either of these possibilities. Moreover, similar associations have been found in populations of widely different origins, and the stratification would therefore have to be similar in all of the populations studied.

It has been suggested that disturbances of blood-group and disease associations might be minimized by looking for the association only within families that segregate for ABO blood types. Thus, for example, an excess of observed over expected O type individuals with duodenal ulcer in O, A segregating families (the expectation being approximately $\frac{1}{2}$ in segregating O $\times$ A matings) would provide evidence for the association between type O blood and duodenal ulcer within families. An observed association within a family could not easily be explained by stratification and would minimize possible environmental and background genetic factors. However, this method has two important drawbacks. First, bias in the ascertainment of families may hamper the proper estimation of the expected proportion of O and A types (see Appendix II on Pedigree Analysis). Second, and more important, the use of *segregating* families limits the possibility of detecting associations to *AO* and *BO* heterozygotes, whereas there is no *a priori* reason for excluding the possibility that the association with the homozygous genotypes, *AA* and *BB*, may be more marked. A number of more or less plausible hypotheses have been presented to explain the physiology of the association between type O blood and duodenal ulcer, but no clear-cut picture has yet emerged. No satisfactory explanations have been offered for any of the other associations (see, however, J. H. Edwards, 1965).

In addition to the diseases listed in Table 5.4, associations have been claimed between A (or "not-O") and salivary-gland tumors, cancer of the esophagus, diabetes mellitus, rheumatic fever, and carcinoma of the cervix, but these associations have not been firmly established.

The disease associations summarized in Table 5.4 certainly establish the fact that there may be significant physiological differences between individuals with different ABO blood types. These may eventually be of considerable clinical interest and may also help in understanding some of the factors influencing the diseases involved. They do not, however, offer much help in explaining the maintenance of the ABO polymorphism, because the implied selective differences are very small, since most of the diseases impose their major effects after the end of the reproductive period. A detailed calculation of the selective differences implied by the association between type O blood and duodenal ulcer, presented in Chapter 6, gives values of only 1 in 10^4 for males and 6.4 in 10^6 for females, which are almost certainly too small to be of much significance in counteracting the effects of incompatibility selection. Many workers believe that duodenal ulcer has increased in incidence greatly during the last 50 years, reducing still further the likelihood of its having had a selective effect during the main period of human evolution. The same is probably true for some of the other diseases listed in Table 5.4. The average age of child-bearing in times of greater infant mortality and shorter life-expectancy is likely to have been appreciably lower than it is now, which would minimize still further the effective selection that could have been due to the association of blood type with these diseases.

Associations between ABO blood types and infections diseases, based on the presence of the ABO antigens on the relevant pathogens, have been claimed but not substantiated, and, in any case, would not, in general, result in any advantage to heterozygotes.

Cross-reacting substances similar to the ABO antigens are found in the cell walls of a number of pathogenic organisms, particularly the enteric bacteria *Escherichia coli*, *Shigella*, and *Salmonella typhymurium*. Such cross-reacting substances may be expected to have some significance for the consequences of infection by one of these organisms. A type-A individual, for example, may be more susceptible to infection by a pathogen carrying an antigen cross-reacting with A on its cell surface, because such an individual will not be able to respond to infection by the production of anti-A which might be needed to combat the infection. Vogel and co-workers (1960) have suggested that present world distributions of ABO blood groups may reflect the past history of the plague, since the responsible organism *Pasteurella pestis* has an antigen that they claim cross-reacts with an antigen called H found in type O individuals (see Section 5.7). They suggest that O is frequent only in areas in which there have been no major epidemics of plague. A similar suggestion was made by these workers relating type A blood and smallpox. This association has, however, not been confirmed in subsequent studies by others (see Azevedo et al., 1964). A lower average titer of antibodies against some strains of *Escherichia coli* in type O than in type A, B, or AB individuals has been reported (Eichner et al., 1963), which fits in with the general hypothesis relating ABO blood groups and infection diseases. It must be emphasized, however, that the type of selection that would be operating if the hypothesis were correct would, in general, act against *heterozygotes*, for the more heterozygous an individual is, the more antigens he carries and so the fewer antibodies he can make in response to infection by organisms containing the relevant antigens. Selection against heterozygotes leads to an unstable polymorphism, so that this type of disease association does not provide an adequate answer to the maintenance of the blood-group polymorphisms in the face of incompatibility selection.

Other reasons for preferential disease resistance of different blood types may not be subject to this problem of interpretation. It is known that sometimes viruses in the course of infecting and multiplying in a host cell incorporate into their coat a portion of the surface of the cell they are attacking. In this way they may acquire some of the antigens of their host. If a virus that has so acquired human antigens attacks a person with a different antigenic constitution, the person can respond with antibodies directed against the acquired antigens, and so may be better protected against the viral infection. Clearly, the greater the heterogenity in human populations with respect to cell-surface antigens that can become incorporated into viral coats, the greater the probability that a virus attacking a person will encounter a "hostile" immune environment. Thus, there may be an advantage of heterozygosity with respect to such antigens. (Lederberg, personal communication.)

Recently a few reasonably well-substantiated observations have been published that show a relation between ABO types and susceptibility to such virus infections as colds or influenza. Tyrrell and co-workers, (1968) for example, found that 30 out of 40 type O volunteers given intranasal inoculations became infected (developed detectable virus or anti-viral antibody), while only 12 out of 27 of type A were infected ($\chi^2 = 7.27$, $P < 0.01$). They found, however, that of those who became infected more individuals of type A than type O actually developed colds.

It is possible that any selective differences among the ABO blood groups that played a part in human evolution were associated with differential resistance to infectious diseases, though these associations may not be easily detected in the modern "civilized" world.

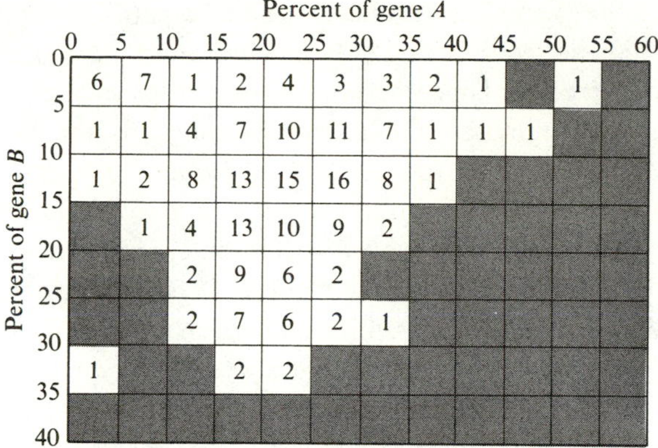

FIGURE 5.5
Quantitative distribution of 215 representative human populations in respect to frequencies of the ABO blood-group genes. (From Brues, 1954.)

If the ABO polymorphism is stable, some selection effectively favoring heterozygotes must counterbalance incompatibility selection.

The ABO polymorphism has been found in a number of primate species, indicating that the polymorphism is probably not transient. Incompatibility selection alone acts against heterozygotes and so, as we have already discussed, would eradicate the polymorphism. There must, therefore, be some compensating selective forces that are adequate to counteract the incompatibility selection and so maintain the polymorphism. These forces most likely favor heterozygotes, either directly if selective values have not changed much, or more likely through selection varying in direction at different times, but on the average favoring the heterozygote. Compensation may also be a contributing factor to the maintenance of the polymorphism. The disease associations so far observed or postulated do not seem to be relevant, though the possibility of significant associations with important infectious diseases and epidemics is not excluded.

It has been pointed out by Brues (1954) that the relatively narrow distribution of ABO blood-group frequencies among the various human races would be hard to explain in the absence of any selective forces. Blood groups and other polymorphisms have been widely used to characterize the differences between human population groups as will be more fully discussed in Chapter 11. A plot of a sample of ABO gene frequencies from different populations is shown in Figure 5.5. The most striking variations are the generally low frequency of *A* and *B* genes among American Indians, the relatively high *B* gene frequency among Asians, and its low frequency among Europeans. No populations are found, however, which have very high *A* or *B* gene frequencies. Using a simple theoretical model for the ABO system, which ignores its interaction with Rh but allows for incompatibility selection and independently varying genotype fitness, Brues has indicated that observed frequencies can only reasonably be explained assuming the existence of appreciable heterozygote advantage. A computer simulation assuming appreciable incompatibility loss, relative viabilities of 0.79 for *OO*, 0.74 for *AA*, 0.66 for *BB*, 0.89 for *OA*, 0.86 for *OB* and 1.0 for *AB*, and allowing for random variations due to finite population size (see Chapter 8), gave a distribution corresponding reasonably well with the observed worldwide distribution of ABO frequencies (Brues, 1963) (see Figure 5.5). While the detailed assumptions on which these results were based are a gross over simplification, and to a fair extent not valid, the qualitative conclusion is clear. The relatively large selective disadvantages assumed for the homozygotes are presumably a function of the relatively large incompatibility effects built into the model, namely loss of 18–25 percent incompatible fetuses. It has been aptly pointed out by Brues that correlations of fitness with blood phenotypes rather than with genotypes will not uncover the nature of the selective forces required to maintain the polymorphism. Their detection will remain a major problem in the absence of suitable immunological or biochemical techniques for distinguishing the genotypes *AO* from *AA* and *BO* from *BB*. Attempts have been made by Morton and others to estimate heterozygote viabilities on the basis of observed segregations of A, B, and O types in families, using the methods of segregation analysis (see Chung and Morton, 1961; and Appendix II). These are, however, based on so many unknown assumptions that it is impossible to judge the real significance of the derived estimates.

In spite of the enormous amount of published work aimed at uncovering the selective forces associated with the ABO polymorphism, no real understanding of such forces has yet emerged. Incompatibility selection is clearly a significant factor though its magnitude is hard to determine. If the ABO polymorphism is stable, which seems likely, both on account of the relatively narrow variation in gene frequencies and its presence in some of the primates, the incompatibility selection must be balanced in some way by selection favoring heterozygotes, but the nature of this balancing selection is also still an open question.

5.7 Other Red-blood-cell Groups

Little or nothing is known about the selective forces associated with blood groups other than the ABO and rhesus groups.

Much less is known about blood groups other than ABO and rhesus. The K_1 antigen of the Kell blood groups, determined by a dominant gene *K* with a frequency of about 95 percent, is sometimes a cause of severe hemolytic disease of the newborn, but at a lower frequency than that caused by either ABO or Rh incompatibility.

Since most observed cases seem to be the result of reactions to prior transfusions, the disease is unlikely to be of much evolutionary significance. An anomalous excess fertility of $K_1(-)$ women, which has no obvious immunological explanation, has been reported by Reed and co-workers (1964), but not confirmed by others. Evidence for an excess fitness of the heterozygote *MN* over homozygotes *MM* and *NN*, in the MN blood groups, has been reported by Morton and Chung, (1959), based on segregation analysis of familial data. Subsequent analysis of more favorable data by the same workers has, however, failed to confirm this excess. In the face of these difficulties, efforts that have been made to estimate the possible level of prezygotic selection associated with the MN system seem rather futile.

One association that is significant and has been amply confirmed, is that between "PTC tasting" (originally considered an "honorary blood group" by Race and Sanger) and different forms of thyroid disease. The ability to taste, as bitter, phenylthiocarbamide (PTC) is determined by a dominant gene *T*, whose frequency in Caucasian populations is about 45 percent. Thus about 25–30 percent of the population are "nontasters," having the genotype *tt*. Adenomaous goiter occurs with a significantly higher frequency in nontasters than in tasters. Among males with this disease the frequency of nontasters is 60 percent and amongst females 37 percent as compared with 30 percent in the population at large (Azevedo et al., 1965). The thiocarbamides are known to be goiterogenic, so that the greater susceptibility of nontasters to adenomaous goiter may be associated with a greater susceptibility to certain goiterogens. The male-female difference is probably associated with the control exerted on the thyroid metabolism by the menstrual cycle. Some evidence for an association between tasters and diffuse goiter has also been reported. Perhaps these opposite selective forces contribute to the balance acting on the PTC polymorphism. The polymorphism is probably stable since it has been shown to be present in several primates.

This miscellaneous series of unrelated facts reflects the poor state of present knowledge concerning the selective factors that are responsible for maintaining the majority of the blood-group polymorphisms.

5.8 The Chemistry and Genetics of Polymorphisms Related to ABO

The genetic control of ABO and related antigens can be interpreted in terms of present knowledge of their chemical structure.

Two polymorphisms are known to be closely related to the ABO system. The first is the "secretor"–"nonsecretor" difference. Secretors secrete into their saliva and other body fluids water-soluble substances with the same antigenic properties as the A and B antigens on the surface of their red blood cells. The substances are detected by their ability to inhibit agglutination of either A cells with anti-A or B cells with anti-B. The difference between secretors and nonsecretors is controlled by a dominant gene *Se*, such that only the recessive homozygotes *sese* are nonsecretors. The second polymorphism associated with the ABO system is the Lewis blood-group system. Individuals carrying the dominant gene *L* have on their red cells either of the antigenic specifications Le(a+) or Le(b+) while *ll* homozygous recessives have neither specificity. Substances with Le(a+) or Le(b+) specificity are also found in serum, saliva, and other secretions, in which the presence of these specificities is also controlled by the *L* gene. Which of the two Lewis specificities is expressed in an individual depends on his ABO and secretor genotypes.

A third rare genetic difference also turns out to be of importance in understanding the overall genetic control of ABO and related antigens. Occasional well-established exceptions to the pattern of inheritance of ABO types have been found. An example is shown in Figure 5.6. An individual (II.2) apparently of type O, whose mate (II.1)

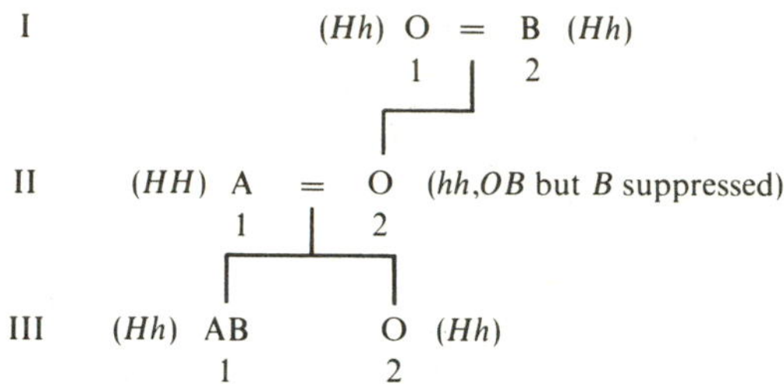

FIGURE 5.6
The inheritance of the Bombay phenotype. The individual with the Bombay phenotype is II, 2. Probable genotypes are given in parentheses. (Adapted from Race and Sanger, 1968.)

was A, had a confirmed AB offspring (III.1). The parents of II.2 were O and B. This individual (II.2) must have carried the *B* gene, but its product, the B antigen, was suppressed. This is interpreted in terms of a recessive "suppressor" gene *h*, such that in homozygotes *hh*, regardless of their ABO genotypes, the A and B antigens are not produced. Thus, II.2 has the genotype *hhOB* but is phenotypically like O. The parent I.2 and offspring III.1 are heterozygous *Hh* and so express A and B antigens normally. The phenotype in which B is suppressed is called Bombay, after the city in which it was originally found.

Some sera are found that identify an antigen, called H, different from A and B, which is present in much larger amounts in O individuals than in A, B, or AB individuals. The Bombay genotype *hh*, lacks this H antigen as well as the A and B antigens and so is phenotypically distinguishable from O. This suggests that production of the H antigen is controlled by the *H* locus and that the H antigen is essential for the expression of the *ABO* locus.

The secretor, Lewis, and Bombay loci have been shown not to be linked. They determine six phenotypes distinguishable by which antigens are secreted and which are on the red blood cells, as shown in Table 5.5. The distinction between A, B, and H antigens is further determined by the ABO genotype. The pattern of phenotypes shown in Table 5.5 gives rise to a strong population association between secretor status and the specificity Le(b+). There is, on the other hand, no association between secretor status and the phenotype Le(a−b−). The frequencies of the *Se* and *l* genes in Caucasian populations are about 0.5 and 0.25, respectively; *h* is always very rare.

TABLE 5.5

The Six Phenotypes Determined by the Bombay, Secretor, and Lewis Loci

Genotype			*Antigens on Red Cells*			*Specificities Detectable in Saliva*		
			ABH	Le(a)	Le(b)	ABH	Le(a)	Le(b)
H·	*Se*·	*L*·	+	−	+	+	[+]*	+
H·	*sese*	*L*·	+	+	−	−	+	−
H·	*Se*·	*ll*	+	−	−	+	−	−
H·	*sese*	*ll*	+	−	−	−	−	−
hh	(*Se*· or *sese*)	*L*·	−	+	−	−	+	−
hh	(*Se*· or *sese*)	*ll*	−	−	−	−	−	−

Source: Based on Ceppellini (1959) and Watkins (1966).
Note: *H*· denotes the genotypes *HH* or *Hh*, etc.

* The brackets indicate smaller amounts of Le(a) here.

It has been shown that the Lewis phenotype is not indigenous to the red cells but is acquired from the soluble Lewis substances found in serum. Thus Le(a−b−) cells can be transformed into Le(a+) or Le(b+) cells by incubation in serum from Le(a+) or Leb(+) individuals (Sneath and Sneath, 1955). The red cell Lewis phenotype is therefore really a reflection of the soluble Lewis substances present in the serum.

Using the serological and genetic data just presented, Ceppellini (1959) recognized that the complex pattern of interactions could be explained by a series of interacting metabolic steps. Thus, on red cells the *H* locus is presumed to control an enzyme that modifies a basic precursor substance to give rise to the H specificity. This is further modified by the *A* and *B* alleles of the ABO locus, to give rise to the A and B specifications: the *A* and *B* alleles require the H substance as a substrate and can not work with the basic precursor substance. For the soluble substances in secretions, *L* is necessary for either of the Lewis specificities. One explanation of the interactions with the secretor locus is that the gene *Se* is necessary for the action of the *H* gene on the soluble substances, and so for the production of the A, B, and H antigens in the saliva. Genes *H* and *L* together with *Se* determine the antigen Le^b while *L* in the absence of *H* gives rise only to the antigen Le^a whether *Se* is present or not. A simplified reaction sequence along these lines is shown in Figure 5.7. An alternative scheme might postulate that the *H* gene product

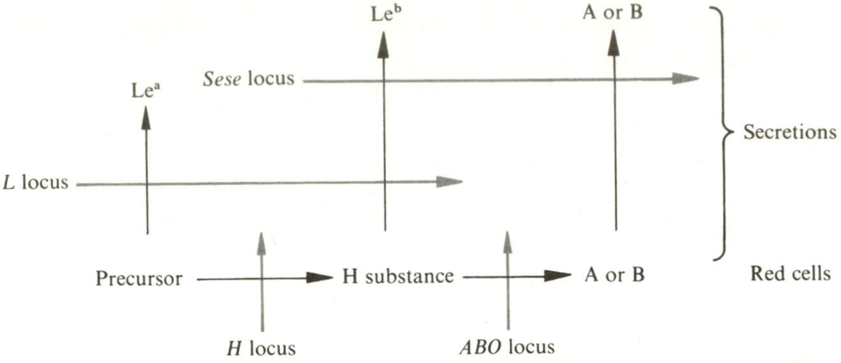

FIGURE 5.7
Simplified reaction sequence for H, AB and Le^a and Le^b substances. The lighter arrows indicate the points of action of the relevant genes.

converts H to A or B and Le^a to Le^b also in secretions, but that its activity there is in some way controlled by the *sese* locus. The complexity of the picture relative to simple metabolic pathways arises from the branching of the paths and the differences between red cells and secretions.

Ceppellini's overall interpretation of the ABO-Lewis-secretor system was strikingly confirmed by a series of elegant chemical analyses of the A, B, Le^a, Le^b, and H specificities worked out independently at about the same time by Morgan and Watkins (see Watkins, 1966). Their work is based on analyzing the substances excreted in body fluids, in particular those obtained from ovarian cysts. The specificities are associated with high-molecular-weight glyco-proteins (proteins attached to chains of carbohydrates) and the chemical structures responsible for the specificities seem to be short chains of carbohydrates attached to large " core " substances. These soluble substances are assayed for their antigenic specificity by their ability to inhibit agglutination reactions of appropriately chosen sera. The initial clues to the chemical structures responsible for the antigenic specificities came from studies on simple substances that inhibited the agglutination reaction in a manner similar to that of the natural substances obtained from body fluids. Thus, the fact that *N*-acetylglucosamine strongly inhibited the agglutinating activity of anti-A sera with A cells suggested that this amino sugar was a primary determinant of the A specificity. The results of many years of elegant work combining this and other methods are summarized in Figure 5.8, which shows the various combinations of D-galactose, *N*-acetylglucosamine, *N*-acetylgalactosamine, and L-fucose that correspond to the H, Le^a, Le^b, A, and B specificities. In each case there is a basic sequence of four sugars, galactose-acetylglucosamine-galactose-*N*-acetylgalactosamine, presumably attached to the core glycoprotein. The various specificities are determined by additions of fucose, galactose, and *N*-acetylgalactosamine to this basic precursor. Any particular glycoprotein presumably may contain many such side chains, which

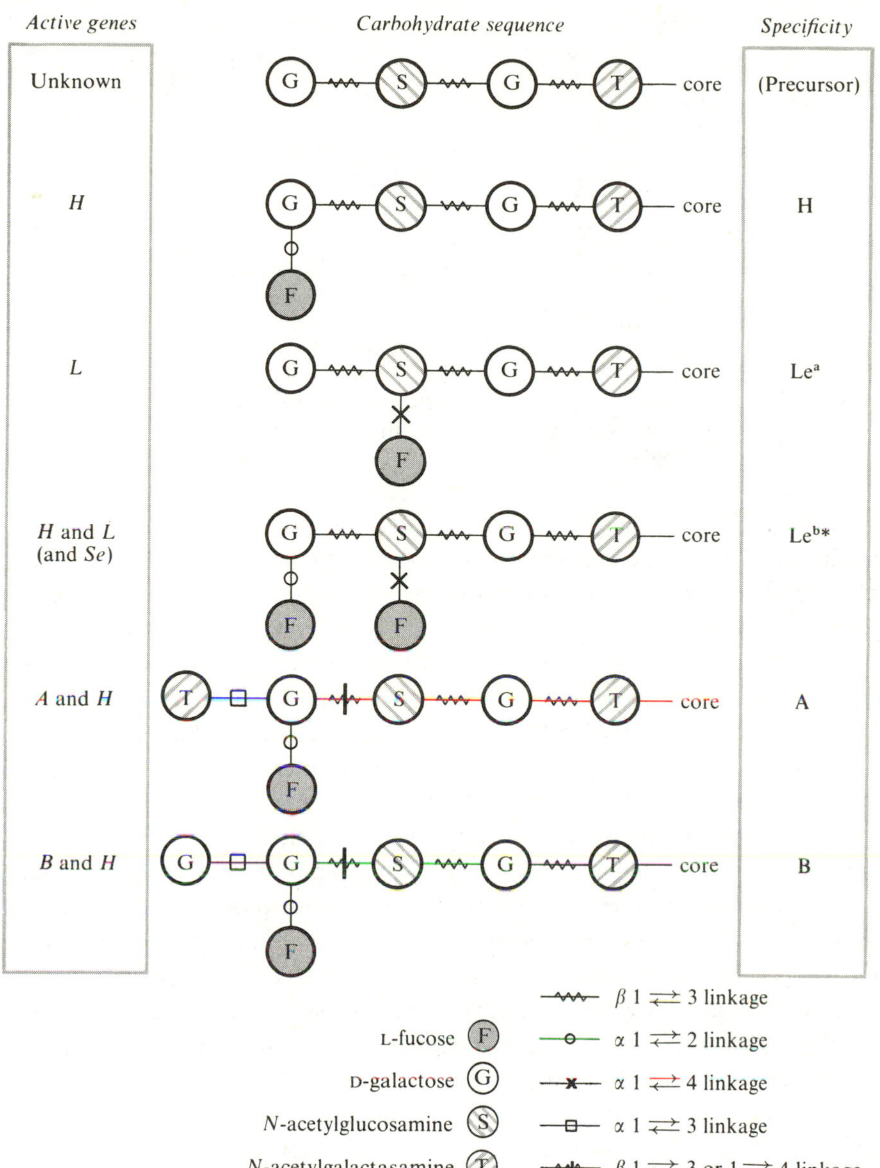

FIGURE 5.8, A

Schematic representation of probable chemical structures responsible for H, Le^a, Le^b, A, and B specificities. The structure marked with an asterisk is not yet confirmed. Genes *A* and *B* can only modify the H substance, and not the precursor, and so are ineffective in the homozygote *hh*. For the soluble substances, gene *H* can probably only act on the precursor in the presence of *Se*, and so is ineffective in the genotype *sese*. (Adapted from Watkins, 1966.)

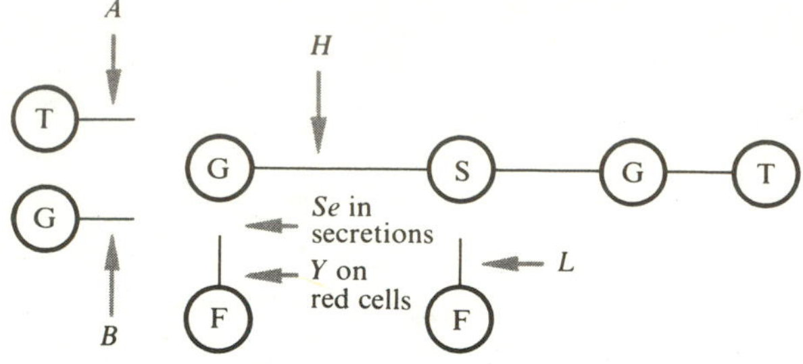

FIGURE 5.8, B
An alternative scheme for the control of the synthesis of ABH, Le^a, and Le^b. *A, B, H, Se, L* are the indicated alleles of the ABO, H, Lewis, and Secretor loci, with the lighter lines and arrows indicating their points of action. *Y* is a gene postulated to act only on red cells, *Se* acts only on secretions. Only *L* can act in the absence of *H*. Gene *h* could be responsible for putting the final GAL on the wrong way.

can include several of the specificities indicated in Figure 5.8, A. The chemical structure of the Le^b specificity is not yet confirmed though most probably it is correct as given. The chemical action of the *Se* gene is not yet known. It may produce a repressor that inhibits the action of the *H* gene on the soluble antigens but not on the red cells, or it may add an as-yet-undetermined chemical group to the soluble antigens that is necessary for the action of the *H* gene. Ginsburg and co-workers (Shen et al., 1968) have shown that the secretor status is correlated with the presence of a soluble enzyme, found in milk, that catalyzes the addition of fucose, and so, when added to the precursor, converts it into H substance (see Figure 5.8, A). This fits in well with the model depicted in Figure 5.8, A. Their study still does not answer, however, the question of whether the *Se* gene product is the enzyme itself or is a controlling substance necessary for the activity of this enzyme. The A, B, and H specificities on red cells are attached to water-insoluble cell-wall glycolypids (sugars combined with fats). Though their chemical structure has not yet been clearly determined, it is presumably the same as that of the water-soluble substances in saliva, serum, and other fluids. The importance of the interaction of *Se* and *H* genes for the water-soluble antigens as opposed to those on the red cells could be related to the different chemical structures of the cores to which the carbohydrates, which give rise to the primary determination of specificity, are attached. It seems likely that the loci for ABO, H, Lewis, and perhaps secretor determine enzymes that are responsible for adding the appropriate sugar residues to polysaccharide chains attached either to the glycolypids or to the glycoproteins. The fact that the genotype $H \cdot Se \cdot L \cdot$ expresses only Le^b on its red cells, or more precisely in its serum, but both Le^a and Le^b in secretions, suggests that in the serum the number of chains available for the joint action of the three genes is limited, so that their action goes to completion leaving no Le^a activity, while in the secretions the number of chains present is in excess, giving rise to both Le^a and Le^b specificities. Alternatively, there may be a set of side chains attached to glycoproteins that cannot accept the activity of all three genes, perhaps because they have a different structure.

An alternative scheme for the synthesis of the various antigens is suggested in Figure 5.8, B. The main differences are that the *Se* gene is postulated to be the structural gene for the enzyme that adds the terminal fucose sugar in secretions, and that another gene, *Y*, is postulated to do the same for red cells. The *H* gene then acts at some earlier stage, perhaps adding the terminal galactose. In this case, the *h* gene perhaps adds this galactose in the wrong configuration. None of the *A*, *B*, *Se*, or *Y* genes can work in the presence of only *h*. Neither *A* or *B* can function in the absence of the terminal fucose.

Recently rare individuals who transmit the combination $A_2 B$ to their offspring have been described. It has been suggested that these individuals carry '$A_2 B$' alleles formed by a rare recombination event in a normal *AB* heterozygote. This might suggest that *A* and *B* are actually mutant alleles in different cistrons controlling distinct enzymes. (See Race and Sanger, 1968; Boettcher, 1966).

Though the story is far from complete, the complex interactions of genotype and phenotype can be plausibly accounted for on the basis of the proposed chemical structures. The complete system demonstrates in a remarkable way how several *unlinked* genes can interact to produce a given detectable antigenic specificity, even when this specificity is not determined by an immediate gene product, namely the protein. It also emphasizes the fact that the "antigenic phenotype" produced by a given genotype may vary according to which cells and tissues are being studied.

It seems clear that the maintenance of the ABO polymorphism cannot be considered separately from the maintenance of the secretor and Lewis polymorphisms or from the fact that the *h* gene is so rare. Disease associations with secretor status have been observed but, as already emphasized, these sorts of associations are unlikely to provide the answer to the selective forces maintaining the polymorphisms. No other indications of selective forces associated with these polymorphisms are yet evident.

5.9 The Complex Genetics of the Rhesus Blood-group System

Inheritance of the many known specificities associated with the original Rh(+), Rh(−) difference can be interpreted in terms of a complex series of closely linked genes.

Soon after the original description of the rhesus antigen, new antigen specificities were discovered, either from other cases of hemolytic disease of the newborn or from cases of transfusion reactions, which were closely associated with the original specificity both in family studies and in random population samples. These new specificities behaved in families as if they were determined by a series of alleles such that a given "allele" carries the genetic determinants for a particular combination of specificities that, in any given family, are always inherited as a completely linked unit. The same specificities may, however, be determined in different combinations by different alleles.

An example of a family showing that a new specificity is closely associated with the original Rh(+) is given in Figure 5.9. A new antibody, found in the mother

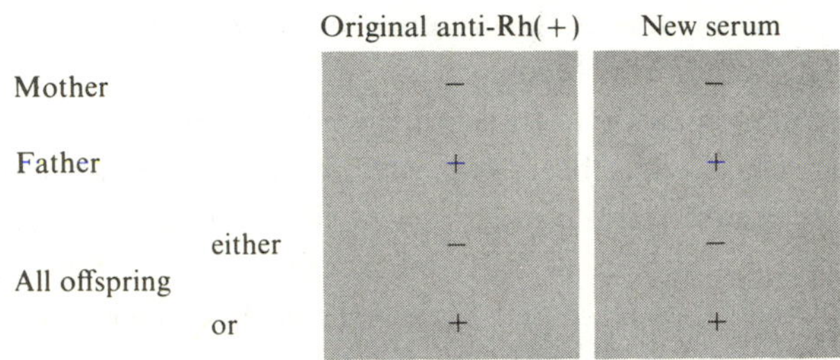

FIGURE 5.9
Demonstration in a family of a new antigen associated with the original Rh(+). The symbols + and − refer to reaction to the indicated serum. All offspring are ++ or −−, none is +− or −+. This suggests that the mother is homozygous −−/−−, the father heterozygous −−/++, and that the genes determining reaction to the two sera are closely linked.

following the birth of an offspring affected by hemolytic disease of the newborn, was different from the original anti-Rh(+), as was shown by the fact that the red cells of some people reacted with one serum but not the other. However, in the family in which it was found, red cells of the father and some offspring reacted to both sera, while those of other offspring reacted to neither. None of the offspring had red cells that reacted to one serum but not the other. Under the assumption that reaction to each serum is controlled by a dominant gene, these red-cell–serum reactions suggest that the mother lacks both genes, and so is −−/−−, that the father is heterozygous for both in coupling, say ++/−−, and that the two genes are closely linked. Further studies confirmed this interpretation and showed that the linkage between the two genetic determinants must be very close, since no cross-overs were found in studies involving large numbers of families. However, families were found in which, for example, a cross −− × ++ would lead to offspring all of whom were either +− or −+. In this case, the genotype of the ++ parent must be +−/−+, in which the genetic determinants for the two specificities occur in repulsion instead of in coupling.

The specificities determined by the main rhesus alleles are shown in Table 5.6. Antibody reactions are given along the rows and allelic combinations along the columns. A "+" indicates that a given allelic combination carries the genetic determinant for a corresponding antigen and a "−" that it does not. The combinations are identified by linkage studies in families as discussed above. Thus in the example

TABLE 5.6
The Main Rhesus Specificities and the Allelic Combinations That Determine Them

	Allelic combinations							
	R_1	R_2	r	R_0	R''	R'	R_z	R_y
	CDe	*cDE*	*cde*	*cDe*	*cdE*	*Cde*	*CDE*	*CdE*
Antibodies								
anti-Rh₀ or D (Rh1)	+	+	−	+	−	−	⊕	⊖
anti-rh″ or E (Rh3)	−	+	−	−	+	−	+	⊕
anti-rh′ or C (Rh2)	+	−	−	−	−	+	⊕	⊕
anti-hr′ or c (Rh4)	−	+	+	+	+	−	−	⊖
anti-hr″ or e (Rh5)	⊕	⊖	⊕	⊕	⊖	⊕	⊖	⊖
anti-d not found	⊖	⊖	⊕	⊖	⊕	⊕	⊖	⊕
Typical European allele frequencies (%)	41	14	39	3	1	1	0.2	very rare

Note: R_1 and *CDe*, etc. are alternative names for the allelic combinations. The antibodies listed in the first column are those used to establish the allelic combinations. Rh_0, D, and Rh1, etc., are alternative names for the specificities detected by these antibodies. R_1, Rh_0, etc., is Wiener's notation; C, D, etc., is the Fisher-Race notation; and Rh1, etc., refers to the notation suggested by Rosenfield and his colleagues (1962). A+ in the body of the table indicates that the corresponding allelic combination (column heading) carries the genetic determinant for the antigen detected by the corresponding antibody (row heading). A− indicates the absence of the corresponding genetic determinant. Thus, for example, the allele R_1 or *CDe* carries the genetic determinants for D, C, and e but not those for E, c and e. The circled reactions (⊕ and ⊖) are those predicted by Fisher in 1944 before the discovery of anti-e or R_z and R_y. The predicted anti-d antibody has never been found. The reactions in the dotted box, which illustrate the complementary nature of anti-C and anti-c, are the ones he used to make his predictions.

given in Figure 5.9 the original Rh(+) was anti-Rh_0 or D and the new antibody was anti-rh′ or C, so that the positively reacting parent was, in the notation of Table 5.6 and ignoring other specificities, *CD*/*cd*. The letters *c* and *d* refer to alternatives not detected by these two antibodies. Here *C* and *D* refer to the genetic determinants corresponding to the antigens detected by the anti-C and anti-D antibodies. The R_1, Rh_0, etc., notation is the one originally introduced by Wiener in which each allelic combination and each antigenic specificity are given a separate designation. The C-D-E notation was introduced later by Fisher and will be discussed below. Anti-Rh_0 (D) was the original rhesus antibody.

The designation of rhesus antigenic specificities and alleles is greatly simplified by the use of single letters for each specificity and a combination of the appropriate letters for each allele.

The specificities listed in Table 5.6 were those that were first discovered within a few years of the original description of the anti-Rh_0(D) and that still account for the majority of allelic combinations found in Caucasian populations.

The encircled reactions in Table 5.6 were predicted by Fisher in late 1943 based on two complementary symmetries he noted in the data available at that time (see also Fisher, 1947):

1. The reactions of the first six alleles with anti-rh′ and anti-hr′ were antithetical—that is, where one was + the other was − (indicated in Table 5.6 by the reactions enclosed by a dotted box).

2. The reactions of alleles R_1 and R'' and of R_2 and R' with the four available antibodies were also antithetical.

On the basis of these observations Fisher predicted first the discovery of two antibodies whose reactions would be antithetical to anti-Rh_0 and anti-rh″, and second that R_z would turn out to be antithetical to r, and third that an allele R_y antithetical to R_0 would be discovered. He suggested using letters C and c to correspond to the respective antithetical specificities rh′ and hr′, and D and E for Rh_0 and rh″, so that the predicted new specificities would be d and e. An allele can then be identified by the combination of antigenic specificities that it determines, so that R_1 corresponds to *CDe*, *r* to *cde* and so on. The predictions concerning the discovery of the antibody, anti-e, the allele R_y(*CdE*) and the remaining reactions for R_z(*CDE*) were soon borne out but the antibody anti-*d* has never been found. Fisher further suggested that the antigenic pairs C and c, E and e, and D and d were determined by alleles at three very closely linked loci, so that new allelic combinations could occasionally be produced by recombination in appropriate heterozygotes. The multiple heterozygote (*CDE*/*cde*, $R_z r$), for example could then give rise to any one of the other six allelic combinations by rare recombination between the three loci. In fact, on the basis of the frequencies of the various combinations and their probable rate of production by recombination he suggested the order of the loci might be *D-C-E*.

A large number of Rh phenotypes can be found in most populations. The frequencies of these phenotypes can be predicted in terms of the frequencies of the known allelic combinations assuming the Hardy-Weinberg law.

Clearly, the number of phenotypes that can be distinguished depends on the number of antibodies available for typing. In some cases the phenotype corresponds to a unique genotype, but mostly there are several genotypes corresponding to any given phenotype. Given the four antibodies anti-C, anti-c, anti-D and anti-E, for example, an individual who is negative to anti-C, anti-D, and anti-E, and positive only to anti-c must have the genotype *cde*/*cde* (*rr*) which is the commonest Rh(−) (that is, D−) genotype. An individual who is C−, c+, D+, E− on the other hand can have either of the genotypes *cDe*/*cde* or *cDe*/*cDe*.

Data on Rh gene frequencies in Sweden, based on tests of 8297 children using six antibodies are shown in Table 5.7. C^w is a specificity that is occasionally found in place of C or c. It was only tested for by anti-C^w on cells that reacted to the anti-($C + C^w$) serum. Anti-e was only used on people who were positive to anti-E, as those negative with anti-E were assumed to be positive to anti-e. Using these sera 15 phenotypes were distinguished. These could be interpreted in terms of the genotypes formed from the twelve allelic combinations constituted by taking any one of C, c, and C^w with D or d, and with E or e, namely *CDE*, *CDe*, *CdE*, *Cde*, *cDE*, *cDe*, *cdE*, *cde*, C^wDE, C^wDe, C^wdE and C^wde. Thirty-six genotypes corresponded to the 15 observed phenotypes. The number of possible genotypes that could be formed is 12 homozygotes $+\frac{1}{2}(12 \times 11)$ heterozygotes, or 78, giving rise to a maximum of 30 potentially distinguishable phenotypes (5 at the first locus determined by combinations of C, C^w, or c namely CC^w (or C^w), C, Cc, C^wc, c; 2 for D+ or − and 3 for E, Ee or e making $5 \times 2 \times 3 = 30$ combinations for all three loci together). The phenotypes not observed are presumably too rare to be found readily even in a sample of more than 8000 individuals. The frequencies of some of these phenotypes can be predicted from the fitted allelic frequencies. Given the allele frequencies, genotype frequencies are obtained, assuming the Hardy-Weinberg law. Thus a homozygote frequency is the square of the relevant allele frequency and a heterozygote frequency is twice the product of the frequencies of the relevant alleles. From these genotype frequencies, we can obtain the expected phenotype frequencies by summing over all genotypes that give rise to the particular phenotype. The data are fitted by starting with an assumed set of allelic frequencies and improving the fit using the method of maximum likelihood scoring (see Appendix I). The fit shown in Table 5.7 is remarkably good. There are only ten genotypes with frequencies higher than one percent; there are only five with frequencies higher than 3 percent—these five genotypes together account for a total of 88 percent of the population. These are constituted from the most frequent alleles found in Caucasian populations namely r (*cde*), R_1 (*CDe*), and R_2 (*cDE*).

Probable rhesus genotypes can sometimes be assigned to phenotypes on the basis of known allelic frequencies.

When one of the genotypes corresponding to a given phenotype is predicted to be much commoner than the others, then this can be assigned as the phenotype's "probable" genotype. Consider for example an individual whose phenotype is cDe.

TABLE 5.7

Rh Frequencies in Sweden: Tests on 8297 Children

Phenotypes										Genotypes		
Reactions with anti-							*Observed*		*Expected*			*Expected*
CC^w	C^w	c	D	E	e		*Number*	*Percent*	*Percent*			*Percent*
−		+	−	−		rr	1,236	14.897	14.596	*cde/cde*	*rr*	14.5962
−		+	+	−		R_0r	123	1.482	1.452	*cDe/cde*	R^0r	1.4174
										cDe/cDe	R^0R^0	0.0344
−		+	−	+	+	r″r	18	0.217	0.225	*cdE/cde*	*r″r*	0.2254
−		+	−	+	−	r″r″	0	0	0.001	*cdE/cdE*	*r″r″*	0.0009
−		+	+	+	−	R_2R_2	256	3.085	2.888	*cDE/cDE*	R^2R^2	2.7892
−		+	+	+	+	R_2r	1,037	12.499	13.392	*cDE/cdE*	R^2r''	0.0985
										cDE/cde	R^2r	12.7612
										cDE/cDe	R^2R^0	0.6196
										cDe/cdE	R^0r''	0.0110
+	−	+	−	−		r′r	32	0.386	0.374	*Cde/cde*	*r′r*	0.3736
+	−	+	+	−		R_1r	2,715	32.723	32.351	*CDe/cde*	R^1r	30.8360
										CDe/cDe	R^1R^0	1.4972
										cDe/Cde	R^0r'	0.0181
+	+	+	+	−		R_1^wr	120	1.446	1.590	*C^{w}De/cde*	$R^{1w}r$	1.5152
										C^{w}De/cDe	$R^{1w}R^0$	0.0735
										C^{w}de/cDe	r'^wR^0	0.0013
+	−	+	−	+	+	r′r″	0	0	0.003	*Cde/cdE*	*r′r″*	0.0029
+	−	+	+	+	+	R_1R_2	1,200	14.463	13.947	*CDe/cDE*	R^1R^2	13.4797
										CDe/cdE	R^1r''	0.2381
										Cde/cDE	$r'R^2$	0.1633
										CDE/cde	R^zr	0.0627
										CDE/cDe	R^zR^0	0.0030

+	−	+	+	+	−	R_zR_2	4	0.048	0.028	CDE/cDE CDE/cdE	R^zR^2 R^zr''	0.0274 0.0005
+	+	+	+	+	+	$R_1^wR_2$	55	0.663	0.685	C^wDe/cDE C^wDe/cdE C^wde/cDE	$R^{1w}R^2$ $R^{1w}r''$ r'^wR^2	0.6624 0.0117 0.0113
+	+	+	−	+	+	r'^wr''	0	0	0.000	C^wde/cdE	r'^wr''	0.0002
+	+	+	−	−		r'^wr	2	0.024	0.026	C^wde/cde	r'^wr	0.0260
+	−	−	−	−		$r'r'$	0	0	0.002	Cde/Cde	$r'r'$	0.0024
+	−	−	+	−		R_1R_1	1,341	16.163	16.681	CDe/CDe CDe/Cde	R^1R^1 R^1r'	16.2861 0.3947
+	+	−	−	−		r'^wr'	0	0	0.000	C^wde/Cde C^wde/C^wde	r'^wr' $r'^wr'^w$	0.0003 0.0000
+	+	−	+	−		$R_1^wR_1$	154	1.856	1.688	C^wDe/CDe C^wDe/Cde C^wDe/C^wDe C^wde/CDe C^wde/C^wDe	$R^{1w}R^1$ $R^{1w}r'$ $R^{1w}R^{1w}$ r'^wR^1 r'^wR^{1w}	1.6005 0.0194 0.0393 0.0274 0.0014
+	−	−	+	+	+	R_zR_1	4	0.048	0.067	CDE/CDe CDE/Cde	R^zR^1 R^zr'	0.0662 0.0008
+	−	−	+	+	−	R_zR_z	0	0	0.000	CDE/CDE	R^zR^z	0.0001
+	+	−	+	+	+	$R_1^wR_z$	0	0	0.003	C^wDe/CDE C^wde/CDE	$R^{1w}R^z$ r'^wR^z	0.0033 0.0001
						Total	8,297	100.000	99.999			99.9999

Source: From Heiken and Rasmuson (1966).

Note: C^w is a specificity which is occasionally found in place of C or c. Two sera were used, anti $(c+C^w)$ and anti-C^w. Only cells positive with the first of these are tested with the second. Similarly only cells positive with anti-E are tested with anti-e, since those negative with anti-E must, because of the "allelism" of E and e ,be positive with anti-e. Genotype and phenotype frequencies are fitted assuming the Hardy-Weinberg law, and assuming the existence of all the 12 possible allelic combinations making C, c, C^w; D, d; and E, e: that is, *CDE, CDe, CdE, Cde*; *cDE, cDe, cdE, cde*; and C^wDE, C^wDe, C^wdE, C^wde. Methods of fitting genotype frequencies are discussed in Appendix II.

He can have either of the genotypes *cDe/cde* (R_0/r) or *cDe/cDe* (R_0/R_0). Assuming random mating and the Hardy-Weinberg Law and using the allelic frequencies given in Table 5.6, the frequencies of these two genotypes are $2 \times 0.03 \times 0.39$ and 0.03^2 or 2.3 percent and 0.09 percent, respectively. Thus the probability of the genotype being *cDe/cde* is more than 20 times the probability it is *cDe/cDe*. The final verification of a genotype always, however, depends on family studies.

More than 25 *antibodies that may be used to identify more than* 30 *rhesus alleles are now known.*

The elegant and relatively simple scheme originally proposed by Fisher, which has formed the basis of most of our discussion so far, has been complicated by the subsequent discovery of new specificities and new allelic combinations that do not all easily fit in with the scheme as first proposed. The new specificities and alleles fall into the following categories.

1. *Alternatives to* D, e, *etc.*: e^s is a modified e almost unique to Africa; D^u is a weak D, C^w, as already mentioned, is an alternative to C and c.
2. *Interaction antigens*: f is a specificity found only in genotypes in which *c* and *e* occur in the same allelic combination, thus the genotype *cDe/CDe* commonly exhibits the f specificity whereas *CDe/cDE* does not. This interaction is analogous to the phenomenon of complementation. Similar antibodies are found for the combinations *Ce*, *CE*, and *cE*. A specificity V corresponding to the combination *ces* is found in Africa. A list of the presently known Rh antigens is given in Table 5.8.
3. *Deletion phenotypes*: individuals are found who react only with anti-D and with none of the other Rh antibodies, suggesting they may be homozygous for a deletion allele *D− −*. Similarly there exist phenotypes interpreted as *Dc−/Dc−* and DC^w-/DC^w- since they show no reactions with any of the E, e, e^s series of antibodies. Finally there is the so-called Rh-null phenotype that shows no reactions

TABLE 5.8
The Known Rhesus Alleles

Allele	*CDE*	*Rh-Hr*	*Allele*	*CDE*	*Rh-Hr*	*Allele*	*CDE*	*Rh-Hr*
Rh1	D	Rh_0	Rh10	V, ce^3	hr^v	Rh19		hr^s
Rh2	C	rh′	Rh11	E^w	rh^{w2}	Rh20	VS, e^s	
Rh3	E	rh″	Rh12	G	rh^G	Rh21	C^G	
Rh4	c	hr′	Rh13		Rh^A	Rh22	CE	
Rh5	e	hr″	Rh14		Rh^B	Rh23	Wiel, D^w	
Rh6	f, ce	hr	Rh15		Rh^C	Rh24	E^T	
Rh7	Ce	rh_1	Rh16		Rh^D	Rh25	LW	
Rh8	C^w	rh^{w1}	Rh17		Hr_0	Rh26		
Rh9	C^x	rh^x	Rh18		Hr	Rh27	cE	

with any Rh antibodies. Phenotypes interpreted as $-C-/-C-$ or $--E/--E$ have not so far been found. As already indicated, it was originally thought that these "deletion" phenotypes were due to the existence of "deletion" alleles, such as $D--$, which lacked completely the genetic information needed for making the C, c or E, e series of antigens. It now seems more likely that individuals with a "deletion" phenotype may be homozygous for a mutation at a gene locus, analogous to the Bombay locus of the ABO system, whose normal product is necessary for the synthesis of the relevant antigens. This locus (or loci) is probably not linked to the Rh locus. Boettcher (1964, 1965) has suggested that the reason no $-$C$-$ or $--$E phenotypes are found is that the D, C, and E antigens are made as part of a reaction sequence D → C, or c, etc. → E, or e, etc. somewhat analogous to that postulated for the ABO and related systems. The "deletion" types are then explained by blocks in the conversion of D → C or of C → E. There is, however, as yet little or no knowledge concerning the chemical structure of the Rh antigens, and certainly no substantial evidence in favor of Boettcher's intriguing model.

4. "*Addition*" *alleles:* individuals are found who, on the basis of family studies, carry an allele *Ccde* determining both C and c specificities.

The total number of different allelic combinations that have been defined in terms of the specificities listed in Table 5.8 and the "addition" alleles is now well above 30.

Though Fisher's original scheme is almost certainly not correct in detail, his notation for the designation of antigenic specificities and allelic combinations greatly simplified the description of the rhesus system, and his prediction of the production of new combinations by genetic recombination, at a time when recombination between alleles had not yet been firmly established, was undoubtedly correct and remarkably farsighted. As indicated earlier, most of the variation found in human populations can still readily be interpreted in terms of Fisher's *C-D-E* notation.

Allelism and close linkage are essentially indistinguishable. A complete understanding of the genetics of the Rh system must await the chemical analysis of the antigen.

There has been much discussion about whether the rhesus system is determined by a series of "alleles" at a single locus, or by a series of alleles at three or more closely linked loci. It may be anticipated that a single-site mutation, namely substitution of a single nucleotide pair in the DNA, can affect a protein in such a way as to change more than one antigenic specificity. This could occur through an effect on the tertiary folding of a protein, which extended over different antigenic sites on the molecule. However, since there are only four alternative nucleotide pairs, that can occupy one such site, the maximum number of combinations of specificities that can occur by mutation at a *single site* is only four. It is clear, therefore, that the large number of allelic combinations observed in the Rh system could not all arise by mutations at a single site. It has been shown, at least in bacteria, that recombinants can occur even for mutations affecting adjacent base pair positions in the DNA. It seems most likely, therefore, that recombinants within the complex rhesus antigenic polymorphism can also be produced.

A possible recombinant within the rhesus system has been reported by Steinberg (1965). Among the offspring of a mating *CDe/cde* × *cde/cde* he observed 4 *cde/cde*, 3 *CDe/cde* and one unexpected type, *Cde/cde*. There was no indication of illegitmacy from data on other polymorphisms segregating in the family. Moreover the members of the family belong to a highly religious sect, the Hutterites, among which Steinberg found not a single *Cde/cde* individual out of 47,444 typed. The *Cde/cde* offspring is most likely therefore, either the product of a crossover between C and D or a mutation D → d or c → C.

Fine-structure genetic studies in microorganisms have clearly established the linear complexity of a single gene, or cistron, and have shown that genetic recombination between mutant genes is independent of the functions they control. Thus mutations in adjacent cistrons may, sometimes, be more closely linked than mutations in the same cistron. The question of whether the rhesus antigen is a polypeptide chain determined by a single cistron, or a complex of chains controlled by a series of closely linked cistrons, or a series of different residues whose specificity is controlled by a set of enzymes, like ABO and Lewis *but* determined by closely linked cistrons, must await a chemical analysis of the rhesus antigens. In this context the definition of an "allele" or gametic combination, is an operational one. It is used to refer to a series of antigenic specificities determined by mutations at genetic sites so closely linked that for all practical purposes they are nearly always inherited as a unit.

The particular allelic combinations found in a given population must be a function of its evolutionary history.

In our earlier discussions of the population genetics of the rhesus system, we ignored all antigens other than D, which is the major one associated with hemolytic disease of the newborn. Much of the genetic variation for rhesus antigens is associated with "allelic complexes" or "haplotypes" mainly *cDe*, *CDe*, and *cDE*, all of which carry the genetic determinant for D, and so are subsumed under the designation *D*, which we used in our earlier analysis of incompatibility selection. Ceppellini and co-workers (1967) introduced the term **haplotype** (from haploid genotype) for the combination of genetic determinants that leads to a set of antigenic specificities which is controlled by one chromosome and so inherited in coupling. Thus each of *cDe*, *CDe*, etc. are haplotypes. This term is equivalent, in our usage, to allelic complexes, or in a loose sense to alleles. However, as we shall see later, it is useful to retain the term alleles, operationally, for sets of differences such as *E*, *e*, and e^s that, apart from very rare exceptions, do not occur together in the same haplotype.

Most discussion of the evolution of the rhesus polymorphism has concentrated on the problem of incompatibility selection against *Dd* heterozygotes and ignored the question of polymorphism with respect to the other antigens and alleles. To our knowledge, only Fisher (1947) in his classic paper on "the rhesus factor" has discussed the origin of other alleles. Following the suggestion that the rhesus antigen differences are determined by mutation at three closely linked sites, Fisher further proposed that the rarer alleles might be derived from the commoner ones by recombination and that they might be maintained at their low frequencies by "recombination-selection" balance. He did not, however, at that time speculate on the origins of the commoner alleles.

The worldwide frequency distribution of the commoner rhesus alleles *cDe*(R_0), *cDE*(R_2), *CDe*(R_1) and *cde*(r) suggests that natural selection has been an important factor in determining the origins of the rhesus system. Thus, there are large differences in allelic frequencies between racial groups of widely differing origins (see Chapter 11), too large, probably, to be readily explained by random genetic drift alone (see Chapter 8). It also seems unlikely that variations in the frequencies of the commoner alleles can be due to mutation pressure in the absence of selection. As already discussed, extraordinarily high mutation rates would be needed to account for the wide disparities in allelic frequencies during the evolutionary life span of the human species. These considerations indicate, therefore, that the one or more mutant genes, which were the origins of the rhesus polymorphism, most probably originally increased in frequency because of some, as yet unexplained, advantage of the newly formed heterozygotes. As already discussed, an analysis of the two-gene *D-d* model suggests that *D* preceded *d* on the evolutionary time scale. Incompatibility in the rhesus system would then be a by-product of other independent selection pressures that initially favored *d* over *D*, and would not be of importance for the initial establishment of the polymorphism.

The alleles *cDE* (R_2), *CDe* (R_1), and *cde* (r) are each connected to *cDe* (R_0) by mutation with respect to a single antigenic specificity. This relation, together with the global distribution of the frequencies of these four alleles, suggests that *cDe* (R_0) may have been the original allele of the system and that R_2, R_1 and r were each derived from it by single independent mutations, a hypothesis at least compatible with the absence of polymorphism and the presence of c – like and D – like antigens on nonhuman primate red blood cells. The other, generally rarer alleles, *CDE* (R_z), *cdE* (R''), and *Cde* (R') may then, as suggested by Fisher, be derived from single crossovers in the respective genotypes, $R_2 R_1$, $R_2 r$, and $R_1 r$, while the very rare *CdE* (R_y) is produced only by crossing-over between these rarer alleles in heterozygotes. This suggested scheme for the evolutionary origin of the major rhesus alleles is illustrated in Figure 5.10. The low frequency of the rarer rhesus alleles (R_z, R'', R', etc.) may either be due to the comparatively recent origin of the primary mutations by which the alleles R_2, R_1, and r arose or, as suggested by Fisher, the frequencies may be held at their low values by the balance between their production by recombination and their elimination by selection. The variations in the frequencies of the commoner alleles presumably are due to differences in their selective values in different populations, either now or at some earlier time. On the basis of our evolutionary scheme the complementary relation between the antigens C, c, E, e, and D and the absence of D is a product of the evolutionary history of the rhesus polymorphism and not necessarily an intrinsic property of the antigens and their genetic determinants.

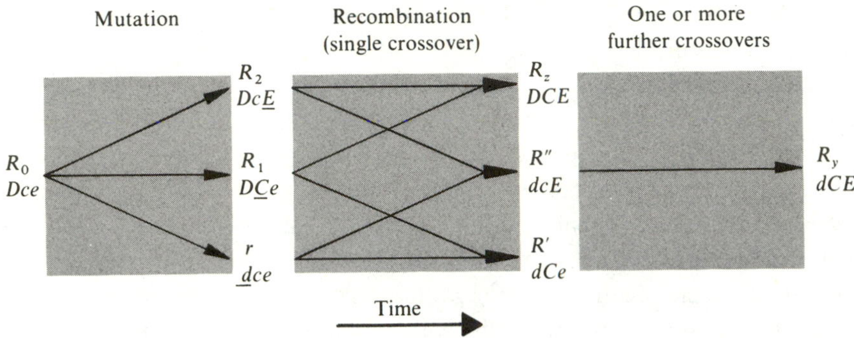

FIGURE 5.10
Suggested evolutionary origin of the major rhesus alleles. (From Feldman et al., 1969.)

This discussion of the rhesus polymorphism has necessarily been over simplified. It ignores, in particular, the existence of many other specificities and alleles and physiological interactions. It seems likely, however, that these complications should not alter the basic qualitative conclusions of this analysis, namely the need for selective forces independent of those associated with incompatibility to explain the origins of the polymorphism and the likely secondary importance of incompatibility selection to the evolution of the rhesus system.

5.10 The Genetic Basis for Histocompatibility Differences

Grafts between donor and recipient pairs combined at random are almost always rejected because the donor and the recipient differ at genetic loci that determine antigenic substances which give rise to graft rejection.

Skin grafts exchanged between pairs of individuals chosen at random rarely, if ever, survive, whereas skin grafts exchanged between identical twins may survive indefinitely. Rejection is an immunological reaction. Thus, a second skin graft between two individuals is generally rejected much quicker than the first in the same way that repeated immunization with an antigen gives rise to a high "secondary" antibody response. Antibodies may also be formed in response to skin grafts and grafts of other tissues and organs, though the production of antibody cannot always be detected at the time of graft rejection. The phenomenon of rejection is under genetic control. The nature of this control has, however, only been elaborated after many years of studies with experimental animals, mainly the mouse, which can be bred easily under controlled conditions and made more-or-less genetically homogeneous by extensive inbreeding, usually by many successive brother-sister matings. These studies have established that graft rejection is determined by genetically controlled antigenic differences, similar to the red-blood-cell antigens that are responsible for transfusion reactions and hemolytic disease of the newborn. The relevant antigens are called **histocompatibility antigens**.

A histocompatibility antigen will elicit graft rejection if the donor has the antigen but the recipient does not.

Such a combination is said to be **incompatible.** Those combinations are **compatible** in which donor and recipient have the same antigens or in which the recipient has at least all the antigens carried by the donor and in addition possibly others that the donor does not have. Compatibility is, therefore, not reciprocal. As for other antigens, histocompatibility antigens are generally determined by codominant alleles. In general, only those antigens controlled by loci that are *polymorphic* are detectable histocompatibility antigens. Thus, which antigens are important for graft rejection in any given population will depend to a large extent on the evolutionary history of the relevant genetic differences within the population.

Intuitively we expect that the probability that two individuals are compatible will be higher for closely related individuals and will decrease as the number of polymorphic histocompatibility antigens increases. At least 12 specific histocompatibility loci have been identified in the mouse and a few more still remain to be worked out, though one, the celebrated H2 locus, seems to be much more important than all the others (Snell and Stimpfling, 1966). In man there are, most probably, about the same number of histocompatibility loci, though, as we shall discuss later, only the ABO locus and a recently discovered antigenic polymorphism on white blood cells called HL-A have so far been identified as histocompatibility antigens. It is thought that the HL-A polymorphism in man may correspond in an evolutionary and physiological sense to H2 in the mouse.

The frequency of compatible parent-child, sib-sib, and random pairs can be predicted given certain simplifying assumptions.

Consider first, for simplicity, a histocompatibility antigen polymorphism controlled by two alleles A_1 and A_2 at a single locus, whose frequencies are equal. If mating is random the three genotypes A_1A_1, A_1A_2, and A_2A_2 occur with frequencies $\frac{1}{4}$, $\frac{1}{2}$, and $\frac{1}{4}$. There are three types of compatible pairs with respect to the antigens determined by A_1 and A_2 : (1) pairs of homozygotes, either both A_1A_1 or both A_2A_2 ; (2) pairs of heterozygotes, both A_1A_2 ; and (3) pairs in which the recipient is a heterozygote and the donor is either type of homozygote A_1A_1 or A_2A_2. In fact, in this simple situation the heterozygote is a universal recipient. The probability that a pair chosen at random from the population are both A_1A_1 is $\frac{1}{4} \times \frac{1}{4} = \frac{1}{16}$, and there is the same probability that such a pair are both A_2A_2. The combined frequency of randomly chosen compatible pairs of types 2 and 3 is simply the frequency of heterozygotes, namely $\frac{1}{2}$. Thus the total frequency of compatible random pairs is $\frac{1}{16} + \frac{1}{16} + \frac{1}{2} = \frac{5}{8}$ or 62.5 percent. Homozygous parents will always be compatible as donors with all their children; heterozygous parents will only be compatible with heterozygous children. The frequency of heterozygotes is $\frac{1}{2}$, and the probability they have a heterozygous child is $\frac{1}{2}$, this being the probability the child receives A_2 from the second parent, having received A_1 from the first, or vice versa. Thus the frequency of parent-child heterozygote pairs is $\frac{1}{2} \times \frac{1}{2} = \frac{1}{4}$. The total frequency of compatible parent-child pairs is, therefore, $2 \times \frac{1}{4} + \frac{1}{4} = \frac{3}{4}$ or 75 percent, which, as expected, is a little higher than the frequency of random pairs. It can be shown that child-to-parent compatibility occurs with the same probability as parent-to-child compatibility.

To calculate the frequency of compatible sib-sib pairs we have to consider each type of mating in turn. The mating $A_1A_2 \times A_1A_2$, for example, occurs with frequency $\frac{1}{2} \times \frac{1}{2} = \frac{1}{4}$, produces offspring in the proportions $\frac{1}{4}A_1A_1$, $\frac{1}{2}A_1A_2$, and $\frac{1}{4}A_2A_2$, and so compatible sibs pair with the same frequency as random pairs from the population, namely $\frac{5}{8}$. The contribution of this mating to the frequency of compatible sib-sib pairs is therefore $\frac{5}{8} \times \frac{1}{4} = \frac{5}{32}$. Calculating the contributions for each mating type separately in this way and then adding gives a total frequency of 78.1 percent, which is just slightly higher than for parent-child combinations. This approach to the calculation of the frequency of compatible pairs with respect to a given polymorphism can readily be extended to more complex genetic situations. The results for a single locus with an arbitrary number of equally frequent alleles are given in Table 5.9 (see also Example 5.2; Newth, 1961; Lunghi, 1965; for review see Elandt-Johnson, 1969). The same data are shown graphically in Figure 5.11. Two main features of these results, which also apply qualitatively to more complex models, should be noted: (1) the probability of compatibility decreases as the number of alleles increases; (2) the probability of sib-sib compatibility is higher than that of parent-child compatibility, which in turn is higher than that for random pairs, and these differences become larger for larger numbers of alleles. It can be shown that the probability of compatibility is, quite generally, *least* when, as assumed here, the alleles have equal frequencies. Any departure from equal frequencies increases the probability of compatibility.

Another important feature of the results given in Table 5.9, which holds quite generally for multiple alleles at a single locus, is that as the number of alleles becomes

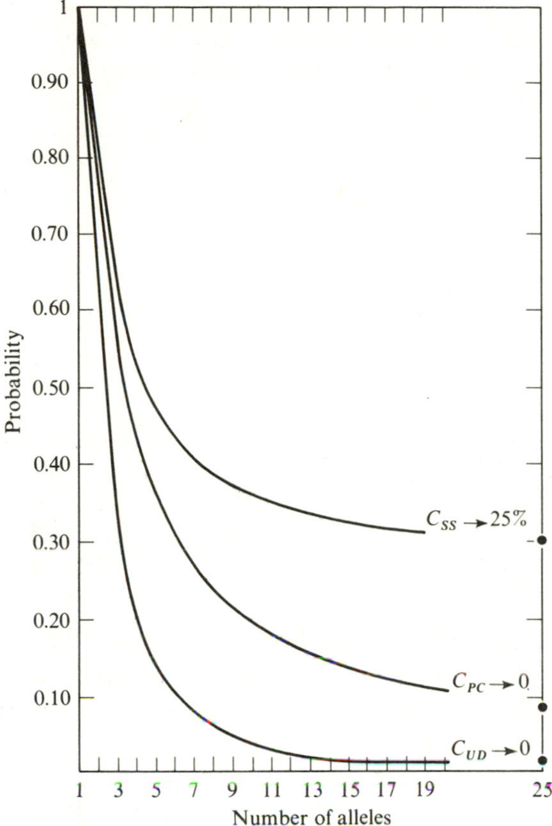

FIGURE 5.11
Probabilities of compatible grafts (C) as a function of the number of alleles (of equal frequency and effect) when the donor is chosen at random among unrelated individuals (UD), among parents or children (PC), or among sibling (SS) of the recipient. See formulas of Example 5.2. (From Table 5.9. Lunghi, 1965.)

very large the random and parent-child compatibility probabilities tend to approach zero, while that for sibs tends to approach $\frac{1}{4}$. In this case essentially all pairs of individuals chosen at random are different and thus clearly incompatible, and hence all matings are of the form $A_1A_2 \times A_3A_4$. Each child, therefore, differs by at least one allele from either parent, and so is incompatible with both of them. There are, however, four types of sibs occurring with equal frequencies of $\frac{1}{4}$, so that the probability of sib pairs being compatible is $4 \times \frac{1}{4} \times \frac{1}{4} = \frac{1}{4}$, as given in Table 5.9.

The probability of compatibility for more than one locus, if they are independent, is obtained from the product of the probabilities for each of the separate loci.

TABLE 5.9

Frequency (in percent) of Compatible Donor-Recipient Pairs for one Locus with n Equally Antigenic Alleles occurring with Equal Frequencies

Number of alleles	*Random pairs*	*Parent–Child*	*Sib–Sib*
n	$(4n-3)/n^3$	$(2n-1)/n^2$	$(n^3+4n^2+2n-3)/4n^3$
2	62.5	75.0	78.1
3	33.3	55.5	61.1
4	20.3	43.7	51.9
10	3.7	19.0	35.4
Large n (approximately)	$4/n^2$	$2/n$	$\frac{1}{4}$

For example, the probability for random pairs with one locus and four alleles is $(4 \times 4 - 3)/4^3 = 13/64$. For five such loci the probability of compatibility would be $(13/64)^5$ or about 0.034 percent. This probability, of course, decreases rapidly as the number of loci increases.

Some limits to the number of histocompatibility loci and alleles can be set by fitting theoretical expectations to observed frequencies of incompatibility.

Attempts have been made in man to use these types of results to estimate the number of histocompatibility loci and alleles, using known frequencies of skin- and kidney-graft survival. This approach, however, faces two major difficulties in the assumptions made to obtain the theoretical results. First, we have assumed that all alleles occur with equal frequency, which will certainly not be the case. As already pointed out, any departure from this assumption will decrease the frequency of incompatible pairs and so lead to an overestimate of the number of alleles. Much more serious is the assumption that all alleles have equal effect and act independently. This is known to be incorrect, at least in the mouse, and most probably also is in man. Some antigenic differences are "strong" and on their own can cause rapid graft rejection. Others are "weak," and a graft across such an antigenic difference may survive for a comparatively long time. Weak antigens may, however, have such cumulative effects that the combination of several of them may lead to rapid graft rejection. Allowing for differences in antigenic strengths and allelic frequencies gives rise to so many alternative models that any given observed frequency of compatibility may be fitted by a large number of different models, aiding little the determination of the real genetic basis underlying graft rejection. Nevertheless the formulas given in Table 5.9 do allow some limits to be set on the possible number of alleles and loci. Some data on the survival of skin and kidney grafts in man are shown in Figure 5.12. The proportion of kidney grafts surviving a given length of time is given in Figure 5.12,A for those between monozygous twins, sibs, parent-child, and random pairs, separating living and cadaver donors. Data are given in Figure 5.12,B on the cumulative mortality of skin grafts between unrelated pairs, parent-child pairs, and sib pairs. Failure of kidney grafts in monozygous twins is presumably due to nongenetic factors, often the recurrence of the disease for which the original transplant was made. The relative ordering of the survivals as a function of relationship of donor and recipient is as expected, though the time scales for skin and kidney are quite different. The first is essentially in days while the latter is in years. The scales were chosen so that the two sets of data would give roughly comparable survival or mortality curves. An important difference is almost certainly the use of **immunosuppressors** in kidney transplantation therapy. These are drugs that suppress the overall immune response and so help to prolong the survival of incompatible grafts. There may also, however, be differences in the antigenic structure of skin and kidney and in the physiology of the grafts, which make, for example, skin grafts more accessible to the immune response than kidney grafts.

The lack of any grafts between unrelated pairs, for whatever organ or tissue, that survive for a long time indicates that a relatively large number of loci and alleles are involved. Since there is no "complete" survival, any time point taken as a cut-off point to distinguish relative success or failure of the graft is arbitrary. It seems likely that strong antigens are more important for short term survival and the accumulated effects of weaker antigens for long-term survival. Thus, fitting the probabilities of survival beyond, say, one year to histocompatibility models such as those already discussed, may give estimates of the number of strong loci and alleles rather than of the total number. This would certainly be significant, since it seems likely that immunosuppressive drugs may be able to cope reasonably with the weaker histocompatibility antigens but not with the stronger ones. Estimates varying from one locus with three or four strong alleles (Simonsen, 1965) to two loci with eight and four alleles, respectively, have been suggested (Serra and O'Mathuna, 1966). If the number of important antigens were really this few, there should be little difficulty in matching donors and recipients for the important antigens, once they have been identified. However, the need for assuming antigens of equal strengths, all of which are equally effective in causing graft rejection, makes these estimates very uncertain. The only real answer to this question lies in the specific identification of each of the histocompatibility antigens, their genetic control, and their relative importance for transplant survival.

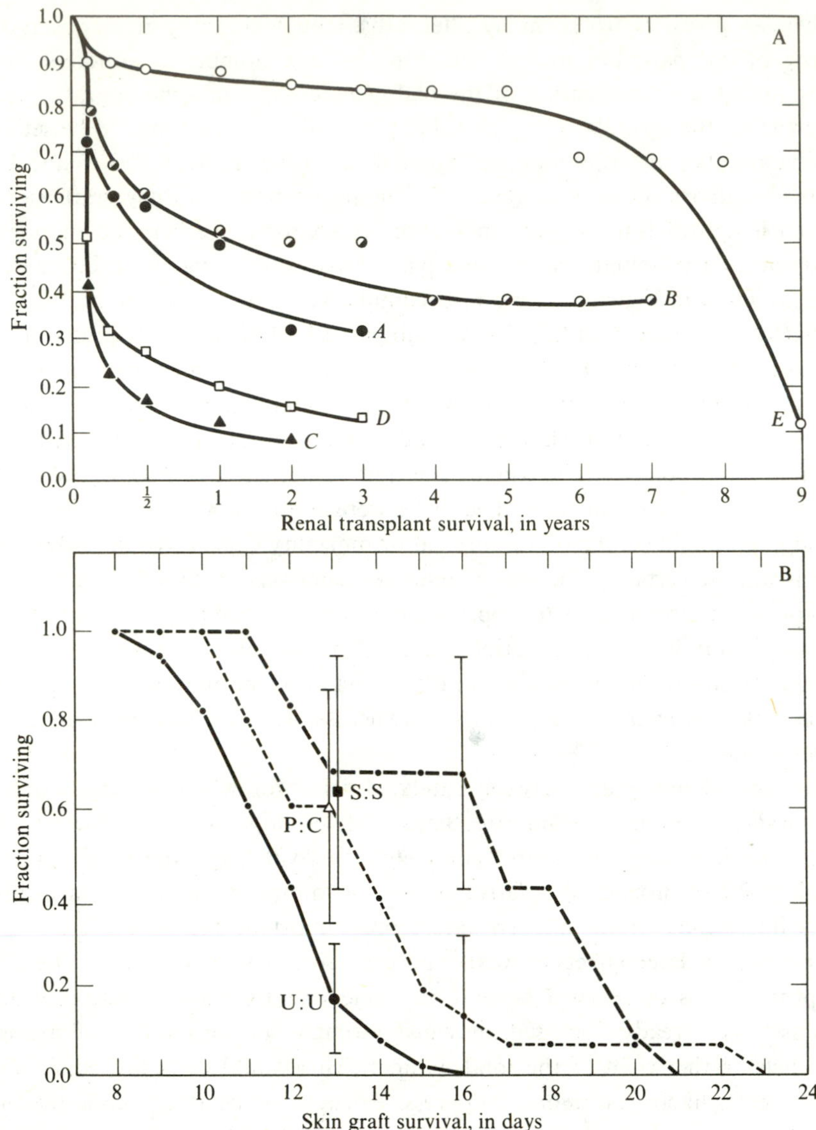

FIGURE 5.12
A: Survival data for renal transplants based on 627 primary transplants. (Courtesy Kidney Survival Data.) *A*: parental donor; *B*: sibling donor; *C*: unrelated living donor; *D*: unrelated cadaver donor; *E*: monozygotic twin donor. (From Ceppellini, 1968.) B: Cumulative mortality curves, with 95 percent confidence intervals at thirteen and sixteen days, of skin grafts classified according to genetic relationship between donor and recipient (S = sibling, P = parent, C = child, and U = unrelated person). Only ABO-compatible grafts are considered. (From Ceppellini et al., 1966.)

There are three main approaches to the problem of overcoming the immunological barrier against transplantation: matching, specific inhibition of the immune response, and nonspecific inhibition of the immune response.

First the identification of the histocompatibility antigens can be attempted, which would permit the matching of donors and recipients who are compatible for all the important antigens. Second, it may be possible to suppress the immune response to specific important histocompatibility antigens, either by the induction of tolerance or by immunological paralysis. Thus, if an animal is challenged with a large dose of a given antigen sufficiently early in its life in relation to the development of the immune system, then it may become "tolerant" toward the antigen, namely, not produce an antibody directed against the antigen if challenged again later with it. Similarly, administration of large doses of antigen to human adults can sometimes "paralyze" their immunological response to the antigen. Last, it is possible to suppress the overall immune response with certain drugs called immunosuppressors or, possibly, with antibodies made against human lymphocytes, the so-called anti-lymphocyte serum. (We should perhaps also mention that artificial organs are being developed that may possibly be inserted into the living body and not elicit an immune response.)

There are disadvantages inherent to each of these three approaches to overcoming the transplantation-immunity barrier, and any solution will probably depend on a combination of them. When the total immune response is suppressed, an individual naturally becomes very susceptible to infectious diseases. It is thus characteristic that deaths of organ-transplant recipients from infection, which are not uncommon, involve organisms, often fungi, that can easily be combated by individuals with a normal immune response. Induction of tolerance is, even in experimental animals, not yet a procedure sufficiently well developed to be ready for clinical application. The routine induction of tolerance to a wide variety of antigens in infants certainly poses many problems. Induction of tolerance in adults probably needs to be accompanied by the administration of drugs whose overall effect may be very detrimental.

The identification of histocompatibility antigens is now progressing rapidly. However, the finding of a compatible graft donor for a given recipient will always remain a major problem. It will certainly be somewhat alleviated by the development, currently underway, of suitable techniques for long-term preservation of whole organs. This will enable the establishment of tissue and organ "banks," built up from material obtained from cadavers, which would be analogous to blood banks. In what follows we shall concentrate on the problems of matching, especially with respect to the HL-A polymorphism.

Some of the antigens on red blood cells may be important histocompatibility antigens.

Undoubtedly, the simplest approach to histocompatibility typing is through the recognition of important antigens on suitable cells and tissues, notably blood cells and skin, by appropriate serological techniques. Such systems are first identified without any reference to histocompatibility and then must be tested, one at a time, for their importance as histocompatibility antigens, most commonly by their effect on skin-graft survival.

The first and most obvious candidates for such an approach are the red-blood-cell groups. The ABO antigens are known to be widely distributed throughout the various tissues of the body. Though it was thought for many years that the red-blood-cell groups were not important for transplantation in man, it has been shown recently that the ABO system is certainly important for histocompatibility (Dausset and Rapaport, 1966; Ceppellini et al., 1966). Data on the survival of skin grafts between unrelated ABO compatible and incompatible pairs are shown in Table 5.10. There is a very significant difference of three days, in the overall mean survival time and it is clear that A_1 into O incompatibility has a major effect on graft survival. It is interesting to note that this is the same combination that is implicated in ABO fetal-maternal incompatibility. Data on the survival of ABO-incompatible kidney grafts confirm directly, for this organ, the importance of the ABO system for graft survival. Further confirmation that the ABO system is a significant histocompatibility system comes from the fact that preimmunization with ABO-incompatible red cells significantly shortens the survival of subsequent ABO-incompatible skin grafts as compared with ABO-compatible skin grafts. There is some suggestion that the P red-cell blood-group system may also be significant for histocompatibility typing.

Antigenic polymorphisms on the white blood cells have now been shown to be important for histocompatibility.

There are a number of lines of evidence that suggest that histocompatibility antigens may be found on the white blood cells. Thus, injection of white blood cells sensitizes against the survival of subsequent skin grafts. Multiple skin grafts lead to the production of antibodies that are directed against white blood cells, suggesting that skin and white cells have important surface antigens in common. More generally, for most sera that react with white cells, the activity can be absorbed out by many different tissues, including skin and placenta. Recently it has been shown that matching for reactions of sera with white blood cells significantly enhances the survival of skin and kidney grafts. Antigens on white cells have also been shown to be important for febrile (fever-inducing) transfusion reactions, which quite often occur with blood fully matched for red-cell groups.

TABLE 5.10

Survival of Skin Grafts Between Unrelated Individuals as a Function of ABO Incompatibility

Donor	Recipient	Graft Survival Time in Days															
		WG[a]	9	10	11	12	13	14	15	16	17	18	19	20	21	23	MST[b]
O	O		2	4	2	3	3	1	1	1							12.12
A	A, AB			1	6	4	3										
O	A, B, AB				1	1	5	2	2								
A_1, A_1B	O	4	3	2	2												9.13
A_2, A_2B	O			1	1	2											
A_1B	A					1											

[a] WG = White graft, a term for a very rapid rejection.

[b] MST = mean survival time.

Source: Ceppellini et al. (1966).

5.11 The HL-A Polymorphism

The recognition of antigenic systems on the white blood cells was at first hindered by a number of technical problems. Sera reacting with white blood cells (mainly lymphocytes or granulocytes) were initially obtained from people who had had many blood transfusions. These sera, not surprisingly, contained a large number of antibodies. In addition, the assays used, which generally depended upon agglutination, were not very reproducible and it was difficult to obtain enough white blood cells to do large-scale absorptions in order to purify the sera. In recent years, the more reproducible cytotoxicity assay, which is based on the ability of some antibodies to kill their specific target cells in the presence of complement, has become the most widely used assay for white-cell typing. In 1958, Payne and Van Rood independently discovered that leukocyte agglutinins could be found in women who had been sensitized during pregnancy by fetal-maternal incompatibility, as in the case of many of the rhesus red-cell-blood-groups. This immediately made available sera, which though still largely multispecific, were much less complex and so were suitable as reagents for the establishment of genetic groups. About 20–30 percent of women who have had two or more children have in their sera white-cell antibodies that will react with their husband's cells and with those of a significant fraction of the population. These antibodies, in contrast to the Rh and ABO antibodies, which are the major cause of hemolytic disease of the newborn, do not in the large majority of cases have any obvious clinical effects on the offspring. Sufficiently large studies have, however, not yet been done to detect incompatibility effects comparable to certain of those claimed for the ABO and rhesus systems. Even with such pregnancy sensitized sera, problems of obtaining enough material for large-scale absorption and their generally low titers, combined with their multispecificity have until recently hindered the definition of antigenic specificities by conventional immunological procedures. Many of the best HL-A typing sera now available have been produced by planned immunization of human volunteers with human white blood cells, using donors carefully chosen to differ only with respect to a desired specificity. Statistical analysis of the pairwise associations of the reactions of a set of primarily multispecific sera, with a panel of random cell donors, at first formed the main basis for the definition of antigenic specifications on the white cells. It is still a very useful procedure for the initial characterization of sera. (See Bodmer et al., 1969 for a review.)

As in the original elucidation of the genetics of the ABO blood groups by Bernstein, the analysis of population associations between white-cell antigens has also proved to be a useful guide for the determination of their genetic control. It is now known that most of the white-cell antibodies produced either by fetal-maternal stimulation or by planned immunization of human volunteers detect antigens controlled by one complex polymorphic system called HL-A. In this section we shall first review some of the statistical approaches used for the definition of antigenic specificities and their genetic control. We shall then discuss the present view of the HL-A polymorphism, the evidence for its importance as a histocompatibility polymorphism, and some speculations on its evolution.

The main statistical tool for the simple analysis of serological specificities is the 2×2 *comparison of the reactions of a set of sera with a random panel of cell donors.*

As shown in Table 5.11, two sera each containing one or more antibodies may be identical, contained one within the other, significantly associated, or, finally, completely independent. The significance of the association between the sera is measured by the usual 2×2 contingency chi square (see Appendix I).

TABLE 5.11

2×2 Table for the Analysis of the Association Between a Pair of Sera Tested on a Random Panel of Cell Donors: a, b, c, d represent the number of observations of the four possible types of reactions $++$, $+-$, $-+$, *and* $--$

		First serum +	First serum −
Second serum	+	a	b
	−	c	d

If $b = c = 0$, the sera are identical.
If $b = 0$, serum 2 is "contained" in serum 1, and vice versa if $c = 0$.
The sera may show a significant association. The sign of the association is that of $ad–bc$. The significance of the association is measured by

$$\chi_1^2 = \frac{n(ad - bc)^2}{(a + b)(c + d)(a + c)(b + d)}$$

and its magnitude by $(\chi^2/n)^{1/2}$, where $n = a + b + c + d$. (See Appendix I.)
The sera may be independent (χ^2 not significant.)

There are two main reasons for an association between the reactions of a pair of sera with a random population sample. Either (1) one or both of the sera contain more than one antibody and at least one of the antibodies is common to both sera, or (2) the sera contain antibodies directed against antigens that are associated in the population.

The first reason for association between serum reactions, the sharing of antibodies, has, as already mentioned, played a key role in the initial identification of many of the antigens of the HL-A system. The analysis of 2×2 associations between serum reactions for antigen definition in the HL-A system was pioneered by Van Rood (1962) and led to his initial description of the antigens 4a and 4b as a simple two-allele polymorphism. The method was further developed by Bodmer, Payne, and co-workers in their definition of the antigens HL-A1 and HL-A2 (Payne et al., 1964). The main principle involved is the recognition of a group of associated sera that share an antibody, which may then be used to define the corresponding antigen. One of the main difficulties in the application of this methodology to the HL-A system turns out to be that of disentangling associations between sera due to shared antibodies from those due to antibodies directed against closely associated antigens.

In view of the large number of comparisons to be made with reactions of, for example, only 30 sera tested on 100–200 people, it is almost essential to use an electronic computer to perform the necessary manipulations. A group of sera each containing an antibody against a common antigenic specificity is recognized by the fact that all sera of the group should show significant positive pairwise associations. The group of sera so defined should then separate people into two categories according to their overall reactions to all the sera of the group, and so define, objectively, an antigenic specificity, using multispecific sera. Many of the antigens now recognized on leukocytes were first identified in this way.

As an example, Table 5.12 shows the pairwise chi squares between the sera that were originally used to define the antigens HL-A1 and HL-A2. Almost all the chi squares within each group are significant and the associations positive, as they are expected to be if the sera within each group share an antibody directed against the

TABLE 5.12

Chi-square Relationships of HL-A1 and HL-A2 Sera

	Serum Number	2	8	14	15	25	30	32	6	7	9	17	28	22
		HL-A1							HL-A2					
HL-A1	2													
	8	**43**												
	14	**69**	**37**											
	15	**42**	**39**	**47**										
	25	**6**	**12**	**10**	**9**									
	30	**24**	**20**	**20**	**25**	**8**								
	32	**18**	**12**	**22**	**8**	2	**15**							
HL-A2	6	**9***	**11***	**18***	**10***	**9***	1*	0.2*						
	7	**11***	1*	**12***	4*	2*	0*	0*	**28**					
	9	**5***	**5***	**14***	**10***	9*	0*	0	**40**	**32**				
	17	1*	1*	3*	**6***	0.2*	0.2	0*	**18**	**19**	**17**			
	28	4*	4*	3*	**13***	**10***	0.6*	3	**40**	**15**	**20**	**16**		
	22	3*	**12***	**9***	4*	0.6*	0	0.2	**24**	**12**	**9**	**19**	**41**	

Source: From Payne et al. (1964).

Note: Boldface type indicates chi-square greater than 5. An asterisk indicates negative association.

appropriately defined antigen. The asterisks indicate negative associations whose interpretations will be discussed a little later on.

The data for HL-A1 are shown in a different form in Figure 5.13, which is a

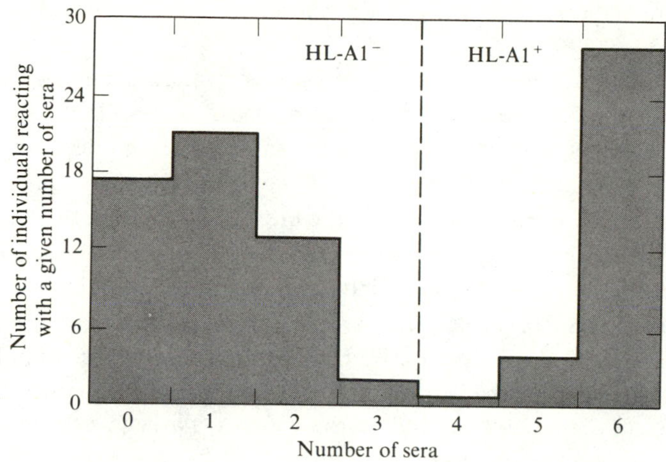

FIGURE 5.13
Definition of the HL-A1 antigen using a group of multispecific sera. (Data from Payne et al., 1964.)

histogram giving the number of people who reacted with a given number of 6 of the sera containing anti-HL-A1 listed in Table 5.12. The distribution is clearly bimodal with a minimum at 3–4 sera out of 6. Individuals reacting to 4, 5, or 6 of the sera are HL-A1+, the remainder HL-A1−. Technical errors account for the fact that some HL-A1+ individuals may not react with all the sera of the group. People reacting to 1, 2, or 3 sera are, presumably, reacting to antibodies in these sera other than anti-HL-A1. The upper limit to the level of misclassification is set approximately by the 3 out of 86 individuals who reacted to 3 or 4 sera. In this example, the definition of the antigen HL-A1 is unambiguous and is not obscured by associations between antibodies. It was confirmed by absorption studies and the later identification of monospecific sera.

The initial identification of the relationships between the genes controlling leukocyte antigens has proceeded from an analysis of their population associations

The best way to define the relationship between the genes controlling the antigens is, of course, by appropriate family studies. When, however, the methods of typing are subject to an appreciable frequency of misclassification as was at first the case for white-cell typing, family studies may prove unreliable. Even a single crossover can be enough to distinguish whether two genes are allelic or not, so that no misclassification can be tolerated in family studies. Associations between the antigens in a population will, however, not be seriously disturbed by moderate levels of misclassification and can, as pointed out in Chapter 2 and discussed below, provide some indications of the genetic relationship between closely linked genes.

A population association between antigens, or, more generally, between two genetically determined traits, may be caused by one or more of the following factors.

1. The traits may be multiple effects of the same gene. An interesting example of this is the association between red-cell ABO compatibility and skin-graft survival.
2. The traits may be the result of epistatic interaction between two or more genes. A striking example is the interaction between the Lewis and secretor loci, such that Le^b is only found on the red cells of individuals with at least one of the dominant alleles at each locus (see Table 5.5).
3. There may be a common selective basis for the maintenance of polymorphism at the two loci. Probably the best example of this in man is the association between G6PD deficiency, thalassemia, and resistance to malaria in certain areas of the Mediterranean (see Chapter 4).
4. Departures from random mating due to inbreeding, assortative mating, or population stratification can lead to nonrandom association between genes (see Chapters 7, 8, and 9). The effects of inbreeding in human populations, are, however, likely to be quite small, and assortative mating with respect to cryptic genetic characters, such as blood-group antigens, is unlikely. Population stratification, on the other hand, particularly as a result of recent racial admixture (see Example 2.2), may be a very significant general source of nonrandom association between un–linked genes. For example, in a truly random sample of the American population a negative association should be expected between the allele R_0 (cDe) of the Rhesus system and the Duffy allele, Fy^a, and a positive association between R_0 and the Kidd allele, Jk^a. This is because R_0 and Jk^a occur with a relatively higher frequency in African than in Caucasian populations while Fy^a has a much higher frequency among Caucasians than Africans. The magnitude of the associations found in the American sample will depend on the proportion of individuals with African ancestry in the sample. The associations will, presumably, eventually disappear if random mating with respect to racial ancestry is established on a more-or-less permanent basis.
5. The last, and from our point of view most important, cause of nonrandom association between traits is allelism or very close linkage. The negative association between the antigens A and B of the ABO system was, of course, the basis for Bernstein's interpretation of the system in terms of three alleles and the association between the antigens S, s and M, N of the MN blood groups was the basis for assigning all these antigens to one system, consisting of the four haplotypes *MS*, *Ms*, *NS*, *Ns*. In an exactly analogous way, the negative associations between the sera

containing anti-HL-A1 and those containing anti-HL-A2 noted in Table 5.12, which indicated a negative association between the respective antigens, suggested that the genes determining HL-A1 and HL-A2 were allelic.

It is this last cause of genetic association that has been emphasized especially by red-cell and white-cell groupers. This is because the other causes can, mostly, be readily recognized or controlled and are in any case likely to have small effects relative to that resulting from the very close linkage between genes controlling antigens that are part of the same system, like, for example, the rhesus blood groups.

As a model situation consider two hypothetical antigenic specificities G and H, for which the relevant gametic types are *GH*, *Gh*, *gH*, *gh*. The presence of *G* (or *H*) in the specification of a gamete denotes the fact that it carries the genetic determinant for the antigen G (or H), while *g* and *h* denote the absence of the respective genetic determinants. If the population frequencies of the gametes *GH*, *Gh*, *gH*, and *gh* are x_1, x_2, x_3, and x_4 (where $x_1 + x_2 + x_3 + x_4 = 1$), then it can be shown that the expected 2×2 association between the antigens G and H in a population is given by

$$\rho = \frac{D\left(2 + \dfrac{D}{q_1 q_2}\right)}{\sqrt{(1 - q_1^2)(1 - q_2^2)}} \sim \frac{2D}{\sqrt{(1 - q_1^2)(1 - q_2^2)}}, \tag{5.3}$$

when D is small, where $q_1 = x_3 + x_4$ = the total frequency of gametes lacking *G*, $q_2 = x_2 + x_4$ = the frequency of gametes lacking *H*, and $D = x_1 x_4 - x_2 x_3$ is a measure of the association between *G* and *H* among the gametes (Chapter 2) and $\rho = (\chi^2/n)^{1/2}$ as defined in the footnote to Table 5.11 (see Examples 5.5, 2.6). In general the association ρ between the antigens G and H in the population is a direct measure of the association D between the genetic determinants of the antigens G and H, respectively. The quantity D will be positive or negative depending on whether the predominant gametes are *GH* and *gh* or *Gh* and *gH*. It was shown in Example 2.6 that in the absence of selective differences between the various genotypes and under the assumption of random mating, D tends to approach zero as the population frequencies of the gametes approach equilibrium, however close the linkage between the genes *G* and *H* may be. The *rate* at which D tends to approach zero, does, however, depend on the closeness of the linkage between genes *G* and *H*. If the frequency with which heterozygotes *GH*/*gh* (or *Gh*/*gH*) produce recombinant gametes *Gh* and *gH* (or *GH* and *gh*) is r, then D decreases by a factor $1 - r$ in each generation. If r is actually zero, then there is, of course, no change in D, while if r is very small D will change very slowly. Fine-structure genetic recombination studies suggest that values of r between 1 in 10^3 and 1 in 10^5 are appropriate for a system of closely linked antigenic determinants.

If $r = 0.001$, D will decrease by a factor of 10 in approximately $n = 2 \times 10^3$ generations or 60,000 years for humans, where n is given by the formula $(1 - r)^n = \frac{1}{10}$

If $r = 0.00001$, then $n = 2 \times 10^5$ generations or six million years. Thus, for very closely linked genes, the time taken to approach an equilibrium, with $D = 0$, in the absence of selection, is comparable to the evolutionary life span of the human species. One explanation, therefore, for a significant association between antigens in the absence of selective differences is that the mutations giving rise to the antigenic differences arose recently enough and are closely enough linked for there to be too little time to reach an equilibrium.

It was originally pointed out by Fisher in 1930 that certain selective interactions between genes can lead to equilibria with D not equal to zero. The quantitative analysis of such situations has been discussed by a number of authors. In general, if the genotypes GH/gh or Gh/gH or both have a fitness which differs from that of gh/gh or GH/GH by less than r, the recombination fraction, equilibria with D not equal to zero will often exist. When r is very small, then clearly a very small selective advantage may be enough to prevent random association at equilibrium. (For a review of this problem see Bodmer and Felsenstein, 1967.)

Nonrandom association of antigens in a population is thus evidence for close linkage between the genes determining the antigens, whether this is due to the recent origin of the relevant mutants or to selective interactions of the type suggested. The analysis of population associations may thus be viewed as a form of linkage analysis for *closely* linked genetic determinants. The sign of D and hence of ρ, will predict which allelic combinations are most prevalent in the population under study.

A series of antigens is said to belong to a "system" if the recombination fraction r, for the genetic determinants of all pairs of antigens, is small.

The effective limit on r is set by the resolution of human pedigree studies for the detection of low recombination frequencies, and so, in most studies, is unlikely to be much less than one percent. Based on the estimated amounts of DNA per chromosome, observed recombination fractions and the average number of nucleotide pairs per cistron, there may, of course, be many, perhaps even hundreds or thousands of, genes between two loci separated by a one percent recombination fraction. It seems likely, however, that antigens belonging to a system are controlled by a block of contiguous genes. Unless such a block is exceptionally large, the appropriate values of r might be much less than one percent, perhaps even as low as 10^{-4} or 10^{-5}. The final answer to these questions will always depend on a detailed chemical understanding of the gene-antigen relationships.

Based on these considerations, however, D, and thus ρ, for most pairs of antigens belonging to a system should be different from zero. The signs of D should help in predicting the most prevalent haplotypes. It is clear that a value of D not significantly different from zero by no means proves two determinants are not part of the same system, since large numbers of observations may be needed to detect small correlations. This is a major limitation to association analysis since a correlation, however small, may indicate that two antigens belong to the same system. Mi and Morton (1966) have analyzed the pairwise associations between a large number of blood factors (blood groups, Gm groups, and secretor status) in a population sample based on families drawn from northeastern Brazil. In their data, the ρ value for the A and B antigens of the ABO system was -0.14, while those for D and E, and D and e, of the rhesus system were 0.18 and -0.05, respectively. The latter value, based on 1991 individuals, was not significantly different from zero at the one percent level. However, only two out of the 10 associations between the five antigens D, E, e, C, and c of the rhesus system were not significant. The third lowest ρ value was 0.12 for the association between D and C, which would have been detected as significant in a sample of about 400 individuals. Thus, in this case, all the antigens would clearly have been assigned to the same system, based only on phenotypic association analysis. The Brazilian sample has a mixed Caucasian, African, and American Indian origin. Mi and Morton emphasize that the associations between systems that must have existed in the ancestral population because of racial stratification are no longer significant. This suggests that there has been enough random mating with respect to racial origin, for a sufficient number of generations, to remove this evidence for racial stratification. Significant associations between white cell antigens along the lines discussed were the original basis used by Dausset, Bodmer, and others for assigning most of them to a single system, HL-A (Dausset et al., 1965; Bodmer and Payne, 1965; Bodmer et al., 1966). However, family studies were subsequently required to elaborate and clarify the system.

The HL-A polymorphism can be described in terms of two "allelic" series of closely linked antigens, the LA and 4 *series.*

The first genetic polymorphism on the leukocytes, called group 4, was described by Van Rood in 1962. At that time it was thought to be a simple two-allele polymorphism but, as has been the case with so many of the red-blood-cell groups, it turned out to be a very complex system, comparable to the rhesus or H2 systems. What was initially thought to be a second major independent polymorphism, LA, was described soon after by Payne, Bodmer, and co-workers (Payne et al., 1964). The independence was based mainly on the fact that the LA and 4 series of antigens did not seem to show significant population associations with each other, and also on incorrect family data showing apparent crossovers between genes controlling antigens belonging to different series.

Subsequently, Dausset, Ceppellini, and others presented both family and population association data which showed that the genes for all the antigens were very closely linked and so that they all belonged to one system, the HL-A polymorphism (see *Histocompatibility Testing* 1967). It now appears that this system of antigens can be described formally in terms of two very closely linked loci with multiple alleles corresponding respectively to the LA and 4 series of antigens.

To illustrate this representation, we shall first trace the history of the LA antigen series. The two antigens HL-A1 and HL-A2 were first described together in 1964 using basically the data shown in Table 5.12. It was then clear—both from the family data and from the negative population association between the two antigens—that they were determined by a pair of alleles. The existence of HL-A1(−)HL-A2(−) individuals indicated the presence of a "blank" (silent) allele, analogous to O in the ABO system. Subsequently two more antigens, HL-A3 and HL-A9 (formerly designated LA4, TO12 or Lc11) were described that behaved as if they were determined by two other alleles of this system, still leaving room for a "blank" allele. The operational criterion for allelism in this context is, as emphasized before, that no individual exists who carries on one chromsome, and so transmits together to his offspring, the genetic determinants for two or more of the antigens. Thus, an individual who has two of the allelic set of antigens, say HL-A1(+)HL-A2(+) must be heterozygous for the corresponding genetic determinants, *HL-A*1/*HL-A*2. No individual should therefore have more than two of the antigens. Data on the expected and observed phenotypic frequencies of a sample of Caucasians with respect to the antigens HL-A1, HL-A2, HL-A3, and HL-A9 are shown in Table 5.13. (These and the subsequent data in this section, unless otherwise indicated, come from unpublished work of Bodmer and co-workers.) The expected frequencies are based on fitting gene frequencies by the method of maximum likelihood (see Appendix I), assuming random mating. The fitted gene frequencies are those given for Caucasians in Table 5.17. As expected, there are no individuals with three antigens and some with none, indicating the presence of a blank allele. The fit is good, giving a chi square of 5.97 with 6 degrees of freedom. The allelism of the genetic determinants for these antigens has now been amply confirmed by family studies (see *Histocompatibility Testing* 1967) which show that no two are ever passed on to offspring together. Recently, more antigens of the series have been described, which almost fill in the blank (see *Histocompatibility Testing* 1970). In time, new antigens will probably be found that fill in the blank completely.

The 4 *series started, as already mentioned, as a two-allele system involving antigens* 4*a and* 4*b.*

A number of other antigens were subsequently described that were clearly closely associated with 4a and 4b. It now appears that 4a and 4b are actually complex in the sense that each is really the sum of a number of constituent specificities. Thus, for example, the antigens HL-A7 (formerly designated 7c or 4d) and HL-A8 (formerly designated 7d) are "included" in 4b, which means, operationally, that all individuals classified as either HL-A7 or HL-A8 are also 4b. There still remain people who are 4b but not HL-A7 or HL-A8, indicating that all the consti-

TABLE 5.13

Expected and Observed Phenotype Frequencies for the LA Series of HL-A Antigens

Phenotype					
HL-A1	HL-A2	HL-A3	HL-A9	*Observed*	*Expected*[a]
−	−	−	−	9	7.2
+	−	−	−	8	9.0
−	+	−	−	25	24.2
−	−	+	−	11	10.6
−	−	−	+	5	9.6
+	+	−	−	7	7.8
+	−	+	−	5	4.1
+	−	−	+	5	3.8
−	+	+	−	6	9.0
−	+	−	+	11	8.3
−	−	+	+	6	4.4
			Total	98	98.0

$\chi_6^2 = 5.97$

[a] The expected phenotype frequencies are based on fitting gene frequencies for each antigen and a "blank," assuming random mating. The fitted gene frequencies are given in Table 5.17.

tuent antigens of the 4b complex have not yet been described. An exactly similar situation exists with respect to 4a. This complexity is most probably due to the fact that cross-reacting antibodies, namely those which have affinity for more than one specificity, are relatively common for antigens within each of these two complexes. Thus, if a serum were found that contained an antibody reacting to either HL-A7 or HL-A8, it would behave as though the sum of these two antigens were a single entity. Until a serum was found that contained an antibody that reacted with only one of the two antigens, there would be no reason to believe that the first described specificity was complex. The frequent occurrence of cross-reacting antibodies for antigens within the 4a and 4b complexes is, presumably, a reflection either of the similarity of the chemical groupings responsible for the set of specificities within each complex or perhaps more likely of their close interrelationship on the cell surface. Undoubtedly, cross-reacting antibodies sometimes can be found between an antigen of the 4a and one of the 4b complex. However, the mere fact that 4a and 4b were initially described as a two-allele system indicates that such cross-reacting antibodies must be much less frequent than those that cross-react within each complex.

A number of antigens such as HL-A7 and HL-A8 belonging to the 4 series have now been described that, when considered as a separate system, behave as if they were determined by a set of alleles analogous to the LA series (Kissmeyer-Nielssen et al., 1968; Dausset et al., 1969; Mickey et al., 1969). Some data on observed and expected phenotype frequencies for four of these antigens (disregarding, for the time being, the LA antigens) is given in Table 5.14. The fit is not as good as it

TABLE 5.14

Expected and Observed Phenotype Frequencies for the 4 Series of HL-A Antigens

Phenotype					
HL-A12	4c*	*HL-A7*	*HL-A8*	*Observed*	*Expected*[a]
−	−	−	−	27	20.5
+	−	−	−	11	17.2
−	+	−	−	19	22.9
−	−	+	−	18	20.9
−	−	−	+	7	9.3
+	+	−	−	11	6.7
+	−	+	−	6	6.2
+	−	−	+	6	3.0
−	+	+	−	10	7.9
−	+	−	+	2	3.9
−	−	+	+	5	3.5
			Total	122	122.0

$\chi^2_6 = 13.8$ $\quad 0.05 > P > 0.025$

[a] The expected phenotype frequencies are based on fitting gene frequencies for each antigen and a "blank" assuming random mating. The fitted gene frequencies are given in Table 5.17.

should be, probably because of some inaccuracies in the typing for HL-A12. This antigen is part of the 4a complex while HL-A7 and HL-A8 are associated with the 4b complex. There is an even larger frequency of the blank allele than for the LA series, indicating that many more specificities of the 4 series remain to be properly defined. Recently, several new specificities in this series have been described (see *Histocompatibility Testing* 1970).

It should be emphasized that only a very small proportion of sera turn out to be useful for typing for the well-defined antigens we have so far mentioned. Many sera are found that are either mixtures or that do not seem clearly to identify a new or an old specificity. This is undoubtedly due to the serological complexities of the system. Many of these sera are probably cross-reacting with the specificities that will be found to fill in the blank. Some cross-reacting antibodies are occasionally found within the LA series. However, it seems clear that the relative ease with which the LA series was worked out compared with the apparent confusion surrounding the 4 series is a reflection of the much higher frequency with which cross-reacting antibodies are found for the latter.

It was originally thought that the loci for the LA and 4 series were unlinked mainly because some of the first described specificities in the two series showed no significant population associations.

All pairs of antigens within each series tend to be associated negatively, as is expected if they are determined by alleles. The chi squares for pair associations of antigens, one member from the LA and one member from the 4 series, are shown in Table 5.15, based on data from a sample of about 140 white Americans in California.

TABLE 5.15

Chi squares for Pair Associations of HL-A Antigens, One Member from the LA and One Member from the 4 Series

		LA series			
		HL-A1	*HL-A2*	*HL-A3*	*HL-A9*
4 Series	HL-A12	4.5†	○	○	4.7
	4c*	○	○	○	○
	HL-A7	○	○	○	○
	HL-A8	27.1	○	○	○

Based on a Caucasian Sample of About 140 People from California.

Note: Dagger indicates a negative association; circles indicate a chi-square of less than 2.5

Only three of the values exceed 2.5 and only one association is really striking, namely that between HL-A1 and HL-A8. On the basis of our analysis of population associations due to nonrandom association of closely linked gentic determinants, we should predict that the genetic determinants other than HL-A1 and HL-A8 should occur in all possible pair combinations, one member from the AL and one member from the 4 series, more-or-less at random. HL-A1 and HL-A8 combinations thus give relatively large values of *D*, the measure of association for pairs of genetic determinants on haplotypes, while, for example the HL-A1 and 4c* combinations agree rather well with the assumption of independence, giving a very low value of *D*.

Very close linkage between the genetic determinants for the two series of antigens LA and 4 has now been amply demonstrated by extensive family studies.

An example showing how haplotypes are determined using family data is shown in Table 5.16. Data on offspring with identical phenotypes have been combined.

TABLE 5.16

Haplotype Determination in a Family of Six Children. 1 indicates presence, 0 absence of the respective antigens. Antigens giving the same pattern in the family (e.g., HL-A2 and HL-A5) are grouped, and children giving the same reactions are grouped.

	Antigens					
	Paternal Backcross		*Maternal Backcross*		*Intercross*	
	HL-A2 HL-A5	*HL-A3*	*HL-A1 HL-A12*	*HL-A9*	4c*	*Assigned Chromosomes*
Father	1	1	0	0	1	A/B
Mother	0	0	1	1	1	C/D
Children (number of)						
(3)	1	0	1	0	1	A/C
(1)	0	1	1	0	0	B/C
(2)	0	1	0	1	1	B/D

PARENTAL HAPLOTYPES

Father	A	HL-A2	HL-A5	4c*
	B	HL-A3		
Mother	C	HL-A1	HL-A12	
	D	HL-A9	4c*	

Two pairs of complementary patterns of antigen segregation in a family are expected on the assumption that all the antigens are determined by closely linked genes. The first pair correspond to backcrosses in which the father has the antigen, the mother does not, and one or more of the children do not. In this case the father must be heterozygous for the corresponding genetic determinant and we call this a paternal backcross. Two complemenatry patterns are expected corresponding to the fact that only one of the two paternal chromosomes is transmitted to any given offspring (see the first two columns of Table 5.16 for HL-A2, 5 and HL-A3). The second pair of complementary patterns corresponds to maternal backcrosses as shown in the next two columns of Table 5.16 for HL-A1, HL-A12 and HL-A9.

It should, in general, be possible to assign the linkage phase of the genetic determinants and so the haplotypes on the basis of such backcross data. Thus for the family shown in Table 5.16 the father transmits HL-A2 and HL-A5 on one chromosome, arbitrarily designated A, and HL-A3 on the other, which is called B. As indicated in the final column of the table the first three offspring must have received the A chromosome from their father and the remainder B. The two maternal chromosomes are C (HL-A1 and HL-A12) and D(HL-A9). The genetic determinants for LA antigens, as expected, do not occur together in the same haplotype. In addition to the paternal and maternal backcross segregation patterns, intercross patterns, namely those for which both parents have the antigen but at least one child does not, may also be found. These should be consistent with the corresponding gene being on one or other of the chromosomes of each parent, as determined from the backcross data. Thus in Table 5.16, 4c* is missing in the child who got the B chromosome from the father and C from the mother and so is presumably determined by genes on the A and D parental chromosomes. (The antigen 4c* is known to be complex and to "include" the antigen HL-A5.)

A few definite crossovers between the LA and 4 loci have now been described. The data suggest a recombination fraction between the loci of about one percent (see *Histocompatibility Testing* 1970).

It is already clear from the data on the LA and 4 series allelic frequencies given in Tables 5.13 and 5.14 that the HL-A system is highly polymorphic, in the sense that there exist a large number of possible genotypes all with fairly small population frequencies. This means that a fairly high proportion of individuals are doubly heterozygous with respect to LA and 4 antigenic determinants. As a result, parental *HL-A* haplotypes can be assigned in a high proportion of families chosen at random, maybe as high as 70 percent. Nevertheless, haplotype frequencies calculated from direct determination using family data, must to some extent be biased because those families in which haplotypes can be sorted out can clearly not be considered a random sample.

Haplotype frequencies can be estimated from population phenotype frequencies.

According to the picture of the HL-A system we have developed so far, all haplotypes must consist of one genetic determinant (or a blank) from each of the LA and 4 antigen series. Thus, given 4 antigens in each system, there are $5 \times 5 = 25$ possible haplotypes and so 25 homozygotes, plus $(25 \times 24)/(1 \times 2) = 300$ heterozygotes, making a total of 325 possible genotypes. It can be shown that in this case the total possible number of phenotypes is $11 \times 11 = 121$. In a random sample of approximately 100 Caucasians, 53 of these phenotypes were, in fact, observed. There are now approximately 14 known antigens at the 4 locus and 9 at the LA locus that can give rise to 8001 genotypes and 3404 phenotypes.The multiplicity of haplotypes, genotypes, and phenotypes certainly complicates the problems of fitting observed data on phenotype frequencies to a specific genetic model. The problem is, however, greatly simplified by an extension of the two-locus two-allele representation of gamete or haplotype frequencies in terms of products of gene frequencies plus or minus a correction term, D (see Example 2.6 and the discussion on association due to close linkage on pages 247–248). Thus it can be shown (see Example 5.5) that the frequency of a haplotype ij, determining antigens i of the LA series and j of the 4 series, can be expressed in the form

$$x_{ij} = p_i P_j + D_{ij}, \tag{5.4}$$

where p_i is the frequency of antigen allele i of the LA series, calculated as in Table 5.13, and P_j is similarly the frequency of the antigen allele j of the 4 series and D^{ij} is the gametic association between the alleles, ignoring all other combinations. This D_{ij} can be estimated directly from the 2 × 2 association table for the i and j antigens (see Example 5.4). A set of haplotype frequencies estimated in this way from a sample of a Caucasian population is given in Table 5.17. The frequencies can be represented in the form of a matrix whose row and column headings are the

TABLE 5.17

Estimated Haplotype Frequencies for a Caucasian Population. The figures in the enclosed square are the frequencies of the haplotypes determining the antigen corresponding to the relevant row and column headings. The figures around the edge are the allele frequencies estimated separately for the LA and 4 series

			4 series				
			HL-A12	*4c**	*HL-A7*	*HL-A8*	*Blank*
			0.164	0.193	0.261	0.104	0.416
LA series	*HL-A1*	0.135	0	0.006	0.011	0.061	0.057
	HL-A2	0.289	0.044	0.020	0.042	0.021	0.162
	HL-A3	0.167	0.051	0.061	0.036	0	0.019
	HL-A9	0.136	0.053	0.019	0.017	0.015	0.032
	Blank	0.276	0.016	0.087	0.020	0.007	0.146

antigens of the 4 and LA series, respectively. The row and column sums are the frequencies of the 4 and LA alleles calculated separately as in Tables 5.13 and 5.14. The most frequent haplotypes are *blank blank* and *HL-A2 blank*, which are the only ones with frequencies of more than 10 percent. As more antigens are found that fill in the blanks, these more frequent haplotypes will become subdivided.

TABLE 5.18

Frequencies of the LA and 4 Antigen Series Alleles of the HL-A System in Three Different Populations

Alleles	*Caucasians* ($n = 111$)	*Western Pygmies* ($n = 115$)	*American Blacks* ($n = 61$)
LA series			
HL-A1	0.15	0	0.05
HL-A2	0.29	0.12	0.18
HL-A3	0.16	0.11	0.11
HL-A9	0.14	0.05	0.20
LA-W[a]	0.12	0.31	0.27
Blank	0.14	0.41	0.19
4 *series*			
HL-A12	0.17	0.09	0.12
4c*	0.10	0.09	0.20
HL-A7	0.19	0.10	0.09
HL-A8	0.12	0.04	0.07
Blank	0.42	0.68	0.52

[a] LA-W is a new LA antigen that helps to fill in the LA blank.

The frequencies of the HL-A antigens vary widely from population to population.

A limited number of studies of the distribution of HL-A antigens in different populations have been carried out. As with rhesus blood-group systems, African populations, in particular, show striking differences from Caucasian populations. Two sera that behave identically when tested on one population, may react quite differently in the other. One possible reason for this is the existence of a second antibody in one of the sera that may react very infrequently with, say, Caucasians, but quite often with people of African descent. A second reason might be that the overall distribution of antigens is sufficiently different, in the two populations, for a serum to show quite different patterns of cross-reactivity when tested on the two types of cells. These complexities mean that it is essential to confirm the characterization of a Caucasian serum on the African population, and, in general, that at least two sera must be used to type for any of the antigens. Some data on LA and 4 allele frequencies in Caucasian and African populations is shown in Table 5.18. The fit of the observed to the expected phenotype frequencies in the pygmy and and American black populations, assuming allelism of the LA and 4 antigens, can be shown to be quite satisfactory, supporting the two-locus multiple-allele structure of the HL-A system in these populations. Corroborating family studies have, however, not yet been done. Since HL-A1 seems to be absent in Africans, the HL-A1 and HL-A8 association found in Caucasians cannot be verified. There is, however, a significant positive association between HL-A3 and 4c* in the pygmies, which is appreciably higher than that found in Caucasians. Most of the information on the HL-A differences between the populations is, however, contained in the allele frequencies given in Table 5.18. There are striking differences in frequency, the most significant, perhaps, being the increased frequency of the blank allele in pygmies as compared with Caucasians. This is accompanied by appreciably lower frequencies of most of the known alleles, which were originally described in Caucasian populations. Only the new specificity LA-W shows a higher frequency in Africans (Bodmer and Bodmer, 1970). This bias in allele frequencies is most probably a reflection of the source of the sera used for typing. As more sera from Africans become available, the high blank frequencies in these populations may, perhaps, be filled in.

The antigen HL-A1 seems to be absent also from Japanese populations and so probably from other Oriental populations (Singal et al., 1969; see also Bodmer and Bodmer, 1970).

There is direct evidence that the HL-A antigens are histocompatibility antigens.

The HL-A antigens are present on a wide variety of tissues including all types of white cells, skin, and kidney, which is consonant with their being "tissue antigens" that may be of importance for transplantation. The best direct evidence that they are histocompatibility antigens comes from studies on the survival of skin grafts exchanged between sibs. When the sibs have identical HL-A types, skin-graft survival is significantly longer than if they differ in HL-A genotype. The HL-A genotypes can be assigned as discussed earlier (see Table 5.16); thus sibs who are identical for HL-A may be identified. Representative data from a single family are shown in Table 5.19, which gives survival of skin grafts exchanged reciprocally between four sibs whose HL-A genotypes were determined. The grafts between the HL-A identical sibs 1 and 3 survived much longer (20 and 22 days) than those between the other pairs, whose mean survival time was 13.5 days. Similar data have been obtained from a substantial number of families.

When lymphocytes of different origin are cultured together, they stimulate each other to enlarge and subsequently divide. This so-called mixed-lymphocyte-culture test was at one time thought to be an *in vitro* mimic of *in vivo* graft rejection. It has been shown, however, by Bach and Amos (1967) that pairs of sibs, which are HL-A identical do not stimulate each other. This shows that HL-A differences dominate the mixed-lymphocyte-culture reaction. Unfortunately, therefore, this test does not provide any information that cannot, in principle, be obtained by HL-A typing.

Data on the cumulative survival of a series of kidney grafts between sibs separated according to whether or not they were matched, that is HL-A compatible though not necessarily HL-A indentical, are shown in Figure 5.14. There is a striking effect of

TABLE 5.19

Survival of Skin Grafts Between Sibs as a Function of HL-A Genotype. Donors and recipients are the same four sibs. Figures in the body of the table are days the grafts survived. Genotypes are indicated in terms of the paternal (A/B) and maternal (C/D) chromosome combinations. Squares indicate grafts between HL-A identical sibs

		Donors and HL-A *Genotypes*			
		1 B/D	2 A/D	3 B/D	4 A/C
Recipients	1	—	14	[22]	13
	2	14	—	13	13
	3	[20]	16	—	13
	4	12	14	13	—

Source: From Ceppellini (1968).

HL-A compatibility on survival that parallels the data on skin-graft survival. Surprisingly, however, the same authors reporting the data shown in Figure 5.14. report survival data for parent-child kidney grafts that give a much less striking difference between survival times of matched and mismatched pairs. Their data show a higher survival of the parent-child grafts than sib-sib grafts at four years after the transplant (60 percent versus 43 percent). This may, in part, be due to the fact that sibs can differ with respect to two HL-A alleles while parent and child always have one allele in common.

Data on survival of grafts exchanged between unrelated individuals are much more difficult to interpret owing both to the multiplicity of possible HL-A phenotypes and to the presumably greater average disparity in other as-yet-unidentified histocompatibility antigens. Nevertheless, there is some indication that kidney grafts between unrelated individuals matched for HL-A antigens fare, on the average, somewhat better than those between mismatched pairs. Some representative data taken from Terasaki's studies are given in Table 5.20. Clinical grading of the quality of a graft from A to F is based on a variety of physiological and histological criteria. A significantly higher proportion of the mismatched grafts have lower grades than the matched ones. However, there are many mismatched grafts that do well and, of course, some matched ones that do not, perhaps because of undetected incompatibilities. It must be borne in mind that all kidney-graft data are based on patients receiving continuous immunosuppressive therapy, which presumably accounts, at least in part, for the survival of grafts across HL-A mismatches. All these data, of course, are on grafts between ABO compatible donors and recipients.

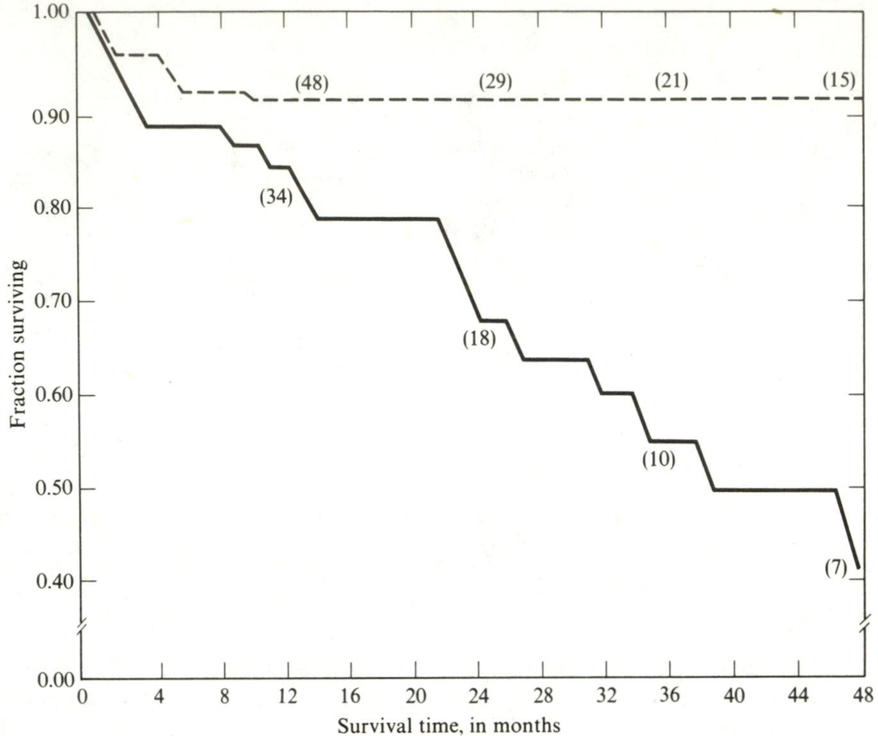

FIGURE 5.14
Survival curves for matched- and mismatched-sibling kidney transplants. The numbers along each curve indicate the number of patients at risk. Survival of matched transplants is plotted by the dashed line; that of mismatched transplants along the solid line. (From Singal et al., 1969.)

TABLE 5.20
Correlation of Overall Clinical Rank of Kidney Grafts Between Unrelated Individuals and HL-A Compatibility

	Clinical Grade of Patients with More Than 4 Months Graft Survival					
	A	B	C	D	F	*Totals*
Matched	6	10	1	0	4	21
Mismatched	4	14	7	3	13	41

Source: Based on Patel et al. (1968).
Note: A test of significance for the difference between matched and mismatched pairs gives $P < 0.02$.

Recent evidence on the chemistry of the HL-A antigens suggests that the LA and 4 series may reside on different molecules.

An understanding of the relationship between the genetic and functional differences of an antigenic polymorphism, such as HL-A, always depends, ultimately, on a determination of the chemical nature of the antigens. The importance of the HL-A antigens for transplantation has greatly stimulated research on their chemistry. The most difficult problem for the analysis of cell-surface antigens is that they reside in components of the cell, the membranes, that are insoluble and so cannot be readily studied using standard biochemical techniques. The first step in their analysis is therefore solubilization, which may be brought about by use of enzymes, detergents, or mild sonication (see, for example, Kahan and Reisfeld, 1969). Recent evidence suggests that the HL-A antigens are predominantly proteins with some associated carbohydrates, though it is not clear which of these two kinds of component carries the antigenic determinants. One group of workers has provided evidence which suggests that the LA and 4 antigens reside in molecules, or molecular fragments, that can be separated by standard biochemical techniques (Mann et al., 1969). This, of course, fits in well with the formal genetic picture of the HL-A polymorphism as a two-locus multiple-allele system. Further work is needed, however, to resolve the questions of how many cistrons are involved and whether the immediate gene products are, in fact, the molecules carrying the antigenic specificities. The fact that the antigens within each of the LA and 4 series behave in a strictly allelic fashion, while associations between antigens of the two series are relatively weak, suggests that if several cistrons are involved, those determining specificities within each of the two series are more closely linked. This interpretation is based on the fact that the association between closely linked genes is, at least to some extent, a function of the recombination fraction between them. As discussed before the smaller this is the longer will be the time needed to achieve random association at the population level between a pair of linked genes. It is, of course, also possible that the pattern of associations observed within the HL-A system is a reflection of a particular pattern of selective advantages for appropriate combinations of antigens. A further possibility is that the mutually exclusive antigens within the 4 and LA series reflect some sort of a metabolic sequence, as has been delineated for the ABO and associated polymorphism, and as was suggested by Boettcher for the rhesus polymorphism.

There are, so far, no clear indications of selective differences associated with the HL-A polymorphism.

Family data on segregation of individual antigens on the whole show no significant departures from expected Mendelian proportions. There is preliminary evidence to suggest that the HL-A polymorphism, or at least closely analogous polymorphisms, may occur in a wide variety of animal species. It has been clearly identified in the chimpanzee and probably also in the rhesus monkey (Balner et al., 1967). The mouse H2 system is also thought to be to some extent analogous to the HL-A polymorphism. The only positive evidence for selection in relation to the HL-A system is that its variation among the human races is narrower than that of most polymorphisms. (Bodmer and Bodmer, 1970; see also Chapter 11.)

The widespread occurrence of such a histocompatibility polymorphism does pose the question of whether there are special selective mechanisms associated with such antigenic differences. Incompatibility selection against sperm or fetus could conceivably lead to an advantage of any new antigenic variant. No clearcut clinical effects of the presence of maternal HL-A antibodies on the fetus have yet been described though the presence of the antigens on sperm has been confirmed. Burnet (1962) has suggested that histocompatibility polymorphisms may have evolved to protect the individual from "infection" by cancerous cells from other individuals. The probability of compatibility between two random individuals is so low that cancerous cells from almost any other individual would be rejected. It must, however, be remembered that we now know that a high proportion, perhaps as high as 30–50 percent of all gene loci are polymorphic (see Chapter 11). The apparent high level of variability associated with the HL-A polymorphism may therefore merely reflect the activity of many closely linked genes.

OTHER WHITE-CELL POLYMORPHISMS

A few other antigenic polymorphisms detected on white blood cells have been described. These seem, however, to be much less complex and have received much less attention than HL-A. A summary of information on the polymorphisms so far described is given in Table 5.21. The 5 system, originally described by Van Rood,

TABLE 5.21
White-cell and Platelet Polymorphisms, other than HL-A

System	*Antigens*	*Alleles and Their Frequencies in Caucasians (in Percent)*
5 (on all white cells and many tissues)	5a, 5b	*5a* (18) *5b* (82)
NA (on granulocytes)	NA1	*NA1* (34)
Zw (or PlA) (only on platelets)	Zw^a, Zw^b	Zw^a (84) Zw^b (16)
Ko (only on platelets)	Ko^a, Ko^b	Ko^a (7) Ko^b (93)
PlE (only on platelets)	PlE^1, PlE^2	PlE^1(97.5) PlE^2(2.5)

Source: Data from Shulman et al. (1964), Van Der Weerdt (1965), Van Rood et al. (1965), and Ceppellini et al. (1967).

seems to occur on many different tissues, as does HL-A, but it has not so far been shown to be significant for transplantation. The NA system (Ceppellini et al., 1965) is specific to the granulocytes. Presence of anti-NA1 in the mother has been shown to be a cause of neonatal neutropenia, a disease in which the fetus has a very significant reduction in the number of granulocytes. Both the NA and 5 polymorphisms have, so far, only been detected by agglutinating sera. The remaining three polymorphisms are specific to the **platelets**. Of the three, Zw has been described independently by two groups of workers, and the other two, Ko and PlE, have each been described only once and thus are not so far fully confirmed (see Shulman et al., 1964; and Van der Weerdt, 1965).

Other polymorphisms and more antigens of presently known polymorphisms clearly will be found. Since the majority of antibodies produced naturally, or by planned immunization, seem to be directed against antigens of the HL-A system, it is perhaps not surprising that the understanding of this system has developed most rapidly.

5.12 The Gm Polymorphism and the Genetic Control of Antibody Structure

The first of the Gm factors, which are detected as antigenic specificities on antibody molecules, was discovered by Grubb in 1956, in the course of a serological determination of gamma globulin levels. The assay he was using is based on the Coombs test.

The Coombs test is an adaptation of the agglutination assay that is widely used in red-cell typing.

The principle of the Coombs test is illustrated in Figure 5.15. Ag is an antigen on the cell surface, and Ab a specific antibody against Ag. Agglutination occurs, in simple terms, if an antibody has at least two antigen binding sites, enabling it to form a bridge between cells by attaching to one cell at one side and another at the other (see Figure 5.15, A). In this way a mixture of cells and antibodies forms a lattice held together by antibody molecules, which is observed as an agglutinated clump. Often, however, an antibody, though it can bind to the antigen on the cell surface, is incapable of causing agglutination. This may be because it only has one effective antigen-binding site (as in Figure 5.15, B), or because the antigen is embedded in the cell surface in such a way that the antibody molecule is not large enough to form a bridge between the antigens on two different cells (see Figure 5.15, C). If, however, we make an antibody against the original Ab *antibody* (C for Coombs reagent in Figure 5.15, B and C), this may be able to attach to the antibody on the

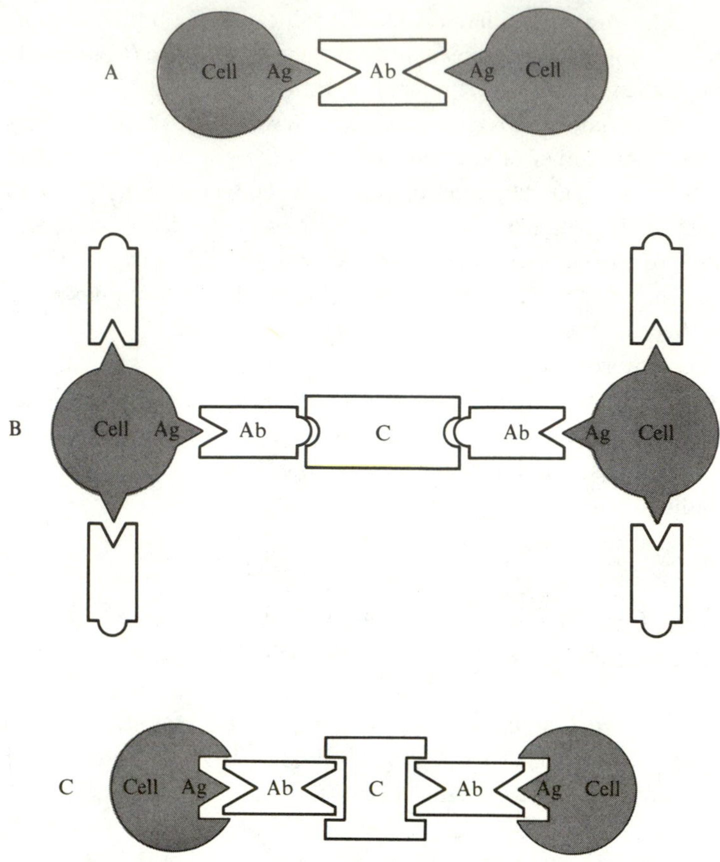

FIGURE 5.15
A: Direct agglutination. B and C: Coombs agglutination—principle of the Coombs test. Ag is the antigenic site on cell surface; Ab is the specific antibody for Ag; C is the Coombs-reagent "bivalent" antibody against Ab. In B the antibody is depicted as having only one site for attachment to the antigen Ag, and so cannot lead to direct agglutination. In C it has two attachment sites, but these are arranged in such a way that the antibody cannot cause direct agglutination.

surface of the cell, causing agglutination. This second antibody acts as though the first antibody were itself an antigen on the cell surface. Ag, for example, might be the rhesus E specificity and Ab an "incomplete" anti-E, as such nonagglutinating antibodies are called. The C antibody, or Coombs reagent, is often an antibody made in rabbits against whole human gammaglobulin, which is the antibody-containing protein fraction of serum. Such a Coombs reagent can be used quite generally for testing for any incomplete antibody coming from a *human* source. A critical part of the test is to wash the cell suspension thoroughly after it has been treated with the primary antibody Ab. Otherwise, if excess antibody remains in

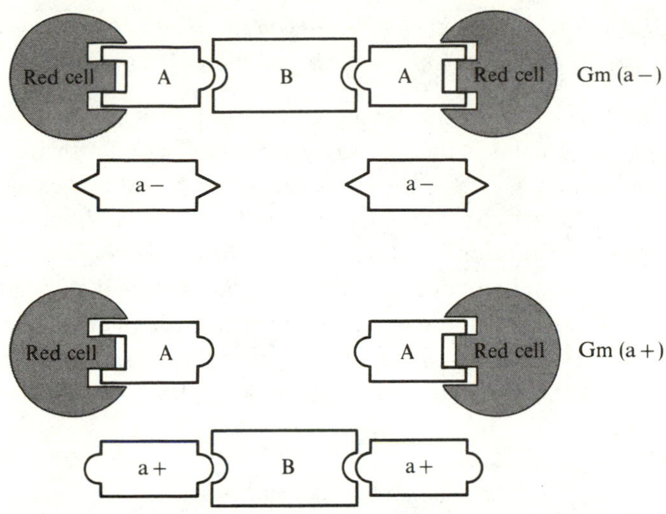

FIGURE 5.16
Schematic illustration of Gm typing by inhibition of agglutination. A is an anti-D antibody that is Gm(a+); B is an anti-Gm(a) antibody; a− are antibody molecules from a Gm(a−) individual; a+ are antibody molecules from a Gm(a+) individual.

solution, it will bind the Coombs reagent in solution, competitively inhibiting the agglutination reaction.

Grubb observed that one of the sera he was testing for its ability to inhibit the Coombs test, itself reacted as if it were a Coombs reagent, agglutinating the red cells that were coated with the primary antibody, which was a rhesus anti-D antibody. The serum that behaved in this way came from an Rh(+) individual, who could not, therefore, have made any anti-D antibody himself.

Grubb found that serum from about 60 percent of normal individuals, which he called Gm(a+), inhibited his unexpected agglutination reaction, that the capacity to inhibit was inherited as a dominant gene, and that the inhibitory fraction of the serum was gamma globulin.

The explanation of the phenomenon that Grubb found is that antibodies, which are themselves gamma globulin molecules, from about 60 percent of the Caucasian population have an antigenic determinant Gm(a). Rare individuals have in their serum an antibody directed against Gm(a), which agglutinates Rh(+) red cells coated with anti-Rh(+) antibody from a Gm(a+) individual. Serum from Gm(a+) individuals competitively inhibits this agglutination while that from Gm(a−) individuals does not. This principle of typing for antigenic determinants of the antibody molecules, which is still the predominant test used for Gm typing, is illustrated in Figure 5.16. It depends on two reagents, the anti-D antibody with Gm(a+) specificity (A in the figure) and the anti-Gm(a) antibody (B in the figure).

Following Grubb's original work, many more Gm specificities have been discovered, all of which behave as part of a single complex polymorphism.

Many of the antibodies for Gm typing come from patients with rheumatoid arthritis. The reason for this is thought to be that rheumatoid arthritis is an autoimmune disease, namely a disease in which people make antibodies against their own cells and cell products. Some Gm-typing antibodies come from patients who have had multiple transfusions and some have been prepared using animals. Occasionally, anti-Gm antibodies are formed by fetal-maternal stimulation. Each new Gm antibody must, as just discussed, be matched with an anti-D carrying the appropriate specificity. A list of the Gm factors that have been described so far is given in Table 5.22 (taken from Ceppellini (1966), who gives a comprehensive

TABLE 5.22
The Factors of the Gm System

Nomenclature		
Original	*Numerical*	*Date of Discovery*
Gm(a)	Gm(1)	1956
Gm(x)	Gm(2)	1959
Gm(b^w)-Gm(b^2)	Gm(3)	1960
Gm(f)	Gm(4)	1965
Gm(b)-Gm(b^1)	Gm(5)	1959
Gm(c)	Gm(6)	1960
Gm(r)	Gm(7)	1961
Gm(e)	Gm(8)	1962
Gm(p)	Gm(9)	1963
Gm(bα)	Gm(10)	1963
Gm(bβ)	Gm(11)	1963
Gm(bγ)	Gm(12)	1963
Gm(b^3)	Gm(13)	1965
Gm(b^4)	Gm(14)	1965
Gm(s)		1966
Gm(t)		1966
Gm(b^0)	= Gm(5)	1966
Gm(b^5)		1966
Gm(c^3)		1966
Gm(c^5)		1966
Gm(z)	Gm(17)	1966
Ro_2	Gm(18)	1965
Ro_3	Gm(19)	1965
Gm(20)	Gm(20)	1966
Gm(g)	Gm(21)	1966
Gm(y)	Gm(22)	1966
Gm(n)		1966

Source: Adapted from Ceppellini (1966).

review of the subject of this section). The original names for these factors reflect, in part, the genetic development of the system. Thus Gm(b), or Gm(5), which was the second to be discovered, behaves, in Caucasians, as if it were controlled by an allele of the gene that specifies Gm(a).

Most of the other Gm(b) specificities (b^0, b^3, b^4, b^5), are highly associated at the population level with Gm(b^1). Following our earlier discussion of associations due to close linkage, this can be interpreted in terms of a predominant b-determining haplotype that includes the genetic determinants for all of b^0, b^1, b^4, b^5, and b^3. Because of the allelism of the genes for Gm(a) and Gm(b) in Caucasians, there are almost no haplotypes in Caucasian populations that determine both Gm(a) and Gm(b). Close associations between Gm(a) and Gm(x), Gm(g) and Gm(z), and between Gm(b) and Gm(f), Gm(y) and Gm(n), are interpreted in a similar way. The resulting set of the most common haplotypes in Caucasians which have been amply verified by family studies, together with their frequencies calculated from population phenotype frequencies, is given in Table 5.23. Four haplotypes account for 98 percent of the total, the remaining, rare haplotypes such as $azb^0b^1b^4b^5b^3$ and *yfg* occurring with frequencies of less than 0.5 percent.

TABLE 5.23
The Major Gm Haplotypes and Their Frequencies in Caucasian Populations. Other haplotypes (for example, $azb^0b^1b^4b^5b^3$ or yfg) occur with frequencies of less than 0.005

Haplotype	*Approximate Frequency*
azg	0.15
axzg	0.13
$yfb^0b^1b^4b^5b^3n$	0.5
$yfb^0b^1b^4b^5b^3$	0.2

Source: Based on Ceppellini (1966); Natvig et al. (1967).

There are striking variations in haplotype frequencies between populations of different racial origin.

As with the other complex polymorphisms, studies on the frequencies of Gm haplotypes in different populations have uncovered striking variations and have contributed substantially to the understanding of the polymorphism itself. The first surprise in the Gm system was that all people of African descent turned out to be Gm(a) *and* Gm(b), and so presumably to be homozygous for a *Gm*(*ab*) haplotype, which is extremely rare in Caucasian populations. Other specificities have served to distinguish to some extent the *Gm*(*ab*) haplotypes occurring in these populations.

The approximate frequencies of the commoner haplotypes found in African and Japanese populations are given in Table 5.24. In each case, the most striking difference from Caucasians is the presence of *ab* haplotypes such as $ayfb^0b^1b^4b^5b^3n$ among the Japanese and $azb^0b^1b^4b^5b^3$ among the Africans. The main African haplotypes are distinguished by the presence or absence of some of the b associated antigens, notably they have c^5 and c^3 in place of b^4 and b^3, and s instead of b^1 and b^4.

TABLE 5.24
Frequencies of the Major Haplotypes Found in Japanese and Africans

Haplotypes	*Approximate Frequencies*
Japanese	
azg	0.5
$axzg$	0.15
$ayfb^0b^1b^4b^5b^3n$	0.1
$azb^0stb^5b^3$	0.25
African	
$azb^0b^1b_4b^5b^3$	0.5
$axzb^0b^1b^4b^5b^3$	0.05
$azb^0b^1c^5c^3$	0.25
$azb^0b^1b^4b^5c^3$	0.05
$azb^0sb^5b^3$	0.1

Source: Adapted from Ceppellini (1966).

Crossing-over is probably an important mechanism for the production of new Gm haplotypes.

As in the interpretation of the Rh system, it seems natural to suggest that haplotypes such as the African $azb^0b^1b^4b^5b^3$ may have arisen by crossing over within the genetic region determining the Gm specificities. This could, for example, have been produced by an $azg/yfb^0b^1b^4b^5b^3$ heterozygote.

Extensive family studies have confirmed the close linkage between the Gm genetic determinants and have been the basis for the definition of the haplotypes listed in Tables 5.23 and 5.24. A few families have, however, been discovered in which crossing-over within the Gm region is clearly demonstrated. Data on one of these are shown in Figure 5.17. Individual II.1 was the propositus of the family. He lacked the b complex that is almost invariably associated with *fy* in Caucasians. His son III.1 had the same unusual phenotype. All other phenotypes in the family could be readily interpreted in terms of the haplotypes commonly found in Caucasian populations. Since II.1 had Gm(x), he must have inherited the *gzax* haplotype

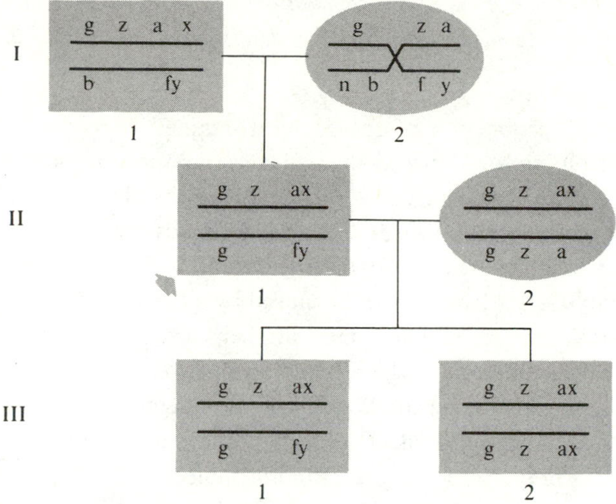

FIGURE 5.17
A Caucasian family demonstrating a crossover between Gm genetic determinants. The arrow indicates the propositus of the family; b stands for the complex of b types usually found together in Causcasians; g fy is the cross over haplotype. Individual I 2 bears the probable crossover. (Adapted from Natvig et al., 1967.)

from his father. Thus the unusual *g fy* haplotype must have come from the mother. She had the phenotype *azg bfy n* and so must have been heterozygous *g za/nb fy*, suggesting that the *g fy* haplotype arose by a crossover in the position indicated in Figure 5.17. The molecular information on the Gm system shows that in order to explain the unusual *g fy* haplotype, as a mutational event, two simultaneous mutations would be necessary, which is extremely unlikely. The number of families that have been screened so far without finding crossovers suggests that the recombination fraction between Gm determinants is no more than 1 in 10^3 or 1 in 10^4.

Three gamma globulin types have been described that are not associated with the Gm system.

The best known of these, which is called Inv(a) occurs with a frequency of about 20 percent in Caucasians and 50 percent in Africans and Mongoloids. Inv(b) behaves as if it were determined by an allele of the gene for Inv(a). The third specificity Inv(l) almost always occurs in association with Inv(a). The independence of the Inv and Gm types has been fully confirmed by family studies and indicates that the two sets of differences must be located on different polypeptide chains.

Two lines of evidence show that individual gamma globulin molecules from a particular person's serum do not carry all the Gm determinants detectable in his serum as a whole.

1. Only a small proportion of anti-D sera obtained from Gm(b+) individuals can be used for typing for Gm(b). This indicates that most of the anti-D antibodies from Gm(b+) people do not carry the Gm(b) antigenic determinant. Similarly not all anti-D sera from Gm(a) or Gm(f) individuals can be used for Gm(a) or Gm(f) typing, though most of it apparently can be.

2. Some people suffer from a disease caused by a tumor of the antibody-producing cells called a myeloma. They have in their serum in high concentration a *homogeneous* gamma globulin, which is the product of the myeloma. It is believed that the myeloma arises from a single antibody-producing cell that can only produce just one type of antibody molecule. On this hypothesis a myeloma protein represents a unique pure example of a gamma globulin molecule. Such proteins have been intensively studied and have been the basis for much of the recently accumulated knowledge concerning the chemical structure of the antibody molecule. As for the Rh antibodies just discussed in paragraph 1, myeloma proteins do not express all the Gm specificities found in whole serum from the individual with the myeloma. The Gm types found on myelomas show associations that are different from, and additional to, the population associations found between individuals typed in the usual way using whole serum. Thus, no myelomas are found that have more than one of the Gm types, a, b, and f. On the other hand, all Gm(x+) myelomas are Gm(a+) while approximately one-half the Gm(a+) myelomas do not have Gm(x). The explanation of these associations depends on a knowledge of the overall chemical structure of the various classes of antibody molecules, which we shall now describe briefly.

There are four main classes of antibody molecules called γM, γA, γG, *and* γD.

Each class of antibody molecules is made up of two types of chains, heavy (H) and light (L). The H chains of the various classes differ, and are used to define them, while the L chains are the same in all four classes. The best-known classes are γM, the so-called 19S or macroglobulin, and γG, which is the usual immune, or 7S, gamma globulin. The symbol S stands for the sedimentation constant, usually determined by ultracentrifugation and related to the molecular weight of the protein.

The four classes of antibody molecules participate to different extents in the various immune phenomena and presumably serve different functions in the body's immune defense system. There are two known types of L chains, called, κ and λ. Approximately 70 percent of the γG L chains are κ. Analysis of the amino-acid sequences of a number of myeloma protein L chains has shown that they consist of a constant half, which has with one exception the same sequence for all chains of

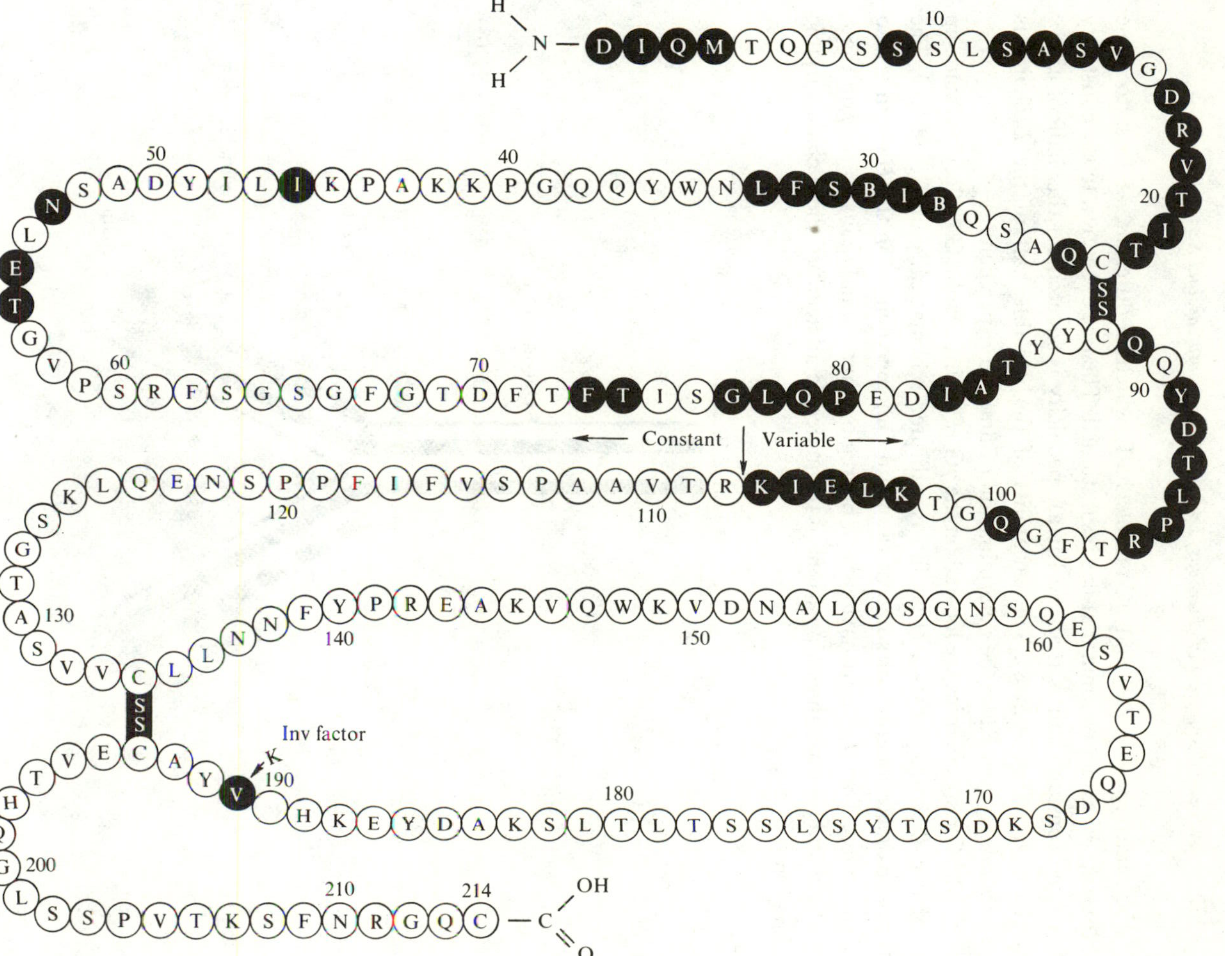

FIGURE 5.18
Amino-acid sequence of a human κ light chain from a protein. See Table 1.1, page 6, for single-letter code of amino acids. The black circles mark variable loci, at which different amino acids have been found in other human κ light chains. "Inv factor" indicates the amino acid position that determines the Inv(a), Inv(b) specificity difference. (From Putman et al., 1967.)

the same type (κ or λ) and a variable half, which is different, at least to some extent, for each myeloma that has been studied. The one exception to the constancy of the constant half is that the Inv(a) specificity is located on this half of the κ chains (see Figure 5.18). The variability of the variable half of the L chains is presumably related to the determination of antibody specificity.

The Gm specificities are located on the γG H chains. Each γG molecule is made up of two L chains and two H chains as indicated schematically in Figure 5.19. The approximate location of some of the main Gm specificities on the H chain is also indicated. The H chains have a variable region that is analogous to that on the L chain and that occupies a position on the molecule adjacent to the L chain variable region.

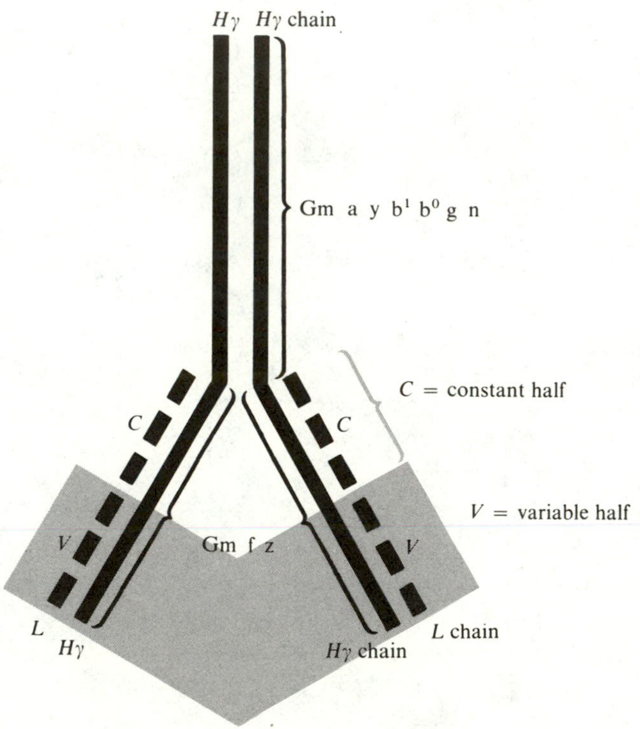

FIGURE 5.19
Schematic structure of γG antibody molecules.

There are four known subclasses of γG *globulins determined by four different* γG *H chains called* γG1, γG2, γG3 *and* γG4.

The differences between the γG H chains are detected serologically using antibodies made by immunizing animals with myeloma proteins. The four subclasses are present in all individuals. Approximately 70 percent of γH chains are γG1,

18 percent are γG2, 8 percent are γG3, and 3 percent are γG4 (see, for example, Natvig et al., 1967). The pattern of Gm types found on myeloma proteins is explained by the fact that any given Gm specificity is always associated with a particular one of the four γG H chains. Thus Gm(a), for example, only occurs on γG1 chains, while Gm(b) only occurs on γG3 chains. Since a myeloma synthesizes just one unique type of antibody molecule, and so one type of H chain, Gm(a) and Gm(b) are not expected to occur together in the same myeloma protein. The heavy chain subgroups or genes associated with the various Gm specificities, which can be determined by typing myeloma proteins for Gm and for the γG H chain subclasses, are indicated in Figure 5.20.

Heavy chain	γG4	γG2	γG3	γG1
Gm specificities	none known	n	y b^0 b^1 b^3 b^4 b^5 s t c^3 c^5	a x y f z

FIGURE 5.20
Association of Gm specificities with γG heavy chain genes.

The simplest interpretation of the genetic control of the γG H chains is that they are determined by very closely linked, probably adjacent, cistrons. The order shown in Figure 5.20 is based on recombination data such as that illustrated in Figure 5.17 and also other criteria. The close linkage of the γG1, γG2, and γG3 genes is based on the inheritance of the Gm specificities in closely linked haplotype combinations, as already discussed. The close linkage of γG4 to the other three is suggested by the similarity of all the four chains to each other, which is why they are grouped into one class determining the γG antibody molecules. A serum sample has been discovered by Kunkel and co-workers (1969) that contains γG heavy chains that seem to be γG3-γG1 hybrid molecules, analogous to the Lepore hemoglobin, which has a β-δ hybrid chain. The most plausible explanation for the origin of this chain is that it arose by unequal crossing-over between the γG1 and the γG3 chains. This is consistent both with the close linkage between these two genes and with the high degree of homology between the two corresponding chains.

The lack of close linkage between the Inv and Gm polymorphisms shows that the L and H chain genes are not closely linked, though it is a reasonable guess that the genes determining the two types of L chains, κ and λ, are themselves closely linked. The overall pattern of genetic control of the γ G globulins is similar to that of the hemoglobins, which are made up of α chains combined with β, γ, δ, or ε chains. In this case, also, the β, γ, δ and ε chain genes are closely linked to each other but are not linked to the α chain gene. A special feature of the antibody molecules, however, is that any one cell synthesizes only the product of *one* allelic gene. Thus, a given single cell from a *Gm(azg)/Gm(yfb)* heterozygote synthesizes γG1 chains that are either a z *or* y f, but not both.

The presence of the Gm(a) specificity in sera from gorillas and chimpanzees but not in those from rhesus monkeys and baboons sets an upper limit to the age of the mutation that gave rise to this specificity.

It must, presumably, have occurred after the separation of the Old World monkeys from the primates (see Chapter 11). It has been shown that Gm(a) is associated with a short polypeptide sequence Asp-Glu-Leu-Thr-Lys, which is present in γG *H* chains from man and higher primates. The corresponding sequence in the Old World monkeys (rhesus monkey and baboon) is Glu-Glu-Leu-Thr-Lys (Wang et al., 1969). From the genetic code (see Table 1.1) the change from Glu to Asp can occur as a result of a mutation in a single nucleotide position, namely:

	Glu		Asp
either	GAA	→	GA*U*
or	GAG	→	GA*C*.

Of the primates and the Old World monkeys that have been studied so far, only the chimpanzee and to a lesser extent the orangutan have shown detectable polymorphism for Gm factors using human typing sera. This does not necessarily mean that the other species are not polymorphic for the gamma globulins, but simply emphasizes the close evolutionary relationship of man and chimpanzee.

The homology between the various gamma globulin chains has led to the suggestion that they may all have evolved from a common ancestral chain.

There is a hierarchy of similarity between the various chains. Thus γG *H* chains (at least in their constant regions) are probably more closely related to each other than they are to γM, γA, or γD *H* chains. Nevertheless, the *H* chains all show significant homologies with each other. Similarly κ and λ chains have many similar amino acids in corresponding positions, though it appears that the κ chains in mouse and man are more similar to each other than either is to the corresponding species λ chains. This suggests that if κ and λ chains have a common evolutionary origin, the split between them occurred, probably, before the evolutionary separation of mammalian species. Heavy and light chains have significant similarities though they are the most different of the total family of gamma globulin polypeptide chains. A schematic illustration of the evolution of the various chains is shown in Figure 5.21. Each step in the evolutionary pathway is presumed to be

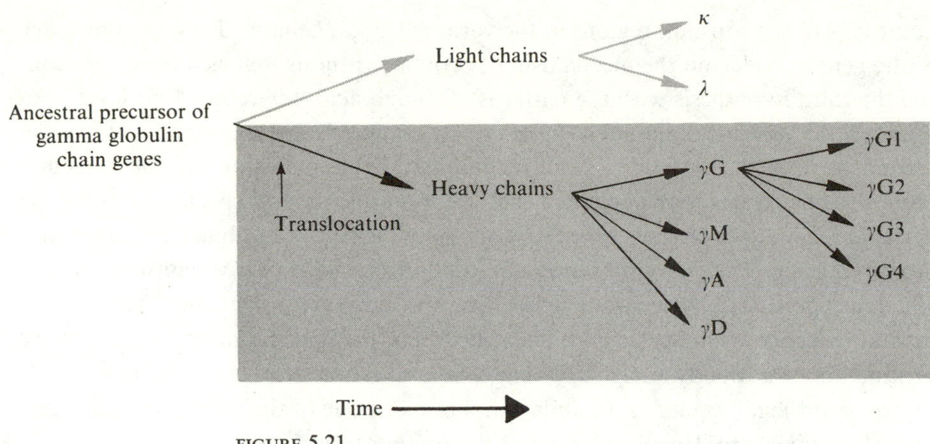

FIGURE 5.21
Schematic illustrations of the evolution of the gamma globulins.

mediated by serial duplication of a preexisting gene followed by divergence due to selection of new mutants. The separation of the *H* and *L* chains presumably occurred by translocation or by duplication of the whole chromosome followed by differential evolution of the relevant parts of the duplicate chromosomes. In the mouse, there is evidence to show that all the genes for the various classes of *H* chain are closely linked (see, for example, Herzenberg et al., 1968). The only direct evidence in humans, of course, involves the Gm specificities found on the γG 1, 2, and 3 *H* chains, though it seems likely that the genes for γA, γM, and γD *H* chains are also closely linked to the γG *H* chain genes.

It has been suggested that the variable and constant regions of both L and H chains may be coded for by different genes.

As we mentioned before, the extreme variability of the variable regions of the *L* and *H* chains is probably the key to the specificity of the antibody molecule. The Gm and Inv polymorphisms are only detectable because they are in the *constant* regions of the molecules. There are three basic types of hypotheses for the origin of the variability: (1) It could be present in the germ line. (2) It could be produced by somatic mutation or recombination during the development of the antibody-producing cell. (3) It could result from "programmed" inaccuracies in the transcription of the gamma globulin genes or the translation of the corresponding messenger RNA's.

If the first hypothesis were true, there would have to be an enormous number of *H* and *L* chain genes. If, in addition, each chain were determined by a *single* gene, the "constant half" of all the genes would have to be the same, indicating remarkable evolutionary conservation. This could, of course, be related to the need for strict homology of the constant regions of the various *L* and *H* chains. Present knowledge of the genetic code and the mechanism of protein synthesis makes it hard to reconcile the third hypothesis with the patterns of amino-acid differences that have been found in the variable region of *L* chains so far analyzed. These arguments leave the second hypothesis, that of somatic recombination or mutation, as the currently preferred one. It has been suggested that there may be several special genes for the variable regions, which have properties allowing them to recombine or mutate with high frequency. Their products are then joined to their respective constant halves, which are produced by separate genes. There is, however, so far, no direct experimental evidence to indicate anything concerning the specific mechanisms for the production of antibody variability. The whole question of explaining the immune response at the molecular and cellular level remains one of the most tantalizing and actively studied problems in biology at the present time.

General References

Ceppellini, R., 1959, "Physiological genetics of human factors." *In* G. E. W. Wolstenholme and C. M. O'Connor, eds., *Ciba Foundation Symposium on Biochemistry of Human Genetics*, 242–261, London, Churchill, 1959. (The first exposition of his genetic theory for the relationship between ABO, Lewis, secretor, and Bombay blood groupings.)

Ceppellini, R., "Genetica delle Immunoglobuline." *Twelfth Annual Meeting of Associazione Genetica Italiana.* 1966. (A comprehensive review of the genetics, population genetics, and biochemistry of the immunoglobins, together with a review of theories of the immune response—in Italian.)

Ceppellini, R., and Nasso, S. and Tecilazich, F., *La Malattia Emolitica del Neonato*, Milano, 1952. (Still one of the best discussions of the problem of hemolytic disease of the newborn—but available only in Italian.)

Histocompatibility Testing 1967 (edited by E. S. Curtoni, P. L. Mattiuz and R. M. Tosi).

Histocompatibility Testing 1970 (edited by P. Terasaki). Copenhagen: Munksgaard. (Two up-to-date and comprehensive symposia on transplantation and the genetics of transplantation antigens.)

Natvig, J. B., H. G. Kunkel, and S. D. Litwin. "Genetic markers of the heavy chain subgroups of human γG globulin," *Cold Spring Harbor Symp. Quant. Biol.* **32**: 173–180, 1967. (A very useful concise paper on the genetics of human gammaglobulins.)

Race, R. R., and R. Sanger, *Blood Groups in Man.* Philadelphia: F. A. Davis, 1968. (The classic work on the subject.)

Watkins, W. M. "Blood-group substances." *Science* **152**: 172–181, 1966. (A review of the biochemistry of ABO and related antigens by one of the major contributors to the field.)

Worked Examples

5.1. *Compensation model for the rhesus blood groups.*

In terms of the simple model for rhesus incompatibility selection discussed in this chapter, and assuming the possible existence of heterozygote advantage unrelated to incompatibility selection, we shall find the conditions for the increase of the D and d alleles when either one or the other is rare. (Based on Feldman et al., 1969.)

The selective scheme for incompatibility selection and heterozygote advantage is given in the following table. The scheme has five independent parameters. The frequencies of the genotypes DD, Dd, dd are u, v, and w, respectively, and random mating is assumed. Only the matings marked with an asterisk, namely ♀ dd × ♂ DD and ♀ dd × ♂ Dd involve incompatibility selection. The parameter l is such that $1 - l$ is the proportion of offspring surviving the incompatibility selection in the mating ♀ dd × ♂ Dd. Compensation is introduced in the mating ♀ dd × ♂ Dd by the fertility factor f, which is the ratio of the average number of zygotes produced by ♀ dd × ♂ Dd matings to the number produced by compatible matings, which are all assumed to have equal fertility. The parameter k, as it stands, includes both the compensation effect in the ♀ dd × ♂ DD mating and the relative viability of offspring

Mating			*Offspring distribution*		
Female	*Male*	*Frequency*	*DD*	*Dd*	*dd*
DD	× DD	u^2	1	0	0
DD	× Dd	uv	$\frac{1}{2}$	$\frac{1}{2}$	0
Dd	× DD	uv	$\frac{1}{2}$	$\frac{1}{2}$	0
DD	× dd	uw	0	1	0
*dd	× DD	uw	0	$1 - k$	0
Dd	× Dd	v^2	$\frac{1}{4}$	$\frac{1}{2}$	$\frac{1}{4}$
Dd	× dd	vw	0	$\frac{1}{2}$	$\frac{1}{2}$
*dd	× Dd	vw	0	$\frac{1}{2}(1 - l)f$	$\frac{1}{2}f$
dd	× dd	w^2	0	0	1
Overall viabilities			$1 - s$	1	$1 - t$

Note: Only the matings with an asterisk ($dd \times DD$ and $dd \times Dd$) involve incompatibility selection.

from these matings, as these two factors cannot be separated in this mating. The viability parameters $1 - s$, 1, and $1 - t$ represent overall relative viabilities associated with the three genotypes *DD*, *Dd*, and *dd*, respectively. These are independent of the incompatibility selection and compensation. It is clear that the model represents a considerable simplification of the physiological situation arising from incompatibility selection. In particular, the viabilities of *Dd* offspring must be functions of the number of pregnancies and the distribution of *Dd* offspring among them. These viabilities will, on the average, decrease rather sharply with increasing numbers of *Dd* children conceived. Thus it could be argued that, since it is unlikely a still birth due to Rh incompatibility will be followed by any live Rh(+) births, f should be restricted to the *dd* offspring of the $dd \times Dd$. In any case, it seems unlikely that $1 - k$ or $(1 - l)f$ are often bigger than 1, as required for overcompensation.

The equations relating the frequencies of the three genotypes in successive generations are then as follows:

$$u' = \frac{(1 - s)(u^2 + uv + \frac{1}{4}v^2)}{T}, \tag{1}$$

$$v' = \frac{uv + uw(2 - k) + \frac{1}{2}v^2 + \frac{1}{2}vw[1 + f(1 - l)]}{T}, \tag{2}$$

$$w' = \frac{(1 - t)(w^2 + \frac{1}{2}vw(1 + f) + \frac{1}{4}v^2)}{T}, \tag{3}$$

where T is a normalizing factor such that $u' + v' + w' = 1$.

When d is very rare, meaning that v is small, then, as approximations, $u = 1 - v$ and w is of order v^2. Ignoring terms in v, $T = (1 - s)$ and thus ignoring terms in v^2 we obtain from Equation 2

$$v' = \frac{v}{1 - s}. \tag{4}$$

Thus if $s \neq 0$, the d gene increases provided $s > 0$; that is, provided only the heterozygote *Dd* is fitter than the *DD* homozygote. When $s = 0$, terms of $0(v^2)$ and so also $0(w)$ must be included in order to determine the fate of the d gene. It can then be shown that

$$v' = v + \tfrac{1}{4}v^2[(1 - t)(2 - k) - 2] + 0(v^3), \tag{5}$$

with the result that d increases provided

$$(1 - t)(2 - k) > 2, \tag{6}$$

or when $t = 0$, $k < 0$. This implies overcompensation for the incompatibility loss in the mating $dd \times DD$.

When the allele D is very rare, meaning that, again, v is small and as approximations $w = 1 - v$ and $u = 0(v^2)$, then to $0(v^2)$ we have

$$v' = \frac{1}{2}\frac{(1 + f(1 - l))}{1 - t} v \tag{7}$$

since, now, $T = 1 - t +$ terms in v.
Thus D increases provided

$$\frac{1}{2}\frac{1 + f(1 - l)}{1 - t} > 1,$$

which can be written in the form

$$f(1 - l) > 1 - 2t, \tag{8}$$

or

$$f > \frac{1 - 2t}{1 - l}, \tag{8a}$$

or

$$t > \tfrac{1}{2}[1 - f(1 - l)]. \tag{8b}$$

When $t = 0$, this condition is

$$f > \frac{1}{1 - l}, \tag{9}$$

and when $f = 1$,

$$t > \frac{l}{2}. \tag{10}$$

Thus from these equations we see that the D gene will not increase in a predominantly dd population unless the heterozygote advantage of Dd, t, or the compensating fertility, f, are large enough to counter the incompatibility selection against heterozygotes Dd. It seems likely, therefore, that d initially occurred in a predominantly DD population in which, for small frequencies of d, the incompatibility selection was negligible.

A balanced polymorphism will be maintained provided condition 8 is satisfied and $s > 0$. This implies heterozygote advantage, with the difference between Dd and dd large enough to counteract incompatibility selection when D is rare. When $s = 0$ a balanced polymorphism will result if condition 6 as well as condition 8 is satisfied. Thus if $s = 0$ and $t = 0$ and there is no intrinsic heterozygote advantage, a balanced polymorphism can exist provided $k < 0$ and $f > 1/(1 - l)$, that is, provided in both the matings $dd \times DD$ and $dd \times Dd$ there is overcompensation for the incompatibility loss.

5.2. *Frequency of histocompatible donor-recipient pairs.*

We shall calculate the frequency of randomly combined, parent-child, and sib-sib pairs that are compatible for a histocompatibility polymorphism involving n equally frequent, equally strong alleles. We assume random mating. Allele A_i, $i = 1, \ldots n$ each determine a corresponding antigen that can, in an incompatible combination, lead to graft rejection. The frequency of each of the n alleles is assumed to be $1/n$.

RANDOM PAIRS

There are three classes of compatible, random, donor-recipient pairs: $A_iA_i \to A_iA_i$, $A_iA_j \to A_iA_j$, and $A_iA_i \to A_iA_j$, where A_i and A_j are any two distinct alleles. The probabilities of getting these combinations are as follows.

$A_iA_i \to A_iA_i$. The frequency of any particular homozygote (assuming random mating and a Hardy-Weinberg equilibrium) is $(1/n)^2$. The probability of picking a particular pair of identical homozygotes, is therefore $(1/n)^2 \times (1/n)^2$. Since there are n types of homozygotes, the total probability of this combination is $n \times (1/n)^2 \times (1/n)^2 = (1/n)^3$.

$A_iA_j \to A_iA_j$. The frequency of any given heterozygote is $2 \times 1/n \times 1/n = 2/n^2$, so that the probability of picking a particular pair of identical heterozygotes is $2/n^2 \times 2/n^2$. The number of types of heterozygotes is $\frac{1}{2}(n^2 - n) = [n(n-1)]/2$, and so the total probability of this combination is

$$\frac{n(n-1)}{2} \times \frac{2}{n^2} \times \frac{2}{n^2} = \frac{2(n-1)}{n^3}.$$

$A_iA_i \to A_iA_j$. There are $n - 1$ heterozygotes of type A_iA_j carrying the particular allele A_i and any of the other alleles A_j $(j \neq i)$. Thus, for a given homozygote, the probability of this combination is $(n-1) \times 2/n^2$. Since there are n different types of homozygotes, the total probability of this combination is

$$n \times \frac{1}{n^2} \times (n-1) \times \frac{2}{n^2} = \frac{2(n-1)}{n^3}.$$

The total probability of compatibility is the sum of these three separate probabilities. That is,

$$\frac{1}{n^3} + \frac{2(n-1)}{n^3} + \frac{2(n-1)}{n^3} = (4n-3)n^3.$$

PARENT-CHILD PAIRS

Homozygous parents. These must each offer compatible grafts to all their children whether or not the other parent is homozygous. The probability that a given parent is homozygous for a particular allele is $(1/n^2)$ and there are n types of homozygotes, so that the probability of this type of compatibility is $n \times 1/n^2 = 1/n$.

Heterozygous parents. A given heterozygous parent $A_i A_j$ (frequency $2/n^2$) passes on just one of his alleles (say A_i) to a child. The child will be compatible only if he has the same heterozygous genotype $(A_i A_j)$ as the parent; that is, if he received the appropriate second allele (A_j) from the other parent. With random mating the probability of this is simply $1/n$. Hence the probability of a heterozygous parent being compatible with his child is $2/n^2 \times 1/n$. There are $n(n-1)/2$ types of heterozygotes so that the total probability of compatibility is

$$\frac{n(n-1)}{2} \times \frac{2}{n^2} \times \frac{1}{n} = \frac{n-1}{n^2}.$$

The total probability of parent-child compatibility is, therefore,

$$\frac{1}{n} + \frac{n-1}{n^2} = \frac{2n-1}{n^2}.$$

An exactly similar argument applies for child-parent compatibility.

SIB-SIB PAIRS

We must consider all possible parental combinations and the frequency with which a random sib pair from any given combination will be compatible. There are six parental combinations to be distinguished, namely matings between homozygotes either like or unlike; matings between a homozygote and a heterozygote who carries the same allele as the homozygous parent; matings between a homozygote and a heterozygote who carries two alleles different from that carried by the homozygous parent; matings between two like heterozygotes; matings between two heterozygotes with one allele in common; and finally matings between two heterozygotes with no alleles in common. The frequencies of the various mating types, the number of such types of matings and the frequency of compatible sib pairs within each type are shown in the accompanying table. The contributions in the last column are the products of the three preceding columns. The frequencies of compatible sib pairs for mating types 1a, 1b, 4, and 6 have already been discussed; the calculation of the others is left as an exercise for the reader. For the number of matings of given type: in 3, $n-2$ is the number of homozygotes $A_k A_k$ that have no

	Mating type	Frequency	Number of matings of given type	Frequency of sib–sib compatible pairs	Contribution to total
1a	$A_iA_i = A_iA_i$	$\frac{1}{n^2} \times \frac{1}{n^2}$	n	1	$\frac{1}{n^3}$
1b	$A_iA_i = A_jA_j$	$2 \times \frac{1}{n^2} \times \frac{1}{n^2}$	$\frac{n(n-1)}{2}$	1	$\frac{n-1}{n^3}$
2	$A_iA_i = A_iA_j$	$2 \times \frac{1}{n^2} \times \frac{2}{n^2}$	$\frac{n(n-1)}{2} \times 2$	$\frac{3}{4}$	$\frac{3(n-1)}{n^3}$
3	$A_kA_k = A_iA_j$	$2 \times \frac{1}{n^2} \times \frac{2}{n^2}$	$\frac{n(n-1)}{2} \times (n-2)$	$\frac{1}{2}$	$\frac{(n-1)(n-2)}{n^3}$
4	$A_kA_j = A_kA_j$	$\frac{2}{n^2} \times \frac{2}{n^2}$	$\frac{n(n-1)}{2}$	$\frac{5}{8}$	$\frac{5(n-1)}{4n^3}$
5	$A_iA_k = A_iA_j$	$2 \times \frac{2}{n^2} \times \frac{2}{n^2}$	$\frac{n(n-1)}{2} \times (n-2)$	$\frac{3}{8}$	$\frac{3(n-1)(n-2)}{2n^3}$
6	$A_lA_k = A_iA_j$	$2 \times \frac{2}{n^2} \times \frac{2}{n^2}$	$\frac{n(n-1)}{2} \times \frac{(n-2)(n-3)}{2} \times \frac{1}{2}$	$\frac{1}{4}$	$\frac{(n-1)(n-2)(n-3)}{4n^3}$
	Total				$\frac{n^3+4n^2+2n-3}{4n^3}$

allele the same as those in A_iA_j; in 5, there are $[n(n-1)]/2$ allele pairs A_k and A_j and $n-2$ remaining alleles different from A_i and A_j; in 6, $[(n-2)(n-3)]/2$ is the number of heterozygotes that can be formed using all the $n-2$ alleles other than A_i and A_j, the factor $\frac{1}{2}$ derives from the fact that the mating pairs can be considered with either heterozygote first. The total frequency of sib-sib compatible pairs is simply the sum of the terms in the last column.

5.3. *The association between a pair of sera with a shared antibody.*

We consider two dispecific sera, one of which contains antibodies anti-X and anti-Y_1, and the other anti-X and anti-Y_2. The expected frequencies of the four corresponding phenotypes, on the assumption that the antigens X, Y_1, and Y_2 are not associated, are shown in the accompanying table, where x, y_1, and y_2 are the

		Anti-1 +	*Anti-1* −	*Totals*
Anti-2	+	$x + y_1y_2(1-x)$ (++)	$y_2(1-y_1)(1-x)$ (−+)	$x + y_2(1-x)$
	−	$y_1(1-y_2)(1-x)$ (+−)	$(1-y_1)(1-y_2)(1-x)$ (−−)	$(1-x)(1-y_2)$
Totals		$x + y_1(1-x)$	$(1-x)(1-y_1)$	1

population frequencies of the antigens X, Y_1, and Y_2, respectively. The probability of being $++$ is the sum of the probability of reacting to X, and that of reacting to Y_1 and Y_2 but not X, which is $x + y_1y_2(1-x)$, since all reactions are independent. The probability of being $+-$ is the product of the probabilities of reacting to Y_1, not to Y_2, and not to X, which is $y_1(1-y_2)(1-x)$. The probability of being $-+$ is similarily $y_2(1-y)_1(1-x)$. The probability of being $--$ is the product of the probabilities of not reacting to any of the antibodies, which is $(1-y_1)(1-y_2)(1-x)$. The cross product for the 2×2 table is

$$\begin{aligned}&[x + y_1y_2(1-x)][(1-y_1)(1-y_2)(1-x)]\\&-[y_2(1-y_1)(1-x)][y_1(1-y_2)(1-x)] = x(1-x)(1-y_1)(1-y_2).\end{aligned}$$

The correlation between the reactions of the two sera, which is the cross product divided by the square root of the products of the marginal totals, is, therefore,

$$\rho = \frac{x(1-x)(1-y_1)(1-y_2)}{\sqrt{(x+y_1(1-x))(1-x)(1-y_1)(x+y_2(1-x))(1-x)(1-y_2)}},$$

which can be rewritten in the form

$$\rho = \frac{x}{\sqrt{\left(x+\dfrac{y_1}{1-y_1}\right)\left(x+\dfrac{y_2}{1-y_2}\right)}}. \tag{1}$$

The association due to a shared antibody is thus always positive. The value of ρ increases as x, the frequency of the shared antibody, increases but decreases as y_1 and y_2, the frequencies of reaction to the unshared antibodies, increase. For given x, ρ tends to approach 1 as y_1 and y_2 tend to approach zero. Thus, as expected intuitively, the association between two sera due to a shared antibody decreases as the frequencies of reaction to the unshared antibodies increase. Given a group of sera all sharing one antibody, anti-X, any pair of them should be associated, though to varying extents depending on the frequencies of reaction to the unshared antibodies. The probability that an individual reacts with all the sera, but does not have the antigen X is $(1-x)\Pi y$, where the product Πy is with respect to the frequencies of reaction to the unshared antibodies in each of the sera. This quantity will tend to approach zero as the number of sera in the group increases. The presence of antigen X is then reliably determined by a reaction with all the sera in a group. In principle, therefore, an antigen can be identified by a group of multispecific sera, all pairs of which have significant ρ values. (See Bodmer et al., 1969.)

5.4. *The association between a pair of antigens due to allelism or close linkage.*

We consider two linked loci each wih two alleles (G, and g, H and h) in a population mating at random. We assume that alleles G and H determine the hypothetical antigens G and H, and that the four gametes GH, Gh, gH, and gh have frequencies

x_1, x_2, x_3, and x_4, respectively. We first calculate the frequencies of the four phenotypes GH, G, H, and – in terms of x_1, etc., or alternatively, in terms of the gene frequencies $q_1 = x_3 + x_4$, $q_2 = x_2 + x_4$ and the gametic association D, defined by

$$D = x_1x_4 - x_2x_3.$$

The frequency of a homozygote, say GH/GH, is the square of the respective gamete frequency, x_1^2, while the frequency of a heterozygote, say GH/Gh, is twice the product of the frequencies of the constituent gametes, $2x_1x_2$ (see example 2.6). The phenotype frequencies are, therefore, as follows.

Phenotype	*Genotypes*	*Frequency*
GH	GH/GH, GH/Gh, GH/gH, GH/gh, Gh/gH	θ_1
G	Gh/Gh, Gh/gh	$\theta_2 = x_2^2 + 2x_2x_4 = q_2^2 - x_4^2$
H	gH/gH, gH/gh	$\theta_3 = x_3^2 + 2x_3x_4 = q_1^2 - x_4^2$
–	gh/gh	$\theta_4 = x_4^2$

Since $\theta_1 + \theta_2 + \theta_3 + \theta_4 = 1$, by subtraction

$$\theta_1 = 1 - q_1^2 - q_2^2 + x_4^2.$$

The expressions for θ_1, etc., can be rewritten in the form

$$\theta_1 = (1 - q_1^2)(1 - q_2^2) + x_4^2 - q_1^2q_2^2 = (1 - q_1^2)(1 - q_2^2) + \tilde{D}, \tag{2}$$

$$\theta_2 = (1 - q_1^2)q_2^2 - \tilde{D}, \tag{3}$$

$$\theta_3 = q_1^2(1 - q_2^2) - \tilde{D}, \tag{4}$$

$$\theta_4 = q_1^2q_2^2 + \tilde{D}, \tag{5}$$

where $$\tilde{D} = x_4^2 - q_1^2q_2^2 = (x_4 - q_1q_2)(x_4 + q_1q_2) = D(D + 2q_1q_2). \tag{6}$$

This follows from the fact that x_4 can be expressed in the form

$$x_4 = q_1q_2 + D.$$

When $D = 0$, then $\tilde{D} = 0$ and the expressions for θ_1, etc., take the form expected if there were independence between the two loci at the population level. The quantity $\tilde{D}$ is the analogue, at the phenotypic level, to D and Equations 2 to 5 are the analogues of the expressions for the gamete frequencies in terms of gene frequencies and D, namely

$$x_1 = (1 - q_1)(1 - q_2) + D, \text{ etc.}$$

The population correlation ρ between the antigens defined as in Example 5.1 is given by

$$\rho = \frac{\theta_1\theta_4 - \theta_2\theta_3}{\sqrt{(\theta_1+\theta_2)(\theta_3+\theta_4)(\theta_1+\theta_3)(\theta_2+\theta_4)}} = \frac{\tilde{D}}{\sqrt{q_1^2 q_2^2(1-q_1^2)(1-q_2^2)}}, \tag{7}$$

or

$$\rho = \frac{D\left(2 + \dfrac{D}{q_1 q_2}\right)}{\sqrt{(1-q_1^2)(1-q_2^2)}}. \tag{8}$$

In order to estimate D, the gametic association, from observed data on phenotype frequencies, we need to express D in terms of θ_1, θ_2, θ_3, and θ_4. From Equations 5 and 6

$$\theta_4 = q_1^2 q_2^2 + \tilde{D} = q_1^2 q_2^2 + D^2 + 2D\, q_1 q_2 = (q_1 q_2 + D)^2,$$

so that

$$D = \sqrt{\theta_4} - q_1 q_2. \tag{9}$$

Now

$$\theta_2 + \theta_4 = (1 - q_1^2)q_2^2 - \tilde{D} + q_1^2 q_2^2 + \tilde{D} = q_2^2,$$

and similarly $\theta_3 + \theta_4 = q_1^2$. Thus, from Equation 9 we have

$$D = \sqrt{\theta_4} - \sqrt{(\theta_2 + \theta_4)(\theta_3 + \theta_4)}. \tag{10}$$

The significance of an estimated D—that is, whether it is significantly different from zero—is measured by the usual 2×2 χ^2 based on the phenotypes (see Appendix I). From the form of the χ^2 it is easily seen that

$$\chi^2 = n\rho^2. \tag{11}$$

This equation enables one to predict the minimum correlation ρ that can, on the average, be detected at a given significance level and for a given number of observations. For example at the 5 percent level $\chi^2 = 3.84$, so for $n = 400$ the minimum ρ that will, on an average, be significant is given by

$$\rho = \sqrt{\frac{\chi^2}{n}} = \sqrt{\frac{3.84}{400}} = 0.098.$$

Mi and Morton (1966) defined the six basic types of logical relationships that can exist between a pair of loci according to the values of D and the gametic frequencies x_1, x_2, x_3, and x_4. Their properties are summarized in the following table.

Type designation	*Gametes*				*Phenotypes*						
	GH	*Gh*	*gH*	*gh*	GH	G	H	–	D	$\tilde{D}$	ρ
Permuted	+	+	+	+	+	+	+	+	D $\neq 0$	$D(D+2q_1q_2)$	$D(D+2q_1q_2)/q_1q_2\sqrt{(1-q_1^2)(1-q_2^2)}$
Segregant	–	+	+	+	+	+	+	+	$-x_2x_3$	$-x_2x_3(x_2x_3+2x_4)$	As above
Complementary	–	+	+	–	+	+	+	–	$-x_2x_3$	$-x_2^2x_3^2$	$-x_2x_3/\sqrt{(1-x_2^2)(1-x_3^2)}$
Codominant	+	+	+	–	+	+	+	–	$-q_1q_2$	$-q_1^2q_2^2$	$-q_1q_2/\sqrt{(1-q_1^2)(1-q_2^2)}$
Subtypic	+	–	+	+	+	–	+	+	p_2q_1	$p_2q_1^2(1+q_2)$	$p_2q_1(1+q_2)/q_2\sqrt{(1-q_1^2)(1-q_2^2)}$
Identical	+	–	–	+	+	–	–	+	x_1x_4	$x_1x_4^2(1+x_4)$	1

Note: The symbol + indicates presence of corresponding gamete or phenotype in a population, – indicates absence; $p_1 = 1 - q_1$ and $p_2 = 1 - q_2$, where p_1 and p_2 are the allele frequencies of G and H.

Permuted is a general positive or negative association.

Segregant ($x_1 = 0$) corresponds to a three-allele locus with one-allele silent, like ABO.

Complementary ($x_1 = x_4 = 0$) corresponds to a two-allele codominant locus like MN.

Codominant ($x_4 = 0$) is a system in which there is no "blank" allele.

Subtypic ($x_2 = 0$ or $x_3 = 0$) is an "inclusion" due to the absence of one of the gametes determining just one of the antigens (either *Gh*, as in the table, or *gH*).

Segregant, complementary, and codominant loci lead to negative associations, as for strict allelism; subtypic and identical loci lead to positive associations. (See Bodmer et al., 1969.)

5.5. *The expression of haplotype frequencies for a pair of multiple allelic loci in terms of gene frequencies and gametic associations.*

Suppose the allelic determinants for the antigens of the first locus have frequencies p_i, $i = 1, \ldots k$ and for the second locus P_j, $j = 1, \ldots, l$ (p_k and P_l may refer to blank alleles). Let x_{ij} be the frequency of the haplotype determining antigen i at the first locus and antigen j at the second. The 2×2 table for the gametic association between i and j, ignoring all other antigens, takes the following form, where the frequencies in the body of the table are those of the corresponding haplotypes:

	FIRST LOCUS $i+$	$i-$	
SECOND LOCUS $j+$	x_{ij}	$P_j - x_{ij}$	P_j
$j-$	$p_i - x_{ij}$	$1 - p_i - P_j + x_{ij}$	$1 - P_j$
	p_i	$1 - p_i$	1

The haplotype $i+j-$ corresponds to the sum of all those gametes with i at the first locus and anything other than j at the second. Similarly $i-j+$ corresponds to the sum of gametes with anything other than i at the first locus, and j at the second locus, while $i-j-$ is the sum of those with anything other than i at the first locus and anything other than j at the second locus. The gametic association between i and j, considering them as alleles of a two-locus two-allele system with the second alleles unspecified, is given by

$$D_{ij} = x_{ij}(1 - p_i - P_j + x_{ij}) - (p_i - x_{ij})(P_j - x_{ij}) = x_{ij} - p_i P_j. \quad (1)$$

Thus x_{ij} can without loss of generality be expressed in the form

$$x_{ij} = p_i P_j + D_{ij}. \quad (2)$$

Equation 2 effectively represents a transformation from the haplotype frequencies x_{ij}, to the allele frequencies p_i and P_j and the pairwise gametic associations D_{ij}. The allele frequencies are readily estimated using standard maximum likelihood techniques (see Appendix I). The D's can be estimated from the 2 × 2 table for the phenotype association between i and j using Equation 9 of Example 5.4. Thus, using Equation 2 just given, haplotype frequencies for the HL-A system can be estimated rather simply, without recourse to the full set of phenotypes.

The total numer of haplotypes is lk and the total number of genotypes, therefore, is $lk(lk + 1)/2$. The number of D's is also lk, but of these only $(l - 1)(k - 1)$ are independent. The total number of independent parameters represented by the D's and the allele frequencies is, therefore,

$$(l - 1) + (k - 1) + (l - 1)(k - 1) = lk - 1,$$

which corresponds to the number of independent haplotype frequencies x_{ij}. This shows the validity of the transformation represented by Equation 2. Assuming that there is a blank allele at each locus, the total possible number of phenotypes, which is the product of the number of phenotypes for each locus, is

$$\left(l + \frac{(l-1)(l-2)}{2}\right)\left(k + \frac{(k-1)(k-2)}{2}\right).$$

(There are $l - 1$ single antigen phenotypes, plus 1 for the blank, plus $(l - 1)(l - 2)/2$ double antigen phenotypes. Triples and higher order combinations do not occur in an "allelic" system.)

6

Genetic Demography and Natural Selection

6.1 Demography and the Measurement of Fitness

The conventional models of population genetics that we have discussed in previous chapters are formulated on the basis of very simple assumptions. In the first place, these models apply to characters whose inheritance can be specified in terms of one or two loci with a definite number of alleles. Time is measured in discrete units of one generation, selection is measured by the probability of survival from birth to maturity and, sometimes, also by the average number of offspring per individual per generation but the age structure of the population is not taken into account in these measurements. Mating is generally assumed to be at random with respect to characters of interest. Many of these assumptions may be reasonable for many investigations and at least produce models that give answers which can be easily interpreted. The goal of the human population geneticist is, however, more ambitious than the results that are obtainable from using such models. Information on man is available in much more detail than specified by these relatively simple models. It is, in fact, probably more completely available for man than for any other organism, thus potentially making him from many points of view the organism of choice for the general study of population genetics.

Conventional population-genetics models ignore the population's age structure, distribution in time and space, and behavioral or socioeconomic characters.

There are three major characteristics of real human populations that are ignored in the usual population-genetics models. First, because a real population has an **age structure**, time cannot be measured in discrete units of one generation. The events in a human life that influence overall reproductivity happen at different rates at different ages, and this must be taken into account in any adequate description of selective differentials. Second, human populations are distributed in space and by socioeconomic characteristics in a way that clearly enforces limits on the mating patterns such that they cannot be at random with respect to these characteristics. Last, many of the characters of interest to the student of human population structure are not simply inherited. They are, in general, complex behavioral attributes often measured by relatively imprecise socioeconomic parameters, which are partly genetically determined and partly dependent on environmental influences. The theory of quantitative inheritance, which will be discussed in Chapter 9, provides the basis for estimating the relative contributions of genotype and environment to such complex characters.

Demography, in its broadest sense, is simply the statistical study of human populations. In so far as such a study is directed at questions of genetic interest we speak of genetic demography—an extension and generalization of population genetics to take account of the detailed information available on human populations. Demography in its more usual, narrower, sense refers to the study of mortality and fertility as a function of age, including socioeconomic and geographic parameters. The measurement of selective differences depends essentially on these characteristics of a population. In this chapter we shall concentrate on those aspects of genetic demography that relate to the measurement of fitness, leaving the discussion of migration and mating patterns and the inheritance of demographic variables to later chapters.

Biological fitness is best measured by the intrinsic rate of increase, which can be calculated if age-specific birth and death rates are known.

The biological fitness of an individual or of a group of individuals with a common characteristic is, ideally, measured by the relative contribution of the individual or the group to the ancestry of future generations. For the simplest population models, which ignore age structure, each generation is assumed to be discrete or nonoverlapping (that is, the population is completely replaced each generation, as in annual plants), and the fitness of an individual is measured simply by the number of his offspring that survive to reproduce in the next generation. However, for a complete specification of fitness in a human population it is necessary to take account of all the many factors and their interactions that contribute to an individual's reproductivity. A complete schematic representation of these factors will be discussed later. They obviously include as major components the probability of having children and the probability of dying, each given as a function of age. The actually achieved reproductivity of a population is a balance between loss by death and by infertility and gain by birth. The demographer measures the rate of the increase or decrease in total size of a population by a quantity r called the **intrinsic rate of increase**, which is calculated from **age-specific birth and death rates** (rates for each of the various age groups in the population). The population's **age distribution** is assumed to remain constant from one generation to the next—that is, the proportion of the population that is of a given age stays the same. The use of the quantity r to describe fitness was first emphasized by Fisher in 1930, who called it the **malthusian parameter** in honor of Malthus' influence on the questions of population growth and the measurement of natural selection. For the *quantitative measurement* of selective differences, this relatively simple information on age-specific birth and death rates, which takes into account the effects of all the various factors contributing to reproductivity, is all that is needed. Only when we are interested in determining the *underlying cause* of a selective difference, do we have to delve deeper into the interrelationships of the factors that influence reproductivity. Our first and major aim in this chapter will be to describe the quantities needed and the calculations for determining the intrinsic rate of increase.

6.2 Age-specific Birth and Death Rates and the Intrinsic Rate of Increase of a Population

It is a classical result due to Lotka, that if the probability of death at a given age and the mean number of children produced at a given age do not change with time, then, whatever the initial age distribution, the population gradually approaches an age distribution that is independent of the initial distribution and depends only on the age-specific birth and death rates. The total population size, furthermore, increases or decreases at a constant rate, which is the intrinsic rate of increase (or decrease).

Because the theory on which this result is based is formulated in terms of the individual rather than the mating pair, it is customary to conduct an analysis by using only females. The theory provides the basis for calculating the intrinsic rate of increase of a population and so for calculating the magnitude of selective differences between genotypes. The only quantities needed for the calculation are the age-specific birth and death rates. We shall illustrate the population model and its consequences by a numerical example and leave the general derivation to Example 6.1 at the end of the chapter.

A numerical example, based on data collected in 1960 *on females in the United States, illustrates the application of the population projection matrix to the determination of the intrinsic rate of increase and the stable equilibrium age distribution.*

We consider, for simplicity, a hypothetical example in which time is measured in 15-year intervals. Let p_0 be the mean proportion of women in the 0–15 age group who survive for 15 years, p_1 the mean proportion of the 15–30 age group who survive for 15 years, and so on. Similarly, let b_1 be the mean number of female children born per woman to the 15–30 age group, and b_2 be the mean number born to the 30–45 age group. The survival probabilities p_0, etc., are usually obtained from deaths tabulated by age. Thus, if, in a given population, D_1 is the total number of deaths of women in the 15–30 group during a 15-year interval and N_1 the number of women in this age group who were alive at the beginning of the interval then

$$p_1 = 1 - q_1 = 1 - \frac{D_1}{N_1}, \tag{6.1}$$

where $q_1 = D_1/N_1$ is the age-specific death rate for females. If B_1 is the number of female children born to the 15–30 group during the 15 years, then $b_1 = B_1/N_1$, since we are concerned only with births of females. Thus the b's and p's can be calculated directly from census and vital statistics data on deaths tabulated by age, births tabulated by age of mother, and number of individuals of a given age. Demographers would not, of course, use a time interval as long as 15 years, since birth and death rates may change quite rapidly during a 15-year period. Moreover, if, because of prevailing death rates, the number of people alive at the beginning of an interval is substantially different from that at the end, it is not clear what value of N_1 should be taken to calculate the death rate. One possibility, for example, is to take N_1 to be the mean of the numbers living at the beginning and at the end of the interval. Much of standard demography is, in fact, concerned with the refinements of adjusting age-specific birth and death rates, based on observed data tabulated over 1–5 year intervals, to smooth out the errors resulting from the grouping of the data into various time intervals. These details are, however, of little or no importance for understanding the basic principles governing the laws of population growth that are needed for the description of fitness differences. We consider 15-year intervals simply for the ease of numerical illustration.

A set of values of b_0, etc., and p_0, etc., for the female population of the United States in 1960 is given in Table 6.1. When arranged in the form given there, this set of values is called the **population projection matrix**, as it provides the basis for predicting the age distribution in any 15-year interval in terms of a given distribution for the previous 15-year interval. The matrix of Table 6.1 is a condensed version of one given by Keyfitz (1966). The numbers 1.37 and 0.465, for example, are the respective mean numbers of female children born per woman of ages 15–30 and

TABLE 6.1
Population Projection Matrix for United States Females, 1960

	Age, in Years			
	0–15	15–30	30–45	over 45
Age-specific birth rates	$b_0 = 0$	$b_1 = 1.370$	$b_2 = 0.465$	$b_3 = 0$
Survival probabilities	$p_0 = 0.992$	0	0	0
	0	$p_1 = 0.988$	0	0
	0	0	$p_2 = 0.964$	$p_3 = 0.880$

Source: Adapted from Keyfitz (1966).

30–45 during the stated 15-year interval. The number of women in these age groups can, most simply, be thought of as the number alive at the middle of the relevant interval. The number of births to women older than 45 years has been neglected. The numbers 0.992, 0.988, and 0.964 are the respective 15-year survival probabilities for the age groups 0–15, 15–30, and 30–45. These can most conveniently be thought of as the probabilities of surviving from the middle of one interval to the middle of the next interval, namely from ages 7.5 to 22.5, 22.5 to 37.5, and 37.5 to 52.5. The lower right-hand diagonal element, 0.88, is the proportion of people older than 45 years who survive for 15 years, or the probability of surviving from age 52.5 to 67.5. This term is included in order to obtain a correct representation of the changes in the number of people over the reproductive age. The application of the projection matrix for one cycle (equivalent to 15 years), starting with a population in which 40 individuals are in the 0–15 age group, 30 in the 15–30 age group and 20 in 30–45 age group, and 10 are older than 45 years is illustrated in Table 6.2. Thus the individuals of ages 0–15 in the subsequent 15-year interval will be those born to women of ages 15–30 years and 30–45 years in this time interval, namely $1.37 \times 30 = 41.1$ and $0.465 \times 20 = 9.3$ respectively. The individuals of age 15–30 in the subsequent interval will be the survivors from the 0–15 age group in this interval, namely, $0.992 \times 40 = 39.68$. Similarly, the 30–45 age group will be the survivors from the 15–30 age group, namely, $0.988 \times 30 = 29.6$, and the over 45 age group will be the survivors from the 30–45 age group plus the survivors from the over 45 age group, $0.964 \times 20 = 19.28$ plus $0.88 \times 10 = 8.8$. The subsequent changes in the population can be predicted by repeating this cycle of operations on the newly derived population (see Table 6.3). One cycle results in a drastic change in the population structure—that is, a leveling of the age distribution and a 50 percent increase in population size—because the age distribution of the initial population is very far from the equilibrium distribution, being weighted heavily towards the younger age groups that are, of course, more fertile. This type of rapid temporary change in population size and age structure may be expected whenever there is a sudden change in the patterns of mortality and fertility.

TABLE 6.2

Application of the Projection Matrix for United States Females, 1960, for one (15-year) Cycle. The projection matrix used is given in Table 6.1. Solid arrows indicate contributions of one age group to the next from survivors, dotted arrows indicate contributions to the births (0–15 age group of the next cycle)

Age Group	*Initial Population*	*Population After One Cycle*		*(in %)*
0–15	40	(1.37 × 30) + (0.465 × 20) =	50.4	34.1
15–30	30	0.992 × 40 =	39.7	26.8
30–45	20	0.988 × 30 =	29.6	20.1
Over 45	10	(0.88 × 10) + (0.964 × 20) =	28.1	19.0
Total	100		147.8	100

The significance of this for changes in the human population structure over the last one or two hundred years will be discussed further later in the chapter. The successive changes in the age structure of this model population over a series of 60-year intervals (each being four 15-year cycles) are given in Table 6.3. The percentages of the population in the various age groups stabilize very quickly and are, in fact, near the equilibrium distribution after only 60 years, or about 2 generations. The rate of population growth also stabilizes rapidly. Thus in the 12 cycles

TABLE 6.3

Successive Changes in Age Structure of a Population Subject to the Projection Matrix for U.S. Females 1960 (Table 6.1).

		Percent in Age Group					
Cycle[a]	*Years*	0–15	15–30	30–45	over 45	*Total Population Size*	*Population Growth per Cycle*[b]
1	0	40.00	30.00	20.00	10.00	100.0	
5	60	28.6	22.1	17.1	31.2	390.5	1.35
9	120	27.9	21.1	16.1	34.7	1198.1	1.32
13	180	27.7	20.9	15.9	35.5	3529.7	1.31
18	240	27.6	20.8	15.8	35.6	10315.6	1.31
Equilibrium		27.55	20.92	15.81	35.72	...	1.307

[a] one cycle = 15 years.
[b] Ratio of population size in given cycle to that in the previous cycle.

from the fifth to the seventeenth, the population grows by a factor of $10315/390 = 26.4$, while the expected increase based on a constant geometric growth rate of 1.31 is $1.31^{12} = 24.9$. The small discrepancy between the actual increase and that expected using the constant asymptotic growth rate of 1.31 per cycle is due to the slightly more rapid population growth during the earlier cycles, when there is still an excess of the population in the younger age groups. The annual rate of growth is $(1.307)^{1/15}$ or about 1.018 for the 1960 United States population. The natural logarithm of this annual rate of growth, $\log_e (1.307)^{1/15} = 0.0175 = r$ is called the **intrinsic rate of natural increase** of the population and is such that once the population has reached the stable-equilibrium age distribution it grows by a factor $e^{rt} = e^{0 \cdot 0175t}$ in t years, where e is the base of natural logarithms. The intrinsic rate of increase r provides the basis for measuring the magnitude of selective differences since it is a measure of the rate of growth of the population that incorporates the effects of all the factors affecting reproductivity. How long a population takes to approach its equilibrium distribution depends, of course, on how far from the equilibrium it is initially. However, as the numerical example just given shows, even when the initial distribution is quite far from the equilibrium, the approach is quite rapid.

An alternative way of obtaining the equilibrium distribution and the corresponding value of r, based on the theory of matrices, is outlined at the end of Example 6.1.

6.3 The Life Table and Lotka's Equation for the Intrinsic Rate of Increase

Given the net reproductive rate R_0 (the number of offspring per person) and intrinsic rate of increase r, the mean generation time T can be defined by the relation $R_0 = e^{rT}$.

The proportion of individuals who survive from birth to age x is given by

$$l_x = p_0 p_1 \cdots p_{x-1} = (1 - q_0)(1 - q_1) \cdots (1 - q_{x-1}), \tag{6.2}$$

where, as before, p_i is the probability of surviving from age i to $i + 1$ and $q_i = 1 - p_i$ is the age-specific death rate. This equation simply states that the probability of surviving from birth to age x is the probability of surviving from birth to the first age group times the probability of surviving from the first to the second age group, and so on, up to x. Consider now the fate of a **cohort** of births numbering N_0, that is, a group of N_0 individuals all born in a given year. The number surviving to age x will be $N_0 l_x$, and the number of births to the survivors of age x will be

$N_0 l_x b_x$, where b_x is the age-specific birth rate. The total number of offspring born to members of this cohort is therefore

$$\sum_{x=0}^{m} N_0 l_x b_x ,$$

summing over all years to the end of the reproductive period, and thus the number of offspring born per individual obtained by dividing this expression by N_0, is

$$R_0 = \sum_{x=0}^{m} l_x b_x ,$$

where m is the maximum attainable age, or the age at which reproduction ceases. The quantity R_0 is often called the **net reproductive rate** and is a measure of the rate of increase of the population *per generation* rather than per unit time. Thus if T is the length of a generation in years, r, R_0 and T are related by the equation

$$R_0 = e^{rT}, \tag{6.3}$$

since e^{rT} represents the proportionate growth of the population in one generation, assuming an equilibrium age distribution, and this by definition is R_0. As we shall see later, T can also be defined, for example, as the mean age at child bearing. The calculation of l_x and R_0 for the 1960 U.S. female population, based on the projection matrix given in Table 6.1, is illustrated in Table 6.4. The cohort is started at

TABLE 6.4
Survival Probabilities l_x and the Calculation of the Net Reproductive Rate R_0 Using the Projection Matrix for the 1960 U.S. Female Population (see Table 6.1)

Age Group	*Representative Age* (x)	*Age-specific Survival Probability* p_x	*Probability of Survival to Age x* (l_x)	*Age-specific Birth Rate* (b_x)	*Number of Births to Survivors of Age x* ($l_x b_x$)
0–15	7.5	0.992	1	0	0
15–30	22.5	0.998	0.992	1.37	1.36
30–45	37.5	0.964	$0.992 \times 0.998 = 0.980$	0.465	0.456
Over 45	—	0.88	$0.980 \times 0.964 = 0.945$	0	0

age 7.5 and so we assume $l_{7.5} = 1$, which is equivalent to assuming that "birth" is effectively at age 7.5. The net reproductive rate is $1.36 + 0.456 = 1.816$, which from Equation 6.3 gives a value of

$$T = \frac{\log_e R_0}{r} = \frac{0.597}{0.0175} = 34.1 \text{ years} \tag{6.4}$$

for the length of one (female) generation. This is somewhat longer than the generally assumed generation length of 29 years. The value given in Equation 6.4 is, however, very approximate because of the coarse age grouping and consequent rounding off of errors in the projection matrix as used. The generation time measured in this way should be approximately the same as the mean age at child bearing. If we assume that the characteristic age of an age group is at its midpoint, the mean age at child bearing for the data on the U.S. 1960 females is, from Table 6.4,

$$\frac{1.36 \times 22.5 + 0.456 \times 37.5}{1.36 + 0.456} = 26.3 \text{ years},$$

which again is subject to the large rounding off errors due to the coarse age grouping. Given T and R_0 Equation 6.3 also, of course, gives r as

$$r = \frac{\log_e R_0}{T}. \tag{6.5}$$

The stable age distribution reached by a population with constant age-specific birth and death rates can be simply represented in terms of l_x, the probability of surviving to age x, and r, the intrinsic rate of increase.

Suppose at the present time the number of new births in a given unit time interval is N_0. If the population has reached equilibrium the number of new births x years ago must have been

$$e^{-rx}N_0$$

since the population will have grown by a factor e^{rx} in x years. Given that l_x is the probability of surviving to age x, the number of these births that survived to the present is

$$N_0 e^{-rx} l_x,$$

and this must therefore be the number of people now living whose age is x. Thus in a stable population the proportion of people at age x is proportional to $e^{-rx}l_x$.

The application of this formula to the data obtained from the population projection matrix for 1960 U.S. females (Table 6.1) is illustrated in Table 6.5. Only the three age groups 0–15, 15–30, and 30–45 are considered. The seventh and eighth columns of this table show the close agreement between the results obtained by using the formula $e^{-rx}l_x$ and the equilibrium distribution obtained by repeated operation of the population projection matrix on the age distribution (Tables 6.2 and 6.3).

TABLE 6.5

Expected Age Distribution Based on the Equilibrium Formula $e^{-rx}l_x$ for the Proportion of People at Age x, using the 1960 U.S. Female Population Projection Matrix (Table 6.1)

		Computation of Age Distribution from Equilibrium Formula					*Computed Age Distribution from Population Projection (Table 6.3)*	
Age Group	*Representative Age* x	rx ($r = 0.0175$)	e^{-rx}	l_x	$e^{-rx}l_x$	*Relative Proportion*	*Relative Proportion Limited to* 0–45 *Yrs.*	*Observed Percent in all Age Groups*
0–15	7.5	0.131	0.877	1	0.877	0.427	0.429	27.6
15–30	22.5	0.394	0.676	0.992	0.670	0.326	0.325	20.9
30–45	37.5	0.656	0.519	0.980	0.508	0.247	0.246	15.8
Total					2.055	1.000	1.000	64.3

Note: The l_x values are taken from Table 6.4 and $r = 0.0175$ is based on the data given in Table 6.3, as described in the text. The observed percentage in age group (last column) is taken from the last row of Table 6.3. These numbers and the $e^{-rx}l_x$ are normalized to unity in the seventh and eighth columns for comparison.

Fertility is more important than mortality in determining stable age distributions.

The age distribution of a population is one of the most readily available demographic parameters, being a direct product of any population census. It is of some interest to consider the effect of different age-specific birth and death rates on the age distribution, assuming it is given by $e^{-rx}l_x$ as derived in the previous section for a stable population. In general, differences in mortality have relatively little effect. Changing patterns of mortality in the last 100 or 200 years have affected both individuals below and those above the reproductive age to about the same extent, and in the same direction, thus tending to eliminate any effect of change on the age structure of the population. A decrease in mortality only of individuals below reproductive age tends to lower the age distribution of the population; a decrease in mortality only of older persons tends to raise the age distribution. Relatively constant increases or decreases in mortality affecting persons of reproductive age have very little effect on the age distribution. Changes in the fertility rates, on the other hand, may have a profound effect on the age distribution. Decreasing fertility raises the age distribution. This has probably been the major factor changing the age distribution of Western industrial societies during the first half of this century.

An example of the effects of different fertility and mortality rates on an age distribution are illustrated in Figure 6.1. This shows the expected stable age distributions for the Swedish population for combinations of the observed mortality rates and fertility rates in the middle of the nineteenth and middle of the twentieth centuries. The differences in mortality have a second-order effect in comparison with those in fertility. Thus 1950 fertility rates give almost the same age distribution

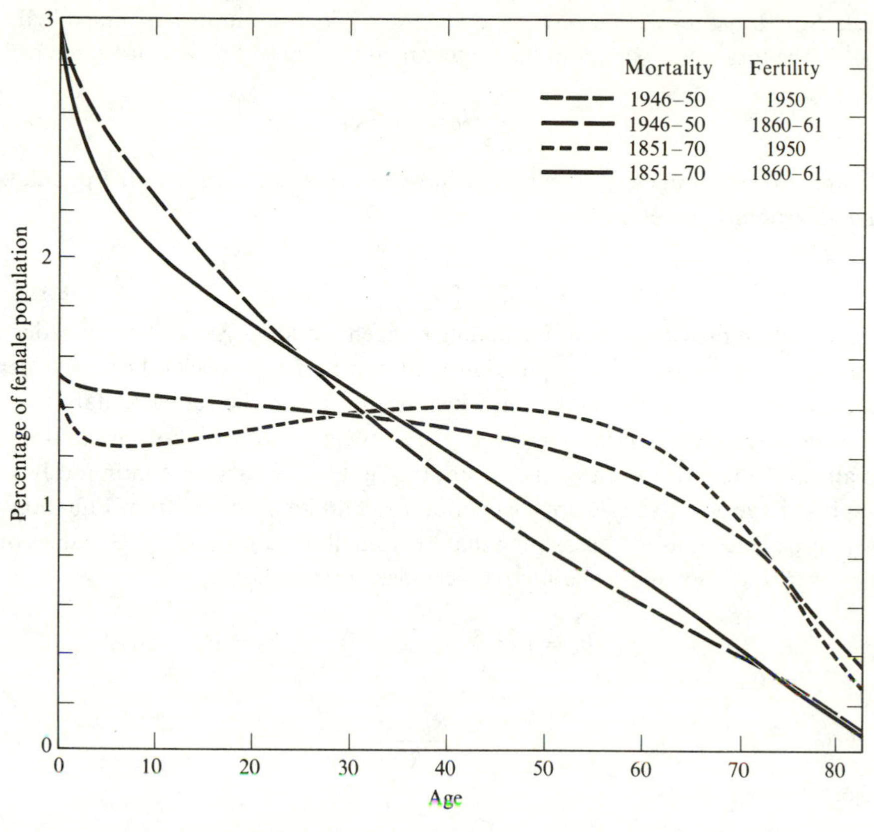

FIGURE 6.1
Stable age distributions, Sweden. (From Coale, 1957.)

with 1946–50 as with 1851–70 mortality rates, and similarly for 1860–61 fertility rates with these same two sets of mortality rates. Transitory changes in fertility will, of course, cause a departure from the distribution expected assuming stability, resulting in "humps and hollows" in the age distribution. Many of the humps and hollows in presently observed age distributions can be traced to unusual birth rates, such as that immediately after the Second World War. (See Coale, 1957, for further discussion.)

The intrinsic rate of increase r can be obtained as the solution of Lotka's fundamental equation

$$\sum_{x=0}^{m} e^{-rx} l_x b_x = 1.$$

The number of births in a unit time interval contributed at the present time by people of age x is

$$N_0 e^{-rx} l_x b_x$$

since b_x is the age-specific birth rate and $N_0 e^{-rx} l_x$ the number of individuals of age x. The total births in a unit time interval, at the present time, must therefore be

$$\sum_{x=0}^{m} N_0 e^{-rx} l_x b_x,$$

summing over all ages. By definition, however, this is N_0 in a stable population and so we must therefore have

$$\sum_{x=0}^{m} e^{-rx} l_x b_x = 1. \tag{6.6}$$

This equation provides a basis for finding r, given l_x and b_x. Lotka's original derivation of the properties of a population with constant age-specific birth and death rates was in terms of this equation, which is considered the fundamental equation of demography. (It is usually given in the "integral" form, appropriate for very small time intervals, when l_x and b_x are thought of as varying continuously with respect to age x.) A simple approximation to r can be obtained from Equation 6.6 by approximating to e^{-rx}, assuming that rx is small. If rx is small for all values of x, then $e^{-rx} \sim 1 - rx$ and Equation 6.6 becomes, approximately,

$$\sum_{x=0}^{m} (1 - rx) l_x b_x = 1 = \sum_{x=0}^{m} l_x b_x - r \sum_{x=0}^{m} x l_x b_x = R_0 - rR_1,$$

giving

$$r = \frac{R_0 - 1}{R_1}, \tag{6.7}$$

where

$$R_0 = \sum_{x=0}^{m} l_x b_x$$

is the net reproductive rate, as defined before, and

$$R_1 = \sum_{x=0}^{m} x\, l_x b_x .$$

The quantity

$$\frac{R_1}{R_0} = \frac{\sum_{x=0}^{m} x l_x b_x}{\sum_{x=0}^{m} l_x b_x} \tag{6.8}$$

is the mean age at child bearing, since

$$l_x b_x \Big/ \sum_{x=0}^{m} l_x b_x$$

is the proportion of children that are born to individuals at age x. Thus R_1/R_0 is approximately equal to the generation time T, as was pointed out earlier. From the data given in Table 6.4,

$$R_1 = 1.36 \times 22.5 + 0.456 \times 37.5 = 47.7,$$

giving, approximately,

$$r = \frac{R_0 - 1}{R_1} = \frac{1.816 - 1}{47.7} = 0.0171,$$

which compares well with the correct value of 0.0175 obtained by iterating the population projection matrix.

6.4 The Use of r for the Measurement of Selective Differences

Evidence for selective differences between genotypes can be obtained from departures from Mendelian segregations and from associations with diseases or with geographic distributions, but most directly from differences in fertility and mortality as indicated by the intrinsic rate of increase.

There are many ways in which evidence for selective differences between genotypes can be obtained, at least in principle. Associations between specific diseases and the genotypes may be looked for, as with the blood groups discussed earlier. Association between the geographic distributions of genotypes and environmental differences may be found, as for sickle-cell trait, G6PD, and thalassemia and malaria. These associations, however, only indicate possible causes of a selective difference and do not give a measure of its magnitude. Departures from expected Mendelian segregation, as discussed for the ABO blood groups, may also provide evidence for the existence of selective differences due to mortality. However, in the absence of any direct observation of a change in gene frequency, the most satisfactory way to detect and measure selective differences is by calculation of the intrinsic rate of increase for the various genotypes, as obtained from their respective age-specific mortality and fertility distributions. Although even this approach is only an approximation to reality, since it is calculated only for females and depends on the assumption of stable age distributions in the population based on constant age-specific birth and death rates, it is the most complete and easily applicable description of selective differences that is at present available. Measuring only mortality or only fertility, as is often done by population geneticists, may be quite misleading especially if the age structure of the population is ignored, and cannot, in general, lead to proper estimates of selective differences. It is always, of course, true that the only forces which can be measured are those *currently acting*, and these may not be representative of the forces that prevailed during the major part of the evolution of man. In addition, measurement of the intrinsic rate of increase, assuming a stable age distribution, is subject to error in times when there are major overall changes in mortality and fertility patterns not related to genotypic differentials. Nevertheless, the demographic background given in some detail in the previous sections of this chapter is essential for the proper measurement of selective differences in human populations and clearly indicates the type of data needed to measure these differences. A lack of appreciation of these needs may severely limit the value of otherwise comprehensive studies on particular genetic traits. The need for a complete specification of reproductivity, rather than just a description of, for example, mortality, was emphasized earlier in our discussion of compensation as a factor possibly needed to explain the maintenance of the sickle-cell-trait polymorphism.

The net reproductive rate R_0, which is the average number of offspring per person, is a satisfactory measure of selection if there are no differences in the generation time T. Two genotypes with the same R_0 but different values of T clearly have different fitnesses, since from Equation 6.5 the genotype with the smaller generation time has a higher intrinsic rate of increase. To relate the intrinsic rate of increase to the usual measures of selection per generation used in the simple population-genetics models discussed in earlier chapters, fitness must be measured by $e^{r_i \overline{T}}$, where $\overline{T}$ is an average generation time for the whole population and r_i is the intrinsic rate of increase of some particular subsection of this population. The relative fitness of two genotypes will then be measured by

$$w = \frac{e^{r_1 \overline{T}}}{e^{r_2 \overline{T}}} = e^{(r_1 - r_2)\overline{T}}, \tag{6.9}$$

where r_1 and r_2 are the two respective intrinsic rates of increase and $\overline{T}$ is either the mean generation time of the population from which the genotypes were drawn or the mean generation time of the two genotypes. Small differences in $\overline{T}$ will not have much effect on the relative fitness. When $(r_1 - r_2)\,\overline{T}$ is small, Equation 6.9 gives, approximately,

$$w = 1 + (r_1 - r_2)\overline{T}. \tag{6.10}$$

There are, unfortunately, so far no genetic traits recognizable at birth (apart from sex) for which enough detail has been published to calculate properly an intrinsic rate of increase.

In some instances knowledge only of the mortality distribution of a trait may be adequate for the calculation of r on the assumption that the fertility distribution of the particular genotype in question is the same as that of the general population.

Let l_x and b_x be, respectively, the probabilities of survival to age x and the age-specific birth rates of the total population, including the genotype or trait in question. Then it can be shown (see Example 6.3) that an approximate measure of the selective effect of the trait is given by

$$R_0^0 - R_0 = \sum_{x=0}^{m} l_x b_x \delta_x, \tag{6.11}$$

where R_0^0 is the net reproductive rate of that part of the population which does not have the trait, R_0 is the net reproductive rate of the whole population, and δ_x is the probability of having died from the trait by age x. Only this latter quantity relates specifically to the effect of the trait. The quantities l_x and b_x can be obtained from standard sources of information on the general population.

The selective difference between blood groups O and A, resulting from the association between them and duodenal ulcers, is 6.4×10^{-6} *for females and* 9.5×10^{-5} *for males.*

Let us consider the application of the above formula to a specific disease. The correlation between blood group O and duodenal ulcer is one of the most-widespread and best-established associations between a disease and a blood group (see Chapter 5). To measure the selective difference implied by this association, we must first consider the fitness of people who get duodenal ulcer. The calculations required for doing this are outlined in Table 6.6. The quantities l_x, q_x, and b_x are taken from a standard life table (that is a table of l_x and q_x values) and from tables of total births per year to people of given age x and of the total number of people

TABLE 6.6

Calculation of the Effect of Duodenal Ulcer on the Fitness of a Human Population (Based on Italian Census and Life Table Data for Females, 1953)

Age Group	l_x	b_x	k_x	q_x	$q'_x = k_x q_x$ $\times 10^{-6}$	$\delta_x = \Sigma_{i=1}^{x} q'_i$ $\times 10^{-6}$	$l_x b_x \delta_x$ $\times 10^{-6}$
15–19	0.937	0.0172	0.0023	0.00285	5.795	5.80	0.94
20–24	0.934	0.192	0.00265	0.00395	8.15	13.95	0.28
25–29	0.930	0.34	0.00196	0.00520	10.2	24.15	7.64
30–34	0.925	0.3792	0.00236	0.00655	15.5	39.65	13.91
35–39	0.918	0.1465	0.00225	0.00900	20.0	59.65	0.08
40–44	0.909	0.0867	0.00338	0.01180	39.7	99.45	7.84
45–49	0.896	0.0122	0.0044	0.01770	77.7	177.15	1.94
							40.57

Note: l_x and q_x come from a standard Italian life table: l_x refers to survival to the mid-point of the relevant age group and q_x is the probability of death in the five-year interval following this mid-point. b_x is the Italian age-specific female birth rate for 1953, given in 5-year intervals, calculated as

$$\frac{\frac{1}{2}(\text{number of births to people of age } x \text{ in a given year})}{\text{total number of people of age } x}$$

k_x is the ratio of the number of deaths from duodenal ulcer among females of age x to the total number of deaths among these females. $q'_x = k_x q_x$ is the probability of dying from duodenal ulcer in the relevant time interval. The other quantities are calculated as indicated:

$$R_0^0 - R_0 = \sum_{i=1}^{n} l_x b_x \delta_x = 40.6 \times 10^{-6}.$$

of age x. The proportion of people k_x dying at age x because of duodenal ulcer is taken from data on deaths due to duodenal ulcer in a given time period and total deaths, obtained from vital statistics, in the same time period. The quantity $q'_x = k_x q_x$ is the probability of dying *at* age x from duodenal ulcer; the probability of dying from the disease *by* age x is, approximately,

$$\delta_x = \sum_{i=1}^{x} q'_i .$$

The other calculations follow as indicated to give $R_0^O - R_0 = 40.6 \times 10^{-6}$. For the life table data given, R_0 $(=\sum l_x b_x)$ is about 1.09, so the effective selective effect of duodenal ulcer in females is to reduce the overall fitness of the population by an amount that is, approximately,

$$\frac{40.6 \times 10^{-6}}{1.09} = 38.5 \times 10^{-6}.$$

The relative incidence of duodenal ulcer in females having type O blood as compared to the other ABO blood types is 1.4 to 1. Thus the proportion of all duodenal ulcers that occur in females of type O is $1.4/(1 + 1.4) = 0.583$, and in females of other types 0.417. The reduction in the selective value of type O females as compared with non-O caused by the association with duodenal ulcer is, therefore,

$$(0.583 - 0.417) \times 38.5 \times 10^{-6} = 6.4 \times 10^{-6},$$

which is a very small selective difference, comparable in magnitude to mutation rates. The male selective difference, calculated in exactly the same way, is 9.5×10^{-5}, which is still much too small to be of any great significance in counteracting incompatibility selection. Two main factors contribute to these low selective values, first, the low overall incidence of duodenal ulcer, and second, the fact that it is primarily a disease affecting people beyond the reproductive age (see T. E. Reed, 1961). It should be pointed out that we have assumed people with duodenal ulcer have a normal age-specific birth rate during their total lifetime. Any reduction in fertility of people who have duodenal ulcer will tend to increase, somewhat, this estimate of selective differences but the effect is not likely to be very large. It is remarkable that such small selective differences associated with a given disease can be measured satisfactorily. This is due to the availability of information on the proportion of deaths at a given age that are due to duodenal ulcer.

Late age of onset of a genetic disease poses special problems for the estimation of fitness, since individuals with the "abnormal" genotype who have not yet manifested the disease are necessarily included among the normal part of the population.

Huntington's chorea is an example of a dominant defect with a mean age of onset (defined by the first appearance of "choreic" movements) of approximately

35 years (Reed and Chandler, 1958). The choreics who are younger than about 30 years are a relatively small fraction, probably less than 25 percent, of all the persons who are heterozygous for the chorea gene. If we assume that heterozygotes have a normal mortality and fertility pattern before they contract the disease, then a plausible model for estimating their fitness in terms of observable parameters can be constructed. Normal fertility (b_x) and mortality (l_x) rates are taken from the general population from which the choreics are drawn. We require, in addition, the probability χ_x that a heterozygote first develops the disease at age x (obtained from the distribution of the age of onset of the disease) and also the age-specific fertility and mortality rates of people who have already developed Huntington's chorea. Ideally these rates should be obtained as a function of the age of onset of the disease, but in practice there will usually not be enough data for such a detailed breakdown, and so it will have to be assumed that these age specific rates (l'_x and b'_x) are independent of the age of onset of the disease. It can then be shown (see Example 6.4) that the number of children born to surviving heterozygotes of age x in the interval x to $x + 1$ is

$$p'(x) = l_x b_x \prod_{i=1}^{x-1} (1 - \chi_i) + l'_x b'_x \sum_{r=1}^{x} \frac{l_r}{l'_r} \chi_r \prod_{i=1}^{r-1} (1 - \chi_i)$$

$$= l_x L_x b_x + l'_x b'_x \sum_{r=1}^{x} \frac{l_r}{l'_r} \chi_r L_r, \tag{6.12}$$

where the first term is the contribution from those heterozygotes who have not yet developed chorea by age x and the second term is the sum of the contributions from those heterozygotes who developed chorea at all ages up to age x. The quantity

$$L_x = \prod_{i=1}^{x-1} (1 - \chi_i)$$

is the probability of not yet having developed the disease by age x. The net reproductive rate of choreics is then

$$\sum_x p'(x)$$

and, from Equation 6.6, their intrinsic rate of increase is the solution of the equation

$$\sum_{x=0}^{m} e^{-rx} \; p'(x) = 1 \tag{6.13}$$

Huntington's chorea provides a useful example of a dominant genetic disease with a relatively late age of onset.

An extensive study on Huntington's chorea has been published by Reed and Chandler (1958) and Reed and Neel (1959). Though the data, as published, are not adequate for a complete estimate of fitness in terms of the intrinsic rate of increase, they do provide information that can easily be used to set upper and lower limits on the fitness of choreics. Thus a simple upper bound for the fitness can be obtained if it is assumed that $b'_x = 0$, that is, that heterozygotes are no longer fertile once they have contracted the disease. A lower bound is obtained if χ_x is the relative probability of death at age x rather than the probability of onset of the disease at age x, in which case $l'_x = 0$ and $l_x L_x$ is simply the probability that choreics survive to age x. In either case, the second term is zero. For the upper bound we are neglecting the fertility of heterozygotes after they contract the disease; for the lower bound we ignore the debilitating effect of the disease on fertility. A third, intermediate, estimate can be obtained if we assume that χ_x is the probability of institutionalization at age x and that fertility stops at institutionalization, being normal (b_x) beforehand. Reed and Chandler (1958) give the distribution of the age at onset, the age at death, and the age at institutionalization of a sample of choreics from Michigan, culled around 1940. Using this information and l_x and b_x values obtained from U.S. Vital Statistics of 1940, we can calculate the three sets of $l_x L_x b_x$ values, from which the net reproductive rates can be obtained by simple summation. The results of these calculations are given in Table 6.7. When the intrinsic rates

TABLE 6.7
Upper and Lower Bounds to the Selective Disadvantage of People with Huntington's Chorea.

	Net Reproductive Rate (R_0)	*Approximate Selective Disadvantage (in Percent)*
U.S. 1940 Total Population	1.0485	—
Choreics		
Assuming fertility stops at onset of disease	0.7854	26.3
Assuming fertility stops at institutionalization	1.0235	2.5
Assuming fertility is normal until death	1.0308	1.8

Source: Calculated using data from Reed and Chandler (1958) and Reed and Neel (1959).

of increase are small and generation lengths are approximately the same for choreics as for normals, the selective disadvantage of choreics is given, approximately, by the difference between the net reproductive rate of the total population and that of choreics. This follows from Equation 6.7, giving

$$r = \frac{R_0 - 1}{R_1} \sim \frac{R_0 - 1}{T}, \tag{6.14}$$

when R_0 is near 1, and so $R_1 \sim T$, and from Equation 6.9, giving the selective difference between types 1 and 2 as

$$w \sim 1 + (r_1 - r_2)T \sim 1 + (R_0^{(1)} - R_0^{(2)}), \tag{6.15}$$

since, from Equation 6.14, $rT \sim R_0 - 1$. Thus the selective disadvantage of type 2 relative to type 1 is approximately $R_0^{(1)} - R_0^{(2)}$, where $R_0^{(1)}$ and $R_0^{(2)}$ are the net reproductive rates of types 1 and 2, respectively. The small disadvantage consequent to the second and third assumptions of Table 6.7 is due to the relatively late age at institutionalization (average about 49 years) and the small effect of Huntington's chorea on mortality before the end of the reproductive age. Using their own data on the fertility of choreics, Reed and Neel (1959) derive an estimate of 19 percent as the selective disadvantage of the Huntington's chorea gene for heterozygotes.

Schizophrenia is a disease with a mean age of onset between 25 *and* 30 *years and thus centered in the midst of the reproductive period.*

The relatively high incidence of the disease (about one percent), its severity, and the fact that its onset tends to be at the prime of life make it one of the major public health problems of our time. There is evidence for a strong genetic component in the determination of the disease, as will be discussed more fully in Chapter 9, though more than one gene is most probably involved.

Studies on the fitness of schizophrenics indicate that the major effect of the disease is through a reduction of the probability of marriage after onset, starting mainly during the period between onset and first admission to a mental institution (see, for example, Goldfarb and Erlenmeyer-Kimling, 1962). Assuming that fertility and mortality are normal before onset or institutionalization, but that fertility is essentially zero afterwards, we can apply Equation 6.11 to the calculation of the fitness of schizophrenics as we did to that of Huntington's choreics.

A reduction in the average age of child bearing, as has occurred over the last 30–40 years, may be expected to have a marked effect on the fitness of individuals having a trait like schizophrenia that has an onset in the middle of the reproductive period. Estimated net reproductive rates, calculated using Equation 6.11, and consequent approximate selective disadvantages of male and female schizophrenics, assuming in turn 1940, 1950, and 1960 United States birth-rate distributions, are shown in Table 6.8 (from Bodmer, 1968). Data on the distribution of ages of male and female schizophrenics at admission to an institution (Rosenthal, 1966) were used to calculate, for each sex, L_x, the probability of not having developed the disease by age x (see Example 6.4). The same L_x values, which differ appreciably between the sexes, were used with the various birth-rate distributions and with a constant value of $l_x = 0.94$, for all reproductive ages, to obtain the estimates given in Table 6.8. Age-specific birth rates for males are not readily available from standard sources of statistics. It was therefore assumed that the b_x values for females

TABLE 6.8

Net Reproductive Rates and Approximate Selective Disadvantages of Schizophrenics Assuming 1940, 1950, *and* 1960 *U.S. Birth-rate Distributions Apply Before Onset of the Disease*

	Net Reproductive Rates[a] ($R_0^{(2)}$)			*Selective Disadvantages*[b]		
	1940	1950	1960	1940	1950	1960
Males	0.56	0.77	0.92	0.86	0.61	0.51
Females	0.76	1.04	1.25	0.40	0.27	0.22

[a] Calculated using Equation 6.12, assuming $b_x' = 0$.

[b] Calculated using Equation 6.18.

Source: Bodmer (1968): b_x values for females for the indicated years from *Vital Statistics of the United States*, 1963, Vol. 1; b_x^* for males was calculated using Equation 6.16.

applied to males aged $x + 2.5$ since on an average males are 2.5 years older than females at time of marriage. Thus, for 5-year intervals, the age-specific birth rate for males b_x^* was given in terms of that for females b_x by

$$b_x^* = \tfrac{1}{2}(b_x + b_{x-1}). \tag{6.16}$$

Selective disadvantages are calculated using a combination of Equations 6.10 and 6.7. Thus from Equation 6.10, the selective disadvantage s is given by

$$s = (r_1 - r_2)\overline{T}. \tag{6.17}$$

If the two generation times corresponding to r_1 and r_2 are both equal to T, then, using Equations 6.17 and 6.7, we have

$$s = (r_1 - r_2)T = \left[\frac{R_0^{(1)} - 1}{R_1^{(1)}} - \frac{R_0^{(2)} - 1}{R_1^{(2)}}\right]T$$

$$\sim \frac{R_0^{(1)} - R_0^{(2)}}{R_0^{(1)}R_0^{(2)}} \tag{6.18}$$

since the generation time T is from (6.8) approximately given by

$$T = \frac{R_1^{(1)}}{R_0^{(1)}} = \frac{R_1^{(2)}}{R_0^{(2)}}. \tag{6.19}$$

Equation 6.18 approximates to 6.15 when $R_0^{(1)}$ and $R_0^{(2)}$ are both near one. For the data in Table 6.8, $R_0^{(1)}$ refers to the normal population and is based on the relevant b_x values with $l_x = 0.94$ for all values of x; $R_0^{(2)}$ refers to the schizophrenics.

The overall fitness of male schizophrenics is much lower than that of females. This may reflect an earlier age of onset of the disease in males, or a tendency for males to be institutionalized at an earlier age because of social pressures, as well as average age differences between males and females at time of marriage. The female values agree reasonably with those obtained by Erlenmeyer-Kimling and Paradowski (1966) in a direct study of the fitness of schizophrenics, while the male values are a little higher. There is an almost two-fold decrease in both sets of selective disadvantages from 1940 to 1960. This is the effect expected from a trend toward an earlier average age of child bearing in the general population. It represents a striking example of the effect of an overall demographic change on the fitness of a particular trait. The model we have used to calculate Table 6.8 is, of course, a simplification, which ignores, in particular, fertility of schizophrenic females after onset, and also that, even before the onset of the disease, a smaller proportion of male schizophrenics than of normal males marry. The increasing use of drugs to mitigate the effects of schizophrenia, may, of course, accentuate still further the decrease in the selective disadvantage of schizophrenia. As the interpretation of the effect of these changes on the frequency of schizophrenia depends on assumptions concerning the genetic basis of the disease, it will be deferred to Chapter 9.

The relationship between IQ and fitness has been accurately measured but IQ is, of course, not a simply inherited character.

The relationship between intelligence and fertility has been discussed for many years, particularly since Galton's conjecture at the turn of the last century that the decline of the ruling classes was a result of their intrinsically lower fertility. In fact, Fisher (1930) founded a theory of the decline of human civilizations on the basis of such observations. Intelligence, usually measured by some form of IQ test, is not, of course, a simply inherited character. Its expression must be under the influence of many genes as well as being strongly molded by the environment. Approaches to the study of the genetics of such characters will be discussed in Chapter 9. At this point we need only emphasize that any differences in reproductivity between people with different IQ's have a *genetic* effect only in so far as IQ is genetically determined. There can be little doubt that socioeconomic factors highly correlated with IQ, such as educational status, have during at least the first half of the twentieth century had a remarkable effect on patterns of reproductivity in industrialized societies, with a generally inverse relationship between IQ and fitness. This pattern is, however, changing rapidly. We do not know how long it existed and we may doubt whether it has had any long-term effect on the average intelligence of the human population, as has been suggested by some writers (see Chapter 9). Bajema in 1963 published a study in which intrinsic rates of increase of a sample of the United States population were calculated as a function of IQ. The data are based on individuals from Kalamazoo born in the years 1916 and 1917, whose fertility was completed at the time they were interviewed. IQ's were measured at an average age of 11.6 years. The intrinsic rates of increase, generation lengths, and derived fitnesses are given for five IQ ranges in Table 6.9. As has previously

TABLE 6.9
The Relationship Between IQ and Fitness

IQ Range m	*Intrinsic Rate of Natural Increase* r_m $\times 10^{-3}$	*Average Generation Length* T_m	*Relative Fitness Measured by* $e^{r_m T}$ *where* T *is the Average Generation Length*
≥ 120	8.9	29.4	1.00
105–119	3.9	28.9	0.87
95–104	0.3	28.4	0.78
80–94	7.5	28.0	0.96
69–79	−10.0	28.8	0.58
Total sample	3.9	28.5	—

Source: Based on Bajema (1963).

been found, the lowest IQ range, which includes mainly those people usually classified as definitely subnormal, has the lowest fitness, due mainly to lower fertility. This estimate is, however, based on only 30 individuals. Apart from this group, the fitness-IQ relationship is U-shaped with high fitnesses at low and high IQ's and lower fitnesses at intermediate IQ's. It is worth noting that the relatively high intrinsic rate of increase of the >120 IQ group is partly offset by their longer-than-average generation length. The observed relationship between IQ and fitness is not the simple inverse one often claimed; it undoubtedly reflects recent changes in overall reproductivity patterns. The implications of a relationship between IQ and fitness for changes in the average IQ of a population, and the problems of interpreting such data will be discussed in more detail in Chapter 9, after we have reviewed the evidence for a genetic component in the determination of IQ.

6.5 The Distribution of Progeny Size

Fertility, as we have already emphasized, is one of the major components of biological fitness. It is measured, crudely, by the mean progeny size after the reproductive period has finished. There are, however, at least two important properties of populations that depend on the variance in the number of progeny, as well as on the mean, (or, more generally, on the overall shape) of the distribution of progeny size.

1. The scope for the action of natural selection and, in fact, the rate at which evolutionary changes take place depend, as we shall discuss later, on the variance in fitness. A major component of this is the variance in progeny size.

2. The effects of random variation in gene frequencies due mainly to finite population size, which will be discussed in Chapter 8, depend on this variance.

In addition, the solutions of some problems in the analysis of human pedigree data depend on assumptions concerning the distribution of progeny size (see Appendix II).

If all women had the same constant chance of bearing children over a given fixed time period, births would be distributed randomly throughout this period and the distribution of their number would be Poisson.

The Poisson distribution is defined by the equation

$$p_r = \frac{m^r}{r!} e^{-m}, \tag{6.20}$$

where $r! = 1 \times 2 \times 3 \cdots r$, ($r! = r$ factorial), e is the base of natural logarithms, and p_r is the probability of r random events, in this case births, occurring in a given time interval. The mean of the distribution is m, its only parameter (see statistical appendix). The principal feature of the Poisson distribution is that its mean m is equal to its variance.

Observed distributions of progeny size cannot, in general, be fitted to a Poisson distribution. They usually have variances that are from 1.5 to 3 times as large as their means. This is hardly surprising since all women clearly do not have the same constant probability of bearing children, which would be the major assumption of an application of the Poisson distribution to this example. Most differences, such as in the probability of marriage, the age at which marriage is contracted, the duration of marriage, and natural fecundity, tend to increase the variance in the number of progeny. Family planning that specifies a desired family size may, however, tend to decrease the variance. Later in this chapter we shall discuss some computer-simulation models that take into account all these various factors.

The negative binomial distribution generally provides a good fit to observed progeny size distributions.

The negative binomial is defined by the equation

$$p_r = \frac{(n + r - 1)!}{r!(n - 1)!} q^{-n}\left(\frac{p}{q}\right)^r, \tag{6.21}$$

where $q = 1 + p$. It depends on the two parameters n and p (or q) and can be shown to be equivalent to a particular type of mixture of Poisson distributions with different means. It presumably gives a good fit because it allows for variation in fertility between women. The mean of the negative binomial distribution is np, the variance is $np(1 + p)$, and the ratio of the variance to the mean is thus $(1 + p)$. The negative binomial tends to approach the Poisson distribution as p becomes small and n large.

A comparison of the fit of the negative binomial and Poisson distribution to a progeny-size distribution obtained from the Msambweni district of Kenya in East Africa is shown in Table 6.10 (data from Brass, 1958). An observed progeny-size

TABLE 6.10
Comparison of Observed and Fitted Distributions of Progeny Size

	Number of Women			
			Expected	
Number of Births	*Observed*		*Negative Binomial*	*Poisson*
0	29		22.4	10.6
1	41		46.4	35.6
2	48		57.1	60.0
3	53		54.1	67.4
4	53		43.6	56.7
5	33		31.5	38.3
6	22		21.0	21.5
7	11		13.2	10.3
8	7		7.9	4.4
9	7		4.5	1.6
≥ 10	3		5.3	0.6
Total	307	Parameters:	$n = 5.37$	$m = 3.37$
Mean	3.37		$p = 0.629$	
Variance	5.49			
χ^2(8 df)			9.1	52.9
			$(0.3 < P < 0.5)$	$P \ll 01\%$

Source: Brass (1958): Data from Msambweni, East Africa.

distribution, to be meaningful, must relate to *completed* fertility. It must therefore be obtained from women who have completed their reproductive period, that is who are generally older than 40-45 years. The "observed" column of the table gives the numbers of women (over age 35) who have had the corresponding number of births shown in the first column. The third and fourth columns show the expected numbers, based respectively on the negative binomial distribution (Equation 6.20) and the Poisson distribution (Equation 6.19). The latter gives a very poor fit, with too narrow a distribution, as expected from the fact that the observed ratio of variance to mean is 1.63. The negative binomial, on the other hand, gives a very good fit to the observed data.

A complication that frequently arises in fitting observed progeny-size distributions is that women who bear no children fall into two categories: those who never married, and those who married but had no children. The first term of the expected distribution applies only to the latter group. When data distinguishing the two categories are not available it is convenient to fit the negative binomial only to the observed numbers of women with one or more births, excluding the zero class. This is done by dividing all the expectations based on Equation 6.21 by $1 - p_0$, to give what is called a **truncated distribution**, and, using the method of maximum likelihood (see Appendix I), to obtain estimates of n and p. Estimates of n and p and the corresponding means and variances for a representative set of countries are shown in Table 6.11. The ratio of the variance to the mean is near 2 for a number of countries that have quite different mean numbers of births. This shows, as we shall discuss later, that the variation in progeny size is not necessarily correlated with large mean progeny sizes.

TABLE 6.11
Estimates of the Negative Binomial Parameters (n and p) and Derived Means and Variances for a Representative set of Countries

Country	*Year*	n	p	*Mean* (np)	*Variance* $(np(1+p))$	$\frac{\text{Variance}}{\text{Mean}}$ $(1+p)$
Brazil	1940	3.61	1.98	7.2	21.3	2.98
Venezuela	1950	2.82	2.13	6.0	18.8	3.13
Ceylon	1946	5.45	1.01	5.5	11.1	2.01
Japan	1950	5.60	0.92	5.2	9.9	1.92
Czechoslovakia	1930	4.20	0.94	3.9	7.7	1.94
Italy	1931	5.01	1.06	5.3	10.9	2.06
England and Wales	1951	3.37	0.60	2.0	3.2	1.60
United States	1950	2.84	0.93	2.6	5.1	1.93

Source: Data from United Nations Demographic Year Book, 1954, 1955, and Brass (1958).

6.6 The Size of Investigations Needed to Measure Selection Intensity

Direct determination of mortality differences as departures from expected Mendelian segregations requires very large bodies of data for the measurement of small selective differences.

Since small selective differences cause very minor departures from the expected Mendelian segregations, very large numbers of observations are needed to detect selective effects in this way. It can be shown that the number of progeny n that must be observed in order to detect a difference between expected ratios of $u_1:1$ and $u_2:1$ is, on the average for a given significance level α, given by

$$n = \frac{\chi_\alpha^2(2 + u_1 + u_2)(u_1 + u_2 + 2u_1u_2)}{2(u_1 - u_2)^2}, \tag{6.22}$$

where χ_α^2 is the value of χ^2 for one degree of freedom at the α percent point (see statistical appendix and Mather, 1951). Suppose, for example, we wish to contrast matings ♀ $AO \times OO$ ♂ and ♀ $OO \times AO$ ♂ in order to detect incompatibility selection, as discussed in Chapter 5. In the absence of selection we expect each mating to give $\frac{1}{2}$ of type A and $\frac{1}{2}$ of type O offspring. Given incompatibility selection of magnitude s, we expect $AO \times OO$ still to give $\frac{1}{2}$A:$\frac{1}{2}$O and so $u_1 = 1$; $OO \times AO$ gives $\frac{1}{2}(1 - s)$A:$\frac{1}{2}$O and so $u_2 = 1 - s$, using the notation of Equation 6.22. We wish to know how many offspring from each type of mating are needed in order to detect a discrepancy between ratios of 1:1 and $1 - s:1$ for any given, possibly quite small, value of s. From Equation 6.22 we have, substituting $u_1 = 1$ and $u_2 = 1 - s$,

$$n = \frac{\chi_\alpha^2(4 - s)(4 - 3s)}{2s^2} \sim \frac{8\chi_\alpha^2}{s^2}$$

when s is small, where now $n/2$ is the number of offspring needed from each mating. When $\alpha = 95$ percent, $\chi^2 \sim 4$. Thus, for example, when $s = 0.01$, $n = 320{,}000$ is the number of offspring needed to detect, on an average, a one percent disadvantage of type A in ♀ $OO \times AO$ ♂ matings at a 95 percent significance level. When $s = 0.1$, then $n = 3200$.

No published study is available that has been done with enough people to detect selective differences of the order of one percent. Few have been done with enough people to detect selective values as large as even 10 percent. The data on the ABO blood groups, which we discussed in Chapter 5, are probably the most comprehensive, and, nevertheless, do not usually give a clear-cut answer to the question of whether there are consistent selective effects. The methods we have used in Section 6.4 can, however, supply, on rare occasions, estimates of very low selection coefficients because they make use of observations made on very large populations, like that of a whole country. As such data may have to be obtained from official statistics, they suffer from the uncertainties associated with such sources of information, as will be discussed later in this section. It is perhaps questionable whether enough *homogeneous* data can ever be collected to detect selective coefficients of about one percent by departures from Mendelian expectations. Even if a selective difference were detected with a large body of data, it would, of course, only pertain to the time of the observations and would not be a complete estimate of fitness, since the data would not take into account the fertility differences that, through compensation, might be a very important factor.

Small selective differences will always be very difficult to determine in human populations. The large numbers of observations needed to detect them make it almost impossible to obtain a sample of each relevant genotype that is properly balanced for the many socioeconomic parameters known to have a considerable effect on reproductivity patterns. Excluding socioeconomic bias or stratification as a factor contributing to any small observed difference in reproductivity between two distinguishable genotypes may, therefore, be difficult. It is a paradox of evolutionary theory that the small differences that may be so effective in molding the future of a species are so difficult to measure directly. The estimation of the larger selective differences due to relatively severe genetic abnormalities poses much less of a problem, though as already discussed very few, if any, studies have obtained the information necessary for the calculation of an intrinsic rate of increase.

Fitness can sometimes be measured by the observed progeny size.

As we have already pointed out, a complete analysis of fitness requires knowledge of mortality and fertility by age for all genotypes concerned. Sometimes, for example when fitnesses are functions of the mating types (as for incompatibility) or gene frequencies (a hypothetical example), further breakdowns of the data may be necessary.

We shall now consider a shortcut method for the estimation of fitnesses, which can usefully be applied in some circumstances: namely, the direct estimation of the average progeny size as a function of genotype, phenotype, or mating. It should be noted that, for these direct estimates of fitness to be valid, they should cover exactly one generation. The number of children born to the **probands** (the individuals whose fitness is being estimated) should be counted when they have reached the same age as the probands. If the fitness is estimated for mating types, the number of progeny should be estimated as the number of progeny who marry and only matings that have occurred sufficiently long ago that all possible progeny have had a chance to marry should be analyzed. This often requires fairly complete records. A survey of mating types limited to those in which both partners are still living, could result in a bias in favor of long-lived individuals.

Estimates of mean progeny size are effectively the same as $2R_0$, where, as before, R_0 is the net reproductive rate. As already discussed, to obtain complete fitness estimates, R_0 values should be adjusted for average generation times.

The size of studies of this kind needed to obtain relative estimates of fitness of a desired sensitivity can be estimated. From Equation 6.45 the selective disadvantage (or advantage) s of a genotype 2 with respect to genotype 1 (the latter considered as having fitness 1) is given, approximately, by

$$R_0^{(1)} - R_0^{(2)},$$

assuming that T is the same for both genotypes and s is small. Suppose the variance of the progeny born to an individual is V and assume it to be the same for the two genotypes. This variance will, as a lower limit, be equal to R_0 (the overall mean progeny size, or the mean of $R_0^{(1)}$ and $R_0^{(2)}$), but usually will be greater by a factor k such that $V = kR_0$, where, as pointed out earlier, k lies, approximately, between 1.5 and 3. An observed difference in fitness

$$s = R_0^{(1)} - R_0^{(2)},$$

estimated from the expectation of progeny over one generation, will be significant at the 95 percent level if s is larger than twice its standard error. That is, we must have

$$R_0^{(1)} - R_0^{(2)} > 2\sqrt{V}\sqrt{\frac{1}{n_1} + \frac{1}{n_2}},$$

where n_1 and n_2 are the total number of progeny counted for each of the two genotypes, so that the variance of $R_0^{(1)}$ is V/n_1 and that of $R_0^{(2)}$ is V/n_2. Taking $n_1 = n_2 = n$, we can estimate a minimum n such that

$$R_0^{(1)} - R_0^{(2)} = s = 2\sqrt{\frac{2V}{n}}.$$

Substituting $V = kR_0$, we have

$$s = 2\sqrt{\frac{2kR_0}{n}}.$$

It follows that for significance at the 5 percent level, the minimum number of progeny per genotype must be

$$n = \frac{8kR_0}{s^2}. \tag{6.24}$$

Thus, for $s = 0.1$ (as, approximately, holds for $\overline{AS}$ versus $\overline{AA}$, or the G6PD deficiency allele),

$$n = 800$$

if $k = 1$ and R_0 is approximately 1, and is proportionately greater if $k > 1$.

For a selection coefficient of order 0.01, n must be larger than 80,000 and is thus greater than can ever be hoped for in ordinary surveys. For a selection coefficient of 0.001, the value of n is such that whole large nations would have to be investigated.

6.7 Evolutionary Advance and the Index of Opportunity for Selection

The rate of evolution depends on the available amount of genetic variation in fitness.

If there is no genetic variation in fitness there can be no differential selection and hence, in general, no significant evolutionary change in the genetic content of a population. Fisher, in 1930, proved a theorem that he called the "fundamental theorem of natural selection." It states that "the rate of increase in fitness of any organism at any time is equal to its genetic variance in fitness at that time." As originally proved the theorem is valid only for certain relatively simple genetic models but, nevertheless, its qualitative importance has no restrictions. The theorem simply says that the rate of evolution by natural selection is directly proportional to the available genetic variation in fitness. The complete proof of this theorem is beyond the scope of this book. We shall, however, give Li (1955) and Crow's (1958) simple interpretation of it in terms of what Crow called the "index of total selection."

The index of total selection is $I = V/\bar{w}^2$, *where* V *is the variance in the number of offspring and* $\bar{w}$ *the mean number of offspring.*

Suppose that a parental generation contains k genotypes with frequencies p_i and relative fitnesses w_i, $i = 1, \ldots k$. We assume the simplest possible hereditary scheme, namely that all individuals of type i produce only offspring of type i, which can, in fact, only be realized with vegetative reproduction or self-fertilization of completely homozygous individuals. The average fitness of the parental population is

$$\bar{w} = \sum_{i=1}^{k} p_i w_i .$$

The relative frequency of type i in the offspring generation is

$$p_i' = \frac{p_i w_i}{\bar{w}}$$

because of selection. The average fitness in the offspring generation is, therefore,

$$\bar{w}' = \sum_{i=1}^{k} p_i' w_i = \sum_{i=1}^{k} \frac{p_i w_i}{\bar{w}} w_i = \frac{\sum_{i=1}^{k} p_i w_i^2}{\bar{w}},$$

so that the relative increment in fitness in one generation is given by

$$I = \frac{\bar{w}' - \bar{w}}{\bar{w}} = \frac{\sum_{i=1}^{k} p_i w_i^2 - \bar{w}^2}{\bar{w}^2} = \frac{V}{\bar{w}^2}, \tag{6.25}$$

where V is, by definition, the variance in fitness of the parental generation (see Appendix I for a definition of statistical terms). The quantity I is called the **index of total selection** and is the instantaneous geometric rate of increase in the fitness since $\bar{w}$ increases by a factor $1 + I$ in each generation. It is, therefore, a measure of the rate of evolution of the population under these highly simplified assumptions, namely that all differences in fitness are completely genetically determined. A real rate of evolution would be much lower, since only a fraction of the differences in fitness would have a genetic basis. Thus I may best be considered as an index of the "opportunity" for selection. It represents the rate of increase in fitness that could obtain if each offspring had a fitness equal to the mean the fitnesses of his parents. Equation 6.25 expresses Fisher's fundamental theorem of natural selection for this special genetic model.

The contribution to I can be broken down into two components, one associated with mortality differences, and the other with fertility differences.

Assume that all mortality up to the end of the reproductive period is prereproductive and that a proportion p_d of individuals (counted at birth) die before the reproductive age and a proportion p_s survive to an age when they can have children, where $p_d + p_s = 1$. Let $p_x(x = 0, 1, 2, \ldots, n)$ be the proportion of people that survive to have x children each so that

$$p_s = \sum_{x=0}^{n} p_x,$$

where n is the maximum possible number of offspring per parent. Fitness can then, most simply, be measured by the number of offspring per parent, x, so that Equation 6.25 is defined in terms of the mean and variance of the number of offspring per parent. Then

$$I = \frac{1}{\bar{x}^2}\left[p_d(0 - \bar{x})^2 + \sum_{x=0}^{n} p_x(x - \bar{x})^2\right], \tag{6.26}$$

where

$$\bar{x} = p_d \times 0 + \sum_{x=0}^{n} x p_x$$

is the mean number of offspring per parent. The variances due to mortality, V_m, and fertility, V_f, are defined as

$$V_m = p_d(0 - \bar{x})^2 + p_s(\bar{x}_s - \bar{x})^2 \quad \text{and} \quad V_f = \frac{\sum_{x=0}^{n} p_x(x - \bar{x}_s)^2}{\sum_{x=0}^{n} p_x},$$

where $\bar{x}_s$ is the mean number of offspring per person surviving to the reproductive age and is given by

$$\bar{x}_s = \frac{\sum_{x=0}^{m} x p_x}{\sum_{x=0}^{n} p_x} = \frac{\bar{x}}{p_s}. \tag{6.27}$$

If we write

$$I_m = \frac{V_m}{\bar{x}^2} = \text{index of total selection due to mortality}$$

and

$$I_f = \frac{V_f}{\bar{x}_s^2} = \text{index of total selection due to fertility,}$$

then it can be shown that

$$I = I_m + \frac{1}{p_s} I_f, \tag{6.28}$$

(see Example 6.5) and also, that $I_m = p_d/p_s$. All of these quantities can be calculated directly from routinely available mortality and fertility tables. Clearly, as p_d decreases, I_m decreases, leaving a larger proportion of the opportunity for selection to fertility differences, as is probably the present situation in most of the developed industrialized societies.

An illustrative calculation of the index and its components is shown in Table 6.12.

TABLE 6.12

Calculation of the Total Index of Selection and its Fertility and Mortality Components (Data from New South Wales, Australia, about 1900)

Number of Children (x)	*Number of Women* (n_x)	*Number of Children* (x)	*Number of Women* (n_x)
0	1110	11	568
1	533	12	422
2	581	13	226
3	644	14	129
4	702	15	57
5	813	16	39
6	855	17	12
7	976	18	5
8	963	19	2
9	847	20	2
10	786	Σn_x	10272

Note:

$$\Sigma x n_x = 63556, \quad \Sigma x^2 n_x = 550046, \quad \bar{x}_s = \frac{\Sigma x n_x}{\Sigma n_x} = 6.2,$$

$$V_f = \frac{1}{\Sigma n_x}(\Sigma x^2 n_x) - (\bar{x}_s)^2 = 15.3, \quad I_f = \frac{V_f}{(\bar{x}_s)^2} = 0.4.$$

The proportion of births surviving to ages 15–20 is

$$p_s = 0.798, \quad \text{so that} \quad p_d = 0.202.$$

$$I_m = \frac{p_d}{p_s} = 0.25, \quad I = I_m + \frac{1}{p_s} I_f = 0.25 + 0.5 = 0.75.$$

n_x is the number of women who are 46 and over—and who are thus considered to be no longer fertile—who have had x children.

Source: Based on Powys (1905).

The data on the distribution of family size comes from New South Wales at about the turn of the last century. The women whose family sizes were counted for this study were all older than 46 years, by which age reproduction has generally ceased. The mean number of children per woman is 6.2, which is $\bar{x}_s$, since only women who survived are included in the calculations, and the variance of the number of children is 15.3. Thus the fertility component of the index $I_f = 15.3/(6.2)^2 = 0.4$. Since the proportion of newborns surviving to age 15–20 is 0.798, which is p_s, the mortality component of the index $I_m = 0.202/0.798 = 0.25$. The total index is, therefore, from Equation 6.28, given by $0.4/0.798 + 0.25 = 0.75$.

Changing fertility and mortality patterns due to major socioeconomic changes have a profound effect on natural selection in human populations.

Examples of values of the total index and the component due to fertility, for various human populations with widely differing overall fertility rates, are given in Table 6.13. Two features of these data need special emphasis:

TABLE 6.13
The Total Index of Opportunity for Selection and its Fertility and Mortality Components for Some Human Populations with Widely Differing Fertilities

			Component of Total Index due to Infertility	
Population	*Mean Progeny per Parent Born*	*Total Index (I)*	$\frac{1}{p_s} I_f$	*Percent of I*
Hutterites (around 1950)	8.8	0.27	0.22	81.5
New South Wales (around 1900)	6.2	0.75	0.50	66.7
United States				
Age 45–49 in 1910	3.9	1.18	0.96	81.4
Age 45–49 in 1950	2.3	1.38	1.27	92.0

Source: Modified from Crow (1961a).

1. The total index is not proportional to the mean number of offspring; it actually increases as the mean number of offspring decreases.

2. Mortality contributes a small fraction to the total index in highly developed industrial societies such as the United States. The major opportunity for selection in such societies is clearly through differential fertility and this is greatly influenced by birth control practice and modern medical care, both of which undoubtedly tend to reduce the variance in fertility. The conditions of highly developed industrial

societies are certainly quite radically different from those that must have prevailed during most of the time the human species has been evolving, in fact, until at most one hundred or two hundred years ago. For most of this time, mortality, largely due to infectious diseases, must have been a major selective agent. This is illustrated in Figure 6.2, which shows the decline in prereproductive mortality and the corresponding decline in fertility in the French population from 1770 to 1950.

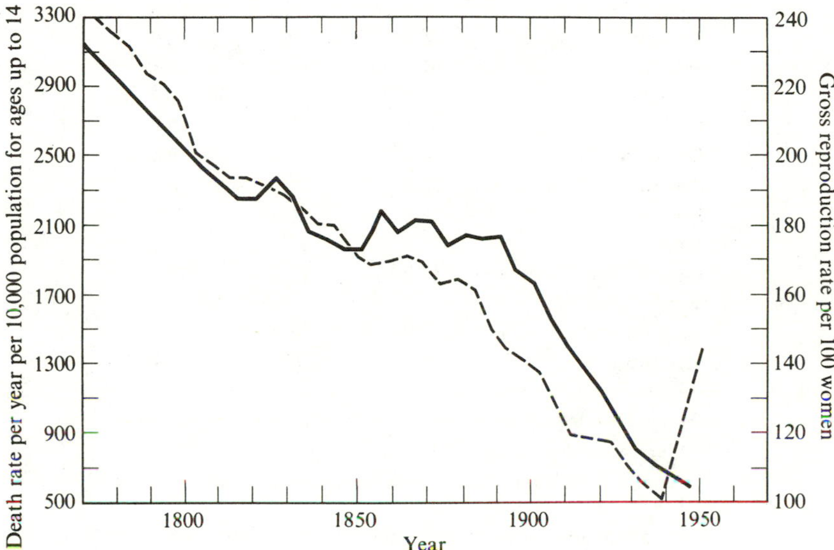

FIGURE 6.2
Change in mortality and fertility in France from 1770–1950. The dashed line is gross reproduction rate; the solid line, death rate. (Based on data from Bourgeois-Pichat, 1951.)

The decline in mortality in France seems to have occurred in two major phases, the first beginning before 1770 and lasting until about 1850 and the second running from 1890 to the present time. Initially, the fertility decline lagged behind that in mortality, leading to an increasing population with a higher proportion of people in the younger age groups. However, in the late nineteenth century and early twentieth century, the continuing decline in fertility caught up with the decline in mortality. This pattern of change during industrialization is quite characteristic. Rapid temporary increases in population size result from a decline in mortality greater than that in fertility, though in France this gap has been less than in other countries. The large increase in the gross reproduction rate in France between 1940 and 1950 is undoubtedly in part the effect of the cessation of World War II.

Present mortality and fertility data for countries at various levels of development, arranged in descending order of crude birth rates, are shown in Table 6.14. These data illustrate the wide divergence in many countries even at the present time from

TABLE 6.14
Crude Live Birth Rates, Infant Mortality Rates, and Death Rates For Various Countries at Various Stages of Development

Country	Live Birth Rates[a]	Infant Mortality Rates[b]	Death Rates[c]
Nigeria (Lagos)	50.1	76.2	12.8
Mexico	45.9	74.4	12.5
Jordan	40.1	63.1	8.1
Ceylon	36.6	57.5	20.3
Chile	35.5	127.7	12.5
United States	24.6	26.4	9.4
Australia	22.6	21.5	8.8
France	18.4	25.2	11.8
Japan	18.2	33.7	7.8
England and Wales	15.9	22.2	11.6
South Africa			
Caucasian	25.0	28.7	8.6
African	46.1	120.6	16.2

Source: Data from United Nations Demographic Year Book, 1961.

[a] Average number of live births per 1000 population, 1955–1959.

[b] Number of deaths in 1958 or 1959 within the first year of life per 1000 live births.

[c] Average number of deaths per 1000 population, 1955–1959.

the low mortality and fertility rates prevailing in, for example, the United States. High birth rates are generally correlated with high infant death rates and high intrinsic rates of increase, and with poor socioeconomic conditions. The contrast between Caucasians and Africans in the Union of South Africa hardly needs comment. Overall death rates are complicated by differences in age distributions and show no obvious trends. Socioeconomic changes have clearly wrought a profound change in the likely future patterns of natural selection in human populations.

6.8 The Fitness Flow Sheet

A fuller understanding of the causes of a selective difference requires a comprehensive specification of the total pattern of reproductivity.

The use of Crow's index of opportunity for selection provides only a crude breakdown of the causes of selection into fertility and mortality components. Many factors, such as age at marriage, the stability of the marriage, the length of the post-partum sterile period, as well as the mortality, need to be taken into account in a more complete description of the total reproductive history.

A schematic representation of these factors in terms of a flow sheet is illustrated in Figure 6.3. The flow sheet is given in the form of a diagram appropriate for writing a computer program to simulate the reproductive process. Time is incremented by one unit ($t \rightarrow +1$) as indicated. Encircled questions denote decisions that have to be made that will, in general, be complex functions both of time and of the previous history of an individual. Thus, for example, the probability of marriage will depend on age, previous marital history if divorced or widowed, and possibly, also, previous reproductive history. (Marriage is used here in a biological sense, and thus the analysis includes births that are considered illegitimate in the legal sense. The probability of conception will depend on the length of time since the previous birth and on the previous number of births, as well as on age. The probability of living will depend on marital history, fertility, as well as age. Each encircled question, therefore, represents a complex of probability distributions describing the relevant aspects of fertility, mortality, and marriage. The final outcome of the process is the number of births, per individual, giving the net reproductive rate. There are four major determinants in this process namely, prereproductive mortality, mortality during the reproductive period, marriage, and fertility. The age-specific birth and death rates used for calculating the intrinsic rate of increase incorporate contributions from all these determinants but do not give any direct indication of their relative magnitudes.

The effects of each of these various determinants may be thought of in terms of a reduction of the maximum reproductivity achievable with optimal conditions. This maximum will be achieved only if there are no still births, no infant deaths, no prereproductive or reproductive deaths, if marriage takes place as early as possible and is not broken by divorce or death of the mate, and if maximum fertility with a minimum post-partum sterile period is maintained throughout the maximum possible reproductive period. Ignoring the possibility of multiple births, the maximum possible achievable number of births per human female, is about 30. The Hutterites are a good example of a group with very high fertility. Their median age at marriage is 22 for women, only about one percent of either sex at age 45 has never married, and birth control is not practiced at all (see Eaton and Mayer, 1954). Including marriages at all ages, whether or not they are fertile, the mean number of children that had been born to women who were 45 years or older in 1950 was 8.9. A more precise definition of the effect of a determinant (its "cost"), such as age at marriage, in reducing the maximum achievable intrinsic rate of increase $r_{\max}$ is given by the reduction in the intrinsic rate of increase obtained when the actual distribution of the determinant is substituted for the optimum, while all other determinants are left at their optimum. This cost can be measured by $r_{\max} - r_{\text{marr}}$, where r_{marr} is the intrinsic rate of increase with the actual age at marriage distribution. These quantities can be calculated by following through the consequences of the model outlined in Figure 6.3. The actually achieved reproductive rate may, at least approximately, be thought of as the maximum minus the sum of the costs

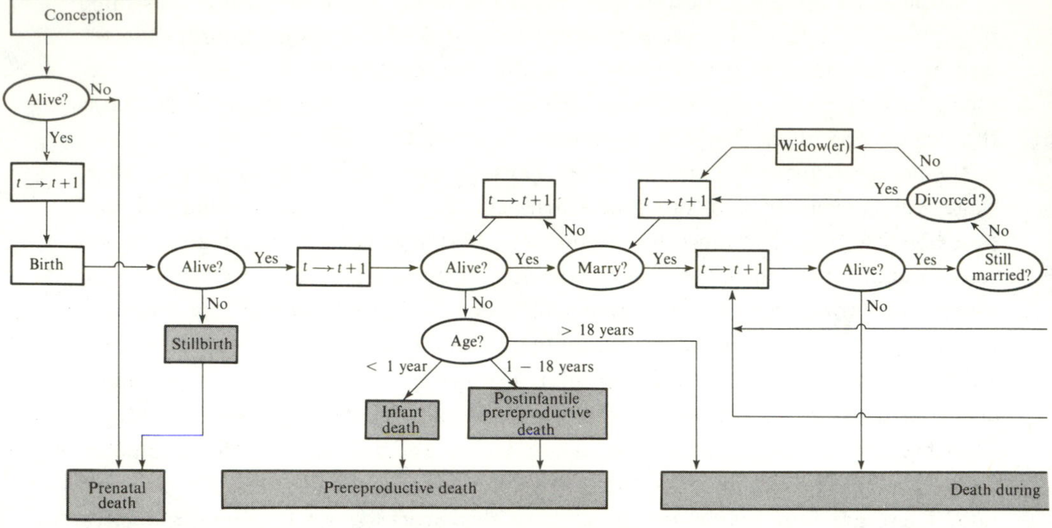

FIGURE 6.3
Fitness flow sheet. The questions in the ovals are answered on the basis of various frequency distributions, which, for example, give the probability of survival from one time interval to the next as a function of age. (Modified from Bodmer and Cavalli-Sforza, 1966.)

contributed by all the individual determinants. We thus obtain a balance sheet of the contributions of the various determinants to the biological "cost of living."

The costs can also be measured relative to some standard set of distributions, for example the overall population mean. Given a knowledge of the various distributions for a particular genotype, the differential cost of each factor can be assessed and thus the underlying causes of a selective difference determined.

An interesting analysis of a model of the Indian population has been given by Ridley and Sheps (1966). These authors have constructed a computer model of a population, based on a simplified version of the sort of fitness flow sheet shown in Figure 6.3. Using a series of observed distributions for the various parameters of their model, they had a computer simulate the life histories of 1000 females and then calculate standard parameters, such as the net reproductive rate, based on these life histories. Their aim was to try to choose a set of parameters that would correspond reasonably to the observed fertility pattern in the Indian population. Their model took into account the age differences between spouses, the probabilities of fetal wastage and infant death, estimates of natural fecundity as a function of age, standard life tables for males and females, estimates of the age of onset of sterility and of the average age at first marriage, and probability distributions for the length of pregnancy and of the post-partum sterile period, each as a function of fetal deaths or live births. They were able to show that a post partum sterile period longer than that generally supposed, and corresponding to the distri-

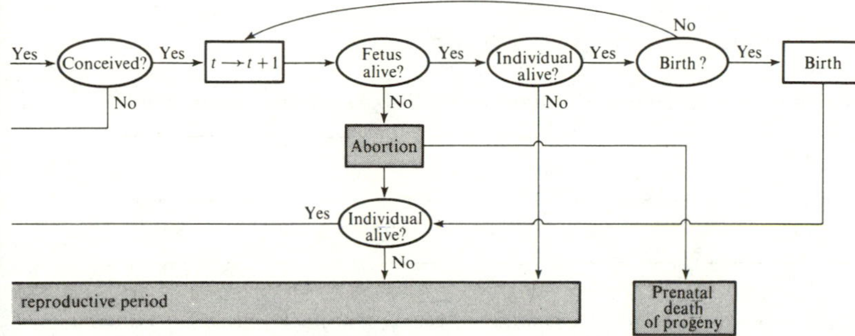

bution of the duration of lactation, was a major factor in giving net reproductive rates comparable to those observed for the Indian population. When the mean length of the post-partum sterile period was 12.9 months, the net reproductive rate was 2.36. When the length of this interval was increased to 20.4 months, the net reproductive rate was reduced to 1.96. An increase in the average age of marriage from 18.2 to 21.3 years reduced the net reproductive rate still further to about 1.5, a figure that corresponds quite closely to that observed in the Indian population.

Computer simulation of the fitness flow sheet can also be used to investigate the effects of a variety of factors on the variation in fertility or reproductivity

Jacquard and Bodmer (1968) constructed a model for the Hutterite population, allowing for all observable distributions, such as those for age at marriage, death rates, and birth rates. The one factor that could not readily be taken into account was the intrinsic variability in fecundity—that is, the variation of the monthly conception rate between individuals. The observed variance of the progeny-size distribution was 13.9; that predicted by the model was only 5.8. The difference of 8.1 is, presumably, a measure of the contribution of intrinsic variation in fecundity to the actual variance in fertility. In another study, using data from the French population, the effect of variation in the age of marriage was assessed by holding it constant at 21 years. Ignoring effects of mortality and marriage instability, the variance in fertility was then found to be reduced from 10.0 to 1.5. Birth control is another important factor leading to a reduction in the fertility variance. Assuming a specific simple model for birth control in which couples aim for a desired family of 2 or 4 offspring and using data from the French population gave the results shown in Table 6.15. It is assumed that couples use birth control for the first two years of marriage and again after having produced a family of the desired size. As the efficiency of birth control, measured by the percentage reduction in the monthly probability of conception, increases, the variance in fertility decreases markedly, and the mean approaches the desired family size. The reduction in the variance, as we have already emphasized, implies a marked reduction in the opportunity for selection.

TABLE 6.15
Effect of Birth Control on the Mean and Variance of the Progeny-size Distribution

	Efficiency of Birth Control							
	70%		90%		95%		99%	
Desired family size	*Mean*	*Variance*	*Mean*	*Variance*	*Mean*	*Variance*	*Mean*	*Variance*
2	6.4	5.5	3.9	2.5	2.8	1.4	2.1	0.49
4	6.7	5.3	4.8	2.4	4.1	1.7	3.7	1.23

Source: Based on Jacquard (1967).

These analyses show how the building of an appropriate model together with computer simulation can isolate one or two out of a number of the many factors affecting reproductivity and its variations as being of major importance in determining an observed fertility pattern. Selective differences between genotypes have not yet been analyzed to this level of refinement, though in a number of cases the probable cause of the difference can be qualitatively identified.

We have already pointed out how socioeconomic changes leading to different fertility and mortality patterns have drastically altered selective forces. Social customs relating to the average age of marriage and child bearing also have very important effects. Thus, for a disease with a late age of onset, such as Huntington's chorea, and where birth control exerts a strong influence on family size, an earlier average age of marriage may very much increase the probability of the desired maximum fertility being achieved before the onset of the disease. This will, of course, tend to relax selection against such disease. The later average age of marriage in men than in women will, for such a disease, lead to there being a stronger selective disadvantage for males than for females. This difference will largely disappear with earlier average ages of marriage. Since, in general, more than 60 percent of babies with Down's syndrome are born to mothers older than 35, a decrease in the average age of child bearing should drastically decrease the incidence of Down's syndrome. These two examples serve to emphasize the need for complete specification of the reproductive history in order to understand selective differences and their interactions with socioeconomic factors.

In many cases it will, in fact, not be adequate to specify individuals only in the fitness flow sheet, since genotypes of mates may interact, as for example with incompatibility selection, in the determination of fertility and mortality. Ideally, the specification of fitness should be in terms of mating pairs and their contributions to the next generation of matings, rather than in terms of individuals. Unfortunately, the hope of obtaining information adequate to satisfy these needs, even for human populations, is still remote.

6.9 Sources of Data

A major limitation to the investigation of fitness differences is the paucity of suitable data.

Complete reproductive histories of very large numbers of people are required for a really adequate study. In this section we shall discuss briefly some of the types of data that are available on a relatively large scale and some of the approaches used for their collection and collation.

There are basically six types of sources of data.

1. *Censuses:* Basic vital and socioeconomic information obtained for a whole population, or a substantial fraction of it, at regular intervals of five or ten years is generally the main source for determining the age structure of a population and provides many data on marriage, migration, and fertility patterns. Most censuses, however, suffer from severe drawbacks as far as genetic studies are concerned. In the first place, since they usually use the household as the basic sampling unit, complete families are included together only if all their members live together at the time the census is taken. Data on an older couple, whose fertility is completed, are thus unlikely to be included with those on their adult children. Because of regulations providing that information given in a census is confidential, it is difficult to relate data for a census to those from another census or from other sources of data. No information is, therefore, generally available from censuses that relates two generations. Since a census collects information only on living persons, it gives no information on mortality patterns. As census data is usually restricted to socioeconomic parameters, such as income, occupation and years of schooling, morbidity information is not generally available. Response errors in large-scale surveys, such as censuses, are likely to be more serious than in small-scale, more-restricted surveys (see Bodmer and Lederberg, 1967).

This may seem a formidable list of disadvantages. But it must be remembered that no other data yet exist on a scale comparable to those from a census, making it worthwhile to expend some effort in extracting useful information from censuses. An effort should be made to influence the census operation at the time of preparation of questionnaires to include special questions of interest to geneticists. The Italian Census of 1962 obtained data on the ages of the fathers and mothers of fertile women in order to permit an investigation of the possible accumulation of sex-linked lethals with grandparental age (see Chapter 3 and 10). The results of this study have not, however, been made available yet.

2. *Vital Statistics: Birth, Marriage, and Death Registrations.* Vital statistics provide the basic source of material for the construction of life tables. The data are, of course, available on a large scale and generally quite accurate. The usefulness of birth, marriage, and death certificates, is however, severely limited by the information collected on them, which is not usually adequate for the construction of pedigrees without extensive collation. Sometimes causes of death, or statements about birth abnormalities accompanying birth and death certificates, can relate to specific genetic abnormalities and thus be useful for fitness calculations. For example, the distribution of age of death for a given disease will likely come from this source of data. Birth-anomaly data collected in this way are not, however, usually very reliable.

3. *Miscellaneous Records.* There are many other records collected on a fairly large scale, for example, by hospitals and from insurance companies. Though these are likely to be biased samples of the population, the information on them may be of use in relation to particular abnormalities, especially if it can be linked to information obtained from other sources.

4. *Special Registers.* Some private companies (for example, insurance companies) and the governments of some nations have special registers for certain classes of abnormalities. Registration of certain infectious diseases is, of course, quite common, but of little genetic significance. However, some Scandinavian countries, notably Denmark, have very complete registries of a wide variety of birth anomalies. An attempt is made to record suitable information on every individual suffering from a particular abnormality. Such registers are, of course, ideal for providing information on well-defined genetic diseases. Recently, testing of blood samples at birth for the detection of phenylketonuria (PKU) has been made compulsory in some states of the United States, notably Massachusetts and California. As a result, detailed PKU registers are being compiled in these states, which should provide valuable information on the fitness of affected individuals but only *after the advent of dietary treatment.* Further extensions of such registers are to be expected and will certainly provide invaluable information for the study of genetic diseases.

5. *Parish books and other records.* In some European countries, parish books recording vital statistics, namely, births, marriages, and deaths, have been accurately maintained for two or three hundred years or even longer. Though the information they furnish is very limited, they do provide a unique source of material for

the study of the inheritance of longevity and fertility. They also provide unique information for determining changing mortality and fertility patterns retrospectively (see, for example, Barrai, Cavalli-Sforza, and Moroni, 1965). The Japanese have a system of collecting and recording demographic and genealogical data within the household, called "Koseki," which have been widely used especially for consanguinity studies (see Chapter 7 and Schull and Neel, 1965).

6. *Special purpose surveys.* Probably the most accurate and complete type of information that can be obtained comes from relatively small-scale special purpose surveys, which are often directed at particular diseases or particular populations. An example of such a survey is the study of Huntington's chorea already mentioned. There are two major drawbacks to such special surveys. In the first place, it may be very difficult to avoid biases in the collection of the data, and in the second place, the expense of special surveys makes it difficult to collect enough data for the determination of small selective differences.

All these various sources of data can be useful, especially if they can be related to each other, which is done by the process called ***record linking****.*

Various criteria are used for deciding whether two records refer to the same individual or to the same family. Identity of the surname is an obvious start, but this is often not enough, even together with a first name, especially if the records used contain clerical errors. Additional information, such as birthdate and birthplace can often be used. If two certificates bear the same name, birthplace, and birthdate, the chances are very high that they refer to the same person. Some names will, of course, be more distinguishing than others, essentially in inverse proportion to their frequency, and empirical scores may be constructed to allow for this. Undoubtedly, the most reliable way of linking records is to assign everyone a number at birth that is to be used on all records. In England this could be the National Health Service Number and in the United States a social security number could be *assigned at birth* and thus used for this purpose. The social security number is already widely used on all conceivable types of records in the United States and would be even more valuable for record linking if it were assigned at birth.

The application of record-linking techniques to genetic studies has been pioneered by Newcombe and co-workers (see, for example, Newcombe and Kennedy, 1962; Newcombe, 1965, 1976). They have linked data from the British Columbia Register of Handicapped Children and Adults to birth, marriage, and death certificates. In this way they are able to construct the complete demographic history of the families that include one or more handicapped children. This provides data such as those on the recurrence risk within families of a particular form of handicap, on the effect of parental age at birth on the probability of having a malformed child, and on the effect of a miscarriage, fetal death, or malformed birth on subsequent fertility. Newcombe and co-workers have shown, for example, that the probability of a birth in the year following a stillbirth is about 2.7 times that following a live birth. This overcompensation disappears, however, in the second year and is followed by a substantial deficiency of births in the third and fourth years following a stillbirth. This leads to an overall probability for the four-year period that is 0.91 that for normal live births (Newcombe and Rhynas, 1962). Barrai and co-workers (1965) have linked parish registries from the Parma Valley in Italy, of baptisms, marriages, and deaths to obtain complete genealogies over extended time periods. Such work should provide data on long-term inbreeding and on the heredity of fertility and mortality.

The scientific and social advantages of "complete" data collection have to be weighed against their possible disadvantages, which are especially easy to imagine if we assume that unscrupulous persons would have access to the data. Some may argue that there is no better way to provide for the rise of a police state than through the complete collation of many different sources of information on every member of a population. Modern electronic computer facilities make this a very immediate prospect whose control has to be faced. Geneticists, social scientists, demographers, and others will certainly welcome the availability of information of such an unprecedented completeness, though the problems of data analysis will then perhaps be even more weighty than present problems of data inadequacy.

The essence of quantitative predictions of evolution by natural selection is the accurate measurement of selective differences. In human populations this requires the use of the demographer's approach as we have outlined it in this chapter. It has been our aim to emphasize the type of information that is needed, the scale on which it must be collected, and the sources from which it can be obtained.

General References

Bajema, C. J., "Estimation of the direction and intensity of natural selection in relation to human intelligence by means of the intrinsic rate of natural increase." *Eugen Quart.* **10**(4): 175–187, 1963. (One of the few complete attempts to estimate overall fitness in man.)

Lotka, A. J., *Elements of Mathematical Biology*. New York: Dover, 1956. (Contains a summary of Lotka's major contributions to demographic theory.)

Keyfitz, N., *Introduction to the Mathematics of Population*. Reading, Mass.: Addison-Wesley, 1968. (A basic reference to the mathematics of demography.)

Worked Examples

6.1. *Derivation of general equations relating the age distributions of a population in successive time intervals.*

General equations are here expressed in **matrix** form to show how Lotka's theorem follows from the well-known theory of matrices. The use of the equations is illustrated by calculating the equilibrium age distribution for the population projection matrix for United States females, 1960 (Table 6.1), by taking successive powers of the matrix.

Let us consider time in terms of, for example, one-year intervals. The individuals at age x years ($x \neq 0$) are the survivors of the group of individuals who were $x - 1$ years old one year ago; the individuals of age 0 are all those born to individuals of age greater than 0. The relationship between the number of individuals of given age in successive years can be simply expressed in terms of birth and survival rates. Thus let

$N_{x,t}$ = the number of individuals of age x at time t,

p_x = the proportion of individuals of age x who survive to age $x + 1$,

b_x = the number of offspring born to each living individual of age x in the time interval x to $x + 1$,

where we assume x ranges from 0 to m, say, m being the maximum age at which children can be produced. It is important to note that p_x and b_x are independent of the time t, meaning that there are no secular changes in the birth and death rates. The relationships between $N_{x,t}$ and $N_{x,t+1}$ are

$$
\begin{aligned}
N_{0,t+1} &= \sum_{x=0}^{m} b_x N_{x,t} && \text{(total births in the time interval)} \\
N_{1,t+1} &= p_0 N_{0,t} && \text{(survivors from age 0 to 1)} \\
N_{2,t+1} &= p_1 N_{1,t} && \text{(survivors from age 1 to 2)} \\
&\ \vdots \\
N_{x,t+1} &= p_{x-1} N_{x-1,t} \\
&\ \vdots \\
N_{m,t+1} &= p_{m-1} N_{m-1,t}.
\end{aligned}
\tag{1}
$$

Since b_x and p_x do not depend on t, these equations can easily be iterated to obtain

$N_{x,1}$ in terms of $N_{x,0}$, and then $N_{x,2}$ first in terms of $N_{x,1}$ and then of $N_{x,0}$, and so on, giving after t cycles $N_{x,t}$ in terms of $N_{x,0}$ for all values of x. The total population size at time t is

$$N_t = \sum_{x=0}^{m} N_{x,t},$$

this being the sum of all the people of different ages in the population. Lotka's theorem, in this form, states that as t becomes large the proportion of people with a given age $N_{x,t}/N_t$ tends to approach a constant n_x, which is independent of the time t, and that the ratio of the total population size at successive times N_{t+1}/N_t tends to approach a constant λ. If $\lambda > 1$ the population is increasing in size, if $\lambda = 1$ it is stationary, and if $\lambda < 1$ it is decreasing. Equations 1 can be written in the form:

$$\begin{pmatrix} N_{0,t+1} \\ N_{1,t+1} \\ \vdots \\ N_{m,t+1} \end{pmatrix} = \begin{pmatrix} b_0, b_1, \ldots, & b_m \\ p_0, 0, 0, \ldots & 0 \\ \vdots & \vdots \\ 0, 0, 0, \ldots, p_{m-1}, & 0 \end{pmatrix} \begin{pmatrix} N_{0,t} \\ N_{1,t} \\ \vdots \\ N_{m,t} \end{pmatrix} = \mathbf{M} \begin{pmatrix} N_{0,t} \\ N_{1,t} \\ \vdots \\ N_{m,t} \end{pmatrix} \tag{2}$$

where **M** represents the array of coefficients of the Equations 1 and is called a **matrix**. The rule for multiplying the matrix **M** by the vector (a **vector** is an array of numbers in a single row or column)

$$\begin{pmatrix} N_{0,t} \\ N_{1,t} \\ \vdots \\ N_{m,t} \end{pmatrix}$$

to obtain the xth term $N_{x,t+1}$ of the vector

$$\begin{pmatrix} N_{0,t+1} \\ N_{1,t+1} \\ \vdots \\ N_{m,t+1} \end{pmatrix}$$

is to multiply term by term the items of the xth row of the matrix **M** by the corresponding elements of the vector

$$\begin{pmatrix} N_{0,t} \\ N_{1,t} \\ \vdots \\ N_{m,t} \end{pmatrix},$$

for example, $N_{0,t+1} = b_0 N_{0,t} + b_1 N_{1,t} + \cdots + b_m N_{m,t}$. If we let $\mathbf{N}_t$ stand for this latter vector, Equation 2 can be rewritten in the form

$$\mathbf{N}_{t+1} = \mathbf{M}\mathbf{N}_t = \mathbf{M}(\mathbf{M}\mathbf{N}_{t-1}) = \mathbf{M}^2\mathbf{N}_{t-1} = \cdots \mathbf{M}^t\mathbf{N}_1, \tag{3}$$

where $\mathbf{M}^t$ represents the result of t successive applications of the matrix **M** to the initial vector $\mathbf{N}_1$.

To illustrate the rules of matrix multiplication, consider the application of Equations 1 to $N_{x,t+1}$ in order to relate $N_{x,t+2}$ to $N_{x,t+1}$ and so to $N_{x,t}$. We have from two applications of Equations 1

$$\begin{aligned} N_{0,t+2} &= b_0 N_{0,t+1} + b_1 N_{1,t+1} + \cdots + b_m N_{m,t+1} \\ &= b_0(b_0 N_{0,t} + b_1 N_{1,t} + \cdots + b_m N_{m,t}) \\ &\quad + b_1 p_0 N_{0,t} + \cdots + b_m p_{m-1} N_{m-1,t} \end{aligned}$$

or

$$\begin{aligned} N_{0,t+2} &= (b_0^2 + b_1 p_0) N_{0,t} + (b_0 b_1 + b_2 p_1) N_{1,t} \\ &\quad + \cdots + (b_0 b_{m-1} + b_m p_{m-1}) N_{m-1,t} + b_0 b_m N_{m,t} \end{aligned}$$

and

$$\begin{aligned} N_{1,t+2} &= p_0 N_{0,t+1} = p_0 b_0 N_{0,t} + p_0 b_1 N_{1,t} + \cdots + p_0 b_m N_{m,t} \\ N_{2,t+2} &= p_1 N_{1,t+1} = p_1 p_0 N_{0,t} \\ N_{m,t+2} &= p_{m-1} N_{m-1,t+1} = p_{m-1} p_{m-2} N_{m-2,t}. \end{aligned}$$

These equations can be written in the matrix form

$$\begin{pmatrix} N_{0,t+2} \\ N_{1,t+2} \\ N_{2,t+2} \\ \vdots \\ N_{m,t+2} \end{pmatrix} =$$

$$\begin{pmatrix} b_0^2 + b_1 p_0 & b_0 p_1 + b_2 p_1 & \cdots & b_0 b_{m-2} + b_{m-1} p_{m-2} & b_0 b_{m-1} + b_m p_{m-1} & b_0 b_m \\ p_0 b_0 & p_0 b_1 & \cdots & p_0 b_{m-2} & p_0 b_{m-1} & p_0 b_m \\ p_1 p_0 & 0 & \cdots & 0 & 0 & 0 \\ 0 & p_1 p_2 & \cdots & 0 & 0 & 0 \\ \vdots & \vdots & & \vdots & \vdots & \vdots \\ 0 & 0 & \cdots & p_{m-1} p_{m-2} 0 & 0 & 0 \end{pmatrix} \begin{pmatrix} N_{0,t+1} \\ N_{1,t+1} \\ N_{2,t+1} \\ \vdots \\ N_{m,t+1} \end{pmatrix} \tag{4}$$

so that

$\mathbf{M}^2 =$

$$\begin{pmatrix} b_0^2 + b_1 p_0 & b_0 p_1 + b_2 p_1 & \cdots & b_0 b_{m-2} + b_{m-1} p_{m-2} & b_0 b_{m-1} + b_m p_{m-1} & b_0 b_m \\ p_0 b_0 & p_0 b_1 & \cdots & p_0 b_{m-2} & p_0 b_{m-1} & p_0 b_m \\ p_1 p_0 & 0 & \cdots & 0 & 0 & 0 \\ 0 & p_1 p_2 & \cdots & 0 & 0 & 0 \\ \vdots & \vdots & & \vdots & \vdots & \vdots \\ 0 & 0 & \cdots & p_{m-1} p_{m-2} & 0 & 0 \end{pmatrix}$$

The rule for matrix multiplication is that the *i*th term of the *j*th column of the product matrix is the sum of the products of the corresponding terms of the *i*th row of the first matrix in the product and jth column of the second matrix. Note that in terms of two general 3×3 matrices the rule gives

$$\begin{pmatrix} a_1 & a_2 & a_3 \\ b_1 & b_2 & b_3 \\ c_1 & c_2 & c_3 \end{pmatrix} \begin{pmatrix} A_1 & A_2 & A_3 \\ B_1 & B_2 & B_3 \\ C_1 & C_2 & C_3 \end{pmatrix}$$

$$= \begin{pmatrix} a_1A_1 + b_1B_1 + c_1C_1 & a_1A_2 + b_1B_2 + c_1C_2 & a_1A_3 + b_1B_3 + c_1C_3 \\ b_1A_1 + b_2B_1 + b_3C_1 & b_1A_2 + b_2B_2 + b_3C_2 & b_1A_3 + b_2B_3 + b_3C_3 \\ c_1A_1 + c_2B_1 + c_3C_1 & c_1A_2 + c_2B_2 + c_3C_2 & c_1A_3 + c_2B_3 + c_3C_3 \end{pmatrix}$$

and is derived from the simple rules of algebra used on page 332. Lotka's theorem now follows from a theorem of matrix algebra well known to mathematicians, which states that for large t, the individual terms of successive powers of a matrix, say $\mathbf{M}^t$ and $\mathbf{M}^{t-1}$, tend to approach a constant ratio λ for all terms of the matrix so that, approximately,

$$\mathbf{N}_t = \mathbf{M}^t\mathbf{N}_0 \sim \lambda\mathbf{M}^{t-1}\mathbf{N}_0 \sim \lambda^t\mathbf{n}, \tag{5}$$

where $\mathbf{n} = (n_0, n_1, \cdots, n_m)$ is a vector whose elements are constant and independent of the time t. This vector $\mathbf{n}$ and the constant ratio λ are properties of the matrix $\mathbf{M}$ and define, respectively, the stable-equilibrium age distribution of the population and the asymptotic rate of growth of the population.

The equilibrium age distribution can be obtained quite rapidly by taking the successive squares of the projection matrix $\mathbf{M}$ to obtain $\mathbf{M}^2$, $\mathbf{M}^4$, $\mathbf{M}^8$, $\mathbf{M}^{16}$, $\mathbf{M}^{32}$ and so on. The number of operations needed to obtain a power of $\mathbf{M}$ near to equilibrium is relatively small. The 16th and 32nd powers of the projection matrix

	M^{16}				M^{32}			
Ages	0–15	15–30	30–45	over 45	0–15	15–30	30–45	over 45
0–15	33.05	43.03	11.57	0	2375.1	3128.7	845.1	0
15–30	24.67	33.05	8.99	0	1802.8	2375.1	641.5	0
30–45	19.10	24.57	6.57	0	1363.1	1795.5	485.0	0
over 45	42.28	56.07	15.28	0.13	3078.4	4055.5	1095.4	0.017

Age groups	*Initial age distribution*	*Age distribution at 33rd cycle*	*Total*	*Percent*
0–15	40	$(2375.1 \times 40) + (3128.7 \times 30) + (845.1 \times 20)$	= 205766.6	27.6
15–30	30	$(1802.8 \times 40) + (2375.1 \times 30) + (641.5 \times 20)$	= 156195.9	20.9
30–45	20	$(1363.1 \times 40) + (1795.5 \times 30) + (485.0 \times 20)$	= 118089.5	15.8
over 45	10	$(3078.4 \times 40) + (4055.5 \times 30) + (1095.4 \times 20) + (0.017 \times 10)$	= 266710.3	35.7
	100		745762.3	100

given in Table 6.1 are shown at the bottom of page 334 together with the derived age distribution for the 33rd cycle, which is, of course, very near to the theoretical equilibrium distribution. The ratios of the individual terms of the 32nd and 16th powers lie between 71.5 and 73, corresponding closely to $(1.31)^{16}$ and thus verifying numerically in this particular instance the rule quoted for the asymptotic behavior of the powers of a projection matrix (see Table 6.3). Leslie (1945) was the first to formalize this matrix approach to the study of population growth.

6.2. *The calculation of the number of years a person is expected to live as a function of his present age.*

The set of l_x values is the basis for calculating the number of years a person is expected to live, which is used by life-insurance companies for the calculation of premiums. The construction of an accurate **life table**, giving the expectation of life at all ages, is one of the main tasks of the actuary and of the demographer. The probability that an individual will die at age x is

$$d_x = q_x(1 - q_{x-1}) \cdots (1\text{-}q_0) = q_x\, l_{x-1} = l_{x-1} - l_x. \tag{1}$$

This follows from Equation 6.2 since this probability is both the probability that an individual lives to age $x - 1$ and then dies at age x and the difference between the probability that an individual survives to age $x - 1$ and that he survives to age x. In a similar way, the probability that an individual dies at age x, given that he was alive at age $n < x$, is

$$\begin{aligned} {}_x d_n &= q_x(1 - q_{x-1}) \cdots (1 - q_{n+1}) \\ &= q_x \frac{l_{x-1}}{l_n} = \frac{1}{l_n}(l_{x-1} - l_x). \end{aligned} \tag{2}$$

The life expectancy of people who have survived to age n, being the mean number of years survived after age n by those who have reached age n, is, therefore,

$$e_n^0 = \sum_{x=n+1}^{m} x({}_x d_n) = \frac{1}{l_n} \sum_{x=n+1}^{m} x(l_{x-1} - l_x) = \frac{1}{l_n} \sum_{x=n+1}^{m} l_x. \tag{3}$$

The expected age at death is $n + e_n^0$, the sum of the present age n and the expected remaining years of survival e_n^0. For the direct application of Equation 3, l_x must be given in one-year intervals. If the intervals are longer, then the l_x values must be replaced by the product of the time interval in years and the average l_x for that interval, which is most simply the average of adjacent l_x values. The table of values of e_n^0, which is the life table, is a complete summary of the mortality pattern of a population assuming a stable age distribution based on constant age-specific birth and death rates.

6.3. *Assuming that a trait's only effect on reproductivity is through a change in mortality (generally to a probability of earlier death), calculate the effective selective difference due to the trait, in terms of the age specific survival probabilities and birth rates of the general population and the death rates, at different ages, of individuals having the trait.*

Suppose that q_x' is the probability of dying from the trait between the ages of x and $x + 1$. Then if k_x is the proportion of all deaths in this age group that are caused by this trait, we must have $q_x' = q_x k_x$, where q_x is the age-specific mortality for the total population. Let q_x^0 be the age-specific mortality for the part of the population that does not have the trait. This means that q_x^0 represents the total mortality other than that due to this trait. Then

$$(1 - q_x) = (1 - q_x^0)(1 - q_x'). \tag{1}$$

Thus l_x^0, the probability of survival to age x among that part of the population that does not have the trait, is given by

$$l_x^0 = \prod_{i=0}^{x} (1 - q_i^0) = \frac{\prod_{i=0}^{x}(1 - q_i)}{\prod_{i=0}^{x}(1 - q_i')} = \frac{l_x}{u_x} \tag{2}$$

from Equation 1, where

$$u_x = \prod_{i=0}^{x} (1 - q_i').$$

The difference between the net reproductive rates of the part of the population that does not have the trait and that of the total population is, therefore,

$$R_0^0 - R_0 = \sum_{x=0}^{m} l_x^0 b_x - \sum_{x=0}^{m} l_x b_x = \sum_{x=0}^{m} \frac{l_x}{u_x} b_x - \sum_{x=0}^{m} l_x b_x;$$

that is,

$$R_0^0 - R_0 = \sum_{x=0}^{m} l_x b_x \left(\frac{1}{u_x} - 1\right), \tag{3}$$

where b_x is the age-specific birth rate of the total population. This is an approximate measure of the selective difference due to the trait. From Equation 6.10 this selective difference is given more precisely by $(r^0 - r)T$, where r^0 and r are the respective intrinsic rates of increase. Using Equation 6.7 for r, when it is small, we have

$$(r^0 - r)T = \left[\frac{R_0^0 - 1}{R_1^0} - \frac{R_0 - 1}{R_1}\right] T \sim \frac{R_0^0 - R_0}{R_0}, \tag{4}$$

assuming $R_0^0 - R_0$ is small and also, as has been derived as an approximation, that $R_1 = R_0 T$. When k_x, and hence q'_x, are very small,

$$u_x = \prod_{i=1}^{m} (1 - q'_x) \sim 1 - \sum_{i=1}^{m} q'_x = 1 - \delta_x,$$

where $\delta_x = \sum_{i=1}^{m} q'_x$, so that $(1/u_x - 1) \sim \delta_x$. Equation 3 then takes the simple form

$$R_0^0 - R_0 = \sum_{x=0}^{m} l_x b_x \delta_x, \tag{5}$$

where δ_x is, approximately, the cumulative probability of death by age x due to the trait.

6.4. *The calculation of the fitness of individuals who contract a disease that has a variable late age of onset.*

We consider only the individuals destined ultimately to contract the disease. With Huntington's chorea, for example, these would be the individuals carrying the gene for the disease. Given that the probability the disease will be first contracted at age x is ψ_x, that normal fertility and mortality (b_x and l_x) prevail before the onset of the disease, and that, after the onset, fertility and mortality change to b'_x and l'_x, we shall calculate the number of children born to survivors at age x.

Let

ψ_x = probability of onset between ages x and $x + 1$,
q_x = probability of dying at age x before onset of the disease,
b_x = age-specific birth rate before onset,
q'_x = probability of dying at age x after onset,
b'_x = age-specific birth rate after onset.

The probability of not contracting the disease by age x is

$$(1 - \psi_1)(1 - \psi_2) \cdots (1 - \psi_{x-1}) = \prod_{i=1}^{x-1} (1 - \psi_i).$$

The probability of surviving to age x before onset is $l_x = (1 - q_1)(1 - q_2) \cdots (1 - q_{x-1})$. Thus, the number of births to individuals of age x who have not yet contracted the disease, is

$$l_x b_x \prod_{i=1}^{x-1} (1 - \psi_i) = l_x b_x L_x,$$

where

$$L_x = \prod_{i=1}^{x-1} (1 - \psi_i)$$

is the probability of not yet having developed the disease by age x. The probability of first contracting the disease at age r is

$$\psi_r \prod_{i=1}^{r-1} (1 - \psi_i).$$

The probability of surviving to age x having contracted the disease at age $r(<x)$ is

$$(1 - q_1)(1 - q_2) \cdots (1 - q_{r-1})(1 - q'_r)(1 - q'_{r+1}) \cdots (1 - q'_{x-1})$$

$$= l_r \frac{l'_x}{l'_r},$$

where

$$l'_x = \prod_{i=1}^{x-1} (1 - q'_i).$$

The number of births to individuals of age x who contracted the disease at age $r(<x)$ is therefore

$$(b'_x)\left(l_r \frac{l'_x}{l'_r}\right)\left(\psi_r \prod_{i=1}^{r-1} (1 - \psi_i)\right).$$

The total number of births to individuals of age x who have already contracted the disease is thus

$$l'_x b'_x \sum_{r=1}^{x} \frac{l_r}{l'_r} \psi_r \prod_{i=1}^{r-1} (1 - \psi_i).$$

The total number of births to individuals of age x of children destined to get the disease is therefore

$$p'(x) = l_x b_x L_x + l'_x b'_x \sum_{r=1}^{x} \frac{l_r}{l'_r} \psi_r \prod_{i=1}^{r-1} (1 - \psi_i), \tag{1}$$

as given in Equation 6.12.

Data on age of onset is usually given in the form of the number (or proportion) of people who develop the disease *at* age x, rather than *by* age x (the probability of developing it by age x is $1 - L_x$). Let a_x be the probability of developing the disease at age x. Then

$$a_x = L_x \psi_x \tag{2}$$

Now, from the definition of L_x we have, also, that

$$L_x(1 - \psi_x) = L_{x+1} = L_x - a_x \tag{3}$$

from Equation 2. Thus

$$a_x = L_x - L_{x+1}$$

for $x = 0$ to m (the maximum achievable age). Since L_0 must be 1, we have

$$\sum_{i=0}^{x-1} a_x = L_0 - L_x = 1 - L_x,$$

and so

$$L_x = 1 - \sum_{i=0}^{x-1} a_x. \tag{4}$$

Equations 3 and 4 allow us to tabulate L_x, and ψ_x, which from Equation 1 are needed for the calculation of $p'(x)$, using observed information on the number of individuals who develop the disease at a given age.

6.5. *The index of total opportunity for selection I, can be expressed in terms of separate mortality and fertility contributions.*

We assume as before that a proportion p_d of persons (counted at birth) die before the reproductive age and a proportion p_s survive to an age at which they can have children, where $p_d + p_s = 1$. Let p_x $(x = 0, 1, 2, \ldots, n)$ be the proportion of parents who survive to have x children so that

$$p_s = \sum_{x=0}^{n} p_x,$$

where n is the maximum possible number of offspring per parent. From Equation 6.26

$$I = \frac{1}{\bar{x}^2}\left[p_d(0 - \bar{x})^2 + \sum_{x=0}^{n} p_x(x - \bar{x})^2\right],$$

where

$$\bar{x} = \sum_{x=0}^{n} x p_x,$$

and further, we define

$$V_m = p_d \bar{x}^2 + p_s(\bar{x}_s - \bar{x})^2 \quad \text{and} \quad V_f = \frac{\sum_{x=0}^{n} p_x(x - \bar{x}_s)^2}{\sum_{x=0}^{n} p_x},$$

where from Equation 6.27

$$\bar{x}_s = \frac{\bar{x}}{p_s}. \tag{1}$$

Now I can be written in the form

$$I = \frac{1}{\bar{x}^2}\left[p_d(0 - x)^2 + \sum_{x=0}^{n} p_x(x - \bar{x}_s + \bar{x}_s - \bar{x})^2\right]$$

$$= \frac{1}{\bar{x}^2}\left[p_d(0 - \bar{x})^2 + \sum_{x=0}^{n} p_x(\bar{x}_s - \bar{x})^2 + \sum_{x=0}^{n} p_x(\bar{x}_s - x)^2\right] \qquad (2)$$

since

$$\sum_{x=0}^{n} p_x(x - \bar{x}_s)(\bar{x}_s - \bar{x}) = (\bar{x}_s - \bar{x})\sum_{x=0}^{n} p_x(x - \bar{x}_s) = 0$$

by the definition of $\bar{x}_s$. Now

$$\sum_{x=0}^{n} p_x(\bar{x}_s - \bar{x})^2 = (\bar{x}_s - \bar{x})^2 \sum_{x=0}^{n} p_x = p_s(\bar{x}_s - \bar{x})^2.$$

Thus from Equation 2 and from the definition of V_m and V_f, we have

$$I = \frac{1}{\bar{x}^2}(V_m + p_s V_f) = \frac{V_m}{\bar{x}^2} + \frac{1}{p_s}\frac{V_f}{\bar{x}_s^2} = I_m + \frac{1}{p_s} I_f,$$

as given in Equation 6.28, since, from Equation 1,

$$\frac{p_s}{\bar{x}^2} = \frac{p_s}{\bar{x}_s^2 p_s^2} = \frac{1}{p_s \bar{x}_s^2},$$

and where, by definition

$$I_m = \frac{V_m}{\bar{x}^2} \quad \text{and} \quad I_f = \frac{V_f}{\bar{x}_s^2}.$$

Note that

$$I_m = \frac{p_d \bar{x}^2 + p_s(\bar{x}_s - \bar{x})^2}{\bar{x}^2} = p_d + p_s\left(\frac{1}{p_s} - 1\right)^2$$

from Equation 1, and so

$$I_m = \frac{p_d p_s + (1 - p_s)^2}{p_s} = \frac{p_d}{p_s}$$

since $p_d + p_s = 1$.

7

Inbreeding

7.1 Consanguinity and Inbreeding

Two individuals are said to be consanguineous if they have *at least one ancestor in common*. The common ancestor must not be too remote or else the concept becomes meaningless, and consanguinity would, in any case, then not be detectable. On the basis of the theory of evolution, all individuals of a species are to some extent consanguineous, since all are descended from remote common ancestors. Consanguinity, and level of inbreeding, must therefore be defined as applying only to relationships established after some evolutionary time point at which, for convenience, everyone is considered to be unrelated. The arbitrariness of the starting point is not, in general, a real problem in the interpretation of inbreeding effects since it is relatively recent consanguinity that is pertinent. In practice, for human consanguinity common ancestors more remote than a great great grandparent are rarely considered. In some human societies more distant consanguinities may have social significance, but from a genetic point of view the connection between two individuals who have one great great great grandparent in common (fourth half-cousins) is, as we shall see, very tenuous indeed.

The progeny of consanguineous parents is, by definition, **inbred**. Genetic effects of inbreeding can be traced to the fact that the inbred individual may carry a double dose of a gene that was present in a single dose in the common ancestor. That is, the gene present in the common ancestor in a single dose may be inherited by *both* consanguineous parents, and their progeny may inherit the gene from *both* parents and thus obtain a *double* dose of it.

A recessive gene carried in a single dose in a common ancestor may remain hidden until it comes to light for the first time in an inbred descendant. Intuitively, we expect, therefore, that recessive traits will occur with increased frequency in the progeny of consanguineous mates. Thus, consanguinity is important in the study of recessive inheritance and of recessive traits.

The frequency with which consanguineous matings occur in a population depends in part on the population structure and also on social customs. Very close consanguineous matings are avoided in human populations, and practically all living societies consider incest to be taboo. The exact degree of relationship at which mating is considered incestuous may differ slightly from one society to another, but, in general, parent-offspring and brother-sister matings are forbidden in all societies. Incest nevertheless occurs, but at a very low rate. For example, from the numbers known to charitable institutions, we can estimate that the number of persons in Italy who are the product of incestuous unions may be in the proportion of one in 100,000 births, but this is likely to be an underestimate. Twenty children conceived in incest, who were one year old or younger, were studied in Michigan by Adams and Neel (1967). These children corresponded to about 10^{-4} of all births and most of them were the result of brother-sister mating, the others of father-daughter matings.

Marriage with a sib's progeny (uncle-niece or aunt-nephew) is incestuous and illegal in some societies. In others it is permissible under specific dispensation; in yet others it is permitted or even considered desirable. A similar variety of societal judgments apply to first-cousin marriages, namely, the unions between progeny of sibs. Discussion of the frequencies of consanguineous marriages will be deferred to Section 7.4 and to Chapter 8.

Genes can be identical by nature or identical by descent.

The understanding of inbreeding is simplified by the distinction between identity of genes by descent and identity by nature, which was pointed out by Malécot (1948).

By definition, an inbred individual is connected through both his father and his mother to the same recent ancestor. He can thus receive two copies of a gene that was carried by the common ancestor. Two such copies are said to be **identical by descent**. They are also identical by nature, that is, physically identical, unless mutation has taken place in one of the lines of descent, which, however, is a very rare event. The probability that mutation has not occurred between the common ancestor and the inbred individual, and therefore that the two genes that are identical by descent are also identical by nature, is $(1 - \mu)^{n_1+n_2}$, where n_1 is the number of generations from the common ancestor to the inbred individual via the father, and n_2 via the mother, and μ is the mutation rate per generation at that locus. This quantity is usually so near to 1 that it will be assumed to be 1 in most of what follows.

An individual who is homozygous for a given gene carries at a certain locus two homologous genes that are physically identical but not necessarily identical by descent. This physical identity is what we call identity by nature. Inbreeding makes an individual homozygous for genes that are identical by descent.

7.2 The Inbreeding Coefficient

The coefficient of inbreeding F is the probability that an individual receives at a given locus two genes that are identical by descent.

The coefficient F was originally defined by Sewall Wright (1921, 1922), who was the main pioneer of studies of effects of inbreeding. The definition that we have given is, however, due to Malécot (1948).

It is easy to compute F from first principles if the common ancestor is a near one. Let us consider as an example an uncle-niece marriage (Figure 7.1) further simplified by the fact that the uncle (B) and his sib (C), who is the parent of the niece,

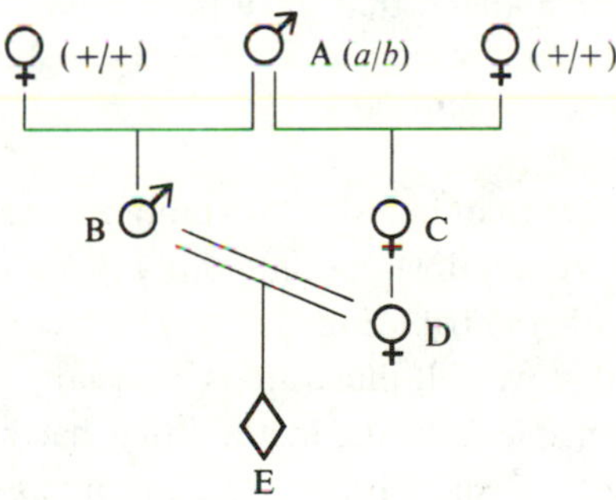

FIGURE 7.1
Half-uncle–niece marriage. Coefficient of inbreeding $F = 1/16$.

have the same father but different mothers (half-uncle–niece). Only genes from the sibs' father (A) can be made homozygous in the progeny (E) of the uncle-niece marriage. Let us call the two alleles at one locus in the sibs' father a and b. Let us call any other allele at the same locus +, and assume that all + alleles are different from each other and from *a* and *b*. We are interested in the probability that progeny E of the uncle-niece marriage represented in Figure 7.1 are homozygous for gene *a* or for gene *b*.

The uncle (B) can be either $a+$, or $b+$, with probability $\frac{1}{2}$ for each genotype. The same is true of C. The niece D will be $a+$ in $\frac{1}{4}$ of the cases, $b+$ in $\frac{1}{4}$, and $++$ in the rest. The probabilities of the various possible genotypes for E are as follows:

GENOTYPES OF D			GENOTYPES OF B: $a+$ ($\frac{1}{2}$)	GENOTYPES OF B: $b+$ ($\frac{1}{2}$)
	$a+$	$\frac{1}{4}$	*aa* in $\frac{1}{4}$ of progeny	—
	$b+$	$\frac{1}{4}$	—	*bb* in $\frac{1}{4}$ of progeny

The probability of homozygous children E is $\frac{1}{4}$ when B is $a+$ and D is $a+$, and this will obtain in $\frac{1}{4} \times \frac{1}{4} \times \frac{1}{2} = \frac{1}{32}$ of all cases. The probability of *bb* E children is the same. The total probability that E has two genes identical by descent is, therefore, $\frac{1}{32} + \frac{1}{32} = \frac{1}{16} = F$.

If the uncle (B) and the niece's mother (C) were *full sibs* the F value would be twice as much, or $F = \frac{1}{8}$, which is the standard value for uncle-niece or aunt-nephew marriages.

Adding a generation to the chain of the relationship halves the probability of identity by descent. Thus in marriages between first cousins, $F = \frac{1}{16}$. The main degrees of relationship and their respective F values are given in Table 7.1. The table also indicates various terminologies that have been used.

Every chain of n generations contributes $(\frac{1}{2})^{n+1}$ *to F.*

Fairly often there is more than one pair of common ancestors in the pedigree of an individual. The effect of multiple consanguinity is additive in the F value with every chain of n generations contributing $(\frac{1}{2})^{n+1}$. The method for taking account of multiple consanguinity that we will illustrate is especially suitable for a computer, but is not different in principle from the method that has been taught for centuries to Roman Catholic priests. Dispensations for consanguineous marriages must be granted before such marriages can be celebrated in the church and therefore priests have been taught how to measure consanguinity, in order to distinguish degrees that are dispensable from those that are not. Most data on consanguinity have actually been obtained from Roman Catholic records.

Figure 7.2 gives pedigrees, in which names have been replaced by numbers, of a consanguineous husband and his wife. The pedigree lists ancestors of the two individuals back to great great grandparents. This makes it possible to detect consanguinities as distant as third cousins. For the measurement of consanguinity it is

TABLE 7.1

The Most Common Types of Consanguineous Matings, Their Symbols and Inbreeding Coefficients: □ = *male,* ○ = *female,* ◇ = *an individual of either sex*

Type	Symbol	*Degrees of Relationship*		*Inbreeding Coefficients (F)*[a]	
		Roman Catholic Usage	*Napoleonic Code*	*Full*	*Half*
Uncle-niece; aunt-nephew	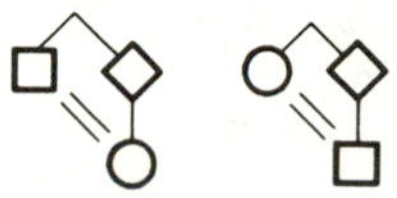	I in II	III	1/8	1/16
First cousins	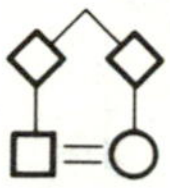	II	IV	1/16	1/32
First cousins once removed ($1\frac{1}{2}$)	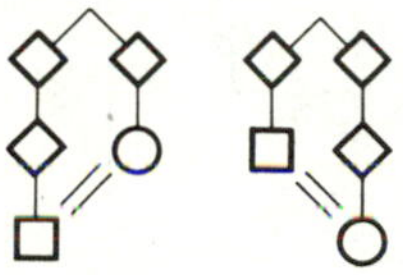	II in III	V	1/32	1/64
Second cousins	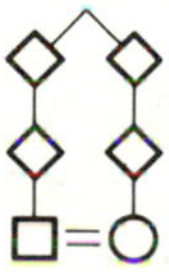	III	VI	1/64	1/128
Second cousins once removed ($2\frac{1}{2}$)	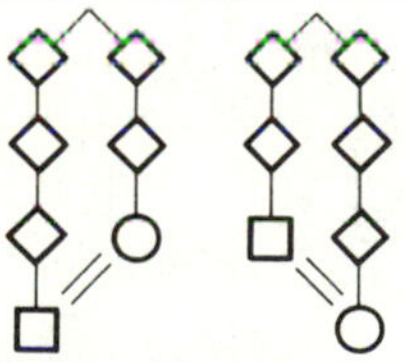	III in IV	VII	1/128	1/256
Third cousins	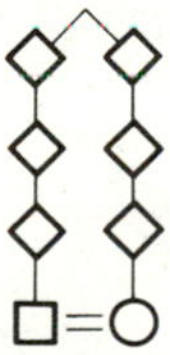	IV	VIII	1/256	1/512

[a] "Full" and "half" refer to the two sibs starting the chains of descent, who are the two top individuals in each pedigree. Full sibs have both parents in common, and half sibs only one parent.

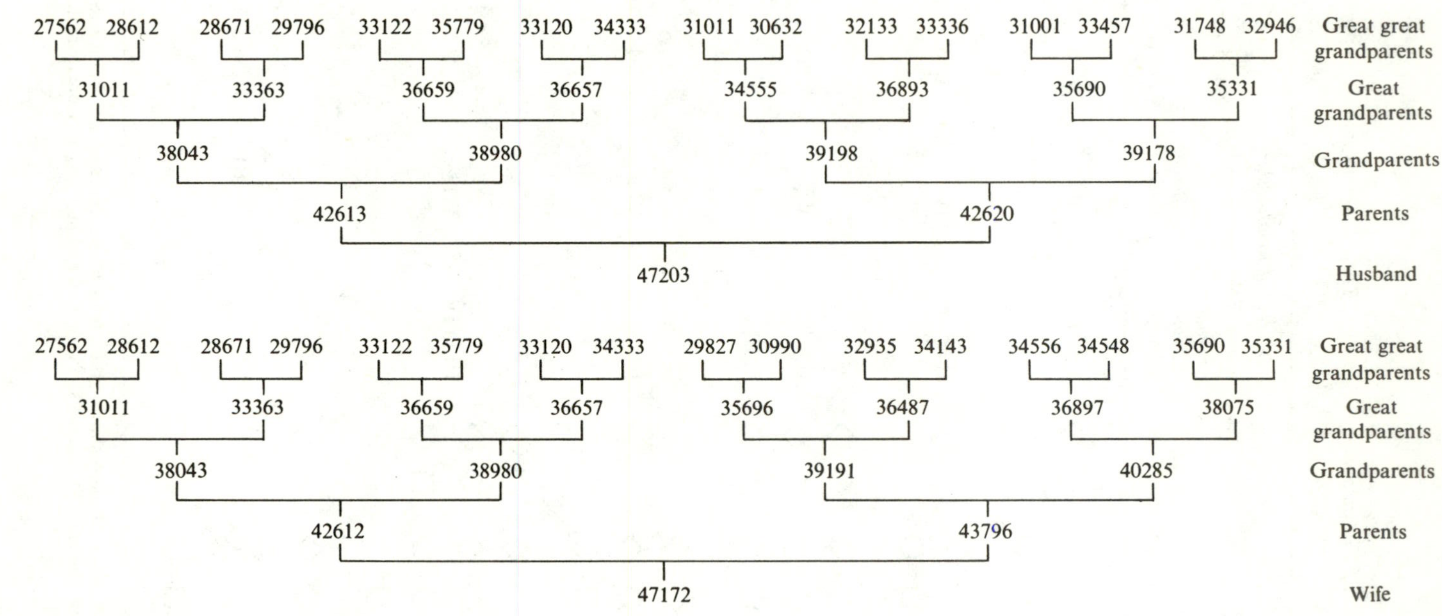

FIGURE 7.2
Pedigrees of consanguineous husband and wife showing their ancestry up to great great grandparents. Names are replaced by numbers to ease computer use. (From Cavalli-Sforza and Zei, unpublished.)

necessary to detect ancestry common to both the husband's and the wife's line. This can be done by testing for identity the numbers (names) of each ancestor in the husband's line with that of each ancestor in the wife's line. Ancestors appearing in both lines are represented in Table 7.2 by the *nearest* common ancestor in any given line. From the analysis of the pedigree of Figure 7.2 given in Table 7.2, it appears that the husband and wife are consanguineous as first cousins, and also as second cousins once-removed, as well as half second cousins once-removed. It will be noted that counting generations as in Table 7.2—namely, starting with consanguineous individuals considered as generation 0—gives the same evaluation as does Table 7.1. The value of F for the whole pedigree is $19/256 = 0.07422$. Figure 7.3 shows the pedigree represented by chains of descent, as is more customary.

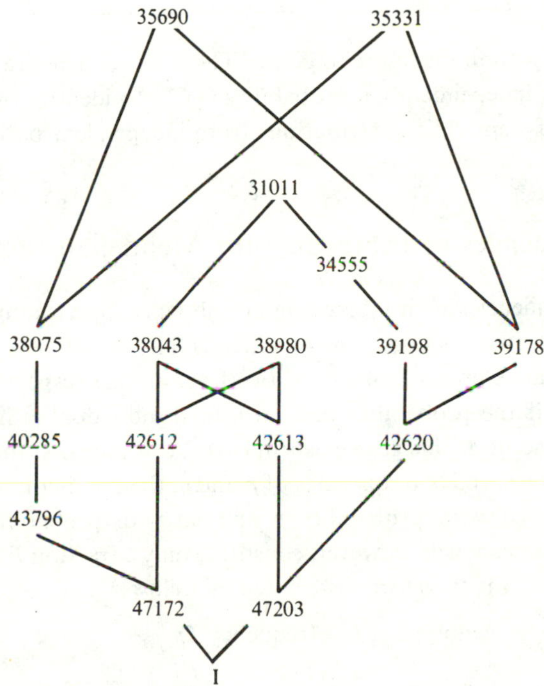

FIGURE 7.3
Same family as in Figure 7.2. I is the individual whose inbreeding coefficient F is computed in Table 7.2; that is, the progeny of the consanguineous husband and wife whose pedigrees are given in Figure 7.2.

A standard procedure, suggested by Sewall Wright (1922) for computing the inbreeding coefficient makes use of pedigrees as shown in Figure 7.3. Here only common ancestors, and ancestors intermediate between husband or wife and the common ancestors are shown. The steps (each corresponding to one generation) are

TABLE 7.2

Identity of Maternal and Paternal Ancestors (from Figure 7.2). Generations (n) are counted from the common ancestors to the consanguineous individuals 47172 and 47203. F is the inbreeding coefficient of their progeny

Maternal Line			*Paternal Line*					
Ancestors		*Generations*	*Ancestors*		*Generations*	*n Corresponding to Each Common Ancestor*		*Contribution to F*
38043	38980	2	38043	38980	2	4(38043)	4(38980)	$(\frac{1}{2})^5 + (\frac{1}{2})^5 = \frac{1}{16}$
35690	35331	4	35690	35331	3	7(35690)	7(35331)	$(\frac{1}{2})^8 + (\frac{1}{2})^8 = \frac{1}{128}$
31011		3	31011		4	7(31011)		$(\frac{1}{2})^8 = \frac{1}{256}$
								$F = \frac{19}{256} = 0.0742$

counted in each path in the figure to give n. The n values here are 4, 4, 7, 7, 7. Each path contributes independently a probability $(\frac{1}{2})^{n+1}$ to identity by descent, where n is the number of steps. The contributions from independent paths are summed to give $F = 19/256$.

7.3 Consequences of Inbreeding for Mendelian Populations

Genotype frequencies with inbreeding are obtained by assuming that the population is split into two fractions, one of which is fully inbred, and the other bred at random. The relative proportions of the two fractions are respectively F and $1 - F$. The quantity F is the probability that an inbred individual is homozygous for a given gene by descent. If this gene exists in two allelic states, A and a, in the general population, with respective frequencies of p and q, then individuals homozygous by descent will be AA with probability p and aa with probability q. Individuals homozygous by descent will, however, constitute only a fraction F of the population, by definition, which is therefore distributed as follows:

$$\begin{array}{rll} \text{genotype } AA: & \text{frequency} & Fp \\ aa: & & Fq \\ \hline & \text{total} & F(p+q) = F. \end{array}$$

In the rest of the population, that is, the fraction $1 - F$, the alleles will be distributed among the three possible genotypes according to the normal rules of random mating. The frequencies of the genotypes are thus as follows:

$$\begin{array}{rl} AA: & (1-F)p^2 \\ Aa: & (1-F)2pq \\ aa: & (1-F)q^2 \\ \hline \text{total} & (1-F). \end{array}$$

The sum of the two fractions, the inbred and the random bred are given in the following tabulation:

Genotypes	*Frequencies*
AA:	$Fp + (1 - F)p^2 = p(p + Fq)$
Aa:	$(1 - F)2pq = 2pq(1 - F)$
aa:	$Fq + (1 - F)q^2 = q(q + Fp)$

This representation of a partially inbred population was first given by Bernstein (1930; see also Wright, 1951).

It is valid not only for the condition for which it was derived here, namely a mixture of fully inbred individuals ($F = 1$) and of individuals mated completely at random ($F = 0$) in the proportions F and $1 - F$, respectively, but also for the progeny of matings having a given inbreeding coefficient F. It is also valid for a population, in which matings have different degrees of inbreeding, with an average value F.

Since F in principle alters the proportions of genotypes, inbreeding in a population can be measured by the departure from Hardy-Weinberg expectations, but in human populations this departure is, in practice, very small (see Example 7.1).

Prolonged inbreeding increases F toward a value of one.

In animals and plants, continued inbreeding is the way to obtain **pure strains**, or, in other words, strains homozygous for as many genes as possible. If the breeder starts with a random mating population with frequencies p and q for alleles A and a, inbreeding with coefficient F will lead to homozygosis for allele A with the probability $Fp + (1 - F)p^2$ and for allele a with the probability $Fq + (1 - F)q^2$. The probability that an individual will be homozygous for one or the other allele will be the sum of these two probabilities, namely $F + (1 - F)(1 - 2pq)$. If the same amount of inbreeding is continued over a number of generations, the proportion of homozygotes will increase as F increases. On repetition *ad infinitum* even of a moderate degree of inbreeding, complete homozygosis ($F = 1$) may eventually be attained. The speed of approach to equilibrium will be higher, the closer the inbreeding. If, on the other hand, there is a mixture of random mating and inbreeding, eventually, even after an infinite number of generations, an average inbreeding coefficient will be reached that is less than one. In experimental organisms, in which inbreeding can be carried out almost at will, it can be shown that a fairly large number of generations may be required for a substantial fraction of the population to become homozygous, even when close inbreeding such as brother-sister mating is practiced at every generation.

TABLE 7.3

Consanguineous Marriages in Human Populations. R.C. = Roman Catholics, disp. = dispensations; α values marked with an asterisk include consanguineous marriages from complex pedigrees not specified in the table

Location	Period Covered	Total No. of Marriages	Percent That Are: Uncle-niece, Aunt-nephew Marriages	First Cousins	1½ Cousins	Second Cousins	α	Source of Data	Author
Argentina (R.C.)	1956–1957	51,391	0.03	0.75	0.10	0.24	0.00058	R.C. disp.	Freire-Maia (1968)
Belgium	1918–1959	2,404,027	0.02	0.49	0.16	0.76	0.0005	R.C. disp.	Twisselmann (1961)
Brazil (R.C.)	1956–1957	212,090	0.06	2.63	0.81	1.32	0.00225	R.C. disp.	Freire-Maia (1968)
Canada									
French population	1885–1895	23,410	0.05	2.03	1.10	3.60	0.0029*	R.C. disp.	Laberge (1967)
	1915–1925	29,178	0.12	2.16	1.27	3.84	0.0032*		
	1945–1955	47,276	0.02	0.55	0.55	1.97	0.0013*		
	1955–1965	50,128	0.01	0.37	0.36	1.36	0.0009*		
total	1959	51,729		0.37	0.24	0.91	0.00045	R.C. disp.	Freire-Maia (1968)
Chile (R.C.)	1956–1957	28,596	0.07	0.80	0.29	0.15	0.00074	R.C. disp.	
Colombia (R.C.)	1956–1957	34,470	0.02	1.25	0.58	1.10	0.00119	R.C. disp.	
Cuba (R.C.)	1956–1957	2,277		0.53	0.26	0.04	0.00054	R.C. disp.	
France									
Loir-et-Cher	1812–1954	212,837		1.35	0.33	1.24	0.0011	R.C. disp. (Bishopric archives)	Sutter and Tabah (1955)
Finistère	1911–1953	243,859		1.02	0.39	2.01	0.0011		
Guinea									
Fouta-Djallon	1955	739		19.08	0.54	6.22	0.026	Survey	Cantrelle and Dupire (1964)
India									
Andra-Pradesh	1963	2,177	7.26	16.62			0.019	Survey	Dronamraju (1964)

Andra-Pradesh	1957–1958	6,945	9.23	33.30			0.032	Survey	Sanghvi (1966)
Italy	1911–1915	1,065,873	0.05	1.62	0.48	1.52	0.0015	R.C. disp.	Moroni et al.
	1916–1920	999,383	0.06	2.17	0.48	1.75	0.0019	(Vatican	(in press)
	1921–1925	1,544,184	0.05	1.85	0.40	1.60	0.0018	archives)	
	1926–1930	1,305,323	0.04	1.51	0.35	1.54	0.0013		
	1931–1935	1,270,328	0.03	1.33	0.32	1.41	0.0019		
	1936–1940	1,463,042	0.02	1.19	0.27	1.25	0.0011		
	1941–1945	1,137,322	0.02	1.14	0.26	1.00	0.0010		
	1946–1950	1,733,270	0.03	1.18	0.36	1.25	0.0011		
	1951–1955	1,522,560	0.02	0.93	0.28	1.02	0.0009		
	1956–1960	1,646,612	0.01	0.77	0.23	0.89	0.0007		
Northern	1640–1699	22,229		0.01	0.03	0.26	0.00007*	R.C. disp.	Moroni (1967b)
Italy	1700–1799	59,509		0.04	0.07	0.30	0.00012*	(Bishopric	
	1800–1899	127,133	0.01	0.32	0.23	1.03	0.00053*	archives)	
	1900–1965	141,508	0.02	0.67	0.25	1.21	0.00077*		
Sardinia	1800–1849	57,745		0.26	0.28	1.66	0.00058*	R.C. disp.	Moroni (1966)
	1850–1899	176,666	0.03	1.09	0.75	2.59	0.00165*	(Bishopric	
	1900–1965	466,189	0.03	1.66	0.56	1.95	0.00167*	archives)	
Japan									
(Average of various places)	various in 1900's	152,790		6.15	1.33	2.28	0.0046		Schull and Neel (1965)
Mexico (R.C.)	1956–1957	28,292		0.17	0.15	0.95	0.00031	R.C. disp.	Freire-Maia (1968)
The Netherlands	1906–1918	572,932	0.04	0.66			0.00005	State	Polman (1951)
	1937–1948	843,005	0.02	0.15			0.00001	archives	
Spain	1911	~138,600	0.07	1.74	0.60	2.43	0.00185*	R.C. disp.	Cisternas and Moroni
	1925	~157.000	0.09	1.87	0.58	2.48	0.00197*	(Vatican	(1967)
	1930	~170,000	0.07	2.00	0.57	2.70	0.00203*	archives)	
US									
R.C.	1959–1960	133,228		0.08	0.02	0.11	0.00009	R.C. disp.	Freire-Maia (1968)
Mormons	1920–1940	132,524		0.61			0.00038	Archives	Woolf et al. (1956)

The amount of inbreeding that takes place among humans is extremely small. The closest consanguineous marriages found in any human population have coefficients of inbreeding of the order of $\frac{1}{8}$ (uncle-niece, aunt-nephew, double first cousins). Except in some very special populations (in India, for example, see Table 7.3) such marriages are extremely rare. The two most common types of consanguineous marriages in most populations are those between first cousins ($F = \frac{1}{16}$) and between second cousins ($F = \frac{1}{64}$). The relative frequencies of these cousin marriages are usually low in most populations, so that the average degree of inbreeding in a human population is always very small and no genetic "purity" could ever be achieved. Further discussion of the effects of continued inbreeding will be found in Chapter 8.

7.4 Average Inbreeding in Human Populations

It is customary to measure the average consanguinity in a population as the average inbreeding coefficient of its individuals. This is defined by the quantity called α, where

$$\alpha = \Sigma\, p_i F_i .$$

Here p_i is the relative frequency of inbred individuals with inbreeding coefficient F_i. This is usually estimated from the marriages rather than from the progeny. Observed values for a fairly large number of human populations are given in Table 7.3. The average inbreeding coefficient α in human populations can be seen from Table 7.3 to be generally less than 1 per 1000.

There are a few areas or populations in which the inbreeding coefficient is higher than 0.01, but they are exceptional. This is true for some isolates, namely relatively small populations that have little or no gene exchange with other populations. Table 7.4 gives especially high values of average inbreeding coefficients found in some such isolates. The inbreeding values observed in small isolates are often as extreme as those of a large South Indian population in which high values are the outcome of preferential uncle-niece mating (see Table 7.3). High values are, however, not necessarily attained even in small isolates. Polar Eskimos, by a careful avoidance of inbreeding, have maintained a low value of $F(<0.003)$ in a very small isolate (Sutter and Tabah, 1956).

These inbreeding estimates take into account only easily detectable consanguinity, which rarely includes relationships more remote than third cousins. The contribution of more distant relationships can sometimes be estimated either directly or indirectly. For direct estimation, more complex approaches are necessary, and these will be considered in Chapter 8. More remote common ancestors contribute less and less to the inbreeding, though they also, get relatively more numerous, the

TABLE 7.4
Special Cases of High Inbreeding Coefficients in Isolates

Isolate and Region		*Period*	*α Values*	*No. of Indiv.*	*Author*
S-Leuts (Hutterites)	South Dakota and Minnesota	1874–1960	0.0216	5450	Mange (1964)
Dunkers	Pennsylvania	1950	0.0254	350	Glass et al. (1952)
Samaritans	Israel and Jordan	1933	0.0434	350	Bonné (1963)
Tristan da Cunha		1938	0.0365	~300	Bailit, Damon, and Amon (1966)
Aeolian Islands		1825–44	0.0033		Moroni (1967b)
		1885–99	0.0092		
		1900–1917	0.0101		
		1918–1949	0.0059		
		1950–1966	0.0031	11,800	
Rama Navajo	New Mexico	last 4 generations	0.0080	614	Spuhler and Kluckhohn (1953)

greater the distance into the past. When more and more past generations are considered, the influence of mutation in introducing new alleles, and thus in making genes that are identical by descent not identical by nature, can no longer be neglected. Migration is also important, and leads to an equilibrium between mutation and migration on one side, and inbreeding on the other. The values of average inbreeding coefficients given in Table 7.3, which include only the closest relationships, may well therefore represent an underestimate of the total inbreeding, as we shall discuss later (Chapter 8).

7.5 Consequences of Inbreeding: Genetic Loads

The increase in homozygosity due to inbreeding leads to a loss of fitness at those loci for which homozygotes are less fit than heterozygotes.

We know that genes which are deleterious when homozygous can belong to either of two major classes:

1. Genes that are not, or are only moderately, deleterious when heterozygous. The fitnesses of the genotypes *AA* and *Aa* are then almost the same and it is the genotype *aa* that is at a disadvantage. Under such conditions the gene frequency of *a* will be kept very low by the balance between mutation and selection.

2. Genes for which the heterozygote is the fittest genotype, and is therefore more fit than either homozygote. Such genes become polymorphic, so that either allele can attain a relatively high frequency. They are often called *overdominant*.

For both types of genes, homozygotes will occur with increased frequency in the progeny of consanguineous matings and thus the average fitness of the population will decrease with inbreeding.

The genetic load is a quantity designed to measure the loss of fitness in a population due to selection.

A **genetic load** is defined to be the relative decrease in the average fitness of a population, with respect to the fitness it would have if all individuals in the population had the genotype that has maximum fitness (Crow, 1958). The concept of *load*, which was first applied to genes maintained in equilibrium by mutation–selection balance, was introduced by Muller in a famous paper in 1950, who called the loss of zygotes due to such loci (class one above) the **mutational load**. In general, if w_{max} is the maximum fitness and $\bar{w}$ the average fitness, the load is defined by

$$\frac{w_{max} - \bar{w}}{w_{max}}.$$

When the optimum genotype is assigned a fitness of one ($w_{max} = 1$), then the genetic load is simply equal to $(1 - \bar{w})$. The genetic load due to overdominant loci is often called the **segregational load**.

Morton, Crow, and Muller, in 1956, developed a method of analyzing the genetic load that was aimed at distinguishing the relative importance of the two classes of genes, deleterious recessives and overdominants, in determining the genetic load. We shall now examine their method of approach because of its interest for the analysis of inbreeding effects, especially in relation to deleterious recessive mutations. Unfortunately, for both practical and theoretical reasons, the original aim of their analysis has not been achieved.

MUTATIONAL LOAD

Let us consider for simplicity a fully recessive lethal gene with frequency q. Homozygotes, whose frequency is q^2 under random mating, die. The resulting loss of individuals per generation is therefore q^2. The load as defined above, is the relative decrease of fitness of the population, taking the fitness of the best genotype as one. In this instance both the normal homozygote (frequency p^2) and the heterozygote (frequency $2pq$) have a fitness of one, while the homozygote for the lethal allele has a fitness of zero. The average fitness of the population is thus

$$p^2 \times 1 + 2pq \times 1 + q^2 \times 0 = 1 - q^2.$$

If the deleterious allele were absent and the population were made entirely of the best genotype, or genotypes, its average fitness would be one. Thus, the decrease of fitness due to the deleterious allele is q^2 and this is, by definition, the genetic load of that population due to the deleterious allele. It is also the loss of zygotes, given that the fitness of the best genotype is one. If the population is at equilibrium we know that $q^2 = \mu$, and thus the genetic load is, in this simple instance, equal to the mutation rate (see Haldane, 1937).

The more general formulation given in Table 7.5 shows that with equilibrium

TABLE 7.5
Mutational Load at One Locus with Two Alleles in a Random Mating Population

			Contribution to Load		
				At Equilibrium	
Genotype	*Frequency*	*Fitness*	*In General*	*If h Is Small*	*If h Is Not Small*
AA	p^2	1	0	0	0
Aa	$2pq$	$1 - hs$	$2pqhs$	neglig.	2μ
aa	q^2	$1 - s$	sq^2	μ	neglig.
Total load (L)			$sq(2ph + q)$	μ	2μ

gene frequencies, the total mutational load L lies between μ for a full recessive, and 2μ for a dominant for which h, the heterozygote fitness factor, is not too small. In fact in the first case $L = sq^2 = \mu$ and in the second $hsq \sim \mu$ and $L \sim 2hsq = 2\mu$, at mutation-selection equilibrium (see Chapter 3). The load at equilibrium is midway between these two extremes, that is, $L = 1.5\mu$, when h is about $(\mu/2s)^{1/2}$ and the loss of a genes from homozygotes is about the same as that from heterozygotes. (See Example 7.2; see Kimura (1961) for a more detailed treatment of this problem.)

In an inbred population with a coefficient of inbreeding F, the proportions of the genotypes are as given in the diagram on page 349, and the mutational load can be calculated as in Table 7.6.

TABLE 7.6
Mutational Load for a Single Diallelic Locus in a Partially Inbred Population

Genotype	*Frequency*	*Fitness*	*Contribution to Load*
AA	$p^2(1 - F) + Fp$	1	0
Aa	$2pq(1 - F)$	$1 - hs$	$2pq(1 - F)hs$
aa	$q^2(1 - F) + Fq$	$1 - s$	$sq^2(1 - F) + sFq$
Total load (L)			$2pq(1 - F)hs + sq^2(1 - F) + sFq$

Reorganizing the terms in the expression for the total load given in Table 7.6, we obtain

$$L = F(sq - 2pqhs - sq^2) + sq^2 + 2pqhs. \tag{7.1}$$

Thus the mutational load is a linear function of the inbreeding coefficient F. It can be rewritten as

$$L = a + bF, \tag{7.2}$$

where

$$\begin{aligned} a &= sq^2 + 2pqhs = sq(q + 2ph) \\ b &= sq - 2pqhs - sq^2 = sq(1 - 2ph - q) \end{aligned} \tag{7.3}$$

In a random-breeding population ($F = 0$) the load (random load) is

$$L_0 = a. \tag{7.4}$$

In a population consisting entirely of inbred individuals ($F = 1$), the load is, from Equation 7.2

$$L_1 = a + b = sq. \tag{7.5}$$

The ratio

$$\frac{L_1}{L_0} = \frac{b}{a} + 1 \tag{7.6}$$

is widely used in applications of the load theory. The value of L_1 is usually calculated using for q the equilibrium gene frequency computed in the absence of inbreeding. The presence of inbreeding reduces gene frequencies, as we shall see later, but since the quantity L_1 is obtained in practice by extrapolation from populations with very low average inbreeding, the approximation introduced by ignoring inbreeding is probably not serious.

Consider first a fully recessive gene, for which $q_e = (\mu/s)^{1/2}$ (Chapter 3) and so $L_1 = (\mu s)^{1/2}$ and $L_0 = \mu$, giving $L_1/L_0 = (s/\mu)^{1/2}$. For a lethal gene $s = 1$, and L_1/L_0 is large, for example, more than 100 if the mutation rate is less than 10^{-4}.

Consider next a gene in which the fitness of the heterozygote is appreciably less than that of the normal homozygote, that is, h is not too small. Then, the equilibrium gene frequency will be approximately $q_e = \mu/hs$ and $L_1 = \mu/h$. The ratio of the inbred load to the random load is $L_1/L_0 = 1/2h$.

Nothing is known about the average value of the quantity h in man. There are, however, estimates for lethals in *Drosophila melanogaster*, giving $h = 0.02$ to 0.03. The extrapolation from Drosophila to man can clearly be questioned. However, accepting this as an average estimate of h, L_1/L_0 would be, on an average, of order 15–25, and therefore lower than for full recessives, but still quite large.

According to this simple theory, the mutational load is expected, from Equation 7.1, to increase linearly with the inbreeding coefficient.

SEGREGATIONAL LOAD

The segregational load also increases linearly with F.

Computations similar to those given for the mutational load can be made when selection is in favor of heterozygotes. The load is again equal to the loss of zygotes when the best genotype, which is now the heterozygote, is given a fitness of one. The application of load theory to overdominant loci is conceptually less satisfactory, because a population cannot contain only heterozygotes. A segregational load can nevertheless be formally defined and calculated as in Table 7.7.

TABLE 7.7
The Segregational Load for a Diallelic Locus in a Partially Inbred Population

Genotype	*Frequency*	*Fitness*	*Contribution to Load*
AA	$p^2(1-F)+pF$	$1-s$	$s[p^2(1-F)+pF]$
Aa	$2pq(1-F)$	1	—
aa	$q^2(1-F)+qF$	$1-t$	$t[q^2(1-F)+qF]$
Total load (L)			$(sp^2+tq^2)(1-F)$ $+F(sp+tq)$

Rearranging terms in the total load L, we find it to be a linear function of F, as follows

$$L = sp^2 + tq^2 + F(sp + tq - sp^2 - tq^2); \tag{7.7}$$

that is,

$$L = a_s + b_s F, \tag{7.8}$$

where

$$\begin{aligned} a_s &= sp^2 + tq^2 \\ b_s &= sp + tq - sp^2 - tq^2 = pq(s + t). \end{aligned} \tag{7.9}$$

The quantity a_s is the random load L_0 or the load in a random-mating population, in which $F = 0$. If the population is at equilibrium, we know (from Chapter 4) that $p = p_e = t/(s + t)$, and $q_e = 1 - p_e$, and therefore the load L_1 in a fully inbred population is

$$L_1 = a_s + b_s = sp + tq = \frac{2st}{s+t}. \tag{7.10}$$

Similarly, since

$$L_0 = a_s = sp^2 + tq^2 = \frac{st}{s+t}, \tag{7.11}$$

the ratio of the inbred load to the random load is

$$L_1/L_0 = 2. \tag{7.12}$$

This ratio will generally be appreciably lower than that for the mutational load, for which it was estimated to range from 10 upwards.

This prediction of a difference in the value of the ratio L_1/L_0 for mutational and segregational loads forms the basis for Morton, Crow, and Muller's approach to the assessment of the relative importance of the two loads.

Other types of genetic loads have been considered: an incompatibility load (due to selection with incompatibility), a substitutional load (due to selection for advantageous alleles, as in transient polymorphisms), and still others (see, for example, Crow, 1958). Because of the difficulties connected with the evaluation and interpretation of these loads they will not be discussed further here. Problems relating to the substitutional load are discussed in Chapter 11.

Mortality data have been extensively used to measure the genetic load.

Losses of zygotes due to adverse selection should, in principle, be countable as actual deaths in a population for which demographic data are available, such as a human population. When counting deaths, we cannot, however, determine which single genes have caused death except in rare circumstances. Thus the effects of deleterious alleles at all loci must be accumulated. In addition, there will be an unknown proportion of deaths due to nongenetic causes, which may be called "accidental."

It is, in most cases, difficult or impossible to classify a single death as genetic or nongenetic. There may certainly be little doubt with deaths among people with serious genetic diseases. Even the fitness of individuals having such diseases, however, depends on the environment, as therapy may in some cases alleviate, or even annul, selection against them. The value of fitness is therefore always a function of the environment.

We shall now consider the estimation of the load due to the combined effects of many loci on the assumption that the corresponding decrease of fitness is entirely accounted for by some fraction of, or perhaps all of, the prereproductive mortality. We further assume (1) that since the various genetic causes of death act *independently*, the probability of surviving the deleterious effects of two loci is equal to the product of the probabilities of surviving the effects of each of them; and (2) that genetic causes of death act independently from environmental causes of death.

These assumptions are critical to the subsequent development of load theory. However, as we shall discuss later, they may be open to serious question.

The probability of death due to deleterious alleles present at a given ith locus is the genetic load at that locus L_i. The probability of survival is thus

$$p_i = 1 - L_i. \tag{7.13}$$

Because of the first assumption, the probability of survival, considering all possible loci, is

$$P = p_1 p_2 \cdots p_n = \Pi p_i = \Pi(1 - L_i), \tag{7.14}$$

where Π indicates the product over all loci. Because of the second assumption, if the probability of surviving all environmental causes of death is P_a, the overall probability of survival is

$$P_s = P_a P. \tag{7.15}$$

Taking logarithms,

$$\log P_s = \log P_a + \log P = \log P_a + \sum \log(1 - L_i), \tag{7.16}$$

where the sum is extended over all loci.

The probability of death, L_i, for each locus is small, especially if we limit our consideration to the mutational loads. We can therefore use the well-known approximation

$$\log(1 - x) = -x, \tag{7.17}$$

which is quite satisfactory for $x < 0.01$. Equation 7.16 then becomes

$$-\log P_s = \sum L_i + X, \tag{7.18}$$

where X equals $-(\log P_a)$, that is, it corresponds to the nongenetic causes of death. As $L_i = a_i + b_i F$, for a given locus i,

$$\sum L_i = \sum a_i + F \sum b_i, \tag{7.19}$$

and so, from Equation 7.18,

$$-\log P_s = X + \sum a_i + F \sum b_i = A + BF. \tag{7.20}$$

Using Equations 7.2 and 7.3, the two constants A and B have the following meaning when we are limiting our consideration to mutational loads:

$$A = X + \sum sq(2ph + q),$$
$$B = \sum sq(1 - 2ph - q), \tag{7.21}$$

where summation is over all relevant loci.

Thus a simple linear relationship is expected between the logarithm of the probability of survival P_s and the inbreeding coefficient F. Given data on P_s for various values of F, the constants of this relationship, the intercept A and the slope B, can be used to estimate two important quantities. The first is the quantity $\sum sq$, which is the sum of the products, over all genes, of the selective disadvantage of the homozygote times the respective gene frequency. If the genes were all fully lethal, it would be equal to the sum of their gene frequencies, otherwise it would be smaller.

The quantity $\sum sq$ cannot be estimated directly, but we can note, from Equation 7.21, that

$$B + A = X + \sum sq \tag{7.22}$$

must be an upper limit, while clearly B itself is a lower limit. Therefore $\sum sq$ must lie between B and $(A + B)$. $\sum sq$ has been called **the number of lethal equivalents.**

If all genes were fully lethal ($s = 1$), the number of lethal equivalents would be the number of lethals carried in a haploid complement, a gamete. If s is on the average, less than one, the number of deleterious genes will be correspondingly higher. The same applies to overdominant loci. Thus, from Equations 7.9

$$a_s + b_s = sp + tq, \tag{7.23}$$

which can be generalized for multiple alleles (see Example 7.3) to

$$\sum s_i q_i \tag{7.24}$$

by summing over alleles at the same locus, each with frequency q_i and selective disadvantages s_i when homozygous.

The second quantity that can be estimated from the relation between $\log P_s$ and F is L_1/L_0, which is the same as $B/A + 1$.

The simple treatment of load theory just given leads to the expectation that B/A will be high for genes maintained by mutation-selection balance and low for those maintained by heterozygote advantage. Thus, in principle, an estimate of B/A based on the relation between P_s and F should tell us whether the load is mainly mutational or mainly segregational. This expectation, however, is based on certain assumptions, many of which have been challenged. The major theoretical objections are as follows.

1. For multiple allelic loci, it has been shown by Crow (1961a) (see Example 7.2) that the load ratio B/A for overdominant loci can be as high as the number of alleles.

2. The assumption of genetic equilibrium may lead to serious error. As shown in Figure 7.4, values of B/A much higher than 2 can be obtained for overdominant loci with two alleles if the gene frequency is far from equilibrium (Levene, 1963). Evolution of disease may make the condition of equilibrium inapplicable, as was especially emphasized by Haldane and Jayakar (1965). Many genes now turning up as deleterious recessives may, in the past, have contributed to resistance to diseases that have now disappeared. Examples of this are thalassemia and sickle-cell anemia. Haldane and Jayakar (1965) said: "We suggest that modern European and American populations are now riddled with genes in the course of slow elimination. ..." Such genes that became very frequent as a consequence of a balanced polymorphism are now subject to slow elimination as mildly deleterious recessives. (See also Chapter 11.)

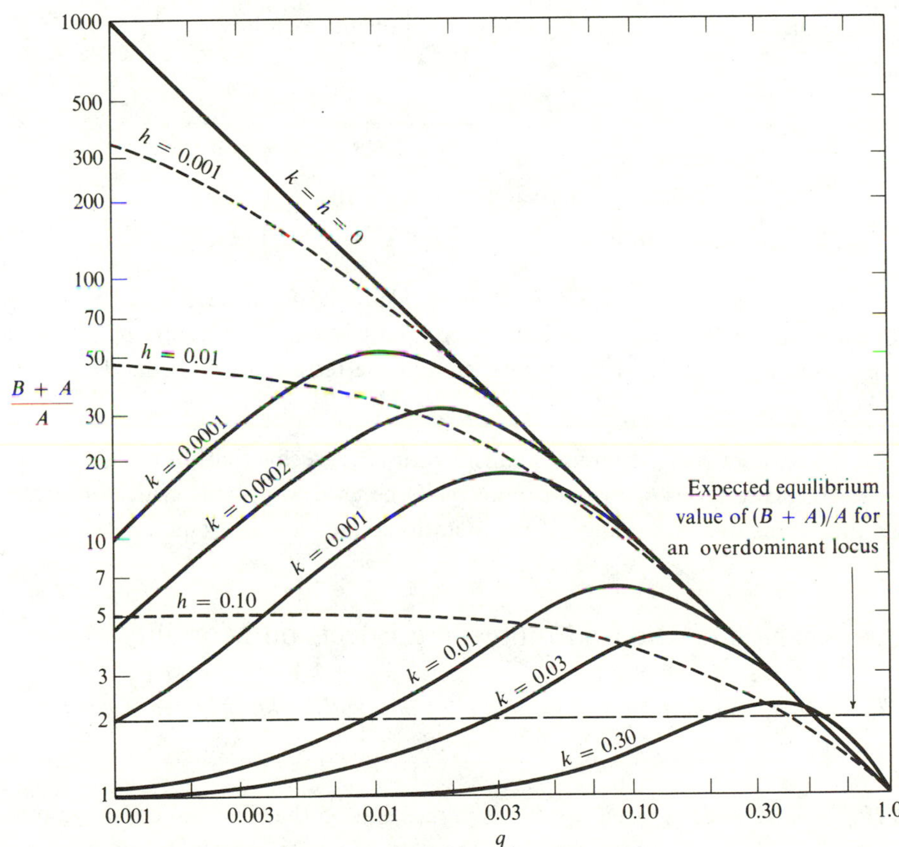

FIGURE 7.4.
Values of $(B+A)/A$ as a function of the gene frequency q for a dominant (deleterious recessive) locus with various h values, and for an overdominant locus with various k values, where $k = s/t$. (From Levene, 1963.)

3. Another possible cause of disturbance, making loads not truly additive, is epistasis, or the violation of the rule that causes of death act independently. Heterotic loci showing nonadditive interactions, whereby all homozygotes are less fit than all heterozygotes, may give high B/A values, up to 2^s, where s is the number of loci concerned (Schull and Neel, 1965). In practice, this makes a B/A ratio of say 10, compatible with a wide variety of proportions in the mutational versus the segregational components. Another type of epistatic model has been introduced by Sved, Reed, and Bodmer (1967), King (1967), and Milkman (1967), which incorporates an upper threshold to the fitness of multiple heterozygotes (see Section 11.7). It can also be shown with this model that the B/A ratio for an effectively overdominant model can take on relatively high values.

4. Other difficulties come from the fact that the "fitness" is rarely examined in enough detail. Conclusions are usually based on data obtained for only one part of the life cycle. Haldane and Jayakar (1965) have considered the confusion that might result for genotypes having fitness components as follows:

	AA	Aa	aa
fertility:	1	1	$1 - k$
mortality (survival):	$1 - l$	1	1
total fitness:	$1 - l$	1	$1 - k$

This fitness model would give a B/A ratio, calculated from deaths alone, of $1/k$. For example, a B/A ratio of 10 would be compatible with the viability of AA being 90 percent of that of the heterozygote Aa and the fertility of aa being 99 percent of that of Aa.

On the basis of these theoretical considerations, it seems that, in the absence of other information, the B/A ratio cannot really be used to provide critical information on the relative importance of the mutational and segregational loads.

7.6 Analysis of Data on Inbreeding Effects on Mortality

The expectation that prereproductive mortality will be linearly dependent on the inbreeding coefficient of an individual is fulfilled.

This expectation has given rise to much research on the effects of consanguinity. In Table 7.8 is given, as an example, the dependence of mortality on F among children of consanguineous marriages in Nagasaki and Hiroshima. Fitting a linear regression to the logarithm of the proportion surviving as function of F, $-\log P_s = A + BF$ (see Equation 7.20), the following equations were obtained:

$$\text{Hiroshima:} \quad -\log P_s = 0.0372 + 0.3957\,F;$$
$$\text{Nagasaki:} \quad -\log P_s = 0.0350 + 0.3193\,F.$$

Averaging the estimates of A and B over the two cities gives $A = 0.0366$ and $B = 0.407$. The number of lethal equivalents is between B and $A + B$—that is, between 0.4 and 0.44.

TABLE 7.8
Percent Mortality of Children in Nagasaki and Hiroshima

	Parental Consanguinity			
City	*Unrelated*	*Second Cousins*	*1½ Cousins*	*First Cousins*
Hiroshima				
Percent mortality	3.55	4.43	7.18	6.12
Number of families	1384	230	192	532
Nagasaki				
Percent mortality	3.42	3.18	4.94	5.25
Number of families	2078	330	263	826

Source: Data from Schull and Neel (1965).

A summary of data from 19 studies on effects of consanguinity on prereproductive mortality is given in Table 7.9. It should be remembered that these studies are highly heterogeneous for cause of mortality and environmental conditions and that there is, probably, some correlation between socioeconomic level and F value. The problem posed by the latter factor will be considered further in Section 7.9, and in Chapter 9, where we discuss the effects of inbreeding on quantitative characters other than mortality. Only one of the 19 investigations whose data were used in

TABLE 7.9
Distribution of A and B Values Computed for Prereproductive Mortality: italic numbers in the body of the table give data on Japanese persons; boldface numbers on Caucasian persons; and roman numbers on persons of other ethnic groups

	B Values						
A values	< 0	0–0.5	0.5–1.0	1.0–1.5	1.5–2.0	2.0–2.5	> 2.5
0–0.1		*1*	*3*				
0.1–0.2	**1**	*1*	*2*	*2*		**1**	
0.2–0.3			**1**		**1**		
>0.3	**1, 1**	**1, 1**	**1**				**1**

Source: Based on data from 19 studies analyzed by Schull and Neel (1965) and by Cavalli-Sforza (1960.)

Table 7.9 deviates significantly (5 percent) from linearity and just one out of 20 would be expected to do so. The agreement with the expectation of linearity is therefore good. There is no obvious correlation between A and B in the data, but the range of variation is very large and the data are scanty. All Japanese values are concentrated in the upper left corner. Values from Caucasians are spread over the full range of B and A values, and so are those from Negroes and Indians, but with higher A values. It should be remembered that A measures the mortality of the noninbred fraction and is therefore to some extent an inverse measure of public health conditions.

Almost everyone carries the equivalent of more than one lethal recessive gene in the heterozygous condition.

As discussed in Section 7.5, data on overall mortality allow the computation of the total number of lethal equivalents, defined by $\sum sq$, where the sum is over all loci. This quantity actually provides a measure of the average number of deleterious genes, measured in terms of lethal equivalents, present in a gamete. When multiplied by 2 this is the average number of lethal equivalent genes carried by any individual in the heterozygous state. As mentioned before, when $s = 1$ for all loci, the number of lethal equivalents is actually the number of genes, since they are then all lethal. As s decreases, the actual number of deleterious genes is roughly the number of lethal equivalents divided by the average value of s.

From Equations 7.21 and 7.22, as pointed out before, it can be seen that the number of lethal equivalents, $\sum sq$, is expected to lie between B and $B + A$. Usually, the difference between B and $B + A$ is found to be small, giving some reliability to the estimates of $\sum sq$ derived from a knowledge of B and $B + A$. The distribution of the estimates of lethal equivalents obtained from 18 of the studies given in Table 7.9 is shown in Table 7.10. The median value of these estimates is 1.1. The

TABLE 7.10
Distribution of Number of Lethal Equivalents in 18 Studies

Equivalents:	<0	0–0.5	0.5–1	1–1.5	1.5–2	2–3	3–4	>4
Number of studies:	2	1	5	4	1	2	2	1

sample is, however, heavily weighted toward Japanese data and so is hardly representative of the world population. (Negative values are, of course, not consistent with the theory and are most probably the result of statistical fluctuations.) The values thus estimated are, as explained before, *per gamete* and should be doubled to be valid for diploid individuals. Previous estimates per individual are somewhat higher than the median value of 1.1 obtained here.

Arner stated in 1908 that "every person on the average contains heterozygously at least one lethal gene which would if homozygous kill an individual between birth and maturity." Actually, Arner's data analyzed by modern techniques give an estimate of 1.97 lethal equivalents per person. Morton, Crow, and Muller (1956) for their calculations of B/A and the number of lethal equivalents made use of data that had been collected much earlier (Arner, 1908; Bemiss, 1858; Sutter and Tabah, 1952, 1955). These yielded an even higher estimated number of lethal equivalents, namely between 3 and 5.

The larger variation in the estimates of B/A from country to country, and even within the same country, needs further analysis and must for the present be accepted with caution. At face value, the present estimate of the number of lethal equivalents per person, based on the median obtained using Table 7.10, is 2.2.

We have seen that the B/A ratio should indicate whether the load is mainly mutational or segregational, but that there are serious difficulties with this indicator. One of the conclusions reached by Morton, Crow, and Muller in their 1956 paper is "... overdominant loci are not making any substantial contribution to B and ... the genetic damage we are measuring is mutational." Their conclusion was based on the high observed B/A ratios, namely 18 and 16 in two French studies by Sutter and Tabah, and 8 and 11 (as approximations) in Arner's and Bemiss' studies. The B/A ratios obtained from later work on other populations are, however, lower. In Japanese data for instance, B/A values range from 1–11. In the studies reviewed by Schull and Neel (1965), the median value is 5.7. This is certainly not sufficiently high to rule out overdominance from making an important contribution to the observed load in prereproductive mortality, even in a straightforward application of the theory in its simplest form. The theoretical objections that have been mentioned show that the situation is too complex for any simple interpretation of the observed B/A ratios to be acceptable at this time.

7.7 Effects of the Inbreeding Level on Equilibrium Gene Frequencies

One interesting fact, noted in the analysis of Table 7.9, is the remarkable variation in A and B values for different countries and peoples. We have already mentioned that A is an indicator of public health conditions. Measuring prereproductive mortality among the noninbred segment of the population, it encompasses many causes of death that are nongenetic, or that are only very weakly correlated with genotype. Genetic causes of death have proved more resistant to modern medicine than infectious and parasitic diseases. It should not be forgotten, however, that even for infectious diseases there may be large genetic components. We have seen this to be the case for malaria. We shall see later that resistance to tuberculosis is at least in part genetically determined. The fact remains that many genetic diseases prove to be the hard core of the pathological conditions affecting man, showing the greatest resistance to treatment. The quantity A has been decreasing with increasing medical care, but within it the proportion of deaths whose cause is strictly genetic has been increasing (see, for example, Sutter and Goux, 1964). The quantity B, which contains almost only deaths caused by recessive genetic traits, is expected to vary much less. We can therefore predict a variation in B/A ratios that parallels the advances in medical treatment. This adds to the difficulties of the interpretation of the B/A ratios.

The quantity B, though it is expected to vary less, actually shows considerable variation between surveys. Japanese studies seem to give lower B values than the most comprehensive European studies. The quantity B is largely dominated by the $\sum sq$ component, or the lethal equivalents. It is thus determined mostly by recessive lethals and semilethals. Differences between surveys in different countries are not likely to be due to differences in s values. It seems likely, therefore, that B differences are due to differences in the q values. These q values, if equilibrium has been reached when $q = \mu/s$, can reflect either mutation rates, or the selective values s, which are hardly likely to change on the average from country to country. An important factor, however, that affects q values is the degree of inbreeding to which a population has been subjected. It is therefore of interest to inquire about the effects of inbreeding on equilibrium gene frequencies.

The effect on equilibrium gene frequencies is noticeable when $\bar{F} > 0.001$.

Some numerical results calculated from the equations derived in Example 7.4 are given in Table 7.11. They illustrate the effects of population inbreeding and of dominance (that is, the h component) with two levels of mutation rates. Even a minor degree of dominance ($h = 0.03$) considerably reduces the effects of inbreeding, so that in practice only the equilibrium frequencies of genes that are almost fully recessive ($h = 0$) are affected even by consanguinity levels as high as $\bar{F} = 0.1$. Even with fully recessive genes, the effect is appreciable only with average F values above 0.001. Populations having such high $\bar{F}$ values are not, however, uncommon. It may be noted that recent inbreeding, as estimated in Table 7.3 from α, is only a fraction of the total inbreeding. The levels of inbreeding (measured by $\bar{F}$ values) found, for example, in Japan may therefore be accompanied by an appreciable lowering of the equilibrium gene frequencies for lethals, thus explaining the lower B values in that country.

Equilibria under inbreeding are reached rather slowly. This is because the rate of increase in gene frequencies depends on the mutation rates, which are small.

The decrease depends on strong selection coefficients (selection against lethals). It is, however, slow because the proportion of individuals subject to selection, namely the recessive homozygotes, is very small. The change in gene frequency is of the order of sq^2 per generation. Therefore, if gene frequencies are near equilibrium in a population in which there is inbreeding, the pattern of inbreeding in the population must have been constant for a long time.

It appears that changes in the traditional pattern of marriage in Japan are either minor or recent. The high frequency of first-cousin marriages is a social custom favored by the practice of arranged marriages, which has probably been in existence for a long time. In contrast, inbreeding levels in Roman Catholic Europe, which are lower than those in Japan, have recently increased as a consequence of social changes. In the centuries before the nineteenth, close consanguineous marriage was exceptional and the α values were, on average, ten to one hundred times lower than they were in the nineteenth century (Moroni, 1967a). From what is known about Roman Catholic legislation of consanguineous marriages, this situation is likely to have prevailed for most of the Christian era. It seems, therefore, that most Christian populations (including practically all Europeans for at least ten centuries) have been subject, on an average, to very low inbreeding coefficients for long periods. Their q values may thus have risen towards those expected for $F = 0$. This may explain the higher B values found in most Caucasian populations. Following the increase in α values in the nineteenth century, they are now, once again, decreasing, because greater mobility is leading to a lower incidence of consanguineous marriages (see Figure 8.27).

TABLE 7.11
Equilibrium Gene Frequencies ($\times 10^{-4}$) *for a Lethal Gene* ($s=1$) *in a Population with Various Levels of Inbreeding*

Mutation Rates	*Dominance*	*Inbreeding Coefficient* (F)				
μ	h	0	0.0001	0.001	0.01	0.1
10^{-5}	0	36.2	31.1	27.0	9.2	1.0
	0.01	9.2	9.1	8.5	4.9	0.9
	0.03	3.3	3.3	3.2	2.5	0.8
	0.1	1.0	1.0	0.99	0.92	0.53
10^{-6}	0	10.0	9.5	6.2	1.0	0.1
	0.01	0.99	0.98	0.90	0.50	0.09
	0.03	0.33	0.33	0.32	0.25	0.08
	0.1	0.10	0.10	0.10	0.09	0.05

7.8 Consanguinity and Detrimental Recessives

The frequency of detrimental recessives is also a linear function of F.

The equations derived for lethality can be applied directly to the study of detrimental recessives, which are mutant genes that have, in homozygotes, a detrimental or debilitating effect, such as causing blindness or deafness, but that are not incompatible with life. The model we can use for these is the following. The number of affected zygotes can be taken as equal to the load L as calculated in Table 7.6, where s is now not the fitness, but the **penetrance** of the gene in the homozygote, and hs the penetrance in the heterozygote. If there are many loci contributing to this **detrimental load** L_D (which is the frequency of affected zygotes) then, as we have seen before when combining the effects on mortality of several genes,

$$L_D = -\log(1 - P_D) = A + BF, \tag{7.25}$$

where P_D is the frequency of affected individuals among people whose inbreeding is measured by F. As P_D is small, we can write

$$L_D \sim P_D \sim A + BF. \tag{7.26}$$

The number of **detrimental equivalents** defined in the same way as for lethal equivalents, namely by $\sum sq$ is, as before, estimated to lie between B and $A + B$.

Estimates of the detrimental load in terms of equivalents are given in Table 7.12. The large variations from country to country must reflect heterogeneity in the criteria used to define genetic morbidity or disability.

TABLE 7.12

Estimates of Detrimental Equivalents and the Constants A and B

Country	*Type of Defect*	*B*	*A + B*	*Ratio of Abnormals among Offspring of First-cousin Matings to That among Noninbreds*	*Author*
France	Conspicuous abnormalities	2.213	2.256	3.76	Sutter (1958)
Italy	Severe defect	0.543	0.573	1.87	Cavalli-Sforza (1960)
Japan	Major morbid conditions	0.654	0.566	1.38	Schull and Neel (1965)
	Minor defects	0.354	0.432	1.26	
Sweden	Morbidity	2.020	2.102	2.26	Böök (1956)
USA	Abnormality	1.164	1.267	2.26	Slatis, Reis, and Hoene (1958)

It is clear in any case that various types of defects caused by detrimental recessive alleles have different probabilities of responding to inbreeding. A breakdown by type of defect is given in Table 7.13. The defects given in Table 7.13 are selected from those that have the highest response to inbreeding. It will be seen that B/A values are fairly high, especially for mental and ear defects.

The last column of table 7.12 gives estimates of the increased risk of genetic disease caused by detrimental recessive alleles in the progeny of first cousins. On an average, the disease incidence is approximately doubled in the progeny of first cousins, as compared with that of noninbred individuals. We would expect, therefore, that the progeny of uncle-niece, aunt-nephew, and other marriages having an F value twice that of first-cousin marriages should have a frequency of disease about twice that of the progeny of first cousins. Data from such marriages could greatly improve our knowledge of the effect of consanguinity on diseases caused by detrimental recessive alleles, but the data are difficult to obtain, except in a few areas where they should be studied more intensively.

TABLE 7.13
Detrimental Load and Equivalents by Type of Defect

				Significance of		
Type of Defect	*A*	*B*	*B/A*	*Heterogeneity Between F Values*	*Regression on F*	*Deviation from Regression*
Mental retardation	0.024	1.399 ± 0.3	58.3	++	++	++
Eye defect	0.058	0.832 ± 0.3	14.1	++	+	−
Ear defect	0.013	0.565 ± 0.2	41.8	++	++	−
Tuberculosis	0.020	0.198 ± 0.2	9.6	−	−	−
Malformation	0.040	0.487 ± 0.3	12.1	−	−	−
Smaller defects	0.081	1.082 ± 0.4	13.3	++	++	−

Source: Survey conducted in Parma, Italy by Conterio and Barrai (1966).

7.9 Retrospective Studies on Consanguinity Effects

Consanguineous marriages tend to be concentrated in areas of low population density. If a character whose correlation with inbreeding is being investigated shows a geographical-social stratification, spurious correlations may easily be generated. This difficulty is not entirely eliminated by the procedures used by some authors of excluding $F = 0$ from the analysis and fitting the regression line only to inbred classes. In fact, the frequencies of the various consanguinity classes also show strong social and geographic stratification. An important problem in studies of consanguinity is therefore the proper choice of noninbred controls.

In a **prospective study**, matings are chosen for study because of their consanguinity. The best way to choose controls for such a study is probably to take married sibs of the consanguineous mates, provided these are not themselves consanguineous with their mate and provided they have not moved to other areas. This choice greatly reduces the probability of bias, but also decreases the sensitivity of the test, because there is a higher probability that the progeny of the sibs of the consanguineous mates are affected by the same diseases as the progeny of the consanguineous mates than that the progeny of the population at large are.

In the retrospective approach one observes the increased consanguinity of parents of individuals who have a disease or abnormality that is caused by a recessive allele.

Another way of studying the effect of consanguinity is by a so-called **retrospective study**, in which the investigator first ascertains which individuals have the genetic disease or abnormality that he is studying and then investigates these individuals for inbreeding. The difficulties already mentioned do not, however, disappear. In a retrospective study, the comparison to be made is between the distribution of the inbreeding coefficients in the selected affected group and that in the general population. Thus a suitable choice of controls is still necessary. They should be chosen to match carefully the sample of defectives surveyed with respect to age, socioeconomic conditions, and place of origin, which may be the most important factor. When the affected are ascertained from hospital data, the part of the population sampled is very difficult to define. Ideally, a complete ascertainment of consanguinity over the

TABLE 7.14
Frequency of First-cousin Marriages among the Parents of Individuals Who Have a Disease Caused by a Rare Recessive Allele

		Frequency of First-cousin Marriages (in Percent)	
Disease	*Area*	*Among Parents of Diseased*	*In the General Population*
Albinism	Europe	18–24	1
	Japan	37–59	6
Alcaptonuria	Europe	21	0.6
Phenylketonuria	Europe	5–14	0.6
Ichthyosis congenita	Europe	30–40	1
	Japan	67–93	6
Microcephaly	Japan	54	6

Source: Based on Serra (1961).

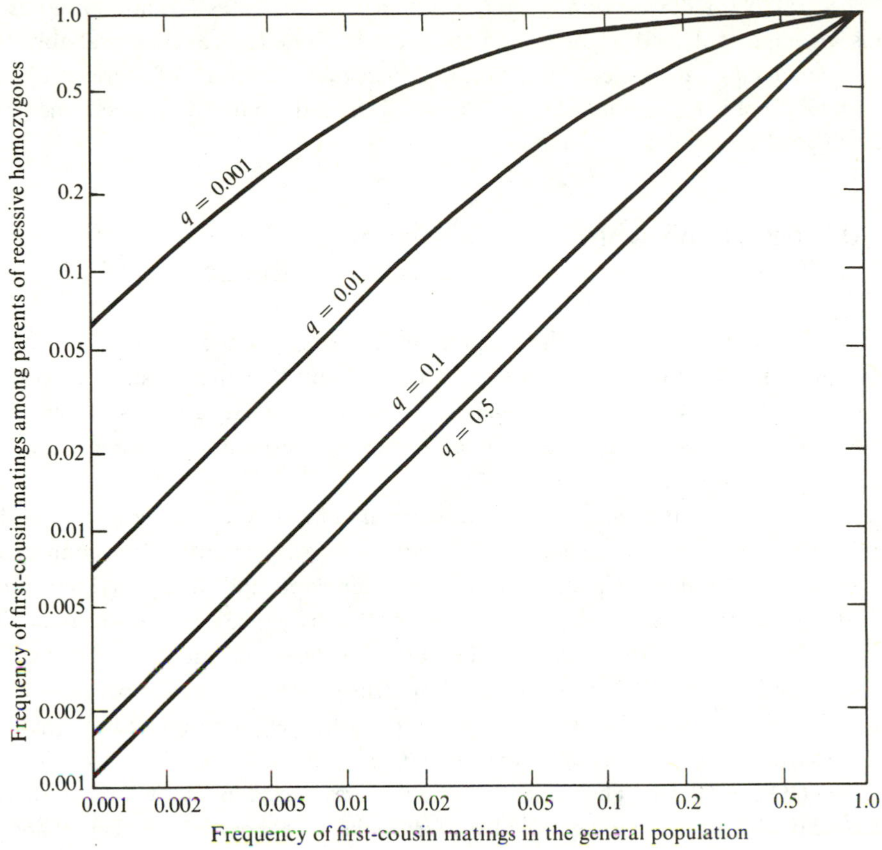

FIGURE 7.5
Frequency of first-cousin matings expected among the parents of progeny displaying recessive phenotypes. The frequency of the recessive gene is symbolized by q.

same area for comparable age classes should be made. If the disease shows a socio-economic stratification, this should be taken into account.

If a genetic disease is recessive, it is very likely that there is more consanguinity among the parents of affected individuals than among the population at large. The expected frequency of first-cousin matings among the parents of recessive homozygotes is plotted in Figure 7.5. The computation is based on an approximation given in Example 7.5 that follows the approach originally suggested by Dahlberg (1948). The increase in the frequency of first-cousin matings relative to that of the general population is extremely sensitive to the gene frequency q, being much greater if q is lower. Table 7.14 shows some actual examples of recessive diseases for which the frequency of consanguinity among the parents of affected individuals is much higher than that of the general population.

The search for consanguinity among the parents of recessive homozygotes is thus an important method of testing for recessive inheritance. It is probably the most convincing one, especially if accompanied by an analysis of segregations in matings. If the heterozygote can be recognized, then the pattern of inheritance can be analyzed even more completely.

7.10 The Number of Genes Determining Mental Deficiency, Deafmutism, and Blindness

We have already seen that the frequencies of defects of the ear, eye, and mind are higher in inbred individuals, which at once shows that the defects are, in part at least, determined by recessive genes. A more complete analysis shows that dominant inheritance and more complex genetic and even nongenetic factors also play an important part.

Among the defects mentioned so far, mental deficiency and illness create the most important social problems. Their frequencies vary greatly from country to country. Institutionalized mental defectives in England and Wales in 1960 numbered 1.3 per thousand of the total population (Penrose, 1963). Not all the mentally ill are, however, institutionalized. The total incidence of mental defectives in England and Wales in 1929 was estimated by Penrose (1938) to be 8.6 per thousand, with the total incidence being made up of approximately 5 percent idiots (having IQ's less than 20), 20 percent imbeciles (IQ's of 20–49), and 75 percent feeble minded (IQ's of 50–69). The frequency of defectives varied greatly with age, reaching a peak of 2.6 percent at age 10-14. " During the school period intellectual deficit is brought sharply into the foreground" (Penrose, 1963). Later, the increase in opportunities for adjustment, possible selective mortality, and other causes may contribute to making the observed incidence lower.

Two bodies of data on mental defect were subjected to segregation analysis (see Appendix II) by Dewey and co-workers (1965). One of them was collected from Wisconsin institutions. The other was the Colchester survey published by Penrose (1938). Mongols, hydrocephalics, and patients suffering from illnesses due to known trauma, neoplasm, or infection were excluded. Only severe mental defect was considered (that is, individuals whose IQ was lower than 50 who had been born to normal parents). Segregation analysis indicated a proportion of **sporadic cases** (those not attributable to Mendelian segregation for a single gene—see Appendix II) as high as 88%. Sporadic cases could be due to unknown trauma, infection, or other environmental cause. They could also be due to chromosomal or other dominant defects that have low penetrance. They might even be due to very low penetrance in heterozygotes for otherwise recessive defects. In any case, they showed no effect of consanguinity, confirming that recessive inheritance was not involved in their determination. The remaining nonsporadic cases (12 percent) segregated in families, in agreement with the proportion of 1/4 expected for recessive inheritance. There was a strong effect of consanguinity on these, as expressed by the equation

$$0.000324 + 0.192F.$$

The A and B values in this equation were estimated by the method given in the previous section. The very high B/A value ($0.192/0.000324 = 593$) is in agreement with the expectation of recessive inheritance.

It is likely that a large number of genes, each with a very low frequency contribute to this mutational load. The number was estimated by Dewey and co-workers (1965) on the basis of the following assumptions. The fitness for severe mental defect is known to be very nearly zero; that is, $s = 1$. Penetrance in the heterozygote is probably very nearly zero. However, even if all the observed sporadic cases were due to low penetrance of the same recessive genes in heterozygotes, they would still lead to a very low h value.

In the study of the effects of consanguinity on mortality, loss of fitness is identified with mortality, which is the character being studied. When the definition of load is extended to detrimental genes, *penetrance* takes the place of fitness in the formulas. If we use the equilibrium between mutation and selection to evaluate mutation rates, fitness must come back into the picture. Formulas can be simplified considerably if we make the assumption that the loss of fitness is expressed as late prereproductive mortality and as decreased fertility, making it possible to count all defectives provided they are identified at a sufficiently young age. Given this assumption, penetrance should be complete, although the fitness may be zero. Then, with complete penetrance in the homozygotes, complete elimination of defectives ($s = 1$), and no penetrance in heterozygotes ($h = 0$), the formulas for A and B given in Equation 7.21 simplify to

$$A = \sum q^2,\ A + B = \sum q \sim B, \tag{7.27}$$

the sum being over all **mimic genes**, namely all genes that have the same effect, for example, that of lowering mental abilities below the threshold used to define mental deficiency.

Assuming that n is the number of mimic genes, all with the same gene frequency q, we have

$$A = nq^2,\ B = nq, \tag{7.28}$$

from which we can estimate,

$$q = A/B,\ n = B^2/A. \tag{7.29}$$

Using these equations Dewey and co-workers (1965) estimated $n = 0.192^2/0.000324 = 114$ as the number of loci contributing to severe mental defect.

The number of mimic genes is bound to be underestimated.

It should be noted that we are talking about a *minimum* estimate. If the gene frequencies at each locus vary as they are sure to do, then, as shown in Example 7.7, the number of loci is underestimated, the more so, the greater the variation in gene frequencies. In Example 7.7, an estimate is given of the possible effect of one source of variation: namely mutation rates. This example shows that variation in mutation rates increases Dewey and co-workers' (1965) estimate of the number of loci contributing to severe mental defect from 114 to approximately 330, which is probably still an underestimate. The major cause of variation in gene frequencies is not likely however, to be the variation in mutation rates, but rather random genetic drift (see Chapter 8). In any case, the total number of loci must be large. Of the great number of genes that must take part in specifying the formation of the central nervous system, many must be important enough that their mutation would cause serious disturbance of mental function.

A similar analysis to that for mental defect was carried out by Dewey and co-workers (1965) for deafmutism. Segregation analysis indicated a lower proportion of sporadic cases, namely 25 percent. The dependence on F for the nonsporadic cases was again very marked, and was given by

$$0.00018 + 0.08\,F.$$

The estimate of the number of mimic genes determining deafmutism was, from Equation (7.9), $0.08^2/0.00018 = 34$, or using the same correction factor (from Example 7.7) as before, 100.

There is more direct evidence for the existence of many mimic genes.

The study of individual cases of inherited defects shows that there are usually many clinical varieties of any given defect. A combination of apparently unrelated symptoms, such as skeletal abnormalities and malformations, pigmentation anomalies, and biochemical abnormalities may accompany, for instance, blindness or deafness. When there is more than one case in a family, the repetition of the combination of symptoms in all affected subjects rules out the possibility that the symptoms are combined simply by chance. Even if the cases are sporadic, the reappearance of a similar combination of symptoms in unrelated subjects suggests a distinct genetic entity.

The analysis of a large number of cases of childhood deafness and blindness from British institutions led Fraser (1965) to formulate a breakdown of causes of these defects as given in Table 7.15. A further criterion can be used for the estimation of the number of loci. The fitness of blind and deaf individuals was practically

TABLE 7.15

Analysis of Aetiology of Childhood Deafness and Blindness. Numbers are percentages of the children studied

	Deafness	*Blindness*
Acquired causes		
Prenatal	6	6
Perinatal	10	34
Postnatal	30	11
Genetical causes		
Dominant	12	17
Recessive	37	16
Sex-linked	2	5
Malformations[a]	3	11
Number of children studied	2355	776
APPROXIMATE NUMBER OF LOCI THOUGHT TO BE INVOLVED IN THE CAUSATION OF CONDITIONS DUE TO SINGLE GENES, INCLUDING ENTITIES NOT REPRESENTED IN THIS MATERIAL		
Dominant	5	15
Recessive	15–25	23–35
Sex-linked	4	9

Source: From Fraser (1965).
[a] Single gene inheritance and more complex types.

zero until the last century, when social care for them started. Their fitness is now believed to be not far from normal, (0.7–0.8) with the limitation that mating is highly **assortative**: the blind often marry the blind, the deaf often marry the deaf. When recessive genes are the cause of the defects, the progeny is often normal, showing that the two parents carry alleles that cause the defect at different loci. There seems to be a higher chance of affected progeny from matings between two bearers of the same defect who are from the same isolated community, as might be expected. Drift alone (Chapter 8) can make some mutant genes much more frequent than others in such communities, and therefore increase the probability that two individuals carry the same recessive gene that causes a defect. Matings between a deaf person from such a community and one from a less isolated community have, however, been observed that have given rise to deaf progeny, indicating that some recessive genes of the sort we have been discussing are more widespread than others. The estimates of number of loci given in Table 7.15 are based on a combination of data from clinical differentiation of defects, gene frequency calculations derived from consanguinity rates among parents of affected children, and results of marriages between affected persons. They are considered by the author to be semiquantitative.

7.11 Estimation of Overall Mutation Rates for Recessive Genes, Taking Account of Inbreeding

In addition to the difficulties of estimating mutation rates already discussed in Chapter 3 is the fact that the prevailing level of consanguinity in the population being studied must be taken into account.

The total mutation rate for recessive genes is approximately equal to the expressed load.

In a population with an average inbreeding coefficient F, the genotype frequencies for a locus with two alleles, one of which is a deleterious recessive, can be written in the form

genotype:	AA	Aa	aa
fitness:	1	$1 - hs$	$1 - s$
frequency:	$p^2(1 - F) + Fp$	$2pq(1 - F)$	$q^2(1 - F) + qF$
loss of a genes:	0	$hs\,pq(1 - F)$	$sq[F + q(1 - F)]$.

Equating the loss of a genes to the gain due to mutation from A to a that proceeds at rate μ, we obtain as an approximation, for one locus, at equilibrium,

$$\mu = sq(F + h + q), \tag{7.30}$$

where terms in q^2F and hqF are neglected on the assumption that all three quantities q, h, and F are quite small (see Example 7.4). If h is zero, and we use for the average inbreeding coefficient F the estimate α obtained from recent inbreeding, the total mutation rate is simply

$$\sum \mu = A + B\alpha, \tag{7.31}$$

where now we are summing over all relevant loci. This is, thus, the total mutation rate over all loci that can mutate to give the defect being studied.

Existing estimates of total mutation rates are probably too high.

An estimate of 192×10^{-5} for the total mutation rate, and thus of 1.7×10^{-5} per locus (obtained by dividing the total mutation rate by the number of loci, which has been estimated to be 114) was given by Dewey and co-workers (1965) for mental defect. This estimate is, however, inflated, by the use of both an estimate for the number of loci that is probably too low and an α value derived from a totally unrelated population. The total mutation rate estimate given by Dewey and co-workers (1965) was in fact derived using a value of $\alpha = 0.006$ obtained from a Japanese population. The inbreeding levels of the populations surveyed for mental defect (Wisconsin and Colchester) are not known accurately, but estimates of them are as much as 10–100 times lower than the estimate for the Japanese population: 0.000406 for a British hospital population, and 0.000048 for the Roman Catholic population in Wisconsin (see Dewey et al., 1965). If these values are used instead of that from the Japanese data, average mutation rates more in line with those discussed in Chapter 3 are obtained.

Taking into account inbreeding removes one source of inaccuracy from estimates of mutation rates to recessive genes, but can hardly remove other sources of inaccuracy, such as the necessity to assume that equilibrium has been established. The difficulties of estimating the total number of pertinent loci makes the estimate per locus especially unreliable. Even for single loci, the difficulty of accepting the assumption of equilibrium must again be stressed. Moreover, when single genes are considered, they are likely to be selected for high gene frequencies because mutations among those with low frequencies are so few as to be almost negligible. Random drift is likely to cause large fluctuations in the frequencies of rare recessive genes. This may explain, at least qualitatively, why some genetic diseases are much more frequent in certain populations, than in others (for example, Tay Sachs disease in Jewish populations, Damon, 1969).

General References

Morton, N. E., J. F. Crow, and H. J. Muller, "An estimate of the mutational damage in man from data on consanguineous marriages." *Proceedings of the National Academy of Sciences* **42**: 855–863, 1956. (The classic paper of the theory of genetic loads.)

Schull, W. J., and J. V. Neel, *The Effects of Inbreeding in Japanese Children.* New York: Harper and Row, 1965. (The most comprehensive survey of the effects of inbreeding in man, with special reference to data from Japan.)

Worked Examples

7.1. *Estimating F from the deviation from Hardy-Weinberg equilibrium due to inbreeding.*

The test for departure from Hardy-Weinberg equilibrium can be carried out using the formula

$$\chi^2 = \frac{(4ac - b^2)^2 N}{(2a + b)^2 (b + 2c)^2},$$

where a, b, and c are the absolute frequencies (that is, the actual numbers of genotypes GG, Gg, and gg) and $N = a + b + c$. In fact, if we write the observed frequencies of the genotypes in a 2×2 table,

	G	g	
G	a	$b/2$	$a + b/2$
g	$b/2$	c	$b/2 + c$
	$a + b/2$	$b/2 + c$	$N = a + b + c$

we may apply the standard formula for a 2×2 chi square (see Appendix I), putting $x = a$, $y = w = b/2$, $z = c$,

x	y	$x + y$
w	z	$w + z$
$x + w$	$y + z$	N

and obtain the result,

$$\chi^2 = \frac{(xz - yw)^2 N}{(x + w)(y + z)(x + y)(w + z)}.$$

(See Li, 1955, Chapter 2.) On the other hand, by substituting

$$\frac{a}{N} = p^2(1 - F) + Fp$$

$$\frac{b}{N} = 2pq(1 - F)$$

$$\frac{c}{N} = q^2(1 - F) + Fq$$

in the first formula, the result

$$\chi^2 = NF^2$$

is obtained (Li and Horvitz, 1953). For χ^2, on the average, to be significantly different from the Hardy-Weinberg expectation ($\chi^2 = 3.84$ at 5 percent probability), N as a function of F must be

$$N = 3.84/F^2.$$

Thus 384 observations will be required to detect a departure from Hardy-Weinberg expectation if F is as high as 0.1, which applies only to exceptionally isolated and small populations. About 40 thousand individuals are required if $F = 0.01$ as is more customary. Thus, estimating F by this procedure is very difficult. With such large values of N, heterogeneities in gene frequencies between subgroups will simulate inbreeding, and other factors (for example, selection) will also alter the genotype frequencies. The method is therefore in practice relatively insensitive with respect to the inbreeding levels encountered in human populations.

Yasuda (1968a) has described a more powerful method of estimating F that uses mating types instead of individuals.

7.2. *The random load as a function of the dominance of a deleterious allele.*

In the system

genotypes:	AA	Aa	aa
fitnesses:	1	$1 - hs$	$1 - s$
frequencies:	p^2	$2pq$	q^2

the total loss of a genes when q and h are small is, approximately,

$$hsq + sq^2,$$

so that, at mutation-selection balance equilibrium,

$$\mu = hsq + sq^2 \qquad (1)$$

(see Chapter 3). If the loss of a genes from heterozygotes and homozygotes is equal, then

$$hsq = sq^2,$$

and so, from Equation 1,

$$\mu = 2sq^2 \text{ or } q = \sqrt{\mu/2s} = h. \qquad (2)$$

The total load is

$$L = sq^2 + 2hsq$$

and so, when q is given by Equation 2, the load is

$$L = 3sq^2 = \tfrac{3}{2}\mu$$

and is half way between the value of 2μ for complete dominants (large h) and μ for complete recessives ($h = 0$). More generally, if

$$hsq = ksq^2,$$

then Equation 1 gives

$$\mu = (1 + k)sq^2 \quad \text{or} \quad q = \sqrt{\frac{\mu}{(1 + k)s}}.$$

The total load is then

$$L = sq^2(1 + 2k) = \frac{1 + 2k}{1 + k}\,\mu,$$

giving $L = \frac{3}{2}\mu$ when $k = 1$. This result should be valid when k is neither very large nor very small. It provides a simple relation between the load L and the equilibrium gene frequency q for loci with intermediate levels of dominance, represented by different values of k. A general treatment of the load for genes with intermediate dominance has been given by Kimura (1961).

7.3. *Segregational load and multiple alleles.*

A complete analysis of the segregational load for multiple alleles is difficult. We consider a simplified example in which all heterozygotes have the same fitness, which is taken to be equal to one, and homozygotes $A_i A_i$ have selective disadvantage s_i, where i is in the range 1 to n. The random load is then

$$L_0 = a_s = \sum_{i=1}^{n} s_i q_i^2,$$

where q_i is the frequency of gene A_i. The quantity b_s (see Equation 7.3) is

$$b_s = \sum s_i q_i - \sum s_i q_i^2 = \sum s_i q_i(1 - q_i),$$

since $h = 0$, where the sums are over all n alleles. The ratio of the inbred load to the random load is

$$\frac{L_1}{L_0} = \frac{a_s + b_s}{a_s} = \frac{\sum s_i q_i}{\sum s_i q_i^2}.$$

If all homozygotes have the same disadvantage s, equilibrium is established when all n alleles have the same gene frequency $q = 1/n$. The random load is then

$$L_0 = \frac{s}{n},$$

and the ratio L_1/L_0 is

$$\frac{L_1}{L_0} = n,$$

which is equal to the number of alleles.

It is easy to show, for two alleles, that, at equilibrium, the load is equal to or less than $s/2$, where s_1 and s_2 are the selective disadvantages of the homozygotes and $s = (s_1 + s_2)/2$. For, at equilibrium, $L_0 = s_1 s_2/(s_1 + s_2)$, and the geometric mean $(s_1 s_2)^{1/2}$ is always less than the arithmetic mean $(s_1 + s_2)/2$ except when $s_1 = s_2$, in which case it is equal to it. Thus

$$s_1 s_2 < \left(\frac{s_1 + s_2}{2}\right)^2$$

and so

$$L_0 < \frac{s_1 + s_2}{4} = \frac{s}{2}.$$

(See Crow, 1961.)

7.4. *Equilibrium gene frequencies for mutation-selection balance with inbreeding.*

The equilibrium frequencies and fitnesses under inbreeding for a locus with two alleles, one of which, a, is a deleterious recessive are

	Fitness	*Equilibrium frequency after selection*
AA	1	$[p^2(1 - F) + Fp]/R$
Aa	$1 - hs$	$(1 - hs)[2pq(1 - F)]/R$
aa	$1 - s$	$(1 - s)[q^2(1 - F) + qf]/R$

where the sum

$$R = p^2(1 - F) + Fp + (1 - hs)[2pq(1 - F)] + (1 - s)[q^2(1 - F) + qF] = 1 - L,$$

and L is the load.

Assuming mutation takes place at the time of formation of gametes, and therefore after selection, the gene frequency after selection and before mutation will be

$$q_s = \frac{pq(1 - hs)(1 - F) + (1 - s)[q^2(1 - F) + qF]}{R}.$$

Mutation will increase q by an amount $\mu p_s = \mu(1 - q_s)$ and so in the next generation, before selection, the gene frequency will be

$$q' = q_s + \mu p_s = q_s(1 - \mu) + \mu,$$

where q_s is as given above.

Table 7.11 is computed from this exact solution. An approximation involving less algebra is used in Section 7.11. It starts from equating the loss of genes a per generation directly to the mutation rate

$$pq(1 - F)(1 - hs) + [q^2(1 - F) + qF]s = \mu,$$

which, for $p \sim 1$ simplifies to

$$\sum sq[F + (1 - F)(h + q)] = \sum \mu$$

(see also Example 7.2).

In the equation of estimation used by Dewey and co-workers (1965), terms in hF and q^2F were dropped, giving the equation

$$\sum sq(F + h + q) = \sum \mu,$$

which, taking q constant for all genes and equal to $\bar{q}$, with $F = \alpha$ and h independent of sq, gives their equation

$$B(\alpha + q + h) \sim \sum \mu.$$

It seems dangerous, however, unless h is very small or zero to assume that h and sq are independent. If h is negligible, and $s = 1$, then

$$\sum \mu = \sum q(F + q) = F \sum q + \sum q^2 \sim A + BF,$$

where F is the population-average inbreeding coefficient (α).

7.5. *The increase in the proportion of first cousins among parents of individuals displaying recessive phenotypes.*

It is easy to predict the expected increase in the proportion of consanguineous marriages among the parents of individuals affected by a recessive trait if we limit our consideration of inbreeding to first cousins, which is a useful approximation in many real situations. If first-cousin matings occur with frequency c in the general population, then in the noninbred fraction $(1 - c)$ the frequency of homozygous recessive individuals is q^2. Remembering that the inbreeding coefficient for first cousins is $F = \frac{1}{16}$, we may compute that in the inbred fraction c the frequency of homozygous recessive individuals is

$$\left(1 - \frac{1}{16}\right)q^2 + \frac{q}{16} = \frac{q}{16}(1 + 15q), \qquad (1)$$

The total frequency of homozygous recessive individuals will therefore be

$$(1 - c)q^2 + c\frac{q}{16}(1 + 15q) = q^2 + \frac{cq(1 - q)}{16}. \qquad (2)$$

Among all the homozygotes, the proportion who have first cousins as parents will, therefore, be given by the ratio of Equation 1 to Equation 2, multiplied by c,

$$c' = \frac{\dfrac{cq}{16}(1 + 15q)}{q^2 + \dfrac{cq(1 - q)}{16}} = \frac{c(1 + 15q)}{16q + c(1 - q)}.$$

This formula was used to construct Figure 7.5, which shows how the increase of consanguinity in the parents of affected recessives can be particularly striking if the inbreeding in the population is low and the gene is rare. (See Dahlberg, 1948.)

7.6. *More exact treatment for retrospective studies of consanguinity.*

If the distribution of the inbreeding coefficient in the general population is known, or is known at least for close degrees of inbreeding F_i having relative frequencies ϕ_i, can we define the mean and variance of the inbreeding coefficient as follows:

$$\alpha = \sum \phi_i F_i,$$

$$\sigma_F^2 = \sum \phi_i F_i^2 - (\sum \phi_i F_i)^2 = \sum \phi_i F_i^2 - \alpha^2.$$

We also need a definition of the incidence, I of the recessive trait in the population. This will be taken as equal to the observed proportion of children born with the trait, or more generally, the frequency of the trait before selection. Among individuals with inbreeding given by F_i, the probability of finding the trait is $A + BF_i$ (see Section 7.8 and Equation 7.26), and thus the frequency of individuals with the trait in the total population is

$$I = \sum \phi_i(A + BF_i) = A + B\alpha.$$

The probability that the inbreeding coefficient of an affected individual is F_i is thus given by

$$\psi_i = \frac{\phi_i(A + BF_i)}{\sum \phi_i(A + BF_i)} = \frac{\phi_i(A + BF_i)}{I}.$$

The average F value among individuals with the trait is

$$\bar{F}_r = \sum \psi_i F_i = \frac{A\alpha + B(\sigma_F^2 + \alpha^2)}{I},$$

since, from the definition of σ_F^2, $\sum \phi_i F_i^2 = \alpha^2 + \sigma_F^2$. Given the incidence of the trait I, the average inbreeding among the affected ($\bar{F}_r$) as well as the average inbreeding (α) and the variance of the inbreeding coefficient (σ_F^2) of the general population, we can estimate A and B values using the equations just given. Further refinements in the methods of estimation, and standard errors of the estimates are given by Morton, (1960) to whom this treatment is due.

One could also carry the analysis further by testing whether each inbreeding coefficient is correctly represented. The distribution function of F_i provides the basis for computing these expectations.

7.7. *Variation in gene frequency as a cause of overestimates of gene number.*

The random deleterious load (that is, the frequency before selection of carriers of the defects in the absence of inbreeding) for a fully recessive gene, assumed to cause no prereproductive mortality, is q^2.

With inbreeding F, the load is

$$L_D = (1 - F)\sum q^2 + F\sum q = \sum q^2 + F(\sum q - \sum q^2),$$

where, as usual, summation is over all relevant loci. Therefore

$$L_D = A + BF,$$

where

$$\begin{aligned} A &= \sum q^2, \\ B &= \sum q - \sum q^2, \\ A + B &= \sum q. \end{aligned}$$

As A is usually very small, $B \sim \sum q$. If deleterious alleles at the ith locus have frequency q_i, with mean $\bar{q} = \sum q_i/n$, and variance

$$\sigma_q^2 = \frac{\sum q_i - \dfrac{(\sum q_i)^2}{n}}{n},$$

then

$$\sum q_i = n\bar{q},$$

$$\sum q_i^2 = n\sigma_q^2 + \frac{(\sum q_i)^2}{n}$$

and so

$$\frac{(B + A)^2}{A} = \frac{(\sum q_i)^2}{\sum q_i^2} = \frac{(n\bar{q})^2}{n\sigma_q^2 + \dfrac{(\sum q_i)^2}{n}}.$$

Hence, B^2/A is equal to n only if the gene frequencies q_i are all equal and σ_q^2 is zero. Otherwise, the estimate of n obtained by taking it equal to B^2/A will be smaller than the true value by a factor

$$1 + \frac{\sigma_q^2}{\bar{q}^2}.$$

This quantity cannot be obtained directly from present data. Its magnitude might, however, be guessed, assuming that the variation of mutation rates is the same for the category of mimic genes considered as that estimated for X-chromosome mutations (Chapter 3). The unselected mutation rates must be considered, and the mean and variance of these was computed to be $M = 3.1 \times 10^{-7}$ and $V = 1.1 \times 10^{-12}$; the latter is computed from

$$V = M^2(e^{V \log m} - 1) = (3.1 \times 10^{-7})^2\,(e^{0.461/0.434^2} - 1) \sim 1.1 \times 10^{-12}$$

(see Example 3.6 and Section 3.8). At equilibrium, with $s = 1, q = (\mu)^{1/2}$ (see Chapter 3). Hence we are interested in the variance of $(\mu)^{1/2}$, which is approximately $\sigma_\mu^2/4\bar{\mu}$ (see Appendix I), or more simply, in the quantity

$$\frac{\sigma_q^2}{\bar{q}^2} = \frac{\sigma^2_{\sqrt{\mu}}}{(\overline{\sqrt{\mu}})^2},$$

which is

$$\frac{\sigma_\mu^2}{4\mu^2} \sim \frac{V}{4M^2}.$$

Assuming that average mutation rates are the same as for sex-linked mutant genes, the correction factor is 2.9. This computation neglects a source of variation in gene frequencies that is probably important, namely that due to random drift (see Chapter 8). There are formulas giving the variation expected at equilibrium for gene frequencies under the balance of mutation and selection, but it is doubtful whether equilibrium conditions apply.

8

Population Structure

Population structure refers to the results of such diverse evolutionary factors as population size, mobility, and nonrandom mating. In predicting consequences of mutation and selection in Mendelian populations we have so far assumed that mating is completely random and that populations are infinitely large. We now discard these assumptions in order to make models that are more realistic. This is done by expanding the concept of **population structure** to include the available knowledge concerning some of the properties of real populations, such as their size and the probabilities of mating between individuals of a given type. In the simple random mating model any two individuals of the same sex have the same probability of mating with a given individual of the other sex, but this clearly is not true of any human population.

We have already seen the consequences of a special type of deviation from random mating: mating between relatives. In the simplest infinite-population random-mating model this event does not take place. But in real populations it does. We have, then, slightly altered the model of an infinite population to take account of the possibility of mating between relatives.

8.1 Random Genetic Drift

Random genetic drift is the random fluctuation of gene frequencies in a population of finite size.

The greatest advocate of the importance of drift in evolution is S. Wright, who contributed many of the early theoretical developments. The mathematics of drift theory is exceedingly complex. Complete solutions of many of the problems have not, so far, been carried out. Recent analytical developments, however, in the field of stochastic processes have helped to refine and generalize some of the earlier solutions and to obtain new ones.

We will, to start with, simplify our problem by idealizing considerably a population of humans and viewing it as that of an organism with nonoverlapping generations, in which all individuals of each generation are born, reproduce, and die at roughly the same time. We can further simplify our problem by considering not the individuals, but rather their genes. This simplification is adequate from the point of view of our present interests, as long as we assume that there is no selection. When there is selection, phenotypes—and therefore individuals—have to be taken into account.

Considering one locus, we therefore have $2N$ genes rather than N individuals, where N refers to the number of individuals forming the group that participates in reproduction, and we take this number to be constant from generation to generation. We assume that there are two alleles at the locus, A and a, with relative frequencies p_0 and $q_0 = 1 - p_0$ at the beginning. Subscripts of p and q indicate the generation number. We are interested in the gene frequencies $p_1, p_2 \ldots p_n$ that can be expected in the successive generations. If there is no mutation and no selection, we expect p_1 to be equal to p_0, and so on for $p_2 \cdots p_n$. The fact that the population is *finite*, however, gives rise to fluctuations in the actual observed gene frequency from generation to generation. We thus expect p_1 to be equal to p_0, on an average, but find that an observed p_1 is similar, but usually not actually identical, to p_0. There is a sampling process involved in determining which genes the individuals of the next generation carry, and this introduces a statistical fluctuation of the gene frequencies.

The formation of the next generation can be viewed as a sampling process.

For creating a new generation individuals form gametes, usually in very large numbers (males of most species form many more gametes than do females). Every gamete receives, at random, one or the other gene that an individual received at any given locus from his father and mother. Only a small fraction of gametes, however, succeed in uniting with gametes of the opposite sex, and develop into adults.

Thus, the next generation is formed by a sample, usually a relatively small sample, of all the possible gametes. Unless gametes containing one allele are favored at some stage over those containing the other, the gametes that are successful in producing zygotes can be considered to be a random sample with respect to alleles A and a. For a given locus, the frequencies of A and a in generation 1 depend on the particular sample of successful gametes, and thus p_1 is the result of random sampling of $2N$ gametes (N male and N female)—that is, of $2N$ genes—from a gene population in which the frequencies of the alleles A and a were p_0 and q_0. The sample size is twice the number of adults in the population.

In this way, p_1 can have any value between 0 and 1 (extremes included), and we can predict exactly the probability that p_1 takes any particular value. The sampling conditions are such that they obey closely the rules of **binomial sampling,** and the distribution of probabilities that thus ensues is the positive binomial, a very well-known probability distribution that is given in Appendix I. Before we continue, however, it is important to realize a fundamental property of drift, namely its cumulative behavior.

Drift is the result of the accumulation of sampling fluctuations generation after generation.

Because of the sampling process, p_1 is not in most real instances identical to p_0, though it is usually similar and, as probability rules show, the expected value of p_1 is equal to p_0 (see Appendix I). In a real population, p_1 is a certain value, probably not too far from p_0, and for the formation of the next generation the gene frequency of A, p_2, depends solely on p_1, and not at all on the p_0 value. The gametes that form the individuals of the next generation are drawn from the individuals of the present generation, among which the gene frequency of A is p_1. Thus, there is again a random sample of $2N$ genes, chosen according to the rules of binomial sampling, but now with the expectation p_1 for A, and so on for the succeeding generations.

We have already mentioned that p_1 may be 0 or 1, and this is also true of $p_2 \cdots p_n$. Clearly, if p_i for any generation i is 0 or 1, the gene frequency in the succeeding generations can change no more, for one allele has become extinct (A if $p = 0$, a if $p = 1$) and the other has become **fixed.** Only new mutation, or migration from an outside population, can restore the lost variation.

In the absence of mutation, migration, and selection, the ultimate outcome of genetic drift is always fixation of one allele and loss of the other (*or others*).

There is a finite, though possibly very small, probability, at every generation, that all gametes forming zygotes carry the same allele. If the process of sampling is continued indefinitely, the probability that fixation occurs is one. There is no definite time by which fixation will certainly have occurred; an average fixation time can be computed, however, which depends on population size, as we shall see. In order to make it easier for those who are unfamiliar with statistics to understand drift, it helps to carry out, mentally at least, a simulation experiment.

Genetic drift may be simulated by a random sampling process.

There are many ways in which a sampling process such as the one described can be simulated. For teaching purposes it is often convenient to make use of beans of two different colors and start, for instance, with equal proportions of, say, white and gray beans and take a random sample of, say, 10 beans. To be exact, sampling should be carried out "with replacement," that is, every bean should be returned to the collection after each sampling, so as not to alter the probability of extracting a bean of a given color at every trial. This is not what happens in reality because sampled gametes are used up, but their number is so large that the probability at subsequent samplings is in practice unaltered. The 10 chosen beans may be taken to represent an adult population of 5 individuals. If, for example, among the individuals there is a p_1 gene frequency of 40 percent, and we decide to let white beans stand for *A* alleles, implying that initially 4 out of the 10 beans were white and 6 gray, then we must take a random sample of 10 beans from a collection in which there are 40 percent white and 60 percent gray. These 10 beans will give us p_2 and so on.

There is a less cumbersome way of carrying out the sampling experiment that requires no material objects and that may serve as an introduction to simulation experiments done by computer. Simulation is a widely used technique for solving problems that do not readily yield to mathematical analysis. The method used for computers will be given here with a simple example.

The procedure of random sampling is imitated by using **random numbers.** These are numbers of one or more digits, in which each digit has the same probability of occurring, and there is no correlation between successive digits or numbers. An extract of a published table of random digits is given in Table 8.1.

We first form a random sample of 10 genes from a population that contains 50 percent *A* and 50 percent *a* genes. To do this we need random numbers of one digit only, and we need as many different random numbers as there are genes to be included in the sample. We can simulate a 50 percent probability of an *A* gene's being chosen by taking a random number between 0 and 4 inclusive as being an *A* gene, and a random number between 5 and 9 inclusive as being an *a* gene. We can use any line or column of random numbers. Taking the third line, say, we have 5, 8, 3, 6, 2, 3, 5, 9, 7, 2. This gives *a*, *a*, *A*, *a*, *A*, *A*, *a*, *a*, *a*, *A*. Therefore, in the first generation we have 4 out of the 10 genes at the locus, or 40 percent, being *A*. This

TABLE 8.1

Sample of Random Digits

7	7	0	0	7	8	8	5	2	6
3	6	9	3	7	1	2	0	7	9
5	8	3	6	2	3	5	9	7	2
0	0	7	7	8	0	3	3	9	1
0	0	0	6	3	2	1	8	6	1
6	6	5	6	2	6	6	7	8	3
5	3	4	7	9	1	2	1	9	2
8	6	8	2	0	9	7	2	7	9
7	5	8	9	9	0	2	9	1	0
8	7	3	0	9	4	3	0	6	9
4	9	3	4	7	2	3	2	4	1
8	2	4	9	0	0	2	4	5	8
6	5	4	6	5	1	2	8	3	6
1	8	7	6	0	4	3	8	8	1
2	9	7	9	3	3	0	5	8	2
8	5	7	2	4	8	1	8	2	5
3	5	7	6	0	0	2	6	1	8
9	9	5	1	1	8	8	6	8	4
5	0	8	8	5	4	3	4	5	3
5	0	3	8	8	0	1	0	1	9

Source: From Fisher and Yates (1953).

TABLE 8.2

Simulation of Random Genetic Drift by Random Numbers. The transformation rule is dictated by the proportion of genes in the preceding generation, so that the expected frequency of random digits representing A (or a) equals the gene frequency of A (or a) in the earlier generation.

Generation	*Random Numbers*	*Rule for Transforming Random Numbers into Genes*	*Corresponding Genes*	*Percentage of A Genes*
0	—	—	—	50
1	5836235972	0,1,2,3,4 = A; 5,6,7,8,9 = a	*aaAaAAaaaA*	40
2	0077803391	0,1,2,3 = A; 4,5,6,7,8,9 = a	*AAaaaAAAaA*	60
3	0006321861	0,1,2,3,4,5 = A; 6,7,8,9 = a	*AAAaAAAaaA*	70
4	6656266783	0,1,2,3,4,5,6 = A; 7,8,9 = a	*AAAAAAAaaA*	80
5	5347912192	0,1,2,3,4,5,6,7 = A; 8,9 = a	*AAAAaAAAAA*	80
6	8682097279	0,1,2,3,4,5,6,7 = A; 8,9 = a	*aAaAAaAAAa*	60
7	7589902910	0,1,2,3,4,5 = A; 6,7,8,9 = a	*aAaaaAAaAA*	50
8	8730943069	0,1,2,3,4 = A; 5,6,7,8,9 = a	*aaAAaAAAaa*	50
9	4934723241	0,1,2,3,4 = A; 5,6,7,8,9 = a	*AaAAaAAAAA*	80
10	8249002458	0,1,2,3,4,5,6,7 = A; 8,9 = a	*aAAaAAAAAa*	70
11	6546512836	0,1,2,3,4,5,6 = A; 7,8,9 = a	*AAAAAAAaAA*	90
12	1876043881	0,1,2,3,4,5,6,7,8 = A; 9 = a	*AAAAAAAAAA*	100

is the probability of *A* genes for the next generation, and we simulate sampling from it by taking digits 0, 1, 2, 3 as being *A* genes, and all other digits as being *a* genes. Random sampling is repeated, using say, the fourth line of random numbers in Table 8.1, 0, 0, 7, 7, 8, 0, 3, 3, 9, 1, and gives 60 percent of *A* genes. This procedure is shown in Table 8.2 and carried out until the twelfth generation, when the gene *a* becomes extinct.

Figure 8.1 shows the vagaries of the gene frequency of *A* in a graphic form. An example of a computer program designed to carry out similar "Montecarlo" experiments is given in Appendix I.

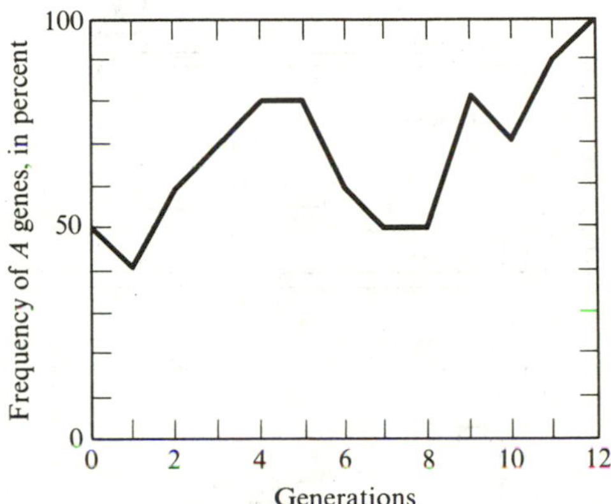

FIGURE 8.1
Variation in the frequency of a gene with time under drift. (Data of Table 8.2.)

The experiment can be repeated a number of times creating a distribution of the gene frequencies found at any given generation. The results of 100 such experiments, done by computer, are given in the form of histograms in Figure 8.2. Already by the fifth generation one allele or the other is fixed in many populations, and the number of populations in which gene *A* is fixed is approximately equal to that in which gene *a* is fixed. This is because we have started with the same number of *A* and *a* alleles.

8.2 Prediction of the Extent of Variation of Gene Frequencies after *n* Generations

It is simple to predict, using the binomial expansion, the distribution of gene frequencies after just one generation. But after two generations, or more, the procedure is complicated by the fact that we must consider all possible outcomes of earlier generations. This can still be easily done numerically and is done in Example

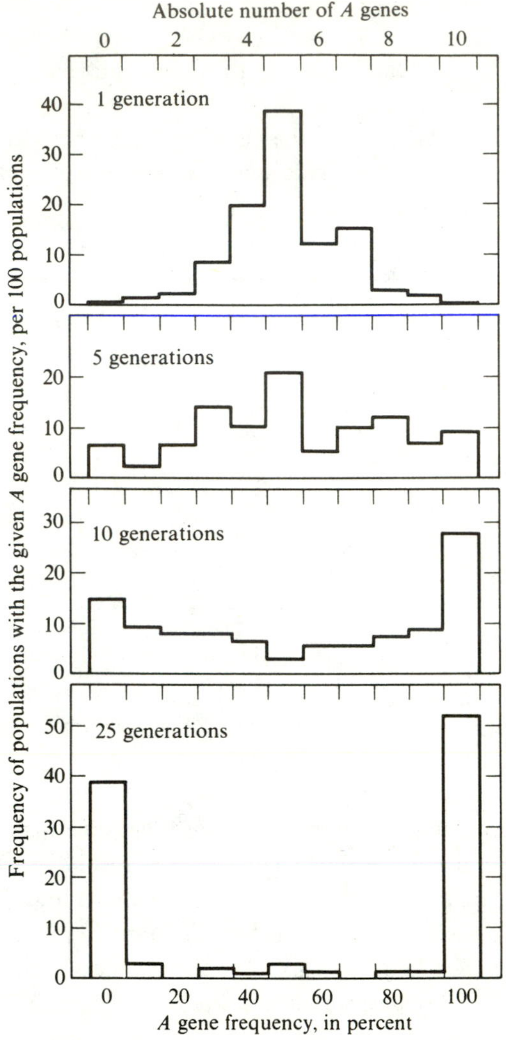

FIGURE 8.2
Results of 100 experiments in which computers were used to simulate random genetic drift for two alleles at a single locus. The distribution of the *A* allele in each experiment at a given generation (t) is given for a selected number of t values. The population is made of 10 genes (that is, 5 diploid individuals) and the initial gene frequency is 0.5 in all 100 experiments.

8.1 for $n = 3$. For a high n, numerical computations become very long, but a computer program can help us to obtain the distribution. An example of distributions calculated with a computer is given in Figure 8.3.

The full theoretical treatment of the prediction of the gene-frequency distribution after n generations has been given by Kimura (1955a, 1964). The equation giving the exact distribution is exceedingly complex. We include here simply its expansion into series:

$$\phi(p, x, t) = 6p(1 - p)e^{-t/2N} + 30p(1 - p)(1 - 2p)(1 - 2x)e^{-3t/2N} + \cdots, \tag{8.1}$$

where p is the initial gene frequency, t is the time in generations, x the gene frequency after t generations, and ϕ the proportion of populations with gene frequency x at time t, given p. This expansion is used in Figure 8.4.

At any time, the mean of this distribution is p, so that *on an average* the gene frequency remains the same, but it will change in any real finite population. Eventually, after an infinitely long time, one or the other allele is fixed in all populations, and the proportion of populations in which A is fixed is p, and that in which a is fixed $1 - p$, in agreement with the constancy of the average gene frequency. The variance of the gene frequencies in the different populations increases with time in a way that can be predicted from the distribution, and also in simpler ways, as will be seen later.

The theoretical distribution is given in Figure 8.4 for $p = 0.5$ and $p = 0.1$, at various times. It may be seen that it tends to flatten into a "rectangular" distribution at $t = 2N$ if $p = 0.5$. As t increases more and more populations have gene frequencies of 0 or 100 percent.

The mean time for fixation is roughly $2N$.

It is clear that the time for fixation depends on the initial gene frequency, p, and on the population size, N. The probability of fixation (or loss) increases with time but becomes one only after an infinitely long time. For large values of t, Kimura has obtained the following approximation for the proportion of populations fixed as a function of t (in generations) and the initial frequency p_0 :

$$P(t, p_0) = 1 - 6p_0(1 - p_0)e^{-t/2N}. \tag{8.2}$$

In Figure 8.5 are given the proportions of populations in which an allele is fixed, as a function of time, using the method of Example 8.1, for various population sizes. The **mean fixation time** in generations has been computed by Ewens (1969), using the diffusion approximation, to be

$$T = 4N[p_0 \log_e p_0 + (1 - p_0)\log_e(1 - p_0)]. \tag{8.3}$$

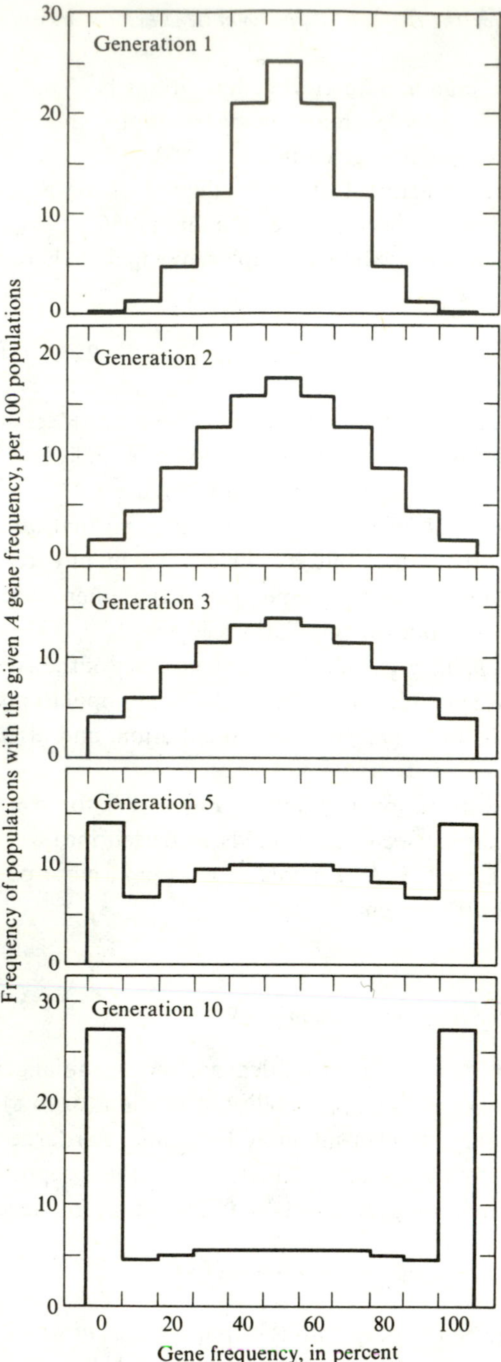

FIGURE 8.3
Theoretical distribution of gene frequencies under random genetic drift in a population of $N = 5$ individuals (10 genes) corresponding to the observed distribution in Figure 8.2. The method of computation is given in Example 8.1.

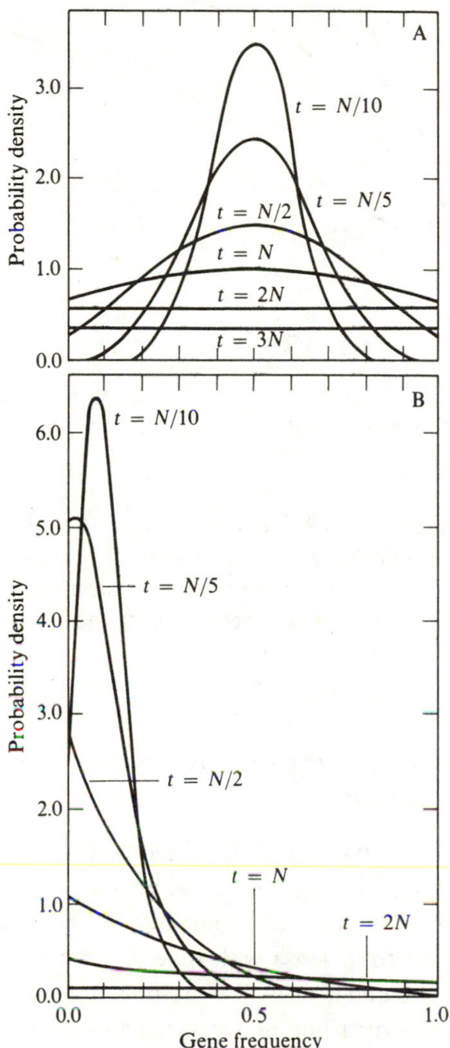

FIGURE 8.4
The process of change in the probability distribution of gene frequency, due to random sampling of gametes in reproduction. It is assumed that the population starts from the gene frequency 0.5 in A and 0.1 in B. The symbol t indicates time (in generations), and N the effective size (number of genes) of the population. (From Kimura, 1955b.)

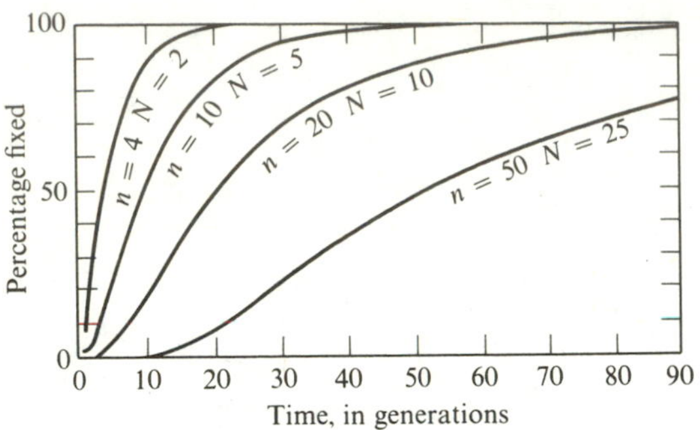

FIGURE 8.5
Proportion of populations in which one allele is fixed under random genetic drift. (See Example 8.1.) The initial gene frequency is 0.5, n is the number of genes, N is the number of individuals in the population ($=\frac{1}{2}n$).

The expression in brackets is 0.69 for $p = 0.5$, and then $T = 2.8N$, which is the maximum value that T can take. For $p = 0.05$, $T = 0.8N$, and for $p = 0.005$, $T = 0.03N$. Mean fixation time becomes shorter as p approaches 0 or 1 because it is then dominated by the shorter time necessary to get to the nearer boundary ($p = 0$ if $p_0 < 0.5$, or $p = 1$ if $p_0 > 0.5$).

The expected value of the gene frequency in populations differentiating under drift remains the initial gene frequency.

If we have k populations, all of equal size (each having N individuals), reproducing under complete isolation, and all starting with a common gene frequency p_0 for a given allele, each population drifts independently, and heterogeneity between their gene frequencies is bound to increase with time. We have seen the general distribution of gene frequencies under drift. We are now interested in deriving the two simplest parameters of this distribution, the mean and the variance. The mean can be obtained as the expected value of gene frequencies. The **expected value** of a random variable is the sum of the products of all the possible values it can assume times their corresponding probabilities. An expected value is denoted by E. It is useful to remember that $E(cx) = cE(x)$ when x is a random variate and c is a constant, and that $E(x + y) = E(x) + E(y)$ when x and y are two random variates (see also Appendix I).

If we have one generation of random sampling, the expected value of the gene frequencies obtained by sampling is equal to the gene frequency of the initial population. This carries on, generation after generation, so that the expected gene frequency of all populations derived by random drift from an initial population having gene frequency p_0 remains at p_0 for all times. Therefore for any generation t, the expected p_t remains

$$E(p_t) = p_0 .$$

It is therefore expected, as shown by Equations 8.1 and 8.2 that the mean of all gene frequencies remains the same under drift, although the gene frequency of any particular population may change dramatically.

For evaluating the variance of gene frequencies under drift it is useful to study first the decrease of overall heterozygosis in independently drifting populations, each mating at random, and therefore each subject to Hardy-Weinberg equilibrium.

8.3 Heterogeneity Between Populations under Hardy-Weinberg Equilibrium and Wahlund's Formula

Suppose we take k populations with different gene frequencies, each in Hardy-Weinberg equilibrium and mix them, without letting reproduction take place. We have

		Genotype frequencies		
Population	*Gene frequency*	*AA*	*Aa*	*aa*
1	p_1	p_1^2	$2p_1q_1$	q_1^2
2	p_2	p_2^2	$2p_2q_2$	q_2^2
...	...	...	...	...
k	p_k	p_k^2	$2p_kq_k$	q_k^2

where $p_i = 1 - q_i$, for the ith population, for all values of i from 1 to k.

We imagine for simplicity that each population is formed by the same number of individuals, If all populations are pooled, the average gene frequency is

$$\bar{p} = \frac{\sum p_i}{k},$$

where the sum, $\sum$, is over all values of i from 1 to k.

We define the variance of the k gene frequencies to be

$$\sigma^2 = \frac{\sum p_i^2}{k} - \bar{p}^2,$$

where we use the divisor k rather than $k - 1$ as we are dealing with a theoretical and not an empirical variance. The average frequency of heterozygotes Aa is

$$\bar{H} = \frac{\sum (2p_i q_i)}{k} = 2\frac{\sum p_i}{k} - 2\frac{\sum p_i^2}{k} = 2\left(\bar{p} - \bar{p}^2 + \bar{p}^2 - \frac{\sum p_i^2}{k}\right)$$

$$= 2(\bar{p} - \bar{p}^2 - \sigma^2) = 2(\bar{p}\bar{q} - \sigma^2), \quad (8.4)$$

using the equation just given for σ^2. Equation 8.4 can be rewritten as follows

$$\overline{H} = 2\bar{p}\bar{q}\left(1 - \frac{\sigma^2}{\bar{p}\bar{q}}\right). \tag{8.5}$$

Similarly, the mean frequencies of homozygotes are

$$AA: \ \bar{p}^2 + \sigma^2,$$

$$aa: \ \bar{q}^2 + \sigma^2.$$

Thus, when populations with different gene frequencies but each at Hardy-Weinberg equilibrium are pooled, there is an overall deficiency of heterozygotes and an excess of homozygotes, by a quantity that is twice the variance of the gene frequencies of the individual populations. The decrease of overall heterozygosity under drift is a direct measure of the heterogeneity between populations drifting independently. We will refer to

$$\frac{\sigma^2}{\bar{p}\bar{q}}$$

computed from the gene frequencies of a group of populations as the standardized, or Wahlund's, variance of gene frequencies.

It is important to emphasize that the decrease in heterozygosity is in *overall* heterozygosity. Since each individual population is mating at random, Hardy-Weinberg proportions are maintained within any given population, subject only to minor perturbations because of its finite size. The effects of random genetic drift are not, therefore, generally detectable as a departure from the Hardy-Weinberg law in any single population, but as an overall deficiency of heterozygotes in the pooled populations, as indicated by the derivation of Equation 8.5.

The inbreeding and kinship coefficients have a close relation to the standardized variance σ^2/pq.

We have seen that the overall frequency of heterozygotes in a set of populations drifting independently is defined by

$$2pq - 2\sigma^2 = 2pq\left(1 - \frac{\sigma^2}{pq}\right),$$

where σ^2 is the variance of gene frequencies between populations, and p and q are the mean gene frequencies.

On the other hand, we know that, under inbreeding, the equilibrium frequency of heterozygotes in any given population is

$$2pq(1 - F),$$

where F is the coefficient of inbreeding. Thus, from the homology of these two expressions, σ^2/pq is a measure of departure of the total population from random mating that is analogous to the inbreeding coefficient. Inbreeding itself may lead to substantial departures from random mating within a single population, while, as already has been pointed out, random genetic drift, in general, does not. The quantity σ^2/pq can be thought of as a form of inbreeding coefficient for a group of populations, as opposed to an inbreeding coefficient for an individual.

Malécot (1948) has defined the inbreeding coefficient F of an individual as the probability that, for a given locus, he carries two genes identical by descent on the two corresponding homologous chromosomes (see Chapter 7). If there is no mutation, F is identical to the coefficient of kinship of the parents. The **coefficient of kinship,** f, between two individuals I and J is defined as the probability that a gene taken at random from I, at a given locus, be identical by descent to a gene taken at random from J at the same locus. It is (see Example 8.2) the same as the inbreeding coefficient of the progeny of the two individuals whose kinship is being measured, apart from a usually negligible correction factor (namely, that due to mutation having occurred in the gametes giving rise to the progeny). The mean kinship coefficient f can be equated to Wahlund's variance, σ^2/pq. When this variance refers to a single population, then σ^2 is the expected variance of the gene frequencies based on repeated realizations of the history of the particular population.

The variance between populations increases, approximately, with time, under drift alone, according to a simple formula.

We can compute the increase of the variance between gene frequencies of independently drifting populations as a function of time. As shown in Example 8.3, the overall frequency of heterozygotes in populations undergoing drift, decreases with time t (in generations) according to the formula

$$\frac{H_t}{H_0} = \left(1 - \frac{1}{2N}\right)^t \sim e^{-t/2N}. \tag{8.6}$$

Here H_t is the mean heterozygote frequency of the populations at time t. We can thus use Equation 8.5 for $\bar{H}$ on the assumption that $\bar{p}$ is the mean and σ^2 the variance of the gene frequencies at time t.

Since

$$\bar{H} = 2\bar{p}\bar{q}\left(1 - \frac{\sigma^2}{\bar{p}\bar{q}}\right),$$

and we have shown that $\bar{p} = p$, the initial frequency, so that $H_0 = 2pq$, we can write, from Equations 8.5 and 8.6,

$$\frac{\sigma^2}{pq} = 1 - \frac{\bar{H}}{2\bar{p}\bar{q}} \sim 1 - e^{-t/2N}. \qquad (8.7)$$

This equation shows very simply the expected way in which the variance σ^2 should increase with time. The mode of increase is shown graphically in Figure 8.6 in

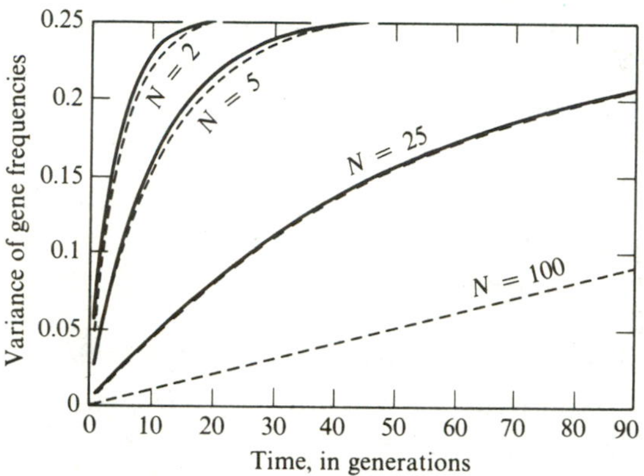

FIGURE 8.6
Increase in variance of gene frequencies σ^2 with time under random genetic drift. Initial gene frequency $p = 1/2$. Dividing σ^2 by pq ($= 1/4$ in this case) gives the standardized (Wahlund's) variance, The symbol N represents the number of diploid individuals. Solid curves are computed by the method given in Example 8.1, and dashed curves by the approximation $pq(1 - e^{-t/2N})$, which is excellent for N greater than or equal to 25.

which both exact values (computed as in Example 8.1) and approximate values computed by using this equation are given. (This result was first derived, in different ways, by Wright, 1931, and Fisher, 1930. See also Crow, 1954.)

When the number of generations t is small relative to N,

$$e^{-t/2N} \sim 1 - \frac{t}{2N},$$

and so

$$\frac{\sigma^2}{pq} \sim \frac{t}{2N}. \qquad (8.8)$$

The rate of increase in σ^2/pq slows down as t approaches N and σ^2/pq approaches its maximum value of one. In fact, when in all populations one or the other of the two alleles has become fixed, there will be p populations made of A alleles only, $1 - p$ of a alleles only, and the variance of the gene frequencies will then assume its maximum value:

$$\sigma^2_{\max} = pq,$$

which is approached gradually as t increases. The Wahlund's variance is one when σ^2 is at its maximum.

These formulas are only approximate but they do indicate in a simple way the kinetics of divergence between populations due to genetic drift. They also indicate that the Wahlund's variance $\sigma^2/\bar{p}\bar{q}$ is a useful and important measure of the differences between populations in that it supplies a simple relationship for the expected variation due to drift. In the early stages of divergence due to drift, the Wahlund's variance is simply proportional to the time in generations divided by the number of genes forming the population.

8.4 Equilibrium Between Drift and Linear Evolutionary Pressures

Usually, when two or more populations of the same species evolve independently in different areas, genetic isolation between them is not complete. An exchange of individuals slows down the differentiation between populations caused by drift. Thus, migration can buffer drift, and the buffering effect depends on the extent of migration.

We will again make recourse to a highly simplified model, which will later be extended to make it more realistic. We will consider an isolated population of constant size N and initial gene frequency p_0, which exchanges a constant number of individuals per generation with an outside population that has a gene frequency $\bar{p}$, which is sufficiently large that random fluctuations of gene frequencies in it may be ignored. The number of immigrants that the population under discussion receives per generation, is given as m. The fundamental property of this model is that fixation of one of the alleles is not necessarily the ultimate fate of the population. A distribution of gene frequencies around a mean will still be found even after infinite time. The final or so-called **steady-state distribution** has an expected gene frequency $\bar{p}$ (that of the population from which the immigrants come) and, for small m, the Wahlund's variance can be shown to be

$$\frac{\sigma^2}{\bar{p}\bar{q}} = \frac{1}{1 + 4Nm}. \tag{8.9}$$

A simple demonstration of a closely analogous result is given in Example 8.5.

The shapes of the distributions at various times, for different Nm values, and

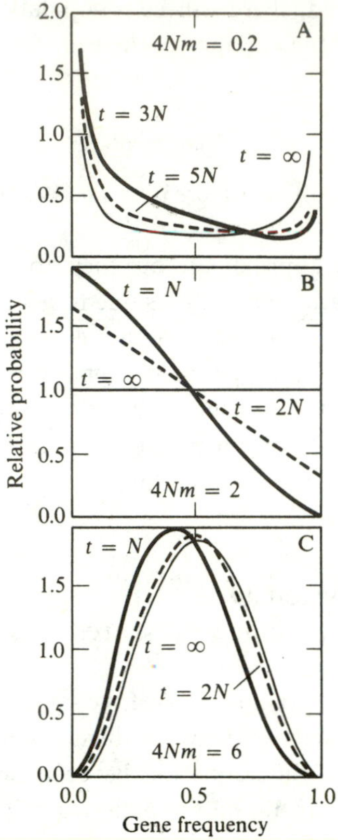

FIGURE 8.7
Asymptotic behavior of the distribution curve for gene frequencies in a finite population with migration. In all three drawings, the gene frequency of the immigrants is assumed to be 0.5 and the initial frequency in the population to be 0.2. The symbol N represents the number of individuals in the population, and m the rate of migration. The same result is obtained for any linear pressure of size m. (From Crow and Kimura, 1956.)

with $\bar{p} = 0.5$, and an initial frequency of $p_0 = 0.2$ are shown in Figure 8.7. (These curves are based on an involved theoretical treatment first given by Crow and Kimura, 1956. Similar approximate results were obtained much earlier by Wright, 1931.) Because of the initial frequency $p = 0.2$, the distributions are skew until $t = \infty$, at which time a steady state is reached. The distributions are then symmetric because the immigrant gene frequency is $\bar{p} = 0.5$. They are U-shaped—showing that many populations are near fixation—if $4Nm < 2$; rectangular if $4Nm = 2$; and bell-shaped otherwise—showing that few populations are near fixation. For $4Nm < 2$ there are populations in which an allele is temporarily fixed, but the arrival of immigrants in the next generation can restore the missing allele. The gene frequency of a temporarily fixed population is, however, constant in time, because a steady-state equilibrium between drift and migration has been reached.

Mutation and, in general, linear systematic pressures have the same effect as migration in counterbalancing drift.

Evolutionary factors whose action on gene frequencies can be described by an equation of the type

$$\Delta p = -b(p - \hat{p}), \tag{8.10}$$

that is, factors which lead to a rate of change that is linear in the gene frequencies, are often described as **linear systematic pressures**. They cause gene frequencies to return to an equilibrium at $\hat{p}$, where b is the coefficient of recall (see Chapter 3). **Migration** is such a linear pressure. If, at every generation, m immigrants come from a population that has a gene frequency $\bar{p}$, Equation 8.17 applies with $b = m$ and $\hat{p} = \bar{p}$ to give $\Delta p = -m(p - \bar{p}) = -mp + m\bar{p}$. **Mutation**, similarly, is a linear pressure, with $b = \mu + \nu$ and an equilibrium frequency $\hat{p}$—that is, $\nu/(\mu + \nu)$. Thus

$$\Delta p = -(\mu + \nu)p + \nu = -(\mu + \nu)\left(p - \frac{\nu}{\mu + \nu}\right). \tag{8.11}$$

Selection is not, in general, a linear pressure. When, however, we consider gene frequencies that are close enough to equilibrium, we may assume that stabilizing selection is approximately linear (see Chapter 4). If s_1 and s_2 are the selective disadvantages of the two homozygotes and the fitness of the heterozygote is 1, then $\hat{p} = s_2/(s_1 + s_2)$, and, near equilibrium, we may use the approximate equation

$$\Delta p \sim (s_1 + s_2)pq(q - \hat{q})$$

to obtain (from the slope at equilibrium)

$$\Delta p \sim -\frac{s_1 s_2}{s_1 + s_2}(p - \hat{p}). \tag{8.12}$$

These factors, when small, can be combined, in the coefficient b of the equation of linear change (8.10) to give, as an approximation,

$$b = m + \mu + \nu + \frac{s_1 s_2}{s_1 + s_2}, \tag{8.13}$$

and

$$\hat{p} = \frac{v + kp^* + m\bar{p}}{\mu + v + k + m}, \tag{8.14}$$

where $k = s_1s_2/(s_1 + s_2)$, $p^* = s_2/(s_1 + s_2)$ and as before, $\bar{p}$ is the gene frequency among the immigrants. Provided that b is not too large, Equation 8.9 given above for the variance of gene frequencies under equilibrium between drift and migration is valid for the more general case of the equilibrium between drift and all linear systematic pressures. Thus at equilibrium the expected value of the gene frequency is $\hat{p}$, and the Wahlund's variance of gene frequencies is

$$\frac{\sigma^2}{\hat{p}\hat{q}} = \frac{1}{1 + 4Nb} \tag{8.15}$$

as long as b (as given by Equation 8.14) is small.

As the mutation rates μ and v are likely to be much smaller than the other two pressures, they can usually be neglected when there is significant selection or migration.

The equilibrium between drift and linear pressures leads to a steady-state distribution of gene frequencies.

Wright (1937) was the first to give the function for the equilibrium distribution of gene frequencies under the joint effect of drift and linear pressures (mutation, migration, and linearized stabilizing selection). The distribution is valid for the steady state in which the increase in the variance of the distribution due to the effect of drift is exactly balanced by the buffering effect of the linear pressure, so that the variance remains constant.

The steady-state distribution of the gene frequencies is a beta distribution (see Appendix I) as follows

$$\phi(p) = Cp^{4Nb\hat{p}-1}(1 - p)^{4Nb\hat{q}-1}, \tag{8.16}$$

where C is a constant making the area of the distribution unity. This distribution has the mean and variance already given. Its shape depends on the values $4Nb\hat{p}$ and $4Nb\hat{q}$. Considering, for simplicity, that only mutation is involved, then $b\hat{p} = v$ and $b\hat{q} = \mu$. The curve given by Equation 8.15 is U-shaped when $4Nb\hat{p}$ and $4Np\hat{q}$ are less than 1—in other words, if $N < 1/4\mu$ or $1/4\nu$. If $4N\mu = 4N\nu = 1$, then the distribution is rectangular; it is bell-shaped if these values are greater than 1 (Figure 8.8).

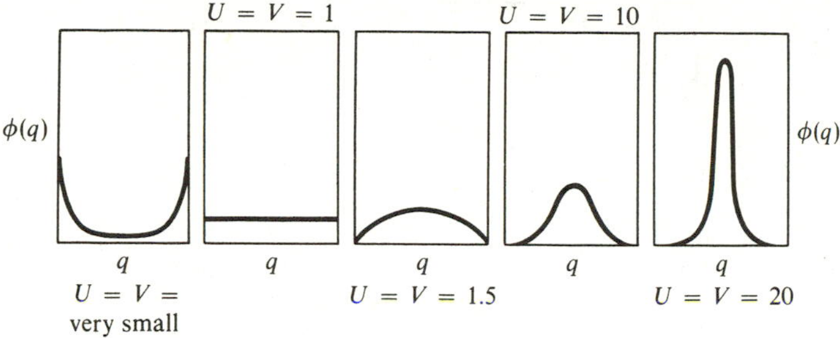

FIGURE 8.8
Steady-state distribution of the gene frequency q under mutation pressures, assuming $\mu = \nu$. The general form of the distribution is $\phi(q) = Cq^{U-1}(1-q)^{V-1} = C[q(1-q)]e^{NV-1}$, where $U = V = 4N\mu = 4N\nu$. (From Li, 1955.)

Drift may affect the balance between selection and deleterious mutations.

It is well known that the likelihood of finding persons who are homozygous for certain deleterious mutant genes is greater in isolated communities, such as those in alpine valleys or on small islands, than elsewhere. This holds for the mutant alleles that cause albinism, deaf mutism, mental deficiency, and other recessive pathological conditions.

It has been shown by Wright (1938) that the joint action of selection against a fully recessive, deleterious gene a that has a fitness value $1 - s$ (in homozygotes aa) of mutation from A to a that proceeds at a rate μ, and of drift leads to a steady-state distribution of the form

$$\phi(q) = \frac{Ce^{-2Nsq^2}q^{4N\mu-1}}{1-q}, \tag{8.17}$$

which is usually extremely skew or J-shaped, when N is not too large. The symbol q represents the frequency of the allele a. As an example, the distributions for a lethal gene ($s = 1$) with a mutation rate of $\mu = 10^{-5}$ and various N values are given in Figure 8.9.

The steady-state gene-frequency distribution for deleterious recessives has a mean that is equal to the value expected under the deterministic model if N is very large, namely $\bar{q} = (\mu/s_1)^{1/2}$. The mean, however, becomes somewhat smaller with decreasing N. Thus it is $\bar{q} = 0.0032$ for $\mu = 10^{-5}$, $s = 1$, $N = 10^6$; with $N = 10^5$, $\bar{q} = 0.003$; with $N = 10^4$, $\bar{q} = 0.002$; with $N = 10^3$, $\bar{q} = 0.0008$; and with $N = 10^2$, $\bar{q} = 0.00026$ (Wright, 1939b). The probability that q may have a large value in a small population may, however, be quite high, even though the mean is small.

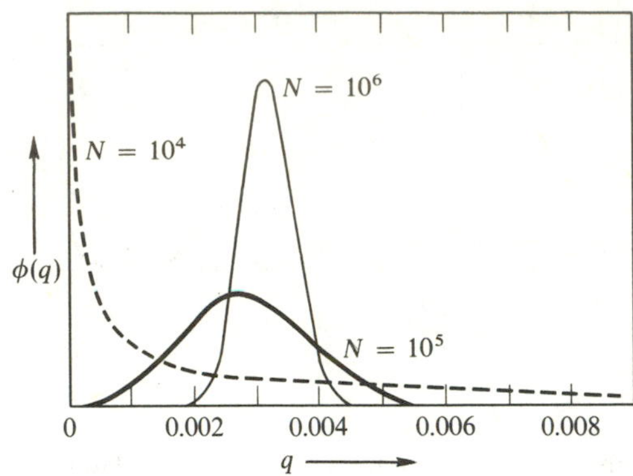

FIGURE 8.9
Distribution of a rare recessive lethal gene, with $\mu = 10^{-5}$. Note that the mean $\bar{q} = (\mu)^{1/2} = (0.00001)^{1/2} = 0.00316$ in large populations and becomes increasingly small as N decreases. (Based on Wright, 1937, 1939; from Li, 1955.)

8.5 The Interaction of Drift and Selection: Some Further Considerations

Additive directional selection increases fixation rates.

Unlike mutation and migration, selection pressures are not generally linear functions of gene frequencies, because selection acts on phenotypes, and so involves squared terms for genotype frequencies in diploid populations. The mathematical analysis of the effect of selection is, therefore, always more complicated than that of mutation and migration. In particular this is true of the interaction of selection with drift, which necessarily requires an indeterministic or stochastic treatment, the mathematics of which is always complex and it is usually impossible to obtain exact solutions with nonlinear pressures. The usual treatment of selection models that assumes the population is infinitely large (as in Chapter 4) is called **deterministic.** A treatment that takes into account statistical fluctuations, such as random drift due to finite population size, is called **stochastic**.

There is one feature common to all selection processes when they are considered jointly with drift and in the absence of other factors. They always culminate in fixation or extinction. This is obviously true of selection, which eventually favors one particular allele. But it is also true of selection for heterozygote advantage. One might think that such selection would keep gene frequencies at some intermediate level near or at the deterministic heterotic equilibrium value. Actually this is not so, for if there is no mutation or migration, drift is an irreversible process. Lost alleles are irreplaceable and so, once an allele is fixed, there is no possibility of reversion (except by migration or mutation). Therefore, even when selection favors heterozygotes, as long as this is the sole factor in a finite population, drift will always ultimately cause fixation or extinction. Selection for the heterozygote may, however, slow down the process.

A full treatment of selection and drift has been given by Kimura (1964) for the simplest case of selection, called **genic or additive selection**, in which selective coefficients for the various genotypes are additive, as in the following set:

AA	Aa	aa
1	$1 - s$	$1 - 2s$

Here the rate of change in gene frequency per generation is approximately (see Chapter 3)

$$\Delta p = spq,$$

with the approximation improving as s becomes smaller. Eventually, the A allele will be fixed in almost all populations. At intermediate times the distribution of gene frequencies can be computed, but it is especially interesting to compute the steady-state rate of decay of heterozygosity. This rate is valid for large values of t when the pattern of change tends to take on its limiting form. The rate is (Kimura, 1954, 1955, 1964)

$$\frac{1 + \frac{2}{5}(Ns)^2 - \frac{2}{1075}(Ns)^4}{2N}$$

and is therefore higher than that due to drift alone, which is $1/2N$.

In a finite population, fixation of a selected allele does actually proceed to completion.

Another aspect of the interaction of selection and drift is that, in a finite population, fixation of an allele selected for can actually occur, while in an infinite population fixation is only approached asymptotically (that is, only reached at an infinite time). Ewens (1969) has given a formula for computing, approximately, the mean fixation time, under additive selection, and under selection in favor of a dominant. The effect of drift is noticeable only when gene frequencies are very near zero or one. The time spent in the rest of the frequency range is largely unaffected by drift. Table 8.3 shows this effect.

TABLE 8.3

Approximate Times (in Generations) Spent in Various Frequency Ranges, Using the Following Fitness Values. Case 1: AA, 1.002; Aa, 1.001; aa, 1.000. Case 2: AA, 1.001; Aa, 1.001; aa, 1.000

	Case 1		*Case 2*	
A Gene Frequency Range	*Infinite Population*	*Finite Population* ($N = 10^6$)	*Infinite Population*	*Finite Population* ($N = 10^6$)
0.1–0.99	13600	13600	105700	85500
0.99–0.999	4600	4500	902300	34500
0.999–0.99999	13800	2690	99×10^6	4000

Source: Adapted from Ewens (1969).

Stabilizing selection (heterosis) may cause retardation of fixation.

Theoretical analysis has confirmed the intuitive prediction that stabilizing selection can act to retard the rate of fixation of alleles under drift, even if it cannot prevent it entirely. One interesting conclusion is, however, that this is not always true. In fact it was found by Robertson (1962) and confirmed by Kimura (1964) that the retardation factor depends on the equilibrium gene frequency. If we take a system with fitness values:

$$\begin{aligned} &1 - s_1 && \text{for } AA, \\ &1 && \text{for } Aa, \\ &1 - s_2 && \text{for } aa, \end{aligned}$$

the equilibrium gene frequency will be

$$p_\infty = \frac{s_2}{s_1 + s_2}$$

for the A gene, and will be 50 percent if $s_1 = s_2$. It is for this value of p_∞ that the retardation effect is at its maximum. When, however, s_1 is different from s_2 and thus $p_\infty \neq \frac{1}{2}$, the retardation is smaller; for p_∞ approximately <0.2 or >0.8, there is actually an acceleration of the fixation process.

Some results given by Robertson (1962) are shown in Figure 8.10. The **retardation factor** in the figure is the ratio between the rate of decay at the given steady state in a population of N individuals subject to heterosis and that of a population of the same size subject to drift alone, which is $1/2N$. The retardation factor becomes less than 1—that is, there is acceleration of fixation, when p is out of range 0.2–0.8.

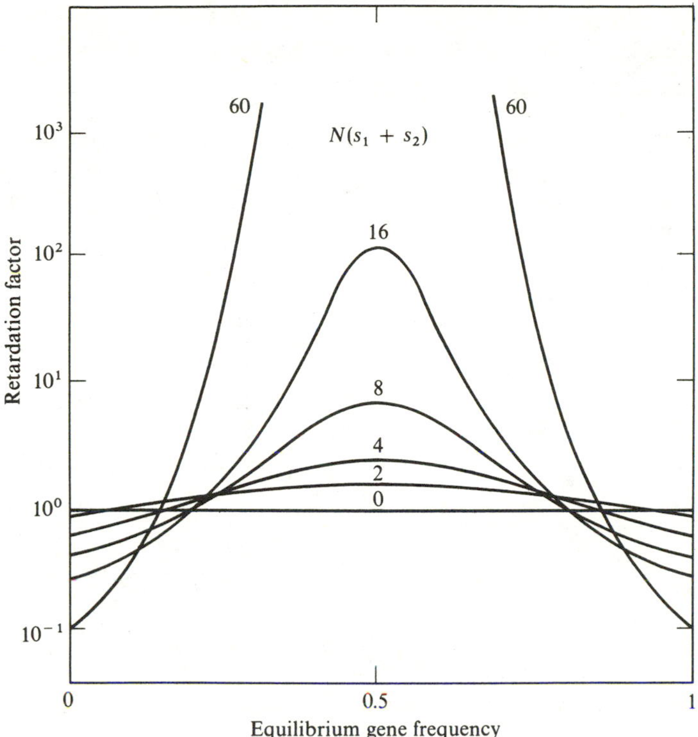

FIGURE 8.10
Graphs showing retardation factor as a function of equilibrium gene frequency for various values of N and $(s_1 + s_2)$, where N is the effective size of the population, and s_1 and s_2 are the coefficients of selection against the two homozygotes. (From Robertson, 1962.)

Other work, both theoretical and Montecarlo or numerical simulation, on this general problem has been published (see Bodmer, 1960; Kimura, 1964). By means of the numerical approach given in Example 8.1, which is exact and can take selection into account, we have essentially confirmed this dependence of retardation on the equilibrium gene frequency. Thus, with a small population ($n = 10$) having an initial gene frequency of 0.5, we have obtained the data given in Table 8.4.

The table shows retardation, as expected, for $p_\infty = \frac{1}{2}$ and also shows that the mean time of fixation increases with s. Retardation is, however, almost absent when selection forces the gene frequency from its initial value of $\frac{1}{2}$ towards a peripheral value. It is worth noting that the expected proportion of A genes in the last two experiments (5 percent and 95 percent, respectively) was not reached, because fixation took place early enough to stabilize the gene frequency at fixation at 10.3 percent and 89.7 percent, respectively.

In another experiment, the initial gene frequency was kept at 0.05. Fixation was then accelerated by the presence of heterosis, unless it was mild (see Table 8.5).

TABLE 8.4

Stabilizing Selection and Drift. Initial gene frequency: $p_0 = 0.5$.

Type of Selection	s_1	s_2	Time to Fixation, in Generations: Mean	Standard Deviation
None	0	0	26.2	19.7
Heterotic ($p_\infty = 0.5$)				
Mild	0.05	0.05	30.4	23.8
Medium	0.2	0.2	56.6	49.9
Strong	0.5	0.5	220.1[a]	141
Heterotic ($p_\infty = 0.05$)	0.19	0.01	27.3	20.2
Heterotic ($p_\infty = 0.95$)	0.01	0.19	27.3	22.2

[a] May be underestimated because of the skewness of fixation and restriction of the experiment because of computer cost.

TABLE 8.5

Stabilizing Selection and Drift. Initial gene frequency: $p_0 = 0.05$.

Type of Selection	s_1	s_2	Time to Fixation, in Generations: Mean	Standard Deviation
None	0	0	7.2	12.8
Heterotic ($p_\infty = 0.5$)				
Mild	0.05	0.05	8.7	16.0
Medium	0.2	0.2	2.9	3.1
Strong	0.5	0.5	1.6	1.5
Heterotic ($p_\infty = 0.05$)	0.19	0.01	6.8	12.0
Heterotic ($p_\infty = 0.95$)	0.01	0.19	2.5	2.3

Geographic variation can be due to selective differences or to drift.

Drift is a source of variation between gene frequencies of a population at different times, or between those of populations having a common origin but now reproducing in partial isolation. In the latter case it can create geographic variation. But it is not, of course, the only factor that can do so.

Whenever geographic variation is encountered, three possible sources of variation should always be considered: (1) differences in local conditions leading to different patterns of selection in different areas; (2) random genetic drift; (3) historical accidents leading to heterogeneity of the populations not yet cancelled by diffusion. This may mean that parts of the area have been populated by an extraneous population, which has not had enough time to mix and homogenize with the rest. Thus equilibrium has not been achieved over the whole area.

The first source of variation can be shown to operate whenever an environmental factor associated with selection is known to exist and can be measured. For example, the geographic variation in the intensity of malaria can be measured, and there are selective differences from place to place. Some changes with time will, however, occur inevitably in the intensity of such a selective factor, especially if it is a disease. Even an endemic disease, such as malaria, does not necessarily have a constant intensity at different times. Other diseases, especially if epidemic in character, change with time and place in a more significant way. If they have a selective effect this is subject to strong changes not only with place, but also with time.

Selective drift is the name given to a special model involving selective conditions that change at random in time and space.

An analysis of the effects of changing selective conditions has been done by Kimura (1954). If we can assume that selective coefficients fluctuate at random from place to place, and from generation to generation, with known mean $\bar{s}$ and variance V_s, then we can compute the distribution of gene frequencies to be expected as a function of time and of the initial gene frequency.

Figure 8.11 indicates what the distribution is when the mean selective coefficient is zero; the mean gene frequency remains unaltered at the initial value (taken to be 0.5). With the passage of time, the distribution becomes flatter and eventually bimodal and then U-shaped. There is no true fixation or extinction, because the population is considered to be infinitely large, but there is **quasi-fixation** or **quasi-loss**, because after many generations the gene frequencies are all very near zero or one (for $t = 100$, the peaks are at 0.0007 and 0.9993).

The similarity between this process and that of random genetic drift is thus very marked, and Kimura has proposed for it the name of **selective drift**. It is useful to form an idea of the remarkable spreading effect that variable selection may have. In the distributions computed in Figure 8.11, V_s is taken to be 0.0025, so that 2/3 of the selective coefficients are in the range -0.05 to $+0.05$ and 95 percent of them are in the range -0.10 to $+0.10$. Under such conditions the process of quasi-fixation is remarkably fast. There are no definite examples of characters that might fit into this scheme. It is, however, possible to imagine that an epidemic disease attacking preferentially an otherwise favored genotype might create a similar selective pattern. Many epidemic diseases show great fluctuations in severity of epidemic waves and could create the fluctuation in selective conditions that is postulated by this theory.

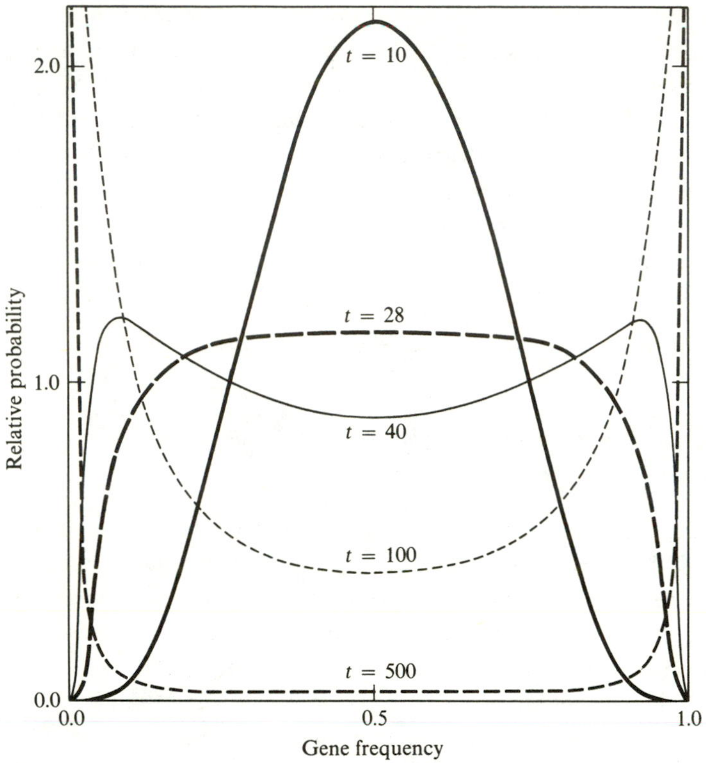

FIGURE 8.11
The process of change in the gene-frequency distribution under random fluctuation of selection intensities. In this illustration it is assumed that the gene is selectively neutral when averaged over a very long period, that there is no dominance, and that $p = 0.5$ and $V = 0.0483$. (From Kimura, 1954.)

8.6 The Fate of Single Mutant Genes

Whenever a mutant gene is produced in a population that is not too large, it quite likely is the only one of its kind in the population. In fact, with a mutation rate of 10^{-6}, and a population of size $N = 10^4$, there will be only one such mutant gene every 50 generations on an average (see Example 8.4). Moreover, the mutation rate is given for a certain locus containing perhaps a thousand nucleotides, and the chance that a new mutation at the same locus repeats exactly a mutation already present may be remote. It is true that the fine structure analysis of mutation in phage T4 indicates that mutant alleles can accumulate at certain mutational sites (Benzer, see, for example, 1961). However, it is still possible that the clustering is due to low (or absent) crossing-over in the accumulation regions. In any case, the repetition of exactly the same mutation is certainly a rarer event than the mutation rate per locus indicates.

It is, therefore, of interest to follow the fate of a new single mutant gene that has an initial frequency of $1/2N$ in a population of size N. The probability that this mutant gene will eventually be fixed is, naturally, a function of its selective value. As the mutant gene (in a random-mating population) will remain for many generations in the heterozygous condition, the selective value of the *heterozygote* will be the important one in the first part of the mutant gene's life, the time during which it is most subject to chance elimination by drift. Later on, if the mutant gene has reached a substantial frequency, homozygotes will appear and only then will the fitness of the individual homozygous for the mutant gene be important. Unless there is clearly heterosis, however, it is only the fitness of the heterozygote that matters in the subsequent computations.

The fate of single mutant genes was first studied by Fisher (1922), and later by Kimura (1962). The probability of ultimate fixation as a function of the initial gene frequency p_0, s (the selection coefficient of the heterozygote under the assumption that there is no dominance), and N takes the form:

$$P(p_0) = \frac{1 - e^{-4Nsp_0}}{1 - e^{-4Ns}}, \tag{8.18}$$

which, for one initial mutant gene ($p_0 = 1/2N$) simplifies to

$$P\left(\frac{1}{2N}\right) = \frac{1 - e^{-2s}}{1 - e^{-4Ns}}. \tag{8.19}$$

This formula is not valid if $s = 0$, in which case we have $P(p_0) = p_0$.

The probability of fixation as a function of s (from Equation 8.19) is illustrated in Figure 8.12 for $N = 10$ and N infinite. This shows how, in small populations, even genes with negative selection coefficients have a chance of being fixed.

The figure also shows that, unless the selective advantage is large, the probability of random elimination is high. This is, of course, especially true in the early generations after a mutation's appearance, and for larger populations. Thus, many mutant genes are lost even though they are advantageous, most being lost soon after they arise.

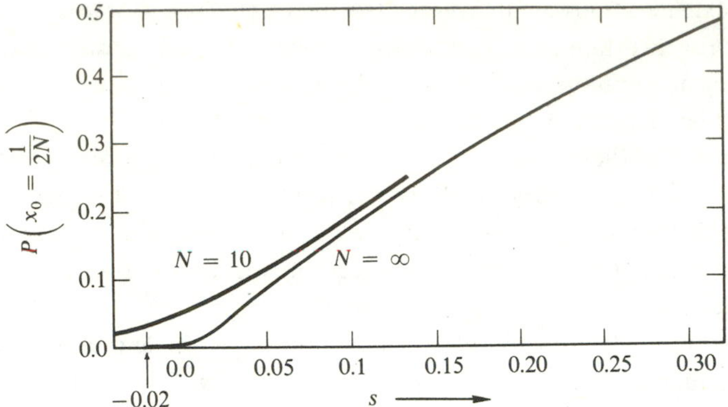

FIGURE 8.12
The probability of fixation for a single mutant gene as a function of its selection coefficients. The function plotted is

$$P = \frac{1 - e^{-4NsX_0}}{1 - e^{-4Ns}}, \quad \text{with} \quad X_0 = \frac{1}{2N}.$$

Two lines are given, one for $N = 10$ and the other for $N \to \infty$. For larger values of s the two lines are effectively superimposed. Note that the initial slope is $2s$ for large N.

8.7 The Number of Alleles That Can Be Maintained in a Finite Population

Drift inevitably causes loss of variation, but we have seen that an equilibrium between drift and new variation, for example from mutation, can be reached. The problem arises then, how many alleles can be maintained in a population if mutation keeps introducing new alleles, and drift eliminates them. Clearly, the number of alleles maintained at equilibrium will be a function of mutation rate and of population size. A model suggested by Kimura and Crow (1964) uses the hypothesis that any mutation forms a new allele. Considering the number of nucleotides involved at a locus, and that each can mutate independently this hypothesis is a reasonable approximation to reality. The assumption that all alleles are "neutral" with respect to selection is, of course, less attractive, and we will later discuss ways in which it may be avoided.

The probability that an individual selected at random is homozygous for one allele at a locus, under equilibrium between mutation and drift, is given approximately by (see Example 8.5)

$$F = \frac{1}{1 + 4N\mu},$$

that is, by the same formula (Equation 8.15) as the kinship coefficient computed before for the same situation. The above equation, however, refers to inbreeding at the individual level due to the finite size of the population. Kimura and Crow (1964) have also shown that, since F is the probability of homozygosity, its reciprocal can be defined as the "effective number of alleles" that are being maintained. In fact, if there were n alleles of equal frequency, $1/n$, the homozygotes would each have a frequency $1/n^2$ and there would be n such homozygotes. The total frequency of homozygotes would be $n \times 1/n^2 = 1/n$. Thus, $1/F$ or the reciprocal of the overall frequency of homozygotes is equal to n, the number of alleles, when these are equally frequent. If alleles have different frequencies the reciprocal will be an underestimate. It has been called the "effective number" of alleles maintained in the population. The relationship between this effective number, population size N and mutation rate is illustrated in Figure 8.13. With mutation rates of order 10^{-6} per locus, the effective number of alleles would be $n = 1.4$ for populations of size 10^5 and $n = 5$ for size 10^6. With mutation rates of order 10^{-5} and population sizes 10^5 and 10^6, it would be 5 and 41, respectively. With a smaller population size, such as $N = 10^3$, $n = 1.04$ and 1.004 for $\mu = 10^{-5}$ and $\mu = 10^{-6}$, respectively, while with $N = 10^4$, $n = 1.4$ and 1.04 for these same mutation rates.

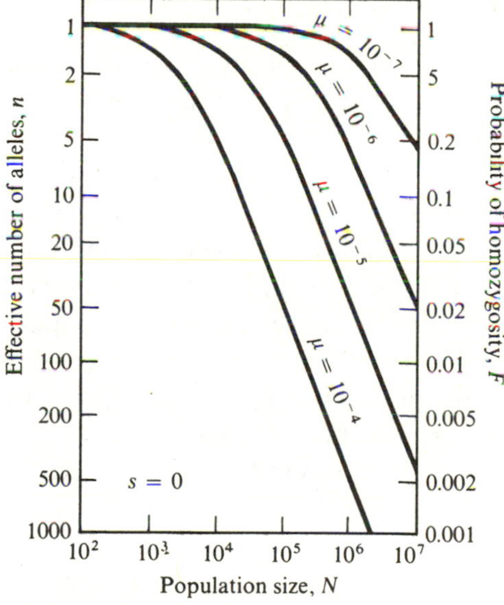

FIGURE 8.13
The probability of homozygosity (F) and the effective number of alleles (n) maintained by a mutation rate (μ) in a population of size N. The mutants are assumed to be selectively neutral and each new mutant allele is of a type not already existing in the population.
(From Kimura and Crow, 1964.)

Overdominance increases the number of alleles maintained only if it is appreciable.

The above estimates are valid for neutral genes. Selection is obviously an important factor in determining how many alleles are in a population. In particular, overdominance may play a role in increasing the number of alleles maintained. It is, therefore, of some interest to consider this possibility theoretically.

With overdominant systems, polymorphisms will be stabilized and so F will decrease. A computation was made by Kimura and Crow (1964) on the hypothesis that all new mutant genes are different, that they are all overdominant, and that all homozygotes have equal fitness $(1 - s)$ with respect to a fitness of 1 for the heterozygotes. With $\mu = 10^{-5}$, $s = 0.001$, and $N = 10^4$, the effective number of alleles is 5, which is very little higher than that predicted for neutral alleles. On the other hand if $s = 0.01$, $n = 8$; if $s = 0.1$, $n = 22$; and if $s = 1$ (the homozygous state is completely lethal), the number of alleles is almost 60.

The question of how many alleles are maintained is closely related to the question of how many polymorphisms can be maintained in a population, which will be discussed in Chapter 11.

8.8 Population Effective Size

The theory of drift as just given is highly simplified. It applies to an organism with nonoverlapping generations and a constant population size of N diploid individuals or $2N$ genes for a given locus. In order to apply it to real situations it is necessary to remove some of the oversimplifications. We shall in particular be concerned with the following: (1) how to take account of changes of N in time; (2) different contribution of the two sexes, and varying distributions of progeny size; (3) how to estimate N with overlapping generations; (4) sampling problems involved; (5) subdivision of the species into several populations that exhibit a certain amount of intermigration that depends on the geographic distribution.

To solve the first three problems, we compute a **population effective size,** N_e, i.e., the population size that would give the same amount of drift in a reproducing population as that in the simplest model so far considered: the model having N constant in time, nonoverlapping generations, and all individuals and each sex expected to make equal contributions to the progeny.

If N varies with time the harmonic mean is the effective population size.

It has been shown by Wright (1939b; see also Crow, 1954) that if at different generations or in different populations, which are indicated by the subscript, population sizes are

$$N_1, N_2, \ldots N_i, \ldots,$$

the effect on the distribution of gene frequencies can be predicted by taking as an estimate of N the harmonic mean of the N_i values (see Example 8.6). Thus, if there are k values N_i at k different times:

$$N_h = \frac{k}{\dfrac{1}{N_1} + \dfrac{1}{N_2} + \cdots + \dfrac{1}{N_k}}. \tag{8.20}$$

The harmonic mean of a set of positive numbers is, in general, smaller than, or at most equal to, the arithmetic mean. It follows that if the population is small during part of its history, the drift effect will be larger than that based on use of the arithmetic mean of the N values at different times.

Bottlenecks in population size and the so-called founder principle are important special examples of variation in population size.

The analysis just given shows that if a population goes through size bottlenecks in its development, these may have important drift effects. In the history of human development, colonization of new areas certainly occurred many times, whether the area was already inhabited or not, and often the number of colonizers must have been small. One extreme case known is that of the six mutineers of the H.M.S. Bounty, whose offspring from eight or nine Polynesian women reproduced, doubling each generation for over five generations.

The often quoted **founder principle** emphasizes the fact that the smallness of the number of founders of a new group can give rise to a large drift effect. It should also be considered that a small number of founders implies that the number of individuals in the population must remain relatively small for several successive generations, even though it will usually increase eventually. Thus, the total drift effect will be higher, the slower the reproduction of the founders. If a population increases from an initial size N to stable size N_t in t generations, with a constant rate of geometric increase r, the contribution to drift in terms of σ^2/pq will be approximately

$$\frac{\sigma^2}{pq} = \frac{1}{2N_0} + \frac{1}{2N_1} + \cdots + \frac{1}{2N_t} = \frac{1}{2N_0}\left(1 + \frac{1}{r} + \frac{1}{r^2} \cdots + \frac{1}{r^t}\right) = \frac{1}{2N_0}\left(\frac{1 - \dfrac{1}{r^{t+1}}}{1 - \dfrac{1}{r}}\right),$$

which tends to approach $(1/2N_0)\, r/(r - 1)$ for large values of t. The expected variance will therefore approach a limit that is $r/(r - 1)$ times larger than that due to the first generation alone. If population growth is slow and thus r is near one, then the inflation of the initial variance may be quite considerable.

As an example, assume that the initial size N_0 is 100 and N_t is 10,000, so that $t = \log_2(10{,}000/100) = 6.5$ for a doubling each generation. In this case $r = 2$ and thus the total effect of the whole growth period will be about twice as high as the initial founder effect, namely the effect there would be if, in the first generation after foundation, the colony had assumed at once its final (relatively large) size. This last extreme case would be possible with various animals, but, of course, not with man, who has a slower reproduction rate.

Knowledge of age-specific reproduction rates is needed to take account of the effect of overlapping generations.

An important complication that makes it impossible to take the census size of a human group as a direct estimate of the effective population size N_e is that human populations have overlapping generations. In fact, since a census estimate includes people from roughly three successive generations, roughly one third of the individuals belong to the same generation, on an average. The exact value depends on the age distribution and other demographic properties of the population.

The most convincing method of computing effective population size with overlapping generations among those published is the one given by Nei and Imaizumi (1966). According to them,

$$N_e = tN_m, \tag{8.21}$$

where t is the mean age at reproduction (in years) and N_m the number of individuals who reach the mean reproductive age per year. We suggest here an improvement that takes demographic information into account more fully when it is available. Instead of taking N_m as the number of individuals reaching the mean reproductive age, we compute N_m as the number of individuals in the reproductive age group, averaging over all reproductive years and weighting by the age-specific fertility. If, at a given time, the number of individuals of age x is N_x, and the number of births per year born of a parent of age x is B_x, then the mean reproductive age is

$$t = \frac{\sum xB_x}{\sum B_x}, \tag{8.22}$$

where the sum is extended over all fertile years. The weighted number of individuals in the reproductive group N_m is

$$N_m = \frac{\sum W_x N_x}{\sum W_x}, \tag{8.23}$$

where W_x, the age-specific fertility, is $W_x = B_x/N_x$. It follows that

$$N_e = tN_m = \left(\frac{\sum x B_x}{\sum B_x}\right)\left(\frac{\sum B_x}{\sum B_x/N_x}\right) = \frac{\sum (xB_x)}{\sum B_x/N_x}. \tag{8.24}$$

This can be computed independently for male and female parents, if desired, and the effective population sizes thus obtained separately for the two sexes.

The validity of this method was tested by a simulation experiment. An artificial population of 50 individuals was constructed, as elaborated in Section 8.15. Age was given in years and time was advanced by yearly increments, thus giving reasonably accurate age distributions. Each individual had five independent diploid loci, all starting with a gene frequency of 50 percent. Ten such independent experiments were carried out, recording demographic data, gene frequencies, and σ^2/pq values among the loci every 50 years. The results of this simulation experiment, with respect to gene-frequency variation as measured by $f = \sigma^2/pq$, are given in Figure 8.14. The mean of fathers' age at reproduction was 38.1 years and that of mothers' 31.42, giving a mean age at reproduction of 34.8. The average N_e obtained by using Equation 8.24 was 18.25 ± 0.24. The median f values at various times are given in Table 8.6. For comparison, f values are also given, which are computed for a population of $N = 18$ with nonoverlapping generations, both by simulation and from Equation 8.7.

TABLE 8.6
Average Values of f, at Different Times, for the Simulation Experiments shown in Figure 8.14 Compared with Values Obtained by a Simulation with Nonoverlapping Generations and from Equation 8.7

Time (*Generations*)	*Overlapping Generations* ($N_e = 18.2$)	*Nonoverlapping Generations* ($N_e = 18$)	*Expected* (*Using Equation* 8.7) $1 - e^{-t/2N}$ ($N_e = 18$)
5	0.11	0.123	0.130
10	0.27	0.256	0.242
15	0.41	0.358	0.341
20	0.56	0.430	0.426
25	0.65	0.469	0.510

As can be gathered from Figure 8.14, the variation between replicates is large. The error of the estimates obtained in simulation experiments is therefore large, but can be reduced by increasing the number of replicates. With the fine subdivision of age distributions (by year) that was desired, these simulation experiments are expensive in computer time and therefore the number of replicates was kept low.

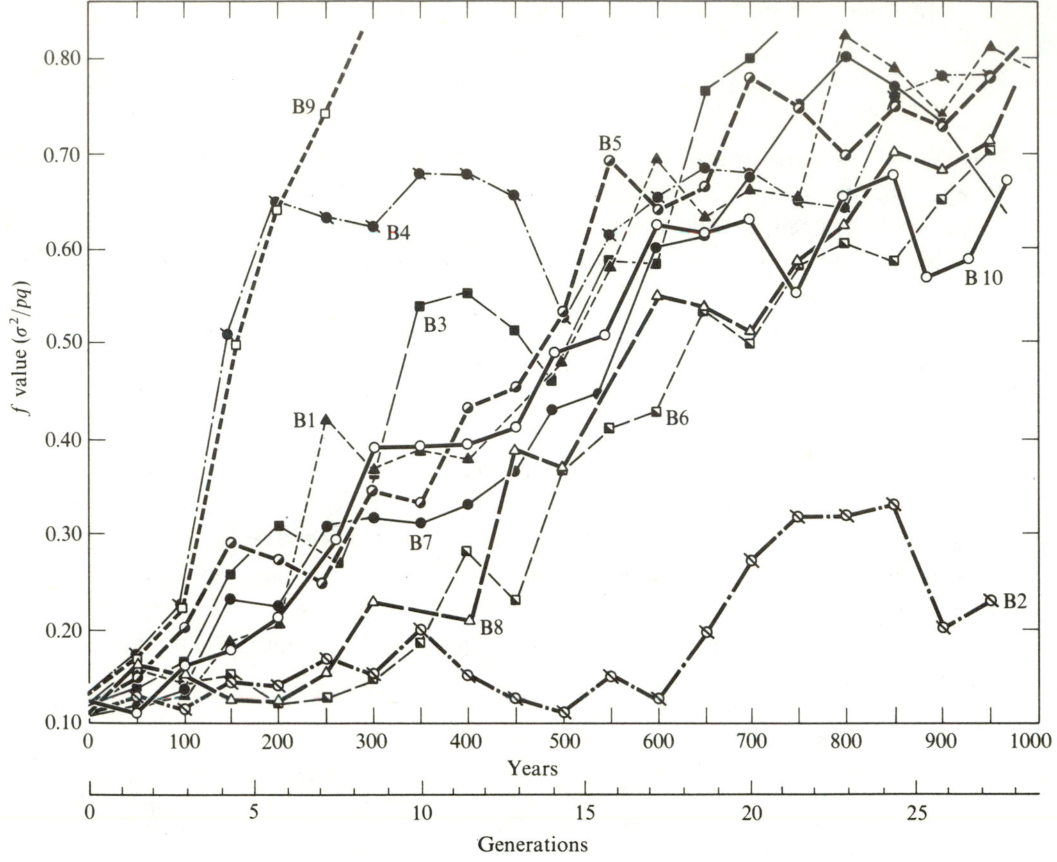

FIGURE 8.14
Simulation of drift in 10 independent populations of 50 individuals, with overlapping generations. The variation of gene frequencies given in the ordinate is the standardized variance σ^2/pq.

It should be added that because the process is stochastic, the error variance increases with time and the earliest observations are, therefore, the most precise ones. It thus seems that the agreement between the two simulations, with overlapping and nonoverlapping generations, as well as that with the theoretical model, can be considered quite satisfactory, in spite of the differences at later times.

Sex ratio, progeny distribution, and variable fertility may affect population effective size.

Unequal contributions of the two sexes cause a decrease in the effective population size. Just as with population sizes that vary from generation to generation (see Equation 8.20) the effective population size N_e for different numbers of males M and females F is a function of the harmonic mean of M and F. Specifically,

$$N_e = \frac{4MF}{M + F},$$

which is easily seen to be twice the harmonic mean of the numbers of males and females.

The numbers in each sex should be counted only among reproducing individuals. Polygamy, therefore, causes a reduction of N_e. With an average of two wives per married male, as is found in some African populations, the reduction is less than a factor of 2. It has been noted by Salzano, Neel, and Maybury-Lewis (1967) that, especially in primitive populations, the number of progeny born to different males is likely to vary greatly. Chiefs and other socially dominant, highly polygamous individuals may have many more descendants than most of the other males. The effect of this on N_e can be predicted, given knowledge of the variance of progeny size, from the following formula, which can be applied separately for the two sexes.

If progeny size is not distributed as a Poisson distribution (see Chapter 6) but has a variance V_k that is higher than the mean $\bar{k}$, a correction factor for N_e can be computed. Approximating from a formula given by Robertson (1961, see also Kimura and Crow, 1963), this correction factor is

$$\frac{1}{1 + \dfrac{V_k - \bar{k}}{\bar{k}^2}} = \frac{1}{1 + \dfrac{V_k}{\bar{k}^2} - \dfrac{1}{\bar{k}}}. \tag{8.25}$$

If fertility (usually measured in terms of progeny size) is inherited with heritability h^2 (the relative genetic contribution to variation in fertility, see Chapter 9), a further reduction has been suggested (Nei and Murata, 1966), as follows,

$$N_e = \frac{N}{(1 + 3h^2)\dfrac{V_k}{\bar{k}^2} + \dfrac{1}{\bar{k}}}.$$

Unfortunately, very little is known about the heritability of fertility, but it is believed to be low (see Chapter 9).

In calculating $f = \sigma^2/pq$ from observed data, the sampling variance must be eliminated from the observed variance of gene frequencies.

When we analyze real data, we compare gene frequencies of different populations at the same time, or, more rarely, of the same population at different times. It is important to define clearly the groups being considered, when samples are taken from them. Ideally, we would like to take the whole population, but this is usually very difficult, and not always necessary. If, however, the samples are too small, the variation between groups may not be significant, most of the observed variation being due to sampling.

If we want to test the significance of variation of gene frequencies between groups (be they villages, tribes, or any other type of grouping), we can estimate how large a sample must be to be significant if we have an idea of the real variation to be expected. With an expected variation of $\sigma^2/\bar{p}\bar{q}$, taking (for simplicity) samples of equal size n of genes from each of k groups, and thus a total sample of $nk/2$ individuals, the heterogeneity is tested by

$$\chi^2 = \frac{n \sum (p_i - \bar{p})^2}{\bar{p}\bar{q}},$$

where p_i is the gene frequency in the ith group. We expect χ^2 to be $k - 1$ because of random sampling alone (see Appendix I). The "true" variation σ^2/pq can then be approximately estimated by subtracting from the observed variation, that due to random sampling, alone, giving

$$\frac{\sigma^2}{pq} = \frac{1}{n}\left(\frac{\text{obs}\ \chi^2}{k - 1} - 1\right). \tag{8.26}$$

This is valid strictly only for genes without dominance, in which case genes can be counted. A correction factor is necessary otherwise. If n varies from group to group it is convenient to use for n the arithmetic mean $\bar{n}$. More satisfactory methods have been suggested by Robertson (1951), and Jayakar and Matessi (in press).

The problem of correcting the observed variance of gene frequencies for the variance due to sampling is partially solved by Equation 8.26. In general, the expected sampling variance of a binomial sample of n is pq/n, but this applies only for sampling with replacement, which is not what we have here. A more exact analysis should take account of the type of sampling, which is without replacement, and of the proportion of individuals sampled. A binomial sampling correction involves subtracting from the observed variance of gene frequencies σ^2, the quantity $\bar{p}\bar{q}/\bar{n}$, where $\bar{n}$ is the average number of genes per group or, equivalently, subtracting $\bar{n}$ from the observed value of $\sigma^2/\bar{p}\bar{q}$. This gives the same correction as in Equation 8.26 above. The f value computed in this way may be negative. Further work will be necessary to test which of various possible formulas might give the most satisfactory results.

The removal of the sampling variance from the observed variation in gene frequencies may sometimes require more elaborate procedures. In particular, one source of variation that may grossly inflate the estimate of variance is the effective duplication of information due to people in the sample from a given village or town being closely related. When several members from the same family are sampled, they may substantially increase the contribution to the χ^2 for gene-frequency variation. One family has only four separate genes to contribute for each locus.

If, for instance, the two parents and several children were available, their contribution to the total information on gene frequencies would have to be evaluated by special procedures. As long as gene-counting methods are used for estimating gene frequencies (see Appendix II) the estimates obtained are reasonably free of the effects of close relationships. There are also less precise but simpler methods for taking these effects into account, either by discarding excessive progeny at random or by determining the number of informative individuals. Thus in families with one parent, the first child contributes information on one gene only, the second on a half gene, the third on one quarter gene, etc.

The size of samples necessary to obtain significant heterogeneity can be anticipated if a rough estimate of the expected variation, in terms of f, can be made. Thus, for $f = 0.01$, on $k = 21$ villages, the average number n of genes sampled per village must be, from Equation 8.26,

$$n = \frac{\dfrac{\chi^2(p = 0.05)}{k - 1} - 1}{f} = \frac{\dfrac{31.41}{20} - 1}{0.01} = 57,$$

where 31.41 is the 5 percent value of χ^2 for 20 degrees of freedom.

8.9 Subdivision of a Species and Models of Isolation

Wright was the first to point out the potential evolutionary significance of subdivision of a species into local units, which are partially isolated and therefore are able to evolve to some extent independently. He has recognized that this may help to speed up evolution in a variety of ways:

1. By favoring adaptation to local environmental niches. Without subdivision, local adaptation would tend to be destroyed by excessive exchange with other neighboring populations.
2. By helping to get different genotypes, having similar but not identical fitnesses, established in different locations, and thus increasing genetic variability. In the absence of subdivision it would nearly always be one and the *same* phenotype, showing optimum fitness, that would get fixed in different places.
3. By favoring the formation of new combinations of genes, some of which may reveal unexpectedly high fitness. The chance "discovery" of epistasis in fitness is certainly favored by subdivision, increasing random genetic drift.

It should be noted that the consequences of subdivision are *not* those of decreasing the *total* population effective size. It has been shown (see Moran, 1962) that, if a population is subdivided into groups having even a very small reciprocal exchange, the amount of drift to be expected for the total population is almost the same as, or only slightly smaller than, if there were no subdivision. The above suggestions made by Wright are all unaffected, however, by this consideration.

The first model in which a population was divided into partially isolated subpopulations—the model that was proposed by Wright (1943) under the name of the "island model"—can only apply in a few circumstances. It does take account of the clustering into colonies (villages, towns, tribes, etc.) but assumes that each cluster exchanges an equal proportion of its genes with every other cluster. Thus it does not take account of the fact that more distant clusters are less likely to exchange individuals with each other. Therefore, further models have been constructed later to take account of the effect of distance.

In the island model, every population exchanges genes equally with every other and there is no effect of distance between populations.

The population is divided into a number of subpopulations, each of size N, that exchange genes with each other, or, more exactly, with an infinite reservoir of genes that has a constant gene frequency. At every generation each subpopulation receives a proportion m of its genes from this infinite pool while a proportion $1 - m$ come from the individuals forming the earlier generation in the same population. This is clearly the same as drift in a single population under migration pressure. The model was given its name because it was believed to be appropriate for a large archipelago of islands, in which each island exchanges genes with all the others, and therefore approaches the requirements of the theory. There is no effect of distance with this model. This, and the necessity to assume equal population size for all the islands, make the applicability of the model to real circumstances questionable even for a group of islands. Its use lies in the simplicity of the formula for the expectation of the variance at equilibrium, which is, as given before,

$$f = \frac{\sigma^2}{pq} = \frac{1}{1 + 4Nm}.$$

For a more realistic approach some consideration of distance is essential. Two types of theoretical models have been suggested: a continuous one, in which population density is constant at any point (Wright, 1943, 1946, 1951; Malécot, 1945, 1966, 1967) and a discontinuous one, in which the population is clustered at the nodes of a lattice (Malécot, 1950; Kimura and Weiss, 1964). Both types exist in at least two versions: a linear one, simulating an approximately one-dimensional distribution of a population (along a coastline, in a narrow mountain valley, along a river or a road, etc.) and a two-dimensional one, representative of the more usual population distribution found in areas where there are no barriers against dispersion in any direction. The discontinuous model, which is also called the stepping-stone model, can be represented in the form of one and two dimensional lattices as in Figure 8.15. Every circle contains the same number of individuals N. In the linear model, migration takes place equally in both directions, with a proportion $m/2$ individuals migrating from each cluster (colony, clan, node, village, town) to the nearest cluster to the left and $m/2$ migrating to the nearest cluster to the right. The contribution of each cluster to itself in the next generation is therefore $1 - m$. In the two-dimensional model each cluster (circle) contributes $m/4$ to every neighboring cluster and $1 - m$ to itself. A three-dimensional stepping-stone model has also been studied theoretically.

Homogeneous migration and the constancy of population distribution are also assumed in the continuous model, in which population density replaces cluster size, and migration is accounted for by the frequency distribution of the distance between birthplaces of parent and child. It is usually assumed that this distribution is normal. A random walk, Brownian movement, or diffusion, all basically yield the normal distribution, which does not seem, however, to be directly applicable to the migration of man as we shall see later. Other "laws" to account for the patterns

Linear stepping-stone model

N N m/2 N m/2 N N

m/2 m/2

Two-dimensional stepping-stone model

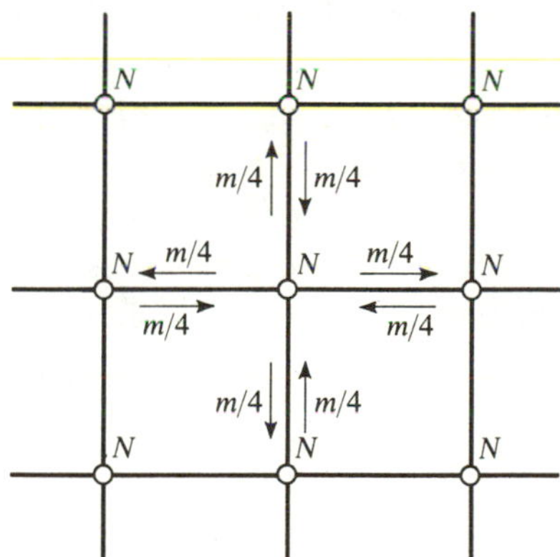

FIGURE 8.15
Population distribution and exchange in one- and two-dimensional stepping-stone models. The symbol m represents the proportion of immigrants per colony and N the population size of all colonies.

of human migration have been introduced, as will also be discussed later in this section.

It is clear that both the continuous and the discontinuous model are still very unrealistic. For one thing, human populations are always highly irregularly distributed over an area, being almost always divided into clusters of different size, separated by varying distances, and connected by roads forming very irregular lattices. Human migration is highly variable. The models of isolation by distance must, therefore, be considered as oversimplifications, but have the advantage of leading to simple and elegant formulas for the expected amount of local variation and of correlations under drift equilibrium. They can help in assessing, to some extent, the effects of the main factors—namely, dimensionality, mobility, and density or cluster size.

There are two ways of expressing concisely the geographic variation in gene frequency expected under a model of isolation by distance.

One expression of the geographic variation is the variance between clusters in the discontinuous model, or between samples taken from different neighboring locations in the continuous model. The other is the correlation between clusters, or between samples taken at a given distance x apart, as a function of x. In either case formulas can be derived in terms of the kinship coefficient f, or of variances of gene frequencies, and it is easy to translate from one to the other, bearing in mind that $\sigma^2/pq = f$.

All models assume a constant population density. This is expressed as the number N of individuals per cluster in the discontinuous model, and by population density δ per unit distance (or unit area) in the continuous one-dimensional or two-dimensional models. Migration is usually considered as isotropic—that is, equal in all parts and directions. Formulas have also been given, in the two-dimensional model, that allow for different mobility in the two dimensions.

The systematic change in gene frequencies is, following Equation 8.10, always assumed to be given by

$$\Delta p = -b(p - \hat{p}),$$

where $\hat{p}$ is the equilibrium gene frequency, b the coefficient of recall to equilibrium, and only balancing linear pressures are considered. The rate of local migration is measured by m. We now consider, in turn, the results expected at equilibrium for the discontinuous and continuous one- and two-dimensional models.

ONE-DIMENSIONAL MODEL—DISCONTINUOUS

According to Malécot (1950), in a linear discontinuous model the mean kinship coefficient between neighboring clusters is, approximately, at equilibrium

$$f_0 = \frac{1}{1 + 4Nb\sqrt{1 + \frac{2m}{b}}}, \tag{8.27a}$$

which for $b \ll m$ simplifies to

$$f_0 = \frac{1}{1 + 4N\sqrt{2mb}}. \tag{8.27b}$$

The correlation between clusters decreases with distance x (number of steps between them) approximately according to

$$\rho(x) = \left(1 + \frac{b}{m} - \sqrt{\left(1 + \frac{b}{m}\right)^2 - 1}\right)^x, \tag{8.28a}$$

which for $b \ll m$ is

$$\rho(x) = e^{-x\sqrt{2b/m}}. \tag{8.28b}$$

The coefficient of kinship between clusters at distance x is

$$f(x) = f_0\,\rho(x).$$

Kimura and Weiss (1964) have given different formulas, using the variance of gene frequencies, which, however, give the same results when $b \ll m$.

The meaning of the symbols f_0 and $f(x)$ should be explained precisely. Use of f with the subscript zero, f_0, refers to the mean kinship coefficient of one cluster. Use of f as a function of x, $f(x)$ refers to the mean kinship of two individuals taken from clusters separated by a distance x. When we consider discontinuous models, the term f_{ij} is a mean kinship coefficient of two individuals taken at random from two different groups, i and j. The quantity σ^2/pq, which corresponds to f_0, is an approximation to the standardized variance of gene frequencies of neighboring clusters (or the limit that the variance in gene frequency approaches when the colonies are very close). More precisely, as already mentioned, it is the variance of a given cluster in replicated experiments. (For the exact relationship between these quantities, see Bodmer and Cavalli-Sforza, 1968.)

The m value (Kimura and Weiss, 1964, use m_1, and Malécot, 1950, and elsewhere sometimes uses $2m$) is the estimate of the **close-range migration**, being the proportion of individuals exchanged with immediate neighbors, while $1 - m$ is the proportion of people autochthonous to each cluster—that is, who do not migrate out of the cluster. There are extensions of the theory for the case in which a *small* proportion of immigrants come from both neighboring and nonneighboring populations. However, for cases in which the migration from a long distance is relatively large the approximation is unsatisfactory. The b value, which is the coefficient of recall to equilibrium, is the sum of mutation, linearized stabilizing selection, and migration not considered in m. This last factor could be long-range migration or migration from an extraneous source that supplies some gene flow into the population being considered. The coefficient b is called m_∞ by Kimura and Weiss, and v or k by Malécot.

It is of interest to compare the results of the linear discontinuous model, using the approximation valid for $b \ll m$, with those of the island model, on the assumption that the only linear pressure, other than migration at rate m, is mutation at a rate μ. For the linear discontinuous model, from Equation (8.27b),

$$\frac{1}{f_0} = 1 + 4N\sqrt{2m\mu};$$

and for the island model, from Equations 8.13, 8.14, and 8.15,

$$\frac{1}{f} = 1 + 4N(m + \mu),$$

where $b = \mu$. Thus in one model, the reciprocal of local variation is linearly dependent on the geometric mean of m and μ, and in the other, on their arithmetic mean. The geometric mean may be much lower (even by two orders of magnitude, if m is large and μ is small) than the arithmetic mean. Thus the inappropriate use of Wright's island model for subpopulations in a straight line, isolated by distance, may significantly underestimate the extent of local variation among them.

ONE-DIMENSIONAL MODEL—CONTINUOUS

In the continuous model, results may depend on the distribution law for migration. Malécot (1967), however, has shown that several different types of migration distributions give the same results as the normal distribution. Following Wright (1943), migration is introduced into the model of isolation by distance as the distribution of distances between the birthplaces of parent and offspring.

The coefficient of kinship between neighbors at equilibrium is where δ is the density

$$f_0 = \frac{1}{1 + 4\sigma\delta\sqrt{2b}}, \tag{8.29}$$

along the line, and σ is the standard deviation of the distance between birthplaces of parent and offspring. This formula is valid for the normal distribution, the exponential distribution and a gamma distribution. It is interesting that the continuous and discontinuous models (see Equation 8.27b) give the same result if $\sigma^2 = m$ and δ (the number of individuals per unit distance) is taken as the size of the colony N in the stepping-stone model (at least as long as $b \ll m$ or $b \ll \sigma^2$). Thus, from Equation (8.28b) we then have

$$f(x) = f_0 e^{-x\sqrt{2b}/\sigma}. \tag{8.30}$$

TWO-DIMENSIONAL MIGRATION

The expectations of the discontinuous (stepping-stone) two-dimensional model are more complicated, and the expressions given by Malécot (1950) and by Kimura and Weiss (1964) for $f(x)$ and f_0 require the numerical computation of integrals, for which the reader is referred to the original papers. There are some discrepancies between the results obtained by these authors. According to Kimura and Weiss the correlation coefficient ρ decreases with distance more rapidly than that in the one-dimensional model, namely

$$\rho(x) \propto \frac{e^{-x\sqrt{2b/m}}}{\sqrt{x}}, \tag{8.31}$$

the approximation being valid for large distances (compare Equation 8.28b). According to Malécot the relationship is still exponential even in two dimensions. It is at the moment difficult to assess the practical importance of the approximations that have to be made to obtain explicit solutions in the two approaches.

In the continuous two-dimensional model, with normal migration, the equilibrium f value is (Malécot, 1950)

$$f_0 = \frac{1}{1 + 8\pi\delta\sigma^2\left(\frac{-1}{\log 2b}\right)}, \tag{8.32}$$

where σ^2 is the variance of the distribution of the distance between parent and offspring birthplaces in one dimension and $2\sigma^2$ is the mean square distance taken over both dimensions assuming isotropic migration (that is, migration equal in the two dimensions) (see Example 8.7). If migration is different in the two dimensions, it is sufficient to replace σ^2 by $\sigma_x \sigma_y$, the two standard deviations σ_x and σ_y being valid for the two separate dimensions.

With equal numbers of individuals in a linear segment and a circle (equating $2\delta\sigma$ in the one-dimensional model with $2\pi\delta\sigma^2$ in the two-dimensional model), f_0 is smaller in the two-dimensional model than in the one-dimensional model.

Distributions other than the normal have been used in the two dimensional model. One that gives the same results as the normal distribution is the *K*-distribution (Malécot, 1967), which, unlike the normal distribution for the two-dimensional model, has no nonzero mode and can have an infinite mode at the origin.

8.10 Demographic Data Relevant to the Analysis of Drift

The difficulties in applying the above drift models to real data will be made clearer if we consider what is known about real populations. Human populations tend to cluster in groups. The size of the clusters may be highly variable and there are great differences in mean cluster size (and variation) depending on socioeconomic conditions.

Migration between clusters also varies a great deal with cultural factors. It depends, in part, on the population density, which is a function of technology of food production, but especially on the economic structure of the population.

The geographic distribution of a population is markedly affected by the physical features of the landscape: mountains, valleys, rivers, plains, forests, etc. This effect is less marked in modern industrial societies, but can be found even there. Every population is, thus, in some sense unique. In addition, every population has had a history of its own. Migration, admixture, and all sorts of historical accidents have taken place and are taking place, which all contribute to the variety of possible situations.

We will consider some aspects of data on population density, cluster size, the measurement of migration, and of the "dimensionality" of the geographic distribution. It is these factors that determine the amount of local variation to be expected because of drift. As we have seen, the continuous and discontinuous models are, to some extent, interconvertible, by setting approximate equivalence relations between population density and cluster size on the one hand, and the different migration estimates on the other.

Population density and cluster size are functions of technological development.

We will leave historical considerations to the chapter on evolution and consider here mostly present-day populations. The technology of food production affects population density most markedly. The most important categories used to classify human populations according to their methods of obtaining food are:

Hunters and Gatherers. There are various populations in the world today that, in part at least, live a life of hunting (or fishing) and gathering without agriculture. People living in the forest, like the Pygmies in Africa, certain Indians in South America, and a few populations in India, Malaya, and Borneo, and also Australian aborigines, are more-or-less in this category. Pygmies live in camps of 10–100 people. A camp is a group of families that are often related. Sometimes new camps are formed quite near to other camps (as many as four camps close to each other have been observed). Pygmy populations occasionally reach sizes up to perhaps 1000 individuals when a number of camps form within 5 or 10 miles of each other. The total size of a Pygmy population among the dozen or so unconnected populations still existing in the African forest, may be as large as 30,000. The population density of one such population is of the order of 0.2 per square kilometer, and has, perhaps, slightly increased over the last several centuries (or even millennia) because of contacts with farmers, with whom the Pygmies exchange some goods. Eskimos number altogether 20,000 over a very large area. They are relatively acculturated, and few live in conditions comparable to those of older times. For the Caribou Eskimos, now almost extinct, the estimate of population density was 0.04 per square kilometer (see Braidwood and Reed, 1957). The Greenland Eskimos were estimated to number 10,000 at the early times of European contact, with a density of 0.06 per square kilometer and the original Aleut population is estimated at 16,000 with a density of 0.6 per square kilometer. In 1940 they numbered 1000 (see Laughlin, 1950). Australian aborigines live (or lived) in extended exogamous families of 20-50 people, each of which is a unit from the point of view of migration and residence. These families formed tribes of an average size of 500 people and the estimate of population density for these is 0.03 per square kilometer, a lower value than that for the Pygmies, which is consistent with their living in a much more arid climate in which less food is available.

Incipient or Primitive Agriculture. This is often accompanied by some food-gathering or hunting activity, as among African Bushmen, who practice a very primitive type of agriculture. In New Guinea, where no game is available, food is obtained mostly by horticulture and by some pig breeding. Tribes in New Guinea that are separated by sharp linguistic and psychological barriers—but that nevertheless do exchange a few individuals—are subdivided into partly exogamous clans. The average number of people per tribe is of the order of 10,000, and the average population density is perhaps 4 persons per square kilometer. Village size is of the order of 100-300. The Mayas in Central America are said to have villages of 500 people and to use an area of 11 square miles. A guess of their density is 20 per square kilometer, but the difficulty in estimating effective population densities here, as elsewhere in forest areas, is that only a small proportion of the land is actually cultivated. These groups use shifting agricultural methods, and do not fertilize the soil, but even so, it would seem that there are factors limiting population size other than the carrying capacity of the land in terms of food production.

Pastoral Nomadism. There are several human groups that specialize in pastoral activity. They usually live among other people, so that their population density is not very meaningful. Three such groups (Braidwood and Reed, 1957), namely, Bedouins, Kirgiz, and Kazakh have densities from 0.06 to 2 per square kilometer. These figures are, however, from very different environments, ranging from desert to steppe. It is difficult to know whether the carrying capacity of the land is fully exploited, though mobile groups may be relatively large (above 1000). In some specialized pastoral groups (for example, the Masai in East Africa), sizes of the population units are, on the average, smaller.

Fully Developed Agriculture. Countries whose population is, or was until a short time ago, mostly occupied in agriculture are few, but, like China and India, they can reach population densities even greater than 400 people per square kilometer over vast areas. Such population densities are probably near the maximum carrying capacity of the land, in the absence of major technological improvements in agriculture.

Population densities per square kilometer in Europe range from those of Iceland (2), Norway (12), and Scotland (67) to those of England (310), Belgium (316), and the Netherlands (350). The European average is 54. In the United States, population densities range from 0.9 (Nevada) to 300 (New Jersey), with an average of 19.

The distribution of cluster size shows great variation. Sociological theories often use statistically unusual measures of variation (Zipf, 1949). With ordinary statistical techniques the distribution of cluster sizes is approximately log-normal.

In highly agricultural areas the clustering of the population is often less extreme (for example, in northern Italy, as compared with southern Italy) with a fairly large fraction of the population living in sparsely distributed houses. The population then approaches a continuous type of distribution. In the majority of human populations, however, clustering is present. In poor rural mountain areas, village size is near to 200-300 on the average (though always variable, in northern Italy and Switzerland, just as in the forests of central Africa). In the Parma Valley in northern Italy the median size of villages varies almost proportionately to population density. The median village size is 300 for those of lower densities—that is, up to 50 inhabitants per square kilometer—475 for a density of 50-95, 680 for 96-140, and 915 for 141-180. The percentage of the population living in isolated houses, which indicates the inverse of the degree of clustering, varies in these density classes from 12 to 66 percent (Cavalli-Sforza, 1958).

8.11 Genetic Migration: The Distribution of Distances Between Birthplaces of Parent and Offspring

The most meaningful measure of migration from the genetic point of view is obtained, as suggested by Wright, by taking the generation as the time unit. This is the basis for using the distribution of distances between birthplaces of parent and offspring to measure migration. This method works only for a continuous model, and is not entirely satisfactory when the population is highly clustered, as most populations are. On the other hand, migratory exchange between nonneighboring clusters is sufficiently frequent to violate the rules of the simplest stepping-stone models. Models in which exchange is limited to neighboring colonies are therefore unsatisfactory. Kimura and Weiss, and Malécot have, however, also given methods for obtaining estimates of m when migration takes place among nonneighboring clusters, though this further complicates their analysis.

There are two distributions for distances between birthplaces of parent and offspring: father-offspring and mother-offspring.

Usually, the mother-offspring distributions show a greater variation and a higher mean, because most populations, especially agricultural ones, are, in anthropological terminology, patrilocal (see Table 8.7). Land is inherited by sons and thus, when two persons from different villages marry, it is the wife who moves from the place where she was born. Among hunters and gatherers, however, patrilocality may not be the customary pattern. This is true of Pygmies, among whom practically no difference is found between the two distributions, and, at the other extreme, patrilocality may not be the pattern in highly industrialized societies.

It is often convenient to study the distribution of the distances between the birthplaces of husbands and wives.

This distribution can easily be obtained from marriage certificates or from census data. Some examples are given in Figure 8.16. The relationship of this to the parent-offspring distribution is not always straightforward. The husband-wife (H-W) distribution sums the movements of husband and wife prior to marriage. If this were all the migration there were, and the two sexes migrated independently, the H-W distribution would be the convolution (statistical sum) of the father-offspring (F-O) and mother-offspring (M-O) distributions. As a rule of thumb, the mean of the H-W distribution would be the sum of the means of those for F-O and M-O (or twice the parent-offspring mean), and the variance would also be the sum of

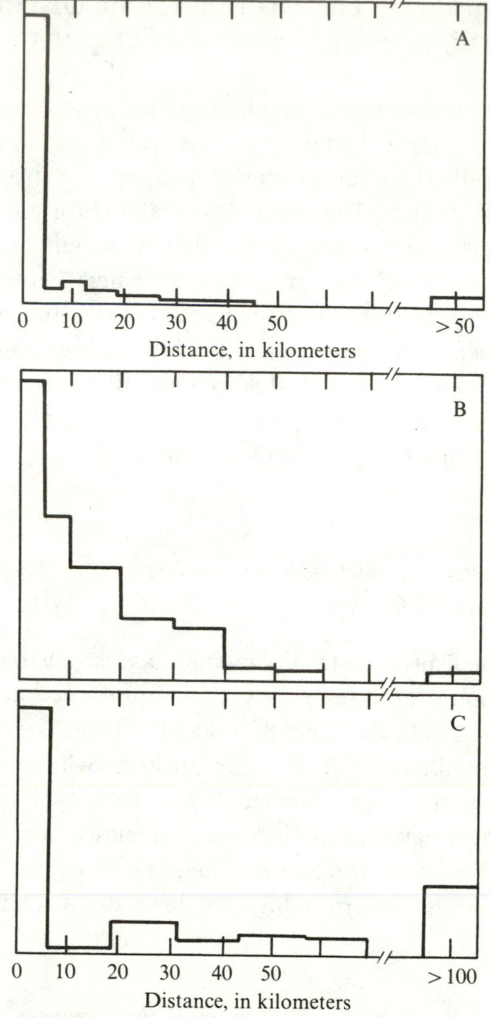

FIGURE 8.16
Distributions of the distances between birthplaces of husbands and wives in three different populations. A: Upper Parma Valley (northern Italy)—population density, 50 persons per square kilometer; endogamy, 55 per cent, village size, 300 persons. B: African rural population (Issongos of the Central African Republic)—population density, 1–2 persons per square kilometer; endogamy, 20 per cent; village size, 100 persons. C: African Pygmies (Babingas of the Central African Republic)—population density, 0.2 persons per square kilometer; camp size, 30 persons; hexogamous.

the two variances. The proportion of husbands and wives born in the same cluster of some defined type, which is often referred to as the **proportion of endogamy**, should be the product of the proportion of fathers and offspring born in the same place multiplied by that of mothers and offspring born in the same place. As an example, from Table 8.7 the F-O and M-O proportions for the upper Parma Valley are seen to be 0.821 and 0.727, respectively. Their product is 0.597, while the observed endogamy in the same area is 0.55. In these data both F-O and M-O migration fit, closely, a gamma distribution.

These rules of thumb can fail if social correlations between marriage partners destroy the independence of the two distributions and increase the variance of the H-W distribution. They also fail if there is migration after marriage and before progeny is born. Then the H-W distribution has a mean and variance that are different from the sums of those of the two parent-offspring distributions. The rules also fail for hunters that keep moving around the same hunting grounds, as is true for Pygmies, and should fail, finally, for nomadic herdsmen for similar reasons. They will be approximately valid for agricultural communities in which most of the migration takes place as an adjunct to marriage itself.

Human migration mostly takes place along predetermined routes.

In evaluating the distance between places geneticists should try, whenever possible, to use road distance and not distances as the crow flies, which are easier to compute, but are perhaps more suitable for crows than for human genetics. Especially nowadays, migration distances are composites resulting from some travel on foot, and other travel that makes use of transportation (horse, camel, canoe, bicycle, automobile or plane). This may account for part of the extreme leptokurtosis (that is, high frequency of extreme values) of the distribution curves, but by no means for all of it. Thus, high leptokurtosis is found in populations that practically never use any of the modern means of transportation (for example, the African Pygmies or an African rural community like the Issongos—see Figure 8.16). In an Italian rural community there was almost no change in migration over the last 300 years, thus apparently negating the importance of improved means of transportation. On the other hand, the increase in individual mobility due to improved means of transportation and higher social mobility is easily seen in data on people in other communities.

There are serious difficulties in fitting distributions to real data.

The discontinuities and irregularities of population distribution create serious problems for the fitting of migration distributions to observed data. One problem is due to clustering of the population in, say, villages. This creates difficulties at the origin of the distribution, where clustering results in a fairly sharp discontinuity.

Even though the distance between husband's and wife's birthplaces is never actually zero, in practice a large fraction of marriages take place between mates whose birthplaces are nearly zero distance apart or at least perhaps within a few hundred yards of each other, namely within a cluster. Moreover, distances between clusters are generally much larger than the average "diameter" of a cluster. Thus, in data from rural sources there may be a gap in the migration distribution until the next village, which may be located several kilometers away. The partition into classes near the origin usually has a strong influence on the fit of theoretical distributions. One way of avoiding this problem is to introduce a separate parameter for the frequency at near zero distance.

Problems also often exist at the tail of the distribution because of large migration distances. Data on such distances may affect very markedly the estimates of the mean and especially of the variance and the mean square distance. One way to avoid such effects is to "censor" the distribution at the upper end and use the total frequency of distances above an arbitrary limit, instead of the set of upper values. The same procedure can be applied at the lower end. Fitting then requires the computation of integrals, which may have to be carried out numerically. An example of this is given in Table 8.7.

Migration may be thought of as being made up of a diffusional component and a gravitational component. The diffusional component is the almost random migratory behavior of individuals, families, or groups of families, which change their residence permanently.

This diffusional component may be due to a permanent change of hunting or grazing grounds, agricultural fields, or industry. This kind of migration is grossly comparable to brownian motion and leads to a normal distribution. In the two-dimensional model, the distribution of distances is given by the function (see Example 8.7)

$$\frac{re^{-r^2/2\sigma^2}}{2\sigma}, \tag{8.33}$$

where r is the distance from the original residence and σ^2 is the variance of r. This distribution is only slightly skew and has not been found to give a good fit in practice. A possible complication is due to variation of individual mobility. If $l = 1/2\sigma^2$ varies from one individual to another according to a gamma distribution ($Ce^{-l}l^{n-1}$) (see Appendix I, and Skellam, 1951) then the distribution of distances becomes

$$\frac{2rn}{(1 + r^2)^{n+1}}, \tag{8.34}$$

which is not too dissimilar from the gamma distribution itself. A further scale

TABLE 8.7
Patrilocal Migration in the Upper Parma Valley

Distances (r)	Father-offspring Distributions Obs.	Father-offspring Distributions Exp.	Mother-offspring Distributions Obs.	Mother-offspring Distributions Exp.	Tests for Difference Between F-O and M-O Distributions
0– 2.5	340	339.8	293	292.4	
2.5– 6.5	11	13.8	18	25.4	
6.5–12.5	8	9.6	21	18.1	
12.5–20.5	10	7.4	16	13.7	
20.5–30.5	6	5.9	16	10.8	
30.5–42.5	4	4.9	7	8.6	
42.5–56.5	4	4.2	5	7.0	
56.5–72.5	7	3.6	6	5.6	
> 72.5	24	24.8	21	21.2	
Total	414	414.0	403	402.8	
χ^2	5.24		6.42		
n	0.0419 ± 0.0066		0.0897 ± 0.0111		$t = 3.71$
k	0.00198 ± 0.00132		0.00686 ± 0.00198		$t = 2.03$
$x_0 = \frac{n}{k}$	21.2 ± 13.8		13.1 ± 3.4		$t = 0.57$
P_0	82.1 ± 1.9		72.7 ± 2.2		$\chi^2_{[1]} = 9.85$

Note: The observed and expected distributions of the distances between birthplaces of father and offspring and of mother and offspring are given. Expected values are computed for a doubly censored gamma distribution; chi square indicates the goodness-of-fit (for 6 degrees of freedom). Parameters of the distribution (n and k) the mean distance ($x_0 = n/k$) and the proportion born in the same village as the parents (P_0) are also given. The unit of distance is 625 meters.

Source: From Cavalli-Sforza (1962a).

parameter is needed for actually fitting this curve to observed data. It gives better fits than the two-dimensional normal distribution, but probably involves too extreme an individual mobility.

The gravitational component in human migration is due to the fact that individuals have, for most of their life, fixed residences to which they return most nights.

This is usually true even of hunters and gatherers, who may stay away for months but usually do have a fixed place around which they gravitate for most of their lives. People may, of course, occasionally change their "permanent residences" as is taken into account by the diffusional component. During their stay-out, individuals explore a certain range, and a spouse may be found in this period, or a child may be born.

Sociologists have long recognized that the gravitational component of migration is proportional to the inverse square of the distance. Individual variation in mobility adds to the variance. Distributions for the gravitational behavior depend on the shape of the function taken to represent the dependence on distance of the exploratory behavior (see also Boyce et al., 1968). This is determined largely by the "attractive" force exercised by clusters of people existing in a given neighborhood.

Migratory behavior of both types, diffusional and gravitational, leads to similar distribution curves, which are closely allied to the gamma distribution. The gamma distribution has a single parameter n, but, for fitting, an additional scale parameter k is necessary, which gives

$$(r) = \frac{k^n}{(n-1)!} e^{-kr} r^{n-1}. \tag{8.35}$$

This has mean distance n/k, variance n/k^2, and mean square distance $n(n+1)/k^2$. An example of a fit of Equation 8.35 to actual data, by censoring at the origin and at the tail, was shown in Table 8.7.

It should be mentioned that some empirical distributions have also been found to be useful, such as

$$e^{-k\sqrt{r}}$$

and

$$re^{-k\sqrt{r}},$$

but they do not lend themselves easily to further analytical treatment. A good fit has also been obtained by using the sum of simple distributions, such as the sum of two or more normals or the sum of two exponentials, $(pe^{-ar} + (1-p)e^{-br})$, which requires three parameters p, a, b. The fitting of sums of these distributions can be done by computer. The sum of normals lends itself more easily to analytical treatment, enabling prediction of migration over two or more generations to be made, as required in some models. The use of sums of distributions is not irrational, considering the heterogeneity of means of transportation already mentioned. Fitting a theoretical distribution to the observed migration data may help to obtain more satisfactory estimates of migration parameters, even of the simplest of them all, the **mean square distance** (Example 8.7). This is equal to the second moment about zero, or to the variance plus the square of the mean (see Appendix I). A direct computation of this quantity from the observed distribution would seem to be the simplest solution, but in practice there are often a few migrations over very long distances that make the error of such estimates extremely large. Omission of just one extremely long migration may easily alter the estimate of the mean square distance by a substantial amount, sometimes even by a factor of 10, showing the enormous error variance of these estimates. Methods of analysis that rely heavily on these types of estimates are, therefore, likely to be subject to very considerable errors. The extremely long migrations can be removed to a separate category (long-range migration), but the cutoff is usually arbitrary. Attempts to represent as complex a phenomenon as human migration by a simple distribution with two or three parameters are likely to give an oversimplified picture and clearly only constitute rough approximations to reality.

The number of dimensions of a geographic distribution is often neither one nor two.

There are some situations in which the number of dimensions over which a human population is spread is unambiguous. A plain, with a well-developed lattice of roads crossing it, is a good example of a two-dimensional distribution. Populations strung out in a long, unbranched valley, along a road through the African forest, or along a shoreline (such as that of Greenland) that has uninhabited land behind it, are good examples of a truly one-dimensional distribution. Often, however, as with populations in branching valleys, or more generally in an inhabited area that has a low number of nodes (branching points) with respect to the number of its population clusters (see Figure 8.17), the number of dimensions is neither one nor two, but is intermediate. An estimate of the number of dimensions can be obtained by counting the number of clusters or inhabitants, at a distance r from an arbitrarily chosen center, or from a center at each cluster, repeated for all clusters. For a one-dimensional distribution, the number of clusters, or of people, found at a distance r, should be independent of r. For a two-dimensional distribution, the number of clusters should increase in proportion to r. An actual example for the Parma Valley, where genetic work to be discussed later was done, is given in Table 8.8. The number of parishes was counted in concentric circles from each parish. For parishes at low altitudes (in the plains) a two-dimensional distribution fits well, but at high altitudes neither a one- nor a two-dimensional distribution fits. For these, the number of parishes found at a distance r increases approximately with the square root of the distance from the center (more exactly, in proportion to $r^{0.54}$) (Table 8.8). This corresponds to a number of dimensions just intermediate between 1 and 2, and the exponent of r plus 1 might be regarded as the number of dimensions.

Situations like that of the Parma Valley make it difficult to choose between the one-dimensional and the two-dimensional models. Similar situations are likely to prevail in all mountainous areas, where higher degrees of isolation, and thus of measurable drift, are more likely to be found.

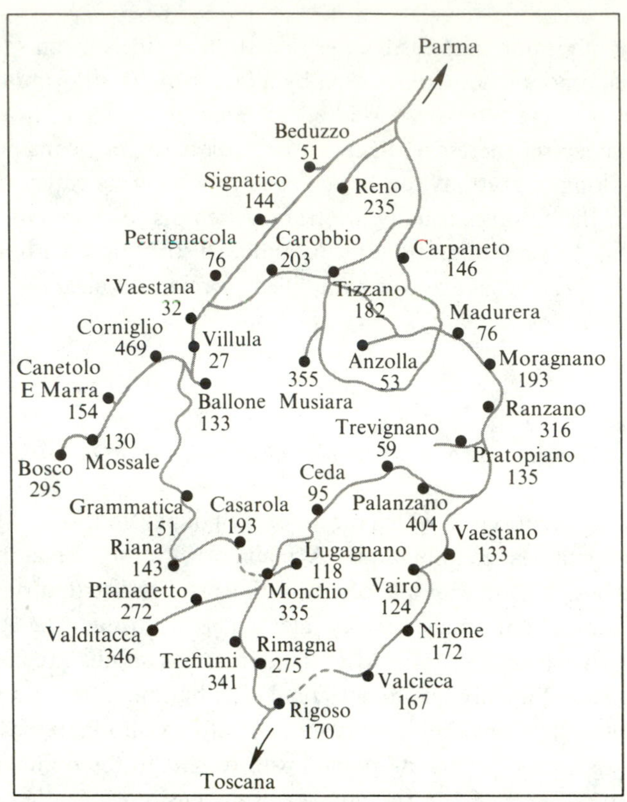

FIGURE 8.17
Map of villages of the Upper Parma valley (northern Italy) showing main connecting roads and village sizes in 1950.

TABLE 8.8
Pattern of Geographical Distribution in the Parma Valley

	Low Altitude (<400 m)		*High Altitude* (>400 m)			
		Expected for		*Expected for*		
Distance (*r*)	*Number of Parishes*	*2-dim Distrib.*	*Number of Parishes*	*1-dim. Distrib.*	*2-dim. Distrib.*	$d = 1.54$
1.5– 2.5	1	3.1	4	15.3	2.8	6.3
2.5– 6.5	32	27.8	47	61.3	25.0	38.9
6.5–12.5	101	88.1	73	91.9	79.3	87.8
12.5–20.5	189	204.0	167	122.5	183.8	157.9
Total	323		291			
$\chi^2_{[3]}$		5.0		21.9	31.7	5.5

Note: Since d is the number of dimensions at distance r the number of parishes is expected to be r^{d-1}
Source: Cavalli-Sforza (1962a).

8.12 Analysis of Geographic Variation

The study of evolution requires the collection of genetic data in time and space. The opportunities for analysis of data collected at different times are usually severely limited, while data on a cross section in space are more readily available. When studying a single species, we can obtain two types of genetic data; data on gene frequencies, and on relationships. Their theoretical analysis can be unified, as we have seen, because the coefficient of kinship and the standardized variance of gene frequencies may, in practice, be equated. Data from such different sources may, however, have entirely different meanings. They should be kept distinct, even though the theoretical analysis may be similar. We will first consider data on gene frequencies.

When geographic variation of gene frequencies is encountered, one of the foremost problems is: can genetic drift explain the observed variation? To answer this problem we need information on the factors that determine the magnitude of variation expected under drift—namely, population sizes and rates of migrational exchange.

The scope of the available data determines the sensitivity of the experiment, and thus the importance of other factors, such as selection, that are detected by departures from the level of variation expected in their absence. When data are available on a large scale, they usually come from a large area and, therefore, a heterogeneous one. For an analysis of drift, variation should be studied with as much detail as possible. When small samples from large populations are examined, the chances of detecting drift, or at least of isolating it from other factors of variation, are small.

The factors other than drift that cause variation, should be mentioned. We have previously mentioned local selective differences, and historical accidents of population migrations as potential sources of heterogeneity. We should also consider the possibility that variation is reduced by stabilizing evolutionary factors, such as those that are included in the coefficient of recall to equilibrium.

Each factor affecting variation has its own mode of action; some of the factors can be isolated from the others.

Drift produces variation that, when measured at the level of gene frequencies as σ^2/pq, is independent of the allele or locus tested.

Local selective differences cause variation that affects only the locus or the loci which are subject to differential selection in the area. It should be noted that, if a selective process is going on in the whole area with an intensity that does not vary from place to place, it should not lead to geographic variation.

Historical accidents of population migration, like local selective differences, may affect each gene differently, for their action depends on the differences in gene frequencies that exist between the immigrant and the local population. Usually, the analysis of expectations under drift is carried out assuming that a steady state has been reached, but this may require a large number of generations. The effects of historical accidents may, therefore, take time to even out. The possibility of hidden gene flow (see Section 8.23) may also have to be considered when different partially segregated populations live in the same area.

Stabilizing evolutionary factors may affect each locus differently. In principle, these factors could include mutation, migration, and balancing selection. It is unlikely, however, that mutation, which is usually of a much smaller order of magnitude than selection or migration, can be measured at this level. Migration should preferably be considered separately, as it is the major force counterbalancing drift and should, as such, affect each locus equally. When possible, it would be rational to distinguish (1) migration from within the area (between clusters within the same larger population), which affects each locus equally and can be considered to be a force equilibrating drift with the same intensity for all loci, and (2) migration from without (namely, from a different overall population than that living in the area under consideration), which can come from populations that have different gene frequencies at some loci, and therefore can affect each locus in a different way. Balancing selection may or may not be present and be of different intensity at each particular locus.

Two main approaches to the analysis of geographic variation are (1) *the comparison of variation at different loci* (*or among different alleles*) *and* (2) *the prediction of the variation due to drift on the basis of demographic knowledge.*

Drift may be separated from the other factors because, when migration counterbalancing it is taken into account accurately, the combination of the two gives rise to an amount of variation that is the same for all loci (and alleles), and can be predicted if adequate demographic knowledge is available. Among the other factors, variation in local intensities of selection, historical accidents, and migration from without all increase the variation at some loci or among some alleles, adding their effects to those of drift, while balancing selection (and, to a probably unmeasurably small extent, mutation) decreases the variation due to drift for other loci or alleles.

It should be added that the estimation of kinship coefficients from genealogy data provides *average* estimates of the variation of gene frequencies at all loci or alleles, and, therefore, does not contribute to that part of the analysis concerning the comparison of different loci or alleles. This can only be done by observing gene frequencies, directly, for several loci and alleles.

It is clear that drift is more likely to be detected when an analysis is confined to a small area, which will probably be more homogeneous and less subject to variations from selective differences or historical accidents. The analysis of **microgeographic variation** is thus more likely to lead to the demonstration of drift effects. Also, it is probably easier to obtain useful demographic information in this kind of survey.

We shall discuss some examples of the analysis of microgeographic variation, as illustrations of the methods than can be used. We will consider first an application of the island model, and then various ways of using the models of isolation by distance. Finally we will introduce both analytical and Monte Carlo simulation methods to take into account migration data as they are usually obtained.

AN APPLICATION OF THE ISLAND MODEL

In some instances, the simple "island model" can be useful. This may be true for the simultaneous investigations of several similar isolates.

Nei and Imaizumi (1966) have collected existing data on ABO blood groups in twelve Japanese isolates, seven in small islands off the Japanese coast, and five in mountainous ranges in the interior. They have analyzed the variances of the three alleles (and also the covariances between alleles, two by two) and, after correction for random sampling of the variances and covariances, obtained an average estimate of $\sigma^2/pq = 0.00224$. (These authors prefer to exclude data from one island that gave somewhat aberrant results, but it is difficult to test if the rejection of this outlier is warranted.) The effective population size, averaged over the twelve isolates, is 1960. In this instance the island model may be adequate, in the sense that these isolates all receive some immigration from the mainland, but are largely independent of one another, as the migration between them must be negligible.

Thus, the formula $\sigma^2/pq = (1 + 4Nm)^{-1}$ is applicable and gives rise to an estimate of $m = 0.06$—that is, 94 percent of offspring stay on the island on which they were born and 6 percent leave, each generation. No direct demographic estimates of m are so far available. They would provide a useful check on the theory.

It is possible that this estimate of m should be a little larger. The deviations of the gene frequencies of each isolate are computed from their general mean, which, as expected, is very close to that for the whole of Japan. As there is a "cline" for ABO gene frequencies in Japan (that is, a linear trend of frequencies along the length of the country), it would be preferable to compute the deviations of gene frequencies from those of the neighboring parts of Japan, rather than from the general average. In the absence of direct information on the origin of the islanders this might be dangerous, however, and the difference would, in any case, be small.

AN EXAMPLE OF DEPENDENCE ON POPULATION DENSITY

In the Parma Valley, inhabitants from a number of parishes were analyzed for ABO, MN, and Rh blood groups. The area was subdivided into various districts corresponding to the administrative units (*comuni*), and for each, a variance of the gene frequencies between parishes was computed. As the results were comparable for the various blood-group alleles (see Cavalli-Sforza, Barrai, and Edwards, 1964), the variation over different loci was pooled. An analysis of the correlation between the variation in gene frequencies, and that of various demographic parameters, showed that the most important parameter was population density. The observed correlation between the variance of gene frequencies (pooled over all genes) and the population density is shown in Figure 8.18.

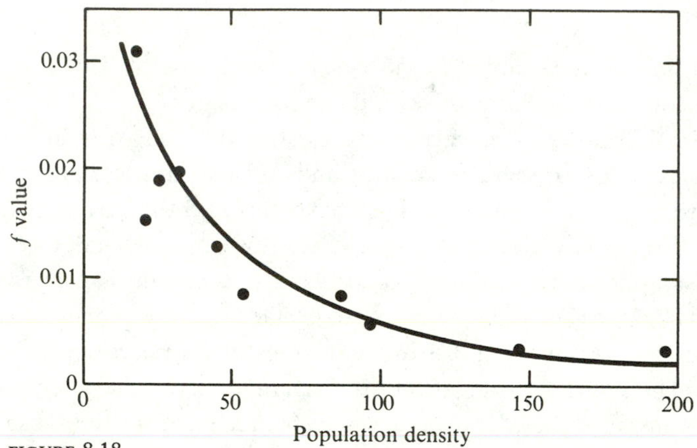

FIGURE 8.18
Relationship between population density, in inhabitants per square kilometer, and f value among parishes belonging to a given administrative unit (*comuni*) in the Parma Valley of northern Italy. (From Cavalli-Sforza, 1969.)

To test the agreement between these data and an isolation model by distance we can fit to the data the curve

$$f = \frac{1}{1 + k\delta}, \tag{8.36}$$

which, if we are using a continuous model, is valid independently of the number of dimensions. Here, k is a coefficient that depends on σ the migration coefficient and b the coefficient of recall to equilibrium (the coefficient b combines all other evolutionary pressures—see Equations 8.29 and 8.32).

The dependence of k on σ and b differs in the one-dimensional and the two-dimensional models. Moreover, δ also has to be estimated differently in the two models. The values given along the abscissa of Figure 8.18 are the customary two-dimensional ones, namely, the number of inhabitants per unit area (here per square kilometer), for which k is estimated to be 1.5. For the one-dimensional model a suitable estimate of density (δ') may be obtained by dividing the mean number of inhabitants per parish by the average spacing between two neighboring parishes. As the average spacing is approximately the same over all the area, we can transform δ into δ' by correcting by an appropriate factor (which depends on what the average spacing is—it happens to be 2.2 kilometers in this area).

As we have already discussed, the expectations under the one- and two-dimensional models are very different, and we have already seen that most of the parishes in the high-altitude area show a geographic distribution that is intermediate between the two types of theoretical distributions. We will examine the meaning of k and the estimate of b that can be obtained from k, assuming in turn, the one- and two-dimensional models.

Population density should be corrected so as to take account of effective population size. The correction factor is 0.4 for this area and can be applied directly to k. Since density should be multiplied by the correction factor, we can simply divide k by it, giving $k/0.4 = 3.7$ for $k = 1.5$. In the two-dimensional model, the k value is to be equated to $8\pi\sigma^2(-1/\log_e 2b)$. The variance σ^2 in this area is approximately 400 square kilometers (from Table 8.7). This is approximate because of the contribution to mean square distance of extremely long migrations, as already mentioned. The resulting value of b is vanishingly small.

For a one-dimensional model, k has to be altered to take account of the change of scale from δ to δ'. With a population density of 44 per square kilometer, the average parish size is 300. With an average spacing between villages of 2.2 kilometers, the linear density is $300/2.2 = 136$ per kilometer. The quantity k has to be multiplied by $44/136 = 0.31$ to take account of the change of scale in density and is, then, $k = 3.7 \times 0.31 = 1.15$. In the one-dimensional model (Equation 8.39),

$$k = 4\sigma\sqrt{2b}.$$

Thus, with $\sigma^2 = 400$ square kilometers and thus $\sigma = 20$ kilometers, $b = 0.00010$.

We know that the distribution is intermediate between a one- and a two-dimensional distribution and so we can only say that b should be smaller than 0.0001. However, we are not considering other features of the data, such as the effect of clustering and the fact that the estimate of migration using σ^2 is unsatisfactory. An approach giving a more satisfactory representation of actual conditions would seem in order and will be discussed in the following sections.

8.13 Analysis of Correlation and Covariation with Distance

The formulas expressing the dependence of covariance and correlation between population clusters on their distance apart, lend themselves to an alternative approach. Pairs of clusters can be grouped according to the distance between them. Within each distance group, the correlation between the gene frequencies of pairs of clusters can be computed. As there is no reason to choose one or the other cluster of each pair for the x (or the y) axis in the correlation diagram, each pair is plotted twice in the correlation diagram, exchanging coordinates. This gives rise to an **intraclass correlation coefficient** (see Appendix I).

Nei and Imaizumi (1966) have used this approach to describe the variation of ABO gene frequencies in Japan from gene-frequency data collected by prefecture. Figure 8.19 shows that the correlation for the frequency of gene A between two

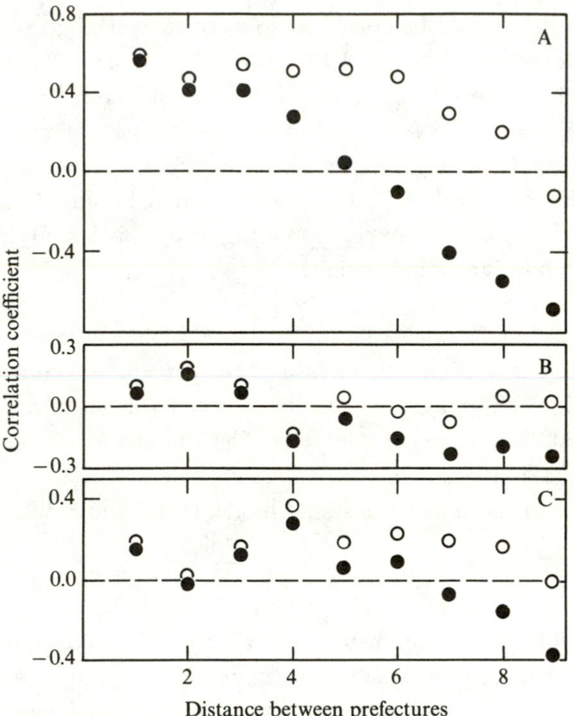

FIGURE 8.19
Correlations of the gene frequencies between different localities for the ABO blood groups. A filled-in circle refers to the intraclass-like correlation and an open circle to the ordinary correlation. A: Gene A. B: Gene B. C: Gene O. (From Nei and Imaizumi, 1966.)

prefectures decreases with distance. The decrease was less clear-cut or not present at all, for B and O. It may come as a surprise that correlations between population clusters can be negative. There is a cline for these frequencies along the length of Japan (see also Section 8.22). In computing the intraclass correlation coefficient each pair of prefectures is plotted twice in the correlation diagram, interchanging which one is plotted along which axis. When a cline exists, a negative correlation—especially over a large distance—can result.

It is possible to correct approximately for the existence of a cline in various ways, such as by fitting a surface to the gene frequencies, computing an expected gene frequency $\hat{p}_{ij}$ for every location, i, j, and then correcting the observed gene frequency at that location by adding the quantity $\hat{p}_{ij} - \bar{p}$, where $\bar{p}$ is the general mean. The method used by Nei and Imaizumi (1966) to obtain the corrected correlation coefficients (see Figure 8.19) was different, though almost equivalent to using this method, and was based on fitting a line (rather than a surface) to the gene frequecies according to their location along Japan. This country is actually long and narrow and therefore has an almost linear configuration.

The dependence of the correlation coefficient for the A gene on distance is not regular, and the B and O genes show little, if any, correlation with distance. This is in part due to the small number of observations, and in part to the grouping into prefectures. There are 45 prefectures and so $45 \times 44/2 = 990$ pairs, which were distributed into nine groups by distance between prefectures, in order to compute the correlations. Correlation coefficients are subject to a fairly large error and the grouping of the population into large units does not allow computation of correlations for small distances, which would be the most informative.

Morton and co-workers (1968) and Azevedo and co-workers (1969) in a series of recent papers, have applied the theory of isolation by distance to two bodies of data collected with a reasonable degree of fine geographic detail. One of them was a survey of ABO blood groups among Swiss conscripts, and the other a fairly exhaustive investigation of a Brazilian population in North East Brazil. Covariances were estimated as a function of distance, by grouping pairs of individuals into classes according to the distances between their birthplaces. The mean kinship coefficient was estimated for each distance class on the basis of various traits. Dependence of the kinship coefficient on distance r was analyzed by fitting, to the data, the semi-empirical equation

$$\phi(r) = f_0 e^{-cr} r^{-d}, \tag{8.37}$$

where $\phi(r)$ is the mean kinship coefficient for pairs of individuals at a distance r apart, f_0 is equal to the kinship at zero distance, d is dependent on the number of dimensions and is zero for one dimension and $\frac{1}{2}$ for two dimensions, and c, if the number of dimensions is clear-cut, can be taken, according to Malécot's continuous theory of isolation by distance, to be equal to $(2b/\delta)^{1/2}$ (see Equation 8.30). Morton and co-workers found that Equation 8.37 did not give an adequate fit to pedigree inbreeding data. Kinship data, based on covariance analysis, could be fitted but the estimate of the dimensionality parameter d was not always as consistent as it should be for different bodies of data from the same geographic region.

The intercept value f_0 should be equal to $(1 + 4Nb)^{-1}$, as before. Unfortunately, the theory does not work well for short distances and in any case the intercept value is especially sensitive to dimensionality. The validity of Morton and co-workers' actual f_0 estimates seems to be questionable. These authors obtained $f_0 = 0.005$ for Alpine isolates, as compared with -0.0004 for random individuals from the whole of Switzerland. It is probable that their f_0 value for Alpine isolates is an underestimate. They have used distances as the crow flies, thus certainly underestimating actual distances between villages in the more remote, mountainous parts of the country, especially for pairs of villages in different valleys. Underestimation of distances near the origin causes a decrease of the value of $\phi(r)$ extrapolated to $r = 0$, which is their basis for estimating f_0. The comment by Morton and co-workers that the upper limit to kinship coefficients in man is 0.02 is in contradiction with the fact that α values (overall inbreeding coefficients) higher than this have been reported (see Chapter 7). Values of α are bound to be lower than kinship coefficients because they neglect remote consanguinity.

Estimation of b, the recall coefficient for linear pressures, from the coefficient c in Equation 8.37 gave values that were so high as to lead Azevedo and co-workers (1969) to remark "we wonder whether long-range migration, perhaps, increasing in recent years, may not be a more significant force than selection." We might also wonder whether the implicit assumption of equilibrium is valid.

The theory of isolation by distance depends on assuming that patterns of migration can be specified in terms of distance alone.

If this is not possible, it may be very difficult to distinguish which part of the migration is accounted for by "distance" and which is not, and so must be absorbed, as long-range migration, into the coefficient of recall b. Under such conditions the coefficient b has only a descriptive value.

Finally, we would like to be able to check that the number of dimensions, as obtained by fitting kinship coefficients to distance, corresponds with that estimated from demographic data. These criticisms illustrate the general proposition that real migration data must be used for measuring the migration pressure that counterbalances drift. The shortcut of using migration distributions defined as a function of distance may be useful as an approximation when no better approach is available.

8.14 Migration Matrices

Migration data are usually obtained from information on the birthplaces of individuals and of their parents (or, less accurately of their spouses) and thus are readily available. A complete statistical summary of such data can be put in the form of a rectangular array of numbers, or matrix (see also Example 6.1), in which birthplaces of parents are given as columns and those of offspring as rows. There is often a difference between father-offspring and mother-offspring migration due to social customs. The two can be pooled to give a single parent-offspring matrix (see Table 8.9). The n_{ij} element of this matrix—namely, the number in its jth row and jth column—is the number of offspring born in cluster i one of whose parents was born in j.

TABLE 8.9

Migration Data from Six Mayan Villages (SA, CDO, etc.) on Lake Atitlan (Guatemala). OV: children born to parents born in outside villages.

Birthplace of Children	Birthplace of Parent: SA	CDO	SAP	SCP	SCL	JAI	*Subtotal*	OV	*Grand Total*
SA	298	0	0	0	0	0	298	16	314
CDO	6	134	0	0	0	0	140	2	142
SAP	0	0	543	2	0	0	545	3	548
SCP	0	1	7	298	0	0	306	3	309
SCL	0	0	0	0	176	3	179	3	182
JAI	0	0	0	0	2	26	28	4	32

Source: From Cann, Barnett, Harris, et al., unpublished data.

Knowing the distance between all pairs of places i and j, we can then construct from these data the frequency distribution of the distance between birthplaces of parent and offspring, and this gives rise to migration distributions, as discussed before. The use of a migration distribution considerably simplifies the presentation of the data. In fact, it allows the whole matrix to be reduced to one or a few parameters, namely, those of the migration distribution function. It is clear, however, that, in this process, much information may be lost, especially if the geographic distribution of the population is irregular.

It is possible, by suitable approaches, not to lose the information contained in the original migration matrix.

Two methods will be considered: calculating the expected gene-frequency variances and covariances from the empirical migration matrix plus other demographic information when it is available, and suitably simulating the population in a computer. Before we go into the details of these approaches, however, we will consider some problems concerning their general validity.

The use of migration matrices is probably the best approach for predicting genetic drift when suitable data are available, but difficulties may be encountered.

A *migration pattern may change with time.* If data from the past were available and the migration matrices for each former generation were known, they could be used to give the final expected variances and covariances. This will seldom be the case however; usually, migration matrices for only the past one or two generations are available. There is an interesting way to check whether an observed migration pattern can be taken as representative of a time period longer than a generation, which we shall discuss later. In fact, if the same migration pattern were to prevail indefinitely, the migration matrix could also be used to estimate the relative proportions of the cluster sizes to be expected. A comparison between the actual sizes of the clusters, and those predicted in this way may show whether the present geographic distribution can be explained on the assumption that the migration pattern has not changed over time.

The observed numbers in the matrix may be small and hence subject to large sampling errors. Pooling suitable rows and columns helps to reduce this source of error.

Prediction may involve heavy numerical computations if the matrix is large. The matrix may be reduced by taking submatrices corresponding to subareas, and examining them independently, or by pooling rows and columns. The pooling process can be carried out by pooling rows and columns that have the lowest values in the main diagonal and the highest values in the symmetrically placed "off-diagonal" position. This should minimize the loss of genetic information.

The theory of migration matrices makes use of the angular transformation of gene frequencies.

The formal theory of migration matrices was developed to predict the expectations of the gene-frequency variances of individual clusters, and the expectations of the covariances. The theory is based on an approximation that makes use of angular values instead of gene frequencies (see Appendix I). It is known that this approximation is quite satisfactory for gene frequencies between 0.05 and 0.95.

The transformation of a frequency p into an angular value θ is defined by

$$\sin\theta = \sqrt{p}. \tag{8.38}$$

It has the property that θ, in contrast to p, has a sampling variance which is independent of p, provided p lies approximately between 0.05 and 0.95. There is a simple relationship between the variance V_θ of the angular value θ and Wahlund's variance σ_p^2/pq, corresponding to a kinship coefficient f, namely,

$$f = \frac{\sigma_p^2}{\bar{p}\bar{q}} = 4V_\theta \tag{8.39}$$

when angular values are expressed in radians, and $f = (\pi/90)^2 V_\theta$ when they are expressed in degrees.

The expected variance of θ in the ith cluster after the nth generation can be shown to be

$$V(\theta_i^{(n)}) = \frac{1}{8}\left[\frac{1}{N_i} + \sum_{j=1}^{k} \frac{1}{N_j} \sum_{r=1}^{n-1} (m_{ij}^{(r)})^2\right], \tag{8.40}$$

where $m_{ij}^{(r)}$ is the ijth term of the rth power of the migration matrix (allowing for migration from the outside) and N_i is the number of individuals in the ith cluster. (See Example 6.1 for a discussion of matrices and how to obtain powers; see Bodmer and Cavalli-Sforza, 1968, for the derivation of these results; see also C. A. B. Smith, 1969.)

Similarly, the expected covariance between two clusters at the nth generation is

$$\mathrm{Cov}(\theta_i^{(n)}\theta_j^{(n)}) = \frac{1}{8}\sum_{l=1}^{k} \frac{1}{N_l} \sum_{r=1}^{n-1} (m_{il}^{(r)} m_{jl}^{(r)}). \tag{8.41}$$

The expected variance of gene frequencies in the total population at the nth generation is

$$V(\bar{\theta}^{(n)}) = \frac{1}{N^2}\left[\sum_{i=1}^{k} N_i^2 V(\theta_i^{(n)}) + \sum_{i=1}^{k} \sum_{\substack{j=1 \\ i \neq j}}^{k} N_i N_j \,\mathrm{Cov}(\theta_i^{(n)}\theta_j^{(n)})\right], \tag{8.42}$$

where

$$N = \sum_{i=1}^{k} N_i,$$

and $\bar{\theta}^{(n)}$ is the mean over all clusters. The expected variance between clusters at the nth generation is

$$V_B^{(n)} = \frac{1}{N}\sum_{i=1}^{k} N_i V(\theta_i^{(n)}) - V(\bar{\theta}^{(n)}) + A, \tag{8.43}$$

where the term A approaches zero with time, if the gene frequencies of the external sources of genes (from which the migrants from outside the area under consideration come) for each cluster are equal.

With the help of a computer, variances and covariances can readily be computed for any generation, and thus the kinetics of the approach to equilibrium can be followed. We can thus determine whether, for any given situation, the time during which the process of drift could have taken place is sufficient for approaching equilibrium.

In this respect, it is important to note that in any finite population corresponding to a given migration matrix with no external immigration, the variance will increase and drift will always lead to fixation. For a steady-state equilibrium to be reached, a balancing force, corresponding to the coefficient of recall to equilibrium, must be present. For a finite area totally cut off from the rest of the world, this coefficient would be the mutation rate plus a "linearized" balancing selection pressure. If, however, there is any influx of genes from populations external to the area being considered or external to the population being considered but living in the same area, then the coefficient of recall to equilibrium should also include this migration pressure. In the infinite, continuous and discontinuous models that we have discussed before, we tend to ignore this pressure, which is effectively included in the b value, but in a finite and more realistic model it cannot be forgotten. The migration-matrix model allows for coefficients of recall that may be different for each cluster. These coefficients are called α_i, one value for each cluster i. If $\alpha_i = \alpha$ is constant with respect to different clusters, it is essentially the same as the recall coefficient b. Allowing for variation between clusters, we can accommodate, for instance, higher migration from the outside for peripheral clusters.

We will use as an illustrative example of the application of migration matrices a small matrix obtained from unpublished material kindly made available to us by Cann, Harris, and co-workers (given in Table 8.9). The data came from five villages inhabited by American Indians and located around the rim of an approximately circular lake (Lake Atitlan in Guatemala). The migration matrix given in Table 8.9 was obtained from preliminary demographic data and is the sum of two matrices, one for father-offspring and one for mother-offspring migration. The stochastic transition matrix, which has rows that sum to 1, is obtained by dividing each value by the row total, as in Table 8.10.

In the last two columns of Table 8.10 are given the population sizes N of the villages (approximate sizes for the first two) and the relative frequency of children born in villages other than those included in the matrix (α_i). The actual numbers of such children were given in Table 8.9 (column OV). The α values given in Table 8.10 are obtained for the first village, for example, as $\alpha_i = 16/(298 + 16)$, etc. The elements of the matrix must be multiplied by $(1 - \alpha_i)$ before taking powers, for use in the variance and covariance formulas (m_{ij} in Equations 8.40 and 8.41).

In what follows we assume that migration from outside the area takes place from populations that have the same gene frequencies, on the average, as those inside. If this is not true, and the actual gene frequencies of the populations from which the migrants come are known, they can be taken into account in the full treatment of the theory.

TABLE 8.10

Stochastic Migration Matrix Obtained from the Migration Matrix of Table 8.9 by Dividing Each Value by Its Corresponding Row Total. Population sizes N and migration from the outside α_i are given in the last two columns.

Birthplace of Children	Birthplace of Parents: SA	CDO	SAP	SCP	SCL	JAI	*Total*	*N*	α_i
SA	1.0	0	0	0	0	0	1.0	3000	0.051
CPO	0.043	0.957	0	0	0	0	1.0	300	0.014
SAP	0	0	0.996	0.004	0	0	1.0	548	0.005
SCP	0	0.003	0.023	0.974	0	0	1.0	309	0.010
SCL	0	0	0	0	0.983	0.017	1.0	182	0.003
JAI	0	0	0	0	0.071	0.929	1.0	32	0.125

The expected variances and covariances of the gene frequencies for the Guatemalan Indian villages, as well as the variance between villages, can be obtained for any generation by using Equations 8.40 through 8.43. These formulas actually give the variance V_θ of the angular transformation of gene frequencies, but, from Equation 8.39, this is simply $\frac{1}{4}f$, where $f = \sigma_p^2/\bar{p}\bar{q}$.

The expected increase in the variance of gene frequencies between villages with time is shown in Figure 8.20. Village sizes have been multiplied by a factor for correcting the total population size to effective population size, which was taken, here, to be 0.3. The matrix of variances and covariances after 300 generations (expressed in angular values) is shown in Table 8.11. At this time equilibrium had not yet been reached.

The equilibrium value of the variance of gene frequencies between villages is about 0.013. The average f value observed for 10 loci (ABO, Rh, MNS, Jk, Fy, Di, Hp, Tf, PGM, PGD) is 0.014 ± 0.009, which agrees very well with the predicted

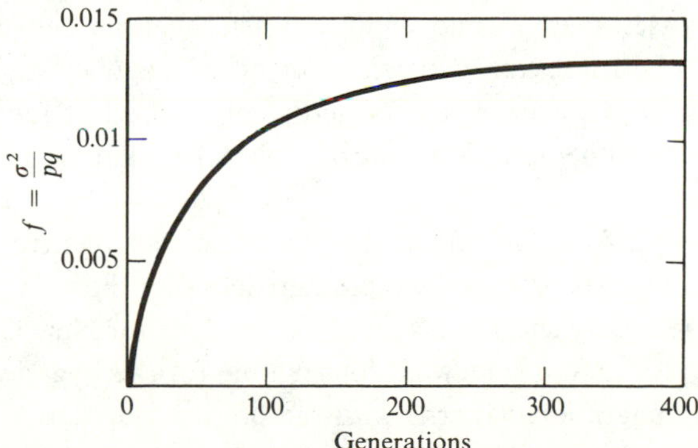

FIGURE 8.20
Increase of f between 6 Guatemalan Indian villages with time computed from an observed migration matrix.

TABLE 8.11

Matrix of Variances and Covariances (Angular Transformations of Gene Frequencies) Between the Six Villages on the Edge of Lake Atitlan, in Guatemala, after 300 Generations

	1	2	3	4	5	6
1	.00042	.00016	.0000003	.000005	0	0
2	.00016	.00393	.000008	.00013	0	0
3	.0000003	.000008	.0166	.00937	0	0
4	.000005	.00013	.00937	.01171	0	0
5	0	0	0	0	.02451	.00835
6	0	0	0	0	.00835	.01428

equilibrium value. The time needed to reach equilibrium is about 200 generations or 4000–5000 years, which is surely longer than these populations have remained stable, or even existed. The major change in f, however, takes place within the first 50–100 generations.

Comparisons of this kind have so far been carried out for 28 parishes of the upper Parma Valley in Italy; 22 colonies of Babinga Pygmies in Central Africa; 13 clans of the Bundi tribe in New Guinea (Malcom et al., unpublished data). In all of these cases, the f values observed from gene-frequency data and those expected on the basis of migration matrices were in good agreement. In the Parma Valley, where observed and expected values showed the greatest disagreement, the observed f value was about 3 times higher than that expected. The analysis is not however, complete. In particular, the migration data so far available do not allow satisfactory computation of the α_i values, which were assumed to be equal. This might contribute to underestimating the expected variation. Estimates of f from pedigree data are nearer to the values expected from the migration matrices.

In all of these cases, there was no significant heterogeneity between the variation at different loci. (This test can be carried out, when angular transformations are used instead of gene frequencies, by Bartlett's chi square test for homogeneity of variances.) Thus, the other criterion for drift being the major source of variation is also met.

Unlike the example from Guatemala, in which there is a relatively high degree of isolation between villages, in all of these other examples the time needed to reach equilibrium was relatively short; only a few generations for New Guinea, about 15 for the Parma Valley, and a somewhat longer time for the Pygmies. A test for the consistency of the migration patterns with the present population distribution in the clusters showed remarkably good agreement, at least in the first two examples, confirming the validity of the migration data.

The theoretical models described earlier correspond to particular forms of the migration matrix. The analysis of these and other theoretical matrices may help in estimating the time needed to reach equilibrium for these theoretical models.

The island model of Wright is given by an infinite matrix, in which all elements of the main diagonal are $1 - \alpha$ (where α is equal to m in the model), while all other elements are 0. The linear stepping-stone model corresponds to $1 - m$ on the main diagonal, and $m/2$ on the elements immediately adjacent to the diagonal, where all elements of a row should be divided by $1 - \alpha$. A "pseudoinfinite" linear stepping-stone model can be imitated by turning the linear pattern into a circular one (see Figure 8.21). A pseudoinfinite square lattice can be similarly generated. Other types of lattices have also been studied. The effect of dimensions is not large, as can be seen by comparing the kinetics of the increase in variance for a linear (circular) and an icosahedral model (see Figures 8.21 and 8.22). The latter model represents roughly 2.5 dimensions. The parameters m and α are the most important factors in determining the rate of approach to equilibrium and the equilibrium f value itself.

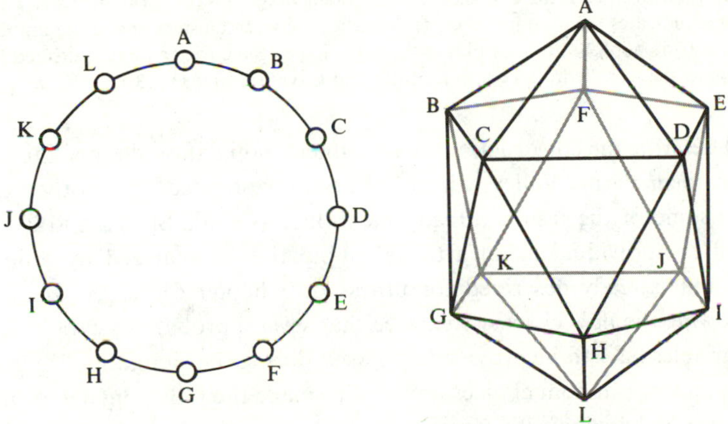

FIGURE 8.21
Spatial representation of population clusters in the circular and in the icosahedral models. Lines connect clusters between which exchange by migration takes place. (From Bodmer and Cavalli-Sforza, 1968.)

8.15 Predicting Drift by Computer Simulation

Early experiments on the simulation of drift by computer were made by Brues (1954, 1963) who imitated the interaction of drift and selection for the ABO blood-group genes and modified input data to obtain a distribution similar to the present world distribution (see Chapter 5). The degree of sophistication of simulation models depends on the available information and computer time.

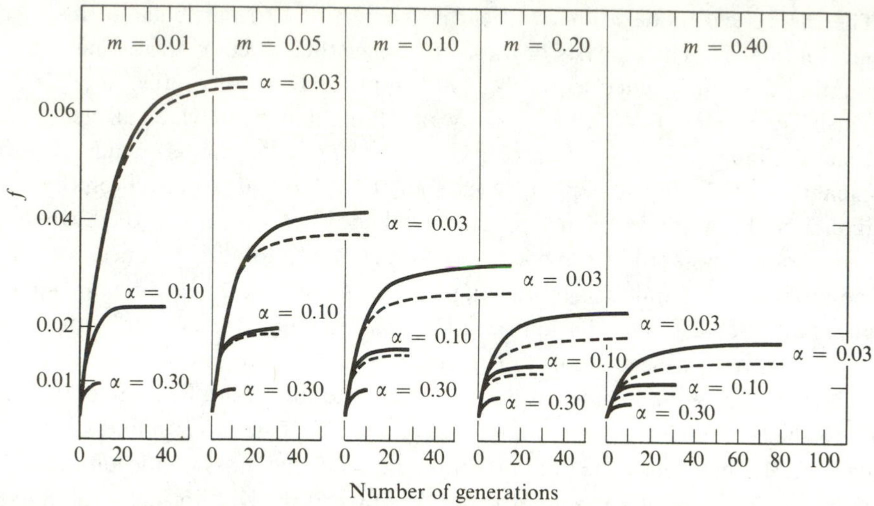

FIGURE 8.22
Increase of f with time in the two theoretical models diagrammed in Figure 8.21. The quantity m indicates the amount of cross-migration between clusters and α the immigration from an outside pool. Solid lines plot data according to the circular model; dashed lines, according to the icosahedron. (From Bodmer and Cavalli-Sforza, 1968.)

We will describe the construction of an artificial population designed to simulate that of the upper Parma Valley, which has been investigated, from other points of view, with some of the results already mentioned (Cavalli-Sforza and Zei, 1967). The life of an individual in an artificial population is analyzed by a flow chart similar to that already described for fitness (in Chapter 6), as shown in Figure 8.23. Following the flow chart, we can see that several probability tables of demographic variables have to be provided and used during the simulation, by means of random numbers. Random choices are used to make the following determinations, in sequence (numbers indicate corresponding items in the flow chart).

1. Probability of death, by age and sex.
2. Probability of marrying, by age (for males).
3a. Probability of marrying woman of a given age (as a function of a male's age).
3b. Probability of marrying woman of a given parish (as a function of a male's residence).
3c. Probability of marrying woman of a given social class (as function of a male's social class).
4. If more than one prospective wife of a given type is available, give each an equal probability.
5. Probability of birth of 0, 1, 2, ... children during a decade, as a function of age of mother and of population rate of growth.

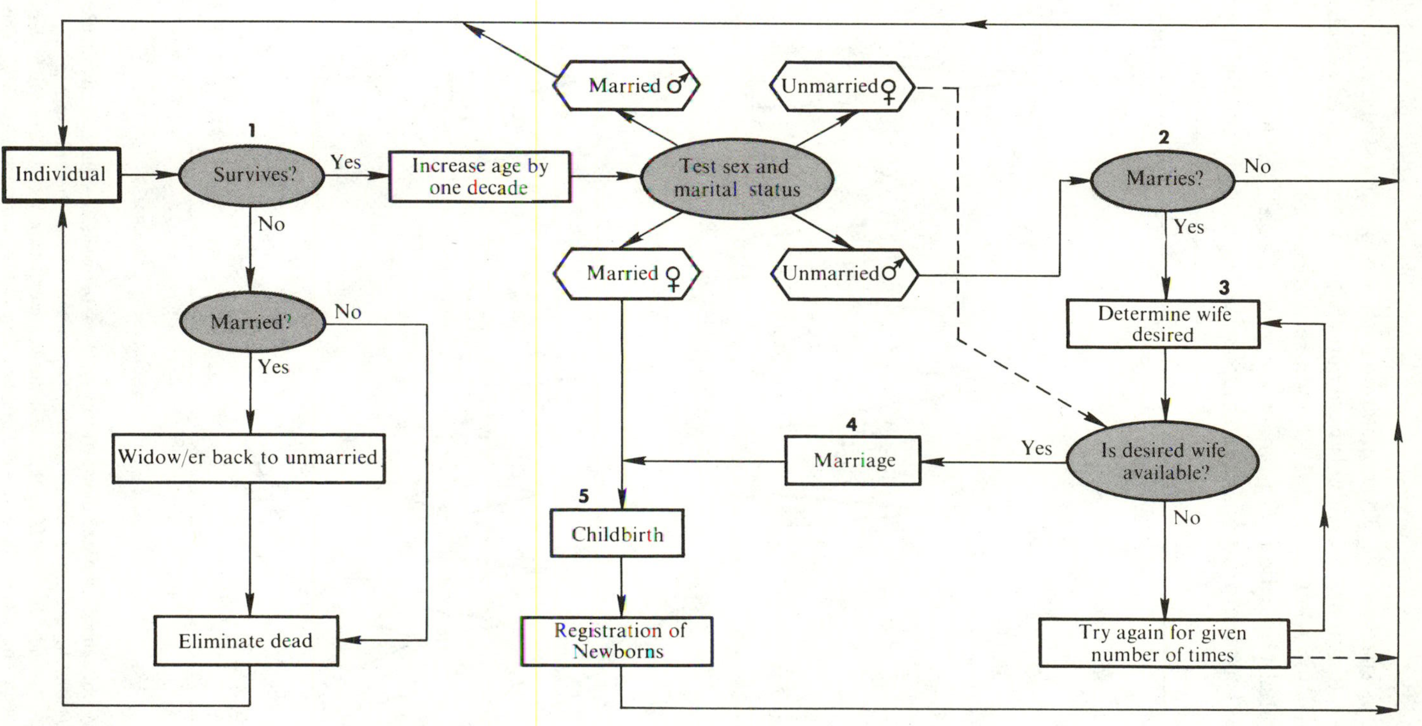

FIGURE 8.23
Flow chart of artificial population for prediction of inbreeding and of drift. Dashed arrows indicate that a portion of flow chart has been simplified. Numbers refer to use of probability tables as specified in text. (Adapted from Cavalli-Sforza and Zei, 1967.)

The probability tables used were, as much as possible, derived from actual data. Information was sparse for 3c. The population itself occupied most of the computer memory (32,000 locations). Individuals were initially given the observed population gene frequencies for ABO, MN, and Rh by a special part of the program, which created the initial population with family groups and ages. The total number of individuals was kept approximately constant from the beginning by suitably regulat-

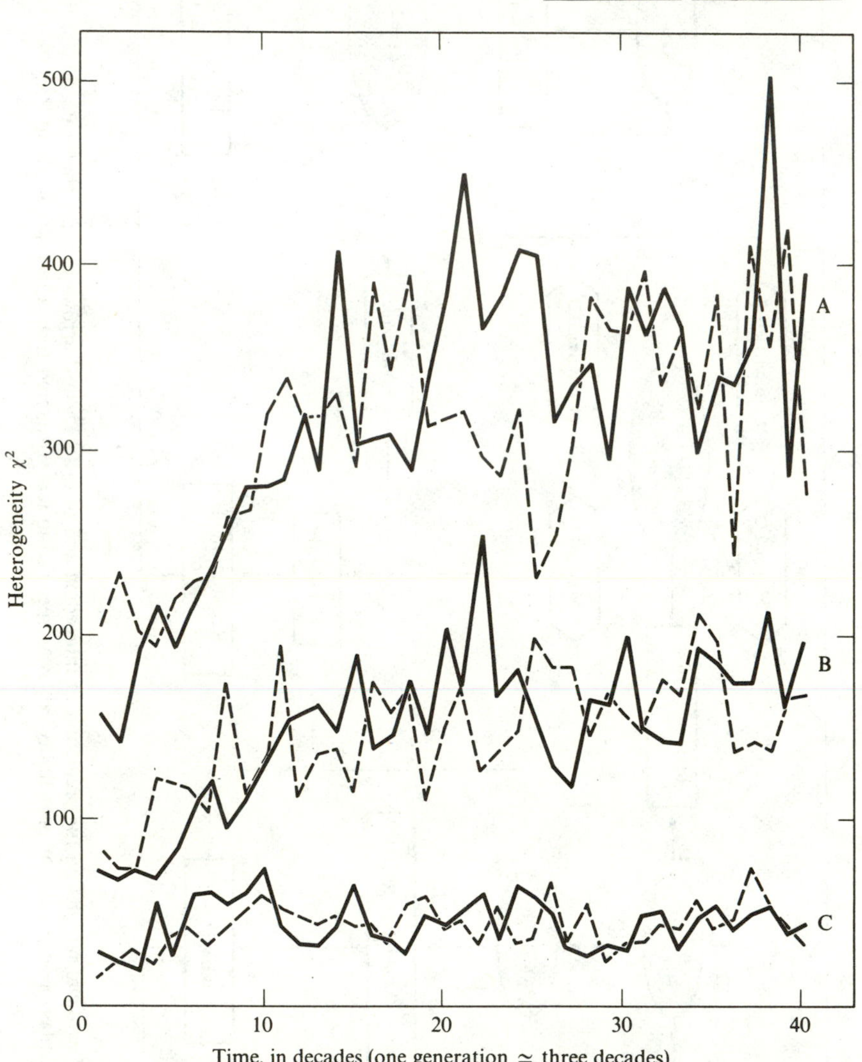

FIGURE 8.24
Heterogeneity chi square for three blood-group systems, between 22 villages in an artificial population simulating the upper Parma Valley. Drift was taken as the only source of variation. The dashed and solid lines represent duplicate experiments. A: The Rh system; seven alleles were considered. B: The ABO system; four alleles were considered. C: The MN system; two alleles were considered. (From Cavalli-Sforza, 1967.)

ing the population growth rate at item 5. The increase in the heterogeneity (measured by chi square) between the blood-group gene frequencies of the 22 parishes represented in the population is shown, for two independent runs of the population, in Figure 8.24.

The computer time required for analysis is proportional to the number of individuals (which was about 5000) and to the reciprocal of the time unit employed. Time is a discontinuous variable in this kind of experiment. In this example a decade was used as the time unit, and so the average length of a generation was approximately 3 units. This, of course, makes the approximation fairly rough from a demographic point of view, but decreases the computer time, which was 2 minutes per cycle on an IBM 7040. A similar method has been used by MacCluer (1967).

The estimate of f derived by computer simulation, 0.01085, compares very favorably with that obtained by the migration-matrix approach using the same migration matrix, which was 0.01160. The two approaches differ mostly in that the latter does not take into account the sampling of migrants from generation to generation. It would seem that the effect of this difference cannot be large. The migration-matrix approach is much more economical in terms of computer time. The only demographic parameters required are the migration matrix and those needed for the computation of effective population size.

8.16 Drift in Time

Glass and co-workers (1952) suggested the possibility of analyzing drift by measuring the difference in gene or genotype frequencies between nonsuccessive generations. This can be done on living people, by considering that representatives of more than one generation are present in human populations and that they can be separated, at least roughly, on the basis of age. Glass and co-workers used data from an isolate formed by a religious sect living in Pennsylvania, the Dunkers. They partitioned the living population into three age groups: 1–27 years, 28–55, and 56 or older. Comparisons between gene or phenotype frequencies for the MN blood group showed a significant chi square between the first and third generations, while those between the first and second generations, and between the second and third were not significant, as might be expected. No such effect was, however, found for the ABO and Rh blood groups.

This method could give a misleading result if there were differential mortality for blood groups with respect to age, but this has not so far been found for these markers (see Chapter 5). Analysis of the age distribution of carriers of the markers can show effects of drift only in highly isolated populations that are very extensively sampled. If the phenotype of every individual in the population is determined and due corrections are made for duplication of genes in each generation resulting from close relationships, the chi square between the first and third generation is expected to be twice the normal expectation for chi square (namely, the number of degrees of freedom), which is only a relatively slight increase. The sensitivity of the test for drift in time could be increased significantly if data on generations that are farther apart were available, because the expectation of chi square between groups n generations apart would be approximately n times the degrees of freedom.

8.17 Inbreeding and Drift

There is a direct relationship between inbreeding and drift. Under random mating in an infinitely large population, no two individuals that mate are related. In a finite population, however all individuals are related to some extent and we have seen that in a population of constant size, in the absence of mutation, the inbreeding or kinship coefficient f approaches one with infinite time. In the presence of mutation or other linear systematic pressures b, f approaches a value $(1 + 4bN)^{-1}$, which is smaller than one. Thus, a finite population is always subject to some degree of inbreeding because of its finite size. It should be noted that, if a population increases in size more rapidly than linearly with time, f approaches a value lower than one, even in the absence of systematic pressure (Malécot, 1948).

There are various approaches to the computation of kinship coefficients from genealogies that can be used to measure the effects of drift.

We have so far mainly considered how to use $\sigma^2/pq = f$ as a measure of genetic drift. An alternative method of measurement, which is based on largely independent sources of data, is to use the mean kinship or the mean inbreeding coefficient, obtained from genealogies, of the individuals forming the population. All the theoretical analyses considered so far could be applied to this alternative method. The mean kinship coefficient can be obtained by taking pairs of individuals at random and, on the basis of their pedigrees, computing the kinship coefficient for each pair, which is the same as the inbreeding coefficient of their progeny. The mean over all pairs would give f.

On the other hand, records on the consanguinity *between mates* exist in the archives, national or religious, of several countries, for example in Japan (through the Koseki) and in countries with Roman Catholic populations. Mates often know if they are fairly closely related, and so a survey of the consanguinity of mates by interview is a possible approach. This kind of information is not, however, always willingly supplied, and consanguinity may often not be revealed. A direct survey in a random sample of the Italian population gave average inbreeding coefficients that were less than half as large as those obtained from Roman Catholic archives.

The two approaches, using the consanguinity of random pairs and that of actual mates, are each based on a knowledge of genealogies. Each can go back for about the same number of generations, which is usually small. This results in a truncation of the information, usually at fairly close consanguinities.

A comparison between random pairs and mates gives reasonable agreement for computed inbreeding coefficients.

Determination of the consanguinity of mates has the advantage, over that of random pairs, that direct use can be made of a greater number of available records. It has the disadvantage that mates form a highly selected sample of pairs. The analysis of all the factors that may bias the sample, and of appropriate corrections, is difficult and often unsatisfactory and will be further discussed later. The study of mates does, however, supply information not obtainable from the study of random pairs: for example, the limitations on population effective size due to effective social stratification and geographic mobility.

The availability of many records of consanguineous matings has made this approach the one mostly followed in practice, though the potential interest in the use of random pairs should be kept in mind. A direct test of the two approaches by Steinberg and co-workers (1966) on two subgroups of the religious sect known as the Hutterites called the S-leuts and L-leuts gave the following results:

Hutterite subgroup	*Observed* $\bar{F}$	F_r^I	F_r^{II}
S–leut	0.0211	0.0248 ± 0.0012	0.0184 ± 0.0005
L–leut	0.0255	0.0311 ± 0.0014	0.0228 ± 0.0006

In this table, $\bar{F}$ is the value computed from actual matings; F_r^I that computed from a random sample of Hutterite individuals, and F_r^{II} that computed from the same random sample after omitting matings between sibs, half sibs, first cousins, and first cousins once removed, which are not approved by the social custom of the Hutterites. The observed values are intermediate between the two expected values, indicating reasonably good agreement. It is interesting that the avoidance of all consanguineous mating between persons more closely related than second cousins has not greatly reduced the average inbreeding coefficient. The Hutterites are, however, a very special group and this result should clearly not be generalized.

On the assumption of randomness (in the sense that consanguineous mating is neither avoided nor favored because of its being consanguineous) we can compute the expected frequencies of various types of consanguineous matings for any given population. We give now an approximate treatment assuming nonoverlapping generations and random mating in a finite population, and will later correct the model to make it more satisfactory. In a population of stable size N, with a Poisson distribution of progeny size, each individual has on the average two children (see Example 8.8), and thus two paternal and two maternal uncles or aunts. Each of these has on the average two children, and so each individual has an average of 8 first cousins, of which 4 on the average are of opposite sex. The chance of marrying a first cousin is therefore

$$\frac{4}{\frac{N}{2}} = \frac{8}{N}$$

since $N/2$ is the total number of individuals of the opposite sex.

It was suggested by Dahlberg (1948) that the **isolate size** N can be estimated from the frequency of first-cousin marriages f_c. Dahlberg's formula, which is $f_c = 4/N$, gives results that are too low by a factor of 2 because he did not consider the variation of progeny size. The isolate size is a rough measure of the effective size of the population from which individuals choose their mates.

This approach can be generalized as follows. If we call d the degree of (even) cousins, then,

$$\frac{2^{2d+1}}{N}$$

is the expected proportion of cousins of desired degree in the population. Malécot (1948) has given the more exact formula

$$p = \frac{2^{2d+1}}{N}\left(1 - \frac{1}{N}\right)^d, \tag{8.44}$$

in which the factor in parentheses is usually very small.

A difficulty with the pedigree approach to population structure and drift is that relatively few individuals are included in most pedigrees.

The analysis of pedigrees in human populations can sometimes be extended as far back as ten or more generations using, for example, parish books (Moroni, 1964). The relative importance of the contribution of different degrees of relatives to the average inbreeding coefficient should be constant, since for degree d,

$$F = (\tfrac{1}{2})^{2d+2},$$

and thus from Equation 8.44, $pF \simeq \frac{1}{2}N$ for each degree of relationship, where p is the frequency of a relationship of degree d. In practice, migration, mutation, and other factors make the contributions of more remote degrees less important.

Most pedigrees cannot be traced for more than two or three generations, and only a few for four or five. Values of F obtained by averaging the inbreeding coefficients of a sample of individuals, or of matings from a population, over so few generations, are likely to be underestimated by an unknown factor. Even when good records are available, there are often gaps in the knowledge of the ancestry that make the estimation of F difficult.

An interesting case that illustrates some of these points, was analyzed by Mange (1964) using data on Hutterites living in South Dakota. This isolate numbered 5450 individuals in 1960, but only about 200 in 1874. Records go back, in part, to 1725 and it was therefore possible to reconstruct genealogies to at least the fourth generation. To estimate $\bar{F}$, individuals were chosen whose pedigrees achieved a certain desired degree of completeness, and the effect of missing ancestors was estimated. It is not clear whether this could bias the data.

Inbreeding coefficients were computed, using all four generations and also using just the first and second generations and just the third and fourth generations. The relative contributions of first- and second-cousin marriages and of third- and fourth-cousin marriages can thus be estimated as follows: from first- and second-cousin marriages $\bar{F} = 0.010 \pm 0.013$; from third- and fourth-cousin marriages, $\bar{F} = 0.012 \pm 0.007$. The total average inbreeding coefficient over the four generations is the sum of these two, namely;

$$\bar{F} = 0.022 \pm 0.014.$$

The decrease in the value of $\bar{F}$ due to not considering earlier generations is unknown.

This population is one of those having the highest observed $\bar{F}$ value (see Chapter 7). It should also be noted that it has considerably increased in size in recent times, while remaining isolated, a fact that helps to explain why $\bar{F}$ has a high value in spite of the relatively large size of the colony. The distribution of values of F for Hutterite families whose records met the criterion of completeness of ancestry used by Mange in his study is given in Figure 8.25.

There is direct evidence for nonrandomness of consanguineous matings.

Other difficulties in using consanguinity data arise from the fact that consanguineous matings are often nonrandom, in the sense that they are either specifically avoided or favored. The determining factors of this nonrandomness are cultural, and vary greatly from one population to another. We have already mentioned a society in which certain consanguineous matings are extremely favored: among the Sudras in Andhra Pradesh (South India) the proportion of matings between uncle and maternal niece may be as high as 12 percent (Dronamraju and Khan, 1960). Elsewhere, these matings are extremely rare (of the order of 10^{-3} to 10^{-4} in European populations) and are, in fact, outlawed in several countries.

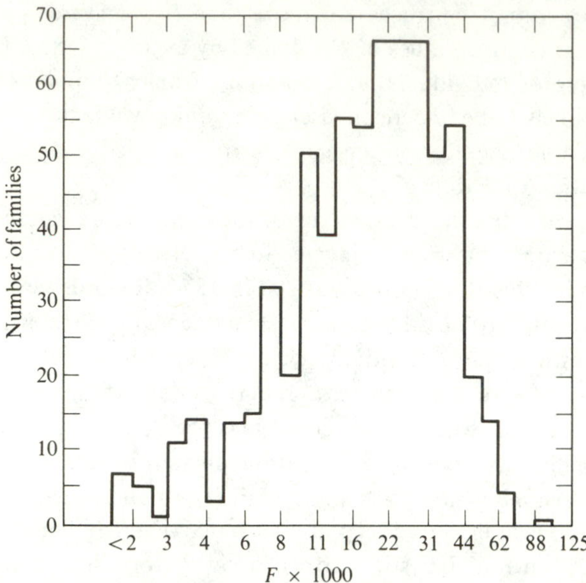

FIGURE 8.25
Distribution of the inbreeding coefficient F (log scale) among Hutterites. Tallied here are 667 families, each family counted once. (From Mange, 1964.)

According to Roman Catholic law, consanguineous matings giving an inbreeding coefficient up to $\frac{1}{8}$ are permitted, but require dispensation. Dispensation was also required from 1550 to 1917 even for third-cousin marriages ($F = \frac{1}{256}$). Later it was restricted to second-cousin marriage ($F = \frac{1}{46}$) and those of closer relations (until 1965), and is now required only for first-cousin and uncle-niece or aunt-nephew marriages. The ease with which dispensations have been granted must have varied with time and place. The frequency of third-cousin marriages and also that of second-cousin marriages, which has always been lower, has remained almost constant throughout the last three centuries in several parts of Italy; the frequency of first-cousin marriages has increased from practically zero to fairly high values during the nineteenth century (Figure 8.26). This change has been attributed to the change in legislation that took place under Napoleon (Moroni, 1967a). The rights of primogeniture were abolished, and those of all children to inherit their father's property nearly equalized. This made it impossible to keep agricultural property from being subdivided, a trend that first-cousin marriages could, in part, reverse. It also seems likely that there must have been a relaxation in the granting of dispensations by Roman Catholic authorities, for before this period, first-cousin marriages were almost nonexistent. Today, permission for a marriage for which dispensation is technically possible is practically always granted by the Roman

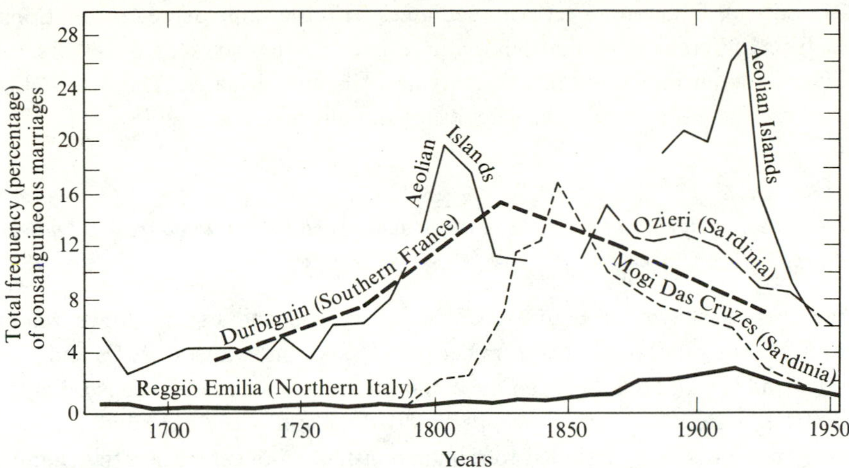

FIGURE 8.26
Secular variation in the total frequency of consanguineous matings (up to and including those between second cousins). (From Moroni, 1964.)

Catholic church. It is possible that consanguineous matings now tend to take place almost at random, in the sense that relationship does not influence the probability of marriage. If this is true, consanguineous matings can provide a simple basis for the measurement of isolate size along the lines suggested by Dahlberg. As Dahlberg himself recognized, however, his original formula does not readily lend itself to estimation and must be corrected in several ways. Improved methods, to be described later, lead to isolate-size estimates that are independent of those obtained from the study of gene frequencies, and which, therefore, provide a check of the randomness of consanguineous matings.

8.18 Estimation of Expected Proportions of Consanguineous Matings

Even if consanguinity itself does not influence the probability of mating, consanguineous matings nevertheless form a highly selected sample of all matings. Overlapping generations, changing population size, correlations between ages of husband and wife, migration, and other factors make the computation of the expected frequencies of consanguineous matings a lengthy one. In addition, there is a great variety of consanguineous matings that are combined under the same general heading and that require different treatments. Even without making a complete enumeration of all possible matings between relatives, as was attempted by Jayakar and Haldane (1965), we can identify four types of matings between first-cousins, 16 of second-cousins (and of first-cousins once removed), and 64 of third-cousins (and of $2\frac{1}{2}$ cousins). Differences between pedigrees with the same degree of relationship arise from different combinations with respect to the sexes of ancestors intermediate between the common ancestors and the actual mates. There are 2^d such ancestors, where d is the degree of consanguineous marriage.

The expected number of relatives of given degree can be estimated from a knowledge of average sibship size.

The mating shown in Figure 8.27 is between second cousins once removed (A and B). The number of ancestors between the common ancestors (E and F) and the husband is i, and that between the common ancestors and the wife is j. Let, now, s be the expected number of sibs per individual and p the expected number of progeny per individual, assumed to remain constant from generation to generation. Then there are (ignoring sex) 2^2p^3s relatives of type A per B individual for the example illustrated in Figure 8.27. This is because B has 2^2 grandparents (D), each of which has on the average s sibs (C), each of which has an average of p^3 great grandchildren (A). In general, there are 2^jp^is relatives of the wife (B) that have the same degree of consanguinity (defined by i and j) as her husband (A), and similarly 2^ip^js relatives of the husband that have the same degree of consanguinity as the wife. The number of progeny p that reproduces is approximately twice the absolute rate of population increase per generation. When family size has variance V and mean p, the expected number of reproducing sibs is (see Example 8.8)

$$E(s) = p + \frac{V}{p} - 1,$$

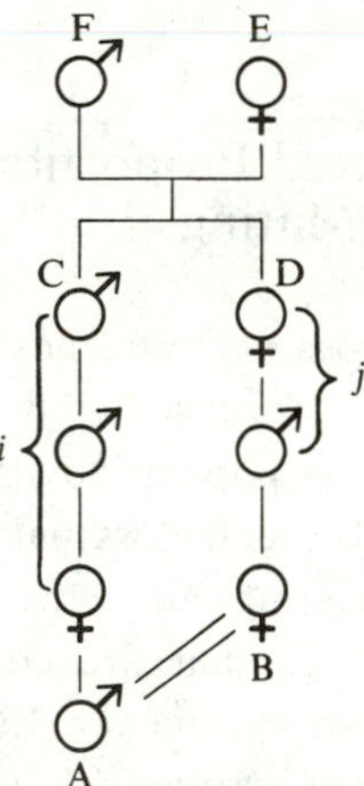

FIGURE 8.27
Pedigree of a consanguineous mating (between second cousins once removed).

which is equal to $p - 1$ for constant progeny size, which was the situation assumed by Dahlberg. If the distribution of progeny size is a Poisson distribution, $V = p$ and thus $E(s) = p$. In practice, $E(s)$ is somewhat larger than p because progeny size usually has a distribution with a larger variance than that of a Poisson distribution (see Chapter 6).

In an actual survey (Mainardi et al., 1962) the total number of living first cousins per individual was found to be 14.1 with a variance of 99.8. The proportion of these who would reproduce is, very approximately, 65 percent. The expected variance (55.5), based on assuming a negative binomial distribution of progeny size (see Chapter 6) and independence of progeny size in successive generations, is significantly smaller than the observed one. This is likely to be the result of correlations in fertility between successive generations. Variation in fertility due to social conditions and the age of the mother, rather than heritable fertility differences, are probably the major cause of this discrepancy.

If the correlation in fertility is r, the expected number of, for example, first cousins for a Poisson offspring distribution is $2p^2(1 + r)$. Thus, the distributions of the number of relatives could in principle be used to estimate correlations in fertility. They cannot, however, be used on their own to separate the parent-offspring correlation of fertility due to social class from that due to biological inheritance.

Correlation of age at marriage has an important effect on the pattern of consanguineous marriage.

It is well known that there is a high correlation for age at marriage. The distribution of age at marriage is skew, but can be fairly well "normalized" by the transformation log $(t - t_0)$, where t is the actual age at marriage and t_0 is the youngest possible age at marriage. Unfortunately, this distribution does not lend itself readily to further analytical computations, but its approximation to a normal was found to be satisfactory for the purpose of this analysis. It is convenient to compute, instead of the correlation coefficient between ages at marriage of husband and wife, the variance of the difference between their ages, which is simply related to the correlation coefficient.

When a general pedigree like the one in Figure 8.27 is considered, the expected age difference between husband and wife is

$$\begin{aligned} M &= (m_j\tau_m + (f_j + 1)\tau_f) - ((m_i + 1)\tau_m + f_i\tau_p) \\ &= (m_j - m_i - 1)\tau_m + (f_j - f_i + 1)\tau_f, \end{aligned} \tag{8.45}$$

with variance

$$S^2 = \sigma_s^2 + \sigma_d^2 + (m_i + m_j)\sigma_m^2 + (f_i + f_j)\sigma_f^2, \tag{8.46}$$

where m_i is the number of male and f_i the number of female intermediate ancestors in the branch leading to the husband ($m_i + f_i = i$), and m_j and f_j the corresponding numbers in the branch leading to the wife. In the actual pedigree of Figure 8.27,

$$m_i = 2,\ m_j = 1,\ f_i = 1,\ f_j = 1.$$

The other quantities in Equations 8.45 and 8.46 are the following demographic parameters: τ_m = mean age at reproduction in males; σ_m^2 = variance in age at reproduction in males; τ_f = mean age at reproduction in females; σ_f^2 = variance in age at reproduction in females; σ_s^2 = variance of the difference in age between sibs; σ_d^2 = variance of the difference in age between husband and wife.

If consanguineous marriages tend to have the same correlation with respect to age as nonconsanguineous marriages, a correction factor can be computed for the probability of a given type of marriage. This factor, when multiplied by the estimate of the probability of a given type of consanguinity obtained from the ratio between the number of relatives and the populations size, gives an estimate that takes into account the correlation between mates for age at marriage. The correction factor can be shown to be

$$I_a = \frac{W}{S\sqrt{2\pi}} e^{-M^2/2S^2}, \tag{8.47}$$

where M and S^2 are given by Equations 8.45 and 8.46 and W is the reciprocal of the average relative frequency of individuals of given age (in years, if this is the unit in which the demographic parameters are expressed), in the reproductive age group (Cavalli-Sforza et al., 1966; see also Hajnal, 1963).

On an average the age correlation makes first-cousin marriages about twice as frequent ($\bar{I}_a \sim 2.0$), marriages between second cousins about 1.7 times, and third cousins 1.4 times as frequent as if there were no correlation between the ages at marriage. Cousins of uneven degree ($i + j$ odd) have I_a less than one, but with increasing remoteness I_a approaches one as in the case of the even cousins.

Dispersal of relatives is another important factor affecting the frequency of consanguineous marriages.

Dispersal of relatives takes place with emigration. It accumulates over generations, and therefore the more remote the consanguinity, the more powerful is dispersal in preventing consanguineous marriage. The functions describing migration that we have already discussed can be used to predict the range within which consanguineous individuals of various degrees can be found. Figure 8.28 gives some idea of the results of such a theoretical prediction, based on approximating the migration distribution by the sum of two exponentials.

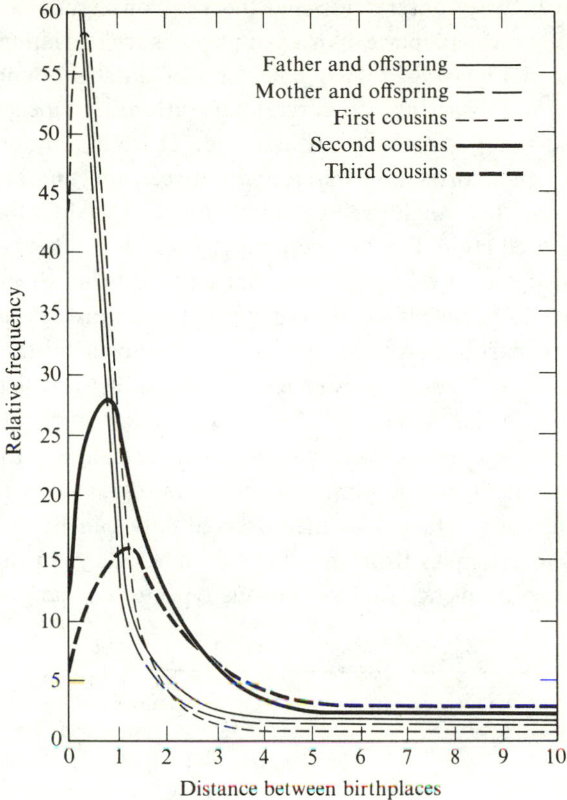

FIGURE 8.28
Theoretical distribution of the distance between birthplaces of various types of relatives. (From Cavalli-Sforza et al., 1966.)

From such distributions, and that of the probability of marriage as a function of distance between birthplaces of husband and wife, the probability of a mating between two consanguineous individuals of given degree of relationship can be obtained. The overall probability of consanguineous matings increases with remoteness, because of the increase in the number of consanguineous individuals, and decreases with remoteness because of dispersal due to migration. Considering even-degree cousins, for which the age effect is small, we find that marriages between second cousins are more frequent than those between either first or third cousins in areas with low dispersal (such as the upper Parma Valley); first-cousin marriages are usually more frequent in areas with high dispersal.

The computation, using theoretical migration functions, of the expected frequencies of consanguineous matings by degree of relationship, gives fairly good predictions. There are, however, several reasons that make it difficult to obtain precise agreement between observation and theory. Consanguinity frequencies are highly variable in time and place. An example of such variation in the Parma Diocese is shown in Figure 8.29. Any collection of data is bound to be heterogeneous. In predicting dispersal over several generations, independence in the migration over successive generations was assumed. There is only one investigation of this point in which a correlation was found between the migratory behavior of parents and children. This might well be largely due to social factors, but it nevertheless affects the prediction. Finally, there may be social ties between consanguineous people, even if they are living apart, that tend to bring them together more often than they would be by chance, and that might therefore affect the probability of consanguineous marriage. An example of this is shown by the Brazilian data analyzed by Morton and co-workers (Azevedo et al., 1969), which, however, are likely to be extreme in this respect.

In view of the difficulties we have mentioned in calculating the expected frequencies of the various types of consanguineous matings, it is perhaps surprising that fairly good fits of the theory to the observed data can be obtained. We will discuss, briefly, two examples from the upper Parma Valley. In the first example, the expected relative frequencies of the various types of consanguineous matings,

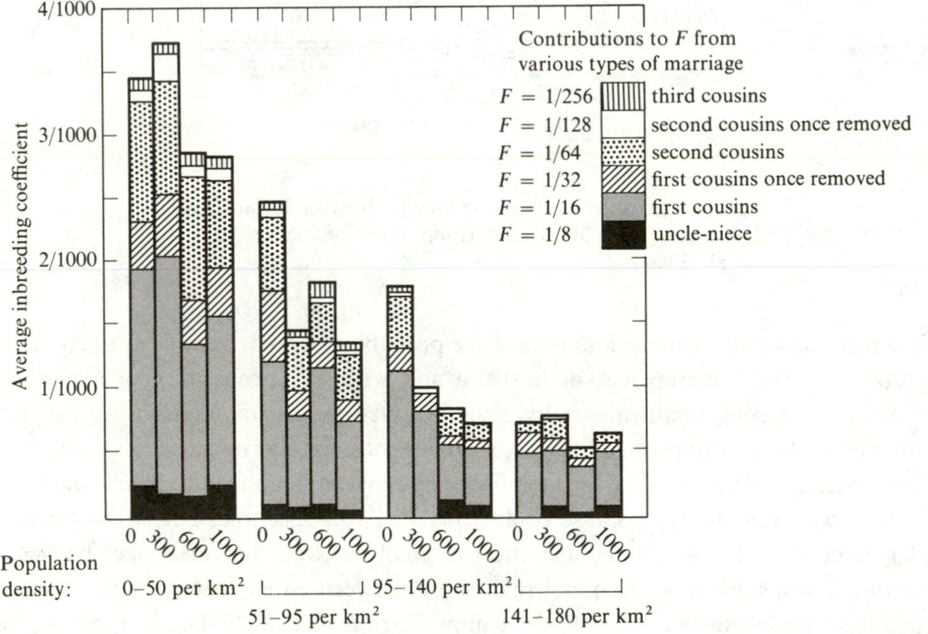

FIGURE 8.29
Average inbreeding coefficient α from Roman Catholic records of consanguineous marriages in the Parma diocese, 1851–1950. Within a given density class, village size is relatively unimportant in determining the α value. Population density, however, is correlated with a five-fold change in inbreeding levels over an otherwise fairly homogeneous area. (From Cavalli-Sforza, 1958.)

taking into account the distribution of the numbers of male and female intermediate ancestors in the pedigree branches leading to husband and wife, were calculated as the products of three quantities: (1) the expected proportion of relatives of the given degree of relationship (calculated as described on page 466, see also Equation 8.44); (2) the age correction factor I_a, as given by Equation 8.47. (3) a factor for the dispersal effect, that was computed using a simplified approach to migration. The relative proportions of different types of consanguineous marriages of given degree (apart from the age factor I_a) were predicted from

$$p_m^{(m_i+m_j)}\, p_f^{(f_i+f_j)},$$

where p_m and p_f are the probabilities that an individual does not emigrate from the isolate—or, more generally, from the "mating range"—for males and females, respectively. The demographic parameters needed for the calculation of I_a ($\sigma^2{}_m \tau_m$, etc.), were estimated by fitting the expected frequencies of the various types of consanguineous matings to their observed values. The results of the overall calculations, as given in Table 8.12, show a reasonable agreement of observation with theory even though the overall chi square for goodness-of-fit is significant. This indicates that the demographic factors taken into account by the three quantities used to calculate the expected frequencies provide an adequate description of the mating patterns in the population. In Table 8.13 the estimates of the demographic parameters obtained from fitting the observed data on the frequency of consanguineous matings to the theory are compared with values obtained from a sample of unrelated married couples living in the same area today. If the actual present-day demographic values are inserted in the equations for computing the expectations, the agreement is not particularly good. The goodness-of-fit seems to be very sensitive even to small variations in these parameters. The values computed from the present-day population cannot, however, be precisely valid for the period for which consanguineous marriages were enumerated. Demographic data over the last 150 years would be needed for this. The differences between the parameters estimated from fitting the consanguineous marriage frequencies and those found by a survey of the present-day sample are, however, small (see Table 8.13).

It is worth considering in more detail the need to use separate parameters for the migration of males and females. We have already commented on the fact that the population in the upper Parma Valley like most human populations, is patrilocal—that is, $p_m > p_f$. In terms of consanguineous marriage frequencies, this fact shows up in the relative frequencies of pedigrees that have different proportions of males and females as the intermediate ancestors. This is seen particularly clearly in the comparison of marriages, such as those between second cousins or third cousins, that have the same expected age difference and to which the same age correction factor applies (see Figure 8.30). Such marriages show regularly decreasing frequencies with increasing proportions of female intermediate ancestors, in agreement with observation (Barrai, Cavalli, and Moroni, 1962).

TABLE 8.12

Observed and Expected Frequencies of Consanguineous Marriages, Upper Parma Valley 1850-1950

Marriage	Number of Male Ancestors in Branch Leading to		Absolute Frequencies	
	Husband	*Wife*	*Observed*	*Expected*[a]
Uncle-niece	–	0	5	6.47
	–	1	4	2.08
Aunt-nephew	0	–	1	0.23
	1	–	0	0.08
First cousins	0	0	109	110.11
	1	0	99	99.55
	0	1	157	142.12
	1	1	96	153.37
1½ cousins: Husband in shorter branch	0	0	21	11.71
	1	0	30	28.67
	0	1	20	11.32
	1	1	55	39.78
	0	2	5	4.11
	1	2	22	12.87
1½ cousins: Wife in shorter branch	0	0	3	2.11
	0	1	14	7.18
	1	0	2	2.06
	1	1	20	8.23
	2	0	2	0.56
	2	1	4	2.25
Second cousins	0	0	41	43.64
	1	0	63	87.57
	2	0	47	10.01
	0	1	93	109.83
	1	1	227	247.03
	2	1	129	125.91
	0	2	60	61.76
	1	2	171	151.91
	2	2	107	87.67
Overall $\chi^2_{[20]} = 88.26$, for goodness-of-fit of expected to observed frequencies	Total frequency		1607	

Source: From Cavalli-Sforza et al. (1964).

The expected frequencies were calculated as described in the text.

The probabilities p_m and p_f, of individuals remaining in the mating range, need to be further qualified. There is no finite mating range, but rather a probability distribution of the distances between birthplaces (or residences) of husband and wife. The estimates obtained from fitting a continuous curve (Table 8.13) correspond fairly well, however, to those obtained by considering the mating range in the upper Parma Valley as the village.

TABLE 8.13
Parameters for the Theory on Frequency of Consanguineous Marriages

	Estimations from Consanguineous Marriages Parma Valley 1850-1950	*Estimations from Demographic Data Obtained from a Sample of Present-day Population Living in Same Area*
Generation time		
Males		
Mean $= \tau_m$	36.10	33.24 ± 0.19
Variance $= \sigma_m^2$	53.32	46.60
Females		
Mean $= \tau_f$	30.95	28.73 ± 0.16
Variance $= \sigma_f^2$	42.59	33.78
Variance of age difference		
Between sibs	53.08	25.78 } 59.09
Between mates		33.31 }
Probability of non-migration per generation		
Males $= p_m$	0.845	0.821 ± 0.019
Females $= p_f$	0.696	0.727 ± 0.070

Source: From Cavalli-Sforza et al. (1964).

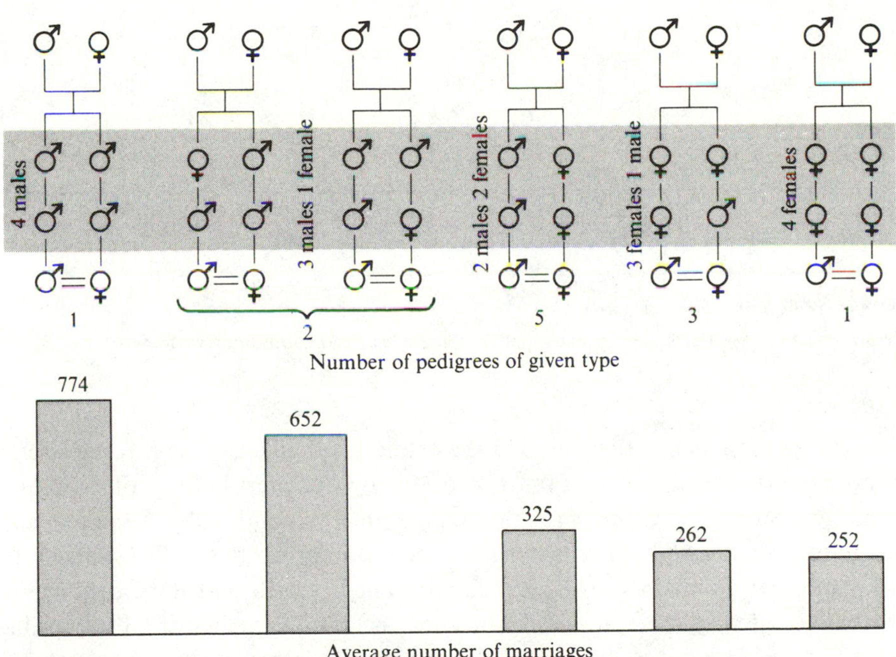

FIGURE 8.30
Frequencies of pedigrees of second cousins, combining data from 1850 to 1951, in the Parma, Reggio Emilia, and Piacenza dioceses, showing the effect of sex of common ancestors on the frequency of the type of pedigree. (From Cavalli-Sforza, 1969.)

A second independent attempt to compute the expected frequencies of consanguineous marriage on the basis of age, migration, and other demographic factors was based on Monte Carlo simulation—that is, counting the consanguineous marriages in the artificial population that had been set up for the analysis of drift. Dispersal was introduced in the form of migration matrices, which is probably the most satisfactory approach even though it does not take account of the apparent "inheritance" of migration habits. The fit to the observed frequencies is not bad, apart from an excess of marriages of uneven degree in the artificial population (see Table 8.14). This is most probably the consequence of the use of an excessive

TABLE 8.14
Consanguinity among 1000 Marriages from an Artificial Population

Multiplicity of Consanguinity in the Marriage[a]	*Uncle–niece*	*First Cousins*	*1½ Cousins*	*First, Twice Removed*	*Second Cousins*	*2½ Cousins*	*Third Cousins*	*Total*
Single								
Full	1	16	17	2	34	36	75	
Half	0	0	4	2	2	5	8	
Total	1	16	21	4	36	41	83	202
Multiple								
Full	1	20	10	0	40	46	133	
Half	0	2	7	0	5	19	15	
Total	1	22	17	0	45	65	148	298
Sum	2	38	38	4	81	106	231	500
Artificial population	0.2	3.8	3.8	0.4	8.1	10.6	23.1	
Real population	0.074	3.41	1.46	0	6.94	—	—	

Source: From Cavalli-Sforza and Zei (1967).

[a] Multiple consanguinities were counted as many times as the single consanguinities they contained.

variance for male generation times in the artificial population. As this parameter cannot be varied at will, but is a function of other parameters in the artificial population, and since time is only roughly represented (with decades as units) and computer costs are high, no attempt was made to improve the fit. It is difficult to make an exact comparison with expectation, but it seems that third cousins are underrepresented in the real population by a factor of 2 or higher. It is possible that some third-cousin marriages may have escaped detection. In view of the accuracy with which Roman Catholic records are kept, this may seem strange, but there may be real difficulties, in practice, in establishing the exact degree of relationship when it is so remote, and certainly some are therefore likely to have been missed.

8.19 Isonymy

Crow and Mange (1965) have suggested an interesting approach to the computation of the mean kinship coefficient in a population that makes use of the frequency of identical surnames, or isonymy. The method is approximate and depends on assumptions that are not always easily tested, disadvantages that are, however, to some extent outweighed by the simplicity of the approach and the opportunity it provides to probe for remote consanguinity to a greater degree than by any other pedigree method.

Suppose there is a given frequency c of first-cousin marriages. One quarter of the couples on the average have identical surnames before marriage because of the inheritance of their grandfather's surname through two sibs, while three-quarters have different surnames. The contribution of first-cousin marriages to the average inbreeding coefficient is their frequency times their F value, which is $\frac{1}{16}c$. If we want to estimate the contribution to F of all first-cousin marriages, on the basis of the frequency ($c/4$) of marriages of first cousins who have the same surnames, we must multiply by four, because only one in four first cousins have the same surname, and then by the F value ($\frac{1}{16}$), or equivalently multiply directly the frequency of marriages of first cousins with the same surnames by $\frac{1}{4}$. Second cousins have the same surnames $\frac{1}{16}$ of the time, and their F value is $\frac{1}{64}$. Thus, taking the frequency of marriages of second cousins with same surnames and multiplying it by $\frac{1}{4}$, we again obtain the contribution of all second-cousin marriages to F. In general, the probability that relatives of any degree (odd or even) have the same surname because of inheritance from a common male ancestor, is always four times the inbreeding coefficient of the particular type of mating (see Table 8.15). This neglects some pedigrees that are rare.

If we assume that all individuals who have identical surnames have them because of common ancestry, then the frequency of isonymous pairs (pairs of individuals with identical surname) divided by four gives directly the F ($=f$) value of the population.

The inbreeding coefficient thus computed goes back as many generations as do the surnames. As usual, kinship and inbreeding coefficients will be identical so long as we neglect mutation, which makes genes that are identical by descent, different by nature.

There are two ways of estimating the mean inbreeding or mean kinship coefficients of the population using identical surnames of couples. We can determine the frequency of isonymous pairs at random from the population, or we can determine it from actually mated pairs. It is likely that, in most cases, the discrepancy between frequencies obtained by these two methods will be small or insignificant.

If the difference is significant, we may partition the total observed kinship coefficient f (obtained from isonymous mates) into a fraction due to random mating f_r and a fraction due to nonrandom mating f_n. The latter may be positive or negative, depending on whether there is a tendency for positive or negative assortative mating with respect to surnames (and hence, presumably, for relationship). The value f_n will differ from zero if people mate preferentially (or avoid mating) with others of the same or related surnames.

The relation between the total inbreeding coefficient f, and its parts f_n and f_r is always of the form

$$1 - f = (1 - f_n)(1 - f_r). \tag{8.48}$$

We can compute f_r by taking all possible pairs of individuals and testing how many are isonymous. The analysis should be designed to eliminate identical surname pairs that are brothers and sisters, and further should limit pairs by age to include only those in the reproductive period. Approximately, but more simply, we can take the frequency of each surname in the population as being q_i and compute $\sum q_i^2$, summing over all surnames. If there are significant differences between the frequency Q_i of a given surname in males and the frequency Q_i in females, it is better to compute $\sum p_i Q_i$.

When this was done by Crow and Mange (1965) for the Hutterites, the random expectation of identical surnames was $\sum Q_i p_i = 79.55/446 = 0.178$. The number of pairs having identical surnames among 446 pairs was 87, which is not significantly different from the random expectation of 79.55. The nonrandom value f_n is, therefore, negligible. Computing f from the random isonymy, we have $f = 0.178/4 = 0.0445$.

Consanguinity between marriage partners whose ancestry was known could be identified. Consanguinity up to and including fourth cousins was thus determined. Direct computation of the average inbreeding coefficient from this data gave $F = 0.0226$, which is about half of the total kinship or inbreeding coefficient evaluated from isonymy. Thus, if identical surnames actually represent common ancestry among the Hutterites, we can estimate that using data on relationships up to and including fourth cousins gives, in this population, total kinship underestimated by a factor of $\frac{1}{2}$. Using data on relationships only up to second cousins the inbreeding coefficient is underestimated by a factor of $\frac{1}{4}$, as compared to the value obtained from isonomy.

LIMITATIONS TO THE METHOD OF ISONYMY

Because the method of isonymy is simple, we should evaluate closely the possible sources of error. In Table 8.15 are given the probabilities of isonymy computed on the simple assumption that pairs of individuals with a relationship of given degree

TABLE 8.15
A Comparison of Observed and Expected Frequencies of Isonymy

Degree of Relationship	Probability of Isonymy (p)	Inbreeding Coefficient		Certain Isonymy Observed[a]
		F	F/p	
Uncle–niece: Aunt–nephew	1/2 = 0.5	1/8	1/4	103/115 = 0.896
First cousins	1/4 = 0.25	1/16	1/4	997/4384 = 0.223
$1\frac{1}{2}$ cousins	1/8 = 0.125	1/32	1/4	247/1653 = 0.149
Second cousins	1/16 = 0.0625	1/64	1/4	774/5764 = 0.134
$2\frac{1}{2}$ cousins	1/32 = 0.031	1/128	1/4	127/1547 = 0.082
Third cousins	1/64 = 0.016	1/258	1/4	167/2184 = 0.076

[a] This column gives the observed frequency of pedigrees in which mates must have the same surname because all common ancestors of the mates are males (based on data from Northern Italy).

have the same probability of marrying, whatever the pattern of male and female ancestry, which determines whether there is isonymy. Also given are the observed frequencies of consanguineous marriages of given degree of relationship showing isonymous mates, based on a large sample of consanguineous marriages collected in three dioceses of Northern Italy. Almost all the expected frequencies are substantially different from the observed frequencies for the reasons that we have already mentioned: differences according to sex dispersal of relatives, difference of age at marriage, and other factors having social causes. It might be argued that mates having identical surnames are more easily identified as consanguineous and, therefore, may be overrepresented in the sample. On the other hand, if this were true, the effect would pertain only to matings in which all common ancestors are males. This is not true, as can be seen from Figure 8.30, which shows that, for any given degree of consanguinity, an increase in the number of females among the ancestry intermediate between the common ancestors and the consanguineous mates is correlated with a decrease in the observed frequency of the pedigree. This correlation corresponds exactly to that expected on the basis of patrilocal migration.

Another important source of error is that there may have been duplication of surnames at the time the names were first introduced. In Europe, surnames arose at various periods, but most arose during the late Middle Ages. They were mostly job names, place names, fathers' names, or nicknames. A certain amount of duplication must have been unavoidable. If the initial duplication were known, its contribution could be removed by an equation corresponding to that for the analysis of f into its components (see Equation 8.48). It is difficult to evaluate the importance of this source of uncertainty. Data from the Parma Valley (Figure 8.31) indicate a rise in isonymy at the beginning of records, and then perhaps a slow decline.

Surnames are inherited, like Y chromosomes, only through the male line, at least in most societies. Thus, they indicate only the kinship deriving from the father. Insofar as the variance of progeny size may be somewhat higher for male than for

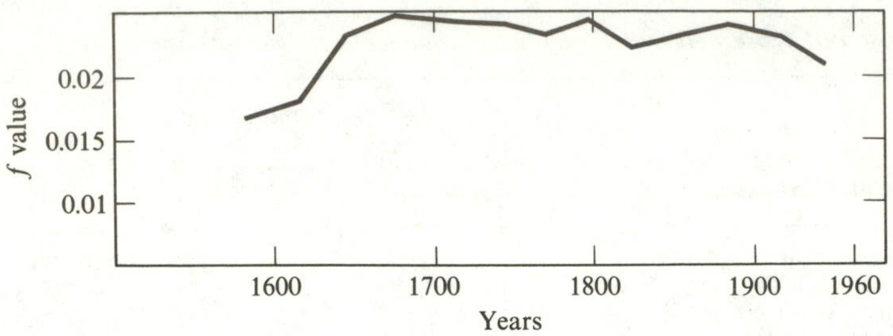

FIGURE 8.31
Observed f value from isonymy data in the upper Parma Valley as a function of time.

female parents (at least in primitive and industrially underdeveloped societies), the estimate of f from isonymy may also be slightly exaggerated for this reason.

Assigning a family's name to an adopted child and giving any name other than the father's to an illegitimate child will increase an estimate of the f value obtained by isonymy. Changing surnames will, however, decrease such an estimate. It seems, on the whole, that isonymy is likely to overestimate the actual amount of kinship by a factor that is difficult to assess.

8.20 A Comparison and Summary of Methods of Measuring the Amount of Kinship and Ascertaining the Effect of Drift

There are, as we have seen, various independent methods of computing kinship coefficients in human populations. They can be usefully grouped as follows.

1. *Methods That Measure Directly the Variation with Respect to Genetic Markers, or Phenotypes That Can Be Distinguished Quantitatively.* Genetic markers are, in general, totally unresponsive to all sorts of environmental forces, in contrast to most common quantitative phenotypic measurements (see Chapter 9). Only genetic markers behave in complete accord with the simple expectations of evolutionary theory, and so we have limited our discussion in this chapter to variation of them.

This first group of methods is based on measuring, directly, gene frequency variances and covariances. They probe into the past as deeply as possible, but are limited to the particular loci or alleles being studied. Averaging over various genes is meaningful only if all show homogeneous variation. If this is not the case, then the interpretation of the average variation may be made difficult by our ignorance of how representative of the whole genome are the genes used in the survey.

2. *Methods That Measure Observed Relationships, Through Pedigrees of Random or Selected Individuals or Through Isonymy.* The extent of the analysis is usually limited to three or four generations for pedigrees, but may cover fifteen generations for surnames. Methods using pedigrees suffer from the problem of truncation, which inevitably reduces the estimated level of inbreeding. Methods using surnames suffer from several uncertainties, as already mentioned. The kinship coefficients observed using these methods are comparable to the mean kinship coefficients obtained from the first group of methods only if the former show homogeneous variation with respect to the various loci.

3. *Estimation of Expected f Values (Kinship Coefficients or Gene-frequency Variances), Using Demographic Data on Migration and Population Sizes, Together with Suitable Models for Predicting Variation.* Of various possible approaches, Monte Carlo simulation is perhaps the most powerful, but is also the most expensive if precision, and, therefore, much repetition, is required. A precise result can be more easily attained by using any of a variety of analytical approaches. Migration matrices give the same result as the Monte Carlo method in a more compact form and less expensively, but can less easily accommodate other parameters whose effects we may want to test (such as variance of progeny size, generation overlap, and a variety of specific selection models). Alternatively, we can resort to other theoretical models, mostly of isolation by distance. These have the disadvantage of being rather inflexible. The parameter that these models are chiefly designed to measure—that is, the coefficient of recall to equilibrium—seems difficult to resolve into its constituent parts of long-range migration, balancing selection, and mutation, and is thus of doubtful value. Perhaps the only exception is the ABO groups, for which the existence of some balance is likely on the basis of other independent considerations, though the exact amount is still highly questionable. We may, finally, mention one measure, to be further discussed later: isolate size in Dahlberg's sense, which occupies a position somewhat all its own, and which, when used alone, is probably the least useful of these methods.

Several complementary approaches to measuring kinship coefficients and ascertaining the effect of drift should be tried whenever possible.

The data on the Parma Valley seem, at present, to be the best relevant example, even though their analysis is unfortunately not yet completed. In particular, the migration matrix presently available is obtained only from marriages, and thus does not allow a full analysis, which it is hoped will be achieved by completing the record linkage of baptisms, deaths, and marriages from parish books. All the f values obtained so far by various different methods are given in Table 8.16. The real f value of the population is probably between 0.01 and 0.03, and more likely

TABLE 8.16
Computation of f Values by Various Methods Based on the Population from the Upper Parma Valley; f_k =values from kinship; f_p values from gene-frequency variation

Method of Computation	f_k	f_p
Consanguineous marriages, real population (1850–1950, including second cousins)	0.00396	—
Consanguineous marriages, computer simulation (up to and including second cousins)	0.00492	—
Same, plus 2½ and 3rd cousins	0.00670	—
Frequency of identical surnames (1930–1950 marriages)	0.0158	—
Computer simulation, variation of gene frequencies	—	0.01085
Migration matrix, expected variation of gene frequencies	—	0.01160
Observed variation of gene frequencies	—	0.0356

Source: From Cavalli-Sforza (1969a).

nearer the former. The observed inbreeding due to marriage between close relatives (up to second cousins) is about 80 percent of what it would be if there were no attempt to avoid consanguineous marriages (0.00396/0.00492). The true value of f is underestimated by a factor of about $\frac{1}{3}$ if only observed marriages between close relatives are used. The discrepancy between the higher and more satisfactory estimates of f values needs further study, but does not seem, in any case, to be large, considering all the possible sources of errors.

In the few cases in which microgeographic variation has been analyzed in sufficient detail, the observed level of variation has been found to be fairly close to that predicted on the basis of random genetic drift alone. Observed differences seem to be well within the error of the available methods, and it is hard to believe that strong evidence for balancing selection can be obtained in this way. In all cases in which the microgeographic variation has been tested at different loci, homogeneity has been found. In all cases in which prediction of the variation due to drift could be made by the methods of group 3 and compared with observed variation, the prediction was correct within a factor of 3. Given all the uncertainties of the assumptions, and the high standard errors of variances, the conclusion seems to be that, at the microgeographic level, no clear-cut evidence of balancing selection has yet been found, and no clear-cut evidence of heterogeneity between loci has been found. By and large, estimates of kinship coefficients from pedigrees tend to be comparable to those from actual gene-frequency variation. Thus, most observed microgeographic variation seems to be the consequence of drift operating at or near the level expected on the basis of predictions made using demographic data.

8.21 The Problem of Isolate Size

The solution of problems of population structure may be more accessible in man than other organisms because of the easier availability of demographic data. Real situations show, however, a wealth of complicating factors that must be taken into account. Overlapping generations are one major source of disturbance. The necessity for examining several populations to compute variances and covariances, and the inevitable introduction of heterogeneity are other major complications.

A central problem in the analysis of population structure is the determination of N, the effective population size. The estimates of the mean kinship coefficient and the variation of gene frequencies depend on this very important parameter. The answer to the most important question of the relative importance of selection and drift also depends on the magnitude of selection coefficients in relation to N. It should be emphasized that, to some extent, N is arbitrary, and depends on the nature of the groups being considered, as will be discussed below.

When analyzing geographic variation between groups, be it for estimation of gene frequencies or kinship coefficient, the unit used for sampling purposes determines the value of N to be used in the analysis. The decrease in effective population size resulting from internal subdivision of a population is sufficiently small, that, in general, it can be neglected. For purposes of comparing the observed variation between groups with that expected under drift alone, the groups can be chosen fairly arbitrarily, The value of N, or any equivalent measure, used for an analysis, is therefore determined by the unit used for sampling. This could be a village, a group of villages, or even a country.

The question arises of whether a special meaning should be attached to Dahlberg's method of estimating isolate size. It will be recalled that Dahlberg (1948) suggested measuring isolate size on the basis of the observed frequency of consanguineous marriages. Under the assumption that f_c, the frequency of first cousins, is expected to be $8/N$ (see Section 8.17), N can be estimated by $8/f_c$. Similar formulas exist for other types of consanguineous marriages. It appears that here isolate size, meaning the effective breeding group, is an independent concept, which escapes the arbitrariness of the choice of the sampling group, already mentioned. In reality, however, there is no clear-cut breeding group. The breeding group should perhaps be viewed as the aggregate of all those groups from which a mate could be chosen by a given individual, each weighted by the probability that the mate will actually come from it. If N_i is the size of each group and p_i the probability of marriage in that group, $\sum p_i N_i$ represents such an estimate. This formulation can easily be transformed to the continuous model. The breeding group thus defined may be useful for understanding social restrictions on marriage. Nevertheless, there remains the difficulty that the frequency of cousin marriages depends, as we have seen, on a number of other factors. Even an oversimplified computation should take account at least of the following factors:

1. The number c of relatives who marry is a function of the growth rate of the population and of the distribution of progeny size. It is, as shown at the beginning of Section 8.18, given by $c = 2^i p^j s + 2^j p^i s$, where p is the average progeny per marriage and s the average number of sibs, which depends on p as shown in Example 8.8. Only if we assume a Poisson distribution and a stable population so that $s = p = 2$, is it correct to assume for cousins of even degree d that $c = 2^{2d+1}$, and thus that c is 8 for first cousins.

2. The age correction factor I_a, which corrects for the correlation in ages between mates, (see Section 8.18 and Equation 8.47), varies with the degree of relationship.

3. The dispersal of relatives due to migration, apart from its dependence on sex, also varies with the degree of relationship. In the simplest model, the dispersal may be taken as P^{2d}, where P is the emigration out of the mating range per generation, assumed to be equal for males and females. At this simple level of approximation, the expected frequency of first cousins is

$$f_1 = \frac{2spI_1P^2}{N} = \frac{8I_1P^2}{N} \tag{8.49}$$

if $s = p = 2$, and that of second cousins

$$f_2 = \frac{4sp^2I_2P^4}{N} = \frac{32I_2P^4}{N} \tag{8.50}$$

if $s = p = 2$.

Here I_1 and I_2 are the relevant age-correction factors. From these two frequencies P and N could be estimated. Thus, from Equations 8.49 and 8.50,

$$P = \frac{1}{2}\frac{f_2 I_1}{f_1 I_2},$$

and thus, for the upper Parma Valley, where $f_1 = 0.0341$ and $f_2 = 0.0694$ and with $I_1 = 2.0$ and $I_2 = 1.7$, we obtain $P = 0.77$, which corresponds fairly well to the average of the estimates of mothers and of fathers born in the same village as their offspring (see Table 8.7). The isolate size N is then estimated from Equation 8.49 by $N = 8I_1P^2/f_1 = 278$, which is somewhat less than, but not far from, the actual mean village size.

It is doubtful however, whether these formulas are reasonable for every situation. The frequencies of first- and second-cousin marriages are the most widely available statistics. However, if only these two frequencies are used to estimate N and P, no degree of freedom is left for testing the internal consistency of the data, and thus one cannot test for other possible sources of departure from randomness.

8.22 Gene Diffusion

Patterns of gene distribution due to gene diffusion arise in a variety of ways. They are especially evident from inspection of maps of the geographic distribution of gene frequencies. Usually, these show complex patterns that are not readily interpreted. Sometimes, however, the patterns can be readily interpreted in simple ecological terms. This is the case, for instance, for hemoglobin S, which is now known to confer relative resistance to malaria infections due to *Plasmodium falciparum* (see Chapter 4). The map of the distribution of the *AS* genotype tends to correspond, with a few exceptions, to that of the infections due to this parasite.

It was the correlation between the geographic distributions of the thalassemia gene and malaria that led Haldane (1949a) to suggest the association between some genetic traits and malaria. Attempts have also been made to correlate, in a similar way, the geographic spread of infectious diseases such as smallpox, plague, and syphilis with the ABO blood-group distributions. The results, however, are controversial (see Chapter 5).

The geographic distributions of some traits, such as skin color, and also of various anthropometric features such as body-fat distribution, size and shape of nose openings, lengths of arms and legs, have been interpreted as adaptive responses to climatological factors. These traits are, however, genetically complex and will not be further discussed here.

Diffusion centers can sometimes be inferred to be in the middle of relatively concentric geographic distributions.

Some types of contemporary geographic distributions in a reasonably homogeneous environment may suggest that a new mutant gene has been caught in the act of diffusing. There are two main ways in which a new advantageous mutant type may arrive in an area in which it was formerly unknown: by fresh mutation and by immigration of an individual or a group carrying the mutant gene.

The spread of an advantageous gene from a single mutation results, in a sufficiently homogeneous environment, in a gene-frequency distribution having concentric "growing rings" with the highest gene frequency at the diffusion center, which is presumably the geographic origin of the mutation. The same is true of a gene brought in by a single immigrant individual or by a small group.

There are several examples of geographic distributions that suggest "a diffusion center" for an advantageous gene. We have already discussed the fact that hemoglobin C may have such a distribution and, that it is likely that the gene that specifies hemoglobin C is on the increase (see Chapter 4). In fact, its geographic distribution itself suggests that it is increasing. On the other hand, we have also seen that the available data do not settle the problem entirely. It remains difficult, however, to imagine that the distribution of the hemoglobin C gene corresponds to a residual pocket of a gene doomed to extinction. Later in this section we shall give a model of migration and selection for this gene. Other red-cell defects connected with malaria that have a restricted distribution may represent similar examples—namely, the "high *F*" gene in Africa and, perhaps, the gene for hemoglobin E in Southeast Asia. The existence of a diffusion center clearly suggests that the mutant gene has arisen and propagated successfully only once, namely, at the center of the diffusion rings.

With genetic traits that have a wider distribution, such as G6PD, thalassemia, or sickle-cell anemia, it is more difficult to accept the idea that the pertinent mutant gene had a single origin. Molecular analysis, in fact, disproves this hypothesis directly for G6PD and thalassemia. A large number of different G6PD molecules are known, and each must have had an independent origin by mutation. No such evidence is available for the S hemoglobin gene, which might indeed have had a single origin. Its geographic distribution in widely different areas (Central and South Africa, Southern Europe, India) is not incompatible with the idea of a single origin together with the assumption that migration was the cause of its spread. The distribution does, however, suggest the possibility of independent recurrence of the same mutation at perhaps three different sites. Thalassemia, of which at least two, and probably more different types exist, must have had several independent origins, some of which gave rise to recognizably different variants of the disease.

Foci of genetic disease provide other instances of diffusion centers. An interesting case of putative concentric diffusion is that of light pigmentation of hair and eyes. The center of this trait is somewhere in South Sweden near the Baltic, and from there the trait, which is inherited in a complex way, seems to have spread almost concentrically, even across the seas. Its selective advantage might have been due to sexual selection, or to some unknown physiological effect.

A model of diffusion for hemoglobins C and S may describe fairly accurately the present distribution of these genes in West Africa.

A rough model for the diffusion of the hemoglobin C and S genes was constructed using a two-dimensional stepping-stone model. The population unit in the model was taken to be the sort of tribe thought to have been the basic population cluster at the time of the beginning of agriculture in Africa: a tribe of 5000 people occupying an area of 10^4 square kilometers. The corresponding population density is 0.5 per square kilometer. A single mutation in one tribe therefore produces a mutant gene that has an initial frequency of 10^{-4}. A deterministic model of diffusion by migration was set up using a square lattice of tribes (as in a two-dimensional stepping-stone model) corresponding approximately to the surface of West

Africa, and a single mutant gene was placed at time zero in the tribe in the center. It was then allowed to increase, having the fitnesses given in the legend to Figure 8.32. The choice of model and parameters is based on our discussion in Chapter 4, where we suggested that gene C is increasing in the presence of gene S. It should be noticed that the interaction of selection with migration puts the gene C at a slight disadvantage if we start with just one C mutant gene in the middle of a population saturated with S genes. A somewhat higher selective value for the genotype AC is thus necessary for the initial increase to take place, as compared with that needed when only a single population cluster is considered and no migration takes place. In fact, migration competes with selection if we start with an initially low frequency of an advantageous gene at one point, and zero frequency for it all around. An exact treatment would have to be stochastic, and this might reduce the importance of such competition between selection and migration. A deterministic treatment such as we have given here is, therefore, very approximate as long as gene frequencies are low. Because of this, and other gaps in our knowledge, the simulation model we have used is a rough approximation, and the results obtained from it can only be considered as suggestive.

The expectations under the model are summarized in Figure 8.32. The C gene, under these conditions, would eventually supplant both the A and S genes. The changes are rapid, if this model is correct. The C gene frequency increases, at present, at a rate of 0.7 percent per generation in the area of maximum change, namely, near the origin. The migration rate needed to approximate the present distribution is fairly high. Present migration-rate estimates are somewhat lower, but they can hardly be compared with those relevant to earlier times. As in almost all historical reconstructions, there are a number of assumptions that have to be made and that cannot readily be checked. The only conclusion that can be reached is that the suggested process does not meet with any major contradictions.

Advantageous mutations may show a constant "wave of advance."

An analytical treatment of the spread of a favored mutant gene was given by Fisher (1937), who considered a linear habitat. If a mutant gene arises at one point and has a relatively high selective value, it will increase locally and also spread. Its advance can be compared to that of a wave, and it can be shown that the rate of advance per generation is constant (in steady-state conditions) and proportional to

$$v = \sigma\sqrt{2s}, \tag{8.51}$$

where σ is the standard deviation of the scattering due to diffusion (the usual measure of migration) and s is the selective advantage (or, equally, $1 + s$ is the fitness of the heterozygote). No dominance is assumed. The length of the wave of

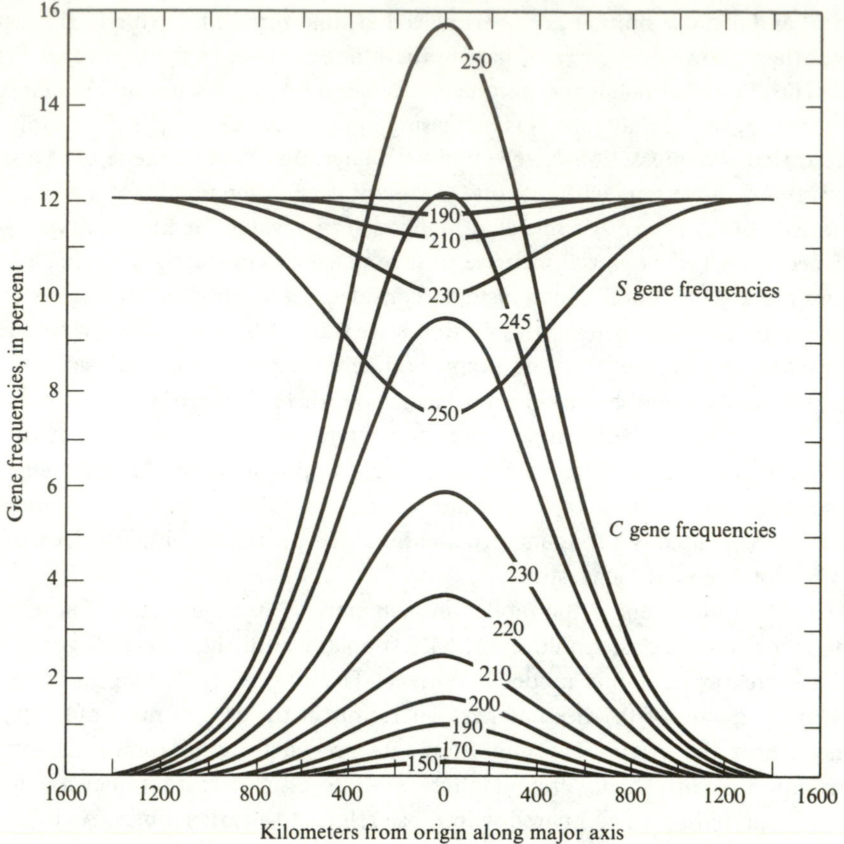

FIGURE 8.32
A model of migration and selection for hemoglobins S and C in West Africa. It is assumed that at time 0, gene *S* is in equilibrium at a frequency of 12.1 per cent and *C* has arisen by a single mutation in a population of 5000 persons located at the origin. Fitness values are: $\overline{AA} = 0.89$; $\overline{AS} = 1$; $\overline{AC} = 0.98$; $\overline{SS} = 0.2$; $\overline{SC} = 0.7$; $\overline{CC} = 1.07$. Migration per generation is 8 percent along the major axis and 4 per cent along the minor axis. Populations are considered to consist of 5000 persons and to be spaced at 100-kilometer intervals in a two-dimensional stepping-stone model. Numbers on curves indicate generations. For simplicity only the situation along the major axis is plotted.

advance is defined in units of $\sigma\,(\frac{1}{2}s)^{1/2}$ so that one unit is, for example, the distance between the position at which the gene frequency is 50 percent and that at which it is 37 percent (forward) or 61 percent (backwards) (from a tabulation by R. A. Fisher, 1937). Using the same table, between the points at which $p = 25$ percent and $p = 75$ percent there are about 4.5 units of wavelength. If $\sigma = 10$ kilometers, as in some rural European populations, and $s = 0.02$ this distance corresponds to a little over 250 kilometers. With these values, the rate of advance is $\sigma\,(2s)^{1/2} = 2$

kilometers per generation and it would take 250 generations, or roughly 6000 years, for the wave to move forward 500 kilometers. Such an explanation is not inconceivable for the evolution of hair and eye pigmentation. Fisher's computation was, however, for a one-dimensional habitat and the analytical treatment for a two-dimensional spread has not yet been done.

A simple simulation to indicate the possible effect of the number of dimensions was done using, once again, a stepping-stone model. This is, however, only a rough approximation to the more refined continuous model used by Fisher. Some results of this simulation are given in Figure 8.33. Because of the discontinuous nature of

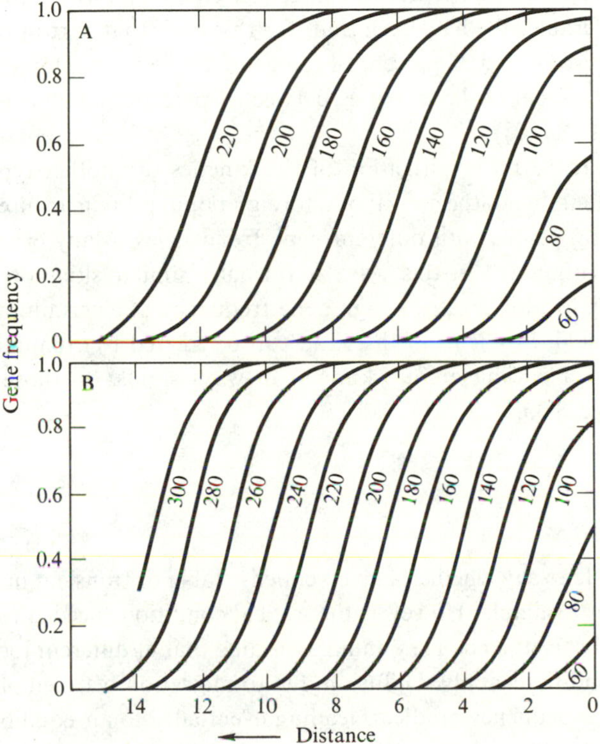

FIGURE 8.33
Wave of advance of an advantageous mutant gene under diffusion in one-dimensional (A) and two-dimensional (B) habitats assuming a stepping-stone model. Migration is isotropic in the two-dimensional model ($m_{lat} = m_{long} = 0.01$), for which only one dimension is shown in the graph, and $m = 0.02$ for the one-dimensional model, so that in both cases the proportion of migrants is the same. The origin of the mutation is at zero abscissa and zero time with an initial gene frequency $p_0 = 0.001$. The selective advantage of the mutant gene (no dominance) is $s = 0.1$. The numbers on the curves indicate generations.

the model, the curves are actually discontinuous and not perfectly parallel. The basic suggestion by Fisher, that a wave of advance of approximately constant shape is formed at a steady state and moves forward at a constant rate, is born out by these numerical results, also for two dimensions.

Clines may be transient.

A cline is a gradient in a phenotype or gene frequency. Diffusion centers form clines that are transient in space, and so the wave of advance considered above is a moving cline. The wave proposed by Fisher is itself a cline, whose total length may be quite considerable. We have seen that, for this case, that part of the cline which is between 25 percent and 75 percent of the gene frequency occupies a distance in a linear diffusion model of 4.5 $\sigma(\frac{1}{2}s)^{1/2}$. Between 1 percent and 99 percent there is a length of about $18\sigma/(2s)^{1/2}$.

There seems to be no investigation of the kinetics of another type of transient cline, namely, that due to the arrival of a foreign population in an already inhabited area occupied by people with different gene frequencies. Many existing clines are likely to have originated in this way. A formally similar situation applies when selection, acting locally, creates a high gene frequency at a certain place, and then selection is relaxed. Problems of this kind can be treated by computer simulation, when the relevant parameters are known, in ways similar to those illustrated in Figures 8.32 and 8.33.

Clines may be stable.

It is impossible to say whether a given cline is stable or transient unless paleontologic evidence is available. However, theoretical conditions needed for the stability of a cline have been studied. They entail assuming that at different locations, different selective conditions apply. Diffusion due to migration between places creates a continuous gene-frequency gradient, leading eventually to an equilibrium between diffusion and selection.

Two different models have been suggested. One of them assumes that selection varies linearly along a linear habitat. The problem is that of predicting the gene frequencies at each point once there has been enough time for diffusion to balance selection. The condition determining a stable polymorphism is that the selective advantage is zero at a point taken to be the origin, increases linearly on one side of the origin, and decreases linearly on the other side. If the rate of diffusion is given, as is usual, by σ^2 per generation, and the selection gradient is g per unit distance, then the value of the gene frequency at equilibrium, as a function of distance, can be found from a table given by Fisher (1950). Fisher also gives, in another table, a transformation (which he called legit) of gene frequencies that gives a variable that is proportional to distance measured in units of $a = [4\sigma^2/g]^{1/3}$. Once again no dominance is assumed.

The second model, analyzed by Haldane in 1948, assumes an abrupt change in the selection coefficient at a boundary line on a plane, such that a recessive phenotype *aa* has selection coefficient $1 + k$ on one side of the boundary, and $1 - K$ on the other. Migration takes place with a coefficient of diffusion σ^2. Haldane calculated that the distance between the points of 25 percent and 50 percent gene frequency of the recessive in the region where *A* is favored is $0.79\sigma/(K)^{1/2}$, while that between the points of 50 percent and 75 percent recessive frequencies in the region where gene *a* is favored is $1.27\sigma/(k)^{1/2}$. The gradient of the gene frequency at the boundary is given by

$$\frac{dp}{dx} = \left[\frac{Kk}{3(K + k)}\right]^{1/2} \sigma. \tag{8.52}$$

For a comparison of the two models we will apply them, as approximations, to an actual example.

In Japan, the ABO gene *A* shows a cline from north east to south west along the archipelago. Along a distance of 2000 kilometers, an almost linear cline from a gene frequency of 0.24 to one of 0.30 is observed (Nei and Imaizumi, 1966). The other alleles, *B* and *O*, also have clines of their own, both decreasing in the direction in which *A* increases. The frequency of one allele, of course, influences that of the others.

Using Fisher's theory, we find that the deviate or legit corresponding to $p = 0.24$ is 0.542, and that corresponding to $p = 0.30$ is 0.401. The difference in the legit scale is 0.141 and so the distance corresponding to one legit is $a = 2000/0.141 = 14{,}200$ km. Fisher showed, as has already been mentioned, that at equilibrium $a^3 = 4\sigma^2/g$ where g is the selection gradient per unit distance. Assuming, therefore, $\sigma^2 = 10$ for Japan (from Yasuda's 1969 data, which probably gives an underestimate in this case), $g = 4\sigma^2/a^3 = 40/(14200)^3$. The predicted difference in selection coefficients between the two extremes, namely, those at the top and bottom of Japan, is, therefore, $2000\ g = 0.3 \times 10^{-7}$, which is a very small difference indeed.

With Haldane's theory, we can use, very approximately, the gradient of gene frequency at the boundary as valid in the range of gene frequencies actually available. Assuming $K = k$, Equation 8.52 gives $dp/dx = \sigma\ (K/6)^{1/2}$. The observed gradient dp/dx is, approximately, $(0.30 - 0.24)/700 = 0.00009$, where 700 is the distance between the two extremes (2000 kilometers) expressed in σ units ($\sigma = 3$). Hence $K = 5 \times 10^{-9}$, and $2K$, the difference between the selection coefficients in the two halves of the plane, is 10^{-8}, a value not very different from that obtained by using Fisher's theory. The two theories give similar results, but their outcome is somewhat surprising. A very small difference in selection coefficients, which is of the order of values for mutation pressure, is enough to maintain the cline. Such a difference would, however, require twenty thousand generations to generate the cline. This number of generations to establish the cline under the estimated difference in selection pressure, is much too high to be acceptable. Probably, the answer is that the cline is not an equilibrium situation, but may be in the process of being generated by stronger selection differences. It may also be a disappearing cline originally due to some early migration. Nei and Imaizumi (1966) note that no significant migrations are known to have taken place in Japan during the last 2000 years. A kinetic computation should be done to see what original difference might still be compatible with the presently available gradient after a time of approximately 2000 years.

To check, independently, for such small selection gradients is practically impossible. It is instructive, however, to see that such very small selective differences can maintain relatively large stable clines.

Another important cline in ABO frequencies is that of decreasing *B* gene frequency from east to west in Asia and Europe. The gradient of gene frequency is of the order of 0.25×10^{-4} per kilometer, and, therefore, comparable to that observed in Japan. Conclusions similar to those regarding the cline in Japan can be deduced. It seems likely that both these clines are the transient consequence of migration, or of selective "accidents."

8.23 Gene Flow

Gene flow is the slow diffusion of genes across a racial barrier.

A special case of gene diffusion is that which takes place, slowly and perhaps in a hidden way, from one population (or race) into another. (The concept of race will be discussed in detail in Chapter 11). Sometimes the barrier is only psychological. This is the case, for instance, with black and white Americans. As is well known, there is an almost complete genetic isolation in the U.S. on the basis of skin color (see Chapter 12). The progeny of a mixed union is conventionally considered to be black, and so is the progeny of a backcross of the hybrid to whites. Further backcrosses to whites may sometime "pass" the racial barrier and thus result in transfer of genes from blacks to whites. Most of the diffusion, however, seems to take place from whites to blacks and not vice versa.

This type of gene diffusion, from one population into another, is often called gene flow. It is a form of racial admixture that takes place at a slow rate, usually because of psychological barriers.

A similar gene flow takes place, for instance, in Africa, this time in the reverse social sense, from the class considered to be socially inferior into the superior one. Pygmies are looked upon as "inferior" by neighboring non-Pygmy tribes, which live by agriculture rather than by hunting and gathering, and are, therefore, economically more advanced. Mixed marriages do occasionally occur but mostly between non-Pygmy males and Pygmy females. The mixed marriages are practically always absorbed into the non-Pygmy culture. Thus almost all of the gene flow is from Pygmy to non-Pygmy. On the other hand, some exchange may have also occurred in the opposite direction, at least in earlier times, as we shall see later.

Estimation of the extent of racial admixture is possible when the gene frequencies of the initial types are known.

The estimation of gene flow is, when gene frequencies of the two populations before admixture are known, only a special case of the analysis of slow racial admixture. If an admixture is formed between two populations A and B, the first contributing a fraction M and the second the other portion $(1 - M)$ to the mixed population, then the gene frequency P_M of an allele in the mixed population will be

$$P_M = MP_A + (1 - M)P_B = P_B + M(P_A - P_B). \tag{8.53}$$

If we know the three gene frequencies P_A, P_B, and P_M, M can be estimated by

$$M = \frac{P_M - P_B}{P_A - P_B}, \tag{8.54}$$

following a method introduced by Bernstein (1931a). Some authors suggest the use of absolute values of the numerator and denominator in this formula, but this may be misleading. In a geometrical representation, the mixed population will have its gene frequencies lying on the straight line joining those of populations A and B, and at distances from them that are proportional to M and $1 - M$, respectively (see Figure 8.34). This is true for any allele, and thus also in the multidimensional space representing all possible gene frequencies.

Gene frequencies are subject to a sampling error that is a function of the gene frequency itself. The process of estimating M and its variance by averaging results from all known alleles is, therefore, complex. Testing the validity of the average estimate by checking whether the various alleles give concordant results is, thus, more complicated than this simple linear approach might suggest.

A complete solution of the problem of estimating the extent of racial admixture requires the simultaneous estimation of many parameters. All the gene frequencies of the parental populations, those of the mixed one, and the admixture coefficient must be estimated simultaneously.

The following is an approximate treatment that is useful for diallelic loci. A complete analysis of multiple allelic systems would have to be much more complex.

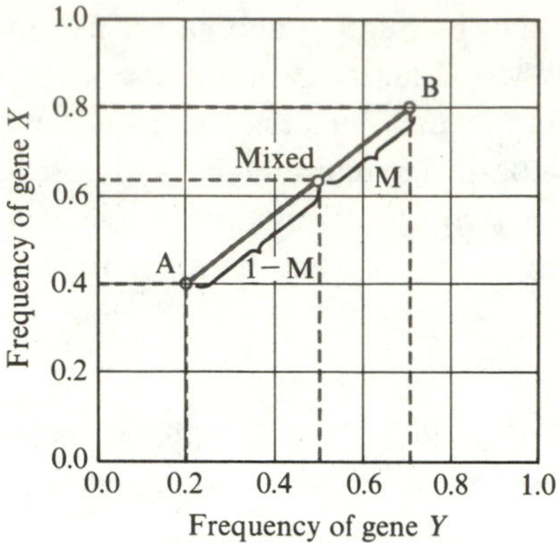

FIGURE 8.34
Gene frequencies at two loci in two populations, A and B, and in a mixed population formed of the proportions M from A and 1 − M from B. If "distances" on the basis of gene frequencies are computed, the mixed population lies between the two parental ones at distances from them that are proportional to the parts in the mixture—that is, to M and 1 − M.

The value of M is estimated from Equation 8.54 for each allele, and then for each M value an error variance is computed from:

$$\sigma_M^2 = \frac{1}{(P_A - P_B)^2}\,[\sigma_{PM}^2 + M^2\sigma_{PA}^2 + (1 - M)^2\sigma_{PB}^2]. \tag{8.55}$$

The variances of the parental gene frequencies, $\sigma^2{}_{PA}$ and $\sigma^2{}_{PB}$, and that of the mixed population, σ_{PM}^2, can be taken as the sampling-error variances estimated by maximum likelihood when only one sample is available, or when there are several samples that do not differ other than because of random sampling. If several samples from any population are available that are heterogeneous, the observed variance between them can be used in the formula for σ_M^2. From Equation 8.55 it can be seen that the error-variance term depends very much on $P_A - P_B$, the difference between the ancestral gene frequencies. The most informative genes are obviously those for which $P_A - P_B$ is largest.

The mean of the M values for separate diallelic systems can be computed as a weighted average from $\overline{M} = \sum WM/\sum W$, where the weight $W = 1/\sigma_M^2$. In addition the heterogeneity between M values obtained from k different loci can be computed, approximately, from

$$\chi^2_{(k-1)} = \sum_{i=1}^{k} \frac{(M_i - \overline{M})^2}{\sigma_{M_i}^2}. \tag{8.56}$$

Alleles conferring malarial resistance may have decreased in frequency because the environments in which black Americans live have changed.

A sample of data on black Americans in Claxton, Georgia, is analyzed in Table 8.17. Error variances were taken from binomial estimates (pq/n) for Claxton blacks and whites and computed from the standard errors of available data for West Africa. Analyzing these Claxton data, Workman and co-workers (1963) concluded that the genes fall into two categories. Most markers gave an admixture rate of about 10 percent of white and 90 percent of (African) black genes in the Claxton blacks. The genes *G6PD*, Hp_1, *Tfd*, and *Hb*, however, showed a higher apparent admixture rate, and so Workman and co-workers suggested that the difference was due to the fact that markers of the second group were affected by selection in favor of the white alleles when blacks moved into the American environment, where malarial infestation was less important or absent.

The analysis in Table 8.17, which is confined to markers for which the African data are most reliable and for which standard errors are small, confirms these conclusions. The chi square for heterogeneity of the seven markers is 581.36 with 6 degrees of freedom, but drops to 18.47 with 4 degrees of freedom if the last two markers, *G6PD* and Hb^s, are omitted. This last chi square is still significant, but the

TABLE 8.17
African-Caucasian Racial Admixture in Black Americans in Claxton, Georgia, Estimated by Using Gene-frequency Data on Claxton Blacks, Claxton Whites, and West Africans. M is the proportion of white genes estimated in Claxton blacks.

	Gene Frequencies		*West African Gene-frequency Data*				
Gene	2090 *Claxton Whites*	1287 *Claxton Blacks*	*Average*	*Standard Deviation*	*No. of Studies*	*Source*	$M \pm \sigma_M$
Fy^a	0.422	0.046	0	0	3	Race and Sanger (1968) and others	0.101 ± 0.0060
R_0	0.037	0.535	0.6051	0.0644	14	Hiernaux (1968)	0.123 ± 0.0205
A	0.246	0.158	0.1468	0.0336	37	Hiernaux (1968)	0.113 ± 0.0113
B	0.050	0.129	0.1431	0.0410	37	Hiernaux (1968)	0.151 ± 0.0110
M	0.508	0.485	0.4644	0.1208	12	Hiernaux (1968)	0.472 ± 0.2365
G6PD	0	0.118	0.1620	0.0670	25	Livingstone (1967)	0.272 ± 0.0133
Hb^S	0	0.043	0.0611	0.0415	72	Livingstone (1967)	0.296 ± 0.0066

Source: Claxton data from Cooper et al. (1963); West African data from a variety of other sources.

big drop in heterogeneity is also clearly significant. Chi square remain significant if the M gene is also left out of the analysis. The M gene on its own gives a highly discrepant estimate of admixture relative to the other markers, but contributes very little information because of its high standard error.

The weakness of the analysis is mostly due to the uncertainty of the origin of black Americans, (see, however, recent historical research summarized by T. E. Reed, 1969b), and the variability of gene frequencies in the probable area of the slave markets in West Africa. In addition, it is unavoidable that gene frequencies have changed somewhat from their original values, due to drift or, in some cases, selection. The opportunities for admixture, and the time available for it, must also have varied widely.

If the average estimate of admixture from the first five genes ($M = 0.113$) is correct, then the gene frequencies of Hb^s and $G6PD$ in Claxton blacks should be 0.144 and 0.054, instead of the observed 0.118 and 0.043. This gives an idea of the change in gene frequency that may have taken place in the ten or so generations during which the African alleles were selected against in the progeny of Africans brought to America as slaves.

Assuming selection against a recessive lethal gene for Hb^s, the change expected in ten generations would be from 0.054 to 0.035 (see Chapter 3). The difference between the expected 0.035 and the observed 0.043 is trivial, and would be smaller considering that the homozygote is usually not completely lethal.

The change in $G6PD$ from the expected 0.144 to 0.118 could be brought about by a selection coefficient of 0.023 in favor of the normal allele, in a nonmalarial or a less heavily infested environment, assuming additive selection. This is, perhaps, surprisingly high, but the data are subject to a large error.

T. E. Reed (1969a,b) has recently analyzed white admixture in black Americans, using the Gm data on four alleles, by maximum likelihood. This analysis confirms that black Americans from Cleveland and Oakland have higher admixture rates (roughly 0.20 to 0.28) than those from Southern Georgia (for which his highest estimate of M is 0.064 ± 0.033). Several hypotheses on the presence of the various Gm alleles among Africans give rise to different estimates, however. It may be added that there are doubts as to the validity of some of the Gm assays used. If these doubts are correct, it is possible that the M value for Gm from South Georgia is higher than 6.4 percent and more compatible with the estimate of 11.3 percent given on the basis of the data in Table 8.17.

One of the best loci for this type of analysis is the Duffy red-cell blood group, as the allele $Fy(a+)$ has zero or almost zero frequency among Africans and a relatively high frequency among Europeans. Estimates of M values obtained by T. E. Reed (1969b) using only Fy are given in Table 8.18. They show the differences in extent of admixture among black Americans from various parts of the United States, which varies from 4 to 26 percent. The difference in white admixture of black Americans from southern and northern states may be the consequence of differential

TABLE 8.18

Estimates of M Derived from Fy^a Gene Frequencies for Black Americans from Various Areas. The frequency of this gene in the African ancestors of black Americans is assumed here to be zero; if this is not true these are maximum estimates (see below); N = number in sample; q = Fy^a gene frequency; S.E. = standard error.

	Negroes		Caucasians		
Region and Locality	*N*	$q \pm$ S.E.	*N*	$q \pm$ S.E.	$M \pm$ S.E.[a]
Northern					
New York City[b]	179	0.0809 ± 0.0147	—	—	0.189 ± 0.034
Detroit	404	0.1114 ± 0.0114	—	—	0.260 ± 0.027
Oakland, California	3146	0.0941 ± 0.0038	5046	0.4286 ± 0.0058	0.2195 ± 0.0093[c]
Southern					
Charleston, S.C.	515	0.0157 ± 0.0039	—	—	0.0366 ± 0.0091
Evans and Bullock Counties, Ga.	304	0.0454 ± 0.0086	322	0.422 ± 0.0224	0.106 ± 0.020

[a] The *q* for Oakland Caucasians (who are of West European ancestry) was used in all estimates: $M = q_n/q_c$.

[b] Two New York City studies were omitted because of selection for dark skin color. The data used here were grouped with both anti-Fy^a and anti-Fy^b. The observed distribution of four Duffy phenotypes differs from Hardy-Weinberg expectation at the 0.025 level of significance.

[c] If the frequency of Fy^a in the African ancestors were 0.02, this estimate would be 0.181.

migration. In countries like Brazil the admixture rates are higher (Saldanha, 1957) and in some groups mating is almost random with respect to skin color (Krieger et al., 1965).

Some gene flow is probably always present, even among minority groups that try to retain a high degree of isolation. Ashkenazim Jews, who lived in Russia, Poland, and Germany for a period that may have been as long as 2000 years, show some effect of local populations on their genetic marker frequencies that is suggestive of gene flow. In principle, selection might also give rise to such an effect (convergent evolution). It seems unlikely, however, that all genes respond to selection *at the same rate*, which would be necessary for selection to mimic gene flow. The computation of gene flow is made difficult, however, here as elsewhere, by uncertainties concerning the original gene frequencies. Averaging over all presently known Jewish groups can give only a rough approximation to the original frequencies among Jews. The only Jewish group that has remained *in situ* (the Samaritans) is extremely small, and since it must have been considerably affected by drift, it could hardly provide the basis for a valid estimate. Similarly, it is impossible to say whether the very high *G6PD* and other deviating gene frequencies of groups such as the Kurdistani Jews (over 60 percent *G6PD* deficiency—see Chapter 4) are the consequence of selection, gene flow, or both, because of the lack of data on the populations from which gene flow might have occurred.

The kinetics of racial intermixture can be predicted on the assumption that admixture rates are constant.

When the number of generations of intermixture is known, we can compute an average rate m of gene flow per generation. Assuming that the gene flow m per generation is constant,

$$M = 1 - (1 - m)^t \tag{8.57}$$

and so

$$\log(1 - m) = \frac{\log(1 - M)}{t}, \tag{8.58}$$

where t is the number of generations and M is the total admixture. Glass and Li (1953) computed an average value for m of 3.6 percent assuming 10 generations of intermixture and a total $M = 30.6$ percent for the amount of Caucasian ancestry in black Americans (based on Rhesus R_0 data). Other black Americans, as mentioned above, show different values of M. Thus, while Cleveland and Oakland blacks have high admixture rates, blacks in Georgia have, on the same basis, a much smaller intermixture ($M = 11$ percent) and so, the admixture rate per generation for them would be about three times as small. It seems unlikely that admixture rates are constant in time, or on a North-South axis. When conditions do not change too rapidly, however, predictions on the basis of observations of admixture rates at one particular time may be valid over a longer period. Roberts and Hiorns (1962) obtained data on the rate of admixture now prevalent in a group of Sudanic Northern Nilotes, which are given in the form of a migration matrix in Table 8.19.

TABLE 8.19
A Matrix of Migration Between Three Groups of Northern Nilotes in Sudan

		Parents		
		Nuer	Dinka	Shilluk
Progeny	Nuer	0.9850	0.0125	0.0025
	Dinka	0.0138	0.9775	0.0087
	Shilluk	0	0.0098	0.9902

Source: Data from Roberts and Hiorns (1962).

"Migration" takes place in the sense that the progeny of mixed marriages usually acquires the ethnic denomination of the village in which it lives. Roberts and Hiorns predict the gene frequencies to be expected in each of the three populations on the assumption that this migration matrix will remain valid in the future. If $\mathbf{q}_0$ and $\mathbf{q}_1$ are vectors of gene frequencies of a series of populations at generations 0 and 1, and $\mathbf{M}$ the migration matrix (as in Table 8.19), then the vector $\mathbf{q}_1$ can be predicted as

$$\mathbf{q}_1 = \mathbf{M}\mathbf{q}_0$$

and after t generations the gene frequencies of the three populations will be

$$\mathbf{q}_t = \mathbf{M}^t\mathbf{q}_0 .$$

The present gene frequencies of the blood-group gene M in the three populations are 0.5750, 0.5760, and 0.5047. At equilibrium, under repeated application of the same migration matrix, the gene frequencies are expected to be identical and equal to 0.5464. The expected kinetics of the change are shown in Figure 8.35.

The extent of admixture can be computed from "genetic distances."

Inspection of Figure 8.34 shows that if we compute "genetic distances" between populations from gene-frequency data (see Chapter 11 for a precise definition of genetic distance), these may lend themselves to a test of whether a given population,

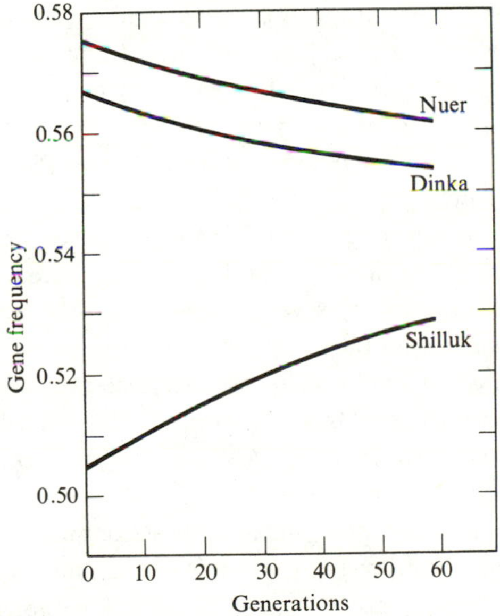

FIGURE 8.35
Predicted approach to equilibrium, through intermixture, of the frequencies of the gene for blood group M in the Northern Nilotes. (From Roberts and Hiorns, 1964.)

M, can be a mixture of two other ones, A and B. Thus if d_{AB} is the distance between A and B, d_{AM} and d_{BM} those between A and M and B and M, then we should have $d_{AM} + d_{BM} = d_{AB}$, and M, the proportion of A in the mixture, should be given by d_{BM}/d_{AB}.

A genetic distance that would satisfy this requirement would be, like the segments in Figure 8.34, given by

$$d_{ij}^2 = \sum \Delta_{ij}^2, \tag{8.59}$$

where Δ_{ij} is the difference in gene frequency between populations *i* and *j*, and the sum is carried out over all alleles and loci.

Naturally, we would like to have a reliable estimate of the error of d_{ij}. All gene frequencies are subject to sampling error, and thus the simple additive relationship of distances cannot be expected to apply exactly. The sampling error of a frequency depends on the frequency itself and a fully satisfactory estimate of the error of the distance defined by Equation 8.59 is not easy to obtain. Other measurements of distance have, therefore, been introduced and will be discussed in Chapter 11. The simple additivity relationship $d_{AM} + d_{BM} = d_{AB}$ is usually not, however, valid for these other measures.

It is still possible to test for additivity of the relationship or, in general, to test the hypothesis that a given population is a mixture of two others by the following procedure. Mixtures in known proportions of M and $1 - M$ of the two putative parental populations are constructed, and the gene frequency of each mixture is computed. The distances between the real, presumably mixed population and these artificial mixtures are then calculated. This was done to test the hypothesis that Western (Babinga) Pygmies are a mixture of an ancestral Pygmy stock, of which the Eastern (Ituri) Pygmies are more "pure" descendants, and of a non-Pygmy stock of Bantu-speaking, farming Central Africans, who may have arrived in Central Africa with the Bantu invasion, perhaps ten or fifteen centuries ago. This hypothesis has been entertained by several authors, mostly on the basis that the Western Pygmies are taller than their Eastern counterparts.

The results are plotted in Figure 8.36. Genetic distance was computed as described in Chapter 11 and its error from the variation between distances computed from each locus. The Western Pygmies are closest (minimum *d*) to a mixture of 55 percent of Eastern Pygmies and 45 percent Central Africans. The distance between the observed population and the hypothesized mixture is, however, still fairly large (almost half that from the Central Africans) and is clearly significant.

Although the hypothesis seems attractive, it cannot be considered as proven. There are many possible reasons for the discrepancy, if the hypothesis is true. The populations taken as putative parents may only correspond poorly with the real ancestors. Also, if the mixture occurred a long time ago, gene frequencies may have changed because of drift or selection. The time postulated by historians for the

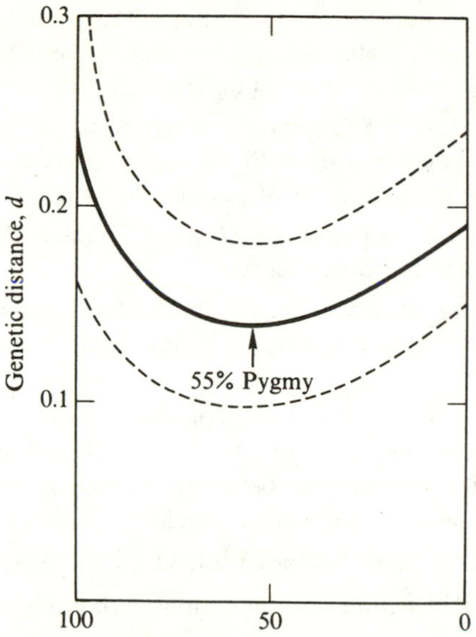

FIGURE 8.36
Theoretical variation of d (genetic distance) between Babingas and mixtures of an Eastern (Mbuti) Pygmy and a non-Pygmy population (Central Africans) as a function of the degree of admixture, which is indicated on the abscissa as the proportion of Mbuti genes. The dashed lines indicate 2σ fiducial belts. (From Cavalli-Sforza et al., 1969.)

Bantu invasion, and, therefore, for the mixture, may seem a little too short for building up all of the observed discrepancy. The possibility that historical events may have been more complex must also be envisaged.

General References

Bodmer, W. F., and L. L. Cavalli-Sforza, "A migration matrix model for the study of random genetic drift." *Genetics* **59**: 565–592, 1968. (A detailed exposition of the underlying theory.)

Cavalli-Sforza, L. L., M. Kimura, and I. Barrai, "The probability of consanguineous marriages." *Genetics* **54**: 37–60, 1966. (A comprehensive theoretical treatment of the problem of predicting the frequencies of different types of consanguineous marriages.)

Kimura, M., and G. H. Weiss, "The stepping stone model of population structure and the decrease of genetic correlation with distance." *Genetics* **49**: 561–576, 1964. (An account of one of the major models of population structure.)

Kimura, M., *Diffusion Models in Population Genetics*. London: Methuen, 1964. (A comprehensive review of genetic drift theory.)

Malécot, G., *Probabilité et Hérédité*. Paris: Press Universitaire de France, 1966. (A collection of Malécot's major works, including those on population structure—in French.)

Wright, S., "Statistical genetics in relation to evolution." *In* Georges Teissier, ed., *Exposés de biometrie et de statistique biologique*. 1–64 Paris: Hermann, 1939. (This paper and the two that are listed immediately below review Sewall Wright's major contributions to the genetic theory of population structure.)

Wright, S., "Isolation by distance." *Genetics* **28**: 114–138, 1943.

Wright, S., "The genetical structure of populations." *Ann. Eugen.* **15**: 323–354, 1951.

Worked Examples

8.1. *Numerical computation of gene-frequency distributions under random genetic drift.*

We will consider a numerical example with a population of $n = 4$ genes (2 individuals) and 2 alleles A and a at one locus. After the first generation, gene frequencies are distributed according to the binomial $P(r, p) = n!p^r(1-p)^{n-r}/[r!(n-r)!]$, where p is the initial gene frequency of A.

No. of A genes	*Gene frequency*	*General*	*With p* = 0.5
0	0%	p^4	0.0625
1	25%	$4p^3q$	0.25
2	50%	$4(4-1)\dfrac{p^2q^2}{2}$	0.375
3	75%	$4pq^3$	0.25
4	100%	q^4	0.0625

In the second generation, the populations with zero or 4 A genes do not change any more, as they can only give rise to the same type of population. Those with one A

gene, in which the gene frequency is $\frac{1}{4} = 25$ percent, can give rise to all possible gene frequencies in the population, with probabilities that are again specified by the binomial, now with $p = 0.25$, and similarly for populations with 2 or 3 *A* genes.

The probabilities of 0, 1, . . . , 4 *A* genes in the second generation, conditional on the gene frequency of *A* in the first generation are, therefore,

A genes in the second generation	*Frequency of the A gene in the first generation.*				
	0%	25%	50%	75%	100%
0	1	0.3164	0.0625	0.0039	0
1	0	0.4219	0.25	0.0469	0
2	0	0.2109	0.375	0.2109	0
3	0	0.0469	0.25	0.4219	0
4	0	0.0039	0.0625	0.3164	1
Sum	1	1	1	1	1

The probability of having zero *A* genes in the second generation is made up of the sum of: (1) the proportion of populations having zero *A* genes in the first generation: 0.0625; (2) the proportion of populations having one *A* gene in the first generation (0.25), times the probability that these have zero *A* genes in the next (0.3164), that is $0.25 \times 0.3164 = 0.0791$; (3) similarly, for populations having 2 *A* genes in the first generation: $0.375 \times 0.0625 = 0.0234$; (4) similarly, for populations having 3 *A* genes: $0.25 \times 0.0039 = 0.0010$. The sum is thus 0.1660. Similarly, for the probability of one *A* gene (25 percent frequency) in the second generation:

$$0.25 \times 0.4219 + 0.375 \times 0.25 + 0.25 \times 0.0469 = 0.2109.$$

The probabilities for the various gene frequencies for a few generations calculated in this way, their means and variances, and the proportion of populations in which one allele or the other is fixed are given below.

No. of A genes	*Gene frequency*	*Generations*					
		1	2	3	5	8	∞
0	0%	0.0625	0.1660	0.2490	0.3587	0.4404	0.5
1	25%	0.25	0.2109	0.1604	0.0904	0.0381	0
2	50%	0.375	0.2461	0.1813	0.1017	0.0429	0
3	75%	0.25	0.2109	0.1604	0.0904	0.0381	0
4	100%	0.0625	0.1660	0.2490	0.3587	0.4404	0.5
Mean gene frequency		0.5	0.5	0.5	0.5	0.5	0.5
Variance of gene frequency		0.0625	0.1094	0.1445	0.1907	0.2313	0.25
Proportion fixed		0.125	0.332	0.498	0.717	0.881	1.0

8.2. *Relation between kinship and inbreeding coefficient.*

Suppose that two individuals I and J have genotypes at one locus that are *ab*, *cd*, and that ϕ_{ac} is the probability that *a* is identical by descent to *c* and similarly for ϕ_{ad}, etc. Then the coefficient of kinship between I and J, $\phi_{IJ} = \frac{1}{4}(\phi_{ac} + \phi_{ad} + \phi_{bc} + \phi_{bd})$. The progeny of I and J can be *ac*, *ad*, *bc*, or *bd* with equal probabilities of $\frac{1}{4}$. The inbreeding coefficient of, for example, an *ac* offspring is, by definition, ϕ_{ac}, and therefore the average inbreeding coefficient of the progeny is

$$F = \tfrac{1}{4}\phi_{ac} + \tfrac{1}{4}\phi_{ad} + \tfrac{1}{4}\phi_{bc} + \tfrac{1}{4}\phi_{bd} = \phi_{IJ}.$$

There is a probability $1 - \mu$ that each of the two gametes forming a zygote does not mutate, if the mutation rate is μ, and, therefore, if mutation is taken into account,

$$F = (1 - \mu)^2 \phi_{IJ},$$

but this correction can be neglected unless we are considering a process lasting many generations.

8.3. *Decrease of overall heterozygosity under drift.*

In a population with gene frequency p_0, the expected gene-frequency variance of a sample of N individuals is given by the binomial

$$V(p_1) = \sigma_{p_1}^2 = \frac{p_0 q_0}{2N}. \tag{1}$$

Here p_1 refers to the gene frequency of the sample, which is the same as the gene frequency of the next generation. From this we can compute the expected frequency of heterozygotes in the next generation as

$$\begin{aligned} E(2p_1q_1) &= \text{expected frequency of heterozygotes} \\ &= 2E(p_1q_1) \\ &= 2Ep_1(1 - p_1) \\ &= 2E(p_1 - p_1^2) \\ &= 2E(p_1) - 2E(p_1^2). \end{aligned} \tag{2}$$

If we add and subtract the quantity $2[E(p_1)]^2$ we have

$$E(2p_1q_1) = 2[E(p_1) - E(p_1^2) + [E(p_1)]^2 - [E(p_1)]^2]. \tag{3}$$

By the definition of the variance, $\sigma_{p_1}^2$ the variance of gene frequencies in the next generation is given by

$$\sigma_{p_1}^2 = E(p_1^2) - [E(p_1)]^2.$$

Since we know that $E(p_r) = E(p_1) = p_0$ at all times, we can write, from Equation 3,

$$E(2p_1q_1) = 2(p_0 - \sigma_{p_1}^2 - p_0^2).$$

Thus, since from Equation 1 $\sigma_{p_1}^2 = p_0 q_0/2N$, and also $p_0 - p_0^2 = p_0 q_0$, we have

$$E(2p_1 q_1) = 2p_0 q_0\left(1 - \frac{1}{2N}\right). \qquad (4)$$

The expected frequency of heterozygotes $2p_0 q_0$ in the earlier generation has thus gone down by the factor $(1 - 1/2N)$. Calling H_1 the expected heterozygote frequency in the next generation, H_0 that in the preceding generation, Equation 4 can be written in the form

$$H_1 = H_0\left(1 - \frac{1}{2N}\right).$$

Applying the same formula for t generations, the expected heterozygote frequency is

$$H_t = H_0\left(1 - \frac{1}{2N}\right)^t.$$

It should be kept in mind that this is the *overall heterozygosity*—namely, the heterozygosity over all independent populations. A particular population with gene frequency p_i may still have a frequency of heterozygotes very near that predicted by the Hardy-Weinberg law, $2p_i(1 - p_i)$.

8.4. *The average number of generations before a mutation appears.*

If a mutation occurs at a rate μ, the expected number of mutant genes produced per generation is $2N\mu$ in a population of size N. The probability that at least one mutant gene will be formed in one generation is, assuming a Poisson distribution for the number of mutant genes produced per generation, $1 - e^{-2N\mu}$. If $2N\mu$ is small ($\ll 1$), then

$$1 - e^{-2N\mu} \sim 2N\mu.$$

If the mutant gene is not initially present in the population and $2N\mu$ is small, then the expected number of generations before it actually appears is the reciprocal of this number, that is,

$$T = \frac{1}{2N\mu}.$$

For example, with $\mu = 10^{-6}$ and $N = 10^4$, $T = 50$.

8.5. *The probability of homozygosis under equilibrium between mutation and drift.*

Kimura and Crow (1964) have considered a model of "neutral" mutations (no selection) in which each mutation produces a different allele. We require the probability that at equilibrium, a random individual is homozygous for a given allele.

Let us call F_t the probability at a given time t. In the next generation, that is, at time $t + 1$, F_{t+1} can be computed as the sum of two independent probabilities: (1) that an individual has two alleles, which are both nonmutated copies of the same gene in the previous generation, (2) that the two alleles are nonmutated copies of genes that in the previous generation were identical by nature, although they came from different individuals. The first probability is $(1 - \mu)^2/2N$ as there are $2N$ genes in the population and the numerator represents the probability that the gene does not mutate while giving rise to two copies of itself in the next generation. The second probability is $(1 - \mu)^2(1 - 1/2N)F_t$, where the first term in the product indicates that the two genes have not mutated in the previous generation, the second that they came from different individuals, and the third the probability that they were identical in these individuals in the previous generation. Thus,

$$F_{t+1} = \frac{(1-\mu)^2}{2N} + (1-\mu)^2\left[1 - \frac{1}{2N}\right]F_t = \frac{(1-\mu)^2}{2N}[1 + (2N-1)F_t].$$

At equilibrium, $F_t = F_{t+1} = F$. Neglecting terms in μ^2, this gives

$$F = \frac{[1 + (2N-1)F](1 - 2\mu)}{2N},$$

and so, solving for F, we have

$$F = \frac{1 - 2\mu}{1 + 2\mu(2N-1)} \sim \frac{1}{1 + 4N\mu},$$

which is the same formula as given in Section 8.4 and elsewhere.

It may be argued that the first of the two probabilities computed in the evaluation of F_{t+1} involves self-fertilization, because this is the only way in which two copies of the same gene can form a new individual in the next generation. Self-fertilization cannot occur in humans, or in most higher animals, but it can be shown that the result does not change much if self-fertilization is avoided (see Wright, 1943; Malécot, 1948).

8.6. *When N varies from generation to generation the effective population size is the harmonic mean of the individual N_i values.*

At every generation, the variance of gene frequencies is given by the binomial

$$\frac{p_{t-1}q_{t-1}}{2N_t},$$

where p_{t-1} and q_{t-1} are the gene frequencies in the previous generation. When N is large, p does not vary much and so we can take it to be approximately constant and equal to p, say. The variance of the gene frequency after k generations will be approximately the sum of the variances at each generation, that is,

$$V = \frac{pq}{2N_1} + \frac{pq}{2N_2} + \frac{pq}{2N_3} + \cdots + \frac{pq}{2N_k},$$

where N_i is the population size at the ith generation.

The total variance just given can be written in the form

$$V = \frac{pqk}{2N_h},$$

where

$$N_h = \frac{k}{\dfrac{1}{N_1} + \dfrac{1}{N_2} + \cdots + \dfrac{1}{N_k}}$$

and is, therefore, the same as it would be if the population size had remained constant and equal to the harmonic mean N_h.

8.7. *Two-dimensional gaussian migration law.*

We are interested in the distribution of the distance between the birthplaces of parent and offspring in a two-dimensional continuum. We assume the distribution in one dimension (x) to be the normal distribution,

$$df(x) = \frac{1}{\sigma_x\sqrt{2\pi}}(e^{-x^2/2\sigma_x^2})\,dx,$$

with mean 0 and standard deviation σ_x, and in the other dimension to be (y)

$$df(y) = \frac{1}{\sigma_y\sqrt{2\pi}}(e^{-y^2/2\sigma_y^2})\,dy.$$

The joint distribution is, assuming independence of migration in the two directions,

$$f(x, y) = f(x)f(y).$$

If $\sigma_x = \sigma_y = \sigma$, the migration is said to be isotropic, and then

$$df(xy) = \frac{1}{2\pi\sigma^2}(e^{-(x^2+y^2)/2\sigma^2})\,dx\,dy.$$

Using the transformation

$$z = \sqrt{x^2 + y^2},$$

it can be shown that

$$df(z) = \frac{z}{\sigma^2} e^{-z^2/2\sigma^2} dz,$$

which is the distribution of the distance z between the birthplaces of parent and offspring in two dimensions. This distribution has mean $\sigma(\pi/2)^{1/2}$ and variance $\sigma^2/(2 - \pi/2)$. The mean square distance $\overline{(z^2)}$ is equal to the square of the mean plus the variance and is thus $2\sigma^2$. It is convenient to use the mean square distance in two dimensions, and take it equal to twice the variance in one dimension.

8.8. *Number of sibs per individual.*

Assume distribution of progeny size (p) per family is given by $\phi(p)$. The frequency of individuals $\psi(s)$ with $p - 1 = s$ sibs is proportional to $p\phi(p)$, because there are, from each family of size p, p individuals each of whom has $p - 1$ sibs. Thus,

Progeny size	*No. of sibs*	*$\psi(s)$ proportional to*
1	0	$\phi(1)$
2	1	$2\phi(2)$
3	2	$3\phi(3)$
...	...	...
p	$p - 1$	$p\phi(p)$

The frequency of s sibs is

$$\psi(s = p - 1) = \frac{p\phi(p)}{\sum p\phi(p)}.$$

The mean value of s is, therefore,

$$\bar{s} = \sum s\psi(s) = \frac{\sum (p - 1)p\phi(p)}{\sum p\phi(p)} = \frac{\sum p^2\phi(p)}{\sum p\phi(p)} - 1.$$

Since

$$\sum p\phi(p) = \bar{p} \text{ (mean progeny size)}$$

and

$$V_p = \sum p^2\phi(p) - \bar{p}^2,$$

we have

$$\bar{s} = \bar{p} + \frac{V_p}{\bar{p}} - 1.$$

If progeny size has a Poisson distribution, $V_p = \bar{p}$ and hence $\bar{s} = \bar{p}$. If $V_p > \bar{p}$, then $\bar{s} > \bar{p}$. If $V_p = 0$, that is, the number of progeny per family is constant, then $\bar{s} = \bar{p} - 1$ (the assumption used by Dahlberg—see Cavalli-Sforza et al., 1966).

9

Quantitative Characters, Polygenic Inheritance, and Environmental Interactions

9.1 The Physiology of Continuous Variation

In the previous chapters we have worked almost exclusively with character differences whose inheritance could be clearly attributed to one or more identified genes. An integral part of Mendel's great contribution to genetics was that he realized that only with such characters would it be possible to elucidate the basic laws of inheritance. Many, if not most, individual variations, both qualitative and quantitative, are not, however, inherited in such an easily discernible way.

A quantitative character is an attribute such as height or weight that is measured on a continuous scale. Only if individuals fall into one or more essentially nonoverlapping ranges of measurement, for example, height of dwarfs as compared with that of normal adults, can a quantitative character be subdivided into qualitatively different categories. Even if we cannot identify simple genetic differences controlling a given quantitative or qualitative character, we may recognize that it has an inherited component. By this we mean that genetically related individuals tend to be more alike with respect to the character than unrelated individuals. Mendel himself realized that the probable explanation for such admittedly vague observations was in terms of the effects of many genes acting together to influence the expression of such a character. In this chapter we shall discuss the approaches that are used to analyze the inheritance of quantitative characters, emphasizing their application to a variety of human data.

Both qualitative and quantitative variation may be associated with an identifiable gene differences.

The effect of a gene substitution may be measured at many different levels. At the ultimate level, namely that of the DNA, all genetic differences are discrete, and thus qualitative. This is, of course, the basic principle underlying Mendelian inheritance. Most gene differences should be recognizable as discrete differences in the amino-acid sequence of the protein gene product. Even at this level, however, it is known from the nature of the genetic code that many nucleotide substitutions occur that do not produce a changed amino acid (that is, they are **synonymous substitutions**).

Proteins differing with respect to one or more amino acids can often be distinguished qualitatively by their physicochemical properties such as, for example, the differences in electrophoretic mobility of A and S hemoglobins (see Chapter 4). In the absence of any qualitative differences between gene products, proteins, especially if they are enzymes, may nevertheless show marked differences in their activity. When the activity of the immediate gene product cannot be identified, the effect of the gene substitution must be measured at a level further removed from that of the protein gene product. It is clear that, in general, as the measurement of a gene effect is based on character differences more and more removed from the primary gene product, the qualitative effect of the gene becomes more and more difficult to detect. This principle is beautifully illustrated by Penrose's data (1952) on the association of phenylketonuria (PKU) with a variety of measurements related to the primary activity of the mutant gene that causes the disease. Recall that PKU is a disease affecting individuals who are homozygous for a recessive mutant allele at the locus controlling the enzyme that converts phenylalanine to tyrosine. Phenylalanine and derivatives of it accumulate in affected individuals and, since these substances are toxic to the brain, cause mental retardation.

The distributions of the amount of phenylalanine in the blood of phenylketonurics and normal people are shown in Figure 9.1,A. This measurement, which is, of course, closely related to the primary activity of the gene, clearly separates two genetic classes of individuals. The set of observations illustrated in Figure 9.1,B concerns IQ. The distributions now overlap slightly, but still discriminate rather well between phenylketonurics and normal individuals. The distributions of head size in Figure 9.1,C show a very considerable overlap. Although there is clearly a very significant effect of homozygosity for the PKU gene on head size, it is not possible to use this character to distinguish phenylketonurics from normal individuals. This measurement is, incidentally, of interest in suggesting that phenylalanine accumulation has an effect during development of the brain. A diet low in phenylalanine

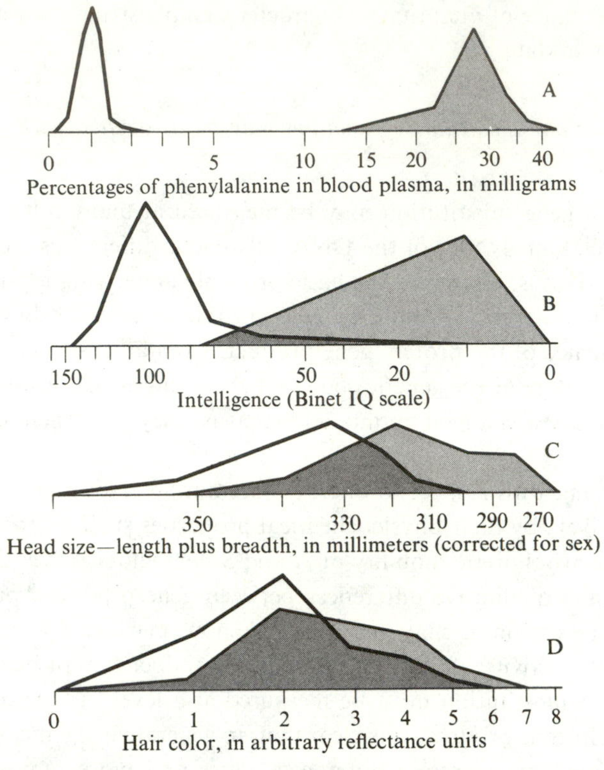

FIGURE 9.1
Frequency distributions: phenylketonurics (right) compared with control populations (left). A: $d/s = 13$, where d is the difference in the means and s is the average standard deviation of the two distributions. B: $d/s = 5.5$. C: $d/s = 2$. D: $d/s = 0.7$. (From Penrose, 1952.)

during development has been shown to mitigate the effects of homozygosity for the PKU gene. Determination of whether phenylalanine accumulation has an effect on mental processes in later life will be an important factor in the decision of whether to continue the phenylalanine-deficient diet for phenylketonurics into later life. The overlap of the distributions for hair color (Figure 9.1,D) is even greater than that for head size, though there is still a significant difference between the two distributions; that is, a significant average effect of homozygosity for the PKU gene on hair color. The main feature of these pairs of distributions that changes as one proceeds from the phenylalanine level in the blood to hair color is the ratio of the difference between the means of the distributions for the two types of individuals and their variances, or standard deviations. This ratio, which is given for each pair of distributions in the legend to Figure 9.1, decreases progressively from 13 for phenylalanine level to 0.7 for hair color. This means that the average effect of being homozygous for the PKU gene on hair color is quite small relative to the total variation in hair color in the control population. This relation between a gene's individual effect on a quantitative character and the total amount of variation with respect to the character is the essential feature distinguishing quantitative from qualitative inheritance.

Phenotypic variation in a quantitative character is the result of a combination of genetic and environmental effects.

The distributions of phenylalanine levels in the blood of phenylketonurics and normal individuals are very clearly separated, but they each show appreciable variability. Differences in dietary habits must be an important factor here, since it is known that a diet with reduced phenylalanine can bring the level of it in the blood of phenylketonurics almost down to that of normal individuals. However, there must also be other genetic factors influencing the phenylalanine level in both normal individuals and phenylketonurics. A special case is, of course, that of heterozygotes for the PKU gene, who are known to have slightly higher levels of phenylalanine in their blood than do normal individuals. In fact, as shown in Figure 9.2, careful

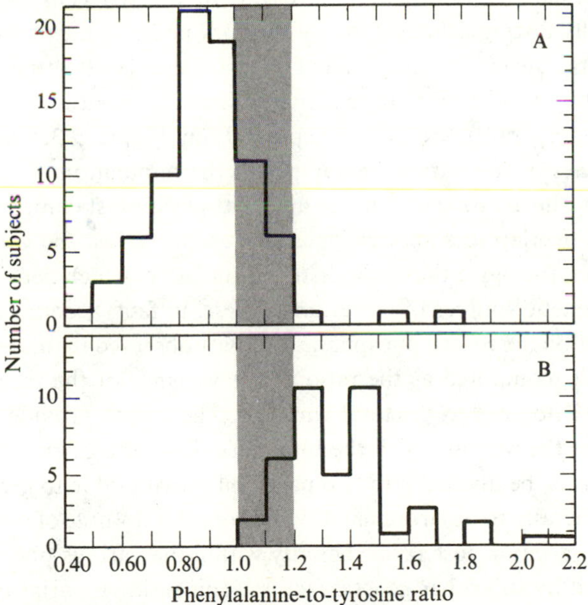

FIGURE 9.2
Distribution of phenylalanine-to-tyrosine ratios in blood taken after overnight fasting of 80 control subjects (A) and 43 phenylketonuric heterozygotes (B). Gray bars indicate a major zone of overlap. (From Perry et al., 1967.)

measurements, under controlled conditions (after overnight fasting), of the ratio of phenylalanine to tyrosine in the blood can separate, with less than 20 percent overlap, normal persons from those who are heterozygous for the PKU gene. This is a good example of the way in which a refinement in a method of measurement can lead to the uncovering of a qualitative genetic difference. The variation in phenylalanine levels observed both for phenylketonurics and normal individuals is the result of both genetic and environmental factors, although homozygosity for the PKU gene clearly has a larger effect on this measurement than any other common genetic or environmental factors.

Genetic polymorphisms are a major source of genetic variability affecting quantitative characters.

It is now known that as many as 30–50 percent of all genetic loci may be polymorphic (see Chapters 4 and 11). Polymorphisms provide a major potential source of the genetic variation affecting quantitative characters.

The way in which variation deriving from a number of identifiable polymorphic genotypes can combine to give a unimodal distribution typical of that for a quantitative character is shown in Figure 9.3. In the red cells are found three electrophoretically distinguishable types of the enzyme acid phosphatase, which are controlled by three alleles at a single locus. Each of the six resulting genotypes has a different distribution of phosphatase activity. These distributions (apart from that for C, which is very rare) are shown separately in Figure 9.3, weighted by the relative frequencies of the various genotypes in the population. Considering only measurements of phosphatase activity, A and B can be almost completely separated while A and BA overlap to a very considerable extent. When the various distributions are summed, they give the outer distribution curve, which betrays no obvious evidence of its genetic sources of variations. These, in fact, account for two-thirds of the total variation in red-cell phosphatase activity observed in this random sample of 275 individuals, computed as the ratio of the variance of the means of the five most common phenotypes to the total variance. These data provide a simple prototype example of the way in which the total variation for a quantitative character can, in simple terms, be divided into two parts, one attributable to genetic variation and the other generally to environmental variation. We do not, of course, know to what extent the variation in enzyme activity within the various phosphatase types may be influenced by still other genetic factors. This residual variation gives, however, an indication of the upper limit of the possible contribution of environmental factors to variation in phosphatase activity, on the assumption that the only genetic variation affecting the activity is the known three-allele polymorphism.

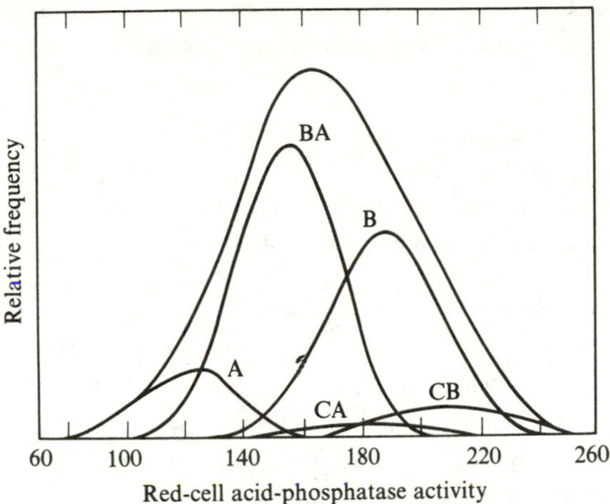

FIGURE 9.3
Distribution of red-cell acid-phosphatase activities in the general population (top line) and in the separate phenotypes (A, BA, B, CA, CB). The curves are constructed from values of the enzyme activities and from the relative frequencies of the phenotypes observed in a randomly selected population. (From Harris, 1966.)

9.2 Genetic Models of Quantitative Variation

The genetic analysis of quantitative characters is based on the construction of models that take into account both the effects of environmental variation and the joint effects of many genes on the given character.

The basis for the genetic analysis of quantitative characters is an extension of the type of simple model just discussed for variation in red-cell acid-phosphatase activity. Many polymorphic genes affecting a character are assumed to be present in the population. Each genotype is assumed to have a given small average effect on the character, the variation within a genotype being due to environmental factors. The total observed variation can then be predicted in terms of the genotype effects, the frequencies of the genotypes, and the magnitude of the environmental variation. The main aim of an analysis is generally to determine the relative contributions of genetic and environmental variations, this being a simple measure of the extent to which a quantitative character is inherited. It should always be kept in mind, however, that this measure refers to a given population in a given environment. The determination depends on the fact that the average differences between individuals should decrease as their genetic relationship increases, and that the extent to which this occurs depends on the relative importance of the genetic and environmental effects.

A major assumption in such an analysis in man is that there is no tendency for the environments of genetic relatives to be similar.

Only on this assumption can the similarity between relatives with respect to a quantitative character be used to measure the relative importance of genotype and environment. It may be possible in working with experimental organisms to control the environment in such a way that there is no correlation with genetic relationship. The social and familial structure of human populations, however, generally causes relatives to have similar environments. As we shall see later, this poses one of the most difficult problems for the study of quantitative inheritance in man.

There are three basic types of quantitative variation, the first of these being variation that can clearly be attributed to single gene differences.

Phenylalanine levels in the blood of phenylketonurics and normal individuals provide an example of variation due to differences at a single locus. In this example, the variation can clearly be subdivided into nonoverlapping, or only partially overlapping, distributions, and then its inheritance can be studied using the standard techniques of Mendelian analysis without recourse to the special techniques needed for the study of quantitative inheritance. Such a situation should, perhaps, be the ultimate goal of all studies on the inheritance of quantitative characters.

The second major category of quantitative variation includes those characters whose distribution in the population is continuous and generally unimodal.

Typical examples are height, weight, and IQ. The distribution of the IQ's of 14,963 Scottish children born in 1926 is shown in Figure 9.4. The data are artificially grouped into ranges of 10 IQ points. The dotted line represents a normal distribution curve fitted to the observed data. Similar distributions, which have a shape fitting closely that of the normal distribution, are obtained for most quantitative characters. The normal curve has a special significance in statistical analysis, which makes it a good candidate for fitting to such data. It can be shown by a proposition known as "the central limit theorem" that any variable that is the sum of a large number of constituent variables tends to follow a normal distribution.

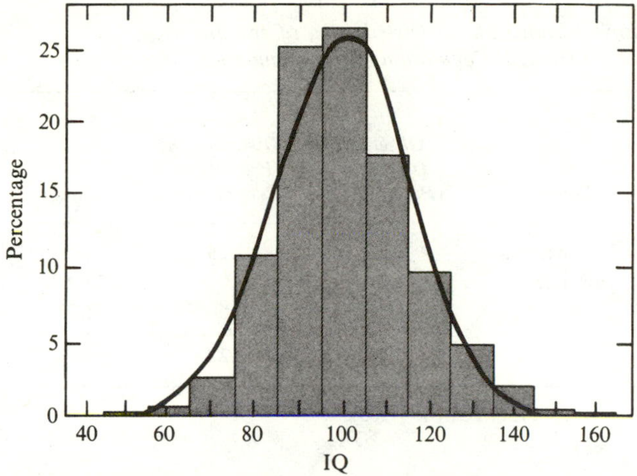

FIGURE 9.4
The distribution of IQ among the 14,963 children born in Scotland on February 1, May 1, August 1, and November 1, 1926. The shaded histogram shows the percentages of the group with IQ's in various ranges of 10 points. This grouping is artificial and is done solely for ease of representation: it does not imply any discontinuity in the values of IQ that children can show. The continuous curve shows the ideal distribution calculated from the observations and representing the statistical population of which the children actually observed are regarded as forming a sample. (Data from MacMeekan; from Mather 1964.)

Thus characters such as anthropometics and IQ, which must represent the sum of a large number of separate effects, both genetic and environmental, should be normally distributed. IQ is defined to have a mean of 100 (in whites) and a standard deviation of approximately 15, regardless of age. It is approximately normally distributed. Thus the IQ's of roughly 95 percent of the population are within the range $100 \pm 2 \times 15$ or 70–130. It is generally true that the extreme left-hand part of the distribution is larger than expected from the normal distribution, suggesting that the individuals who are classed as being mentally retarded are in some sense out of the normal range of IQ variation. As discussed in the chapter on inbreeding, it seems likely that many of these individuals have specific recessive diseases causing mental retardation and are, therefore, genuine outliers. Illustrative data from Penrose's classic book *The Biology of Mental Defect* (Third Edition, 1963) are shown in Table 9.1. The classification into feebleminded (IQ: 50–70), imbecile (IQ: 20–50), and idiot IQ: <20) is, of course, arbitrary, but it serves as a useful guide to define the severity of mental defect. The observed proportion of feeble minded is very close to that expected from the normal distribution. But the proportions of imbeciles and especially of idiots are very much higher than expected from

TABLE 9.1
Lower Limits of the Distribution of Intelligence in the General Population (Age Group 10-14 Years)

Category	Observed Defectives (Percentage)	Normal Distribution Prediction (Percentage)	IQ Range
Feebleminded	2.26	2.23	50–70
Imbecile	0.24	0.04	20–50
Idiot	0.06	0.00001	< 20
Total	2.56	2.27	

Source: From Penrose (1963).

the left-hand part of the normal distribution. It would be interesting to know whether the same is true at the upper end of the scale, since this would clearly suggest the existence of genetic outliers with superior intellect. However, the arbitrariness of the IQ measurement in this range makes such a simple analysis essentially meaningless. The analogues of geniuses and idiots for the character of height are, of course, giants and dwarfs, for both of which, in many cases, a simple genetic etiology is well established.

Children of "subnormal" parents are more likely to be "subnormal" than the children of normal parents.

Data illustrating the general tendency toward familial concentration of mental retardation are shown in Table 9.2. When both parents of an imbecile or idiot child

TABLE 9.2
Classification (in Percentages) of Sibs of Imbecile or Idiot Children According to Whether Parents are Normal, Consanguineous, or Subnormal. These figures should be compared with those in Table 9.1.

Parents	Normal (> 85)	Dull (70–84)	Feeble-minded (50–69)	Imbecile (< 50)	Total Offspring
Normal	90.4	5.2	1.6	2.8	1897
Consanguineous normal	79.6	2.0	4.1	14.3	49
One or both subnormal	54.6	19.7	15.2	10.5	513

Source: From Penrose (1963).

are normal, the IQ distribution of their other children is like that of the normal population with, as usual, an excess in the very low IQ ranges. When, however, one or both parents are subnormal, the distribution is quite different: there are more children in all the lower IQ categories, suggesting a general tendency for such parents to have offspring with lower IQ's. In contrast, consanguineous parents show a significant increase only in the imbecile or idiot categories, which lends weight to the suggestion that much imbecility and idiocy is caused by simple recessive genetic defects. Similar data have been obtained in a recent very extensive study carried out by Reed and Reed (1965). While matings involving one or more subnormal parents show the typical inheritance pattern for a quantitative character that has a significant genetic component, consanguineous normal parents show a pattern suggesting predominantly simple recessive inheritance of imbecility or idiocy. The quantitative interpretation of such data is one of our major tasks.

Some of the problems in such investigations are well illustrated by a famous study on gifted children and their performance as adults, carried out by Terman (see Terman and Oden, 1959), who was one of the first researchers to advocate and construct IQ tests. About 1500 children who were between 7 and 14 years old in the early 1920's and who had IQ's higher than 135 were the subject of an extended follow-up study. Their mean IQ was 150. That of their childen was 133, indicating a very marked tendency for the children of these "superior" individuals to have an IQ comparable to that of their parents. More than 85 percent of the originally selected children completed at least one year of college, indicating, perhaps, that high IQ has at least some correlation with academic ability. Of the spouses chosen by Terman's group of gifted children, more than 75 percent had completed one or more years of college, showing a strong tendency toward assortative mating with respect to intellectual ability. It is clear, therefore, that what the data from Terman's study can tell us about the genetics of IQ is limited because there is a strong tendency for gifted persons to share a common environment, namely, that of the middle- to upper-middle-class university graduate. The data do not enable us to sort out to what extent the similarity in IQ between parents and offspring is due to a common environment rather than to the similarity of their genotypes.

Non-Mendelian "all-or-none" attributes are the third major category of quantitative variation.

There are many attributes, for example, most of the commoner congenital malformations such as cleft palate and anencephaly, which are "all-or-none" attributes but which nevertheless are not inherited in a simple Mendelian fashion. They must be considered quantitative in the sense that their expression is, generally, influenced by the action of many genes and by the environment, just as is that of unimodal continuously distributed characters. The simplest framework for thinking about such discrete attributes is that of a **threshold model**, first introduced by S. Wright (1934) in a study of the number of toes in guinea pigs. It is assumed that there is an underlying continuous variable x, say, whose inheritance is determined in exactly the same way as was discussed for unimodal continuous variates. All individuals with a value of x greater (or less than) some threshold value T are assumed to have the attribute. This simple device relates a continuous distribution to a discrete difference. The theory of the inheritance of such characters is the same as for any other quantitative character, but with the added complication of specifying the relationship between the distribution of x, the underlying continuous variable, and p, the proportion of people who have the attribute (see Figure 9.5).

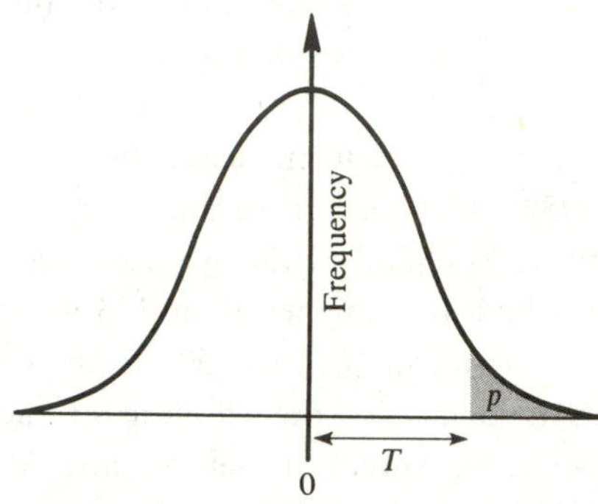

FIGURE 9.5
Threshold model. All individuals with a value of x greater than T are affected. The proportion of affected individuals is the area under the distribution curve beyond T.

Congenital malformations and a number of other important human diseases are "threshold" characters.

Threshold characters range all the way from those that show hardly any evidence of familial concentration, such as breast cancer, anencephaly, and spina-bifida, to those that can almost be explained in terms of a single-gene hypothesis with reduced penetrance, such as diabetes mellitus and schizophrenia.

The overall incidence at birth of the commoner congenital malformations (excluding Down's syndrome and other chromosomal abnormalities) is given in Table 9.3. Their total frequency of 1.2 percent makes them a very important source of illness in present-day western industrialized societies, which have very low infant-mortality rates (see Chapter 6). These incidence figures do not include the total frequency of genetic defects or chromosomal abnormalities which, though

TABLE 9.3
Population Incidence of Common Congenital Malformations (Excluding Down's Syndrome and Other Chromosomal Abnormalities)

Malformation	*Approximate Incidence per* 1000 *Live Births*
Anencephaly	3
Spina-bifida	3
Heart malformations	1
Harelip ($\pm$ cleft palate)	1
Clubfoot	1
Pyloric stenosis	3
Congenital dislocation of the hip	1
Total	12 (1.2 percent)

Source: Based on Carter (1965).

individually rare, together must have a comparable total incidence. The proportions affected of relatives of persons with some of these malformations, given in terms of the population incidence, are shown in Table 9.4. Thus, for example, the frequency of harelip among monozygotic twins of individuals with harelip is 500 times the population incidence; the frequency among sibs of affected individuals is 35 times the population incidence. Clearly, in each of the examples shown, with the possible exception of anencephaly and spina-bifida, there is a strong familial concentration, suggesting that important genetic factors are at play. However, in none of the examples does the incidence among sibs rise above 4–5 percent, and it should be noted, for comparison, that the expected incidence of a simple recessive defect among the sibs of a person having the defect is 25 percent. Though there is no clear-cut evidence for specific environmental factors as causative agents for these malformations, there are some striking correlations between their incidence and certain socioeconomic and demographic parameters. In the mouse and other animals certain chemicals (teratogenic agents), administered to pregnant females markedly increased the incidence of some congenital malformations. The thalidomide story is a gruesome reminder that the same phenomena are certainly to be expected in man. Even excesses of vitamins A and K are known to be teratogenic in animals. In a study on the incidence of anencephaly and other malformations in Scotland, John Edwards (1958) showed that the frequency of anencephaly differed from 0.9/1000 among professional people to 3.6/1000 among skilled workers. The malformations studied by Edwards also showed significant variation between localities and 30–50 percent seasonal variation. The incidence of many congenital malformations is different for different birth orders and varies between the sexes. The

TABLE 9.4

Proportion Affected of Relatives of Persons with Some Common Malformations, Relative to the Incidence in the General Population

		Congenital Dislocation of the Hip			*Pyloric Stenosis*		
	Harelip (± Cleft Palate)	*All Patients*	*Female Relatives of Female Patients*	*Talipes Equinovarus*	*All Patients*	*Female Relatives of Female Patients*	*Anencephalus and Spina Bifida*
Population incidence (approximate)	0.001	0.001	0.0018	0.001	0.003	0.001	0.005
Monozygotic twins[a]	500	500	300	325	150	—	—
Sibs[a]	35	40	35	20	20	100	8
First cousins[a]	7	4	3	5	4	12	—
Second cousins[a]	3	1½	2 2	2	1½	3	2

Source: From Carter (1965).

[a] The numbers in the second and later rows are the ratios of the incidence in twins, etc., relative to the population incidence given in the first row.

sex ratio (males to females) for pyloric stenosis is 5 : 1; for clubfoot 2 : 1 and harelip 1.5 : 1. All these factors greatly complicate any simple interpretation of the inheritance of these important diseases.

9.3 Interaction of Genotype and Environment

Interactions between the genotype and environmental factors may be very important for the interpretation of quantitative inheritance.

So far, in discussing the joint effects of genotype and environment on a quantitative trait, we have made the implicit assumption that their effects are **additive**—that is, that if one genotype has a trait greater than another in any one environment, it will also be the "superior" genotype in all other environments. This assumption, which is characteristic of most quantitative inheritance models, is often wrong. The logical relationships that may exist between two genotypes (populations or species, if reference is to mean values) in two environments were discussed by Haldane (1946) in an illuminating paper entitled "The Interaction of Nature and Nurture." Two individual genotypes labeled A and B are measured for a quantitative trait in two environments X and Y, and the four corresponding values of the trait are numbered from 1 to 4 in order of magnitude. There are $4! = 24$ ways of arranging 4 items in a sequence. However, if one arbitrarily designates AX as the largest measurement, there are only 6 distinguishable logical arrangements. These are illustrated in Table 9.5. Arrangements 1a and 1b represent the conventional assumption, namely, that $A > B$ in both environments. However for 1a, both values of A are better (larger) than the highest of B—that is, A is always better than B; in 1b environment X is better than Y, though within them A and B retain the same relative ranking. (Here we assume that larger trait values are better.) The second arrangement represents a type of interaction in which A is better than B in environment X, but worse in environment Y, although environment X is better than Y. Haldane points out an analogy between this type of interaction and wild (B) and domesticated (A) species in natural habitats (Y) and man-made habitats (X). All fare better in man-made habitats (which offer protection) than in the wild, though wild types do relatively better in the wild than domestic species. Interactions of types 3 and 4 represent situations in which the environments have opposite effects on the two types of individuals. For interactions of type 3, A is still always better than B though while X is better for A than Y, the reverse is true for B. Haldane here draws the analogy between mentally retarded (B) and normal (A) individuals in normal (X) and special (Y) schools. The last types of interaction, 4a and 4b, represent specialization such that A and B are each optimally adapted to their respective environments X and Y. This is the situation generally expected for evolutionary adaptation

TABLE 9.5

Relationships of the Measurements of a Quantitative Trait on Two Genotypes A and B in Two Environments X and Y. Measurements are numbered 1 to 4 in order of magnitude. AX is always assumed, arbitrarily, to be the largest.

Type		X	Y	
1*a*	A	1	2	$A > B$ in X and Y
	B	3	4	
1*b*	A	1	3	
	B	2	4	
2	A	1	4	
	B	2	3	$A > B$ in X, $B > A$ in Y; $X > Y$
3	A	1	2	$A > B$ in X and Y but $BX < BY$ and $AX > AY$
	B	4	3	
4*a*	A	1	3	$A > B$ in X; $B > A$ in Y
	B	4	2	
4*b*	A	1	4	
	B	3	2	

Source: Modified from Haldane (1946).

of individuals to their respective environments. The adaptation of sickle-cell heterozygotes to malaria represents a specific human example of this form of interaction.

An additive model always implies a logical relationship of the first type, in which the relative ordering of the genotypes is the same in all environments.

Suppose that the phenotypic value of an individual for a quantitative trait in a given environment is the sum of an individual contribution and an environmental contribution. The expected pattern of observations will then be

GENOTYPE	ENVIRONMENT X	ENVIRONMENT Y
A	$a + x$	$a + y$
B	$b + x$	$b + y$

where a and b are the respective contributions of genotypes A and B, and x and y the contributions of the environments X and Y. All values are assumed to be positive. In addition, since we assume without loss of generality that AX is the largest, we must have $a > b$ and also $x > y$. This immediately ensures that $AX > BX$ and $AY > BY$. We get interactions of type 1a or 1b according to whether AY is greater or less than BX—that is, according to whether $a - b$ is greater or less than $x - y$. The distinction between 1a and 1b thus depends on whether genotypic differences are greater or less than environmental differences. It is for this reason that Haldane labels the first the "eugenical" and the second as the "environmental" viewpoint. In both cases, however, A is greater than B and X is greater than Y.

9.4 The Basic Model of Polygenic Inheritance

In this section we shall derive some of the consequences of the simplest genetic models of quantitative inheritance and illustrate their application to human data.

Experiments with self-fertile plants made homozygous by continued inbreeding clearly illustrate the principle of partitioning the total phenotypic variation into a portion due to environmental variation and one due to genetic variation.

The first geneticist to draw clearly the distinction between genetic and environmental variation was Johannsen (1909). He worked with bean plants, and since these can be self-fertilized, he was able, by repeatedly self-fertilizing them, to obtain strains that were effectively homozygous (see Chapter 7). He then showed that whatever the seed weight of a parent plant derived from such an inbred strain, its offspring by self-fertilization reproduced the same variation as that present in the homozygous strain from which the parent was taken. He thus demonstrated, as is, of course, expected, that the variation within such a strain is entirely environmental. He was able to distinguish environmental and genetic variation in a simple way by obtaining homozygous, genetically identical, strains which cannot be done in man. The results of a cross between two such homozygous lines illustrate in a simple way how observed phenotypic variation can be partitioned into genetic and environmental components. An example of data on corolla length in the tobacco plant is shown in Figure 9.6. The variance of the parents and that of the F_1 is expected to be the same, being the environmental variance V_E. The F_1 is, of course, genetically uniform, though not homozygous, since all plants must be homozygous at every locus at which their parents are identical, and heterozygous at every locus at which they are different. The F_2 (progeny of F_1 plants self-fertilized) has a distribution with a larger variance than V_E. This must be attributed to segregation

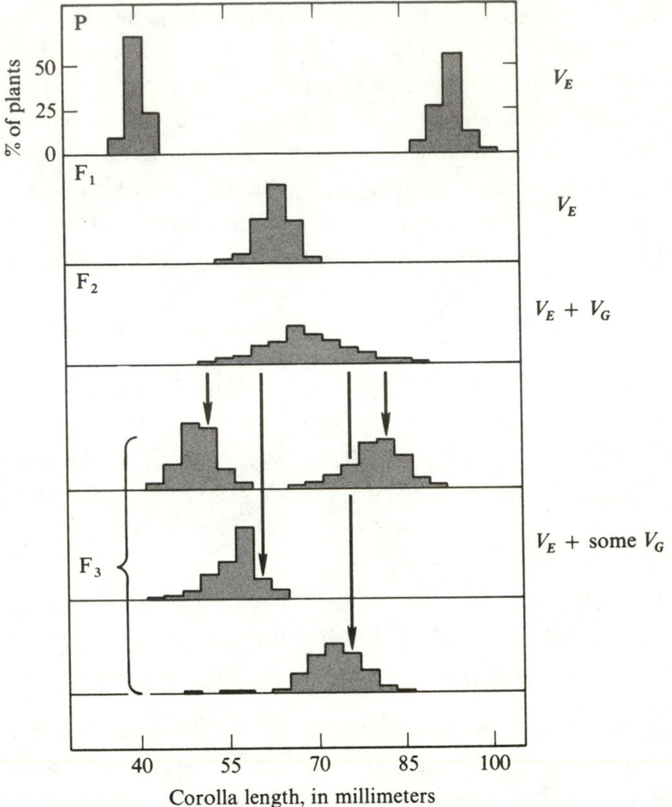

FIGURE 9.6
The inheritance of corolla length in *Nicotiana longiflora*. For ease of presentation, the results are shown as the percentage frequencies with which individuals fall into classes, each covering a range of 3 millimeters in corolla length and centered on 34, 37, 40, etc., millimeters. This grouping is quite artificial and the apparent discontinuities are spurious: corolla length actually varies continuously. The means of F_1 and F_2 are intermediate between those of the parents. The means of the four F_2 families are correlated with the corolla length of the F_2 plants from which they came, as indicated by the arrows. Variation in parents and F_1 is all nonheritable, and hence is less than that in F_2, which shows additional variation arising from the segregation of the genes concerned in the cross. Variation in F_3 is, on the average, less than that of F_2 but greater than that of parents and F_1. Its magnitude varies among the different F_2's according to the number of genes that are segregating. (From Mather, 1949; and East, 1916.)

at those loci at which the F_1 plants are heterozygous. Assuming that the environmental and genetic effects are independent of each other, we can write the variance of the F_2 distribution as

$$V_E + V_G,$$

where V_G is the variance due to genetic segregation. Recall that the variance of the sum of two independent variables is the sum of the separate variances (see Appendix I). The value of V_G can be estimated as the difference between the F_2 variance and the average of the F_1 and parental variances. The ratio $V_G/(V_E + V_G)$ is a measure of the proportion of the F_2 variance that is contributed by genetic segregation, and is thus a measure of the relative genetic (as opposed to environmental) determination of the character. The relation between this quantity, the **degree of genetic determination**, and that commonly called "heritability" will be discussed later. Some results of an F_3 generation (F_2 self-fertilized) are shown in Figure 9.6 in relation to the mean values of the F_2 plants from which they were derived. The mean of the F_2 plants and F_3 plants is approximately the same, while the variance of the F_3 (obtained by self-fertilizing a given F_2 plant) lies between V_E and $V_E + V_G$. This is expected, because in choosing F_2 plants with a given measurement we have reduced the genetic variance by restricting ourselves to those genotypes that can achieve the chosen F_2 value.

The German physician Weinberg, who contributed much to human population genetics, was probably the first to attempt to build, in 1909, a detailed mathematical model for the inheritance of quantitative characters. Though his work in this area is now little known, his approach followed in general terms the same lines as those of later investigations. Fisher, in 1918, published a paper whose content was the forerunner of much of the modern approach to the analysis of the inheritance of quantitative characters. He was presumably unaware of Weinberg's earlier work, which certainly followed the same general approach, though Fisher's analysis was much more sophisticated and complete.

The genetic variance can be divided into two components, as originally shown by Fisher in 1918, *one resulting from differences between homozygotes, the* ***additive genetic variance****, and the other resulting from specific effects of various alleles in heterozygotes, the* ***dominance variance****.*

In order to construct a simple quantitative model we first consider the effect on a quantitative trait of genetic variation at one locus with two alleles A_1 and A_2. We assume, for convenience, that the origin of the measurement in question is midway between the average values of the two homozygotes A_1A_1 and A_2A_2, and that the average measurements of the three genotypes A_1A_1, A_1A_2, and A_2A_2 are as indicated in Figure 9.7. These averages are taken over all possible environments and

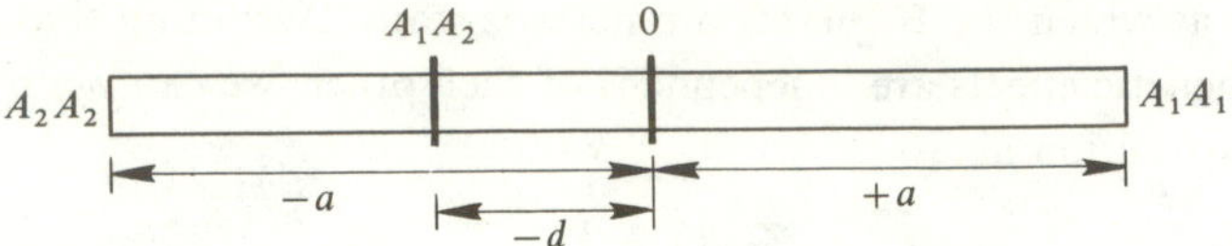

FIGURE 9.7
Model for quantitative effects of variation at one locus with two alleles A_1 and A_2. The origin was chosen, for convenience, to be at the midpoint between the homozygotes. The position of the heterozygote may be on either side of the origin, or at it, depending on the sign and magnitude of the "heterozygote effect" d.

with respect to variation at all other relevant loci. Consider now the expected contribution of this locus to the genetic variance of the F_2 derived from a cross between two homozygous lines, carrying different alleles at this locus. The frequencies of the three genotypes in the F_2 are $\frac{1}{4}A_1A_1$, $\frac{1}{2}A_1A_2$, and $\frac{1}{4}A_2A_2$, and so the mean measurement of the F_2 is

$$\sum p_i x_i = \tfrac{1}{4}a - \tfrac{1}{2}d - \tfrac{1}{4}a = -\tfrac{1}{2}d,$$

where p_i is the frequency of each class and x_i its phenotypic value, and a and d are defined as in Figure 9.7.

The contribution of this locus to the variance of the F_2 is, therefore,

$$\sum p_i x_i^2 - (\sum p_i x_i)^2 = \tfrac{1}{4}a^2 + \tfrac{1}{2}d^2 + \tfrac{1}{4}a^2 - (-\tfrac{1}{2}d)^2 = \tfrac{1}{2}a^2 + \tfrac{1}{4}d^2.$$

Suppose, now, that there are many such loci, each contributing independently to the genetic variance of the F_2; we can then write the genetic variance of the F_2 as

$$\tfrac{1}{2}\sum a^2 + \tfrac{1}{4}\sum d^2 = V_A + V_D, \tag{9.1}$$

where summation is over all the various loci. The term $V_A = \frac{1}{2}\sum a^2$, which is a function only of the difference between homozygotes, is usually called the **additive genetic variance**. When $d = 0$, the heterozygote is exactly intermediate between the two homozygotes and genes A_1 and A_2 are said to be "additive." Traditionally d is called the **dominance deviation** and $V_D = \frac{1}{4}\sum d^2$ the **dominance variance.** Dominance here refers to deviations of the heterozygote from the mean of the two homozygotes. The total phenotypic variance of the F_2 is

$$V_{F_2} = V_A + V_D + V_E,$$

where V_E is the environmental variance. Using the same genetic model, the contribution of such a set of loci to the genetic variance of other sorts of crossses (for example, the F_3 or backcrosses to the parents) can also be derived in terms of V_A and V_D. Thus, if enough data of this sort are available, the quantities V_A, V_D, and V_E can be separately estimated. These provide a basic description of the genetic and environmental components of variation with respect to a quantitative character.

It can be shown (see Example 9.1) that the sum of the variances of the two backcrosses (to each of the parental types) is

$$V_B = V_A + 2V_D + 2V_E. \tag{9.2}$$

Thus, for example, given observations of V_{F_2}, V_B, and $V_{F_1}(=V_E)$ we would estimate

$$V_E = V_{F_1},\ V_D = V_B - V_{F_2} - V_{F_1}, \quad \text{and} \quad V_A = 2V_{F_2} - V_B. \tag{9.3}$$

9.5 Skin Color as an Example of Polygenic Inheritance

Matings between Negroes and Caucasians can give useful information on polygenic inheritance of skin color.

The theory we have just described for crosses between inbred lines has been applied by Harrison and Owen (1964) to the study of skin color inheritance in matings between Negroes and Caucasians. The rationale underlying its application here is that African and European populations were effectively separated for a long enough period of time to make it quite likely that they are each homozygous for the different major genes controlling their difference in skin colors. The population studied comprised a community of West African Negroes and English Caucasians in Liverpool, the offspring (F_1) of matings between the Negroes and Caucasians, the offspring of matings between the F_1 and either Negroes or Caucasians (backcrosses), and the offspring (F_2) of matings among the F_1. The use of a reflectance spectrophotometer and of a standard location on the human body at which skin color was determined gave objective and reliable measurements of skin color. The mean percentages of light reflected at different wavelengths (the reflectance) by the skin of the Negro, Caucasian, hybrid, and backcross groups are shown in Figure 9.8. As expected for a quantitative character that is, in part, genetically determined, the F_1 curve falls roughly midway between the Caucasian and Negro curves, and the curves of the backcrosses fall between these curves and that of the F_1. It is very clear from these data that different results will be obtained from an analysis depending on which measurement is used. Thus, the F_1 value is nearer the midpoint of Caucasian and Negro values, suggesting absence of dominance, at 545 mμ than at either 425 mμ or 685 mμ. On the other hand, the variances of the Caucasians (V_C), the Negroes (V_N), and the $F_1(V_{F_1})$ are all quite different for each of these measurements, while our theory tells us that on simplest assumptions they should all be approximately equal to the environmental variance.

These discrepancies may be due only to the fact that the scale of measurement used does not adequately represent the underlying action of the genes. Harrison and Owen found that taking the antilog of the reflectance at 685 mμ gave measurements

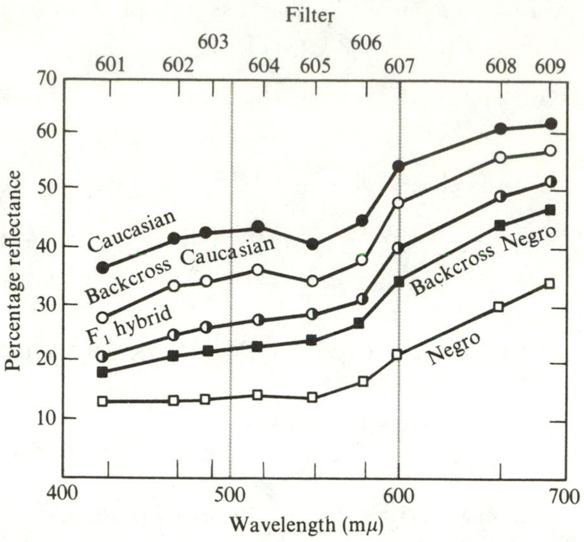

FIGURE 9.8
Mean reflectance curves for skin color of West African Negro, English Caucasian, and various hybrid groups. (From Harrison and Owen, 1964.)

that were reasonably consistent with the simple expectations of the model of quantitative inheritance we have just outlined. The data on means and variances of this measurement are given in Table 9.6. The environmental variance is estimated by the mean of V_C, V_N and V_{F_1}, giving $1000\ V_E = 1.24$.

If there is genetic variation for skin color within the Negro population or the Caucasian population, which is certainly very likely, then this variation is, of course,

TABLE 9.6
Means and Variances of Skin Color Measurements Antilog of (Reflectance at 685 mμ) for Various Matings

	No. of Observations	*Mean*	*Variance* (× 1000)
Caucasian	105	$0.421 = \overline{C}$	$1.09 = V_{\overline{C}}$
Negro	106	$0.225 = \overline{N}$	$1.05 = V_{\overline{N}}$
F_1 hybrid	94	$0.334 = F_1$	$1.59 = V_{F_1}$
Negro backcross	26	0.304	$1.71 = V_{BN}$
Caucasian backcross	30	0.382	$2.00 = V_{BC}$
F_2 hybrid	14	0.346	$1.99 = V_{F_2}$

Source: Data from Harrison and Owen (1964).

included in V_E. If the amount of such variation were large relative to the true environmental variation, and also quite different in the two populations, then we would not necessarily expect $V_N = V_C$. However, it is at least clear that this source of variation is likely to be small relative to the difference between Caucasians and Negroes. The simplest way to estimate the additive and dominance variances V_A and V_D would be to use the observed value of V_{F_2} together with Equation 9.3. The F_2 variance estimate is, however, based on only 14 observations and is therefore subject to a very high sampling error. The sampling variance of a variance V (assuming a normal distribution) is $2V^2/n$. Thus, from Equation 9.3,

$$V_A = 2 \times 1.99 - 1.71 - 2.00 = 0.27,$$

with standard error

$$\sqrt{2\left(4 \times \frac{1.99^2}{14} + \frac{1.71^2}{26} + \frac{2.00^2}{30}\right)} = 1.65.$$

The ratio V_D/V_A can, however, be estimated in another way, based on differences between the means rather than the variances, on the assumption that the variability of the effects of individual genes is negligible or follows a certain pattern. The formula to be used is

$$\frac{V_D}{V_A} = \frac{1}{2}\left[\frac{\frac{1}{2}(\bar{C} + \bar{N}) - \bar{F}_1}{\frac{1}{2}(\bar{C} - \bar{N})}\right]^2 \tag{9.4}$$

(see Example 9.2). Using the data of Table 9.6, we obtain from Equation 9.4

$$\frac{V_D}{V_A} = \frac{1}{2}\left[\frac{\frac{1}{2}(0.421 + 0.225) - 0.334}{\frac{1}{2}(0.421 - 0.225)}\right]^2 = 0.00625,$$

suggesting that the skin-color genes have essentially no dominant effect. Now, using this value of V_D/V_A, the estimate $1000\ V_E = 1.24$, based on the mean of V_C, V_N, and V_{F_1}, and Equation 9.2, to give

$$\frac{3.71}{1.000} = (V_A + 2V_D + 2V_E),$$

we obtain $1000\ V_A = 1.215$ and $1000\ V_D = 0.008$. Thus, the additive genetic variance for skin color in crosses between Negroes and Caucasians, due to the genes by which they differ, is comparable to the variance within each of these groups. On the assumption that the relevant genes have more-or-less equal effects, we can obtain a *rough* estimate of the number of skin-color genes by which Negroes and Caucasians differ. Thus from the definition of V_A and the expression of a variance in terms of a sum of squares and the square of the mean (see Example 9.2 and Appendix 1), we have

$$2V_A = \sum a^2 = \sigma_a^2 + k\bar{a}^2; \tag{9.5}$$

while from the definition of $\bar{C}$ and $\bar{N}$,

$$\tfrac{1}{2}(\bar{C} - \bar{N}) = \sum a = k\bar{a}. \tag{9.6}$$

Thus from Equations 9.5, and 9.6, if $\sigma_a{}^2$ can be ignored,

$$\frac{[\frac{1}{2}(\bar{C} - \bar{N})]^2}{2V_A} = k. \tag{9.7}$$

Substituting the values of $\bar{C}$ and $\bar{N}$ from Table 9.6, and the estimate of V_A already obtained, gives

$$k = \frac{(0.098)^2}{2 \times 0.001215} \sim 4,$$

suggesting that a relatively small number of genes may be responsible for the major differences in skin color between Negroes and Caucasians. This estimate is, however, subject to a high standard error. An analysis of the same problem by Stern (1960), using different data and methods, came to similar conclusions.

As we have emphasized throughout our discussion of these data, there are many simplifying assumptions that have to be made in order to obtain a simple answer. The main ones are:

1. That the model for crosses between inbred lines is appropriate. This implies that the genes responsible for the major differences between Negroes and Caucasians are homozygous in the original populations and that genetic variation within each population is small relative to environmental variation, or at least is comparable in the two populations.

2. That the scale of measurement is so chosen as to eliminate interactions between genes and between genotypes and the environment. Harrison and Owen found that the reflectance at 545 mμ and the log of the reflectance at 425 mμ, as well as the antilog of the reflectance at 685 mμ (see Table 9.6) satisfied some standard criteria used to test scaling, which will not be described here. All three measurements gave very similar estimates for k and for V_D/V_A and $V_{\bar{N}}/V_{\bar{C}}$, suggesting that scaling was not a major problem with their data. Their analysis indicates that a combination of simple model building with appropriately collected data can give valuable information on the genetic determination of a quantitative character. The next step is to use more refined measurements, especially biochemical measurements, in order to identify specifically and separately the various major genes affecting skin-color differences between Negroes and Caucasians.

It should be emphasized that this treatment of the problem is possible because we can make one important assumption: that Negroes and Caucasians have "fixed" most of the differences for skin color as a consequence of a selective process. One group can be considered to have all the "white" alleles, the other all the "dark" alleles at any locus. They are thus "pure lines" for skin color, an assumption that is only approximately true. When, however, we consider populations that have not been separated, but are mating at random, the treatment has to be quite different, as we shall see in the next section.

9.6 Partitioning the Genetic Variance in a Random-mating Population

It is also possible to partition the genetic variation of random-mating populations into two components: one due to additive effects and one due to dominance. These, however, have a slightly different meaning than before. We assume, as before, one locus with two alleles A_1 and A_2 that have quantitative effects as illustrated in Figure 9.7. In a random mating population if p is the gene frequency of A_1, the genotypes A_1A_1, A_1A_2, and A_2A_2 occur with frequencies p^2, $2pq$, and q^2, following the Hardy-Weinberg law, where $q = 1 - p$. The mean of the population is therefore

$$m = ap^2 - 2pqd - aq^2 = a(p - q) - d2pq \tag{9.8}$$

since

$$p^2 - q^2 = (p - q)(p + q) = p - q.$$

The variance due to segregation at this locus is given by

$$\begin{aligned} &p^2 \times a^2 + 2pq \times d^2 + q^2 \times a^2 - m^2 \\ &= a^2(p^2 + q^2) + 2pqd^2 - [a(p - q) - 2pqd]^2. \end{aligned}$$

This can be written in the form

$$\begin{aligned} &2pq[a^2 + 2ad(p - q) + d^2(1 - 2pq)] \\ &= 2pq[a + d(p - q)]^2 + 4p^2q^2d^2, \end{aligned} \tag{9.9}$$

for $1 - 2pq - (p - q)^2 = 1 - p^2 - q^2 = 2pq$, since $p + q = 1$. Thus, if there are many such loci each acting independently the total contribution to the genetic variance can be written as

$$V_G = \sum 2pq[a + d(p - q)]^2 + \sum 4p^2q^2d^2 = V_A + V_D, \tag{9.10}$$

where

$$V_A = \sum 2pq[a + d(p - q)]^2 \tag{9.11}$$

and

$$V_D = \sum (2pqd)^2, \tag{9.12}$$

and summation is over all polymorphic loci affecting the character in question. The quantities V_A and V_D are the additive and dominance variances. Clearly, $V_D = 0$ when $d = 0$ at every locus—that is, when all genes are nondominant. In this case $V_A = \sum 2pqa^2$, where a, as before, is one-half the difference between homozygotes. Both V_A and V_D depend on the gene frequencies. The terms of V_D are always at a maximum when $p = q = \frac{1}{2}$ while, in general, the terms of V_A are at a maximum when $p = q = \frac{1}{2}$ only if, in addition, $d = 0$. Note that when $p = q = \frac{1}{2}$ for all loci, Equation 9.10 is equivalent to Equation 9.1, which gives the variance for the F_2 between two inbred lines. This is to be expected, since such an F_2 is equivalent to a Hardy-Weinberg population with all gene frequencies at one-half.

In a random-mating population the correlations between individuals of given genetic relationship with respect to a quantitative character can be expressed in terms of V_A and V_D and the environmental variance V_E.

This requires the computation of covariances between relatives, which can be done using an approach similar to that already used for variances, and which is given in detail in Example 9.3.

All covariances take the form

$$W = lV_A + mV_D, \tag{9.13}$$

where l and m depend on the relationship. The covariances for some of the relationships that can easily be studied in human populations are given in Table 9.7. Of these, only the covariance between full sibs includes a contribution from the dominance variance V_D, so that when $V_D \neq 0$ the covariance between sibs is greater than

TABLE 9.7

Covariances Between Relatives in a Random-mating Population. Half sibs have one parent in common. The midparent-offspring covariance, which is the covariance between offspring and the mean of the two parents, is the same as that between offspring and a single parent.

Relationship	Covariance	Coefficient of V_A	Coefficient of V_D
Parent-offspring (also midparent-offspring)	$W_{O/P}$	$\frac{1}{2}$	—
Full sibs	$W_{S/S}$	$\frac{1}{2}$	$\frac{1}{4}$
Half sibs		$\frac{1}{4}$	—
Uncle-nephew		$\frac{1}{4}$	—
First cousins		$\frac{1}{8}$	—

that between parent and offspring. The covariance decreases as the degree of relationship decreases. The coefficient of V_A is the coefficient of relationship.

The correlations between relatives are obtained by dividing the covariances by the total variance, which is

$$V_A + V_D + V_E,$$

where V_E is, as before, the environmental variance. One exception is the correlation between midparent and offspring. Here the correlation is

$$\frac{\text{cov(midparent-offspring)}}{\sqrt{\text{var(midparent)} \times \text{var(offspring)}}} = \frac{\frac{1}{2}V_A}{\sqrt{\frac{V_A + V_D + V_E}{2} \times (V_A + V_D + V_E)}}$$

$$= \sqrt{2} \times \text{parent-offspring correlation}$$

It should be emphasized that the covariances do not include any contribution from the environmental variance, only on the assumption that there is no interaction between genotype and environment. This means, in other words, that there is no tendency for relatives to have a similar environment. It is only under this assumption that the total covariance between relatives is due only to the genetic correlation between relatives. This assumption, as we have already discussed, is a very serious limitation to the application of this simple theory to human data.

The estimation of V_A, V_D, and V_E from data on correlations between relatives goes back to Fisher in 1918, who first derived the results given in Table 9.7 and so with Weinberg, as mentioned before, laid the basis for reuniting Mendelism with data on the inheritance of quantitative characters. Lists of formulas for variances, covariances, and correlations in crosses and in random-mating populations are given by Mather (1949), who uses the symbols D (equivalent to $2V_A$) and H (equivalent to $4V_D$). Our notation is close to the one originally introduced by Fisher.

From Table 9.7 the correlation between parents and offspring is

$$r_{P/O} = \frac{\frac{1}{2}V_A}{V_A + V_D + V_E}, \tag{9.14}$$

and that between sibs is

$$r_{S/S} = \frac{\frac{1}{2}V_A + \frac{1}{4}V_D}{V_A + V_D + V_E}. \tag{9.15}$$

Each of these correlations has a maximum value of $\frac{1}{2}$ when V_D and V_E are both zero.

Fingerprint patterns can be quantitated by counting the density of ridges, as indicated in Figure 9.9. There is considerable variability in the total ridge count, which is the sum of the ridge counts on all fingers. Thus, for example, for a group of 825 British males the mean count was 145 and the standard deviation 51. Holt

FIGURE 9.9
The method of ridge counting of finger prints for arches, whorls, and loops. The white line runs from the center of a pattern to the Y-shaped junction, where three sets of ridges meet, which is called the triradius. The number of ridges crossed by this line constitute the ridge count for the pattern on a finger. Note that whorls have two triradii and hence two ridge counts; and loops have one triradius and one ridge count. Arches have the triradius and the center of the pattern or have no center or triradius at all: in either event no line can be drawn and the ridge count of arches therefore is always zero. The counts of loops and whorls depend on the size of the patterns and the thickness of the ridges. (Photo by L. Razavi.)

(1961) found that the parent-offspring correlation was 0.48 and the sib-sib correlation 0.50 ± 0.04. The correlation between parents (who are, of course, generally not related genetically) was 0.05 ± 0.07, which is not significantly different from zero. This indicates that almost all the observed variation in the population can be attributed to additive nondominant genetic variation and little, if any, to environmental effects.

At least 70–80 *percent of the variation in blood pressure may be genetically determined.*

Extensive data on the correlation between relatives with respect to blood pressure was collected by Miall and Oldham (1963). The measurements they analyzed were blood-pressure scores that were corrected for age and sex. The data they obtained for correlations between sibs, and between parents and offspring, are given in the first two rows of Table 9.8. Using Equations 9.14 and 9.15 for systolic pressure we have

$$r_{P/O} = 0.237 = \frac{\frac{1}{2}V_A}{V_A + V_D + V_E} = \frac{\frac{1}{2}V_A}{V_P}; \tag{9.16}$$

thus

$$\frac{V_A}{V_P} = 2r_{P/O} = 0.474.$$

TABLE 9.8
Blood-pressure Correlations Between Sibs, and Between Parents and Offspring and the Corresponding Variance Partition. $V_P = V_A + V_D + V_E =$ *the total phenotypic variance.*

Correlations and Variance Ratios	*Systolic Pressure*	*Diastolic Pressure*
$r_{P/O}$	0.237	0.183
$r_{S/S}$	0.333	0.265
V_A/V_P	0.474	0.366
V_D/V_P	0.384	0.328
V_E/V_P	0.142	0.306

Source: Data based on observations of 612 families from Miall and Oldham (1963).

Also, since

$$r_{S/S} = 0.333 = \frac{\frac{1}{2}V_A + \frac{1}{4}V_D}{V_A + V_D + V_E}, \tag{9.17}$$

and, subtracting Equation 9.14 from Equation 9.15,

$$(r_{S/S} - r_{P/O}) = \frac{\frac{1}{4}V_D}{V_P}, \tag{9.18}$$

we have

$$\frac{V_D}{V_P} = 4(0.333 - 0.237) = 0.384.$$

Finally,

$$\frac{V_E}{V_P} = 1 - \frac{V_A}{V_P} - \frac{V_D}{V_P} = 1 - 0.474 - 0.384 = 0.142.$$

Thus we see that only 14 percent of the total variation seems to be environmental and that the dominance variance constitutes almost 45 percent of the total genetic variance. Exactly similar calculations indicate that for the diastolic pressure 30 percent of the total variation is environmental and 47 percent of the genetic variance is due to dominance. The two sets of data are thus fairly consistent in indicating a very significant dominance effect, while the environmental variance is somewhat higher for diastolic than for systolic pressure. Miall and Oldham in analyzing their data seem to ignore the difference between $r_{P/O}$ and $r_{S/S}$, and thus to ignore the dominance effect, and as a result estimate that only 20–30 percent of the variation is genetically determined. Assuming that our simple model is valid, as they apparently do, there seems to be no good reason for rejecting the high estimate of the proportion of the genetic variation due to dominance. This is, after all, the only simple *genetic* explanation for values of $r_{S/S}$ being higher than values of $r_{P/O}$. More plausible explanations for the difference might, once again, lie in the fact that environmental correlations between sibs may be stronger than those between parents and offspring. If the genetic interpretation is correct it suggests that there may be relatively few genes involved in determining blood pressure, since a consistent dominant effect of many genes seems unlikely. In this case specific genetic segregation of major genes affecting blood pressure should be looked for. Their existence has been suggested by some authors but the data obtained so far are inconclusive.

The proportion of the total phenotypic variance that is due to additive genetic variation

$$h^2 = \frac{V_A}{V_A + V_D + V_E} \tag{9.19}$$

is generally called the **heritability**. This should be distinguished from the degree of genetic determination, which in this case would be $(V_A + V_D)/(V_A + V_D + V_E)$. The heritability is a measure of the amount of genetic variability excluding that expressed by heterozygotes. When practising selective breeding, animal and plant breeders generally aim to produce individuals of improved economic quality, having, for example, higher milk or meat production. Their goal is then, usually, to produce a superior stock whose quality will not deteriorate very much when its individuals are mated with each other. Thus, they depend on making individuals homozygous for genes which improve their economic qualities. The heritability h^2 is a measure of the amount of genetic variability available for the plant and animal breeder to use as a basis for selective breeding.

The correlations between sibs and between parents and offspring for characters such as stature, arm span, length of forearm, and IQ are generally close to 0.5,

TABLE 9.9
The Resemblance Between Relatives for Some Characters in Man

	Correlation Coefficient	
Character	*Parent–offspring*	*Full Sib*
Stature	0.51	0.53
Span	0.45	0.54
Length of forearm	0.42	0.48
Intelligence	0.49	0.49
Birth weight	—	0.50

Source: From Falconer (1960).

giving *prima facie* evidence of very little environmental effect and genetic determination mainly by additive nondominant genes. Some examples of such correlations are shown in Table 9.9. As always, however, these estimates are subject to a large number of reservations. Epistasis, or nonadditive interaction between genes, may lower the correlations. In addition to the problem of correlated environments, assortative mating with respect to the characters in question will increase the observed correlations between relatives above the levels expected from our simple model, which assumes random mating. Fisher, as his classic paper of 1918 shows, was well aware of this and took it into account in his analysis, though his approach is too complex to be reproduced here. Our simple model assumes random mating with respect to genotypes, which implies, of course, random mating with respect to phenotypes. It is, on the other hand, well known that in human populations there is often a strong correlation between husband and wife for IQ and for anthropometric characters such as stature, as will be discussed in the next section.

9.7 Assortative Mating and Its Effect on Quantitative Characters

Assortative mating is a deviation from random mating in which like individuals preferentially mate with each other (positive assortative mating, also called homogamy) or unlike individuals preferentially mate with each other (negative assortative mating, or disassortative mating).

Assortative mating must be distinguished from other types of deviation from randomness in mating, such as inbreeding, though this shows some formal similarity to positive assortative mating and can be, to some extent, treated similarly, as we shall see. It should also be distinguished from **sexual selection**, in which some phenotypes—male or female—are favored, resulting in differential mating. The main difference between assortative mating and sexual selection is that some genotypes are favored in sexual selection and are thus selected for, bringing about changes in gene frequencies and mean phenotype frequencies with time.

With assortative mating, in principle, gene frequencies do not change.

If the probabilities of a given mating are different from those expected under random mating, the expected frequencies of genotypes do not follow the Hardy-Weinberg law. If each genotype has the same probability of mating as any other, the gene frequencies in the progeny will be the same as those in the parents.

It should be noted that, especially in the presence of strong assortative mating, the situation in which each genotype has the same probability of mating may be difficult to achieve in practice. If, say, an individual of genotype *AA*, which happens to be rare, wants to mate only with another *AA* individual, he may have difficulties in finding a mate. This may decrease the probability of mating. The result will be sexual selection and a change in gene frequencies. It may be difficult to have assortative mating that is completely free from sexual selection. In theory, however, the possibility exists and we will limit ourselves here to considering it, in the interests of simplicity. A treatment of some more general models entailing sexual selection has been given by Scudo and Karlin (personal communication).

As assortative mating is a deviation from random mating, the reader might wonder why we didn't include it in Chapter 8, in which we discuss population structure. Our consideration there, however, was limited to single gene traits. Although there are almost no examples of assortative mating in man for simple genetic markers, there is evidence of it for quantitative traits. It is for this reason that we consider it here.

Exceptions to the above rule that there is no assortative mating for simple genetic markers may be found in populations with a high degree of social stratification. Recent immigrants with a different ethnic background may sometimes have gene frequencies different from those of people who have been *in situ* for a longer time. This may create a **secondary assortative mating**, due to **social homogamy**—that is, a tendency to marry within the same social class. This is likely, however, to disappear with time, if sufficient social mobility exists to cancel the initial social stratification in genetic markers. An example of such social stratification is revealed by the correlation between blood groups and the geographic origins of surnames in Australia, to be discussed in Chapter 11.

Positive assortative mating for a recessive phenotype increases its incidence.

Another example of assortative mating for single gene differences is found for some types of genetic defects such as deafness. It appears that marriages between a deaf and a normal person are unlikely to take place and tend to be unsuccessful, but that marriages between two deaf persons are common and tend to be successful (Fraser, 1965). Education and opportunities for the deaf have increased the fitness of deaf people almost to that of normal persons. If deaf people married only other deaf people and had the same probability of having progeny as normal persons, selection against deafness would be reduced to zero. Although gene frequencies would change only very slowly, as a consequence of new mutation, the frequency of the deaf phenotype would increase.

Predictions of the effects of assortative mating have been made by assuming that a fraction r of the population mates assortatively and the rest, $(1 - r)$, at random. For a recessive phenotype "a" this would mean that a proportion r of marriages are either A × A or a × a. If the frequency of the recessive phenotype at any one time is R_t, there will be R_t marriages a × a and $(1 - R_t)$ A × A in the assortatively mating fraction (r), while in the rest $(1 - r)$ there will be R_t^2 a × a and $2R_t(1 - R_t)$ A × a marriages. The frequency of recessives in the next generation can then be shown to be (see Example 9.4)

$$R_{t+1} = (1 - r)q^2 + \frac{r[q^2 - R_t(1 - 2q)]}{(1 - R_t)}, \tag{9.20}$$

where q is the recessive gene frequency. Under these conditions, the gene frequency remains constant from generation to generation. The kinetics of the change of R_t are shown in Figure 9.10; it is assumed that at time 0 the frequency of R_t is that under random mating.

The equilibrium value $\hat{R}$ of the recessive phenotype is

$$\hat{R} = \tfrac{1}{2}[1 + q^2 - rp^2 - \sqrt{(1 + q^2 - rp^2)^2 - 4q^2}] \tag{9.21}$$

and is equal to the gene frequency only if $r = 1$. Under these conditions, heterozygotes disappear at equilibrium. If r is less than one, the equilibrium genotype frequencies are as given in Figure 9.10. Only complete homogamy brings about the disappearance of the heterozygote at equilibrium. The approach to equilibrium is slower, the lower the gene frequency. If assortative mating is weak, however, much of the process takes place in the first few generations, or even the first generation, starting from genotypes in random mating equilibrium.

The application to the problem of deafness is complicated by the fact that there are a number of mimic genes, mostly recessives, that determine deafness as already discussed in Chapter 3. If deafness were due to a single gene, under assortative mating its incidence would rise from the present one of, say, 0.00014 (corresponding to a gene frequency of 0.012) to at most 0.012. This would be valid under the assumptions that the present frequency is determined by random mating, that there is now full assortative mating, and that deaf people have normal fitness. These assumptions are not exact, but are perhaps, not very far from the truth. The increase would, however, be slow. In the first generation of assortative mating, for instance, deafness would reach a frequency about twice the original one, and the rate of increase would decrease thereafter.

The fact that there are many recessive mimic genes will diminish the rate of increase of deafness due to present mating patterns. According to Dewey and co-workers (1965), there are $n = 35$ genes that cause deafness; their average gene frequency is 0.002. We have already discussed this estimate and its possible sources of error (Chapter 7). What matters here is the variation in the frequency of each gene, for if one were very frequent and many were rare, the single gene model would be nearly correct. If all the genes had equal frequencies and deaf people married among themselves without assortment for the genetic type of defect, then the increase of deafness in the next generation due to assortative mating would be much smaller.

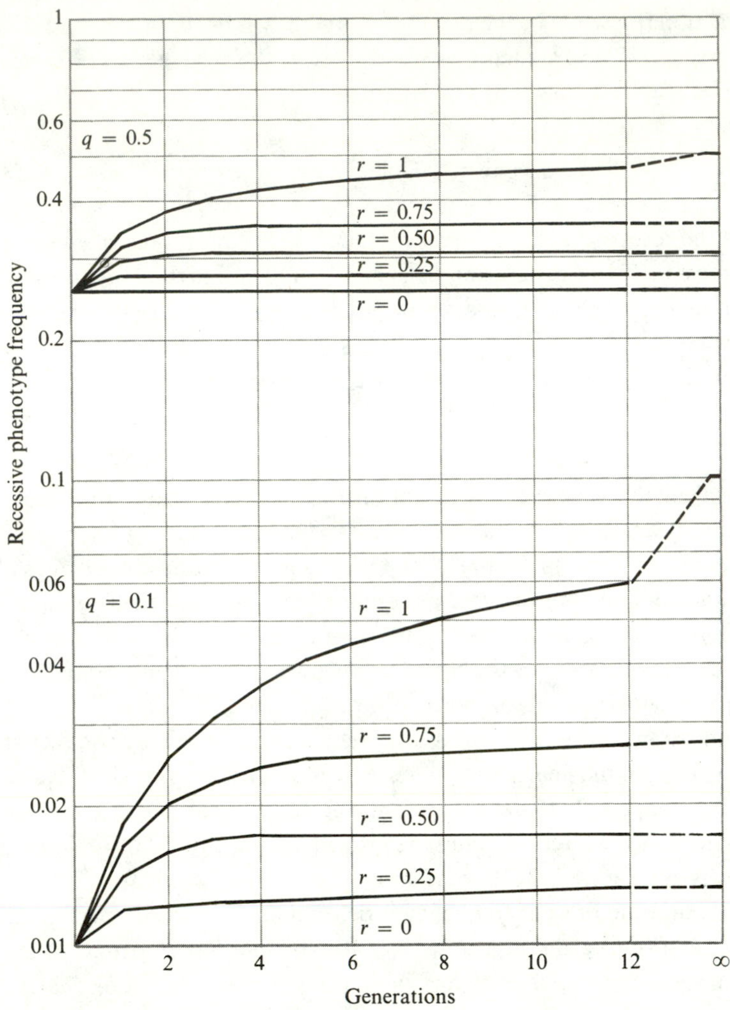

FIGURE 9.10
Change in recessive phenotype frequency with time under assortative mating of degree r. The gene frequencies q, which do not change during the process, are taken to be 0.5 and 0.1.

In fact, only marriages between deaf individuals carrying the same genetic defect will produce deaf offspring, and these constitute only about $1/n$ of all marriages between deaf persons. Thus, the rate, as noted by Crow and Felsenstein (1968), will be roughly 3 percent higher, rather than doubled after the first generation. The real increase lies somewhere between these two extremes, probably nearer to the former.

There is weak assortative mating for physical characteristics.

Since Pearson's work at the beginning of this century, it has been clear that several anthropometric traits show some, usually weak, correlations between matings. An abstract of available data summarized from a more complete tabulation by Spuhler (1968) is given in Table 9.10. Some representative values of the mean correlation coefficients are: + 0.09 for head circumference, + 0.19 for eye color; + 0.20 for stature; + 0.24 for hair color. Spuhler has investigated, in a sample of 734 married couples in Ann Arbor, Michigan, the correlation between fertility and the similarity of each couple for each of 29 traits. None of the correlations was significant, considering the number of traits examined, and, therefore, as far as this sample goes, there was no direct effect of similarity between couples.

TABLE 9.10
Homogamy for Physical Characteristics. Number of studies by correlation coefficients.

	Correlation Coefficients						
Characteristic	< 0	0–0.1	0.1–0.2	0.2–0.3	0.3–0.4	0.4	*Total*
Weight	—	1	2	3	1	—	7
Stature	1	6	8	7	4	1	27
Chest circumference	—	2	5	—	—	—	7
Sitting height	1	0	3	3	—	—	7
Head circumference	2	3	1	2	—	—	8
Cephalic index	2	12	5	3	—	—	22
Facial index	4	7	3	—	1	—	15
Nasal index	3	2	1	2	—	—	8
Hair color	—	—	2	2	1	—	5
Eye color	1	1	1	1	—	1	5

Source: From a tabulation by Spuhler (1968).

There are some objections to the interpretation of such data on assortative mating that make it difficult to accept available data without further study.

1. Socioeconomic status has a known correlation with some of these traits, in particular, for instance, with stature. Such a correlation may be in part a direct effect of socioeconomic status on the phenotype; for instance, in the example of

stature, through the amount and quality of food. There is a negative correlation between stature and progeny size. It has not been sufficiently explored whether, and what part of the correlation, between socioeconomic status and stature is due to the negative correlation between socioeconomic status and progeny size.

It is known that there is a strong correlation for socioeconomic status between husband and wife. Social homogamy seems to be strong in almost all societies. It may, therefore, be that a large part of the correlation between mates with respect to stature is due to their correlation for socioeconomic status. As stature is correlated with a number of other physical traits for example, weight, and chest circumference, there are expected to be, and are indeed found (Schreider, 1964) socioeconomic status differences for all these traits. In agreement with the idea that much of the correlation between mates for physical traits is due to social stratification plus social homogamy, populations whose social structure presumably differs from the Caucasian one, such as the Ramah-Navajo Indians and others (Spuhler, 1968), show no assortative mating for physical traits.

If the effect of socioeconomic status on a trait is only, or partly, phenotypic, and thus does not correspond to genotypic differences, as is possible, this creates additional, and to some extent unpredictable, complications for a theoretical analysis.

2. Many of the traits studied show secular trends. Stature, for instance, has been changing at the rate of 0.5–1.0 standard deviations per generation. There is known to be a strong correlation for age at marriage, of the order of 0.8. Therefore, mates will show a correlation for stature simply because they tend to be born at approximately the same time. A secular trend, especially when it is pronounced, is likely to be entirely or almost entirely phenotypic. Therefore, the correlation between mates due to the secular trend does not really measure assortative mating. It should be eliminated by partial regression, or equivalent methods, from the correlation between mates used in the theoretical analysis of genetic variances.

An example of the effect of a secular trend on assortative mating was shown by Beckman and Elston (1962) in an analysis of assortative mating for fertility. Examining a random sample of 477 married couples residing in the city of Uppsala, they found a high correlation in fertility between the pairs. This was measured by the number of sibs of the wife and the husband. The data are presented in Table 9.11. The chi square for independence is 16.97, showing what is apparently highly significant assortative mating for fertility. If mates in this study are reclassified according to their birth year, the association disappears almost entirely. It could hardly disappear completely, since associations due to socioeconomic status and differential fertility with socioeconomic status would still be present.

The use of observed correlations between mates for predicting effects on genetic variances is, therefore, complex. Secular trends should be removed from the correlation (they usually have not been.) Also the fraction due to nongenotypic correlations between socioeconomic status should be removed, but this is very difficult to ascertain.

Assortative mating for polygenic traits inflates the genic (additive) portion of the variance.

The effects of assortative mating on a polygenic trait were studied by R. A. Fisher (1918) in his famous paper, in which he gave the methods for predicting the "correlation between relatives on the supposition of Mendelian inheritance." Other aspects of the problem were covered by S. Wright in (1921b). A recent paper by Crow and Felsenstein (1968) gives an exposition of these theories from an elementary point of view. We will limit ourselves to citing their main conclusions.

A simple exact model assumes n genes, unlinked (or more specifically in linkage equilibrium), no dominance and no epistasis, and allows for variation of gene effect α and gene frequency p. Assortative mating is measured by the correlation coefficient between mates, r. The variance of the effect of each locus is

$$\sigma_i^2 = \sum p_k \alpha_k^2 - \left(\sum p_k \alpha_k\right)^2, \tag{9.22}$$

the sum being over all alleles at the locus. An equivalent number of genes n_e is computed from

$$n_e = \frac{\sum_{ij} \sigma_i \sigma_j}{\sum_i \sigma_i^2} \tag{9.23}$$

The average inbreeding coefficient at equilibrium under assortative mating is computed as the weighted mean of all loci, using as weights the $\sigma_i{}^2$ values. It can be shown to be given by

$$\hat{f} = \frac{r}{2n_e(1-r)+r}. \tag{9.24}$$

This is a small quantity under the conditions that must be prevalent in man, namely, of r being low and n_e being likely to be large.

The additive genetic variance, at equilibrium, is inflated by the increase in homozygotes according to the formula

$$\hat{V}_A = \frac{V_A}{1 - r\left(1 - \dfrac{1}{2n_e}\right)}, \tag{9.25}$$

where V_A is the value for random mating. For a large number of genes

$$\hat{V}_A \sim \frac{V_A}{1-r}. \tag{9.26}$$

After one generation of assortative mating, the additive variance is $V_A(1 + r/2)$.

This would mean, if all the assortative mating for stature were genotypic, an increase of 10 percent in the additive variance after the first generation and of 22 percent at equilibrium, assuming that $r = 0.2$.

When dominance and environmental variation are introduced, the simplifying assumption is made that V_D, the variance due to dominance effects, is not affected by assortative mating. The formulas for $\hat{V}_A$ are still valid, but now r must be replaced by the quantity

$$A = \frac{rV_A}{V_P} = rh^2,$$

where h^2 is the heritability, and V_P is the total phenotypic variance. For equilibrium, $\hat{A} = r\hat{V}_A/\hat{V}_P$, and

$$\hat{V}_P \sim V_A\left[\frac{1}{1 - \hat{A}\left(1 - \frac{1}{2n_e}\right)}\right] + V_D + V_E. \tag{9.27}$$

If we take n_e to be large,

$$\hat{V}_P \sim \frac{V_A}{1 - \hat{A}} + V_D + V_E. \tag{9.28}$$

Thus the effect of assortative mating is to increase the additive variance by a factor that is approximately $1/(1 - \hat{A})$.

The effects on correlations between relatives can now be predicted in the following way. The **genic value** z of an individual can be considered as the phenotypic value it would have if there were no environmental variation and no dominance

TABLE 9.11
Effect of a Secular Change in Fertility on the Correlation in Fertility Between Husbands and Wives

CORRELATION BETWEEN WIFE AND HUSBAND IN RESPECT TO NUMBER OF SIBS, POOLING ALL PERIODS OF BIRTH. EXPECTED FIGURES IN PARENTHESES

Wife No. of Sibs	*Husband No. of Sibs* 0–3	4–	*Total*
0–3	173 (150.9)	105 (127.1)	278
4	86 (108.1)	113 (90.9)	199
Total	259	218	477

TABLE 9.11 (*continued*)

CORRELATIONS WHEN WIVES AND HUSBANDS ARE CLASSIFIED ACCORDING TO PERIOD OF BIRTH AND SIZE OF SIBSHIP

Wife		*Husband Born in*						
		–1899		1900–1919		1920–		
		No. of Sibs		*No. of Sibs*		*No. of Sibs*		
Born in	*No. of Sibs*	0–3	4–	0–3	4–	0–3	4–	*Total*
1899–	0–3	16	11	1	1	0	0	29
	4	18	28	0	1	0	0	47
1900–1919	0–3	4	1	34	40	1	1	81
	4	1	7	28	42	2	0	80
1920–	0–3	0	0	21	11	96	40	168
	4	0	0	4	12	33	23	72
Total		39	47	88	107	132	64	477

CHI SQUARES AND PROBABILITY LEVELS TO TEST CORRELATIONS BETWEEN WIFE AND HUSBAND IN RESPECT TO NUMBER OF SIBS WHEN THE PARTNERS ARE GROUPED ACCORDING TO AGE

Age Categories	χ^2	*P*	*No.*
1. Both partners born before 1900	2.73	$0.1 > P > 0.05$	75
2. Both partners born 1900–1919	0.48	$0.5 > P > 0.3$	144
3. Both partners born after 1920	2.42	$0.2 > P > 0.1$	192
4. Partners of unlike birth periods	10.38	$0.005 > P > 0.001$	68
Sum of 1–3	9.87	$0.005 > P > 0.001$	409
Sum of 1–4	16.97	$P < 0.001$	477

Source: From Beckman and Elston (1962).

effect. A parent of phenotype x (taking both x and z on a scale having a mean of zero) has an expected genic value z, given by

$$z = \frac{\hat{V}_A}{\hat{V}_P} x, \tag{9.29}$$

where $\hat{V}_A/\hat{V}_P$ is the regression coefficient of z on x because V_A is the common variance (the covariance) of x and z, and V_p is the total variance (see Appendix I). Because of r, the phenotypic correlation between mates, the other parent has an expected phenotype of $x' = rx$ and thus an expected genic value

$$z' = rx' = r\frac{\hat{V}_A}{\hat{V}_P}x. \tag{9.30}$$

The mean genic value of the two parents is $(z + z')/2$ and the regression coefficient of the mean genic value of the parents on the phenotype of one parent is, therefore,

$$\frac{\hat{V}_A}{\hat{V}_P}\left(\frac{1+r}{2}\right). \tag{9.31}$$

If there is no other cause of correlation between parents and offspring than their genic value, which is expected to be identical in parent and offspring, then the value given in Equation 9.31 is also the correlation between parent and offspring under equilibrium of assortative mating of degree r; namely,

$$r_{P/O} = \frac{1+r}{2}\frac{\hat{V}_A}{\hat{V}_P}. \tag{9.32}$$

Fisher found, in a similar way, the expected correlation between more remote ancestors, and also that between sibs and between descendants of sibs. The correlation between ancestors and their descendents after n generations can be shown to be

$$\frac{1+r}{2}\frac{\hat{V}_A}{\hat{V}_P}(1+\hat{A})^{n-1}. \tag{9.33}$$

The variance between full sibs is approximately the same under assortative mating as under random mating, because it depends only on the heterozygosity of the parents, which is not significantly reduced by assortative mating if the number of genes is large. From this consideration, Fisher computed the sib-sib correlation to be

$$r_{S/S} = \frac{1}{2}\frac{\hat{V}_A}{\hat{V}_P}\left(1 + r\frac{\hat{V}_A}{\hat{V}_P}\right) + \frac{1}{4}\frac{V_D}{\hat{V}_P} = \frac{1}{4}\frac{\hat{V}_A + V_D}{\hat{V}_P}\left[1 + \frac{\hat{V}_A}{\hat{V}_A + V_D}(1+2\hat{A})\right], \tag{9.34}$$

where, as before, $\hat{A} = r\hat{V}_A/\hat{V}_P$. He also obtained the correlation between uncles (or aunts) and nephews (or nieces) and that between first cousins.

Fisher used this approach to analyze data on the inheritance of stature, collected by Pearson and Lee (1903) among English University students. The observed correlations found by Pearson and Lee were

marital correlation coefficient: $r = 0.2803$;
correlation between parent and offspring: $r_{P/O} = 0.5066$;
correlation between sibs: $r_{S/S} = 0.5433$.

Fisher, using these correlations, found that the fraction of the total variation due to inheritance must be large. In fact, from Equation 9.32,

$$\frac{\hat{V}_A}{\hat{V}_P} = \frac{2r_{P/O}}{1+r} = 0.7913 = c_1 c_2, \quad \text{say,} \tag{9.35}$$

and also, by definition,

$$\hat{A} = \frac{r\hat{V}_A}{\hat{V}_P} = 0.2219.$$

From the correlation between sibs (Equation 9.34) he obtained the quantity, which he called c_1, defined by

$$c_1 = \frac{\hat{V}_A + \hat{V}_D}{\hat{V}_P} \tag{9.36}$$

using the formula

$$c_1 = 4r_{S/S} - c_1 c_2(1 + 2\hat{A}), \tag{9.37}$$

which is easily derived from Equation 9.34 for $r_{S/S}$, and Equation 9.35, which defines $c_1 c_2 = \hat{V}_A/\hat{V}_P$.

From Pearson and Lee's data, c_1 was estimated to be 1.031. The value of c_1 should not be higher than one. This led Fisher to assume that the quantity which makes c_1 in general smaller than one, namely the environmental variance, must be very small, and, to a first approximation, negligible.

Partitioning the total variance further, the fraction of the genetic variance due to additive genic effects (often called "fixable" because it can in principle, be fixed by selection) is equal to

$$c_2 = \frac{\hat{V}_A}{\hat{V}_A + \hat{V}_D} = \frac{\dfrac{\hat{V}_A}{\hat{V}_P}}{c_1} = 0.7913. \tag{9.38}$$

A fraction of this, equal to $\hat{A} = r\hat{V}_A/\hat{V}_P$, is at equilibrium under assortative mating, due to the increase in additive variance because of assortative mating, while the residual $1 - \hat{A}$ is the fraction of the additive genetic variance that would obtain if mating were at random.

The final partition of the total variance for stature was, thus;

Environmental $1 - c_1 = V_E/\hat{V}_P$ 0%
Nonenvironmental (genetically determined) $c_1 = 1 - V_E/\hat{V}_P$
  Nonadditive (due to dominance) $c_1(1 - c_2) = V_D/\hat{V}_P$ 21%
  Additive $c_1 c_2 = \hat{V}_A/\hat{V}_P = \hat{h}^2$
    Expected under random mating $c_1 c_2(1 - \hat{A})$............ 62%
    Due to assortative mating $c_1 c_2 \hat{A}$........................ 17%

where V_E is the environmental variance.

Fisher applied the term "variance due to genotypes" to the sum of the fractions 62 percent and 21 percent—that is, the fraction that would be genetically determined under random mating (now usually called the genotypic variance).

It is difficult to believe that the environmental variance is actually zero, though it may have been especially low for the fairly homogeneous sample that was analyzed. We know today from other lines of evidence, that there are environmental effects on stature.

There is strong assortative mating for measures of intelligence.

Some of the reservations already mentioned for the interpretation of marital correlations may also hold for correlations for IQ, but it seems likely that at least a large fraction of this correlation is primary, and that it is not an indirect consequence of other reasons for assortment. Thus, it is clear that coeducation at universities favors meeting and marriage between people of higher IQ. The association between husband and wife for educational attainment is very strong as shown in Table 9.12.

Burt and Howard (1956) have given an analysis of the variation in IQ along the lines suggested by Fisher. The correlations they found, corrected for several technical sources of error, are

between mates: $r = 0.3875$;
between parent and child: $r_{P/O} = 0.4887$;
between sibs: $r_{S/S} = 0.5069$.

Taking these correlations at face value and carrying out the analysis as before, using Equations 9.35 to 9.38, we obtain

$$c_1 c_2 = \frac{\hat{V}_A}{\hat{V}_P} = \frac{2r_{P/O}}{1 + r} = 0.7044,$$

$$\hat{A} = r \frac{\hat{V}_A}{\hat{V}_P} = 0.2730,$$

$$c_1 = \frac{\hat{V}_A + V_D}{V_P} = 4r_{S/S} - c_1 c_2(1 + 2\hat{A}) = 0.9386,$$

TABLE 9.12

Expected and Observed Percentage of White Women, 35-44 and 45-54 Years Old, Married and Husband Present, of Three Educational Levels Choosing Husbands of Specified Education[a]

Education of Wife and Husband	*Percent of Wives 35–44*		*Percent of Wives 45–54*	
	Expected	*Observed*	*Expected*	*Observed*
WIFE—COLLEGE 4+[b]				
Husband				
College				
4+[b]	10.7	59.9	8.8	50.2
1–3	9.9	15.7	8.3	16.4
High school				
4	29.3	16.3	17.6	16.8
1–3	21.7	4.9	20.8	8.7
Elementary				
8	14.7	2.1	22.4	5.3
under 8	13.7	1.1	22.1	2.6
	100.0	100.0	100.0	100.0
WIFE—HIGH SCHOOL 4				
Husband				
College				
4+[b]	10.7	9.8	8.8	9.9
1–3	9.9	12.3	8.3	12.4
High school				
4	29.3	41.5	17.6	33.4
1–3	21.7	20.6	20.8	20.6
Elementary				
8	14.7	10.4	22.4	15.7
under 8	13.7	5.4	22.1	8.0
	100.0	100.0	100.0	100.0
WIFE—ELEMENTARY: UNDER 8				
Husband				
College				
4+	10.7	0.5	8.8	0.5
1–3	9.9	1.4	8.3	1.2
High school				
4	29.3	7.5	17.6	4.4
1–3	21.7	14.8	20.8	10.9
Elementary				
8	14.7	17.5	22.4	18.5
under 8	13.7	58.2	22.1	64.5
	100.0	100.0	100.0	100.0

Source: From Kiser (1968).

[a] The expected proportions among white women 35–44 were based upon the educational distribution of white males 40–44 years old, regardless of marital status, in the United States in 1960. For white women 45–54 the expected proportions are based upon the educational distribution of white males 50–54 years old. The observed proportions were derived from U.S. Bureau of the Census, *Women by Number of Children Ever Born* (1964), Table 26.

[b] Numbers refer to years of college, high school, or elementary education.

which may be used to obtain the following breakdown of variance components:

Variance component	*Relative contributions to the total*	
	Face value (our analysis)	*Corrected (Burt and Howard's values)*
additive random	51.2%	47.8%
assortative mating	19.2%	17.9%
dominance	23.4%	21.9%
environmental	6.1%	12.4%

Burt and Howard carried out their analysis in a different way. It is interesting that their estimate of environmental variation, derived from an entirely different source (variation between twins reared apart), was 12.4 percent, which may still be compatible, within the limits of error, with the environmental variance that we obtained by computation. They corrected their figures to make the environmental variance equal to that between monozygous twins reared apart.

These results show that assortative mating, when sufficiently strong, does have a considerable effect in increasing the genic variance. A hidden assumption behind the computations is that equilibrium for assortative mating has been reached. This may be reasonable, because equilibrium is often reached in a relatively short time. In general, it should never be forgotten that the real situation is oversimplified in all these models, in order that the theoretical treatment can lead to useful predictions.

9.8 Inbreeding Effects

The effect of inbreeding on the mean of a quantitative character can give some indication of the average dominance of genes affecting it.

Consider a single locus with two alleles A_1 and A_2 such as was used for the model of Figure 9.7. If the gene frequencies of A_1 and A_2 are p and $q = 1 - p$, respectively, and the population inbreeding coefficient is F, then the frequencies of the three genotypes A_1A_1, A_1A_2, and A_2A_2 can be written as

$$p^2 + Fpq,\ 2pq(1 - F),\ \text{and}\ q^2 + pqF,$$

respectively (see Chapter 7). With means a, $-d$, and $-a$ for the three phenotypes, the mean of the inbred population is

$$m_F = a(p^2 + Fpq) - d(2pq)(1 - F) - a(q^2 + Fpq)$$
$$= a(p - q) - 2dpq + F(2pqd). \tag{9.39}$$

Thus, summing over a number of independent loci, the mean of the inbred population is

$$m_F = m + F\sum(2pqd), \tag{9.40}$$

where m is the mean of the population in the absence of inbreeding (see Equation 9.8). Thus, if $\sum d2pq$ is not zero, m_F changes linearly with F, and so a regression of m_F on F provides a test for the presence of dominance. The size of the change depends on the average size of d. If heterozygotes tend to be nearer the "higher" genotype A_1A_1, d will be negative (see Figure 9.7) and the mean of the inbred group will decrease as the extent of inbreeding increases. This applies particularly to characters exhibiting heterosis—that is, characters for which the mean of the heterozygote is higher than that of either homozygote. Thus Equation 9.40 provides a simple analytical basis for the phenomenon of inbreeding depression. An analysis of a quantitative character based on this approach was used by Barrai and co-workers (1964) for data on chest girth and stature of males born in the Parma Province (Northern Italy) between 1892 and 1911. The mean measurements, corrected for their correlation with a series of associated socioeconomic and geographic parameters, are shown as a function of the inbreeding coefficient in Figure 9.11. The linear decrease in mean chest girth with F is very significant, suggesting that alleles increasing chest girth tend to be dominant over those decreasing it. The data on stature do not show significant linear variation with F. Their peculiar variation (in particular, the peak in stature for the progeny of third cousins) may be connected in some way with associations between levels of inbreeding and social class, and hence stature, though the statistical corrections to the primary data were intended to remove these effects. The results show that great care must be taken in allowing for such stratifications in the analysis of human population data, and that the methods of correction used are not always successful.

Schull and Neel (1965) have done an extensive study on the effects of inbreeding on Japanese children. They collected a large amount of data on anthropometric measurements (weight, height, head girth, chest girth, etc.) of noninbred individuals, and of partially inbred individuals, most of whom were the offspring of first-cousin marriages. There was a significant reduction in the mean of the anthropometric measurements of all inbred individuals; for the offspring of first-cousin marriages this generally was of the order of 0.5 percent of the outbred value. As in the analysis by Barrai and co-workers, allowance was made for socioeconomic factors by correcting for these using a multiple-regression analysis. Most of the anthropometric characters are, of course, highly correlated with overall size and thus with each other. Taken at face value, the results suggest that genes affecting overall size tend to be dominant, in agreement with the results on chest girth obtained by Barrai and his colleagues. A number of other characters, some concerned with neuromuscular

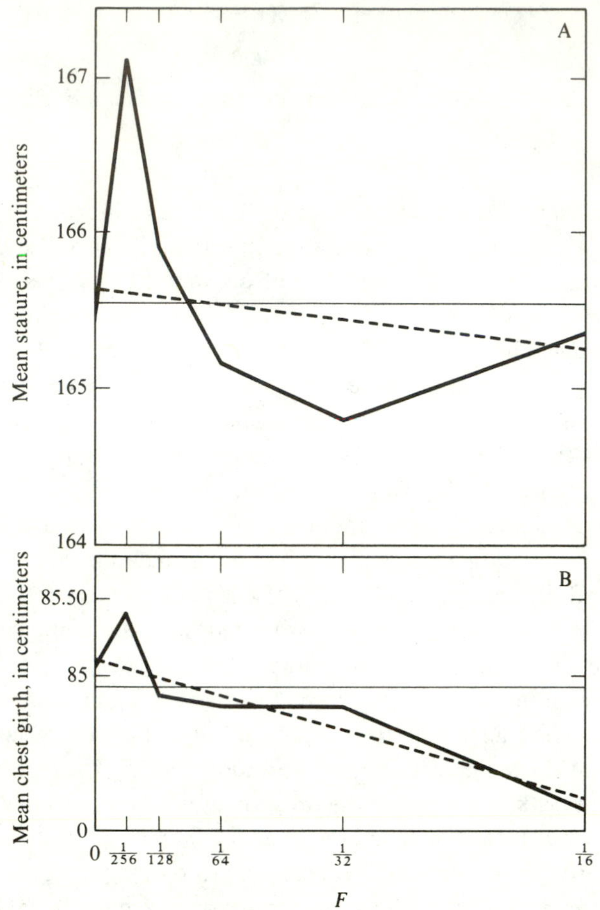

FIGURE 9.11
Data on quantitative characters of males born in the period 1892–1911 in the Parma Province of Northern Italy. A: Regression of corrected means of stature on *F*. Chi square for departure from linearity is 191; $P < 1$ per cent. B: Regression of corrected means of chest girth on *F*. Chi square for departure from linearity is 2.0; $P < 90$ per cent. (From Barrai et al., 1964.)

ability and some with academic performance, also showed a decline with inbreeding. Once again, we can come to the same conclusion concerning the overall dominance of genes affecting these traits in a positive direction. Though Schull and Neel took great care, as already mentioned, to allow for socioeconomic stratification of inbred marriages, there remains a lingering doubt that the observed inbreeding depressions are still influenced in some way by environmental factors that might not be entirely corrected for.

9.9 Heritability of Threshold Characters

An estimate of the heritability of a threshold character can be obtained from a comparison of its incidence in the general population with that in relatives of persons having the character.

We have already pointed out that all-or-none attributes, such as congenital malformations, that are not simply inherited may be converted into quantitative characters by using the concepts of the threshold model (see Figure 9.5). The study of the inheritance of a threshold character then becomes the study of the inheritance of the continuous variate underlying it. The technical problem that needs to be solved is that of obtaining information on variances and covariances of the underlying variate from observations on the incidence of the attribute, which are the only available data. We have already discussed data showing the increased incidences of many diseases among the relatives of affected individuals as compared with the incidence in the general population (see Table 9.4). This, of course, provides the primary evidence that there is an important genetic component in the determination of these diseases. We shall now show how such information can sometimes be used to obtain quantitative estimates of the relative importance of genetic and environmental variation with respect to a threshold character.

Following the approach to this problem set forth by Falconer (1965), we make use of some of the concepts used by plant and animal breeders to predict the outcome of selection experiments. The basis for the calculations is illustrated in Figure 9.12. The upper curve in this figure shows the distribution of the underlying continuous variate, or liability as Falconer calls it, in the general population. Its shape is assumed to be normal. The vertical solid line indicates the position of the threshold T, so that the incidence in the general population q_g is the area under the tail beyond the threshold. The lower curve shows the distribution of the liability in relatives of affected people. Its shape is also assumed to be normal and with the same variance, but its mean position is moved somewhat towards the threshold, as expected if liability is in part genetically determined. This results in a higher proportion of individuals lying above the threshold and so in a higher incidence q_r among the relatives. The two incidences q_g and q_r are the only observable quantities. The dotted vertical line through G indicates the position of the mean liability of the general population. For simplicity we consider all measurements to be made in terms of the standard deviation of the liability. In other words, we assume that both distributions have variance unity. The liability distribution among the relatives can be thought of as the liability distribution expected among given relatives of a random sample of people from the general population, whose liability is above the threshold. Thus, for example, for parents and offspring, the curve represents the expected liability distribution among children of parents who have a liability greater

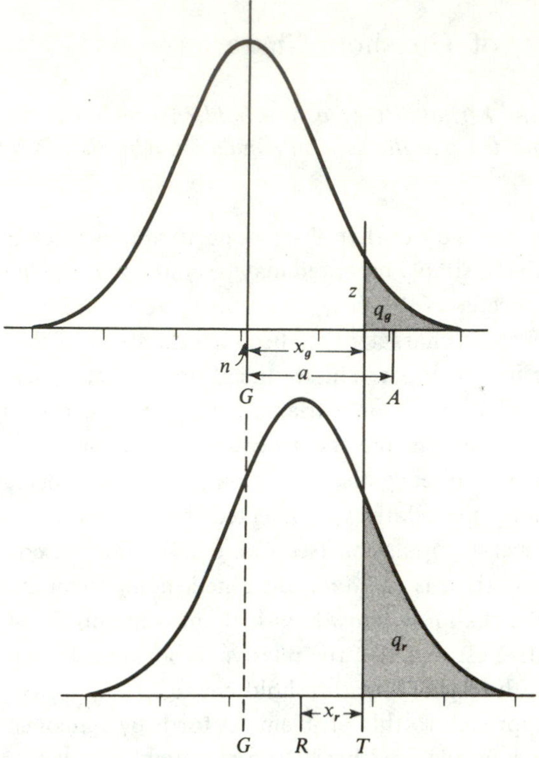

FIGURE 9.12
The inheritance of liability to diseases. Two distributions representing the general population (upper curve) and the relatives of affected individuals (lower curve) compared with reference to the fixed threshold *T*. *G* is the mean liability of general population. *A* is the mean liability of affected individuals in the general population. *R* is the mean liability of relatives. *q* is the incidence—that is, the proportion of individuals with liabilities exceeding the threshold. *x* is the deviation of the threshold from the mean—that is, the normal deviate. *z* is the height of the ordinate at the threshold. *a* is the mean deviation of affected individuals from the population mean ($=z/q$). *n* is the mean deviation of normal individuals from the population mean [$=z/(1-q)$], subscript *g* refers to the general population, subscript *r* to the relatives.

than *T*. This is exactly the same situation that is faced by an experimental breeder in his selection experiments. If the experimenter breeds from all individuals with a liability above *T*, what is the expected liability among the offspring? In other words, what is the gain in liability due to the selection process or, as it is sometimes called, what is the **selection response**. The gain is, of course, a function of the heritability of liability, assuming there are no genotype-environment interactions. If A is the mean liability of affected individuals, that is, those in the tail of the general population, and R the mean of their relatives (for example, offspring), then the realized gain in liability is $R - G$. The difference between the mean of the general population G and that of the "selected" individuals A, $A - G$ is often called the **selection differential**. The ratio of these two differences

$$b = \frac{R - G}{A - G} \tag{9.41}$$

is the regression of relatives on affected propositi with respect to liability. This regression coefficient is the expected slope of the line that would be obtained by plotting a number of $R - G$ values against $A - G$ values. The regression coefficient of one variable, say R, the mean liability of relatives, on another, say A, the mean liability of affected propositi, is given by

$$b = \frac{\text{cov}(A, R)}{V(\text{affected})} \tag{9.42}$$

where cov stands for covariance, assuming the variances of affected individuals and their relatives are the same (see Appendix I).

If there are no genotype-environment interactions then cov(A, R) is simply the genetic covariance between relatives, as calculated above, while V (affected) is the total phenotypic variance ($V_A + V_D + V_E$). Thus, assuming our simple model of independently acting genes, on the basis of which the covariances given in Table 9.7 were calculated, the value of b for parents as propositi, and offspring as the relatives, would be

$$\frac{\frac{1}{2}V_A}{V_A + V_D + V_E} = \tfrac{1}{2}h^2,$$

where h^2 is the heritability. More generally,

$$b = rh^2, \tag{9.43}$$

where r is the coefficient of relationship, or the inbreeding coefficient of the offspring of matings between relatives of given degree. This equation is only strictly valid when the dominance variance $V_D = 0$, or for those relationships whose covariance does not involve V_D, which includes almost all those readily available for analysis in human populations except sibs. Thus, if we can express b, as given by Equation 9.41, in terms of the incidences q_g and q_r, we can obtain an estimate of the heritability. The distance from the mean of a distribution to the threshold, for example, $x_g = T - G$, is a function of the proportion q_g in the tail of the distribution. This distance can readily be determined from conventional tables of the normal distribution. For example, when $q_g = 2.5$ percent, it is well known that x_g is approximately 2 (more precisely 1.96). The distances $x_g = T - G$ and $x_r = T - R$ can thus be determined from a knowledge of the two incidences q_g and q_r, and thus the numerator of the expression for b,

$$R - G = (T - G) - (T - R) = x_g - x_r.$$

is also determined. The quantity

$$A - G = a,$$

which is the difference between the mean of the selected individuals with liabilities above the threshold and the mean of the general population, is also a function of q_g, which can readily be determined from tables of the normal distribution. It is a well-known result of selection theory (see Example 9.5) that

$$a = \frac{z}{q_g}, \tag{9.44}$$

where z is the height of the normal distribution at T, the threshold. Given x_g, z is simply

$$\frac{1}{\sqrt{2\pi}} e^{-(x_g{}^2/2)},$$

which is tabulated in many standard statistical tables. For example, if $q_g = 2.5$ percent, then $x_g = 1.96$, from which $z = 0.0586$ and so

$$a = \frac{z}{q_g} = \frac{0.0586}{0.025} = 2.34.$$

From Equations 9.41, 9.43, and 9.44, we have

$$h^2 = \frac{b}{r} = \frac{x_g - x_r}{ra} = \frac{q_g(x_g - x_r)}{rz}. \tag{9.45}$$

Falconer gives as an appendix to his first paper on this method of analysis of threshold characters a useful table of values of x and a for given incidences q. Thus given q_g and q_r, the corresponding value of b, as given by Equation 9.41, can be obtained from tables of the normal distribution or from Falconer's tables. A simple graphical representation of the relationship between the population incidence, the incidence in first-degree relatives, for which $r = \frac{1}{2}$ and thus from 9.43 $b = \frac{1}{2}h^2$, and h^2 is given in Figure 9.13. It is clear that the larger the ratio q_r/q_g, the higher the estimate of heritability. The spread of q_r/q_g values is much greater for low general population incidences q_g, showing in simple terms that the effect of a given level of heritability on q_r/q_g is much greater for lower values of q_g. In other words, it is

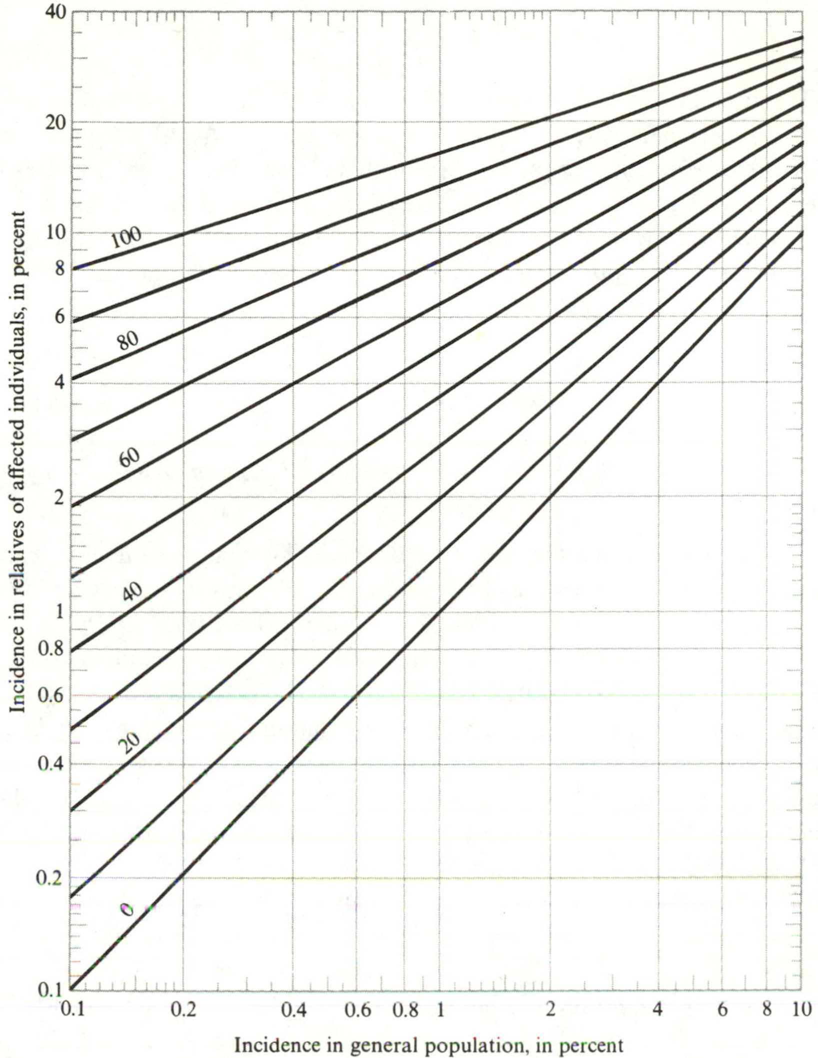

FIGURE 9.13
Graph for estimating the heritability of liability for a threshold trait from two observed incidences, when the relatives are sibs, parents, or children. The numbers on the lines are the heritabilities, h^2, in percent. (From Falconer, 1965.)

more difficult to detect by this method heritable components of characters that are relatively frequent in the general population.

As an example of the threshold-model approach to estimating heritability, take the data on harelip given in Table 9.4. The incidence of harelip in the general population is $q_g = 0.001$, from which we find $x_g = 3.1$ and $a = 3.7$. The incidence in first degree relatives is $q_1 = 0.035$, giving $x_1 = 1.8$. Thus, from Equation 9.43,

$$h^2 = \frac{3.1 - 1.8}{\frac{1}{2}3.7} = 0.7,$$

which value can also be obtained using Figure 9.13. The h^2 values obtained using the data on second-degree ($r = \frac{1}{4}$) and third-degree ($r = \frac{1}{8}$) relatives are 0.65 and 0.75, respectively. The reasonable agreement between these three estimates is certainly reassuring, and lends some weight to the rather high heritability of harelip. In particular, the fact that the value from sibs, which would be inflated if $V_D \neq 0$, is not higher than the other values suggests the absence of significant dominance effects.

Apart from the usual questions of the validity of a simple additive genetic model with no dominance and with no genotype-environment interactions, this method of estimating the heritability of threshold characters is subject to some other special sources of error.

1. The assumption that the liability is normally distributed with equal variances in both the general population and the population of relatives of affected individuals is not generally valid. That the distribution of the liability in the general population is normal is a reasonable assumption in the absence of obvious major genes affecting a liability. It can, however, easily be shown that the distribution among relatives cannot be the same as that in the general population. This source of error can be handled by a method to be suggested later in this section. It seems especially serious in applications of the method to monozygotic "identical" twins. Thus, for the data on harelip in Table 9.4, the incidence q_r in co-monozygous twins is 0.5, giving $x_r = 0$. Using the same values of x_g and a that we used before, we have $h^2 = b = 3.1/3.7 = 0.84$, since $r = 1$. This estimate is appreciably higher than the estimates obtained from the other relatives. It is possible that the heritability estimates of threshold characters obtained from identical twins are generally too large. An alternative method gives, however, a heritability estimate that is in agreement with this one. Although the assumption of equal variances does not seem, in general, to be a serious source of error, it can to some extent be corrected for by more complex formulas.

2. The second major source of possible error is the choice of the "control," or general, population. It is, of course, very important that the control sample represents the population from which the relatives of affected propositi are drawn. If, for example, the data on relatives come from a special local survey while the data on the general population come from some form of national survey, the general population incidence is not necessarily the one that is relevant for the particular sample of relatives chosen for study. An appropriate control sample might in this instance be taken from the same geographical area, or might even be taken from relatives of a sample of unaffected individuals from the same area. If the control

sample necessarily excludes relatives of affected people, this would bias slightly the incidences estimated from it. However, so long as the general incidence is low this effect can safely be neglected.

3. A third difficulty, which is common to all such studies, is due to intrinsic heterogeneity in the population under study. For example, if the disease has a variable age of onset, as diabetes mellitus and schizophrenia do, the sample of relatives, which, of course, includes people of all ages, is bound to be heterogeneous with respect to the expected incidence of the disease. The same is true if there are differences in incidence of a disease between the sexes or if a disease incidence has been changing with time, perhaps due to improved medical care. If there is such heterogeneity, it is important to obtain heritability estimates from as homogeneous a sample as possible, using, for example, only relatives of a given age or sex. In this way variations in the heritability with age and sex may be detected and allowed for.

Diabetes mellitus is an example of a threshold character with a variable age of onset.

Diabetes mellitus is a disease with quite variable expression, whose familial incidence is well documented and, according to some authors, was noted in India in the early seventeenth century. The reported incidence of the disease increases markedly with age, from about 0.1 percent among persons 25 years old or younger to 3–4 percent among persons 60-70 years old. The incidence is, to some extent, a matter of the criteria used to define the disease, which may vary from an almost complete lack of insulin activity to a mild but consistently elevated blood-sugar level which is of little clinical consequence. Clearly the incidence increases as the stringency of the criteria used to define the disease decreases. The rationale behind using less stringent criteria is the hope of identifying at an earlier age those individuals ultimately destined to be affected. Tests of tolerance to glucose loading have been devised with this aim in mind. These have not, so far, however, been very helpful in the definition of the prediabetic state. If the disease were due to a single gene, the variable expression and age of onset would surely lead to incomplete penetrance of the gene's effect. In other words, not all individuals with the supposed diabetic genotypes would (at the time of observation) have diabetes. A number of authors have suggested single-gene inheritance, particularly inheritance due to a single recessive gene with incomplete penetrance, as the genetic basis for diabetes (see, for example, Neel et al., 1965). However, as was pointed out very clearly by J. Edwards (1960), unless all the variability in penetrance is ascribed to environmental effects, there is no great distinction between the model of a single gene with incomplete penetrance and a hypothesis of multifactorial inheritance. The goal is the *clear-cut* identification of specific genes affecting the incidence of the disease. Diabetes is generally much more severe in cases that have an early age of onset.

This suggests that the so-called "juvenile" diabetes who acquire a severe form of the disease before they are 20 years old, may be a separate genetic entity. Some people have, in fact, suggested that the severely affected juvenile diabetics are homozygous for a diabetic gene, while the more mildly affected individuals are the heterozygotes (see Simpson, 1962). However, none of these hypotheses fits the observed data without an appeal to a considerable level of impenetrance. Thus, the estimated proportion of juvenile diabetics born to normal parents, assuming recessive inheritance, is only 0.067. Compared with the expected frequency of 0.25, this would suggest a penetrance of only $0.067/0.25 = 0.268$, a value that is still much too low to be accounted for only by the variable age of onset (see Barrai and Cann, 1965).

It seems reasonable to consider diabetes as a threshold character. The blood-sugar level and insulin activity are good analogues of the liability variable, though other factors also may, of course, contribute to the diabetic liability. Falconer (1967) has applied his method of analysis to diabetes, separating age groups on the assumption that sibs of affected propositi are, in general, in the same age range as the propositi. The variation in heritability estimates with age for three different bodies of data is shown in Figure 9.14. The heritability seems to decline from

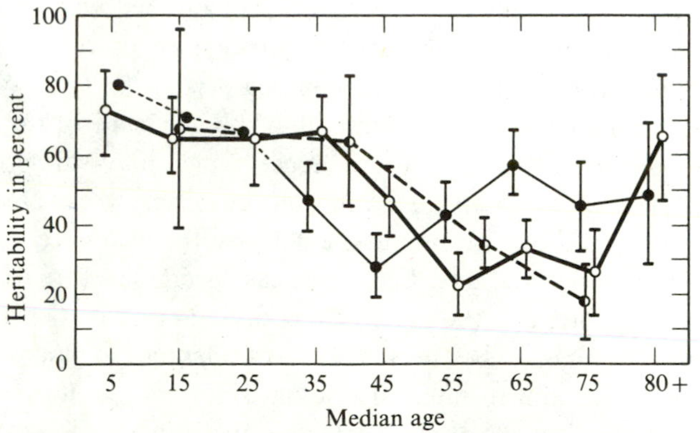

FIGURE 9.14
Diabetes mellitus: changes in heritability with increasing age. The heritability of liability estimated from the sib correlations. The fine line plots data on Canadian males; the heavy solid line, data on Canadian females; the heavy dashed line, data on both sexes in Birmingham, England. The vertical lines extend to plus-or-minus one standard error. (From Falconer, 1967.)

60–80 percent for younger groups (less than 40 years old) to 20–40 percent for the older groups (up to 70 years old). This is certainly consonant with the idea that the genetic etiology for younger people may be different from that for older people. When data for all ages are combined, heritability estimates from parents and offspring are comparable to those for sibs, and the combined estimates are in the range of about 30–40 percent. The older people, of course, dominate this estimate since young diabetics are a relatively small proportion of the total. A multifactorial hereditary basis for diabetes with a threshold for manifestation, taking into account the possibility that different genes may influence early and late onset, of the disease, seems at the present time to be the most satisfactory hypothesis.

A slightly different but related model makes possible the direct estimation of the phenotypic correlations for threshold characters.

A model that is preferable to the one given by Falconer with respect to the objection to the assumption that there is the same distribution of liability among relatives of the propositi as in the general population, goes back to Pearson. It has been revived recently and elaborated by J. Edwards (1960, 1969). This model assumes that inheritance is multifactorial with an abrupt threshold for manifestation of the disease and that the correlation between two relatives can be expressed using a bidimensional normal distribution, which is illustrated in Figure 9.15. One axis, say x, indicates the liability of an individual (the propositus) and the other, y, the liability of a relative of given degree of the propositus. The frequencies of pairs

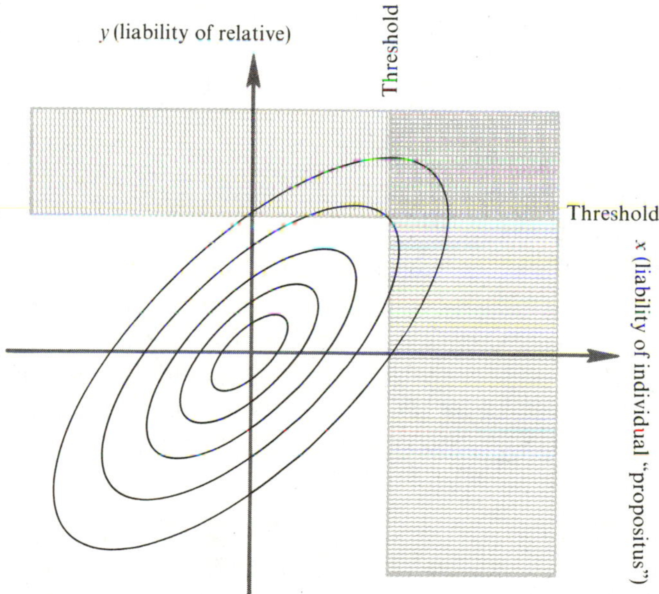

FIGURE 9.15
Bidimensional normal distribution theory for the threshold model. The liability of the propositus is indicated by the x axis; that of relatives by the y axis.

of x and y values would be given in an axis, not indicated in the figure, which is perpendicular to the x and y axes. The joint distribution of the frequencies would be described by a bell-shaped "correlation" surface with a peak at the center and skewed along a major axis at 45° with respect to x and y. The ellipses indicate lines on this surface that correspond to equal frequencies. Shaded areas beyond the threshold lines indicate affected individuals. The proportion affected in the general population (q_g) is given by the volume under the correlation surface corresponding to the area shaded by horizontal bars (for y) or by vertical bars (for x). The relatives of an affected individual (the propositus) that are affected are situated in the area shaded both vertically and horizontally. This distribution may be shown by a 2 × 2 table:

		PROPOSITUS		
		Normal	*Affected*	*Total*
RELATIVES	*Affected*	$q_g(1 - q_r)$	$q_r q_g$	q_g
	Normal	$1 - 2q_g + q_g q_r$	$q_g(1 - q_r)$	$1 - q_g$
Total		$1 - q_g$	q_g	1

The computation of q_g, given the position of the threshold, can be done with a table of normal integrals as already described. The computation of the proportion of affected relatives of affected individuals (q_r/q_g) requires special tables (the 2 × 2 correlation tables calculated by Pearson), which permit the evaluation of ρ, the correlation between relatives, given q_r and q_g. The quantity ρ is equal to rh^2 (the product of the relationship coefficient and the heritability), and it is this which we wish to estimate. Figure 9.16 allows the direct evaluation of ρ from data giving q_g and $k = q_r/q_g$. J. Edwards (1969) suggests an approximation, which is easy to evaluate, namely,

$$\rho = \frac{0.57 \log k}{-\log q_g - 0.44 \log k - 0.26}. \tag{9.46}$$

He previously had suggested another approximation,

$$q_r = \sqrt{q_g},$$

which is less satisfactory.

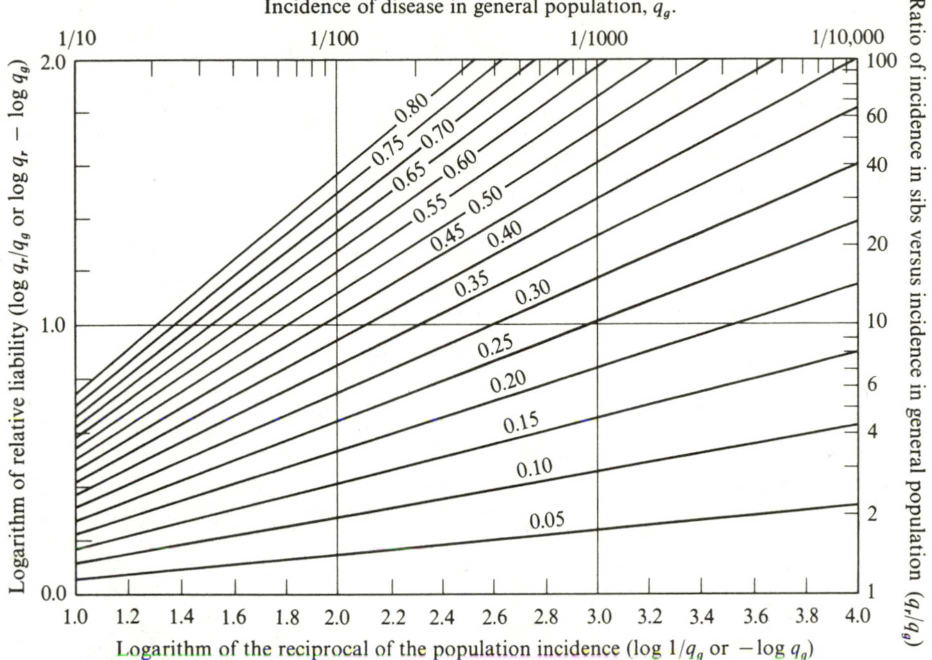

FIGURE 9.16
Graph relating ρ(the phenotypic correlation on Pearson's model)to the population incidence of a threshold trait q_g and the relative liability ($k = q_r/q_g$) of persons related to an affected individual.

Using the harelip data, $q_g = 0.001$, $q_r/q_g = 35$ for first-degree relatives, giving $\rho = 0.43$ (from Figure 9.16 or Equation 9.46). Assuming no dominance, the expected value for this correlation coefficient is 0.5. The heritability estimate is thus $0.43/0.5 = 86$ percent, which agrees well with the estimate obtained before from monozygous twins. From second-and third-degree relatives, $\rho = 0.17$ and 0.09 (which may be compared with the expected values of 0.25 and 0.125), giving heritability estimates of 68 percent and 72 percent, respectively These heritability estimates are thus somewhat higher than those obtained by Falconer's method, but still in fair agreement with them.

An interesting diagram collecting information for several diseases was constructed by Newcombe (1964). In this diagram (Figure 9.17) the disease incidence in the general population q_g is plotted against the ratio $k = q_r/q_g$, where q_r refers to the disease incidence in sibs of affected individuals. The lines in the diagram correspond to three hypotheses: simple dominant inheritance; simple recessive inheritance; multifactorial inheritance with a threshold and with various degrees of heritability. The lines for the multifactorial model are obtained from the tetrachoric correlation coefficient, taken as 0.5 for 100 percent heritability. In the original

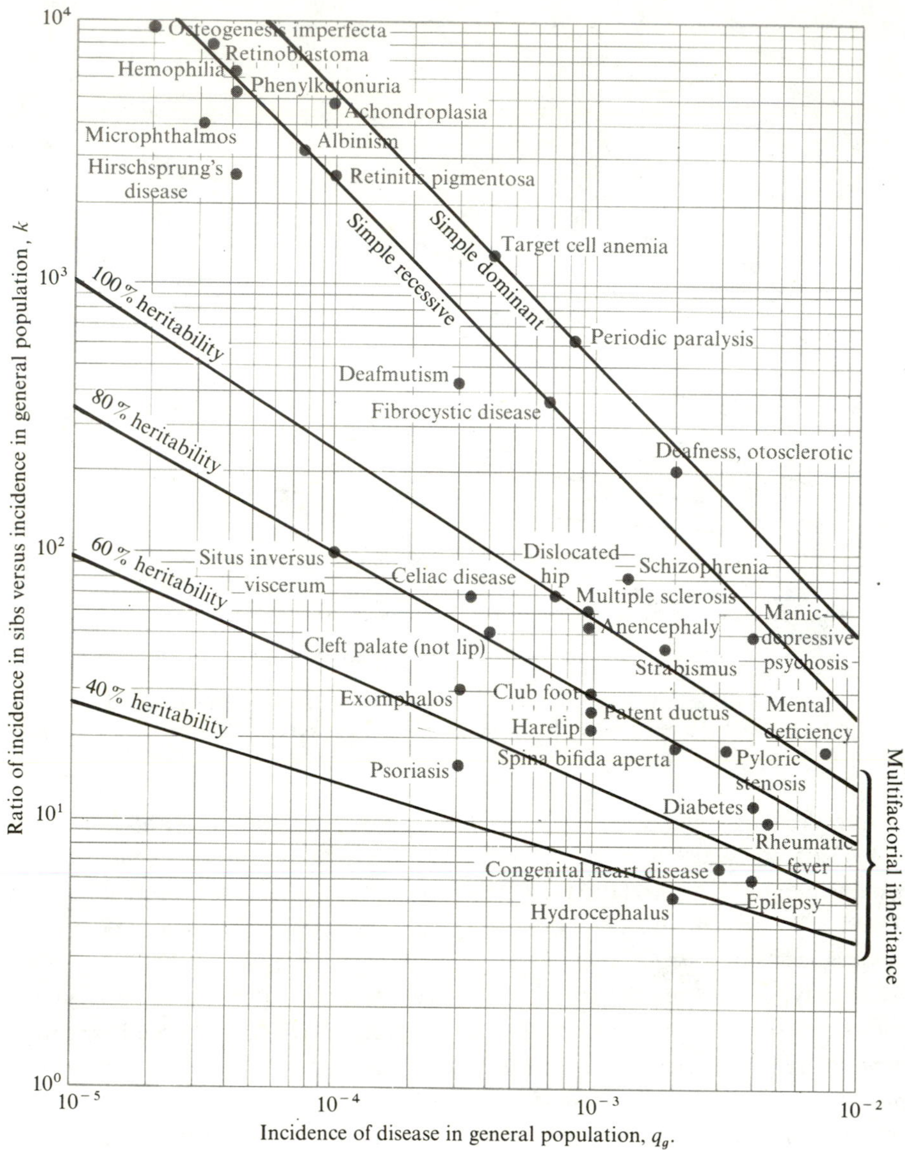

FIGURE 9.17
Relation between disease incidence (q_g) and relative incidence in sibs of affected individuals (k) for a number of diseases. The lines indicate the expected relationships for simple dominant, simple recessive, and multifactorial inheritance with different heritabilities. (Modified from Newcombe, 1964).

diagram by Newcombe there was only one line for multifactorial inheritance, and this corresponded to the approximation suggested earlier by Edwards, namely,

$$q_r = \sqrt{q_g}.$$

The lines labeled "simple dominant" and "simple recessive" correspond to the expectations for characters determined in a straightforward way by single dominant or recessive genes. For a recessive gene, the ratio $k = q_r/q_g$ is, approximately, $\frac{1}{4}/q^2 = 1/(4q^2)$, where q is the gene frequency and so $q^2 = q_g$ is the disease incidence, ignoring all matings other than $Aa \times Aa$ on the assumption that q is small. For a dominant gene, again assuming the gene frequency q is small, and thus assuming that $Aa \times aa$ is the only significant mating, the ratio is $\frac{1}{2}/2q$ or $k = 1/(4q)$, where $2q$ is the approximate incidence. Most of the diseases illustrated in Figure 9.17 fall into two distinct groups, those clearly compatible with simple Mendelian inheritance, which lie close to the two upper lines, and those suggestive of additive multifactorial inheritance with various degrees of heritability. Among those most clearly intermediate between the two is schizophrenia, which will be further discussed in the last section of this chapter.

9.10 The Biological Basis of Twinning

Identical or monozygous multiple births are the only source of humans that have identical genotypes.

Francis Galton, the famous British biometrician of the nineteenth century, was one of the first to emphasize the significance of twins for studies of human inheritance. Since that time they have been studied very extensively with a view to determining the relative importance of genetic and environmental effects on a wide variety of diseases and other attributes, many of which are behavioral. Comparisons are generally made between monozygous, or monozygotic (MZ), and dizygous, or dizygotic (DZ), twins, with the latter coming from separate fertilizations and thus being genetically comparable to sibs. This is undoubtedly an oversimplification, as there must be special factors, especially in relation to personality development, which exert a greater joint influence over monozygous twins than over dizygous twins. A study of monozygous twins separated soon after birth should (as was first pointed out by Muller in 1925), to some extent mitigate these effects and help to confirm which similarities between twins cannot be readily attributed to their similar environment. It is also important to realize that, in general, *differences* between monozygous twins must be due to environmental factors. Thus, a study of such twins should provide a test of those theories of personality development that appeal heavily to the importance of differences in environmental stimuli. An important limitation of twin studies is that they indicate only the degree of genetic determination of a character and can give no information as to its mode of inheritance.

There are four biologically different types of monozygous twin pairs.

Monozygous twins are presumably always derived from a single fertilized egg. The immediate products of cleavage, after fertilization, are not differentiated and thus each has the potential to develop into a normal individual. Monozygous twins presumably arise, therefore, from separation and subsequent independent development of early cleavage products. A substantial fraction of monozygous twins develop with separate chorions, amnions, and placentas, giving rise to the four types of pregnancies illustrated in Figure 9.18. Dizygous twins, which are the result of double ovulation and simultaneous or almost simultaneous fertilization, always develop in separate amnions and chorions, but may share a common placenta.

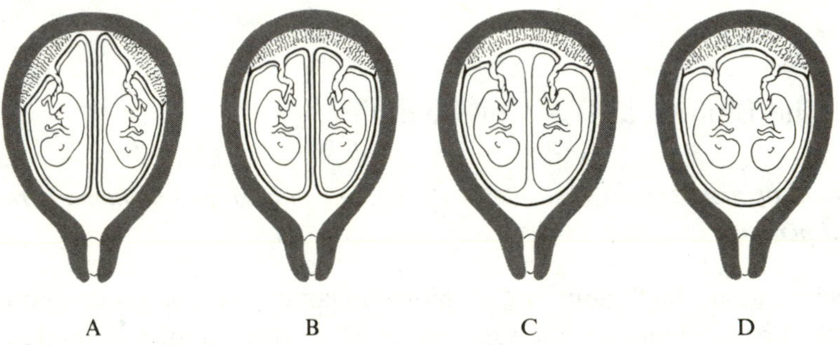

FIGURE 9.18
Diagrams of twin pregnancies enclosed in the uterus. A: Monozygous or dizygous twins with separate amnions, chorions, and placentas. B: Monozygous or dizygous twins with separate amnions and chorions, and fused placenta. C: Monozygous twins with separate amnions, and sharing a chorion and a placenta. D: Monozygous twins sharing a single amnion, chorion, and placenta. (A-C after Potter, 1948; D after Stern, 1960.)

About one out of every eighty babies born (to Caucasians) is a twin, and about 30 *percent of these are monozygous.*

The total frequency of human twin births is generally between 1 and 1.5 percent. The frequency of twin births shows some racial variation, with Japan having an unusually low rate of 0.65 percent. The birth rate of twins is also a function of socioeconomic factors and of parental age. Although demographic data do not distinguish the two types of twins, it is important to distinguish variations in birth rates of monozygous and dizygous twins since the two types are biologically different.

A simple method (suggested by Weinberg in 1901) *of estimating the relative proportion of twin pairs that are monozygous is based on the total proportion of twins of like sex and on the distribution of sexes among twins.*

Monozygous twins must always, of course, be of like sex, while among dizygous twin pairs the sexes are presumably combined at random. Thus, the excess of twin pairs of like sex over that expected from a random combination of sexes should provide a basis for estimating the relative proportions of the two types of twins. If the proportion of male births is y (which is usually slightly larger than 0.5, see Chapter 10) and of female births is $x = 1 - y$, then the expected proportion of dizygous twin pairs that are of unlike sex is $2xy$. If, further, the proportion of all twins that are dizygous is d, then

$$d = \frac{\text{proportion of all twin pairs that are of unlike sex}}{2xy}$$

$$\sim 2\,(\text{proportion of all twin pairs of unlike sex}), \text{ if } x \sim \tfrac{1}{2}. \tag{9.47}$$

A major survey of twins is being undertaken in Denmark with the aim of recovering as much information as possible on all surviving twins born in the period 1870–1901 (Harvald and Hauge, 1965). The overall proportion of dizygous twin pairs of unlike sex in the data collected so far is 2525/6893 = 0.367. Using Equation 9.47 the estimated proportion of monozygous twins is calculated as $1 - 2 \times 0.367$ or 0.266, which corresponds very well with the directly ascertained frequency of 0.255 based on zygosity diagnosis (see below).

Table 9.13 gives some data on twinning rates computed using Equation 9.47 for various races, and interracial crosses (see Chapter 11 for a discussion of the strict definition of race). Birth rates of monozygous twins are not appreciably different from race to race. The overall differences between races are thus mostly due to differences in birth rates of dizygous twins. Negroids have the highest rates, Caucasoids have rates that are intermediate, and Mongoloids have the lowest. It is interesting that interracial crosses show no paternal effect on the birth rate of dizygous twins, but do show a strong *maternal* effect: the observed rate is almost equal to that of the race of the mother (Table 9.13).

An example of data showing the variation of twinning rates with maternal age is given in Figure 9.19. There is very marked variation in the DZ rate, while the MZ rate is fairly stable. In general it seems that the DZ rate is much more susceptible to environmental influences than the MZ rate. Maternal age is, of course, highly

TABLE 9.13
Twinning Rates per Thousand Births by Maternal Race

Race of Mother	No. of Births	Observed Rates MZ	Observed Rates DZ	Expected[a] DZ
Caucasoid	50,570	3.7	6.6	7.0
Oriental	90,518	4.5	2.5	2.4
'Part-Hawaiian'	23,547	5.5	3.2	2.4
Caucasoid-Oriental	7,533	5.0	2.1	3.2
U.S. White, 1922–36	—	3.8	7.4	—
U.S. Negro, 1922–54	—	3.9	11.8	—
Japan, 1926–31	—	4.2	2.7	—
Japan, 1956	—	4.1	2.3	—
Nigeria Negro	—	5.0	39.9	—
Antigua Negro	—	3.9	11.5	—
Johannesburg Negro	—	4.9	22.3	—
Salvador, Brazil	—	4.0	13.6	—

Source: From Morton, Chung, and Mi (1967).
[a] Adjusted for maternal age, year, and birth order.
$Y = 7.0 - 4.6M_1 - 1.5H_1$.

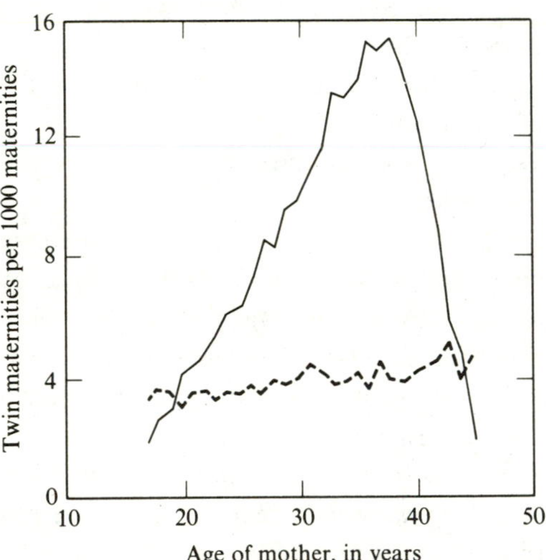

FIGURE 9.19
Twin maternities per 1000 maternities for given age of mother. Solid line plots births of dizygous twins: dashed line, of monozygous. Italian data, 1949–54. (From Bulmer, 1958.)

correlated with parity. Both factors have been shown to influence the DZ twinning rate independently. Twinning rate increases steadily with parity at all ages, but peaks at 30–40 years for all parities. There are up to now, no good explanations for these strange correlations.

The frequency of triplets is approximately the square, and that of quadruplets the cube, of the twinning rate.

This empirical rule, known as Hellin's rule fits observed data remarkably well. Triplets and quadruplets may all come from different zygotes, all from the same zygote, or from an intermediate combination. Attempts have been made to derive Hellin's rule from theoretical considerations in terms of the probability of double ovulation and the probability of a fertilized zygote splitting into two, but so far they have not met with any noticeable success (Allen, 1957; Bulmer, 1958). Triplets and quadruplets could, of course, in principle, also be useful for quantitative inheritance studies but their incidence is too low for them to be of any real practical use. Recently, pituitary and other hormones introduced for the treatment of some types of sterility have lead to striking cases of multiple births.

There may be a small genetic component to the twinning rate, particularly for dizygotic twins.

The data on the inheritance of a tendency toward twinning is largely inconclusive. Certainly, exceptional families have been reported with a striking recurrence of multiple births. On the whole, however, twinning rates among relatives of twins are at most slightly higher than the rate in the general population. Thus, Weinberg early this century concluded that the DZ twinning rate was higher in relatives of the mothers of twins than in those of the fathers. Further evidence in favor of a maternal effect has been given by Waterhouse (1950) and others. Greulich in 1934 found an approximately two-fold increase in the DZ twinning rate of relatives on both sides. Other studies have also suggested some increase for MZ twins. However, as pointed out by Bulmer (1960) these estimates are very sensitive to variations in the average ages and parities of the related groups. Certainly, factors are known that increase the rate of multiple ovulation, such as the hormones already mentioned. Strains of mammals, sheep, for example, with unusually high twinning rates are also known. Maternal effects on the DZ twinning rate, which have been observed for both interracial matings and matings in which both persons are members of the same race, are at the moment the best evidence for some inheritance of twinning.

The objective diagnosis of the zygosity of twins depends on a detailed analysis of genetic similarity.

The striking outward similarities between monozygous twins, even as compared with dizygous twins (whose similarity, after all, is no greater than that of other sibs) often allow a reasonably clear-cut classification of twin pairs as monozygous and dizygous. Clearly, however, care must be taken not to allow such an admittedly subjective procedure to bias the zygosity determination of twins. The only true criterion is identity of the genotype and this cannot, of course, be checked directly. However, the number of known polymorphisms is now large enough that if twins are classified with respect to all or most of them, the probability of dizygous twins being identical for all of them is so small that it may be discounted. Identity for all the polymorphisms may be used to diagnose monozygosity. This is a satisfactory and objective, if laborious, method. Other quantitatively inherited attributes such as fingerprint ridge counts or patterns can also be used, in an empirical way, to aid in diagnosis. Another direct test for monozygous twins is the survival of skin grafts from one to the other. As was pointed out in Chapter 5, there must be a large number of polymorphic loci for histocompatibility antigens, which are responsible for graft rejection. Thus the probability that any two individuals other than monozygous twins, even dizygous twins or other sibs, are identical with respect to all relevant loci is vanishingly small, so that the extended survival of a skin graft is an efficient, though generally impractical, test for the probable identity of twin genotypes.

The general principle of zygosity diagnosis using blood groups and other polymorphic markers is as follows:

1. The probability of sibs, that is, dizygous twins, being identical is calculated separately for each locus. The nature of this calculation depends on the information available on the parental types. For example, if one parent was *MN* and the other *N* and the twins *NN*, the probability of identity for this locus would be $\frac{1}{4}$, this being simply the probability that two sibs from the mating $MN \times N$ are both *MN*. If no information on the parents is available, we must calculate the probability of identity in terms of the expected frequencies of different mating-type combinations in the population. Thus, given the gene frequency of *M* as $\frac{1}{2}$ and assuming random mating, we may compute the frequencies of the matings that can produce *MN* offspring:

$$\tfrac{1}{4}MN \times MN,\ \tfrac{1}{4}MN \times N,\ \tfrac{1}{4}MN \times M, \text{ and } \tfrac{1}{8}\ M \times N.$$

Since the respective probabilities of a pair of *MN* sibs are $\frac{1}{4}$, $\frac{1}{4}$, $\frac{1}{4}$, and 1, the total probability of producing identical sibs is

$$\tfrac{1}{4} \times \tfrac{1}{4} + \tfrac{1}{4} \times \tfrac{1}{4} + \tfrac{1}{4} \times \tfrac{1}{4} + 1 \times \tfrac{1}{8} = \tfrac{5}{16}.$$

Tables of such identity probabilities for the common blood groups, constructed by using the gene frequencies observed in England, are given by Race and Sanger (1968) in their standard work on blood groups. Use can be made of data on quantitative traits if the distribution of twin-pair differences is known for both monozygous and dizygous twins. For example, suppose the proportion of monozygous twins whose height difference lies in the range 0.5–1 inch is 0.4 and that of dizygous twins is 0.2. Then, for a pair of twins whose difference is in that range, the relative probability of being dizygous is 0.2/0.4 = 0.5.

2. The product of the individual identity probabilities for polymorphisms, quantitative characters, and sex, obtained as in (1) gives the probability P_I of the twins being identical, given that they are dizygous.

3. If the relative proportion of twins in the population that are monozygous is m, then the overall probability of the twins being monozygous given they are identical (nonidentical twins are obviously dizygous), is

$$P_M = \frac{m}{m + (1 - m)P_I}. \tag{9.48}$$

This is because the proportion of twins that are dizygous but identical is $(1 - m)P_I$, and thus the total proportion of twins that are identical is $m + (1 - m)P_I$, of which only the fraction P_M as given by Equation 9.48 are monozygous. For example, data on a pair of female twins gave identity probabilities as follows: Sex 0.5; blood groups: ABO 0.48, MNS 0.47, Rh 0.50, P 0.85, Secretor 0.54, Kell 0.95, Lutheran 0.80, Kidd 0.85; and fingerprint ridge count 0.77. By multiplying together all of these separate probabilities, we obtain $P_I = 0.013$.

Given the proportion of monozygous twins $m = 0.3$, Equation 9.48 gives, for the probability that the twins are monozygous.

$$P_M = \frac{0.3}{0.3 + 0.3 \times 0.013} = 0.987.$$

A few twin pairs have been found that must be chimeras—that is, they each have mixtures of two genetically different types of cells.

About 90 percent of dizygotic twin calves have mixtures of two genetic types of blood, due presumably to anastomoses (vascular connections) of the blood vessels of the calves. Such chimeras are much rarer in man, though a few have been described. A **chimera** is defined, in general, as an individual formed of genetically different types of cells that come from different zygotes. A **mosaic** is formed from a single zygote, but has a mixture of types of cells because of somatic mutation, nondisjunction or aberration of chromosomes, or somatic crossing-over. Human mosaics also are rare, and are usually more easily observed with cytogenic studies than by an examination of blood groups or other polymorphisms. The blood groups of one set of chimeric twins and of the members of their family are shown in Figure 9.20. Even though each of the twins has almost equal numbers of the two types of cells and thus is not readily distinguishable from the other by this criterion, the genotype of II.2 can clearly be ascertained from the genotypes of her offspring. Thus, since III.1 and III.2 are both *MM* and III.3 is heterozygous for Rh *E* and *e*, II.2 must have the genotypes *MN* and *cDE/cde*. The fact that all three offspring have O type blood is also compatible with II.2 being OO. This twin, although OO genetically, has only anti-B antibodies in her serum. The same is true of three other

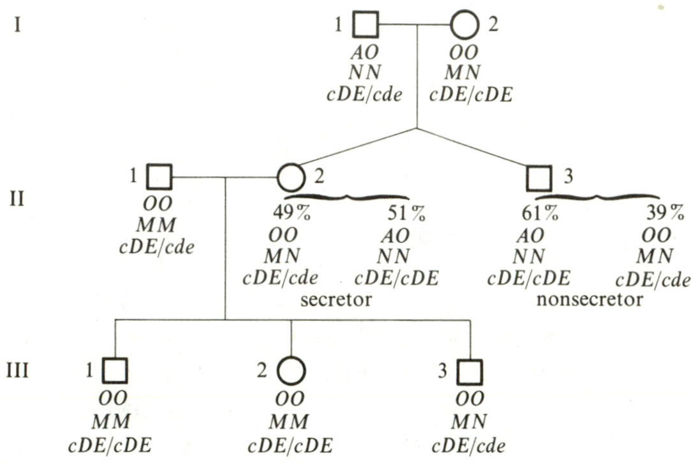

FIGURE 9.20
Blood groups of the W family. II.2 and II.3 are chimeric twins. The genotype of II.2 is ascertained from those of her offspring. Both twins had only anti-B in their serum. (From Race and Sanger, 1962.)

chimeric twins who are OO, but have "grafted" A cells. This is most probably due to the induction of tolerance to the A antigen by the presence of A red cells at an early stage of fetal development. There are some indications that the white cells of II.3 are also mixed, based on recognition of a small proportion of granulocytes with the characteristic female "drumstick" nodules on their nuclei. This would suggest the possibility that skin grafts exchanged between the chimeric twins should survive. This idea follows from the fact that it is likely that the white cells carry most of the histocompatibility antigens, so that a chimeric twin may become tolerant to its cotwin's histocompatibility antigens. This phenomenon is known to occur in chimeric cattle twins. Undoubtedly, there is a small chance that chimeric twins could be mistaken as monozygous if they were not carefully studied.

The systematic use of simple visual criteria for twin diagnosis is almost as efficient as the use of blood groups and other polymorphisms.

It may not always be possible to do zygosity diagnosis using blood groups, especially in large-scale twin surveys, such as that being carried out in Denmark. Simple questionnaires asking about the degree of similarity between twin pairs have been found at least 90–95 percent efficient for the diagnosis of monozygosity. Typical questions asked relate to eye color, hair color and texture, height, weight, body build, and the tendency to be mistaken by parents, close friends and casual friends, in addition to the twin's own diagnosis of zygosity. Interestingly enough, in one survey, as many as 24 percent of monozygous twins thought they were dizygous while 19 percent of dizygous twins thought they were monozygous and 10 percent were not sure. It seems that these opinions may have been influenced by the physician's incorrect diagnosis at birth, based on whether there were common chorions (see Figure 9.18) (see Nichols and Bilbro, 1966).

These methods of diagnosis generally underestimate slightly the proportions of dizygous twins, since it seems that few dizygous twins are similar enough to be mistakenly classified as monozygous but a significant percentage of monozygous twins may differ enough to be misclassified as dizygous. Race and Sanger, however, comment, "for several years Mr. James Shields of the Genetics Unit of the Maudsley Hospital has been sending us samples of blood from twins. We find that the blood groups practically never contradict the opinion of such a skilled observer of twins."

9.11 The Use of Twins for the Study of the Genetic Determination of Traits

The use of twin studies for the measurement of the degree of genetic determination of a continuously varying character is based on a comparison of the variance of the differences between monozygous twins pairs and the differences between dizygous twin pairs.

The distributions of the differences in height between 50 pairs of monozygous twins, 52 pairs of dizygous twins, and 52 pairs of sibs who are not twins are shown in Figure 9.21. The means of these three difference distributions are 1.7 centimeters, 4.4 centimeters, and 4.5 centimeters, respectively. As is expected if determination of height has a significant genetic component, the monozygous twin pairs on the average, differ less than the dizygous twin pairs. This greater variation between dizygous twin pairs is a measure of the extent to which sibs differ with respect to genes affecting height. Distribution curves similar to those shown in Figure 9.21 are obtained for a wide variety of continously varying anthropometric and also biochemical attributes, as well as for IQ.

Simple quantitative estimates of the degree of genetic determination for such characters can be obtained by a comparison of the means of the squares of the

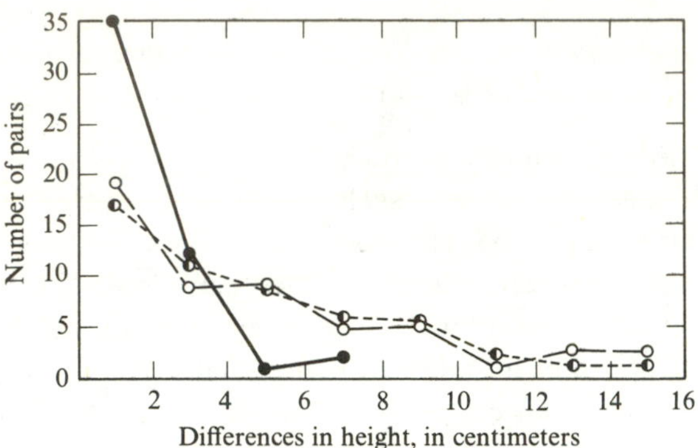

FIGURE 9.21
Curves of distribution of differences in standing height of 50 pairs of identical twins (plotted along the solid line), 52 pairs of nonidentical twins (plotted along the line of longer dashes), and 52 pairs of sibs (plotted along the line of shorter dashes). (After Newman, Freeman, and Holzinger, 1937, from Stern, 1960.)

differences between members of monozygous twin pairs (V_{MZ}) and the means of the squares of the differences between dizygous twin pairs (V_{DZ}). These quantities are each equal to twice the mean of the variance *within* a pair, the average being taken either over all monozygous or over all dizygous pairs. If we assume that the environmental influences on the two types of twins are equivalent and that there are no genotype-environment interactions, then V_{MZ} should be an estimate of the portion of V_{DZ} that is due to environmental variation. Thus, the quantity H, defined by

$$H = \frac{V_{DZ} - V_{MZ}}{V_{DZ}}, \tag{9.49}$$

is a measure of the proportion of the variance of the differences between dizygous twins that can be attributed to genetic variation. It is, therefore, in this limited sense, a measure of the degree of genetic determination of the character in question. In order to assess the significance of variation between H values obtained from different sources, it is useful to have an expression for the variance of H. This is the same as the variance of V_{MZ}/V_{DZ}. This ratio, being the ratio of two variances, has the distribution of the F statistic used to test the significance of the ratio between two variances (see Appendix I). The distribution of F has a mathematical form known as a "β distribution of the second kind" which depends on the expected value of the variance ratio and on the numbers of observations from which the two variances were calculated. It can be shown that

$$\text{Var}\,(H) \sim \frac{2F^2 n_2^2(n_1 - 1)(n_1 + n_2 - 4)}{n_1^2(n_2 - 3)^2(n_2 - 5)}, \tag{9.50}$$

where F is the observed value of V_{MZ}/V_{DZ} and n_1 and n_2 are the respective number of monozygous and dizygous twin pairs on which measurements were made.

Some estimates of *H*, with their standard errors, are given in Table 9.14 for a number of morphological characters. They are taken from an extensive twin study carried out by Osborne and De George (1959) on a series of approximately 60 monozygous and 40 dizygous twins. The aim of their study was to define as completely as possible the genetic basis for morphological variations between adults of both sexes. A large number of types of measurements were made of which just a few are represented in Table 9.14. The *H* values for stature, arm length, and foot length are very high and are approximately equal in the two sexes. These three length measurements are each strongly correlated with overall size and so, of course, with each other. By contrast the *H* values for weight are not significantly different from zero, though that for females is suggestively higher than that for males. This fits in with the intuitive expectation that weight is more subject to obvious environmental variation than are the three length measurements. The differences between the sexes might be attributed to females working harder to maintain their ideal weights and thus tending to lessen the environmental variation. Other twin studies have, on the whole, given *H* values for weight that are appreciably larger than those obtained by Osborne and De George, who suggest that this difference may be a function of the ages at which the twins were studied. At younger ages, say, before they are 20–25 years old, weight correlations may be higher since they then probably reflect growth rate rather than final weight, and this may be less subject to environmental variation. However, as is clear from the standard errors shown in Table 9.14 the sampling errors of the estimates are large, and this may be enough to explain the different results obtained by different investigators. The sampling errors for the male values are particularly high because they are based on only ten male dizygous twins. The *H* values for hip circumference show a significant difference between the sexes, which is easily explained in terms of secondary sexual differences and a closer control over weight by the females. These effects of sex may have been accentuated

TABLE 9.14
H Estimates from Monozygous and Dizygous Twin Data

Characteristic	*Male*	*Female*
Stature	0.79 ± 0.21	0.92 ± 0.03
Weight	0.05 ± 0.94	0.42 ± 0.25
Arm length	0.80 ± 0.20	0.87 ± 0.06
Foot length	0.83 ± 0.16	0.81 ± 0.08
Hip circumference	0.19 ± 0.93	0.66 ± 0.15
Cephalic index (head breadth/head length)	0.90 ± 0.10	0.70 ± 0.13
Masculinity-feminity index	0.78 ± 0.21	0.85 ± 0.06

Source: Data from Osborne and De George (1959).
Note: H and its variance (and so its standard error) are calculated using Equations 9.49 and 9.50, respectively.

by the age group studied by Osborne and De George. The ages of the twins were in the range 24–27 years, a range in which females are probably highly motivated to control their weight. Heritability of a character may be a function of age and sex, and the genes contributing to its determination may be different at different ages and in the two sexes. At younger ages growth rate is important, while at intermediate ages the strength of secondary sexual characteristics may markedly influence environmental variability. These are, of course, intuitively obvious facts. What needs to be emphasized is the influence of age and secondary sexual characteristics on esimates of the degree of genetic determination. Thigh circumference showed no significant heritability in Osborne and De George's data. It is, of course, strongly correlated with weight and also somewhat more subject to measurement error. The cephalic index (head breadth/head length), which has been used widely by anthropologists in the characterization of skulls of different racial groups, shows a very high heritability. As it is a measure of head shape rather than size, it may give quite different results from either head length or breadth. In fact Osborne and De George found high H values for head breadth, but values so low as to be insignificant for head length. It seems likely that many of the outward physical attributes by which we recognize each other, for example, facial expressions and relative body build can only be expressed in terms of combinations of simple measurements—that is, indices such as the cephalic index. The study of the heritability of these attributes depends, therefore, on the discovery of the correct indices. Osborne and De George made an interesting attempt to construct an index of masculinity-femininity. It was based on an objective point-scoring system of a variety of morphologic attributes that are generally considered to be secondary sexual characteristics, such as the unevenness of the male calf as compared with the smooth curves of the female calf. As can be seen from Table 9.14, this rating has a relatively high heritability in both sexes.

The variation between twin pairs is often expressed in terms of the mean of the absolute differences between the pairs rather than the variance of the differences. These two quantities are closely related. For two independent measurements having the same normal distribution, it can be shown that the mean of the absolute difference between them, which for two observations is the range, is

$$\frac{2}{\sqrt{\pi}} = 1.13$$

times the standard error of the parental normal distribution. Thus, in the twin data, the square of the mean difference between twins should be approximately proportional to the variances of the differences. The H index of genetic determination then takes the form

$$H_t = \frac{M_1^2 - M_2^2}{M_1^2}, \tag{9.51}$$

where M_1 and M_2 are the mean observed absolute differences for dizygous and monozygous twins, respectively. The average IQ difference found by Newman,

Freeman, and Holzinger (1937) between 50 monozygous twin pairs was 5.9 points and that between 52 dizygous twin pairs 9.9 points (47 non-twin sib pairs had an average difference of 9.8 points). The H value here is, therefore,

$$\frac{(9.9)^2 - (5.9)^2}{(9.9)^2}$$

or 0.64. Comparable values have been found in a number of other twin studies. Some interesting data on the differences between twin and sib pairs with respect to the age at menarche are shown in Table 9.15, summarized by Tanner (1960). Tanner

TABLE 9.15
Mean Difference in Months in Age at Menarche of Related and Unrelated Women

	From Petri (1935)		*From Tisserand-Perrier (1953)*	
Relationship	*No. Pairs*	*Diff. Months*	*No. Pairs*	*Diff. Months*
Identical twins	51	2.8 ± 0.33	46	2.2
Non-identical twins	47	12.0 ± 1.62	39	8.2
Sisters	145	12.9	—	—
Unrelated women	120	18.6	—	—

Source: From Tanner (1960).

used this measurement, which identifies a definite stage in the development of the female, because he was interested in the genetic determination of growth rates. The estimated H value for the data from Petri (1935) is 0.9 and that from Tisserand-Perrier (1953), 0.93, indicating that there is a very high heritable component for age at menarche, even though it is known to be significantly affected by nutrition and other socioeconomic factors. Averages for age at menarche vary considerably among different populations, and the overall average has actually been declining in Europe at a rate of approximately four months per decade. This rate of decline is clearly too fast to be ascribable to genetic changes.

Though the H statistic is the simplest and most convenient way to summarize twin data on continuously varying characters, its interpretation is subject to many limitations.

Even under the assumptions of equivalent environments for monozygous and dizygous twins and the lack of genotype-environment interactions, the use of H to measure the extent of genetic determination is nevertheless subject to two further limitations. The first is the limited extent of environmental variation allowed for by restricting consideration only to twins, and the second is the limitation of the extent of genetic variation to differences between sibs. In order to illustrate the latter we consider the expression of V_{MZ} and V_{DZ} in terms of the additive variance (V_A), the dominance variance (V_D), and the environmental variance (V_E). The variance of the difference between two variables x and y is given by

$$V(x - y) = V(x) + V(y) - 2\operatorname{cov}(x, y). \tag{9.52}$$

If x and y are the measurements in the two twins of a pair, then for both monozygous and dizygous twins

$$V(x) = V(y) = V_A + V_D + V_E,$$

the total phenotypic variance. For monozygous twins $\operatorname{cov}(x, y) = V_A + V_D$, the total genetic variance. This follows from the fact that $\operatorname{cov}(x, y)$ is the covariance between the genetic effects, which are identical in monozygous twins, and from the fact that the covariance between identical variables is the same as the variance. Cov(x, y) for dizygous twins is the same as that between sibs, which is $\frac{1}{2}V_A + \frac{1}{4}V_D$ (see Table 9.7). Thus, from Equation 9.52,

$$V_{MZ} = 2(V_A + V_D + V_E - V_A - V_D) = 2V_E \tag{9.53}$$

and

$$V_{DZ} = 2(V_A + V_D + V_E - \tfrac{1}{2}V_A - \tfrac{1}{4}V_D) = V_A + \tfrac{3}{2}V_D + 2V_E, \tag{9.54}$$

and so, by the definition of H, we have

$$H = \frac{V_A + \frac{3}{2}V_D}{V_A + \frac{3}{2}V_D + 2V_E} = \frac{\frac{1}{2}V_A + \frac{3}{4}V_D}{\frac{1}{2}V_A + \frac{3}{4}V_D + V_E}. \tag{9.55}$$

Thus, if V_E were a representative environmental variance, H would always underestimate the degree of genetic determination, $(V_A + V_D)/(V_A + V_D + V_E)$. Since, however, we have no good way of estimating the effect on values of V_E obtained from twin data of limiting environmental variation to difference between twins, we cannot readily relate H values from twin studies to estimates of the degree of genetic determination from studies on random samples of other combinations of relatives from Mendelian populations.

Departures from random mating due to positive assortative mating with respect to a character under study lead to a positive correlation between the parents, and therefore increase the correlation or covariance between sibs. This, from Equation 9.51, tends to decrease V_{DZ}, but not V_{MZ}, and thus tends to decrease H. Assortative mating with respect to many quantitative attributes is certainly well documented (see Section 9.7), and may be an important factor tending in general, to increase heritability estimates. There are two plausible factors that might make H an overestimate of the genetic differences between twins. One is a tendency for the environmental variation between monozygous twins to be less than that between dizygous twins. This certainly may be an important factor for some behavior attributes, but the difference would be expected to be very small. The second factor that might tend to inflate H values is the existence of sampling biases in the collection of twin data. We have already discussed diagnosis of zygosity. If this is done objectively, then misclassification of relatively different monozygous twins as dizygous should be negligible.

The main factor that can bias the collection of data on normal (nondiseased) twins is differential survival of twin pairs. Clearly, only data from those pairs both of whom have survived to the time of contact will be collected. Even if registries of twin births are maintained, there may still be biases for characters that cannot be detected at birth and that are associated with differential survival. The older the twin pair, the less likely it is that both have survived. Differential survival between the sexes has a marked effect on the sex ratio among ascertained twin pairs. Thus, Osborne and De George note that the aggregate ratio of male-to-female twin pairs in six different studies is 918 : 1373, which represents a highly significant deficiency ($\chi_1^2 = 98.2$) of male pairs. Other factors may affect the ratio such as the greater reluctance of male pairs to be studied. Availability of twin pairs for study, generally in the same overall locality, is also a factor that may bias their selection as well as their common willingness to be studied. All these biases are likely to favor similarity between ascertained twin pairs, though to what extent it is almost impossible to say. Ascertainment biases in twin studies should be minimized by avoiding as far as possible those factors that might lead to nonrandom sampling of twin populations.

We can at least say that a significant H value is likely to be an indication of enough genetic polymorphism with respect to genes affecting the character in question to give rise to variation between dizygous twins (that is, sibs) which is significant, relative to the observed environmental variation between monozygous twins. However, nothing at all can be said with respect to the probable number of such polymorphic genes or their relative dominance.

The study of threshold characters in twins is based on a comparison of the frequencies of concordance in monozygous and dizygous pairs.

Twin pairs are said to be **concordant** with respect to a threshold character if they both have it or both do not. The concordance frequency is the proportion of concordant twin pairs among all those which include at least one with the trait. A significantly higher concordance frequency in monozygous than in dizygous twins is considered evidence for a significant genetic component to the determination of the trait. The significance of the difference can be tested simply by a 2×2 contingency chi square as illustrated for the data on epilepsy shown in Table 9.16. These data are taken from the extensive Danish twin study by Harvald and Hauge, to which we

TABLE 9.16
Twin Concordance for Epilepsy

Twins	*Concordant Pairs (Percent)*	*Discordant Pairs*	*Total Pairs*
MZ	10 (37)	17	27
DZ	10 (10)	90	100
Total	20	107	127

$$\chi_1^2 = \frac{127(10 \times 90 - 10 \times 17)^2}{20 \times 107 \times 100 \times 27} = 11.7 \qquad P \sim 0.1\%$$

Source: Data from Harvald and Hauge (1965).

have already referred. The discordant pairs are those with only one twin affected. The data on like-sexed and unlike-sexed dizygous pairs were combined since they showed no significant differences in concordance.

An index for estimating the degree of genetic determination of threshold characters from twin data, which was originally suggested by Holzinger in 1929 and is widely quoted, is

$$H = \frac{q_M - q_D}{1 - q_D}, \tag{9.56}$$

where q_M and q_D are the respective concordance frequencies in monozygous and dizygous twins. The quantity defined by this equation, seems, however, to be quite arbitrary. It is difficult to see what relation, if any, it can have to the usual estimates of the degree of genetic determination of continuously varying characters, which are based on a variance analysis. The appropriate method of approach to is make use of Falconer's or Edward's analysis of threshold characters. Given an estimate of the incidence of the trait in the general population, heritability estimates can be obtained separately from monozygous and dizygous twins using Equation 9.45. As we pointed out earlier, estimates from monozygous twins may be biased because of significant departures from normality of the distribution of monozygous twin liability. Differences between estimates from monozygous and dizygous twin pairs can give an indication of the relative magnitude of the dominance variance, since, by definition, the regression coefficient b is the ratio of the covariance between twins to the total genetic variance. Thus, for monozygous twins we estimate $(V_A + V_D)/(V_A + V_D + V_E)$, and for dizygous twins $(\frac{1}{2}V_A + \frac{1}{4}V_D)/(V_A + V_D + V_E)$. Here, as before, V_E refers only to the environmental variation between members of the same twin pair.

Whether estimates of the general population incidence of a trait are available from twin studies depends on the way the studies have been carried out. In the

Danish twin study, all twins born are located and then pairs, one or both of which have the trait, are identified. The trait incidence is then given approximately by the proportion of dizygous twins that are affected. For a very rare disease this approach may not be practicable as the size of the initial twin sample needed to obtain a sufficient number of pairs with one or more affected may be prohibitively large. A search may then be made for affected individuals who have twins. The proportion of their twins that are affected is then the concordance frequency. In this case the twins are ascertained through the fact that one of them has the rare trait being studied. Unless a further study on the incidence of the trait in the population from which the twins were drawn is undertaken, this approach provides no indication of the general population incidence. The method of ascertainment may seriously bias the twin sample, often in favor of twins both of which are affected. This is because the probability of the twins being ascertained is the probability that at least one of the two affected is ascertained, which is approximately twice the probability of ascertaining a twin pair with only one affected (see Appendix II). On the assumption that $V_D = 0$ or that V_A/V_D is given, an estimate of the degree of genetic determination using the threshold model can be obtained from a combination of data on monozygous and dizygous twins even in the absence of an estimate of the general population incidence. Following Figure 9.12, let q_g, x_g, and a be, respectively, the incidence, the deviation of the threshold from the mean, and the mean deviation of the affected individuals from the mean of the general population. Then from Equation 9.45 we have for monozygous twins

$$\frac{V_A + V_D}{V_A + V_D + V_E} = \frac{x_g - x_M}{a} \tag{9.57}$$

and for dizygous twins

$$\frac{\frac{1}{2}V_A + \frac{1}{4}V_D}{V_A + V_D + V_E} = \frac{x_g - x_D}{a}, \tag{9.58}$$

where x_M and x_D are the deviations of the thresholds from the means for monozygous and dizygous twins (based on the concordance frequencies q_M and q_D). Taking the ratio of these Equations 9.57 and 9.58, we have

$$\frac{4V_A + 4V_D}{2V_A + V_D} = \frac{x_g - x_M}{x_g - x_D}. \tag{9.59}$$

Given $V_D/V_A = k$, say, Equation 9.59 allows us to estimate x_g in terms of x_M and x_D. Thus, substituting $V_D = kV_A$ into Equation 9.59 gives

$$\frac{4(1 + k)}{2 + k} = \frac{x_g - x_M}{x_g - x_D} \tag{9.60}$$

or

$$x_g = \frac{1}{2 + 3k}\left[4(1 + k)x_D - (2 + k)x_M\right].$$

When $k = 0$, and thus $V_D = 0$, we have

$$x_g = 2x_D - x_M, \tag{9.61}$$

and when k is very large, and V_A is much less than V_D, Equation 9.60 approaches

$$x_g = \tfrac{4}{3}x_D - \tfrac{1}{3}x_M. \tag{9.62}$$

Since x_M is generally less than x_D, the value of x_g given by Equation 9.62 is usually smaller than that given by Equation 9.61 and thus the resulting population incidence q_g is higher. As can be seen from Figure 9.13, the higher q_g, the lower the heritability estimate for a given incidence among the relatives. Thus, in general, Equations 9.61 and 9.62 give upper and lower limits to the estimate of the degree of genetic determination given by Equations 9.57 and 9.58. However, as we saw earlier, since for most quantitative characters determined by many genes it seems likely that V_D is small relative to V_A, Equation 9.61 usually gives a satisfactory estimate. The heritability estimate obtained in this way is analogous to taking the ratio

$$\frac{V_{DZ} - V_{MZ}}{V_{DZ} - \frac{1}{2}V_{MZ}}, \tag{9.63}$$

for continuously varying characters, which from Equations 9.53 and 9.54 is an estimate of $V_A/(V_A + V_E)$ when $V_D = 0$. Let us take as an example the data on epilepsy given in Table 9.16. For this, $q_M = 0.37$ and $q_D = 0.1$, from which we obtain $x_M = 0.33$ and $x_D = 1.28$ using Falconer's table. Assuming $V_D = 0$, Equation 9.61 gives

$$x_g = 2.23,\ q_g = 0.013,\ a = 2.58. \tag{9.64}$$

Thus from Equations 9.57 and 9.58 we have

$$\frac{V_A}{V_A + V_E} = \frac{2.23 - 0.33}{2.58} = 0.73.$$

If we assume $V_A = 0$, and use Equation 9.62, we obtain $x_g = 1.59, q_g = 0.056$, $a = 2.015$ and so, from Equation 9.57,

$$\frac{V_D}{V_D + V_E} = \frac{1.59 - 0.33}{2.015} = 0.63,$$

which is, actually, not much lower than the estimate obtained using the assumption that $V_D = 0$. The population incidence of epilepsy estimated from the incidence in dizygous twins was about 0.02. Using this value for q_g gives $x_g = 2.05$ and $a = 2.42$. Equations 9.57 and 9.58 then give

$$\frac{V_A + V_D}{V_A + V_D + V_E} = \frac{2.05 - 0.33}{2.42} = 0.71 \tag{9.65}$$

and

$$\frac{\frac{1}{2}V_A + \frac{1}{4}V_D}{V_A + V_D + V_E} = \frac{2.05 - 1.28}{2.42} = 0.32. \tag{9.66}$$

Dividing Equation 9.65 by Equation 9.66, we obtain

$$\frac{4(V_A + V_D)}{2V_A + V_D} = 2.2,$$

and so

$$\frac{V_D}{V_A} = \frac{2 \times 2.2 - 4}{4 - 2.2} = 0.22.$$

There is, thus, an indication of a slight effect of dominance, though this has very little effect on the estimate of the degree of genetic determination, which is 0.71, as opposed to 0.73 on the assumption that $V_D = 0$. The various estimates of this quantity are reasonably consistent and indicate that there is a major genetic component in the determination of epilepsy. The estimate of the dominance variance using data from monozygous and dizygous twins is not influenced by the limitations on V_E, at least not in the absence of significant genotype-environment interactions.

Data from the Danish Twin Registry on concordance frequencies for a number of different diseases are given in Table 9.17. As before, data on all types of dizygous

TABLE 9.17

Twin Concordances and Derived Degrees of Genetic Determination for a Number of Different Diseases. All comparisons other than those for cancer and acute infection are highly significant.

	Percent Concordance[a]		*Limits of Genetic Determination*[c]	
Disease	*MZ*	*DZ*[b]	*Upper* ($V_D = 0$)	*Lower* ($V_A = 0$
Cancer at same site	6.8 (207)	2.6 (767)	0.33	0.23
Cancer at any site	15.9 (207)	12.9 (212)	0.15	0.1
Arterial hypertension	25.0 (80)	6.6 (212)	0.62	0.53
Mental deficiency	67 (18)	0 (49)	Both 1	
Manic-depressive psychosis	67 (15)	5 (40)	1.05	1.04
Death from acute infection	7.9 (127)	8.8 (454)	−0.06	−0.06
Tuberculosis	37.2 (135)	15.3 (513)	0.65	0.53
Rheumatic fever	20.2 (148)	6.1 (428)	0.55	0.47
Rheumatoid arthritis	34 (47)	7.1 (141)	0.74	0.63
Bronchial asthma	47 (64)	24 (192)	0.71	0.58

Source: Data from Harvald and Hauge (1965), and based on the Danish Twin Registry.
[a] The numbers in parentheses are those on which the relevant percentages are based.
[b] Data on all *DZ* twins are combined.
[c] Upper and lower estimates are obtained using Equations 9.61 and 9.62, respectively.

twins have been combined. The only comparisons between the two types of twins that are not significantly different are for cancer and for deaths from acute infections. Upper and lower heritability estimates were calculated using Equation 9.61 assuming $V_D = 0$ and Equation 9.62 assuming $V_A = 0$. The two limits differ by, at most, about 10 percent. The heritability estimates of 1 for mental deficiency and 1.04 to 1.05 for manic-depressive psychosis must, of course, be overestimates since not all monozygous twins were affected. These inflated values are probably due to the normality assumption mentioned before. The negative estimates for deaths from acute infections and low estimates for cancer are not significant. It is, however, consistent with *a priori* expectation that cancer at the same site should have a higher heritability than cancer at any site. Even cancer at the same site is not necessarily a homogeneous etiological entity. The inheritance of certain, possibly rarer, forms of cancer may easily go undetected if the data for a variety of different forms, many showing no hereditary predisposition, are pooled. The contrast between data on deaths from acute infections and those on deaths from tuberculosis is striking. Both are infectious diseases and so *a priori* might be strongly influenced by, for example, the closer personal contact that usually prevails between monozygous than between dizygous twins. The very low heritability of acute infections and relatively high heritability of tuberculosis suggests that this is not an important factor. The high heritabilities for arterial hypertension, rheumatic fever, rheumatoid arthritis, and bronchial asthma are in line with other comparable studies in suggesting that there are important genetic components for these diseases.

Of course, all these estimates are subject to the same reservations that apply to the estimation of heritabilities of threshold characters from other relationships than twins and also to the problems of twin studies. A striking effect of age on the heritability estimates from twins for diabetes is illustrated by the data given in Table 9.18, which are also taken from the Danish Twin Study. We have already discussed

TABLE 9.18

Twin Concordances and Heritabilities for Diabetes Mellitus as a Function of Age. The calculations and symbolism are as described in the footnotes to Table 9.17.

	Percent Concordance		*Limits of Genetic Determination*	
Healthy Twin	*MZ*	*DZ*	*Upper* ($V_D = 0$)	*Lower* ($V_A = 0$)
$\geq$ 30 years	49	10	0.88	0.79
$\geq$ 50 years	55	11	0.93	0.86
$\geq$ 70 years	73	31.5	1.12	1.04
Total	47 (76)	9.7 (238)	0.86	0.77

Source: Data from Danish Twin Registry and Harvald and Hauge (1965)

the complications inherent in a genetic analysis of diabetes and the apparent change in its heritability with age. The main problem for twin studies is, that since there is a distribution of the age of onset of the disease, if observations on twins are made at an early age there is a high probability that only one twin will at that time have developed the disease even if both are ultimately destined to do so. The concordance frequencies for both monozygous and dizygous twins thus increase markedly as the minimum age of twins included in the study increases. The heritability estimates also increase, though the values of 1.12 and 1.04 for twins older than seventy years must, of course, be overestimates. The total data are dominated by the younger twins and show a heritability corresponding more to that of the youngest ages included. These heritability estimates, as expected, are higher than those shown in Figure 9.14, which were obtained from data on sibs. They also show an *increase* in heritability with age, where the other data showed a decrease. Apart from sampling errors, which certainly must be appreciable, the most plausible explanation for this difference is a much smaller level of environmental variation between twins, especially monozygous twins, even at later ages. This might suggest that the environment at early ages may be an important factor in influencing the later development of diabetes.

Probably the most important limitation of twin studies is the lack of any control over the similarities and differences of the environment of a twin pair.

Lack of control over the environment is, of course, a problem with all studies on quantitative inheritance in humans. It is, however, especially aggravating with twin studies because of the inherent difficulties in comparing the environmental variation between twins with that applicable to unrelated individuals chosen at random, and also with that between sibs who are not twins. As we have previously emphasized, one of the major difficulties in human populations is the fact that the environments of relatives, especially sibs, are likely to be highly correlated. We should, therefore, distinguish at least two types of environmental variation, that within families, say E_1, and that between families, say E_2. The variation between twins, dizygous or monozygous is only a part of E_1. By restricting variations to individuals born to the same parents at the same time, the effects of parental age, parity and secular trends are eliminated. In addition, development in the same uterus at the same time may well eliminate a major source of variation that is present between sibs who are not twins. There are, however, some factors that tend to emphasize differences between twins. These are mainly related to asymmetrical development before and after birth. Intrauterine competition can sometimes lead to marked differences in birth weight, which may be accompanied by size differences after birth and into later life. Often, in childhood one twin of a monozygous pair tends to dominate their relationship. This seems to occur among at least 60–70 percent of all monozygous twin pairs. There is good evidence to suggest that the socially dominant twin is likely to be the one who weighed more at birth. Such asymmetries of development *in utero* and afterwards are presumably caused by competition for resources and tend to accentuate environmental differences. Morphological asymmetries between twins may be caused by different embryonic and fetal positions in the uterus resulting primarily in differential gravitational effects during development. Thus, while the frequency of left-handedness is about 6 percent among single births, it is at least two to three times higher among monozygous and dizygous twins. Shields (1962) suggests that there may be a tendency for the left-handed member of a twin pair to be the smaller. Some early data from Dahlberg (1926) illustrating this point are as follows:

Both twins R	*1R 1L*	*Both L*	*Total*
89	29	6	124

The proportion of left-handed twins is (29 + 12)/248 = 16.5 percent. A test for random coincidence of right and left-handedness gives $\chi_1^2 = 1.9$, with $P > 10$ percent. Thus the association between right-handedness and left-handedness in twins is not significant, though it is presumably a tendency toward opposite-handedness that increases the frequency of left-handedness in the twin pairs. There is no evidence for a higher concordance in monozygous than in dizygous twin pairs, which is in keeping with the lack of any genetic component indicated by the data just presented. Some data on the familial incidence of handedness are shown in Table 9.19. While on the whole these data confirm the lack of any genetic component, they do suggest the existence of a maternal effect. Thus, left-handed female × right-handed male matings produce a significantly higher proportion of left-handed

TABLE 9.19
Comparison of Handedness of Children According to Parental Matings

	R♂ × R♀	*R♂ × L♀*	*L♂ × R♀*	*L♂ × L♀*	*Total*
Number of matings	14	9	17	13	53
Right-handed children (R)	36	15	41	26	118
Left-handed children (L)	4	9	6	5	24
Total	40	24	47	31	142
Percent left-handed	10	37.5	12.8	16.1	16.9
Heterogeneity	$\chi_2^2 = 8.82$, $P < 0.05$				

Source: From Falek (1959).

offspring than other matings. This could either be a psychological effect, as suggested by Falek (1959), or a genuine maternal effect connected with the early *in utero* development.

A possible explanation of the apparent paradox that some degree of inheritance of handedness (though of *maternal* type) is found from family studies, and none in twin analysis, is the following. The discordant monozygous twin pairs may all, or almost all, be due to the phenomenon known as mirror-imaging, which is probably connected with the different positions twin pairs must assume *in utero*. Mirror-imaging is observed in twins for several characteristics and relates to finding the same type of asymmetry anomaly in opposite sides of two twins. If most, or all, of the discordance in handedness is due to mirror-imaging, then the apparent non-heritability in twins is explained. This also explains, moreover, why left-handedness is higher in twins. The observed frequency of left-handedness, considering only concordant monozygous pairs, corresponds, in fact, almost exactly to that observed in the nontwin general population.

It is interesting to note that handedness is one of the few traits for which **maternal inheritance** (that is, an effect of the maternal phenotype or genotype on the immediate offspring that does not carry over to later generations) has been found in man. If it were shown that this carries over to later generations, it might be considered to be **cytoplasmic** or **extranuclear inheritance**–that is, due to genetic determinants in the cytoplasm. For a character that may depend on the orientation and distribution of organelles in the cytoplasm, an interpretation in terms of maternal or cytoplasmic inheritance, or both is not inconceivable. One other trait known to show maternal effects in man is birth weight (see Morton, 1955), and there are readily understandable reasons for the observed correlation.

A tendency for monozygous twins to have more similar environments than those of dizygous twins may result from the generally greater similarities in character and temperament of monozygous twins, making them more likely to do things together more of the time and to seek similar environments. An interesting expression of these tendencies is illustrated in Table 9.20, which shows the response of dizygous and monozygous twins to the question of whether they were satisfied with their "twin state." Of the monozygous twins, 77.9 percent were satisfied, while only 41.9 percent of the dizygous twins were, the difference being highly significant ($\chi_1^2 = 18.8$, $P < 0.001$).

The similarities between monozygous twins that are due to their having similar environments can, to some extent, be determined by comparing monozygous twins brought up apart and brought up together.

A small proportion of monozygous twin births are separated at birth or soon after. This provides a unique natural experimental situation for comparing the

TABLE 9.20
Classification of Twins According to Whether They are Satisfied with Their Twin Status

Classification	MZ	DZ	Total
	Individuals		
Satisfied	53	36	89
Not satisfied	7	21	28
Indifferent	4	28	32
No answer	4	1	5
	68	86	154

Source: From Conterio and Chiarelli (1962).

performance of two identical genotypes in different families and so, generally, in quite different environments. Muller in 1925 was perhaps the first to realize fully the significance of observations on separated twin pairs and to urge the collection of more data. Probably less than one in a thousand twin pairs are separated at birth or soon after. Thus, since the frequency of monozygous twin births is only about 0.4 percent, the collection of data on separated twins is no easy task. Three major studies have been published: Newman, Freeman, and Holzinger (1937), which utilized 19 separated pairs: Shields (1962), 44 pairs: Burt (1966), 53 pairs. We shall concentrate mainly on the Shields study giving just a brief survey of its main findings. The original account is fascinating reading and creates a vivid picture of the problems of analyzing, in depth, human similarities and differences.

Shields obtained his twin samples by means of a questionnaire advertised through a television program of the British Broadcasting Corporation. It is an amusing commentary that the incentive for Shields' twins to cooperate was their contribution to science, and that for Newman, Freeman, and Holzinger's twins was a free visit to the Chicago World's Fair! Shields' appeal was for monozygous twins, especially if they had been brought up apart. Of about 5000 responses, approximately one in 62 was from separated monozygous twins and 76 percent of these (44 pairs) turned out to be suitable for analysis. Among the twins investigated there were twice as many female as male pairs, paralleling the sex bias in twin studies that we discussed earlier. Control pairs of monozygous twins were chosen from "reared-together" pairs that had responded to the appeal, matching as much as possible for age and sex. In using data from a study such as this, we should ask if there are obvious ways in which the sample studied (the 44 pairs of separated monozygous twins) might not be representative of the general population. In particular, it is likely that the separated twins come from poorer than average socioeconomic backgrounds as reflected in the reasons for which they were separated. The time of separation of the twins and the extent of their contact in later life varied to some extent. Only four pairs were separated after five years of age and thirty pairs were separated before twelve months. The commonest reason for separation was the ostensible inadequacy of the mother due to physical conditions (including death) or illegitimacy. Only 8 pairs of twins were reunited before twenty years of age. The extent of contact depended on the relationship between the families that adopted the twins. Fourteen pairs were taken into unrelated families: thirty pairs were brought up by different branches of the same family. One pair of separated twins came to know each other only through Shields' investigation, and were reunited for the first time, at the age of 36, after having been separated when they were 16 months old. At the other extreme were 4 pairs who, most of the time, attended the same school. It is clear that the environmental variations due to these four separations are not by any means as great as those between a random pair of individuals. Also, intrauterine and very early postnatal similarities cannot be eliminated. With such a relatively small number of twins, each pair must to some extent be considered as a unique case history, and as a result much interesting information is given by Shields in an anecdotal rather than a quantitative form. For example, a small number of examples of relatively large differences in weight and IQ were traced to specific differences in the home environment, such as a poorer versus a more affluent home, or to particular diseases such as duodenal ulcer and disseminated sclerosis, and in one example of a large IQ difference, to epilepsy in only one of the twins.

The zygosity diagnosis was confirmed in most cases by blood grouping as well as by Shield's personal evaluations combined with the usual type of questionnaire. In addition to the conventional anthropometric measurements on height and weight, detailed IQ and personality tests were given the twins and detailed case histories emphasizing behavioral and psychological traits were obtained. A summary of the distributions of the differences in height between the separated (S) and control (C) monozygotic twins and a sample of about 30 dizygous twins, some of whom were brought up apart, is shown in Figure 9.22. In all cases the distributions for C and S are much more similar than either is to that for dizygous twins, the tendency being towards small differences between both C and S monozygous twin pairs. By analogy with the H statistic (Equation 9.49) as a measurement for genetic determination from the comparison of monozygous and dizygous twins, Newman Freeman, and Holzinger suggested the statistic

$$E = \frac{V_S - V_C}{V_S} \tag{9.67}$$

as a measure of the proportion of the variation between the S pairs that is due to their having been brought up apart. Here V_S and V_C are the variances of the differences between the S and C pairs. A comparison of the estimates of H and E obtained by Shields, by Newman, Freeman, and Holzinger and by Burt for height, weight, IQ and personality rating is shown in Table 9.21. Newman, Freeman, and Holzinger

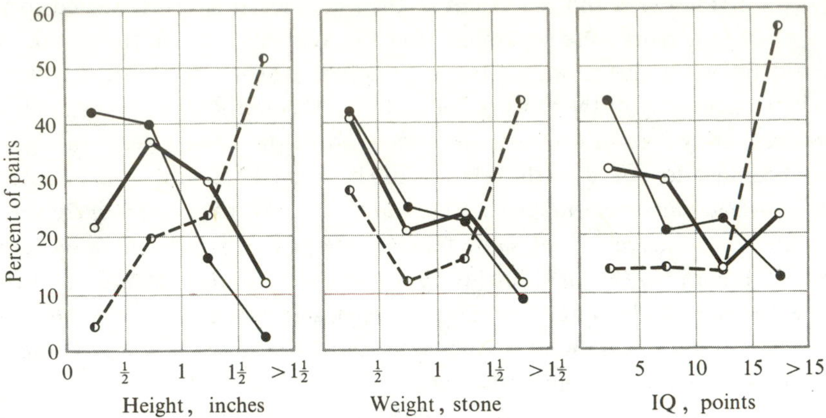

FIGURE 9.22
Distributions of height, weight, and IQ differences between twins compiled to provide information about the genetic-versus-environmental components of determination of traits. Data on monozygous pairs of twins that were brought up separately are plotted by the heavier solid lines. Data on monozygous pairs brought up together are plotted by the lighter solid lines. Data on dizygous pairs are plotted by the dashed lines. (Data from Shields, 1962.)

studied 69 monozygous twin pairs, 50*C* and 19*S*. Burt 148 monozygous pairs, 53*S* and 95*C*. Though the IQ and the personality tests used were different, the *H* estimates are, on the whole, in remarkably good agreement. Weight generally seems to be less heritable than height, and personality rating less than IQ. The relatively low heritability of personality rating is most probably due to the low precision and to some extent arbitrariness of the tests. The *E* values are rather erratic and hard to interpret. A negative value indicates *S* twins are more similar than *C* twins. Certainly the experimental error is large, both because of the sampling errors due to the small numbers of twins and because of the inevitable biases in the environments of the *S* twins. A comparison of the mean difference between twins and of the intraclass correlations between the twin pairs for height, weight, and IQ is shown in Table 9.22. The intraclass correlation *r* is a measure of similarity between twins that has often been used. It can be shown (see Appendix I, Equation 93) that

$$1 - r = \frac{V}{S^2} \tag{9.68}$$

where V is the mean of the square of the differences between members of a twin pair, r the intraclass correlation and S^2 the total variance of all the observations. Using subscripts MZ and DZ for the two types of twins, the quantity

$$H' = \frac{r_{MZ} - r_{DZ}}{1 - r_{DZ}} \tag{9.69}$$

TABLE 9.21

Estimates of Hereditary (H) and Environmental (E) Determination of Traits, from Monozygous Twins Brought up Apart and Together and from Dizygous Twins: A Comparison of Data

	H Values			*E Values*		
	Shields (1962)	*Newman, Freeman, and Holzinger* (1937)	*Burt* (1966)	*Shields* (1962)	*Newman, Freeman, and Holzinger* (1937)	*Burt* (1966)
Height						
Females only	+0.89	—	—	+0.67	—	—
Males only	—	—	—	+0.89	−0.54	+0.33
Both sexes	—	+0.81	+0.93	—	−0.64	+0.33
Weight						
Females only	+0.57	—	—	—	—	—
Both sexes	—	+0.78	+0.83	−0.62; +0.68	+0.27	+0.39
IQ						
Dominoes and vocabulary	+0.53	—	—	—	—	—
Binet	—	+0.68	—	—	—	—
Other	—	—	+0.86	−0.04	+0.64	+0.40
Personality						
Extroversion	+0.50	—	—	−0.33	—	—
Neuroticism	+0.30	—	—	−0.36	—	—
Woodworth Mathews Neur. Ques.	—	+0.30	—	—	−0.06	—
Other	—	—	—	—	—	+0.95
Educational attainment	—	—	+0.90	—	—	+0.95

is approximately the same as the quantity H defined by Equation 9.49, as long as S_{DZ}^2 and S_{MZ}^2 are not too different. H and H' have both been used interchangeably in the analysis of twin data. The data of Table 9.22 show rather good agreement among the three studies, as was noted before. In every case the mean difference between S twins is less than that between dizygous twins and, except for Shields' female weights, the intraclass correlations are also greater for S twins. The small difference between the r values of S and C twins for height, relative to the much lower r value for dizygous twins is quite striking. The same is true for Shield's data on IQ, though in Newman, Freeman, and Holzinger's data the S IQ correlation is near that of the dizygous twins. However, the sampling errors of these correlations are undoubtedly fairly large. At least for these three measurements, the similarity between monozygous twins is, apparently, only slightly decreased by their early

TABLE 9.22

Mean Differences and Correlations Between Monozygous Twins Brought Up Together (C) and Apart (S) and Dizygous Twins: A Comparison of Data

	Height			Weight			IQ		
	Brought Up Together		*Brought Up Apart*	*Brought Up Together*		*Brought Up Apart*	*Brought Up Together*		*Brought Up Apart*
	MZ(C)	DZ	MZ(S)	MZ(C)	DZ	MZ(S)	MZ(C)	DZ	MZ(S)
NEWMAN ET AL. (1937)									
Mean difference	1.7 cm	4.4 cm	1.8 cm	4.1 lb	10.0 lb	9.9 lb	5.9	9.9	8.2
Correlation	+0.932	+0.645	+0.969	+0.917	+0.631	+0.886	+0.881	+0.631	+0.670
SHIELDS (1962)									
Mean difference	1.3 cm	4.5 cm	2.1 cm	10.41 lb	17.3 lb	10.5 lb	7.4	13.4	9.5
Correlation (males)	+0.98	—	+0.82	+0.79	—	+0.87	—	—	—
Correlation (females)	+0.94	+0.44	+0.82	+0.81	—	+0.37	+0.76	+0.51	+0.77
BURT (1966)									
Correlation	+0.962	+0.472	+0.943	+0.929	+0.586	+0.884	+0.925	+0.453	+0.874

separation, relative to the differences between dizygous twins. This emphasizes the importance of a genetic component in the determination of size and IQ.

As we have already mentioned, much of the evidence for similarities and differences between the twins studied, and for specific environmental influences, especially with respect to personality traits, is contained in the details of the individual case histories.

Thus, in one case a relatively large personality difference between two male *S* twins, causes Shields to comment that "the harsh attitude of Alfred's stepmother, together with the effects of child neglect in the first two years, would seem to be an adequate explanation for the big difference between the twins in personality and behavior. Alfred was consistently more anxious and paranoid than his more fortunate twin brother." Many such personality differences could be traced to plausible environmental factors, especially in the early home background, and some could be traced to the effects of physical factors such as diseases, and one was determined to be related to partial deafness. However, the degree of personality difference did not by any means always match the extent of environmental differences. Thus, Ken, one of an *S* pair, was brought up in a more-or-less normal home while his twin, Dick, was brought up by a mother having chronic mental illness and a very deaf father. While Ken "seemed to have more social poise and to be a warmer person than Richard, who was inclined to be bored and sulky," nevertheless Shield's final comment on the pair is "it is remarkable that Dick shows so few signs of having been brought up in such an unfavorable environment and that the twins are as alike as they are."

The leadership of one twin of each pair in the *C* group seems to be a significant factor in their mutual development, and this factor may persist into later life. The leader tends to be more extroverted and more successful, at least financially. Significantly, most of the twins who made the initial contact with the study were the dominant members of a pair. As mentioned above, a major fact in determining who is leader seems to be bigger size, at birth and later, due mainly, perhaps, to asymmetries in the early embryonic and fetal development of the twin pairs. Shields suggests that this asymmetry of the leadership relations in *C* twins, which, of course, tends to accentuate differences between them, may be a reason why more personality differences are not found between the two groups of twins. No such relation can, of course, develop between *S* twins, though their overall environments differ more in other respects.

The time of first menstruation, as we noted earlier (Table 9.15) seems to be highly heritable. There was in the female monozygous twins a significant association between intelligence and early menstruation. In the combined data on *S* and *C* groups the first twin to menstruate was the more intelligent 27 out of 34 times ($\chi_1^2 = 10.6$, $p \sim 0.01$). This effect was not correlated with general growth rate, since the twins who menstruated first did not tend to be taller or heavier.

There was both concordance and discordance with respect to overt psychiatric disorders. One pair of *S* twins was concordant for epilepsy; in another pair, only one twin had epilepsy, but the other had an electroencephalogram pattern that showed some of the characteristics associated with epilepsy. In other pairs, both *S* twins have had nervous breakdowns at approximately the same age, both were unruly as children and later became heavy drinkers and, according to Shields, classifiable as "borderline psychopaths." Striking differences between *S* twins were also noted. For example, in one pair one twin developed homosexual attachments and had attacks of amnesia. In another, one twin made a suicide attempt and subsequently suffered from severe attacks of depression. Shields suggests that "illnesses of a probably hysterical kind are conspicuous in the discordant pairs while anxiety states count for several concordant pairs." Obsessional tendencies were often present in both members of a pair. Jule-Neilsen and Harvald (1958) have studied electroencephalogram patterns in a group of twins brought up apart and find "practically complete concordance as regards both normal qualities and abnormalities." An example of tracings of brain waves for three pairs of *S* twins is shown in Figure 9.23.

We have discussed at some length the "anecdotal" material on twins to emphasize the difficulties in getting precise, quantitative, and unbiased information on personality and behavioral traits. The information is complex, and there is no easy way to reduce the many undoubtedly significant observations to meaningful numbers. The smallness of the samples also contributes to decreasing the chance of finding statistically significant conclusions.

Behavioral characteristics that have to some extent been quantitated are drinking and smoking habits.

A summary of data on the cigarette-smoking habits of Shields' female twins is shown in Table 9.23. Twins were classified as alike if both were either smokers or

TABLE 9.23
Concordance for Smoking Habits in Female Twins

	Smoking Habits		
Twins	*Alike*	*Unlike*	χ_1^2
Monozygous			
S	23	4	
C	21	5	
Total	44	9	0.005
Dizygous	9	9	6.25, $p \sim 1\%$

Source: Data from Shields (1962).

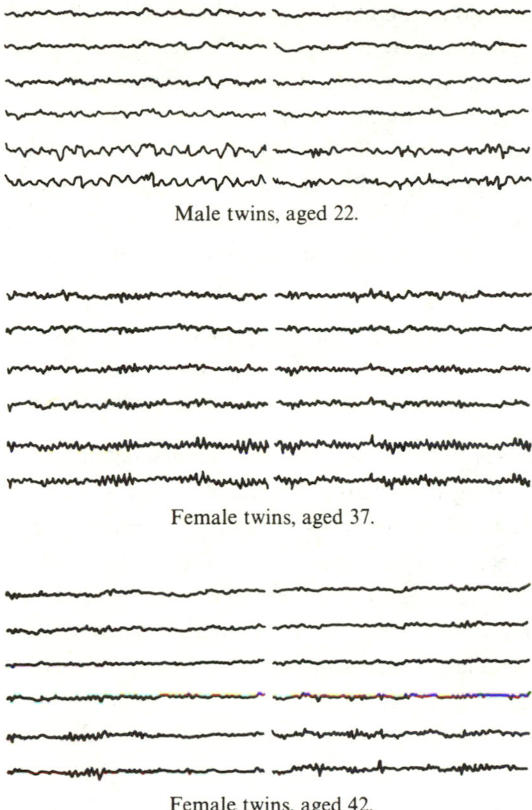

Male twins, aged 22.

Female twins, aged 37.

Female twins, aged 42.

FIGURE 9.23
Electroencephalograms of three pairs of monozygous twins brought up apart. The left and right sides of the figure compare the tracings of the pairs of twins and show practically complete concordance for all features of the tracings. In all cases standard lead: left frontal-left ear, right frontal-right ear, left occipital-left ear, right occipital-right ear. (From Juel-Nielsen and Harvald, 1958.)

nonsmokers. The data in the lower part of the table are for *S* and *C* twins combined and show monozygous twins to be significantly more similar in being smokers or nonsmokers than dizygous twins. In fact, the habit of smoking seems to be distributed at random in dizygous pairs. The upper part of the table shows the remarkable fact that there is no difference between *S* and *C* twins with respect to their being smokers or nonsmokers. Undoubtedly the similarity between the monozygous twins must have a genetic basis, perhaps through similarities in personality traits connected with the habit of smoking. Other twin studies, though not including separated twins, have given analogous results. The similarities extend to the amount of tobacco consumed and to the tendency to inhale. An analysis of alcohol drinking habits among Swedish twins by Kaij (1957) showed significantly greater concordance among monozygous than among dizygous twins. A similar analysis in Italy by Conterio and Chiarelli (1962) failed to show an effect. This is perhaps because the uniformly high consumption of alcohol in Italy leaves too little scope for individual variation. The difference may really be due to the type of alcohol consumed, for example wine as opposed to distilled spirits. Consumption of coffee showed a very significantly higher concordance rate in the Italian monozygous twins.

Shields ends his study with the following two propositions: (1) *family environments can vary quite a lot without obscuring basic similarity in a pair of monozygous twins*; (2) *monozygous twins, even if reared together, can differ quite widely.*

The relative emphasis given to these two statements depends on whether one views twin studies as providing information on genetic effects or environmental effects. Both viewpoints are valid and important. There has been a tendency in the past, even among the researchers who have carried out the studies, to minimize the results of twin studies because of their possible biases and also because of the uncertainties concerning the extent of environmental variation. However, it seems that different studies, especially those on separated twins, agree rather well in the average extent of genetic determination indicated for a variety of characteristics. It is our feeling that twin studies do give a useful and valid indication of nature-versus-nurture effects when there is significant genetic polymorphism for a quantitative trait. Analysis of monozygous pairs whose members were separated early is an especially powerful tool and it is unfortunate that these pairs are rare. (Unfortunate in the sense of the scientist's need for empirical data. It is, of course, fortunate for all of us that the basic values of human societies prevent any sort of scientific experimentation that would interfere with personal choices regarding how and where a child should be reared.) The main limitation of twin studies is that they can give no information on the possible basis for the genetic determination.

Information on adopted children can sometimes be useful for indicating the relative importance of genotype and environment.

Studies on adopted children are, of course, closely analogous to studies on separated twins. Their main drawbacks are that the biological parents of the adopted children are not usually available for study and that most adoption agencies naturally make an attempt to match the background of the foster and biological parents. Typically, an attempt is made to compare foster-parent–adopted-child correlations with control parent–offspring correlations. The former should be essentially a measurement of the effect of the environment as represented by the foster parent. An example of such data, given in Table 9.24, is taken from two

TABLE 9.24
Correlations (r) of Children's Intelligence with That of True Parents and with That of Foster Parents

Parents	Children and Foster Parents		Children and True Parents	
	r	N	r	N
DATA OF BURKS (1928)				
Father's (MA)[a]	0.07	178	0.45	100
Mother's (MA)	0.19	204	0.46	105
DATA OF LEAHY (1935)				
Father's (Otis score)[a]	0.15	178	0.51	175
Mother's (Otis score)	0.20	186	0.51	191

Source: From Neel and Schull (1954).
[a] MA and Otis score refer to the type of intelligence test used.

different studies of adopted children. The control groups were selected to match as closely as possible the group of foster children. The correlation between IQ of foster parent and child is in each case lower than that between true parent and offspring. This, of course, suggests that the effect of the environment on IQ is not enough to replace the effect of heredity. As in the twin studies, only information relating to environmental variation is obtained. We shall see later, when discussing the genetics of schizophrenia, how such studies can be very helpful when there is an indication that a character may be strongly influenced by the home environment.

9.12 The Problem of Nature versus Nurture and the Limitations to the Concept of Heritability

Most traits of social importance are determined by a large number of genes and strongly affected by environmental factors. This severely limits the possibilities for genetic prediction where there is great interest in being precise.

This is not true of man alone. Most economically important characters in domesticated animals and plants have the same limitations. The insight and tenacity of breeders has led to extraordinary results even in the absence of genetic theory, through their application of the principle of artificial selection—namely, that like breeds like. But the development of a genetic theory of quantitative characters, coupled with the application of rigorous principles of experimental design, has led to a considerable increase in the rate at which desirable characters are propagated and augmented.

The theory developed by breeders helps, among other things, in predicting the long-term results of crosses and of artificial-selection experiments. The predictions are often inaccurate, but still useful. The inaccuracy is due to the difficulty of estimating the importance of dominance, of interaction between genes (epistasis), of linkage between genes with minor effects, and of linkage between such genes determining the trait and others that determine important characters such as fertility, and finally of the **pleiotropy** of these genes (that is, the capacity of a gene to influence simultaneously several characters). All these difficulties are shared by the human geneticist. He has, however, two great handicaps that the animal or plant breeder does not have. The human geneticist cannot perform the crosses that could be informative, and can only hope to find them—if indeed they occur at all. He is also handicapped in having no control over the environment such as the breeder can have. The environments in which people live vary in an almost infinite number of ways. Some of these we can observe and measure, and thus bring under statistical control. But many, perhaps the most important ones, are unknown or very difficult to measure, and the lack of any sort of control over them may be an ever-present source of error.

Among the techniques that breeders have developed to cope with their prediction problems are the measurement of heritability and of the degree of genetic determination of a character. These concepts can, in part, be carried over to human populations, as we have seen. We have tried to keep the two concepts distinct. By **degree of genetic determination** we mean the fraction of the total variance that is determined genetically, and by **heritability** the fraction of total genetic variation that can be fixed in a selection experiment.

The genetic (sometimes called genotypic) fraction of the variance is made up of four major portions.

$$\text{Genetic variance} = V_A + V_{am} + V_D + V_I .$$

The **additive portion of the variance** V_A is, roughly speaking, the portion due to the difference between the homozygotes for the various genes, summed over all genes. Thus, if there were two genes with two alleles each, *A*, *a* and *B*, *b* determining a quantitative character, the difference between *AA* and *aa*, weighted by the respective genotype frequencies plus the same for *BB* and *bb* would form the basis for computing the additive portion of the variance. (This is only approximately true if there is dominance). It should be added that the frequencies of the genotypes to be considered are those that would obtain under random mating. Additivity refers to the assumption that the genes act additively, meaning that their effects are simply summed. This is the major portion of the genetic variance that can be fixed under selection, for the population may eventually be made entirely of individuals all of whom are *AA*, *BB*, etc., and then the additive variation will have disappeared.

The **variance due to assortative mating** (V_{am}) arises from the fact that, under positive assortative mating, homozygous genotypes tend to be more frequent than they would be under random mating. They thus inflate the additive variance by increasing the frequency of individuals that bear the extreme expressions of a trait. Knowledge of the intensity and nature of assortative mating can help in separating it from the additive portion, with which it is otherwise confounded.

The **dominance variance** V_D is due to the fact that the heterozygote for a pair of genes may have a phenotype not exactly intermediate between the two homozygotes. It affects mostly the correlations between collateral relatives such as sibs and cousins and not those between direct descendants, parent-offspring, etc. Inbreeding effects reveal the average *direction* of dominance. Thus, the inbred progeny will have a mean lower than that of noninbreds if the mean of heterozygotes, say *Aa*, for a gene tends to be higher than the means of the respective homozygotes, *AA* and *aa* (and vice versa). It is, however, possible that there is a high V_D value and no effect on the means of inbred progeny. This happens when positive deviations in heterozygotes at some loci balance exactly negative deviations at other loci.

The **variance due to genic interaction (epistasis)** V_I is that fraction arising from the fact that different loci may not act simply additively. For instance, if the effect of allele *A* increases the mean by two units and that of allele *B* by three, the presence of both may have an effect that is quantitatively different from the sum of the two. Models have been devised from which, using appropriate breeding designs, V_I can, in principle, be estimated. Though this fraction of the variance may be very significant, it is difficult, if not impossible, in human populations to collect the information necessary for its estimation. Part of this variance is fixable by selection.

The total phenotypic variance V_P is the genetic variance plus the environmental variance. The degree of genetic determination, which is the genetic variance divided by the total variance, is larger than the heritability, which is $(V_A + V_{am})/V_P$ (sometimes V_A/V_P), unless V_D and V_I are both zero. It should be remembered that many of the various estimates that are given the name of heritability estimates, especially in studies on twins, have somewhat different meanings.

The greatest complications regarding variance in human genetics arise in the estimation of the environmental variance. This has several components that careful experimentation in practical breeding could control and analyze separately, but that cannot easily be separated, and certainly not all separated, in human populations. It is of interest to list some of these components. Thus we can write

$$V_E = V_{ind} + V_{fam} + V_{soc} + V_{rac} + V_{GE}.$$

The **variance between individuals within families** V_{ind} is included in all variances, but changes in a subtle way from one to another. The environmental variance between monozygous twins cannot be the same as that between dizygous twins, as we have discussed, because monozygous twins tend to choose more similar environments. The environmental variance between twins cannot be the same as the environmental variance between non-twin sibs, which includes, among other things, the variance due to birth order. Also, V_{ind} is probably not the same in families of different sizes.

The **variance between families within socioeconomic strata** V_{fam} inflates the covariance between parent and offspring, but not V_{DZ} (the variance between DZ twins) and related variances. Some idea of its importance is given by the correlation between foster-parents and adopted children. The matching process that is often carried out by adoption agencies probably removes, from such a statistic, part of the variance between socioeconomic strata. The variance between monozygous twins reared apart is probably the best source of an estimate of the combination of V_{ind} and V_{fam}.

The **variance between socioeconomic strata** V_{soc} is commonly believed to be removed by partial correlation or by analysis within socioeconomic strata. Probably, most socioeconomic strata classifications are too crude to give a proper account of this component. In any case, it usually is not removed from the published correlations, and has effects similar to V_{fam}.

Cultural differences between families or between social groups may be maintained by socio-cultural inheritance and may have effects especially on behavioral traits. They may thus create correlations between relatives that are very difficult to distinguish from those due to genetic determination. These differences are likely to be highest in the comparison between racial groups. The isolation in different environments, which has permitted the development of genetic differences between races (see Chapter 11) has also created a parallel, but probably largely independent, development of cultural differences. It seems therefore likely that the **variance in environmental conditions** (including sociocultural ones) **accompanying racial group differences**, V_{rac}, is large. It is clear that V_{rac} can remain high even in the highly equalizing environment of North America. The problem of interpreting racial differences in a quantitative character will be dealt with in Chapter 12 when we discuss the mean difference in IQ between black and white Americans.

The **variance due to genotype-environment interaction** V_{GE} is the most complex of all. It arises when given genotypes show different phenotypes in different environments. A striking example comes from an experiment on rats (Cooper and Zubek, 1958). Animals were passed through a maze test, and those showing the fastest, or slowest, learning abilities were bred. In this way, after several generations of selection, a line of "bright" rats that could learn to perform the task much more quickly and with fewer errors was obtained, and also a line of "dull" ones that performed much less successfully in the maze than had the original rats. When the environment in which the rats were grown was changed, the difference between the dull and bright rats disappeared and they all performed equally. This result was consistent regardless of whether the environment was improved or worsened.

Clearly, the capacity to respond was conditioned by growth in the environment in which selection had occurred. The isolation of V_{GE} from strictly genetic or environmental variation in man is an unsolved problem.

It is apparent that heritability and degree of genetic determination are, in general, inadequate yardsticks for measuring the relative importance of nature versus nurture.

Heritability and degree of genetic determination can be changed by a change in environmental conditions. They give measurements that are valid, with some reservations, only for the particular population examined and at the time of analysis. They allow few useful predictions to be made with respect to new, untested environments. As an example, we know that the heritability of stature is high. It is believed that environmental variation is very small, and nevertheless, in the last hundred years we have witnessed a change in average human stature of more than one standard deviation. This change is almost certainly the consequence of secular changes in the environment, but measures of heritability are hardly informative in this respect. Thus large environmental effects may be observed even where heritability measurements apparently indicate a very small environmental component. Heritability tells us only about the ratio of the *prevailing* individual genetic differences to the prevailing individual environmental differences and cannot, in general, be extrapolated to other populations or other environments. Heritability may remain constant and yet there may still be major changes in the environment affecting the average phenotype. In principle, we could change the environment for each particular genotype, and thus so modify the phenotype as to reduce, or increase, the apparent heritability at will. It may, of course, be extremely difficult to find the right techniques for achieving this result!

In spite of all the above reservations, measurements of degree of genetic determination and heritability can serve a useful purpose in telling us something about the relative importance of genetic and enviromental differences. The real advance in genetic knowledge will occur, however, only when we are able to analyze the existing genetic variation in depth, isolating single genetic components that can be shown to behave in a Mendelian fashion. This will be achieved by associating the variation with biochemical or other differences detectable by laboratory techniques. Such an approach has the added advantage of giving clues to the physiological action of the genes and to their modification by the environment. It may also be the case that simple genetic techniques, associating the study of continuous variation with that of the transmission of standard genetic markers, can lead to a more precise genetic definition of quantitative variation. In principle, continuous genetic variation can be so dissected into its elements, given a sufficient knowledge of standard genetic markers, that linkage between the quantitative character and the markers can be established. This requires, however, the study of a large number of individuals. It is still uncertain whether present knowledge of genetic markers in man is sufficient to carry out such a research program. Animal experimentation especially by Thoday and his colleagues (see, for example, Thoday, 1966) has proved the validity of this approach. Some of the traits that are under polygenic control are sufficiently important, from a social point of view, that their study may be justified, even if the expense is relatively large.

9.13 The Effect of Natural Selection on Quantitative Characters

Natural selection acts on phenotypes, not genotypes. It is, however, only the resulting genetic changes that influence the future constitution of a population. To interpret the effect of natural selection on quantitative characters, we must, therefore, translate the effects on the phenotype into those on the genotype, and then translate back again from the selected genotypes to their phenotypes.

There are three basic ways in which natural selection can act with respect to a quantitative character—stabilizing, directional, and disruptive.

If individuals with intermediate values for a given quantitative trait have the highest fitness then selection is said to be **stabilizing** (or normalizing or balancing, or optimal). When fitness increases as the value of the character increases (or decreases) selection is said to be **directional**. Lastly, if intermediate values are relatively disadvantageous in terms of fitness, and the extremes have high fitness, selection is said to be **disruptive**. These three basic types of relationship between the values of a quantitative character and fitness are illustrated in Figure 9.24. The solid lines on the first row simply represent the normal distribution of the character among the parents. The dashed lines show the hypothetical relationship between the value of the quantitative character x and the fitness. The lower distributions show, schematically, the expected distribution of x after one generation of selection. In the first case of stabilizing selection, outlying values of x are at a relative disadvantage, and the offspring distribution should have the same mean, but a smaller variance. Directional selection should simply increase the mean of the offspring distribution and disruptive selection should increase the *variance* and leave the mean unchanged. In each case the extent of the response, of course, depends on the relationship between phenotype and genotype, mainly the degree of genetic determination of the quantitative character x. Clearly these three modes of selection are idealizations that represent the main possible forms of relationship between fitness and a quantitative character. In practice, selection is usually a combination of these basic modes.

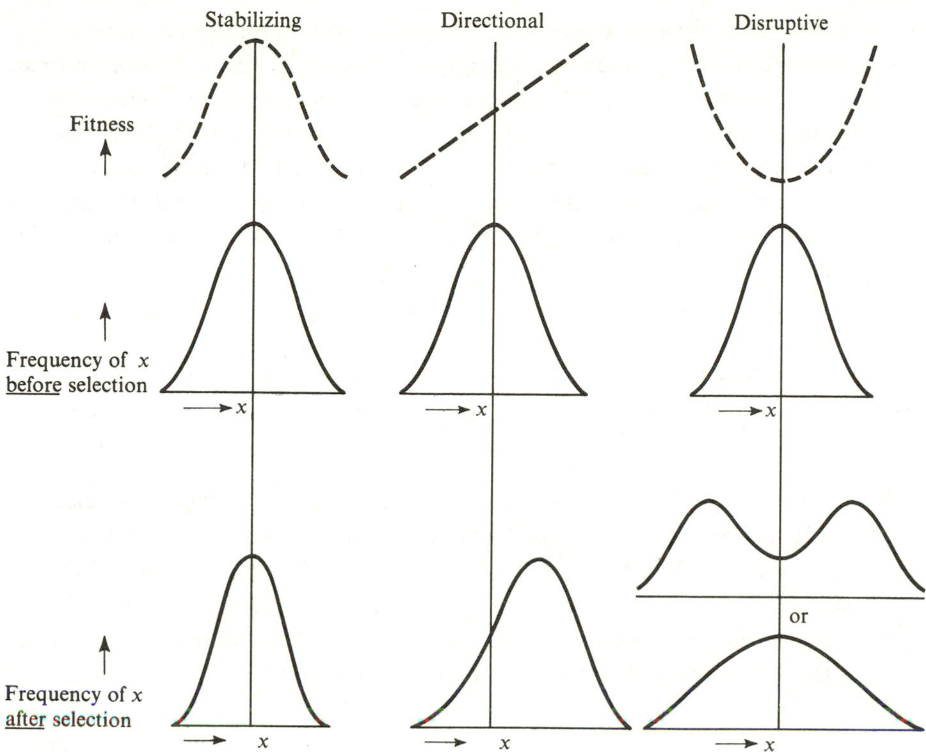

FIGURE 9.24
Schematic representation of the three types of selection for quantitative characters and their effects. Solid lines show the frequency distributions of the quantitative character x. Dashed lines show fitness as a function of x.

Much natural selection for quantitative characters is believed to be stabilizing.

If the mean value of x for optimal fitness is not at the population mean, then there will be a combination of stabilizing and directional selection until the population mean is adjusted to the position of optimal fitness. Directional selection clearly cannot continue indefinitely. Perhaps only fitness itself is strictly subject to directional selection, by definition, though selection for other quantitative characters may be approximately directional at any given time. Most artificial selection as practiced by plant and animal breeders is directional selection in which, most simply, all individuals above a given threshold value of x are selected as parents for the next generation. It is a common experience that the limits to selection for economic characters may, to some extent, be set by the poor viability of the extreme selected phenotypes. This is a reflection of stabilizing *natural* selection for the character opposing the directional *artificial* selection. Disruptive selection is the most unusual of the three types. It is, essentially, selection for diversification with respect to the character and, as Mather (1955) and Thoday (1960) have emphasized, may contribute to the maintenance of polymorphisms. Perhaps the only human attributes naturally subject to disruptive selection are certain secondary sexual characteristics. Indeed we might well argue, as did Darlington (1958), that the evolution of secondary sexual differences is, to some extent, the result of such disruptive selection aimed at improving the efficiency of the bisexual mating process. Of course, in many cases, the same genes may not control secondary differences in the two sexes.

Because of its importance for experimental breeding programs, directional selection has been the subject of much more theoretical discussion than have the other two types of selection.

The simple theory for predicting a response to directional or threshold selection (which is often called culling, in experimental breeding) is the same as that used for the analysis of threshold characters. Thus Equations 9.41, 9.43, and 9.45 effectively state that, if a is the difference between the mean of the parents after selection as compared with that before selection and $g(=R-G)$ is the difference between the mean of the offspring and the mean of the unselected parents, then

$$g = ah^2, \tag{9.70}$$

where h^2 is the heritability. This is because the regression of offspring on midparent is h^2, rather than $\frac{1}{2}h^2$. Thus if we know the heritability, h^2, of a character, Equation 9.70 predicts the change in the mean of the population in terms of the imposed selection, represented by the difference, a, between the means of the selected and unselected parents. It is clear that the response to a given level of selection increases with the heritability of the character, and is zero if the heritability is zero.

The application of these results to human stature, which has been increasing steadily during this century in most western industrial societies, shows that natural selection could be responsible only for a small fraction of the change.

From data on Italian conscripts, Conterio and Cavalli-Sforza (1957) estimate the mean increase of stature as 0.1 centimeter per year or about 3 centimeters per generation during this century. There are, of course, many factors, both genetic and environmental, that could be causing this increase. General improvement in living conditions, particularly with respect to nutrition and disease, are likely to be the most important. Thus, in the Italian data (based on males only) there is a very significant positive correlation of stature with socioeconomic group, with average heights ranging from 170.2 centimeters for professional and clerical workers to 167.5 centimeters for the lowest income groups. In addition there is a negative correlation with family size, with a decrease in average stature of the children of roughly 0.4 centimeter per extra child. There are two ways, possibly contrasting, in which natural selection could be responsible for the increase in stature. On the one hand, taller people may have a higher fitness (viability or fertility) which would mean that there is simply directional selection for stature at the present time. On the other hand, medical progress may be relaxing selection for diseases that are in some way genetically associated with tallness. To investigate the second possibility, Conterio and Cavalli-Sforza analyzed data on the heights of brothers of persons who died before reaching the age of twenty. These data were obtained by linking records on conscripts (which in Italy include height and chest-girth measurements) with death certificates, which contain information on the cause of death. Two causes of death out of eight showed a barely significant association with stature. The brothers of individuals who died of tuberculosis were 2.07 centimeters shorter, and brothers of individuals who died as a result of premature birth, 1.9 centimeters taller. Let us make the extreme assumption that $h^2 = 1$, or, in other words, that there is complete additive genetic determination of height and also an absolute correlation of death due to prematurity with stature. Parents of children who die prematurely should then, on the average, be $2 \times 1.9 = 3.8$ centimeters taller than the population mean, since the regression of offspring on one parent is $\frac{1}{2}h^2 = \frac{1}{2}$ in this case. Thus, from Equation 9.70, with $h^2 = 1$, their children should also be 3.8 centimeters taller. Deaths of babies born prematurely have decreased from 7 percent at the beginning of the century to 3 percent today. Assuming this difference of 4 percent represents an average of the increased contribution of taller parents to the present generation, this factor contributes 0.04×3.8, or about 0.16 centimeters per generation to the increase in mean stature of the population. This is clearly a negligible amount compared with the effects of other environmental factors and might well be counter-balanced by the increased survival of individuals susceptible to TB, who on the average, are 4 centimeters shorter than the population mean.

In another study by Conterio and Cavalli-Sforza (1960) differences in height of Italian conscripts at 20 years of age were determined as a function of survival and marriage. The difference in height between individuals born in 1890–1891 who were still living in 1957 and those who had died by that time was 1.3 centimeters (significant at the 1 percent level). The difference in height between the married and single was only 0.06 centimeter and was not significant. No data on fertility were available. If we assume 100 percent heritability for stature, the maximum effect of the selection due to mortality would be simply to increase stature by 1.3 centimeters per generation. Assuming 62 percent heritability for stature (from Fisher's analysis of English data), the expected increase would be 0.62×1.3 or 0.8 centimeters, roughly $\frac{1}{4}$ of the total increase. Another factor that may decrease this estimate will be considered later. This is certainly much more significant than the effect of reduced mortality, but still not enough to explain the observed rate of increase. In fact, as we shall discuss later, this effect may well be counterbalanced by the negative correlation between height and family size. From Figure 9.25 (or Falconer's tables) we can, given the mean difference between selected and unselected parents, and the mean and variance of the distribution of unselected parents calculate the proportion eliminated due to the selection favoring an increase of the quantitative character. We assume that a threshold is chosen such that the mean of the individuals above it is increased by a given amount, and we calculate the proportion of individuals above the threshold. The variance in stature of the Italian males is 41.8 cm^2. The difference in stature between survivors and nonsurvivors as a proportion of the standard error is, therefore, $1.3/(41.8)^{1/2} = 0.22 = a$ in the notation of our analysis of thresholds. This corresponds to an elimination of about 11.5 percent of the population and a threshold height for survival of 157.25 centimeters. The general relation between *a*, called the selection intensity, and the proportion selected, assuming culling, is shown in Figure 9.25.

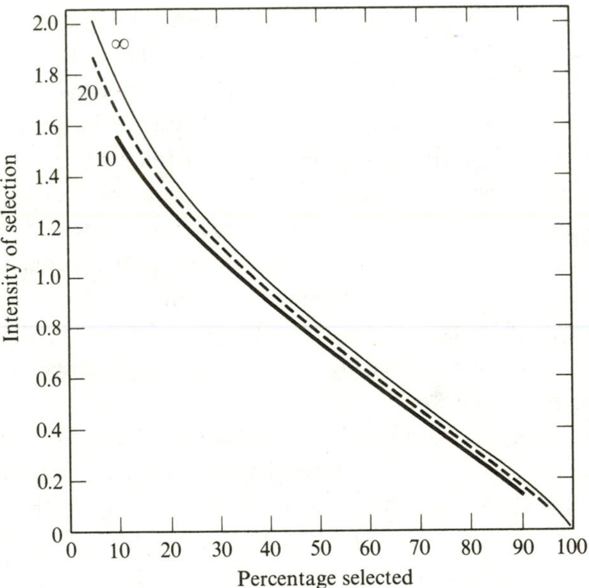

FIGURE 9.25
Intensity of selection in relation to proportion selected. The intensity of selection is the mean deviation of the selected individuals, in units of phenotypic standard deviations. The upper curve refers to selection out of a large total number of individuals measured: the lower two curves refer to selection out of totals of twenty and ten individuals, respectively. (From Falconer, 1960.)

A model of directional selection with intensity increasing gradually with the trait gives a more satisfactory method of analysis.

It is, of course, ridiculous to assume that there is a sharp threshold in stature below which all people die before reproducing and above which all survive. A more plausible directional selection model, at least with respect to natural as opposed to artificial selection, would assign some smooth regularly increasing curve relating fitness to height, whose minimum approached zero, maximum approached one, and which was centered on the approximate threshold. A convenient analytical shape for such a curve is provided by the probit transformation, which is the integral of the normal distribution, and so, in general, is defined by

$$P(x) = \frac{1}{\sigma\sqrt{2\pi}} \int_{-\infty}^{x} e^{-(y-\mu)^2/2\sigma^2}\,dy = [N(\mu, \sigma)]_{-\infty}^{x} = [N(0, 1)]_{-\infty}^{(x-\mu)/\sigma}, \qquad (9.71)$$

where $N(0, 1)$ is the normal distribution function with mean zero and variance one. This function, which cannot be expressed in simple algebraic form, is tabulated as the probit or error function. It is the integral of a normal (gaussian) distribution with mean μ and standard deviation σ, taken between the limits indicated outside the square brackets. The function tabulated as the integral of the normal curve has mean zero and standard deviation one, and therefore when standard tables are used, the upper limit must be transformed from x to $(x - \mu)/\sigma$ (see Appendix I).

A representative example of how such a selection scheme might work for stature, from the Italian male data, is shown in Figure 9.26. The selection curve is defined by two parameters μ and σ. The average position of the fitness curve is determined by μ, the mean of the normal distribution function under the integral sign in Equation 9.71, and its spread or slope is determined by the standard error σ. When $\sigma = 0$, this model becomes just the simple culling model. It is always easiest to transform the basic distribution (in this example, the distribution of height), into a standard normal distribution with mean zero and variance one, as was done in Equation 9.71. This is done simply by working in terms of $(x - 165)/6.5$, where 165 centimeters is the mean and 6.5 centimeters is the standard deviation of the original distribution of heights. It can then be shown that the proportion surviving selection is

$$s = [N(0, 1)]_{\mu/\sqrt{1+\sigma^2}}^{\infty}, \qquad (9.72)$$

which is the same as for a culling model with the threshold $\mu/\sqrt{1 + \sigma^2}$ instead of μ. The mean of the selected individuals can be shown to be

$$m = \frac{\exp[-\mu^2/2(1 + \sigma^2)]}{s\sqrt{2\pi(1 + \sigma^2)}}, \qquad (9.73)$$

where s is given by Equation 9.72 and "exp" stands for e (the natural base of logarithms) to the power indicated in brackets. Since selection is now described in

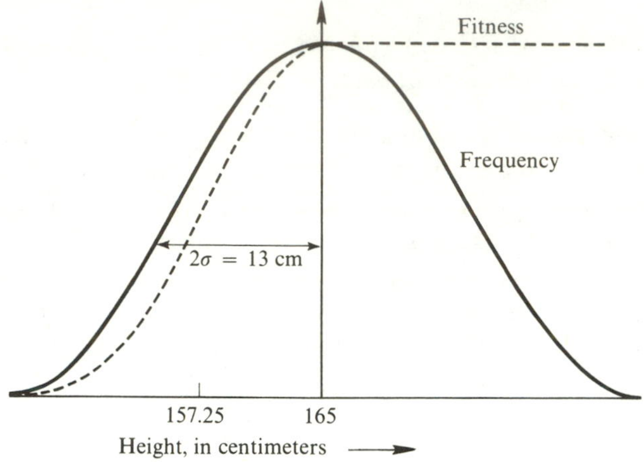

FIGURE 9.26
Directional selection for fitness according to the probit threshold model. The solid curve is the normal distribution of stature. The dashed curve gives the relation between stature and fitness. Its shape is that of the probit transformation, which is the integral of the normal distribution. (Equation 9.71). (Based on data from Conterio and Cavalli-Sforza, 1960.)

terms of two parameters, we can no longer estimate s given only the mean m of the selected individuals, as was true for the culling model. However, it can be shown that the variance of the selected individuals is

$$v = 1 - m\left[m - \frac{\mu}{1 + \sigma_2}\right]. \tag{9.74}$$

The two equations 9.73 and 9.74, with s defined by 9.72, allow us to estimate μ and σ^2, the parameters of the fitness curve, given the observed values of v and m. Then by Equation 9.72 we can estimate the proportion eliminated by selection. Though it is not possible to obtain explicit analytical solutions for μ and σ in terms of m and v, they can easily be obtained numerically. Using the observed variance of survivors, 39.68, which is $39.68/41.8 = 0.95$ in standardized units, the estimates of μ and σ^2 needed to give $m = 0.22$ give a survival probability of around 50 percent, which is rather too low. This estimate is, of course, very much influenced by the estimate of the variance of the survivors, which is subject to a large sampling error. It follows from these results that for this selection model, more individuals need to be eliminated for a given change in mean than with simple culling. This corresponds with the intuitively obvious result that culling should be the most efficient way of changing the mean for a given level of elimination.

If the proportion surviving is known, then this, rather than the observed variance, can be used to estimate μ and σ^2 (by Equation 9.72). The numbers of dead and living observed by Conterio and Cavalli-Sforza were 274 and 499 respectively, from which $s = 0.646$. The corresponding estimates of μ and σ^2 (using Eq. 9.72 to find $\mu/(1 + \sigma^2)^{1/2}$ and then Eq. 9.73 with $m = 0.22$ and $s = 0.646$ to obtain $(1 + \sigma^2)^{1/2}$ are -0.098 and 5.85 in standardized units. These give, from Equation 9.74, an expected standardized variance $v = 0.9185$ and so an actual variance of $v \times 41.8 = 38.4$ cm^2, which certainly does not differ significantly from the observed value of 39.68. The mean absolute position of the fitness curve ($\mu \times 6.5$ cm) is 6.4 cm below the population mean, so that the heights corresponding to 2.5 percent and 97.5 percent survival are 37 cm below and 25 cm above the population mean, respectively. This model thus seems to fit the data reasonably well. Certainly the culling model gives too low a level of elimination, as well as an unreasonably sharp distinction between survival and death. The extent to which the observed difference in height between survivors and those exposed to risk is passed on to the next generation is, of course, a function of the heritability, as we have already discussed.

The estimate, given on page 605, that the observed mortality differential for stature may account at most for $\frac{1}{4}$ of the observed secular increase is not affected by the use of a graded as opposed to a threshold model of selection. This is due to the fact that the estimate was based on a direct comparison of survivors with the original population. The analysis shows, however, that only the graded model predicts the observed variance and percentage of survivors. There are, however, two factors that may make this estimate of the contribution of genetic differences to secular changes in stature too high. Mortality is correlated with socioeconomic group, and so is stature. The correlations with stature may be entirely phenotypic. If the heritability estimates do not take proper account of the purely phenotypic effects of socioeconomic status on height, which are very difficult to evaluate, the estimate of the contribution of differential mortality to the stature increase may be excessive. There is also another factor involved, namely that the height difference correlation with mortality was observed at 67 years of age, while only mortality before the end of the reproductive period is relevant. Thus the height difference relevant for predicting the change over one generation is that occurring up to about fifty years of age. Since an appreciable fraction of the total mortality is likely to occur between ages 50 and 67, this will reduce significantly the potential effect of mortality selection on the change in height over one generation.

Undoubtedly, as already emphasized, the major part of the differences in height observed from one generation to the next must be environmental.

We have already seen that mortality differences cannot generate much (in any case no more than $\frac{1}{4}$) of the observed secular increase in height. Fertility differences have not been analyzed by the same prospective approach, except for the fraction due to mating differentials. This, as far as males are concerned, makes no contribution to the trend, as will be shown later. A retrospective study of fertilities, from the correlation between height and progeny size, would, if anything, predict a decrease of stature, because, as we have mentioned, the correlation is negative. It is dangerous, however, to draw conclusions from these retrospective data. The decrease in average stature with increasing progeny size might be in part, or entirely phenotypic, and might even conceal an actual genotypic increase, though this last hypothesis seems unlikely.

There is direct evidence of environmental influences on height.

The comparison between immigrants to North America and their relatives who did not emigrate shows a dramatic effect. The difference is such that it might well account for most of the observed secular trend. To quote a recent example, Hulse (1968) investigated the heights of Italian-Swiss from Canton Ticino born in Switzerland and born in California. There is a difference of almost 4 centimeters in height at maturity between California-born and Switzerland-born. Although the standard error of the difference is not given, it must be of order ± 0.7, and therefore the difference is highly significant. On the other hand, there is no difference between emigrants and their brothers who stayed home, excluding selection due to migration itself. There are also interesting effects on cephalic index and other measurements, which decrease while stature increases. The factors at play are not entirely understood. Food has been considered a major factor. Decreased incidence of disease and decreased progeny size, may be others. The possibility remains, however, that presently unknown environmental factors are involved, which may be the same as those determining the observed secular trend in developmental rates. Increased stature, is, in fact, accompanied by earlier development.

A suggestion repeatedly advanced to account for increased stature, namely heterosis, deserves some comment. The hypothesis has been suggested by several authors, especially in the past, that increased human migration favors increased heterozygosity and therefore an increase in stature due to heterosis. This hypothesis is contradicted by several facts. The change in overall heterozygosity in man cannot be a large one. Values of F above 0.02 are rare, and thus, even if they decreased to zero, they could cause only a minor change in the frequency of heterozygotes. Direct estimation of the effects of inbreeding on stature show small effects, if any. The increase in stature has, moreover, occurred also in areas where there has been no substantial increase in outbreeding, as, for instance, in the Parma Valley in Italy.

Mortality, at least up to the end of the reproductive period, and fertility are necessarily subject to directional selection.

The striking increases in life expectancy during this century were discussed in Chapter 6. These, clearly, must be due almost entirely to improved living conditions, both nutritional and hygienic, as well as improved medical care. However, a rough idea of the possible genetic contribution to this increase can be obtained using the simple culling theory. Considering only the survival of individuals beyond five years of age, the standard error of the average age at death is approximately 21 years, assuming a mortality distribution corresponding to white U.S. males in the year 1902. If we assume a threshold at approximately 25 years of age, the proportion surviving is 0.915. This assumption is, of course, an oversimplification implying that there is no reproduction before age 25 and that reproduction continues until death. Strictly speaking, we should use a fitness curve like the probit, which corresponds to the relative probability of reproduction at different ages. The mean deviation in age at death of those who survive to age 25 is, from Falconer's Table, 0.17 or, in years, $0.17 \times 21.1 = 3.6$ years. This, therefore, is the amount that must be multiplied by the heritability to obtain the expected gain in survival beyond 5 years due to genetic causes. Heritability estimates from parent-offspring and sib correlations have proved to be very unreliable, possibly because of the great difficulties in avoiding sampling biases. The Danish Twin study, from which we have already drawn much information, has, however, provided a heritability estimate based on twin data. Considering only twins both of whom survived to 5 years, the mean difference in age at death for monozygous twins was 14.5 ± 0.94 years, and for dizygous twins, 18.6 ± 0.84 years (difference significant with $P < 0.01$; see Harvald and Hauge, 1965) and the estimate of H, according to Equation 9.49, was 0.29. Using this as the heritability estimate, the estimated genetic change per generation in the average age at death after reaching 5 years is $0.29 \times 3.6 = 1.05$ years. The life expectancy of the white American five-year-old male has increased by 9.6 years from 1900 to 1953, which corresponds to a change in the average age of death of about 6 years per generation. The possible genetic contribution to the change in life expectancy clearly falls far short of that realized. Apart from the assumption of a sharp threshold of survival, there are at least two further assumptions in these calculations that may seriously affect our estimate. The first is the implicit assumption that the age at death is normally distributed. It is, in fact, markedly skewed toward the older age groups. The second is that the heritability estimate from twins is valid for the whole population. As we discussed above, this is certainly not true, as the environmental variation between twins is quite restricted. Each of these assumptions tends to lead to an overestimate of the expected gain, reemphasizing the fact that the realized change must be largely environmental. In spite of all the approximations in these examples, they are at least instructive in illustrating the principles of the action of natural selection on quantitative characters, and they also indicate that even the maximum possible rates of genetic change are very slow.

Fertility could, in principle, be treated in the same way as mortality. There is, however, almost no satisfactory data on the heritability of fertility. What evidence there is, suggests it is low. Fisher, in his famous book *The Genetical Theory of Natural Selection*, estimated the correlation between the fertility of daughters and their mothers, using data from the British Peerage. On the average there is a regression of 0.21 children born per number of children in the mother's family. Fisher thought that much of this correlation might be genetic, and not due to a traditional continuity or "social inheritance." His attitude was based on the fact that the regression of the number of children born to granddaughters on that of their grandmothers was almost exactly half as great (0.1065). On the assumption of biological inheritance, or organic inheritance, as Fisher calls it, a halving of the regression at every generation would be expected. These data indicate a heritability of fertility of about 40 percent. This does not necessarily mean that the inherited fertility is physiological. It might just as well be psychological, and nevertheless inherited. It must, in fact, be largely psychological, for the average number of children, even among the British peerage, was below the potential maximum number. Some form of birth control must therefore have been in operation. The distinction between social and biological inheritance (the latter conceived in its broadest sense—that is, including psychological aspects) is exceedingly difficult. This is true of any society that practices birth control, and even primitive groups like the Bushmen, Pygmies, or Yanomama Indians have been found to practice it.

It is to be expected that additive genetic variation with respect to fitness should be low, since it will be consumed rapidly while it exists, in order to approach the maximum possible fitness, so long as the population as a whole departs from this maximum. It seems likely that, in general, heritability for quantitative characters subject to strong directional selection should usually be fairly low.

Perhaps the most clear-cut example of a human character subject to stabilizing selection is birth weight.

A typical example of the relationship between birth weight and the logarithm of the mortality before 28 days of age is shown in Figure 9.27. The data were taken by Mather (1964) from a classic study by Karn and Penrose (1951) of the relation between birth weight, gestation time, mortality, parental age, and parity. The curve for survival, which is simply the complement of that for mortality, is unimodal as expected for normalizing selection. It is, however, not quite symmetrical and has a mode at 8 pounds, whereas the mean birth weight is only 7.16 pounds. The total elimination due to the neonatal mortality is 4.8 percent.

Birth weight is highly correlated with gestation time and with parity, though hardly at all with parental age. The regression of birth weight on parity is 0.17 pounds increase per birth. The correlation with gestation time is the basis for defining prematurity as having a birth weight less than 5.5 pounds. Karn and Penrose show that a parabola provides a good fit to the relationship between birth weight and the log of the ratio of survivors to nonsurvivors. This effectively assumes that the birth-weight distributions of both survivors and nonsurvivors, and thus that of the

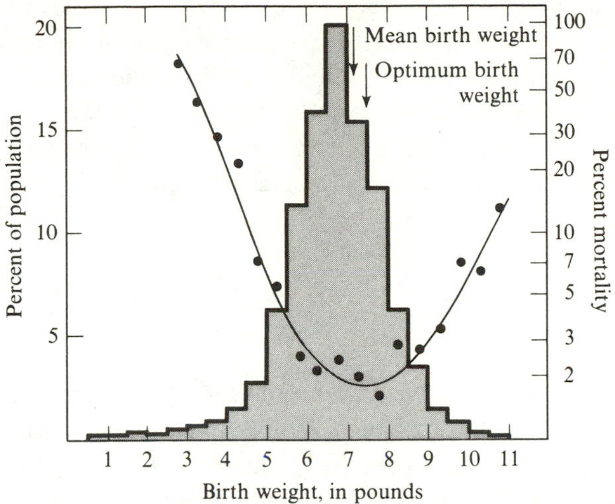

FIGURE 9.27
The distribution of birth weight among 13,730 children and the rates of early mortality of the various birth-weight classes. The histogram shows the proportions of the population falling into the various classes in respect of the birth weight. The plotted curve is the curve of mortality in relation to birth weight, the values actually observed for the classes of birth weight being represented by the points to which the curve is an approximation. The percent mortality is set out as a logarithmic scale for ease of representation. The optimum birth weight is that associated with the lowest mortality. (Based on data from Karn and Penrose, 1951.)

total population, are approximately normal. They estimate that for babies weighing between about 5.9 pounds and 10 pounds at birth, the mortality rate is less than average while for those outside these limits it is greater than average. Very similar results have been obtained by other workers with data from widely differing populations. This suggests that the parabolic shape of the survival curve, as defined by Karn and Penrose, and optimal survival being that of babies weighing between 7.5 and 8 pounds at birth, are fairly general properties of human populations.

A relatively simple theory for the effects of stabilizing selection can be constructed if it is assumed that the shape of the fitness curve, relating the probability of survival to the quantitative measurement, is that of the normal distribution.

In this case, if the original distribution of the quantitative character is normal, so is that of the selected population. Assuming maximal survival is unity, the normal distribution used for the fitness curve has to be multiplied by a factor which brings its peak height to unity. Alternatively, the height of the peak can be used, as suggested by O'Donald (1969), as a further parameter representing mortality not connected with the quantitative variation under study. If the population distribution before selection is normalized to have mean zero and variance one, and the fitness curve is $f_0 e^{-(x-\mu)^2/2\sigma^2}$, which has a maximum value of f_0 when $x = \mu$, then it can be shown that the mean of the selected population is $\mu/(1 + \sigma^2)$, and the proportion selected is given by

$$s = \frac{\sigma f_0}{\sqrt{1 + \sigma^2}} \exp[-\tfrac{1}{2}\mu^2/(1 + \sigma^2)]. \tag{9.75}$$

The larger σ^2 is, the smaller the deviation of the mean of those selected from the mean of the original population, and the higher the proportion surviving. For true normalizing selection $\mu = 0$ and the mean is unchanged, while s depends only on σ. We consider, for simplicity, $f_0 = 1$. Let us try the fit of this theory to Karn and Penrose's data on birth weight. They find the standard error of the birth-weight distribution to be 1.3, so that μ, in standardized units, is $(8 - 7.16)/1.3 = 0.645$. To estimate σ^2 we equate $\mu/(1 + \sigma^2)$ to the observed difference in birth weight between all births and surviving births, which was 0.08 pounds or $0.08/1.3 = 0.0615$ in standardized units. Thus, $1 + \sigma^2 = 0.645/0.0615 = 10.5$, so $\sigma^2 = 9.5$ or $9.5 \times 1.3^2 = 16.1$ pounds^2. Substituting $\mu = 6.45$ and $\sigma^2 = 9.5$ in Equation 9.75 (with $f_0 = 1$) gives $s = 0.95$ as the proportion of survivors. This agrees remarkably well with the observed level of survival (95.2 percent), suggesting that our "nor-optimal" selection model is appropriate for this type of data and that the assumption $f_0 = 1$ is reasonable in this case. The value of μ has surprisingly little effect on the survival probability s as long as σ^2 is large, since then s is largely determined by the magnitude of σ^2. Osborne and De George (1959) quote birth-weight differences of 207 grams for monozygous and 340 grams for dizygous twins, which, using Equation 9.51, gives an H estimate of 0.63. Thus, the maximum change in birth weight per generation expected from genetic causes is 0.08×0.63 or only 0.05 pounds, which is very little compared with the standard deviation of the birth-weight distribution. This is expected since the selection on birth weight is essentially stabilizing.

Mate selection has a stabilizing effect on stature, at least for males.

An interesting and rather striking case of normalizing selection for stature was found by Conterio and Cavalli-Sforza (1960) in their analysis of the data on Italian conscripts. While there was hardly any difference, as already mentioned, in the mean height of married and unmarried men, the *variance* of the heights of the married men was very significantly less ($P = 1$ percent) than that of the bachelors. This suggests that very tall and very short men have a harder time finding wives, which intuitively seems plausible. Since the means are not different, we have $\mu = 0$. It can be shown that using our "nor-optimal" selection theory, the expected variance of the distribution of selected individuals is $\sigma^2/(1 + \sigma^2)$, which is, of course, less than the original variance of unity. Equating the observed ratio of selected variances to unselected variances, $35.8/42.3 = 0.833$ to $\sigma^2/(1 + \sigma^2)$, gives the estimate $\sigma^2 = 5$. Substituting this into Equation 9.75 with $\mu = 0$ and $f_0 = 1$ gives

$$s = \frac{\sigma}{\sqrt{1 + \sigma^2}} = \sqrt{\frac{5}{6}} = 0.91. \tag{9.76}$$

Thus, the observed reduction in the variance of heights of married as compared to that of unmarried men implies an elimination of 9.1 percent of the men, where, of course (biological) celibacy is genetically equivalent to elimination.

The action of selection on IQ seems, at present, to be partly directional and partly stabilizing.

Fears that the apparent negative correlation between IQ and fertility might be leading to a gradual decline in average IQ have been widely expressed ever since Galton pointed out this relationship for the English ruling classes. Assuming that IQ is affected only by additive genes, that the negative correlation is with respect to overall fitness and not just fertility, and that there are no environmental changes opposing a decline in IQ, our simple selection theory would, of course, be in agreement with these fears. The fact is that there is no good evidence for any decline in IQ, so far as this can be reliably measured over a long period of time. The best and most widely quoted data come from a survey of the IQ's of eleven-year-old Scottish children in 1947 compared with a similar survey done 15 years earlier. There was no evidence for a decline in IQ. In fact, the mean score increased by an average of about one IQ point during this time. Other earlier surveys did indicate a decline in IQ of as much as one or two points per generation. Clearly the comparison of IQ tests done at different times and ages, to this level of precision, is very difficult. A great deal must depend on the test and the way it is given, a subject that will not be discussed here. A striking example of these problems is discussed by Pettigrew (1964) in a most illuminating article on the problems facing the black people of the United States. He quotes data showing that the average IQ of blacks was at least 3 points higher when the test was administered by blacks than when it was administered by white examiners. This, of course, emphasizes the fact that the variation in any measurement has to be related to the range of environmental circumstances over which the measurements are made.

Is average IQ decreasing?

In view of the disagreement between observation and the simplest predictions with respect to a secular decline of IQ, we must question closely, the premises underlying the predicted decline and see where they fail. It should be emphasized that since height and also weight show practically the same negative correlation with family size as does IQ, the same arguments used to predict the decline in IQ also predict a decline in height and weight. We discussed above the positive correlation between height and increased life span and its possible effect on an overall increase in height, which was certainly very small relative to the observed secular trend. As mentioned then, a complete discussion of the effects of selection should relate height or IQ to overall fitness, measured as discussed in Chapter 6. This must take account of both mortality and fertility. The negative correlation of height with fertility is almost certainly enough to counterbalance the positive correlation of height with viability. Certainly if there is concern, for genetic reasons, that IQ may be declining, the same concern should apply to height and weight.

In Chapter 6 we discussed briefly Bajema's data relating fitness, as measured by the intrinsic rate of increase, to IQ. (see Table 6.9). The relation is not linear, but is, in fact, more or less U-shaped with relatively high fitnesses at low and high IQ. It is remarkable that much of the literature on the relation between fertility and IQ ignores the fact that families with no children are excluded from analysis. This introduces an ascertainment bias exactly analogous to that in the ascertainment of matings between heterozygotes for a recessive through their recessive children (see Appendix II). The fertility distribution obtained is only that relating to matings producing at least one child, and ignores infertility as well as the probability that mating occurs at all. This fact was very clearly pointed out by Higgins, Reed, and Reed (1962), whose data show that more than 30 percent of persons whose IQ is 70 or lower have no children, as compared with only 10 percent of persons in the IQ range 101–110 and 3–4 percent of those whose IQ is higher than 131 (see Figure 9.28). This contrasts sharply with the typical data obtained from studies on the relation between IQ and family size, an example of which is shown in Figure 9.29. Here the relative family size, even in the lowest IQ range, is apparently almost three times that in the highest range.

There can be little doubt of the significant negative correlation between IQ and family size as illustrated in Figure 9.29, which many studies have shown is about −0.3. What is cause and what is effect is hard to say. One of the most complete attempts to analyze this correlation is the study on the IQ of Scottish children, to which we have already referred. This showed an average decrease of about 2 IQ points for each extra child. Many socioeconomic factors, particularly crowded living conditions, were obviously, in part, responsible for this decline. The one overriding correlation was with occupation of the head of the household, or more

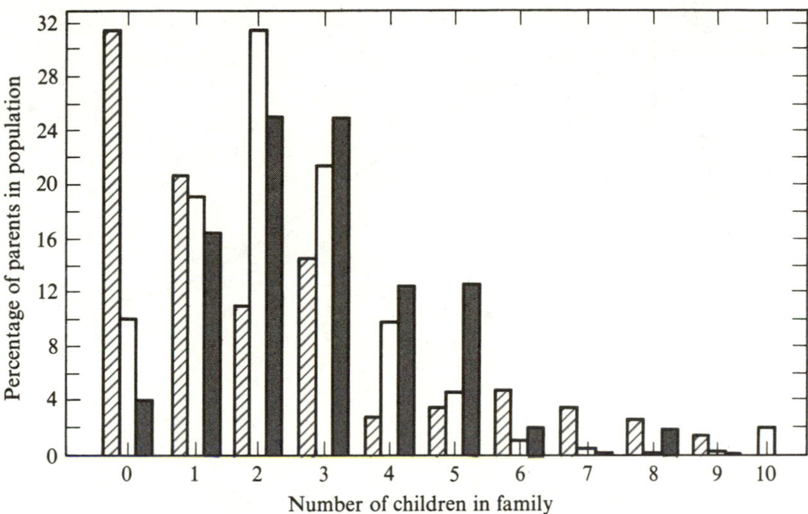

FIGURE 9.28
Distribution of size of family with three groups of average IQ's of parents, 70 and below (striped bars), 101–110 (white bars), and 131 and above (gray bars), measured in percentage. (From Higgins, Reed, and Reed, 1962.)

directly, with socioeconomic level. An interesting further indication, however, of other possible important environmental components to the negative correlation was given by data on twins. It was found that the mean IQ of twins was 5 points lower than that of nontwins. This unusually large effect was not correlated with differences in the sizes of families with twins and without twins, or with any of the other variables, such as socioeconomic level and parental age, which were included in the analysis. Thus, the difference between twins and their apparently more fortunate single sibs remained with all tested subclassifications. This effect may be correlated with incompletely understood physiological consequences of twinning on the *in utero* environment. However, perhaps the most likely explanation is the inevitable reduction in the attention parents can pay to each of two children born at the same time, from that which would be paid to a single child, especially when they are very young. In principle, this suggestion could be verified by comparing the average IQ's of twins brought up apart and together. However, apart from the small number of such twins, which produces high standard errors of the estimates, environmental differences would be completely confounded with the possible disadvantage to twins of having to share the attention of a single set of parents. Shields, in fact, found his separated twins had an average IQ about 6 points lower than that of the controls, an effect he ascribes entirely to the differences in average socioeconomic levels of the parents of the two groups. Perhaps another way to test for this effect of parental attention is to look at the differences in IQ between sibs, brought up in families of given size, as a function of the interval between their births.

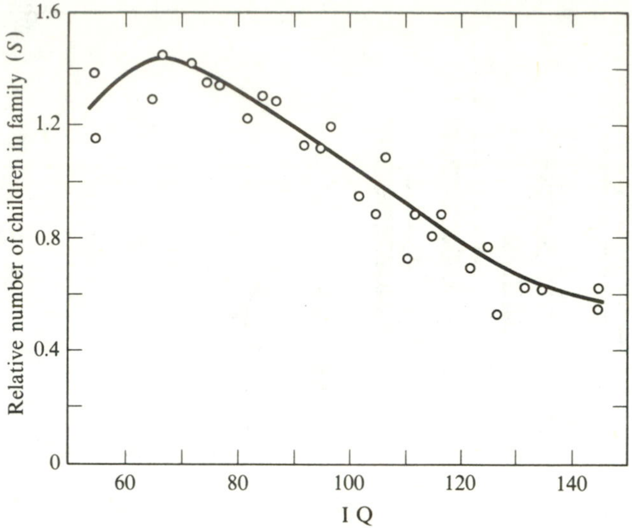

FIGURE 9.29
The relation of size of family and the average IQ of children in it. S is the family size relative to the mean family size for the whole sample. (From Mather, 1964.)

Children born closer together in time might be expected to have lower IQ's than children born farther apart. Biological effects on intelligence of lighter birth-weight (which characterize twins) cannot, however, be excluded. This discussion just serves to emphasize the problems of environmental correlations with IQ. If the effect of family size were entirely environmental, the negative association with IQ would clearly be of no consequence to the IQ of the subsequent generation. Simple nutritional factors may certainly be one component in the correlation between IQ and socioeconomic status. Thus, Pettigrew (1964) quotes a study by Hurrell and co-workers which claimed to show that fortifying the diet of pregnant mothers from lower socioeconomic groups with vitamin B and iron increased the average IQ of their children at three years of age by 5 points, and at 4 years of age, by 8 points, which, if reproducible, is a truly remarkable effect. Eighty percent of the mothers in this sample were black.

As we have already emphasized, any assessment of the effect of differential fitness with respect to IQ on the IQ of succeeding generations depends on a knowledge of the heritability of IQ.

The twin and other studies that we have discussed suggested a fairly high heritability for IQ. This is quite a general finding as indicated by the summary shown in Figure 9.30 of the results of 52 different studies on genetic correlations with respect to IQ compiled by Erlenmeyer-Kimling and Jarvik (1963). All the data are consistent with a large heritable component for IQ. It is clear, of course, that such a statement must be relative to the average range of environmental variation encompassed by each of the studies. Thus, for example, the effects of homozygosity for the PKU gene are severe in an environment in which no account is taken of the special dietary needs of the phenylketonurics, but may be negligible when these are allowed for. In the same vein, even if the negative correlation of IQ with fertility existing at the present time does have some genetic consequences with respect to IQ for the next generation, it is clear that the most direct way to improve overall IQ now is to improve the relevant environment.

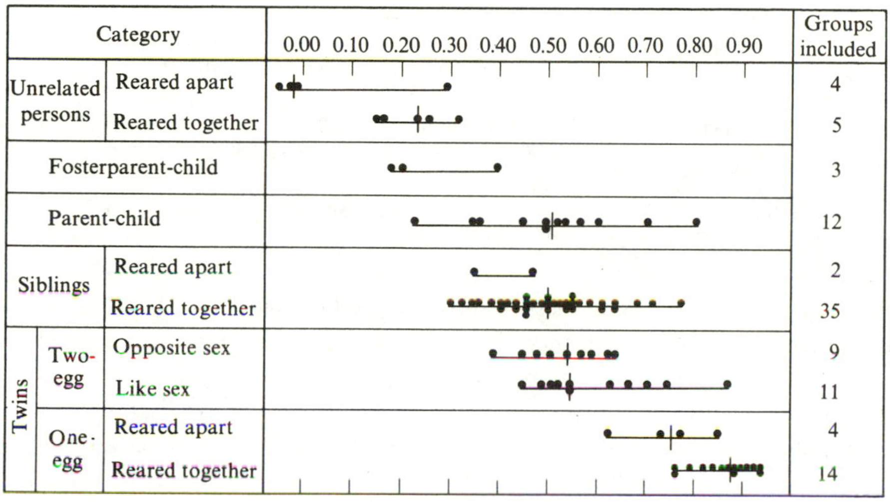

Category			0.00 0.10 0.20 0.30 0.40 0.50 0.60 0.70 0.80 0.90	Groups included
Unrelated persons		Reared apart		4
		Reared together		5
Fosterparent-child				3
Parent-child				12
Siblings		Reared apart		2
		Reared together		35
Twins	Two-egg	Opposite sex		9
		Like sex		11
	One-egg	Reared apart		4
		Reared together		14

FIGURE 9.30
Correlation coefficients for "intelligence" test scores from 52 studies. Some studies reported data for more than one relationship category; some included more than one sample per category, giving a total of 99 groups. More than two-thirds of the correlation coefficients were derived from IQ's, the remainder from special tests (for example, Primary Mental Abilities). Midparent-child correlation was used when available, otherwise mother-child correlation. Correlation coefficients obtained in each study are indicated by dark circles; medians are shown by vertical lines intersecting the horizontal lines that represent the ranges. (From Erlenmeyer-Kimling, and Jarvik, 1963.)

Bajema's data on fitness as a function of IQ can be used directly to calculate the expected IQ of the "selected" population, as indicated in Table 9.25. We assume, for simplicity, that before selection the population mean IQ is 100 with a standard deviation of 15 points. From this we obtain the proportion of people with given IQ. Multiplying these proportions by the relative fitnesses and renormalizing gives the proportions after selection. The resulting mean IQ is 101.01, actually a slight increase. This is due, mainly, to the higher estimated fitness of the > 120 IQ group

TABLE 9.25
Change in Mean IQ Due to Natural Selection

1 IQ Range	2 Mean Value[a]	3 Proportion in Range[b]	4 Relative Fitness[c]	5 Proportion after Selection[d]
≥120	126	0.092	1	0.106
105–119	112.5	0.279	0.87	0.284
95–104	100	0.258	0.78	0.235
80–94	87.5	0.279	0.96	0.313
< 80	74	0.092	0.58	0.062
		1.000		1.000

Mean IQ before selection (Σ column 2 $\times$ column 3) = 100
Mean IQ after selection (Σ column 2 $\times$ column 5) = 101.01

[a] Approximate mean IQ's for ranges given.
[b] Proportions in IQ ranges, calculated from a normal distribution with mean 100, and standard error 15.
[c] Relative fitnesses according to Bajema (1963).
[d] Relative proportions in IQ ranges after selection. They are calculated as column 3 $\times$ column 4/Σ column 3 $\times$ column 4.

offsetting the relatively high fitness of those in the IQ range 80-94. Thus, taking the best data presently available, there is, in fact, no case for predicting a decline in IQ. Undoubtedly, the relationship between IQ and fitness is a changing one. If it was, indeed, negative at one time, it is hard to believe it can have been so for long. The rapidly changing conditions of modern industrial society, especially the so-called demographic revolution of the nineteenth century (see Chapter 6), must have created the opportunity for new patterns of selection that may, indeed, be quite transient. Certainly as discussed in Chapter 6, the present trend is toward a reduction in the variance of fitness and a corresponding evening-out of the socioeconomic correlations with fitness. This emphasizes more than ever the need to concentrate on environmental improvements, which can achieve rapid results, rather than on altering hypothetical genetic selection patterns.

9.14 The Genetics of Schizophrenia

We conclude this chapter with a discussion of the evidence for a genetic basis for schizophrenia and some speculations on the mechanisms that maintain the disease at a relatively high frequency in a population. Schizophrenia is used as a final example because it illustrates so well the problems and pitfalls of quantitative genetics. (A number of papers that provide an excellent review of this subject will be found in a recent book edited by Rosenthal and Kety, 1968).

About half of the patients in mental hospitals in the United States (that is, 250,000 *persons) are classified as schizophrenics, with the overall incidence of the disease being around one percent.*

As with many common and yet complex diseases, the problem of diagnosis of schizophrenia is not straightforward, and those diagnosed as schizophrenic may be a heterogeneous group. Nevertheless, the criteria for diagnosis are reasonably definite. In broad terms, to quote a well known authority on schizophrenia, its major characteristic is "no response or inappropriate response to other people and their environment." As we discussed in Chapter 6 when analyzing the effect of the declining age at marriage on the fitness of schizophrenics, the disease has a variable age of onset, with a mean below 30 years of age, the age of onset being somewhat earlier for males than females. The incidence and severity of the disease, combined with the fact that it strikes at the prime of life, make it one of the most important diseases to understand at the present time.

Hypotheses about the genetic control of schizophrenia have ranged from theories of determination by a single dominant gene to the idea that there is no genetic basis for it at all.

There has been a great deal of discussion about the possible hereditary basis for schizophrenia ever since it was first described. Some authors have attempted to explain its obvious familial concentration in terms of single dominant or recessive genes with imperfect penetrance. Others have been impressed with the overriding importance of the home environment in the development of such a behavioral disorder and have therefore minimized the possibility of a genetic basis for the disease. The answer most probably lies between these two extremes.

A summary of data on the incidence of schizophrenia in the relatives of affected individuals, taken from Gottesman and Shields' (1967) illuminating discussion, is shown in Table 9.26. Concordance in monozygous twins is uniformly higher than in dizygous twins, though it does vary widely from study to study. Gottesman (1968) pointed out that there is probably a relationship between severity of the disease, and monozygous twin concordance. Increasing severity is associated with higher concordance rates as would, in any case, be expected for a disease with a variable age of onset (diabetes provides another example). Thus, differences in the diagnosis of schizophrenia could readily account for differences in observed concordance rates. The incidence among parents and sibs of an affected person shown in Table 9.26 is around 10 percent. Because of the variable age of onset of the disease, the ratios are corrected for the observed age distribution of the schizophrenic probands. A review by Zerbin-Rudin, quoted by Slater (1968), of data from 25 independent investigations on the incidence of the disease in relatives of affected persons gives a figure of only 3.8 percent for the expected proportion of parents of schizophrenics

TABLE 9.26
Heritability of the Liability to Schizophrenia

Relatives of Affected Persons	Investigator	Incidence: Numbers	Incidence: Percentages	Falconer's h^2 (in percent): $q=1\%$	Falconer's h^2 (in percent): $q=2\%$
Same-sex twins	Slater (1953)				
	MZ co-twins	28/41	68	105 ± 8	104 ± 8
	DZ co-twins	11/61	18	106 ± 14	94 ± 15
	Gottesman and Shields (1966)				
	MZ co-twins	14/28	50	87 ± 9	85 ± 3
	DZ co-twins	4/34	12	86 ± 21	72 ± 23
	Kringlen (1966)				
	MZ co-twins	28/64	44	82 ± 6	79 ± 7
	DZ co-twins	12/100	12	86 ± 12	73 ± 13
Parents, sibs, and children	Odegaard (1963)				
	Age corrected	84/832	10	79 ± 4	64 ± 5
Sibs	Erlenmeyer-Kimling *et al.* (1966)				
	Observed	131/2007	6.5	61 ± 3	45 ± 4
	Age corrected	131/1260.5	10	80 ± 2	66 ± 3
Aunts and uncles (second-degree relatives)	Odegaard (1963)				
	Age corrected	81/1749	4.6	96 ± 8	61 ± 9

Source: From Gottesman and Shields (1967).

who themselves have the disease. However, this is appreciably biased on the low side by the method of ascertainment of schizophrenia in the parents. The incidence in sibs, as indicated in Table 9.26, ranges from 8–12 percent, and seems to be only slightly increased if one parent is schizophrenic. The incidence among children with one schizophrenic parent also is about 12 percent. The raw incidence among the children of two schizophrenics (based on five studies—see Erlenmeyer-Kimling, 1968) is 25 percent, but the age-corrected incidence ranges from 35–44 percent, depending on the method used for correction. None of these figures are, of course, compatible with simple dominant or recessive inheritance. As we discussed before, there does not seem any point in introducing the concept of incomplete penetrance, as opposed to polygenic inheritance, unless there is other clear-cut evidence for the involvement of one major gene. The distinction between a major gene with modifiers and polygenic inheritance, which, of course, may only involve a relatively small number of genes, cannot readily be made on the incidence data alone. As emphasized by Gottesman and Shields, schizophrenia is clearly a case for the application of the threshold model.

The data shown in Table 9.26 on the incidence of schizophrenia among the relatives of schizophrenics provide the basis for calculating estimates of heritability based on Equation 9.45. Heritabilities, calculated in this way, for various categories of relationship, are shown in the last two columns of Table 9.26. These have been calculated on the basis of both a population incidence of 1 percent and one of 2 percent. The estimates are all uniformly high. As expected for any quantitative character, they are higher for twins than for persons of any other relationship. The agreement of the heritability estimates from monozygous and dizygous twins suggests that the genes involved are acting additively (see Equations 9.65 and 9.66). Though these data are certainly consistent and suggestive, they do not remove the lingering doubts concerning the influence of a stressful home situation, including the presence of a schizophrenic, on the development of schizophrenia. A much-quoted environmental correlation is the apparently high incidence of schizophrenia in the lower socioeconomic groups, the difference ranging up to eight-fold. However, two recent studies have shown that when schizophrenics are classified according to their father's socioeconomic status, this correlation disappears (Goldberg and Morrison, 1963; Dunham et al., 1966), The most likely explanation is that there is a preschizophrenic state before the disease can be overtly diagnosed, which is detrimental and which leads to a decline in socioeconomic status. Thus, if this is the case, schizophrenia is the cause of the low socioeconomic status and not the effect. Kohn (1968), however, disputes the interpretation of this data, maintaining that socioeconomic status may be a causative factor in schizophrenia.

Probably the most convincing evidence that the environment does not predominate in the causation of schizophrenia comes from studies by Heston (1966) and Kety et al. (1968) on adopted children. Heston surveyed the records of all persons born between 1915 and 1945 to schizophrenic mothers who were in the Oregon State Mental Hospital. Fathers were not available for study, but none were known to be schizophrenic. Subjects were chosen for study only if they were separated from their mothers within three days of birth and had *no* subsequent contact with the mother or her relatives. Controls were selected from the same adoption agencies to which the subjects were sent, matching for sex, type of adopting family, and length and date of stay in foundling home. There were 30 males and 17 females in the experimental group, and 33 males and 17 females in the control group. The results are based on a follow-up study of these persons, with independent blind assessment of schizophrenia by two psychiatrists. A summary of the data obtained is given in Table 9.27. The higher incidence of schizophrenia and other behavioral disorders in the experimental group is quite striking in spite of the small numbers. The age-corrected schizophrenia rate in the experimental group is 16.6 percent-which, perhaps, is a little higher than the incidence among offspring of schizophrenic parents reported in other studies. This may be because the selection of subjects probably was biased toward severe cases. As discussed above, higher twin concordance rates are found for increasing severity of the disease in propositi. What is remarkable in these data is not only the increased incidence of schizophrenia, but also the increased incidence of other behavioral abnormalities. Thus,

TABLE 9.27
Incidence of Schizophrenia and Other Related Attributes in Adopted Children Whose True Mothers Were Schizophrenic and in Matched Controls

Attribute	*Experimental Group* (30♂, 17♀)	*Control Group* (33♂, 17♀)
Mean IQ	94	103.7
IQ < 70	4	0
Schizophrenia	5	0
More than once in penal or psychiatric institution	11	2
Psychiatric or behavioral discharges from the army	8 of 21	1 of 17

Source: From Heston (1966).

Heston classified nine of the males, excluding the schizophrenics, as having "sociopathic" personalities, as revealed by their repeated arrests for assault, battery, thefts, and the like. No such arrests were found in the controls.

The study in Denmark by Kety and co-workers (1968) was based on a combined screening of the adoption register of the State Department of Justice and the psychiatric register of the Institute of Human Genetics, University of Copenhagen, for the years 1924–1947. The 33 schizophrenic index cases were chosen from 507 persons who had been reared by adopting, rather than true, parents and who had been identified as having been admitted, for any reason, to a psychiatric facility, An equal number of controls, matched as carefully as possible for time of adoption. sex, age, age at transfer to the adopting parents, and socioeconomic status of the adopting family, were chosen from the pool of approximately 5000 persons who had been reared by adopting parents and for whom no admission to a psychiatric facility had been recorded. Psychiatric disorders were then identified in the biological relatives of subjects and controls. A summary of the incidence of schizophrenia found by Kety and co-workers among the biological and adopting close relatives of subjects and controls is given in Table 9.28. The increased incidence among the biological relatives of subjects, compared with the incidence among the subjects' adoptive relatives, or among the relatives of the controls, is quite striking. The observed incidence of schizophrenia of about 10 percent among the biological relatives of the schizophrenic persons reared by adopting families agrees reasonably with already discussed estimates that were not based on data pertaining to adoptions.

These studies certainly represent the most clearcut, controlled evidence for a very significant genetic component to the determination of schizophrenia, and are fine examples of the careful use of studies on adopted children. The level of heritability indicated by the data of Tables 9.27 and 9.28 is comparable to that obtained

TABLE 9.28
Distribution of Schizophrenia Spectrum Disorders Among the Biological and Adoptive Relatives of Schizophrenic Subjects and Controls. Numerators = number with schizophrenia, or inadequate personality. Denominators = number of identified relatives.

	Biological Relatives	*Adoptive Relatives*
TOTAL SAMPLE OF 33 SUBJECTS AND 33 CONTROLS		
Subjects	$\frac{13}{150}$	$\frac{2}{74}$
Controls	$\frac{3}{156}$	$\frac{3}{83}$
p^a (one-sided, from exact distribution)	0.0072	N.S.
SUBSAMPLE OF 19 SUBJECTS AND 20 CONTROLS SEPARATED FROM BIOLOGICAL FAMILY WITHIN ONE MONTH OF BIRTH		
Subjects	$\frac{9}{93}$	$\frac{2}{45}$
Controls	$\frac{0}{92}$	$\frac{1}{51}$
p^a	0.0018	N.S.

Source: From Kety et al. (1968).

[a] These are the probability or significance levels for the comparison of subjects with controls. N.S. means "not significant".

from the studies summarized in Table 9.19. This suggests that much of the correlation observed between relatives when schizophrenic persons live together in the same family, may in fact be genetic.

The severe reduction in fitness of schizophrenics was discussed in Chapter 6. The fitness of male schizophrenics at the present time was estimated to be about 0.5, and that of females 0.8. How then can we explain the continued persistence of schizophrenia at a relatively high frequency against such strong selection? Advocates of a single-gene theory have suggested that if the genetic basis is that of a recessive allele, then the heterozygote for this gene must be at an advantage. The genetic situation, as we have emphasized, must be more complex than this. A two-gene theory has been suggested by Karlsson (1968). Even if we accept a polygenic basis for the disease, we must explain how the contributing genes, which on their own do not lead to schizophrenia, are maintained polymorphic. Heterozygote advantage, of course, is always a possible explanation. Such a hypothesis for the maintenance of schizophrenia was suggested by Ceppellini and co-workers in 1963, and is essentially the viewpoint favored by Gottesman and Shields (1967). It is interesting that Heston suggests, anecdotally, that nonschizophrenic relatives of schizophrenics may have "intellectually desirable" attributes. Erlenmeyer-Kimling and Paradowski (1966) have some preliminary evidence suggesting that such individuals may have higher than average fitness. Perhaps, as indicated by Gottesman and Shields, the genes contributing to the causation of schizophrenia are just a sample of the large number of polymorphic genes that must exist which affect behavioral attributes. Understanding how they are maintained in the population poses the same problems as the maintenance of any other polymorphism, save for the known selection against them when they occur together in a schizophrenic individual. This selection is, however, likely to be slight since, if there are even only three or four such genes, the relative proportion of individuals carrying any one of the genes who are schizophrenic is likely to be quite small. This also implies that a relaxation of selection against schizophrenia, such as was discussed in Chapter 6, will have very little immediate effect on the incidence of the disease. It might even reduce the elimination of genes that are in some sense advantageous.

As was mentioned at the beginning of this chapter, the genetic variation for quantitative characters must be due to the large number of polymorphic genes, which we know must exist, though at the present time we can only identify a minute fraction of them. The approaches we have discussed in this chapter attempt to define the relative contributions of genotype and environment to genetically complex quantitative and qualitative characters. They are useful if they help to clarify thinking on the causation of such characters. The ultimate goal of genetic understanding, however, is always the identification of major distinct genes affecting a character, and so, in fact, the elimination of the need to apply the statistical concepts of quantitative genetics as we have developed them here.

General References

Edwards, J. H., "The simulation of mendelism," *Acta Genetica et Statistica Medica* **10**(1–3): 63–70, 1960. (This paper and the one listed immediately below contain discussions of the basis for using the threshold model to estimate heritabilities.)

Falconer, D. S., "The inheritance of liability to certain diseases, estimated from the incidence among relatives." *Annals of Human Genetics* **29**: 51, 1965.

Falconer, D. S., *Introduction to Quantitative Genetics.* Edinburgh: Oliver and Boyd, 1960. (A good general text covering much of the subject matter of this chapter.)

Fisher, R. A., "The correlation between relatives on the supposition of mendelian inheritance." *Transactions of the Royal Society of Edinburgh* **52**: 399–433, 1918. (A classic paper that laid the modern foundations for the variance approach to the analysis of quantitative characters.)

Haldane, J. B. S., "The interaction of nature and nurture." *Annals of Eugenics* **13**(3): 197–205, 1946. (An elegant and succinct discussion of the interaction between genotype and environment.)

Mather, K., *Biometrical Genetics.* London: Methuen, 1949. (A useful text on variance analysis in quantitative genetics.)

Newman, H. H., F. N. Freeman, and K. J. Holzinger, *Twins: A Study of Heredity and Environment.* Chicago: University of Chicago Press, 1937. (One of the three major studies on separated identical twins.)

Shields, J., 1962, *Monozygotic Twins Brought up Apart and Brought up Together*, London: Oxford University Press. (Another one of the three major studies on separated identical twins.)

Worked Examples

9.1. *The variance of backcrosses for a polygenic character.*

As a further example of the type of analysis possible when crosses between pure lines are available, consider the variation within progeny of backcrosses of the F_1 to either parent. We will first consider a single locus. As before, a is one-half the difference between the phenotypes of the homozygotes A_1A_1 and A_2A_2 and d is the difference between the phenotype of A_1A_2 and the mean of those of A_1A_1 and A_2A_2. The genotype frequencies in the progeny of the two backcrosses (B_1 and B_2) are

$$\text{for } B_1\text{: } \tfrac{1}{2}A_1A_1 \text{ and } \tfrac{1}{2}A_1A_2\text{ ; for } B_2\text{: } \tfrac{1}{2}A_1A_2 \text{ and } \tfrac{1}{2}A_2A_2;$$

and the respective means, therefore,

$$\tfrac{1}{2}(a-d) \text{ and } -\tfrac{1}{2}(a+d)\ .$$

The corresponding variances are

$$\tfrac{1}{2}a^2 + \tfrac{1}{2}d^2 - [\tfrac{1}{2}(a-d)]^2 = \tfrac{1}{4}a^2 + \tfrac{1}{4}d^2 + ad$$

and

$$\tfrac{1}{2}a^2 + \tfrac{1}{2}d^2 - [-\tfrac{1}{2}(a+d)]^2 = \tfrac{1}{4}a^2 + \tfrac{1}{4}d^2 - ad,$$

which, when summed, give

$$\tfrac{1}{2}a^2 + \tfrac{1}{2}d^2.$$

Thus, summing over all loci, the genetic variance among individuals, combining data from both backcrosses, is $V_A + 2V_D$, where $V_A = \frac{1}{2}\sum a^2$, and $V_D = \frac{1}{4}\sum d^2$, as before. The total variance from backcrosses is, therefore,

$$V_B = V_A + 2V_D + 2V_E,$$

since each backcross variance includes a contribution V_E. The difference between the variance of the two backcrosses is

$$V_{B1} - V_{B2} = 2da,$$

and, summing over all genes, $2\sum da$. If r is the correlation between d and a values for individual genes, σ_d and σ_a are the standard deviations of d and a values, and n is the number of genes, then

$$\sum da = \frac{1}{n}(\sum d)(\sum a) + r\sigma_d\sigma_a$$

from the definition of the correlation coefficient (see Appendix I). Thus, if r is zero,

$$V_{B1} - V_{B2} = 2\bar{d}\sum a = \bar{d}(\bar{P}_1 - \bar{P}_2),$$

where

$$\bar{d} = \frac{1}{n}\sum d.$$

9.2. *Estimating V_D/V_A from the average dominance, using the skin-color data on Negroes and Caucasians.*

Recalling the one-locus, two-allele model of Figure 9.7 and the fact that the origin was taken to be halfway between the homozygotes, we have

$$\tfrac{1}{2}(\bar{C} + \bar{N}) - \bar{F}_1 = \sum d, \tag{1}$$

where summation is over all skin-color loci that have different homozygous alleles in Caucasians and Negroes, and $\bar{C}$, $\bar{N}$. $\bar{F}$ are the Caucasian, Negro, and backcross means (see Section 9.5).

$$\tfrac{1}{2}(\bar{C} - \bar{N}) = \sum a, \tag{2}$$

so that

$$\left(\frac{\sum d}{\sum a}\right)^2 = \left[\frac{\frac{1}{2}(\bar{C} + \bar{N}) - \bar{F}_1}{\frac{1}{2}(\bar{C} - \bar{N})}\right]^2. \tag{3}$$

We shall show that, under certain assumptions,

$$\left(\frac{\sum d}{\sum a}\right)^2 \sim \left(\frac{\sum d^2}{\sum a^2}\right) = 2\,\frac{V_D}{V_A} \tag{4}$$

from the definition of V_D and V_A, so that from Equation 3 we have, approximately,

$$\frac{V_D}{V_A} = \frac{1}{2}\left[\frac{\frac{1}{2}(\bar{C}+\bar{N}) - \bar{F}_1}{\frac{1}{2}(\bar{C}-\bar{N})}\right]^2.$$

Put

$$\sum d^2 = \sigma_d^2 + k\bar{d}^2; \sum a^2 = \sigma_a^2 + k\bar{a}^2,$$

where $\bar{d} = \sum d/k$, $\bar{a} = \sum a/k$, k is the number of loci at which Negroes and Caucasians differ, and σ_d^2 and σ_a^2 are the variances of d and a at the various loci. We have

$$2\,\frac{V_D}{V_A} = \frac{\sum d^2}{\sum a^2} = \frac{k\bar{d}^2 + \sigma_d^2}{k\bar{a}^2 + \sigma_a^2} = \left(\frac{\bar{d}^2}{\bar{a}^2}\right)\frac{k + \dfrac{\sigma_k^2}{\bar{d}^2}}{k + \dfrac{\sigma_a^2}{\bar{a}^2}}.$$

Thus the estimates of V_D/V_A from Equation 4, tend to coincide if σ_d^2 and σ_a^2 are both zero; that is, all genes concerned have exactly the same phenotypic effects when considered individually (as could be the case if they were exact duplicates of one another). They also coincide if the coefficients of variation of the additive effects $(\sigma_a/\bar{a})$ and of the dominance effects $(\sigma_d/\bar{d})$ of different loci are the same, or, in other words, if d is proportional to a at each locus. If the variation in d is higher than that in a, as is perhaps likely, then use of Equation 4 will overestimate V_D/V_A.

9.3. *Computation of the covariances between relatives in random-mating populations.*

We first calculate, for the genes at one locus, the covariance between parents and offspring in a random-mating population. To do this we need to know the frequency with which parents of given genotype have offspring of given genotype. Since mating is at random, the offspring frequencies can be obtained by combining alleles A_1 and A_2 at random with the parental gametes in proportion to the gene frequencies p and q, as is done in the following table, in which a, d, and $-a$ stand for the phenotypic values.

				Offspring	
			p^2	$2pq$	q^2
			A_1A_1	A_1A_2	A_2A_2
			a	$-d$	$-a$
Parents p^2	A_1A_1	a	p^3	p^2q	—
$2pq$	A_1A_2	$-d$	p^2q	pq	pq^2
q^2	A_2A_2	$-a$	—	pq^2	q^3

Thus an A_1A_1 parent produces offspring A_1A_1 and A_1A_2 in the relative proportions p and q, an A_2A_2 parent produces pA_1A_2 and qA_2A_2, and an A_1A_2 parent produces

$$(\tfrac{1}{2}A_1 + \tfrac{1}{2}A_2)(pA_1 + qA_2) = \tfrac{1}{2}pA_1A_1 + \tfrac{1}{2}A_1A_2 + \tfrac{1}{2}qA_2A_2\,.$$

The frequencies in the body of the table are obtained by multiplying the frequency with which a given parent produces a given offspring by the population frequency of the parent genotype, which, of course, is given by the Hardy-Weinberg law. Thus, for example, the frequency of A_2A_2 offspring from A_1A_2 is $2pq \times \tfrac{1}{2}q = pq^2$.

The mean of the offspring distribution is

$$a(p^3 + p^2q) - d(p^2q + pq + pq^2) - a(pq^2 + q^3) = a(p^2 - q^2) - 2dpq = m,$$

the mean of the parent distribution. This is expected from the Hardy-Weinberg law, which ensures that the overall offspring distribution is the same as that of the parents. The covariance is the sum of the products of parent and offspring values multiplied by the frequencies of the combinations, corrected by subtracting the product of parental and offspring means. (See Appendix I). Thus, the covariance between parent and offspring is

$$\begin{aligned} &p^3(a^2) + p^2q(-ad) + p^2q(-ad) + pq(d^2) \\ &\quad + pq^2(ad) + pq^2(ad) + q^3(a^2) - m^2 \\ &\quad = a^2(p^3 + q^3) + 2\,pqad(q - p) + pqd^2 - [a(p - q) - 2pqd]^2 \\ &\quad = a^2pq + 2pqad(p - q) + pqd^2(p - q)^2 \\ &\quad = pq[a + d(p - q)]^2 \end{aligned}$$

since

$$p^3 + q^3 - (p - q)^2 = 2pq - p^2(1 - p) - q^2(1 - q) = 2pq - pq(p + q) = pq$$

and

$$pq - 4p^2q^2 = pq(1 - 4pq) = pq[(p + q)^2 - 4pq] = pq(p - q)^2.$$

Summing over all loci, which are assumed as usual to act independently, we obtain as the overall covariance between parent and offspring

$$W_{\mathrm{P/O}} = \sum pq[a + d(p - q)]^2 = \tfrac{1}{2}V_A$$

from Equation 9.12. The covariance between any pair of relatives can be obtained in a similar way, though the algebra is somewhat tedious. In every case it takes the form

$$lV_A + mV_D,$$

where l and m depend on the relationship. The covariances computed in this way for the relationships that are more easily studied in human populations are given in Table 9.7.

9.4. *Positive assortative mating for a recessive character*, a.

We assume that a proportion r of matings are either A × A or a × a (with frequencies equal to those of the dominant and recessive phenotype, $1 - R_t$ and R_t respectively at generation t), while a proportion $(1 - r)$ occur at random. Calling D_t and $2H_t$ the frequencies of the homozygous dominant AA, and the heterozygote Aa at time t ($D_t + 2H_t + R_t = 1$), the frequency of AA among phenotypes A is $D_t/(1 - R_t)$ and that of Aa, $2H_t/(1 - R_t)$. The progeny from assortative matings is then,

	Phenotypes	Genotypes	Frequency of mating	Progeny *AA*	Progeny *Aa*	Progeny *aa*
Total	A × A	$AA \times AA$	$\frac{D_t^2}{(1-R_t)^2}$	$\frac{D_t^2}{(1-R_t)^2}$		
	$(1 - R_t)$	$AA \times Aa$	$\frac{4D_tH_t}{(1-R_t)^2}$	$\frac{2D_tH_t}{(1-R_t)^2}$	$\frac{2D_tH_t}{(1-R_t)^2}$	
		$Aa \times Aa$	$\frac{4H_t^2}{(1-R_t)^2}$	$\frac{H_t^2}{(1-R_t)^2}$	$\frac{2H_t^2}{(1-R_t)^2}$	$\frac{H_t^2}{(1-R_t)^2}$
	a × a R_t	$aa \times aa$	1			1

The progeny frequencies from the r fraction are

$$AA: \quad (1 - R_t)\left[\frac{D_t^2}{(1 - R_t)^2} + \frac{2D_tH_t}{(1 - R_t)^2} + \frac{H_t^2}{(1 - R_t)^2}\right] = \frac{(D_t + H_t)^2}{1 - R_t}$$

$$Aa: \quad (1 - R_t)\left[\frac{2D_tH_t}{(1 - R_t)^2} + \frac{2H_t^2}{(1 - R_t)^2}\right] = \frac{2H_t(D_t + H_t)}{1 - R_t}$$

$$aa: \quad \underset{\text{from } A \times A}{\frac{H_t^2}{1 - R_t}} + \underset{\text{from } a \times a}{R_t}$$

The gene frequency of A in the progeny of the assortatively mating fraction is

$$p = \frac{(D_t + H_t)^2}{1 - R_t} + \frac{H_t(D_t + H_t)}{1 - R_t} = D_t + H_t,$$

and therefore equal to that in the parental generation. In the random-mating portion also, the gene frequency does not change, and therefore there is no overall change in gene frequency.

Substituting gene frequencies in the results for the progeny frequencies, in the fraction that is mating assortatively and summing the contributions from the random-mating and the assortative portions, we have, for the next generation,

	From random mating		*From assortative mating*
AA:	$(1-r)p^2$	+	$\dfrac{rp^2}{1-R_t}$
Aa:	$(1-r)2pq$	+	$\dfrac{2rp(1-p-R_t)}{1-R_t}$
aa:	$(1-r)q^2$	+	$\dfrac{r(1-p-R_t)^2}{1-R_t}$

The frequency of recessives in the next generation is thus

$$R_{t+1} = (1-r)q^2 + r\frac{(q-R_t)^2}{1-R_t}.$$

At equilibrium, putting $R_{t+1} = R_t = \hat{R}$, we obtain the formula for $\hat{R}$ given in the text.

The basis of this example is due to Wright (1921b). Further details can be found in Crow and Felsenstein (1968).

9.5. *The selection differential for culling from a normal population.*

We assume, without loss of generality, that the quantitative character x is normally distributed with mean 0 and variance 1. The proportion of individuals with x values above a threshold T is

$$q = \frac{1}{\sqrt{2\pi}}\int_T^\infty e^{-x^2/2}\,dx.$$

The mean x value of these "culled" individuals is

$$a = \frac{1}{q} \times \frac{1}{\sqrt{2\pi}}\int_T^\infty xe^{-x^2/2}\,dx.$$

Now

$$\int_T^\infty xe^{-x^2/2}\,dx = [-e^{-x^2/2}]_T^\infty = e^{-T^2/2}.$$

Thus

$$a = \frac{1}{q} \times \frac{e^{-T^2/2}}{\sqrt{2\pi}} = z/q,$$

where z is the height of the normal curve at the threshold T. This equation is equivalent to Equation 9.44, which was used in Falconer's approach to the study of threshold characters. The difference between the mean of the culled individuals, and that of the general population, a, which is the selection differential, is expressed in terms of the proportion culled, q, and the normal distribution ordinate at the threshold, z.

10

The Sexual Dimorphism

The sexual dimorphism is undoubtedly the most important of all human polymorphisms, and provides the most direct example of a simple Mendelian segregation.

The evolution of sexual reproduction must have depended almost exclusively on the advantage conferred on the species, rather than on the individual. Muller said in 1932 "only the geneticist can properly answer the question 'is sex necessary?'" The existence of sex is one of the most significant factors influencing our biological and social organization. Colombo (1957) goes so far as to say that "the sex ratio at birth is the cornerstone of our social organization and biological life. It is the fundamental condition of monogamous marriage and of a social structure and value system based on the family cell; it also affects the heredity, distribution and selection of the genetic patrimony of the species."

10.1 Sex Determination

The mechanism of sex determination has been the subject of much speculation throughout history (for a general review see Crew, 1933). Its elucidation depended on the interpretation of the meiotic behavior of chromosomes in terms of Mendelian segregation. Mendel himself mentioned an unusual 1 : 3 sex ratio in plants in one of his letters to Nägeli, but never discussed the general problem of the genetic determination of sex, perhaps because he was concerned with the genetics of plants and not that of higher animals.

The ultimate verification of the sex determination mechanism in man came only with the development of suitable techniques for looking at human chromosomes, at the end of the 1950*'s.*

Typical karyotypes (chromosome pictures in which homologous chromosomes are paired and the pairs arranged according to their size) of a normal human male and female were shown in Figure 1.7. The X chromosome is a moderately large metacentric (having its centromere in the middle) chromosome, probably a little smaller than chromosome 6, the largest of the so-called "C group" chromosomes. The Y chromosome is one of the smallest of the human chromosomes and is acrocentric (has its centromere near one end) and so clearly lacks most of the genes present on the X. Enough homology between the X and the Y must exist, however, for them to pair regularly at meiosis, form bivalents, and thus ensure the regular distribution of X and Y chromosomes to the gametes. This is, of course, the basis for expecting that the proportion of males to females should be a Mendelian 1 : 1 ratio, since every cross is a backcross female XX = male XY. Though few if any known mutant differences can be assigned to the Y chromosome, it must at least carry the primary male determining genes. Secondary sexual differences may be determined by autosomal genes (that is, genes not on the X or the Y) that interact with the sex-determining genes. An inherited character that is expressed predominantly in one sex, such as the autosomally inherited pattern of baldness in the male, is called **sex limited**. The sex hormones play a major role in sexual differentiation during early embryonic development and may be the major factors in sex-limited inheritance.

Testicular feminization provides an interesting paradox for sex-linked inheritance.

Persons having testicular feminization are intersexes (that is, individuals having some of the characteristics of both sexes) with a normal XY karyotype and well-developed female secondary characteristics. Most of them regard themselves as females. They may have undescended testes, and lack normal internal female genital organs, apart from a small "blind" vagina. Their "outward" phenotype is clearly female (a famous photographic model and twin airline stewardesses with the syndrome have been reported) and many of them are happily married, though all of them are infertile. The frequency of the syndrome is estimated to be about 1/65,000 males born (see McKusick, 1962). An example of a pedigree showing the

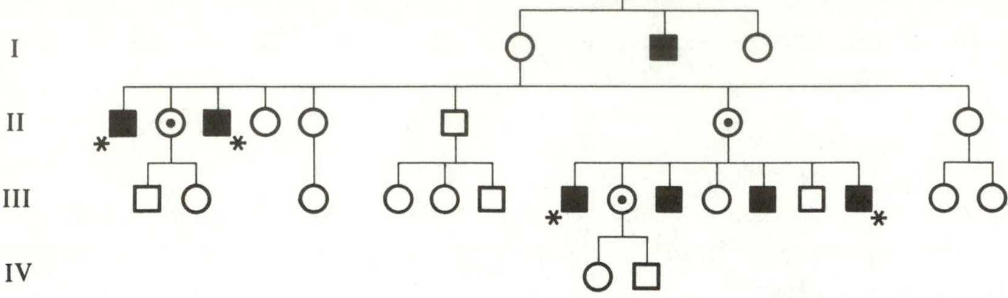

FIGURE 10.1
Pedigree of the testicular feminization syndrome. Note that autosomal dominant inheritance with male limitation explains the *prima facie* pedigree evidence. Circles with dots centered in them indicate women who are hairless (by examination). Black squares indicate men who have the testicular feminization syndrome. Asterisks mark cases of the syndrome that have been confirmed surgically. (From McKusick, 1962; adapted from Schreiner, 1959.)

inheritance of testicular feminization is given in Figure 10.1. The pattern is that typically expected for an X-linked recessive. Carrier females can, as indicated in this pedigree, often be identified because they have almost no female secondary hair, though they are otherwise normal. The inheritance pattern can, however, equally well be explained by an autosomal dominant gene whose expression is essentially limited to the male. Only studies on the linkage of this character with other X-linked genes can determine which model of inheritance is correct. There is so far no detectable linkage with hemophilia, color blindness, or the Xg blood group, but the data are inconclusive.

10.2 The X Chromosome Map

The X is the only human chromosome for which there exists even a partial linkage map.

As was mentioned in Chapter 3, up to sixty genetic differences in man have been identified as being X-linked. This contrasts sharply with the relatively small (though rapidly increasing) number of autosomal linkages detected up to now (see Appendix II). The difference is undoubtedly due to the ease with which a gene can be assigned to the X chromosome because of the characteristic pattern of inheritance of X-linked genes. In small pedigrees chance variations may, however, show an apparent X-linked pattern of inheritance for an autosomal gene.

Genetic mapping of the X chromosome is greatly facilitated by the fact that the male is effectively haploid for the X, so that his phenotype reflects directly its genetic content. This, together with the existence of large numbers of X-linked mutant genes, has led to extensive investigation of the genetic map of the X-chromosome. The main limiting factor is the low frequency of most of the X-linked mutant genes.

There are four X-linked polymorphisms that have been pivotal for X-mapping studies: color blindness, G6PD deficiency, the Xg blood group, and the Xm serum group.

Color blindness has a gene frequency in Caucasians of about 8 percent; G6PD deficiency in malarial, or formerly malarial, areas, reaches a gene frequency higher than 50 percent in some populations (see Chapter 4). There is also a G6PD electrophoretic variant that has a frequency as high as 35 percent in Negroes, though it is rare in Caucasians. The Xg(a+) antigen on red cells is determined by a dominant gene that has a frequency in Caucasians of about 60 percent. The Xm(a+) antigen is a serum protein detected by precipitation with a specific rabbit antiserum. The corresponding gene has a frequency of about 26 percent in Caucasians in the United States.

The relatively high gene frequencies of these polymorphisms make it fairly easy to find suitable doubly heterozygous females for linkage studies on either (1) pairs of these polymorphisms or (2) one of the polymorphisms and a rare X-linked disease. When available, the phenotypes of the father of a doubly heterozygous female determine her linkage phase (coupling versus repulsion, see Figure 10.2); the phenotypes of her male offspring provide a direct sample of her gametes, and thus a direct count of recombinants versus nonrecombinants. (The genotype of the husband is not relevant as he only contributes a Y chromosome to his male offspring.) Doubly heterozygous females are usually ascertained through their having sons with two different X-linked traits, either alone or in combination. The general problem of how the method of ascertainment of a pedigree affects its interpretation is discussed in Appendix II. Analysis of autosomal linkage using pedigrees is greatly complicated by the fact that the genes first must be assigned to their chromosomes, which, if linkage is loose, may require very large bodies of data. Moreover, the linkage phase of double heterozygotes cannot usually be directly ascertained. These problems are discussed further in Appendix II.

Six pedigrees that provide information on the linkage between color blindness (deutan type) and G6PD deficiency are shown in Figure 10.3. They were ascertained through testing black American school boys for color blindness and then testing the families with color-blind probands for G6PD deficiency. The male children giving information on the recombination frequency between the two loci are those in sibships in which both color blindness and G6PD deficiency are segregating. These are encircled in Figure 10.3. Only one (marked with an asterisk in the figure)

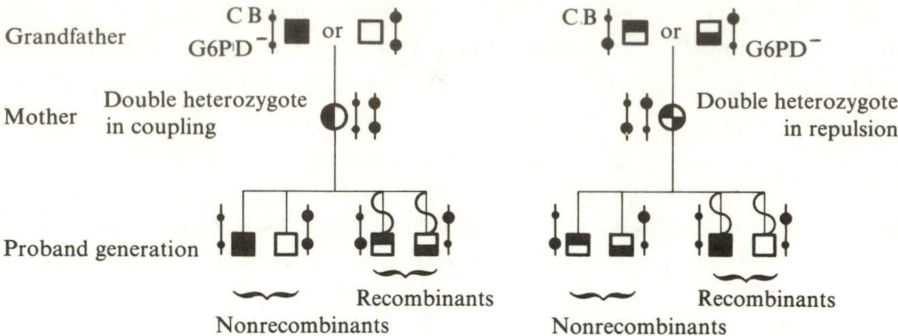

FIGURE 10.2
The grandfather method for determining linkage of common X-borne recessive traits — color blindness (CB) and glucose-6-phosphate dehydrogenase deficiency ($G6PD^-$). Males showing one trait are ascertained. Then all males in the sibship of said probands are tested for the second trait. The families of those sibships in which both traits are segregating are subjected to further study. Most mothers of such sibships are double heterozygotes. (The mothers are tested for both traits to exclude the relatively rare situation of a mother homozygous for one or the other gene or both.) The coupling phase in the doubly heterozygous mothers must be known and is determined by testing their fathers, the maternal grandfathers of the probands. The recombination fraction is the proportion of males in the proband generation who are recombinant individuals. (From McKusick, 1962.)

out of 23 is a recombinant, indicating close linkage between these two loci. More complicated statistical procedures can be used to find the best estimates of the recombination fraction from such data, taking into account, also, pedigrees in which the grandfather is not available (see Appendix II).

Two recent versions of the genetic map of the X chromosome are shown in Figure 10.4 (see also Berg and Bearn, 1968; and Race and Sanger, 1968). There are two main regions of the chromosome for which meaningful recombination data have been obtained. The first is a relatively tight cluster involving G6PD, color blindness, and hemophilia A. The second involves looser recombination fractions, but nevertheless detectable linkage between the Xg blood group, ichthyosis (dry scaly skin), ocular albinism, and, possibly, angiokeratoma (vascular skin lesions). The Xm locus falls somewhere between these two regions (Figure 10.4). The version of the X map shown on the right in the figure is a direct computer output, incorporating data on hemophilia B, which is very loosely linked with hemophilia A. It assumes a total map length of 250 centimorgans (one percent recombination = one centimorgan, for closely linked loci), based on chiasma counts during meiosis and on the relative DNA content of the X chromosome. Results from pedigrees involving deletions of the short arm of the X suggest that the Xg locus is near the distal end (away from the centromere) of this arm. In view of the large size of the X chromosome it is not surprising that many pairs of X-linked markers do not give recombination fractions significantly less then 0.5.

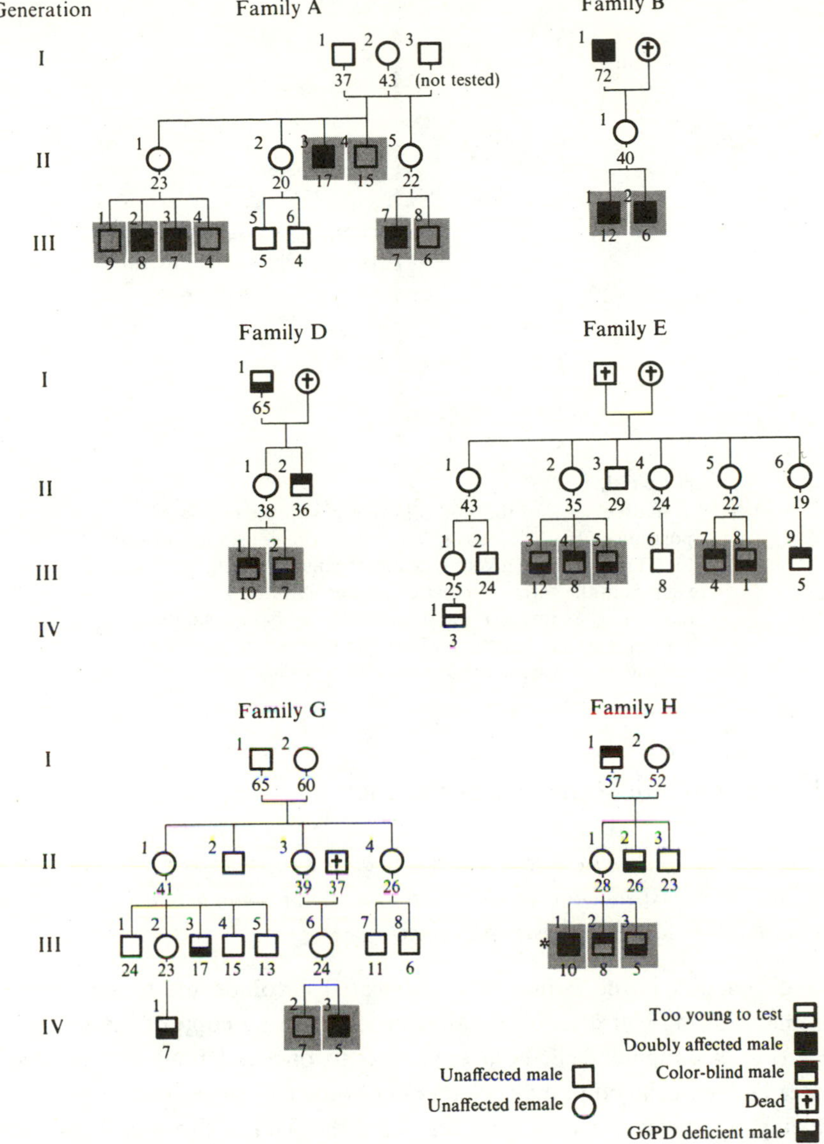

FIGURE 10.3
Linkage of color blindness and glucose-6-phosphate dehydrogenase deficiency. Among 3648 Negro school boys, color blindness was found in 134 (3.7 percent). All males (totaling 236) in the sibships of 106 of the color-blind probands were tested for glucose-6-phosphate dehydrogenase deficiency. In ten sibships both traits were discovered. Six of these families having the deutan type of color blindness are diagrammed. Only one definite instance of recombination was discovered (in family H), out of 22 relevant offspring, which are indicated in the diagram by the gray squares. Numbers placed above and to the left of symbols are pedigree numbers used to identify individuals in a pedigree. Numbers below symbols are ages at time of testing. (See Proter, Schulze, and McKusick, 1962; adapted from McKusick, 1962.)

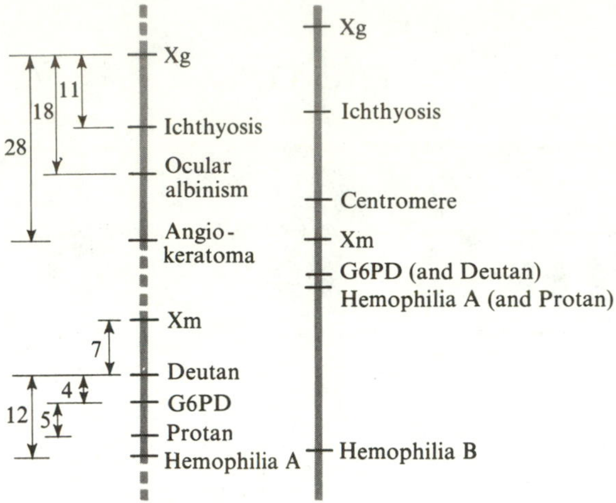

FIGURE 10.4
Tentative maps of the X chromosome with the estimated position of the Xm locus in relation to previously positioned loci. Distances are given in centimorgans. The map at the right is in the form of direct computer output and is the most likely estimate of the position of the loci considered, assuming a map length of 250 centimorgans and no interference. (From Berg and Bearn, 1968.)

10.3 Sex-chromosome Abnormalities

Nondisjunction gives rise to a wide variety of sex chromosome anomalies, some of which have been instrumental in identifying the relative importance of the X and Y chromosomes for sex determination in man.

Nondisjunction is the failure of a duplicated chromosome to separate at the centromere during cell division, so that instead of one copy of the chromosome going to each daughter cell, both copies go to one and none to the other (see Chapter 1). Nondisjunction may occur either at the first or second meiotic division, producing, whenever it occurs, gametes that either lack a chromosome or contain an extra one. If nondisjunction occurs during, or soon after, fertilization, it generally produces a mosaic containing mixtures of genetically different cells, some of which have a normal chromosome complement, some with an extra chromosome (said to be trisomic for that chromosome), and some with a chromosome missing (monosomic).

The expected consequences of nondisjunction for the X and Y chromosomes in either the first or second divisions of meiosis are illustrated schematically in Figure 10.5. In the male, first-division nondisjunction gives rise to XY and O sperm, and

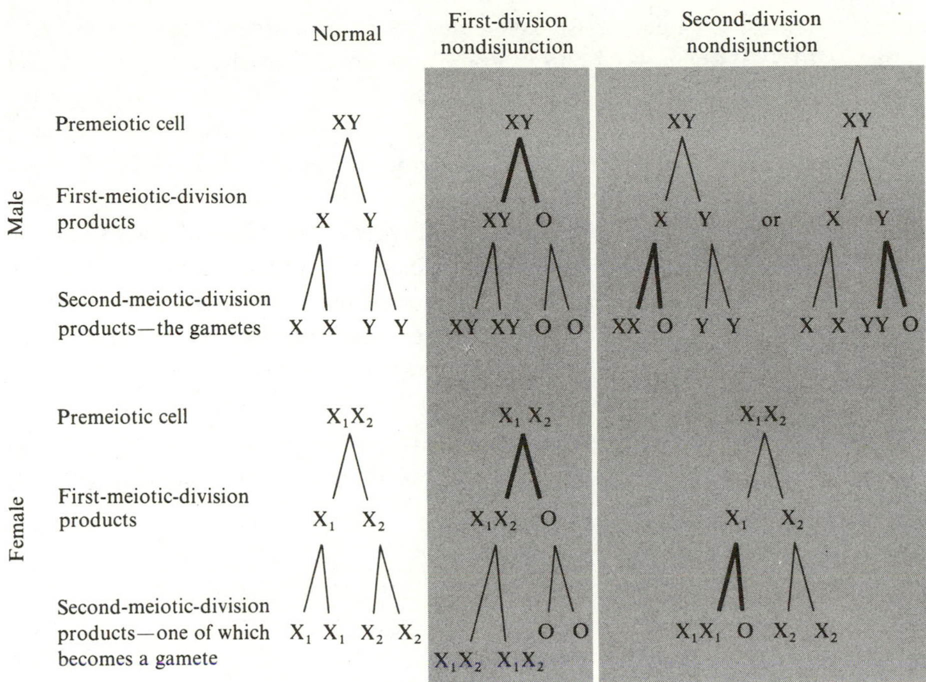

FIGURE 10.5
Patterns of meiotic nondisjunction. Only the X and Y chromosomes are being followed. Heavy lines indicate the divisions in which nondisjunction occurs. The subscripts on the X symbols are simply identifiers to indicate which copy of the X is found in the gametes. In the female only one of the products of meiosis, presumably determined at random, ends up as an egg nucleus. The remaining three become polar bodies.

second-division nondisjunction to XX, O, and YY sperm. In the female, nondisjunction at either first or second division produces eggs that are XX or O. Following second-division nondisjunction, the chromosomes would be identical by descent, if there were no crossovers, and following first-division nondisjunction the chromosomes are copies of homologues. Aneuploid gametes, when fertilized by normal X- or Y-carrying gametes, give rise to the well-known sex-chromosome abnormalities XXY (Klinefelter's syndrome), XO (Turner's syndrome), and XXX and XYY, which were mentioned in Chapter 3. Persons having the chromosome complement YO have never been observed and the lack of an X chromosome is thus presumed to be lethal.

X-linked markers can sometimes be used to determine whether the abnormal gamete that gave rise to a particular Turner's or Klinefelter's individual originated in the father or the mother, and at which meiotic division. Some hypothetical examples are illustrated in Figure 10.6 using the Xg blood group. In Figure 10.6,A, since the X chromosome in the offspring must have come from the father, the nondisjunctional O gamete must have come from the mother. The reverse is true in Figure 10.6, B, where the *X* chromosome must come from the mother. In neither example is it possible to tell in which meiotic division nondisjunction occurred, since O gametes can be produced in either first or second division (see Figure 10.5). The Xg(a+) Klinefelter's offspring in 10.6, C must be XaX − Y since nondisjunction in the mother (first or second division) would produce X−X− gametes, and thus an Xg(a−) Klinefelter's offspring. The paternal nondisjunction must be at the first division, for only then can an XY gamete be produced. In Figure 10.6, D, first-division paternal or maternal nondisjunction gives rise to an Xg(a+) Klinefelter's offspring. The offspring must, therefore, be the product of second-division nondisjunction for the maternal X chromosome.

Assuming that all the types of events listed in Figure 10.5 occur with equal frequency, we nevertheless expect to find more of some nondisjunctional types than others. Thus, for example, though O gametes would be expected in equal frequencies

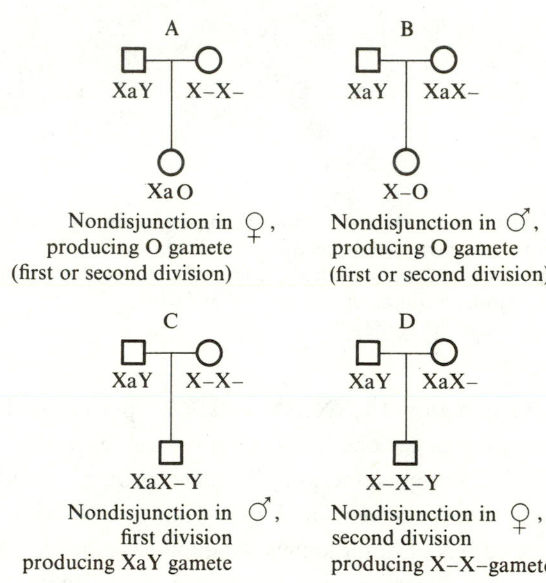

FIGURE 10.6
Determination of the origin of a sex-chromosome nondisjunction—some hypothetical examples using the Xg blood group, which is determined by genes on the X chromosome, as a system of markers. The symbol Xa refers to the chromosome carrying the Xg(a +) genetic determinant and X- to the one lacking it. The father's and mother's phenotypes in A and B determine the offspring's chromosome complement. The nondisjunctional phenotypes of the offspring in B and D make it evident that the mothers must be heterozygous XaX-.

from nondisjunction in the male and in the female, only one-half of those produced by the female give XO offspring, the remainder being YO, which are lost because they are not viable. Thus half as many XO offspring are expected from nondisjunction in the female than the male. Superimposed on this difference, is the fact that it can be shown that the frequency of matings in which the paternal origin of an X (in an XO offspring) can be determined is much lower than the frequency of that in which the maternal origin can be established. In a survey of 56 mother/father/XO combinations, out of the 21 in which the parental origin of the X could be determined, 20 were maternal, indicating that the O gamete must have been produced by nondisjunction in the father (see McKusick, 1962). A further complicating factor may be that nondisjunction involving loss of the Y chromosome is more frequent than that involving loss of the X. It has even been suggested that "loss" of the Y chromosome may often occur after fertilization, before or during the first cleavage, so that only XO cleavage products are produced. This may be an explanation for the fact that only XO among the known nondisjunctional phenotypes (which include trisomy for chromosomes 21, 13 or 15 and 17 or 18 as well as XXY) shows no increasing incidence with increasing parental age (see Penrose, 1963).

The fact that persons with Turner's syndrome (XO) are basically female while those with Klinefelter's syndrome (XXY) are basically male, demonstrates the male determining capacity of the Y chromosome.

The same phenotypes for XO and XXY are observed in mice as in man, whereas Bridges' classical work on sex determination in *Drosophila* showed that there, the XO is a male while XXY and even XXYY are females. In *Drosophila* it is apparently the balance between the number of X chromosomes and autosomes that determines sex, the Y chromosome having no obvious effect. A wide variety of sex-determining mechanisms exists in plants and animals, ranging from single gene differences to the X−Y chromosome mechanism in mammals, and even ploidy (males being haploid and females diploid, for example) in some species of bees and wasps (see Crewe, 1933). We shall discuss the evolution of sex determination toward the end of this chapter.

An enormous variety of sex-chromosome abnormalities in man has been described, including mitotic nondisjunction at an early cleavage stage, forming mosaics, and also such bizarre chromosome complements as XXXXY and XXXXX. Many persons having abnormal chromosome complements are infertile, and nearly all show features characteristic either of *Turner's syndrome* (females with gonadal aplasia, short stature, web neck, wide-set nipples), if they have no Y, or of *Klinefelter's syndrome* (males with small underdeveloped nonfunctional testes and prostate, and scanty secondary sexual hair, if they have at least one Y, together often with a distinctly sub-normal IQ. Rare mosaics, such as XX/XY, may show combinations of the two syndromes. The relative prominence of the various features depends, at least in part, on the relative frequency with which the different cell types occur in the mosaic individual.

The overall frequency of sexually aneuploid males is estimated to be about 0.2 *percent, and of females* 0.16 *percent.*

Sex-chromosome abnormalities constitute about one-quarter to one-fifth of all chromosomal anomalies observed at birth (Court-Brown, 1967). The majority are either XXY, XO, or XYY, though it is not easy to get reliable unbiased estimates of their frequency at birth. Moreover, as discussed in Chapter 3, there is evidence for a remarkably high frequency of sex-chromosome abnormalities among embryos that are spontaneously aborted early in pregnancy, perhaps even as high as 5 percent. (A sample of 50 induced abortions reported by Carr (1967) showed no chromosomal anomalies, while among 200 spontaneous abortions the overall incidence of chromosome anomalies was about 22 percent.) Since there is considerable mortality of embryos with these anomalies, the frequency of the abnormalities at birth may grossly underestimate the frequency with which they are produced during meiosis. The relative frequencies of males having at least two X chromosomes in a registry of abnormal karyotypes is shown in Table 10.1. About 80 percent are XXY, 10

TABLE 10.1
Abnormal Sex-chromosome Complements of Males Having at Least Two X Chromosomes in the Edinburgh Registry of Abnormal Karyotypes (up to March, 1966)

Sex Chromosomes	*No. of Subjects*
XX	5
XXY	166
XXYY	4
XXXY	5
XXXXY	3
XY/XXY	25
XX/XXY	3
XXY/XXXY	2
XXXY/XXXXY	2
XY/XXY/XXYY	1
XX/XY/XXY	1
XO/XY/XXY	1
Total	218

Source: From Court-Brown (1967).

percent are XY/XXY mosaics, and the remaining 10 percent include, in approximately equal frequencies, a variety of more complex combinations. The distribution of the frequencies of some of these types, separated according to the source of the data is given in Table 10.2. There are more XY/XXY mosaics among newborn babies and a noticeably high frequency of XXY among subfertile males and those with endocrine abnormalities. Particularly striking is the fact that most of the more complex combinations, such as XXXY, and even an XY/XXY/XXYY mosaic, were found among the mental defectives and not at all in the newborn babies. This suggests that these abnormalities may contribute substantially to the social load of mental defectives. The frequency of about 70 percent XXY among newborn male babies with at least two X chromosomes was obtained from a survey of 10,500 consecutive male births and so is an unbiased estimate of this overall relative frequency.

In a well-known study of mentally subnormal male patients in a state hospital in Scotland, most of whom were admitted after repeated convictions for criminal offences, Jacobs and colleagues (1965) found twelve out of 197, or 6.1 percent to have chromosomal abnormalities, nine of which were sex-chromosome anomalies. Of these nine, seven were XYY, one was XXY, and one was XXYY. It was this remarkable concentration of XYY persons that sparked the recent interest in the XYY and other related abnormalities, and their relationship to a propensity toward criminality. Jacobs found that the XYY males in her original study had a mean height of 73.1 inches, while the XY's had a mean height of 67 inches, suggesting that the extra Y chromosome was responsible for a significant increase in height. Later studies have, on the whole, tended to corroborate Jacobs' initial findings, though there seems to be a fair amount of variability in the relative frequencies of XYY and other sex-chromosome abnormalities among the inmates of penal institutions. A major problem in the interpretation of the data is the lack of any reliable estimate of the frequency of XYY individuals in the "normal" population and also of information about what attributes associated with the XYY phenotype might in themselves lead to a predisposition toward criminality. A review by Court-Brown (1968) of 15 surveys of hospitalized mentally subnormal males, totalling more than 13,000 subjects, produced the estimate that 9.4 out of 1000 subnormal males have abnormal sex-chromosome complements. This is approximately five times as many as are observed among the newborn. The frequency among the males with IQ's higher than 50 was 17.6 per 1000, while that among males having IQ's lower than 50 was only 6.4 per 1000. Court-Brown points out that this association of sex-chromosome anomalies with relatively high grade mental subnormality could account almost entirely for their increased incidence among criminals. That is, the frequency in penal institutions of males with sex-chromosome anomalies is comparable to that found among mentally subnormal individuals with IQ's higher than 50.

In the absence of prospective studies, it is not known what proportion of the

TABLE 10.2

The Distribution of the Different Sex-chromosome Complements of Abnormal Males Having at Least Two X Chromosomes, according to the Source of Ascertainment

Source	*Total Subjects*	XXY		XXYY		XX		XXXY, XXXXY		XY/XXY		XX/XXY		*Other Mosaics*[a]	
		No.	%	*No.*	%	*No.*	%	*No.*	%	*No.*	%	*No.*	%	*No.*	%
Newborn babies	21	14	66.7	1	4.8	0	—	0	—	6	28.6	0	—	0	—
Subfertile males	39	36	92.3	0	—	2	5.1	0	—	1	2.6	0	—	0	—
General patients	43	33	76.7	0	—	2	4.6	2	4.6	5	11.6	1	2.4	0	—
Endocrinology patients	32	29	90.6	0	—	1	3.1	0	—	2	6.2	0	—	0	—
Mental hospitals	24	20	83.3	0	—	0	—	0	—	4	16.7	0	—	0	—
Mental defectives	55	33	60.0	3	5.4	0	—	6	10.9	4	7.3	2	3.6	7	12.7

Source: From Court-Brown (1967).

[a] The other 7 mosaics included XXY/XXXY (2), XXXY/XXXXY (2), XY/XXY/XXYY, XX/XY/XXY and XO/XY/XXY.

XYY males born end up in penal institutions. Assuming the frequency of XYY among the institutionalized criminals is k-fold higher than that in the general population, and that the overall frequency of such criminals is x, then a proportion kx of all newborn XYY will be expected to end up as institutionalized criminals. Even taking $k = 15$, based on a 2 per 1000 newborn incidence of XYY and a 3 percent incidence in criminal institutions, and taking x as high as 1 percent, this leaves 85 percent of the XYY individuals born not represented among the criminals. The estimates of the frequency of XYY among newborns vary over an almost tenfold range and if, as Court-Brown puts it, XYY individuals are "unusually prone to nonparticipation," that is, do not readily attend clinics or other services used as a basis for collecting data, the estimates of their frequency among the normal adult population may be seriously biased on the low side. One report of an XYY with an IQ of 125 exists. Some attention has been paid to the significance of an XYY karyotype in court in criminal cases. Being an indication of an inborn predisposition toward criminality, it is considered a mitigating factor. Before such evidence should be accepted, much more information is clearly needed on the XYY phenotype. Moreover, it is not clear what should be the legal significance of a "predisposition." After all, there must be many other factors, both genetic and nongenetic, that are beyond the individuals' overt control and that may dispose him toward criminality.

10.4 Y-linkage and Partial Sex Linkage

A Y-linked gene, located on the Y chromosome, should show strictly male-to-male transmission.

There have been a few unconfirmed reports over the years suggesting that certain specific traits, notably "porcupine" skin and webbed toes, might be Y-linked (see Stern, 1957). However, only one genetic character, hairy ears, which is found with a relatively high frequency in India, seems to be reasonably well substantiated as being Y-linked (Dronamraju, 1960). In the mouse there exists a well-defined Y-linked histocompatibility locus that is responsible for the rejection of male skin grafts by females within inbred lines. The striking pattern of inheritance expected for Y-linked genes makes it unlikely that any such known inherited differences would be missed. The Y chromosome at least carries genes that are critical for the development of male sexuality.

As many as 2–3 percent of normal males have Y chromosomes that are either unusually long or unusually short, and mostly the former (Court-Brown, 1967). In one case, the presence of a long Y in only 3 out of 4 brothers allowed detection of false paternity, later confirmed by detailed blood-group studies (Nuzzo et al., 1967). The existence of polymorphism for the Y chromosomes suggests, of course, that its total content is not critical for the development of a normal male.

The X and Y chromosomes are paired short end to short end during meiosis. Pairing between homologous chromosomes during meiosis seems to be an essential prerequisite for the orderly distribution of the chromosomes to the gametes. At least a small section of the X and Y chromosomes must therefore be homologous. Genes occurring in this region should show "ordinary" linkage with sex. The consequences of a crossover in the homologous region for a doubly heterozygous male $AX/A'Y$ are illustrated in Figure 10.7. It has to be assumed that the differential segments of the two chromosomes carry the essential sex-determining genes and so may effectively be labeled X and Y. Crossingover in the homologous region leads to the production of recombinant AY and $A'X$ gametes and should result in "partial" linkage with sex, as observed for pairs of linked autosomal markers. Haldane (1936), in one of the first attempts at linkage analysis in man, claimed to find some evidence for such **partial sex linkage**. A later reevaluation of the available data by Morton (1956a), however, failed to confirm Haldane's conclusion (see also Appendix II). It is likely that evolution has maximized the difference between X and Y, and thus minimized the extent of homology, leaving only the smallest possible region to satisfy the mechanical needs of meiosis. The small extent of homology between X and Y may, in fact, predispose them to meiotic nondisjunction and so account, in part, for the apparently high relative incidence of sex-chromosome aneuploidy.

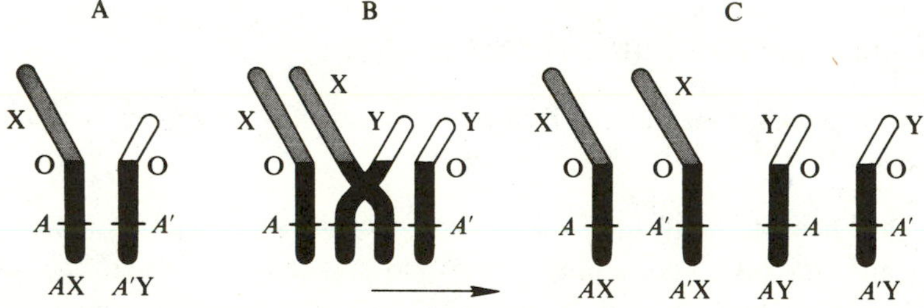

FIGURE 10.7
Diagram of hypothetical crossing-over between the homologous segments of the sex chromosomes. A: X chromosome to the left, Y chromosome to the right. Black = homologous segment; gray = differential segment of the X chromosome; white = differential segment of the Y chromosome; O = differential end point: A and A' = a pair of alleles. B: Four-strand stage with crossover. C: The four chromosomes resulting from crossing-over at the four-strand stage. The sex-determining genes are presumed to be carried on the differential parts of the X and Y chromosomes. (Adapted from Stern, 1960.)

10.5 Dosage Compensation and X-inactivation

A quantitative unbalance for autosomes leads to gross phenotypic abnormalities, but since all X-linked genes are present in two doses in the female and only in one in the male, this is, obviously, not true for the X chromosome.

Persons who are heterozygous for autosomal recessive mutations that cause enzyme deficiencies, such as the PKU mutation that gives rise to a defective phenylalanine hydroxylase enzyme, generally have half the enzyme activity of normal homozygotes. Enzymes controlled by X-linked genes, however, such as G6PD, generally show the same level of activity in males and females. There must, therefore, exist a mechanism that can compensate for the X-gene-dosage difference, and that guards against metabolic imbalance due to the presence of different numbers of X-linked genes in males and females.

It is now a widely accepted hypothesis that dosage compensation in mammals is due to the fact that only one of the two X chromosomes of the female is metabolically active in any given cell at a given time.

The history of the development of this hypothesis starts with the discovery in 1949 by Barr and Bertram that most nondividing nerve cells of the female cat have a small dark-staining body in their nucleus that is never present in the cells of a normal male cat. This seemingly trivial observation was later extended to many other cell types in many other mammalian species, man in particular. The presence of this **Barr body**, or **sex chromatin**, in female, but not male, cells provided a way of sexing a population of cells before the more-recent development of mammalian cytogenetics lead to the specific identification of the human X and Y chromosomes. Already at this time, it had been recognised that individuals with Turner's syndrome, though basically female, lacked the Barr body, while the male-like Klinefelter's syndrome was associated with the presence of one Barr body. It is now established that individuals with n X chromosomes may have cells with up to a maximum of $n-1$ Barr bodies. Surveys for sex-chromosome abnormalities are, in fact, often done by screening for Barr bodies in cells obtained from buccal smears (from the inside of the mouth) as this is a much simpler procedure than karyotyping. The karyotype is then determined only for individuals who have cells with unusual numbers of Barr bodies, mainly females with none and males with one or more.

The sex chromatin or Barr body is an inert X chromosome.

Following the development of cytogenetic techniques for mammals, a careful survey of the origin of the Barr body throughout the cell cycle revealed that it derived from one of the X chromosomes. Chromosome regions that stain heavily at some stage in the cell cycle, as does the Barr body, are called **heterochromatic** and are known in many organisms to be relatively "inert" from a genetic point of view. The identification of the heterochromatic Barr body with an X chromosome, suggested that one of the X chromosomes in a cell with a Barr body is regularly heterochromatic, and so probably inactive. This correlation was further strengthened by the observation, using tritiated thymidine autoradiography, that one X chromosome always tends to replicate later than the other, and this is the one destined to form the Barr body. Late replication during the cell cycle is a characteristic of heterochromatin. (See Ohno, 1967, for a recent review of the material of this section.)

These observations were united by Lyon (1961) into her inactive X hypothesis. This states specifically that at some stage early in embryogenesis one of the X chromosomes, chosen at random, of each cell of the female is inactivated and that the same X remains inactive in all the descendents of any given cell. This leads to the expectation that females heterozygous for X-linked mutants should be phenotypic mosaics, being mixtures of mutant and nonmutant cell lines. The patchiness of female mice heterozygous for X-linked recessive genes that determine coat color was, in fact, one of the observations that led Lyon to formulate her hypothesis. It has subsequently been confirmed for a number of human X-linked genetic markers, notably G6PD, that heterozygous females contain mixtures of two types of cells corresponding to inactivation of one or the other X chromosome. The time of inactivation during development is not precisely known. The fact that even very small pieces of skin give rise to clones of both types suggests that it cannot be very early, otherwise there would be expected to be much more spatial heterogeneity in the distribution of the two types of clones. However, the time of first appearance of sex chromatin suggests that inactivation may occur when the blastula has reached the two- to three-thousand cell stage. Very rare cases of female heterozygotes for X-linked mutations, such as that producing hemophilia A, show phenotypes approaching that of the affected male. These probably represent situations in which, by chance, one whole tissue has been populated predominantly with cells in which the mutant X-linked gene is active.

All genes of the "inactive" X are not necessarily inactive, as shown in mice by L. Russell (1963), though it seems likely that at least the majority are. Also, only a proportion of female cells have a readily detectable Barr body.

The comparatively mild effects of X chromosome aneuploidy compared with autosomal aneuploidy (in particular, the monosomic state) are easily explained by dosage compensation and the inactive X hypothesis. The relation between the number of X's and the number of Barr bodies indicates that at least the majority of cells even in X aneuploids have only one active X. The major consequences of the XO, XXY, XXX, etc., states are, therefore, probably expressed rather early in embryonic life—before X inactivation has taken place.

10.6 The Sex Ratio

The observed sex ratio at birth is significantly different from the expected 1 : 1 *ratio.*

Note that we use the term **sex ratio** for the ratio of the number of males to 1 (or 100) females and **sex proportion** for the proportion of individuals who are males.

Though the departure from equality is not large, being approximately 106 : 100 in the white population of the United States, the amount of data on the sex ratio that is available for analysis is generally so large that even quite small differences are significant. There are two major mechanisms that can lead to variations in the sex ratio from the expected 1 : 1.

1. There may be deviations from 1 : 1 in the sex ratio of fertilized zygotes caused either by different rates of production of X- and Y-carrying sperm (meiotic drive) or by differences in the efficiency with which the two types of sperm effect fertilization (gametic selection).
2. There may be differential mortality of males and females after fertilization.

The first mechanism can be thought of as differential survival of gametes, as opposed to that of zygotes. There is abundant evidence for sex differential mortality at nearly all ages, but the evidence concerning departure from a 1 : 1 sex ratio at fertilization is not conclusive. The sex ratio at fertilization is called the **primary sex ratio**, that at birth the **secondary sex ratio**, and that among mature adults, the **tertiary sex ratio**. The time at which the tertiary sex ratio is computed is, of course, arbitrary and must be defined. Since the sex ratio varies with age, due to differential mortality, the total sex ratio of a population is, among many other factors, a function of the age distribution of the population.

Since the mortality rate of males is slightly higher than that of females at most ages, the sex ratio decreases with increasing age.

Changes in overall mortality rates over the last 100–200 years have, of course, affected the change in ratio with age. Some data on male and female mortality rates of the French population from about 1770–1950 are shown in Figure 10.8 as a function of age. These show that at nearly all times and ages male mortality has been higher than female mortality. Overall mortality has decreased most for the younger ages (see Chapter 6), though at these ages the sex differential in mortality has not changed much. There has, however, been an increase in sex differential mortality at older ages (more than 35 years), accompanied also, of course, by a substantial overall mortality decline. Much of this differential may be due to the apparently greater susceptibility of the male to heart disease and cancer, which are two of the major causes of death of persons beyond the reproductive age. The general effect of the overall decline in mortality at younger ages, even without any appreciable change in the sex differential, is to slow down the rate of decrease in the sex ratio with age, at least up to the reproductive years. Thus in 1805 the proportions of males and females surviving to age 25 were approximately 0.746 and 0.762, respectively, and in 1950 the corresponding proportions were 0.932 and 0.946. Assuming, at both times, a sex ratio at birth of 103 : 100, the sex ratio at age 25 in 1805 is expected to be $103 \times 0.746 : 100 \times 0.762$, or 100.8 : 100. The expected sex ratio for the same age in 1950 is given by $103 \times 0.932 : 100 \times 0.946$, or 101.5 : 100,

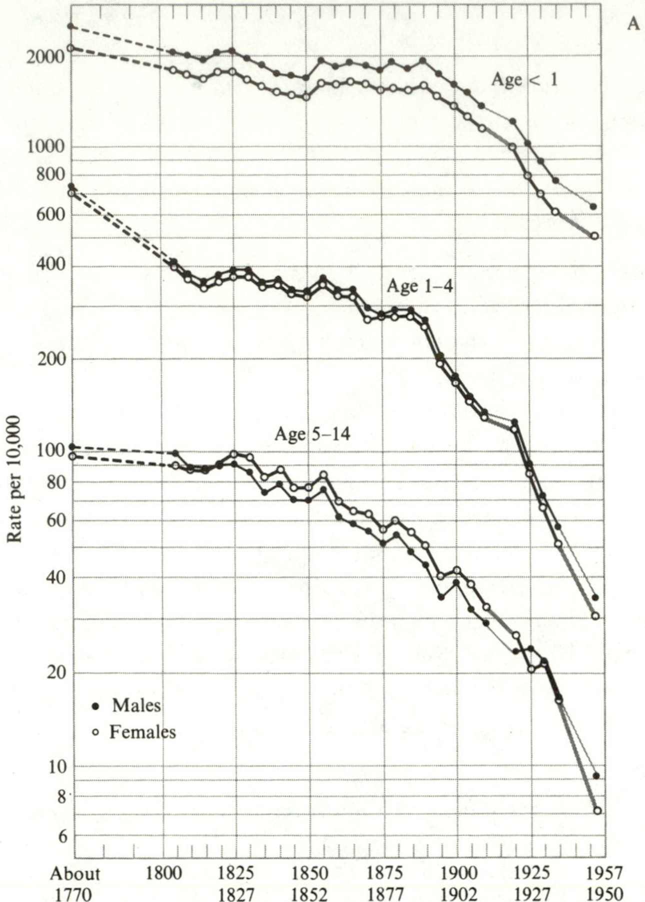

A
Age < 1
Age 1–4
Age 5–14
Rate per 10,000
2000
1000
800
600
400
200
100
80
60
40
20
10
8
6
• Males
○ Females
About 1770
1800
1825 1827
1850 1852
1875 1877
1900 1902
1925 1927
1957 1950

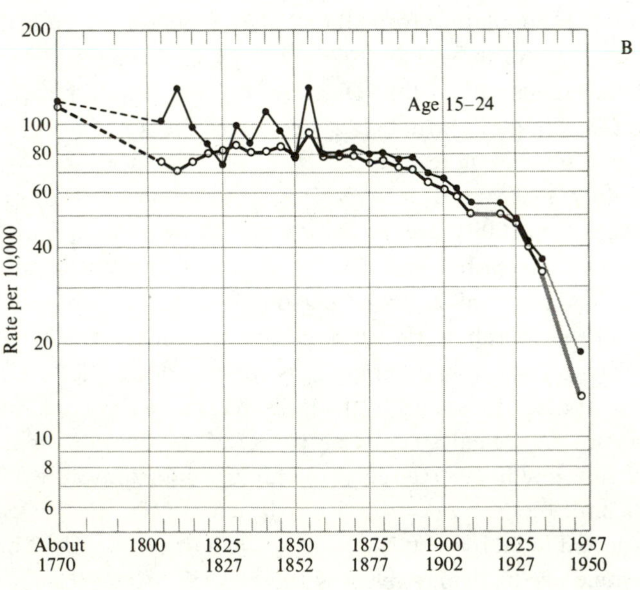

B
Age 15–24
Rate per 10,000
200
100
80
60
40
20
10
8
6
About 1770
1800
1825 1827
1850 1852
1875 1877
1900 1902
1925 1927
1957 1950

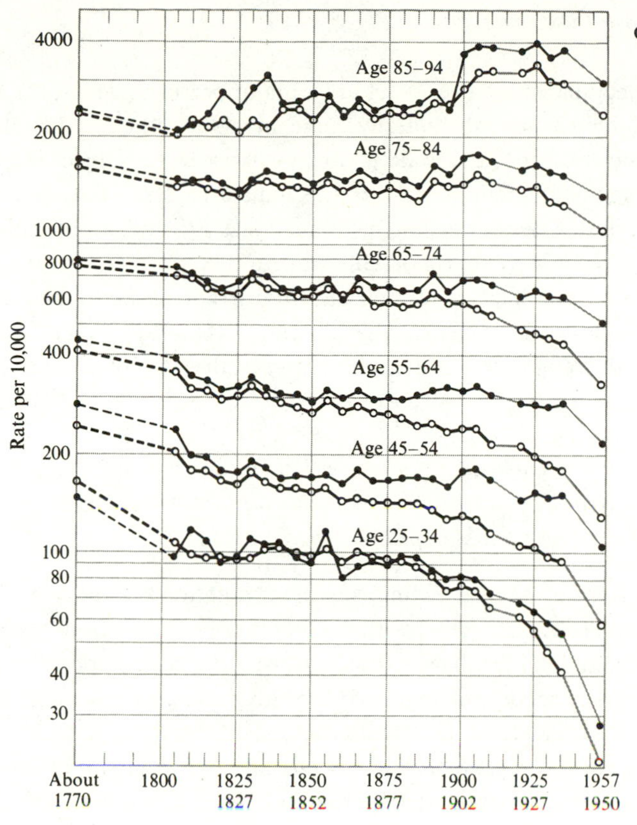

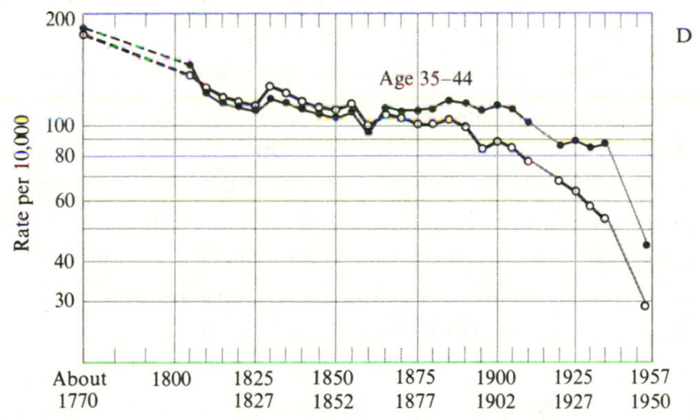

FIGURE 10.8
A: Mortality rates per 10,000 live births in the French population for age groups from <1–14 years. The rate for the less-than-one-year group is per 10,000 live births; the rates for the older groups are per 10,000 persons. Black circles plot data on males; white circles, those on females. B: Mortality rates per 10,000 persons for the age group 15–24 years. Black circles plot data on males; white circles, those on females. C: Mortality rates per 10,000 persons for age groups older than 24 years. Black circles plot data on males; white circles, those on females. D: Mortality rates per 10,000 persons for the age group 35-44 years. Black circles plot data on males; white circles, those on females. (From Bourgeois-Pichat, 1951.)

a small but detectable difference. Wars are a major perturbing factor in sex differential mortality, increasing substantially male mortality but not female mortality. In primitive societies, too, the male is always likely to be subject to more mortal dangers. The main differential risk to the female is death during or soon after childbirth, the latter being due, in most instances, to infection. This must be a major source of sex differential mortality in primitive peoples, and was, until quite recently, an important factor even in the industrialized developed countries. In general, in nonindustrial societies, the sex ratio seems to change from an excess of males at birth to an excess of females between the ages of 15 and 20.

It was thought for some time that the primary sex ratio was quite high, and was reduced to nearly a 1 : 1 *at birth by large mortality differentials in utero, but more recent data do not suggest a high primary sex ratio.*

Until the advent of nuclear sexing, sex determination of embryos aborted early was very unreliable. This, together with the fact that early abortions are undoubtedly not a random sample of fertilized zygotes, must have been responsible for some of the spurious earlier reports of high primary sex ratios, even as high as 160 : 100. The relatively high incidence of XO among early abortions is itself an indication of this bias. A careful review of data on morphological sexing of fetuses aborted early in pregnancy by Tietze (1948), however, gave a sex ratio of 107.9. Carr's (1965, 1967) study of chromosomes in early spontaneous abortions ($<$154 days after the last menstruation) gave 104 female to 85 male (excluding 11 X-0 specimens), a ratio that is not significantly different from 1 : 1.

Central sources of vital statistics on stillbirths are probably the only sources of data that are numerous enough to assure the significance of relatively small differences in sex ratio. Their main drawback is that they are available only for births after seven months of gestation, since before that time a stillbirth is generally classified as an abortion and doesn't require a birth certificate. One of the most extensive surveys of the sex ratio of abortions and stillbirths is that reported by McKeown and Lowe in 1951, based mainly on stillbirths notified in Birmingham, England for the years 1936–1949. The sex proportions (percentage of males) as a function of cause of stillbirths and duration of gestation are illustrated in Figure 10.9. The most striking departure from equality is among the fetal malformations, which show a marked deficiency of males. Fetal malformations also show a significant increase in sex proportion from the seventh to the ninth months, while the other causes of stillbirths show no significant variation in this respect. The proportion of stillbirths due to ill-defined or unknown causes remains at 25–30 percent from the seventh to the ninth months. The proportion of stillbirths due to the other causes, however, changes markedly. Thus, from the seventh to the ninth month, the proportion due to fetal malformations decreases from 20 percent to 12 percent, that due to diseases in

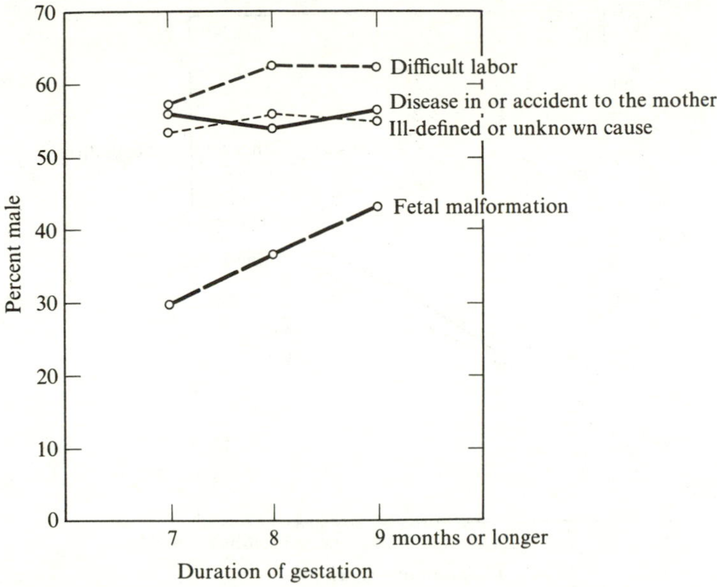

FIGURE 10.9
Sex ratio of stillbirths related to duration of gestation and abnormality. Birmingham, 1936–1949. (From McKeown and Lowe, 1951.)

or accidents to the mother decreases from 42 percent to 17 percent, and that due to difficult labor rises from 6 percent to 45 percent. The change in the overall sex proportion of stillbirths from the seventh to the ninth month reflects, therefore, both changes in the relative frequency of the causes and changes in the sex proportions within each cause category (including, of course, malformations). The major contributing factor is undoubtedly the change in the sex proportions of stillbirths due to fetal malformation. A breakdown of this sex-proportion variation according to three major categories of malformation is illustrated in Figure 10.10. Two major types of malformation, anencephalus and spina bifida, account for practically all the variation, showing a sex-proportion increase from 22 percent at the seventh month to 39 percent at the ninth month. An analysis of the sex-differential mortality of anencephalus and spina bifida from the seventh to the ninth month suggests that the sex proportion of those malformations present *in utero* at 7 months is about 26 percent and that there is a slightly greater loss of females than of males during the seventh to eighth month of gestation. There are two important lessons to be learned from this epidemiological analysis. The first of these regards the complexity of the factors that may interact to produce a given sex ratio; and the second, how these can be meaningfully analyzed to isolate the major relevant factors.

A survey by McKeown and Lowe (1951) of a number of studies on stillbirth sex proportions indicates a consistent proportion of about 57 percent at nine

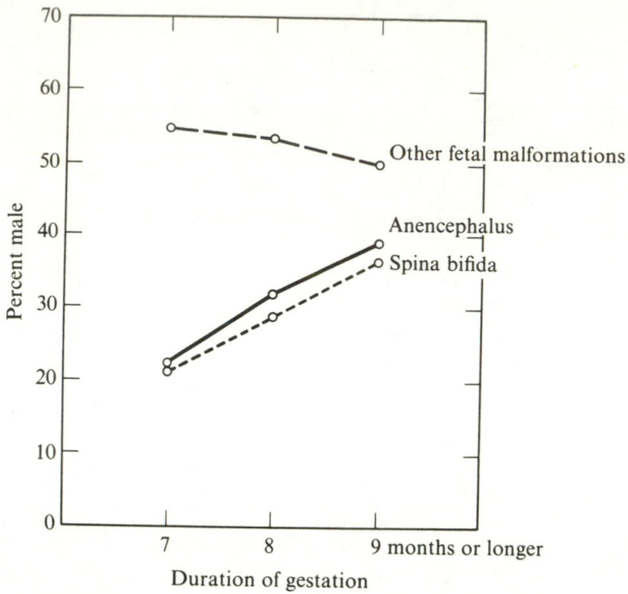

FIGURE 10.10
Sex ratio of stillbirths attributed to fetal malformations related to duration of gestation, Birmingham, 1936–1949. (From McKeown and Lowe, 1951.)

months (as compared to the live-birth sex proportion of about 51.5 percent) and an increase from the seventh to the ninth month. In their own, and also in some other reliable bodies of data, the sex proportion of stillbirths at the seventh month is about 50 percent, which means that the sex ratio of normal fetuses *in utero* at that stage cannot be very far from a 1 : 1. A simple model for the amount of sex differential mortality *in utero* needed to get from a given primary to a given secondary sex proportion can readily be constructed. Let the primary and secondary sex proportions be $y_1 : x_1$ and $y_2 : x_2$, respectively, where $y_1 + x_1 = y_2 + x_2 = 1$. Then, given that the proportions of males and females surviving to term are $1 - m$ and $1 - f$, respectively,

$$y_2 = \frac{(1 - m)y_1}{(1 - m)y_1 + (1 - f)x_1}. \tag{10.1}$$

If, for example, $y_2 = \frac{1}{2}$, then we have, from equation 10.1,

$$\frac{1 - m}{1 - f} = \frac{1 - y_1}{y_1}. \tag{10.2}$$

If we now assume a relatively high primary sex ratio, such as 130 : 100, $y_1 = 0.565$ and so $(1 - m)/(1 - f) = 0.77$. Thus, if, for example, $f = 0.1$, which implies a 10 percent loss of fertilized female zygotes during gestation, $m = 0.31$; if $f = 0.3$, which is perhaps a more realistic value, $m = 0.54$. In either case, the excess mortality of males over females has to be very large. There certainly does not seem to be any compelling evidence to substantiate such high primary sex ratios, and the corresponding extremes of sex-differential *in utero* mortality they would imply.

The secondary sex ratio decreases significantly with increasing parental age.

The decrease in the secondary sex ratio with increasing parental age, could have three primary components: (1) an effect of father's age, (2) an effect of mother's age, and (3) an effect of parity, or birth order. An examination of the effect of variation in each of these three variables, in turn, while keeping the other two constant, reveals their relative importance. Using multiple regression analysis, Novitski and Kimball (1958) showed that the only clearly significant factor was birth order. There was, however, also an unexplained second-order interaction between birth order and father's age. The relation between parental age, birth order, and sex proportion is illustrated in Figure 10.11. The cause of the decrease, which though significant is quite small, is unknown. It seems most likely to be related to changes in *in utero* sex-differential mortality with birth order, similar to those observed for anencephaly and spina bifida. As mentioned in Chapter 9 these congenital malformations are known to show striking variations in incidence with a number of socioeconomic parameters that might well interact with birth order. Once again, we see how difficult it is to interpret variations in the sex ratio.

Genetic control of the sex ratio at birth could occur either through control of the primary sex ratio or through subsequent sex-differential mortality.

A single gene causing males to produce almost only female offspring has been described in *Drosophila*. It is associated with a chromosomal abnormality that interferes with male meiosis and spermatogenesis in such a way as to produce only X-bearing functional sperm (Sturtevant and Dobzhansky, 1936). A striking case of abnormal segregation ratios for an autosomal gene in the mouse, the T-locus, has been extensively investigated by Dunn and colleagues (1956). Males heterozygous for certain mutant t-alleles at this locus, which are lethal when homozygous, transmit the mutant allele to their offspring much more frequently than the normal allele. The extent of this segregation distortion varies from one t allele to another. The only clear-cut case of segregation distortion, or meiotic drive, in man involves the D-21 translocation that is implicated in a certain proportion of cases of Down's syndrome. Here D refers to a human D-group chromosome and 21 to the chromosome that, when trisomic, causes Down's syndrome. Normal carrier males, with

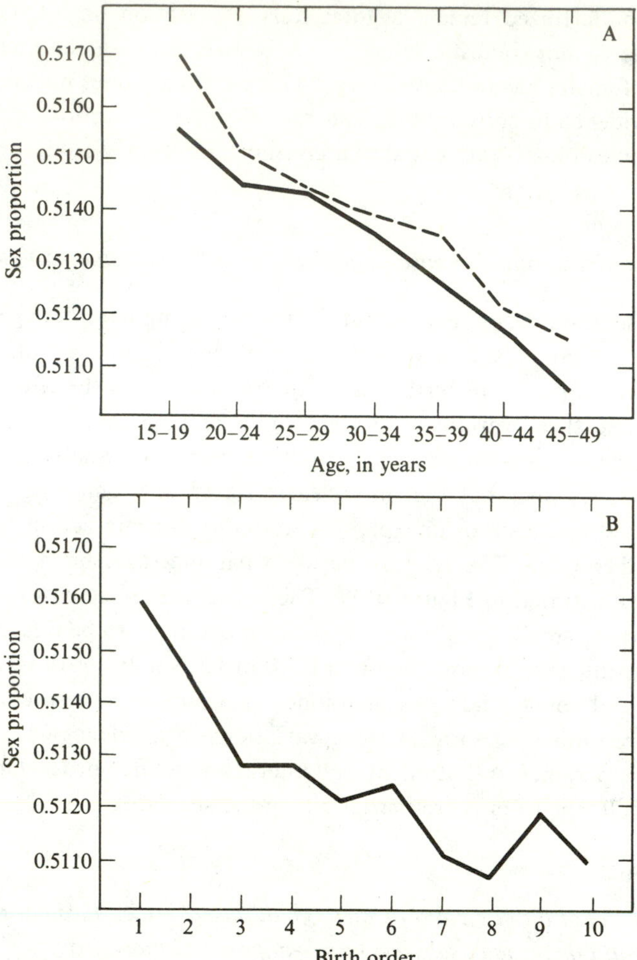

FIGURE 10.11
A: A plot of the sex proportion at birth of live offspring as a function of the age of the mother (solid line) and as a function of the age of the father (dashed line) without correcting for the correlation between the ages of the spouses. B: A plot of the sex proportion at birth of live offspring as a function of the birth order. (A and B From Novitski and Sandler, 1956.)

the constitution D-21, D, 21 rarely transmit the D-21 combination to their offspring, since the frequency of D-21, D, 21, 21 Down's syndrome among their offspring is is about $\frac{1}{20}$ of that expected, but when the female parent is the translocation carrier it is approximately $\frac{1}{3}$, as expected (zygotes with only one chromosome 21 presumably die; Penrose and Smith, 1966). These examples clearly show that it is possible to have genetic variation affecting a segregation ratio. There are, however, no unequivocal examples in man of genes that affect the sex ratio. There are one or two striking examples in the literature of unisexual pedigrees extending over several generations, which, if valid, do suggest the existence of sex-ratio genes like those in *Drosophila* (see Stern, 1960). A statistical study by Slater in 1943 suggests that sex ratios tend, to a very slight extent, to run in families but no further studies have been carried out to confirm this observation.

Some indication of a possible disturbance in the primary sex ratio, possibly due to meiotic drive resulting in overproduction of Y sperm, comes from the observation of an unusually high sex proportion in one population, the Koreans (somewhat above 0.53). There is the important additional information that Korean fathers, in interracial crosses in Hawaii (Morton et al., 1967), also have a high sex proportion among their progeny (0.5308), just significantly different (at the 5 percent level) from the rest of the population. Korean mothers show no difference, ruling out effects of sex-differential mortality *in utero*.

The most extensive studies on genetic variation of the human sex ratio are based on statistical analyses of the secondary sex proportion in pooled sibships.

Such analyses, of course, cannot distinguish variation in the primary sex ratio from that due to differences in sex-differential mortality *in utero*. If there is no variation in sex ratio between sibships, the distribution of sex ratios in families of given size should fit precisely the binomial distribution. The expected proportion of families of size n with r males and $n - r$ females is then

$$\frac{n!}{r!(n-r)!}p^r(1-p)^{n-r},$$

where p is the observed overall proportion of males (see Appendix I).

Genetic variation *between* sibships leads to a distribution whose variance is larger than that of the binomial. It can be shown that the expected variance in the number of males in families of size n is $np(1-p) + n(n-1)\sigma_p^2$, where p is now the mean proportion of males and σ_p^2 the variance in this proportion between families (see Example 10.1). Presumptive evidence for nonzero σ_p^2, and so for genetic variation in the sex ratio, can thus be obtained by comparing the observed variance in sex ratio with that expected assuming a binomial distribution.

Variation in the sex ratio *within* families, for example as a function of birth order, leads to an expected variance that is less than the binomial variance. It can be shown (see Example 10.1, b) that the expected variance in the number of males for families of size n is $np(1-p) - n\tilde{\sigma}_p^2$, where now $\tilde{\sigma}_p^2$ is the variance between p values *within* families. Thus, variation *within* families reduces the expected variance, while variation *between* families increases it. As A. W. F. Edwards (1960a—see also 1962) has pointed out, these two effects may well cancel each other out, so that a comparison of the observed variance with the binomial will be uninformative unless there is a significant deviation, in one or the other direction. Since there is known to be variation in sex ratio with parity, it could be argued that a strictly binomial variance is evidence for variation between families in an amount just large enough to cancel the variation within them. However, such negative evidence is clearly not very convincing.

Most statistical analyses of sex-ratio data are essentially in agreement with a binomial variance.

One of the most frequently analyzed bodies of data on the sex ratio was collected by Geissler (1889) who, as a registrar in Saxony, collected information on more than four million births over the period 1876–1885. A plot of the observed frequencies of males and females in families of twelve children compared with the binomial expectation is shown in Figure 10.12. The expectations were obtained by calculating the terms of the binomial for $n = 12$ and $p = 0.5147676$, which is the overall observed sex proportion in Geissler's data. The fit seems to be very good, though the small deviations from expectation are, in fact, significant.

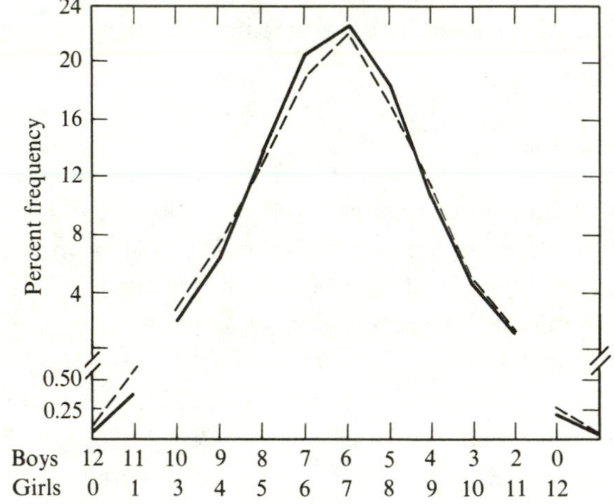

FIGURE 10.12
Percent frequencies of various sex ratios in sibships of twelve. The solid line plots the expected frequencies; the dashed line, the observed frequencies. Note that two different scales were used for the percentage of frequencies. (Data from Geissler, 1889; figure from Stern, 1960.)

A careful reanalysis of Geissler's data by A. W. F. Edwards (1958) suggested an approximate variance of the sex proportion between families of 0.0025. There still, however, remains an unexplained negative correlation between the sexes of successive births. Analyses of other bodies of data have given conflicting results. One based on the families of 5477 Swedish Protestant ministers, fits the binomial distribution precisely with no evidence for any variation in sex ratio between families (Ewards and Fraccaro, 1958). Two other relatively large bodies of data have shown significant, though unexplained, positive correlations between the sexes of successive births, in contrast to the negative correlations found in Geissler's data (see Edwards, 1962).

Family planning has no effect on the sex ratio.

As long as the sex of each birth is independently determined, no degree of birth control, even if directly related to the desire for male or female offspring, can change the overall population sex ratio.

Family planning in relation to sex can, however, affect the distribution of the sequences of births, and thus the apparent correlation between successive births.

Suppose, for example, that all couples continued to produce children until they had just one male offspring, and then stopped. The only possible sequences of the sex of offspring, writing M for male, F for female, would be M, FM, FFM, ... $(F)_n$M with probabilities p, $(1-p)p$, $(1-p)^2p, \ldots (1-p)^n p$, respectively, where p is the probability of a male birth, This distribution is called the geometric distribution (see Appendix I). Note that

$$\sum_{n=0}^{\infty} p(1-p)^n = p\sum_{n=0}^{\infty}(1-p)^n = p\frac{1}{1-(1-p)} = 1.$$

The mean family size is given by

$$\sum_{n=0}^{\infty}(n+1)(1-p)^n p = p\frac{1}{p^2} = \frac{1}{p},$$

since

$$\sum_{n=0}^{\infty}(n+1)(1-p)^n = -\frac{d}{dp}\left(\sum_{n=0}^{\infty}(1-p)^{n+1}\right) = -\frac{d}{dp}\left(\frac{1}{p}\right) = \frac{1}{p^2}.$$

As there is just one male in each family, the mean sex proportion being the mean number of males divided by the mean family size, is $1 \div 1/p = p$, the probability of a male birth, as expected. This family pattern, however, clearly imposes a strong correlation on the sexes of successive births.

Analysis of birth-sequence data suggests the existence of significant birth-control effects. Data on the sex sequence of the first two children, separated according to whether there were only two or more than two children in the family, are shown

TABLE 10.3
Frequency with which Procreation is Continued after the Birth of Two Children, by Sequences According to Sex

Sequence of First Two Children	*Complete Families*	*Incomplete Families*	*Total*	*Proportion Continuing Procreation*
MM	4,400	9,152	13,552	0.6753
MF	4,270	8,327	12,597	0.6610
FM	4,633	8,249	12,882	0.6404
FF	4,218	8,619	12,837	0.6714
Total	17,521	34,347	51,868	0.6622

For testing difference between complete and incomplete families
$\chi_3^2 = 42.86^{***} \ll 0.01\,\%$

Source: From A. W. F. Edwards (1966).

in Table 10.3. The two patterns differ very significantly, though only by small amounts. Parents whose first two children are FM are least likely to continue having children, while parents whose first two children are both male or both female are most likely to continue. This pattern seems to persist with larger families. When the sequence of the latest pair is FM, then the family is least likely to continue procreation (A. W. F. Edwards, 1966).

In spite of all the complicating factors influencing sex-ratio data analysis, Edwards is of the opinion that a positive correlation does exist, at least in some cases, between the sexes of the successive children. No clear-cut evidence, however, exists for between family variation in the sex ratio. For some reason it seems that the sexes of successive children born alive are slightly more likely to be the same than different. The opposite possibility was predicted by Renkonen and others (1962), who suggested that sperm carrying the Y chromosome, may somehow immunize the mother against Y-determined antigens and so make her more likely to reject subsequent fertilization by a Y-bearing sperm. There is, however, no experimental evidence for this phenomenon (see McLaren, 1962). The sperm's phenotype seems to be almost exclusively a function of the male donor's diploid genotype. This presumably explains, at least in part, why segregation-ratio distortion is probably a rare phenomenon.

Disturbances in the secondary sex ratio related to mother-child ABO blood-group combinations have been reported (Cohen and Glass 1956, 1959; Allan 1959). These ABO effects are, however, most likely to be a function of interaction between ABO types and *in utero* survival (see Chapter 5). Attempts to demonstrate segregation of A and B antigens on sperm from an AB individual have not given clear-cut results. It seems likely that A and B substances may be adsorbed onto sperm from the seminal fluid in an analogous way to adsorption of Lewis substances onto red cells from the serum (see Race and Sanger, 1968).

It has been claimed, though not confirmed, that X- and Y-bearing sperm from rabbits and cattle, can be partially separated by electrophoresis (Gordon, 1957). Possibly a difference in density resulting from the smaller size of the Y chromosome could provide the basis for physical separation of X- and Y-bearing sperm. Sooner or later this purely technical problem will be solved permitting, through artificial insemination, the determination, at will, of the sex of a desired child (see also Chapter 12).

Recessive X-linked lethal genes should affect the secondary sex ratio, since they eliminate males while having little or no effect on the viability of heterozygous females.

The net effect of recessive X-linked lethal genes, therefore, should be to decrease the secondary sex ratio. However, since recessive lethals are exposed to selection in males as if they were dominant, they are maintained in the population in much lower proportions than are recessive lethal genes on the autosomes (see Chapters 3 and 4). The effect of recessive X-linked lethals on the sex ratio is, therefore, likely to be very small.

The frequency of females heterozygous for a recessive X-linked lethal, H_1, is equal to half the frequency of the heterozygous females in the previous generation, H_0, plus the genes formed by new mutations in the gametes at the rate μ_m for males and μ_f for females. Thus

$$H_1 = \frac{H_0}{2} + \mu_m + \mu_f. \tag{10.3}$$

At equilibrium, $H_1 = H_0 = H$, and, so, from the above,

$$H = 2(\mu_m + \mu_f). \tag{10.4}$$

Assuming equal mutation rates in the two sexes, females carrying lethals should, therefore, have an equilibrium frequency that is four times the mutation rate. This must be the *total* mutation rate for all genes that can mutate to lethals on the X chromosome. Homozygous females occur with negligible frequencies. Males carrying the lethals, M, occur with half the frequency of heterozygous females, plus the new mutant genes formed in female gametes. Thus

$$M = \frac{H}{2} + \mu_f. \tag{10.5}$$

They will therefore, at equilibrium, be three times as frequent as the mutation rate μ, assuming $\mu_m = \mu_f = \mu$. These results are based on the assumptions of complete penetrance and complete recessiveness (in females) for the lethals. Otherwise, equilibrium frequencies would be higher.

It has been suggested that the excess of male over female deaths (for example, among stillbirths) might be due to sex-linked lethal mutations. Little is known about the causes of death among stillbirths, mainly because the tissues of the dead fetus undergo extensive maceration before they become available for analysis. However, enough is known, for example, about purely maternal causes, such as the effect of mother's age, and about socioeconomic effects, to make it unlikely that much of the stillbirth mortality can be due to purely genetic causes. The excess of male over female deaths among the stillbirths gives, therefore, only an upper limit to the effect of X-linked lethals on the sex ratio. The male proportion among stillbirths, for example in Italy (1930–1952) was 0.5513, compared with 0.5135 among live births. Stillbirths constituted 3.105 percent of all births (see Cavalli-Sforza, 1962a and b). From Equation 10.5, assuming the sex-ratio difference between live births and stillbirths is all due to X-linked lethals, the mutation rate should be $\frac{1}{3}$ of $0.03105 \times (0.5513 - 0.5135)$—this being the number of males eliminated by sex-differential mortality—that is, 0.00039. The X chromosome is about 5 percent of the whole genome. Thus, assuming that mutability is the same in all chromosomes, the total genomic mutation rate to lethals per generation would have an upper bound of 0.0078. Allowing for the possibility of incomplete penetrance would increase this estimate.

The contribution from recessive lethals to the total variance of the primary sex ratio is bound to be small. Offspring from mothers heterozygous for a lethal should have a sex proportion of $\frac{1}{3}$ instead of $\frac{1}{2}$. Their frequency is four times the mutation rate, and thus the variance due to the presence of families segregating for recessive lethals can be computed from

$$(1 - 4\mu)(\tfrac{1}{2})^2 + 4\mu(\tfrac{1}{3})^2 - [(1 - 4\mu)\tfrac{1}{2} + 4\mu(\tfrac{1}{3})]^2 \sim \frac{\mu}{9},$$

ignoring terms in μ^2. With a mutation rate as computed above, the contribution to the variance is $0.00039/9 \sim 0.00004$. The observed variance given by Edwards was 0.0025, and thus sex-linked lethals can be responsible for only a very small fraction of the total variance.

Sex-ratio variation with the age of the mother, and with those of the maternal grandparents, may provide an estimate of the accumulation of mutant genes with age.

Changes in the frequency of production of X-linked lethals with age, expected, for example, if mutation rates increase with age, might be expected to result in detectable changes in the secondary sex ratio as a function of age. In view of the many factors already discussed that can lead to a correlation between parental age and sex ratio, Cavalli-Sforza (1962a and b) initiated a study on the relation between a grandfather's age at the birth of his daughters and the sex ratio of the offspring produced by his daughters. The daughters carry any X-linked recessives produced by their fathers in the heterozygous state (and so themselves are unaffected), so that, on the average, the mutant alleles will be transmitted to 50 percent of their sons. If the mutation rate increases with parental age, then the secondary sex ratio should decrease with increasing age of the maternal grandfather. This approach to searching for age effects on X-linked lethal mutation rates has, unfortunately, produced inconsistent results. An initial analysis on a 1/200 random sample of the Italian population collected by a special questionnaire gave no significant correlation between the sex ratio of live births and maternal grandfathers' age at birth of the daughter. It did, however, give a significant correlation, in the expected direction, between the sex ratio of stillbirths and grandparental age (increase in sex ratio with increasing grandparental age) (Cavalli-Sforza, 1962a). An analysis by Krehbiel (1966), based on Minnesota fetal-death certificates, also gave a significant effect of grandparental age on stillbirths. The two estimates of increased mutation rate with age were in reasonable agreement. However, a larger sample than either of these obtained from Italian birth certificates for the year 1960 gave no significant correlations and was actually in significant disagreement with the earlier investigations (Cann and Cavalli-Sforza, 1962a). This last result is in line with the negative finding of age effects on mutation rate for two sex-linked recessives in man, which we have already mentioned (see Chapter 3). It seems possible that the geographic stratification of the sex ratio among stillbirths observed in Italy, which is probably due to socioeconomic stratification, in conjunction with socioeconomic stratification of age at marriage, may be responsible for the discrepancies. As no way to correct for these effects has yet been devised, the problem still remains, at the present time, an open one.

Ionizing radiations may influence the sex ratio, probably by inducing sex-linked recessive lethals.

A problem of considerable practical importance is genetic damage due to ionizing radiations. The average exposure of the human population is sufficiently small to cause no great concern, but some segments of the population are subject to heavier-than-average exposure (see also Chapter 12).

The expectation is that the sex ratio of the offspring of irradiated mothers would be lower than that among the offspring of mothers who had not been irradiated, because of the induction of sex-linked recessive lethals, which would cause some of the male fetuses to be inviable. The sex ratio among children of irradiated fathers might be altered in either direction. The possibility of radiation-induced sex-chromosome aberrations may also complicate the picture. An extensive study was carried out by Neel and Schull (1956) on the survivors of the Hiroshima and Nagasaki atomic bomb explosions. They failed to find statistically significant effects on the sex ratio (or on the other genetic indicators employed) of the progeny of exposed individuals.

10.7 Natural Selection and the Sex Ratio

The action of natural selection on the sex ratio cannot be treated in the same way as its action on other genetic differences, since the selective consequences of the production by an individual of offspring in a given sex ratio must take into account the prevailing sex ratio in the population, and since all matings must involve a male and a female, whatever the relative viabilities of the two sexes might be.

The selective value attached to production of offspring with a given sex ratio must be a function of the reproductivity of the offspring. Unless there exists some genetic variation in the sex ratio, however, natural selection can, of course, have no effect on the sex ratio.

The advantages of sexual reproduction are, in general, advantages conferred on whole populations, and not on individuals. Presumably, those populations that have evolved a sexual dimorphism are the ones most likely to survive and succeed. The genetic changes within the population may to a large extent have evolved independently of the selection among populations. Restricting our attention, for the time being, to selection for a given sex ratio, there are therefore two levels at which the problem of the action of natural selection on the sex ratio can be discussed. First, what is the explanation for the attained sex ratio *within* any given population, and second, what types of mechanisms might be responsible for populations with a given sex ratio being more successful than those with a different sex ratio. It seems likely that with respect to the sex ratio, selection within a population is more important and we shall discuss this first.

Fisher in 1930 laid the basis for the analysis of the action of natural selection on the sex ratio within populations. Formal mathematical developments of Fisher's approach were later given by Bodmer and Edwards (1960) and Kolman (1960). The fact that males and females contribute equally to the next generation, whatever the prevailing sex ratio, must be an essential feature of any such theory. Fisher's approach depends on taking into account the "parental expenditure" of effort involved in rearing offspring to maturity.

This expenditure is a function of the time and energy that parents spend in rearing their offspring to maturity, and so depends at least in part on the mortality of offspring during the period when the expenditure is made. There may be a differential expenditure of effort on different genotypes. Thus, if, for example, one genotype consistently tends to die early, the expenditure involved in rearing an offspring of this genotype to maturity would be greater than average, though the average expenditure per child conceived would be less than average. A mating tending to produce an excess of such genotypes would, on the average, produce fewer mature offspring. The fitness of a mating, which is measured by the number of offspring surviving to reproduce, is thus inversely proportional to the expenditure required to rear offspring to maturity. Given that there may be a differential expenditure with respect to the sexes, the fitness attached to production of a given sex ratio is a function of both the relative fitnesses of individual male and female offspring and, the relative parental expenditures in rearing them. We shall follow here the approach of Bodmer and Edwards (1960).

Consider first the selective advantage attached to reproduction with a given primary sex proportion. Suppose that the mean primary sex proportion in a population is X, and in a given part of the population is x. Let the proportion of male offspring surviving to reach sexual maturity (that is, surviving throughout the period of parental expenditure) be M in the whole population and m in that part of the population producing offspring in the sex proportion x, and let the corresponding proportions of females be F and f.

Now consider the average fitness of an individual at the moment when parental expenditure on his or her behalf has just ceased. Because the total number of genes contributed to the next generation by all males equals that contributed by all females, the average fitness of a male is inversely proportional to the number of males at that stage in the whole population, or proportional to $1/XM$. Similarly, the average fitness of a female is proportional to $1/(1-X)F$. Thus the average fitness of the offspring of an individual whose whole progeny has the sex ratio x is given by the mean of these values weighted by the proportions of the two sexes living at the end of the period of parental expenditure, and so is proportional to

$$\frac{\dfrac{xm}{XM} + \dfrac{(1-x)f}{(1-X)F}}{xm + (1-x)f}.$$

If we now write $xm/[xm + (1-x)f] = x'$, the sex proportion at the end of the period of parental expenditure, or effectively the tertiary sex proportion, and $XM/[XM + (1-X)F] = X'$, then the above fitness is proportional to

$$x'(1 - X') + (1 - x')X'.$$

We now find the expression for the average parental expenditure required to raise one child to the end of the period of parental expenditure in the 'x' part of the population. Suppose, for simplicity, that the expected total expenditure on a child is proportional to the probability of its surviving to the end of the period. This assumes that children dying before the end of the period incur a negligible expenditure compared with those who survive. This is likely at least, in man, since most deaths occur in pregnancy, at a time v..en the parental expenditure is small. Let the expected expenditure on a male child surviving to maturity relative to that on a female child surviving to maturity be h to $1 - h$. Then the expected relative expenditure on a male child is mh and on a female $f(1 - h)$. The average expenditure per child conceived is therefore proportional to $xmh + (1 - x)f(1 - h)$. Dividing this by the probability of a child's living to the end of the period of parental expenditure, the average expenditure required to raise one child to the end of that period, as a function of the primary sex proportion x, is proportional to

$$\frac{xmh + (1 - x)f(1 - h)}{xm + (1 - x)f}.$$

Substituting for x' as before gives $x'h + (1 - x')(1 - h)$.

The fitness per unit of parental expenditure in the 'x' part of the population is, therefore, proportional to

$$R = \frac{x'(1 - X') + (1 - x')X'}{x'h + (1 - x')(1 - h)}. \tag{10.6}$$

This expression is a measure of the relative selective advantage attached to reproduction with particular sex and parental expenditure ratios. Note that when $x' = 1 - h$, $R = 1$ for all values of x'.

The rate of change of R with respect to the individual tertiary sex proportion x' is

$$\frac{dR}{dx'} = \frac{1 - h - X'}{(x'h + (1 - x')(1 - h))^2} \tag{10.7}$$

Thus, if X' is less than $1 - h$ large values of x' are at an advantage and X' moves towards $1 - h$ in the next generation; if X' is greater than $1 - h$ small values are at an advantage, and X' again moves towards $1 - h$.

The population is therefore at a stable equilibrium when the mean sex ratio at the end of the period of parental expenditure, X', is $1 - h$, and all sex ratios are then equally advantageous, and we then have $X'h = (1 - X')(1 - h)$, which gives "Fisher's Law," that at equilibrium, the expenditures for a child of each sex are equal.

It is important to note that this equilibrium refers to the sex proportion X' at the end of the period of parental expenditure, and not to the primary sex proportion or the sex proportion at birth. The selective advantage R also depends only on the tertiary sex proportions x' and X', so that allowing for variation in x' takes into account both variation in the primary sex proportion and in the differential mortality during the period of parental dependence. For most species it is likely that the major part of any differences between the two sexes in parental expenditure depends on differential mortality, and not on any differential demands that the young make on their parents. The quantity h, which is a measure of the latter more-restricted part of the differential expenditure, may therefore be expected to remain constant and near to the value one-half. We have thus shown that, subject to the assumptions of our model, natural selection tends to maintain the sex proportion within a population at the end of the period of parental expenditure near the value one-half. This is in agreement with the observed numerical equality of the sexes in man in the age group 15–20 years.

Assuming complete genetic determination of the ability to produce with a given sex ratio, Bodmer and Edwards have shown that the population variance in the sex proportion V, changes very slowly and that the rate of change in the sex proportion depends directly on V.

More specifically, if the equilibrium tertiary sex proportion is $1 - h$: h, then it can be shown, as an approximation that (see A. W. F. Edwards, 1963)

$$[X' - (1 - h)] = [X - (1 - h)][1 - V(2h(1 - h)]^{-2}. \tag{10.8}$$

writing $Y = X - (1 - h)$, the departure of the population's tertiary sex proportion from its equilibrium value, and iterating Equation 10.8 gives

$$Y_n = Y_0\left[1 - \frac{V}{(2h(1 - h)^2}\right]^n \tag{10.9}$$

or, if V is small,

$$Y_n = Y_0 \exp\left[-\frac{V_n}{[2h(1 - h)]^2}\right] \tag{10.10}$$

as an approximation of the expected change in sex proportion after n generations. The rate of approach to equilibrium is $V/[2h(1 - h)]^2$ and has a minimum of $4V$ for given V, when $h = \frac{1}{2}$. In this case, from Equation 10.10, Y_n will decrease by a factor $e = 2.718$ in approximately $\frac{1}{4}V$ generations. For example, on these assumptions if $V = 0.0025$, as estimated by Edwards from Geissler's data, it would take 100 generations or about 2000 years to change the human tertiary sex proportion from 0.5200 to 0.5074. This is, moreover, likely to be an underestimate both because V may well be less than 0.0025 and because the model of completely determinate inheritance of the sex ratio maximizes the possibilities of genetic change. Changes in the sex ratio of human populations due to natural selection are undoubtedly much too slow to be detected in our lifetime. They are, in any case, masked by the effects of changing hygienic conditions, family sizes, sex-differential mortality, and stillbirth rates.

Interpopulation selection with respect to the sex ratio, if it exists, probably depends on mechanisms completely different from those we have discussed thus far, which apply only to intrapopulation changes.

Three mechanisms for interpopulation selection with respect to the sex ratio have been suggested.

1. The reproductive potential of most sexually reproducing populations is limited by the number of females they contain. Thus, for a given total population size, the intrinsic rate of population increase might be expected to increase with decreasing sex ratio.

2. If encounter between the sexes occurs at random, the maximum probability of encounter between members of opposite sex is for a sex proportion of one-half. Attempts have been made to construct quantitative models having fixed total resources for reproduction, whose utilization must be maximized, while assuming random collisions between gametes as the basis for reproductive unions. The development of ways to recognize the opposite sex for mating purposes would seem to be a more efficient way to mate, and the one that is generally found in nature. A more interesting problem is why male and female gametes differ so greatly in size. This aspect of sexual differentiation is related to the developmental needs of the fertilized zygote coupled with maximization of the probability of a sexual union achieving fertilization. Economic arguments are probably relevant here. They suggest that the greatest reproductive efficiency must have been achieved through specialization of reproductive functions leading to sexual differentiation (see Kalmus and Smith, 1960; and Scudo, 1967).

3. A further mechanism that has been suggested (Kalmus and Smith, 1960) to make a 1 : 1 sex ratio advantageous for a population is that this maximizes the effective population size for given total population size, and thus minimizes drift and inbreeding effects (see Chapter 7).

The fact that in polygamous species one male can provide adequate reproductive material for many females suggests that if interpopulation selection occurs, then mechanism (1) must be more important than (2) or (3) (see A. W. F. Edwards, 1960b). Since in such species, (which include man, at least under primitive conditions) the sex ratio at birth does not seem to differ widely from 1 : 1, it follows that the intrapopulation selective mechanism must predominate over interpopulation selection, forcing all populations to approximate equality of the sexes.

10.8 The Evolution of the Sexual Dimorphism

Sexual reproduction, as pointed out by Muller and Fisher in the 1930'*s, greatly accelerates the rate of evolution by facilitating the accumulation in a single individual of different, independently occurring, advantageous mutations.*

This feature of sexual dimophism is undoubtedly responsible for its evolution. The advantages of sexual reproduction relate to the population as a whole. Given an asexual and a sexual population competing on equal terms, the more rapid response of the latter to prevailing selection pressures made possible by its greater potential rate of evolution would ensure its replacing the asexual population. Fisher went so far as to suggest that sex was the only major attribute whose evolution depended almost entirely on inter-, rather than intrapopulation selection. In order to understand why evolution in sexual populations is accelerated, it must be contrasted with evolution in asexual populations. Two different advantageous mutations in an asexual population can only be incorporated into a single individual, and so into the population, if one of the mutations occurs in a descendent of an individual in which the other occurred. In a sexual population, on the other hand, the two mutations can be brought together in the same individual by recombination. The relation between the rate of formation of individuals carrying two advantageous mutant genes by double mutation in a single line and that by recombination between single mutant genes depends on the frequencies of the single mutant genes in the population. The rate of formation by recombination is, however, nearly always faster by a factor of at least two. When multiplied over many loci, the rate of formation of new combinations by recombination must be orders of magnitude greater than the rate achievable by sequential mutation. Muller (1958) has aptly characterized this difference as a contrast between evolution in series and evolution in parallel. A schematic diagram of this contrast, taken from a paper by Crow and Kimura (1965) and following an earlier suggestion by Muller (1932), is illustrated in Figure 10.13. In the asexual population, the accumulation of the mutant genes takes place in the sequence $A \rightarrow AB \rightarrow ABC$, each step involving essentially complete incorporation of the new combination into the population. Mutational events out of this sequence, such as C, B, or AC, are lost. In the sexual population, on the other hand, the single mutant genes A, B, and C are recombined stepwise into the final product ABC, without the need to wait for complete sequential replacement of one combination by another.

We shall now analyze a simple two-gene model that illustrates the different rates of evolution in sexual and asexual population. We consider a haploid population, since haploidy must have been the condition under which sexual reproduction evolved, diploidy being essentially a consequence of sex. Let a and b represent the prevailing alleles at two loci and A and B the corresponding new advantageous

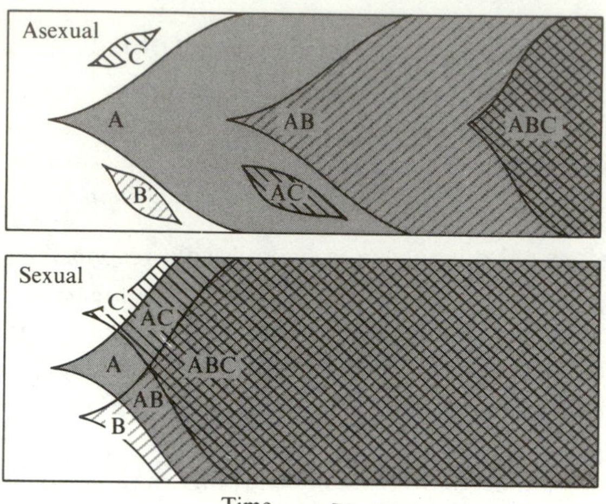

FIGURE 10.13
Evolution in sexual and asexual polulations. (From Crow and Kimura, 1965.)

alleles. For simplicity we assume complete symmetry with respect to A and B, so that genotypes Ab and aB both initially occur with equal frequencies x_0 and have selective advantages $1 + s$ relative to 1 for ab. We assume further that the mutation rates $a \to A$ and $b \to B$ are both μ. Consider now the situation in which the double mutant genotype has not yet arisen and genotypes Ab and aB both have frequency x_0. We shall calculate the approximate average number of generations needed to produce *one* individual of type AB, assuming in turn sexual and asexual reproduction and using an approach originally suggested by Muller (1964). Under both assumptions, since x_0 is presumably very small, and the frequency of ab nearly 1, almost all matings involving genotypes Ab and aB are $Ab \times ab$ and $aB \times ab$. The approximate frequencies of Ab and aB after n generations are therefore, $(1 + s)^n x_0$. If the population size is N and the recombination frequency between the loci r, then the average number of AB individuals produced by recombination in the ith generation, assuming random mating, is

$2x_0^2(1+s)^{2i}$	$\times \frac{1}{2}r$	$\times N$
(frequency of mating $Ab \times aB$)	(probability of AB offspring)	(population size).

The number of AB individuals produced in n generations is the sum of the above expression over all values of i from 0 to n. We require the value of n such that this sum is 1. It can be shown (see Example 10.2; and Bodmer, 1970) that the appropriate value of n is given by the equation

$$n = \frac{1}{2}\frac{\log\left(1 + \frac{s(2+s)}{Nrx_0^2}\right)}{\log(1+s)} - 1. \tag{10.11}$$

The average number of *AB* individuals produced by mutation in the *i*th generation is

$$2x_0(1+s)^i \times \mu \times N,$$

since mutation can occur either in *Ab* or *aB*. Summing, as before, over *i* gives the number of generations $\bar{n}$ required on the average to produce one *AB* individual by mutation. It can be shown (see Example 10.2) that $\bar{n}$ is given by

$$\bar{n} = \frac{\log\left(1 + \dfrac{s}{2x_0 \mu N}\right)}{\log(1+s)} - 1. \qquad (10.12)$$

This treatment, as pointed out by Muller (1964), is only approximate, and works only if $(1+s)^i$ is not too large. Crow and Kimura (1965) have given a more precise analysis. Our treatment also ignores the effects of random drift, in particular with respect to the probability of survival of new genotypes. From Equations 10.11 and 10.12 we see that

$$\bar{n} \gtrless 2n,$$

depending on whether

$$\frac{s}{2x_0 \mu N} \gtrless \frac{s(2+s)}{Nrx_0^2}$$

or, as an approximation,

$$x_0 \gtrless \frac{4\mu}{r}, \qquad (10.13)$$

assuming s is small. Thus, in particular, when linkage is loose and r is nearly $\frac{1}{2}$, and if $x_0 > 8\mu$, the rate of formation of *AB* individuals is at least twice as fast in a sexual population as in an asexual population. Both n and $\bar{n}$, as expected, decrease as s and N increase. The value of $\bar{n}$ also decreases as μ increases. The value of n, on the other hand, *increases* as r, the recombination fraction between the loci, decreases. Thus, as might be expected, tight linkage between the loci reduces markedly the rate of evolution by recombination. If there is initially just one mutation of each type, and consequently only one of each of the genotypes *Ab* and *aB* are present in the population, then $x_0 = 1/N$. In this case, from Equation 10.12, $\bar{n}$ is independent of N, and n, from Equation 10.11, is given by

$$\frac{1}{2}\left[\frac{\log\left(1 + N\dfrac{s(2+s)}{r}\right)}{\log(1+s)} - 1\right]$$

and increases with increasing population size N. Now $\bar{n} > 2n$ if, as an approximation,

$$N < \frac{1}{8\mu}. \tag{10.14}$$

Recombination thus has a more favorable effect the *smaller* the population size. This is because the frequency of the mating $Ab \times aB$ depends on the square of the genotype frequencies, which increase as N decreases. If, for example, $x_0 = 1/N$, $s = 0.01$ and $\mu = 10^{-6}$ then $\bar{n}$ is, approximately, 520 generations. When $N = 10^5$ then n is 365, and when N is 10^6, n is 480, and is still less than $\bar{n}$ though not by much. Actually μ here refers to mutation to a specific site within the gene, and so is more likely to be of the order of 10^{-7} to 10^{-8}. The population size, on the other hand, refers to that group within which mating can effectively occur at random, and this, even for bacteria, is not likely to be as large as 10^7 and is probably much smaller for most other organisms. Thus, in the majority of cases, $x_0 > 8\mu$, and so the rate of formation of individuals having the double mutant genotype in sexual populations is at least twice the corresponding rate in asexual populations. This result was predicted by Fisher in 1930, though characteristically without any accompanying analysis. The greater rate of evolution by recombination accumulates geometrically with increasing numbers of genes, resulting in enormous advantages for organisms having many genes.

Once the double mutant genotype (AB) has formed, provided there are no selective interactions between the loci, it evolves independently of the single mutant genotypes (Ab and aB) in a sexual population. In this case, the rate of increase of the double mutant genotype is the same in asexual and sexual populations. The advantage of sex is thus only for the initial rate of production of the type AB and not for its rate of increase in the population once it has arisen. Maynard-Smith (1968; see also Crow and Kimura, 1969) has used this fact to counter the ideas originally put forward by Fisher and Muller concerning the advantage of sexual reproduction. He suggests a situation in which Ab and aB were originally at a disadvantage with respect to ab and were maintained at low frequencies by a mutation-selection balance. Changes in the environment then so improved the fitnesses of Ab and aB that they came to be advantageous with respect to ab. In this case, if the original selective disadvantages of the genotypes Ab and aB were t and a mutation-selection-balance equilibrium had been reached, the frequency of the double mutant genotype before the change would be approximately μ^2/t^2 in both sexual and asexual populations, and there would be no apparent advantage of sexual reproduction. However, with $\mu = 10^{-7}$ to 10^{-8} and t as low as 0.01, μ^2/t^2 is, at most, 10^{-10}, making it very unlikely that the double mutant genotype even exists in any population, let alone has reached an equilibrium state. The analysis just given for the rate of formation of the double mutant genotype is, therefore, still quite relevant to this situation. Where mechanisms exist that can maintain *Ab* and *aB* at subpolymorphic frequencies, recombination is, of course, even more favorable as a mechanism for producing the double mutant genotype. The same also applies to hybridization between populations in one of which *Ab* is rare and *aB* common, and vice versa in the other.

Muller (1964) and Crow and Kimura (1965) in their analysis of sexual and asexual rates of evolution ignored the length of time needed for recombining two mutant genes into the same genotype in a sexual population. They therefore assumed that all the favorable mutant genes that are produced within the time needed for one new mutation to be incorporated into an asexual population (essentially Equation 10.12 with $x_0 = 1/N$) could successfully be incorporated into a sexual population. They then used this number as the relative advantage of sexual over asexual reproduction. It would seem from our analysis that their treatment is an oversimplification that must, in general, overestimate the advantage of sexual reproduction, especially for organisms with large numbers of genes.

THE INTERACTION BETWEEN SELECTION AND LINKAGE

In Chapter 5 we pointed out that pairs of genes that interact selectively may be held together on a chromosome if they are linked closely enough. Recombination disrupts associations between mutually advantageous genes while increasing the rate of formation of new advantageous combinations. This duality of the recombination mechanism was pointed out long ago by Fisher, Darlington, and others. On the basis of this duality, Muller emphasized that the advantage of sexual reproduction would be lost if nonadditive selection interaction between genes were common. As has already been emphasized, genes that are very closely linked behave essentially as if they were a section of an asexual organism, as Fisher pointed out in 1930 (see Equation 10.11, when $r \to 0$, $n \to \infty$). The advantages of both sexual and asexual systems can therefore, to some extent, be achieved according to the chromosomal arrangement of the genes of a sexual organism. The evolution of pairing between homologues, and subsequent recombination within the chromosome, must have been an essential part of the evolution of the sexual mechanism since pairing followed by recombination is required if full advantage is to be taken of the increased opportunities for recombining mutant genes of different origin.

The duality of the effects of recombination is nicely illustrated by the following model, taken from Bodmer and Parsons (1962). Suppose *Ab* and *aB* are disadvantageous with respect to *ab* and are each maintained at equilibrium frequencies μ/t by mutation-selection balance. Suppose further, however, that genes *A* and *B* interact in such a way that *AB* has a fitness $1 + \alpha$ in comparison with *ab*. The increase in the frequency of *AB* is counteracted by recombination, which leads to a breakdown of *AB* to the disadvantageous types *Ab* and *aB*. When *AB* is rare, almost all matings are of the type $AB \times ab$, and the opportunity is thus provided for loss of AB by recombination at a rate r. The genotype AB increases in frequency only, therefore, if its selective advantage, $1 + \alpha$, is great enough to counteract recombination breakdown or if, approximately,

$$(1 + \alpha)(1 - r) > 1 \qquad (10.15)$$

or

$$r < \frac{\alpha}{1 + \alpha}.$$

This result has been proved rigorously by Bodmer and Felsenstein (1967). It provides the simplest example of a model illustrating an interaction between linkage and selection. Given selective interaction such that AB is fitter than ab, but Ab and aB are not, AB only increases in frequency if the two genes are closely enough linked. In particular from Equation 10.15 the recombination fraction must, approximately, be less than the selective advantage of AB over ab. Now suppose that genotype AB has not yet been formed and consider the question: under what conditions will its rate of formation be greater by recombination than by mutation? The rate of formation of AB by mutation is

$$\frac{2\mu}{t} \times \mu = 2\frac{\mu^2}{t}.$$

The rate of formation by recombination is

$$\frac{2\mu^2}{t^2} \times \tfrac{1}{2}r = r\frac{\mu^2}{t^2}.$$

Thus the rate of formation by recombination is higher than that by mutation if

$$\frac{r}{t^2} > \frac{2}{t}$$

or

$$r > 2t. \qquad (10.16)$$

Conditions 10.15 and 10.16 limit r to the range

$$\alpha > r > 2t \qquad (10.17)$$

Only if expression 10.17 holds does sexual reproduction favor the production of the type AB and also allow it to increase in frequency.

A mixture of sexual and asexual reproduction, as is found in many plants but only in a few animal species, might seem to offer the best compromise. A new advantageous interacting genotype could then reproduce asexually at will without recombinational breakdown into less favorable combinations. The sexual mechanism would be held in reserve to produce new genetic changes that might be needed as the environment changed. It seems, however, from the few animal species that have dual systems, that the temporary advantages of asexuality are not enough to be worth the modification of an exclusively sexual system, especially if, as might be the case for man, this has evolved to an extent at which it infringes on individual selective values. Perhaps the balance between asexuality and sexuality would be too unstable. The former, due to its temporary advantage might oust the latter, only to destroy the potential of the species for future adaptation.

Diploidy, other than for the purpose of gamete formation, is not an inevitable consequence of sexual reproduction, though it is widespread among sexual organisms.

Three main reasons have been suggested for the advantage of diploidy over haploidy.

1. In Chapter 3 it was shown that in diploids, the equilibrium mutational load for genes conferring some deleterious effect on the heterozygote is 2μ in a random-mating population, where μ is the average mutation rate. In a haploid population this load is approximately μ, and so is, in general lower than in a diploid population. Thus diploidy does not protect against the overall deleterious effects of mutation because, though the heterozygous condition mitigates against the severer consequences of mutations, diploidy allows them to reach higher gene frequencies. However, if an organism is suddenly converted from haploidy to diploidy with random mating, much of its mutation load immediately disappears because there will usually be a normal homologous gene to counteract the effects of deleterious mutant genes, The load will, of course, eventually reach its new higher equilibrium value of approximately 2μ. Thus, though a change from haploid to diploid offers no long-term advantage in terms of the mutational load, it does offer an *immediate* advantage (see Muller, 1958). The change in the load is, however, very small even if it is halved.
2. Diploidy does, however, offer a consistent advantage with respect to mutation through its protection against the effects of *somatic* mutation. The risks of somatic mutation to the survival of complicated organ structures in higher organisms might be appreciable in a haploid, but almost negligible in a diploid (Muller, 1958).
3. Many phenomena, such as heterosis, that maintain genes at high frequencies in a population depend on diploidy. In general, diploids can maintain a larger genetic variance in a population and, hence, following Fisher's fundamental theorem of natural selection, should be able to evolve at a correspondingly greater rate.

Once sexual differentiation has been established, even if relatively ineffectively by a single gene difference, further-interacting genes that increase the efficiency of the differentiation will be selected for.

These, however, must be kept together as a block and so, as already mentioned, must be preserved from the effects of recombination. These requirements for sexual differentiation presumably underlie the evolution of the X-Y sex-determination mechanism. Just enough homology has been preserved between X and Y to ensure regular synapsis and the orderly distribution of X and Y to the germ cells following meiosis. Homology with respect to sex differential genes on the X and Y chromosomes would lead to recombination between these genes and a consequent breakdown of sexual differentiation. The relative inactivity of the Y is presumably the result of evolution toward a minimization of other differences between the sexes than those needed for the maximal efficiency of sexual reproduction. It seems likely that the evolution of the X-Y difference occurred relatively early in mammalian evolution. A number of authors have commented on the evidence for homology of the X chromosome between mammals, which, is, perhaps, an expected consequence of early evolution of the X-Y system. Thus, G6PD is known to be X-linked in man, horse, donkey, and European wild hare. Hemophilia is X-linked in man, dog, and probably also the horse (see Ohno, 1967).

General References

Bodmer, W. F., and A. W. F. Edwards, "Natural selection and the sex ratio." *Annals of Human Genetics* **24**: 239–244, 1960. (One of the first mathematical treatments of the subject.)

Crew, F. A. E., *Sex-Determination*. London: Methuen, 1933. (A useful short text.)

Crow, J. F., and M. Kimura, 1965, "Evolution in sexual and asexual populations." *American Naturalist* **99**: 439–450. (A discussion of the evolutionary advantages of sex.)

Edwards, A. W. F., "Genetics and the human sex ratio." *Advances in Genetics* **11**: 239–272, 1962. (A useful review.)

McKusick, V. A., *On the X Chromosome of Man*. Washington, D.C.: American Institute of Biological Sciences, 1962. (Contains much basic information on X linkage, X chromosome abnormalities and the problem of X inactivation.)

Ohno, S., *Sex Chromosomes and Sex-linked Genes*. Berlin: Springer-Verlag, 1967. (A stimulating survey of the problems surrounding the function and evolution of the X chromosome.)

Worked Examples

10.1. *Construction of a model for the effect of variation in the probability of a male birth on the variance in the sex ratio.*

We consider the following two statistical problems.

a. Suppose we draw n successive independent samples whose outcome is either success or failure, with the probability of success of the ith sample being p_i. What is the expectation and variance of the number of success? Sampling in this way (not to be confused with the Poisson distribution) is called Poissonian after Poisson, who first considered this problem.

The expectation of success at the ith sample is p_i and the variance, from the variance of the binomial distribution (see Appendix I), is $p_i(1-p_i)$. Since the samples are independent, the overall mean and variance are simply the sums of their separate components and so are given by

$$m = \sum_i p_i = np \tag{1}$$

and

$$v = \sum_i p_i(1-p_i) = \sum_i p_i - \sum p_i^2, \tag{2}$$

where p is the mean of the p_i. The variance of the p_i is, by definition,

$$\frac{1}{n}\sum_i (p_i - p)^2 = \sigma_p^2, \text{ where } \sum_i (p_i - p)^2 = \sum_i p_i^2 - 2p\sum_i p_i + np^2 = \sum_i p_i^2 - np^2.$$

Thus, substituting in Equation 2, the variance of the number of successes is given by

$$v = np - \sum_i (p_i - p)^2 - np^2$$

or

$$v = np(1-p) - n\sigma_p^2. \tag{3}$$

This is *less* than the binomial variance by an amount $n\sigma^2$, unless $p_i = p$ for all i when $\sigma_p^2 = 0$. If we translate samples into births, and successes into males, then

Equation 3 gives the variance in the expected number of male births in families of size n, when the probability of a male birth varies with parity.

b. Consider now the following sampling problem, called Lexian after the German economist Lexis. We take N samples consisting of n trials whose outcome is either success or failure, such that, at the ith sample out of the N, the probability of success for each of the n trials is p_i. What is the mean and variance of the number of successes per sample of size n?

The mean from the ith sample is np_i. from the binomial distribution. Summing over the N independent samples, the overall mean is

$$\sum_i np_i = nNp,$$

where p is the mean of the Np_i values. The mean per sample is, therefore,

$$m = np.$$

From the binomial distribution, the mean square deviation of the ith sample from np_i is simply the binomial variance $np_i(1 - p_i)$. However, since the overall mean is np, the contribution of the ith sample to the variance is

$$np_i(1 - p_i) + (np_i - np)^2.$$

The variance per sample is given by

$$v = \frac{1}{N}\left[\sum_i np_i(1 - p_i) + \sum_i (np_i - np)^2\right]. \tag{5}$$

From Equation 2 and the definition of σ_p^2,

$$\sum_i p_i(1 - p_i) = Np(1 - p) - N\sigma_p^2,$$

where

$$\sigma_p^2 = \frac{1}{N}\sum_i (p_i - p)^2,$$

is the variance of the p_i. Substituting into Equation 5 gives

$$V = np(1 - p) + n(n - 1)\sigma_p^2. \tag{6}$$

This variance is *greater* than the binomial variance unless all the p_i are equal. Once again, translating trials into births, successes into males, and now also samples into sibships, we see that the variance in the number of males per sibship of size n is greater than the binomial variance if the probability of male births varies from one sibship to another.

A third modification of simple binomial sampling obtains if trials are made with a fixed correlation between the results of successive trials. This would be the case, for example, if having a male offspring predisposed against the next offspring being a male and similarly for females. The variance would be increased relative to the binomial variance if the correlation between successive trials were positive and it would be decreased if the correlation were negative (see A. W. F. Edwards, 1960a).

10.2. *Rates of evolution in sexual and asexual populations.*

We assume the situation described in Section 10.8, namely two loci in a haploid organism with common alleles a and b and rare alleles A and B. The selective advantage of Ab and aB over ab is $1 + s$, their initial frequencies are x_0, the recombination fraction between the loci is r and N is the population size.

Then, as before, the average number of individuals having the genotype AB produced by recombination in the ith generation, assuming random mating is

$$2x_0^2(1 + s)^{2i}\tfrac{1}{2}rN. \tag{1}$$

We require the value of n such that the sum of expression 1 for i going from 1 to N is 1. This is the solution of the equation

$$Nrx_0^2\sum_{i=0}^{n}(1 + s)^{2i} = 1. \tag{2}$$

Since

$$\sum_{i=0}^{n}(1 + s)^{2i} = \frac{(1 + s)^{2n+1} - 1}{(1 + s)^2 - 1},$$

this being a geometric series, Equation 2 gives

$$(1 + s)^{2n+1} = 1 + \frac{s(2 + s)}{Nrx_0^2}. \tag{3}$$

Taking logarithms of both sides gives the following equation for n, the average number of generations required to produce one AB individual by recombination:

$$n = \frac{1}{2}\left[\frac{\log\left(1 + \dfrac{s(2 + s)}{Nrx_0^2}\right)}{\log(1 + s)} - 1\right], \tag{4}$$

which is Equation 10.11.

The average number of AB individuals produced by mutation in the ith generation is

$$2sx_0(1 + s)^i\mu N$$

since mutation can occur either in *Ab* or *aB*. Summing, as above, over *i*, the number of generations, $\bar{n}$, required on the average to produce one *AB* individual by mutation is given by

$$2x_0 \mu N \sum_{i=0}^{\bar{n}} (1 + s)^i = 1. \tag{5}$$

Since

$$\sum_{i=1}^{\bar{n}} (1 + s)^i = \frac{(1 + s)^{\bar{n}+1} - 1}{s},$$

Equation 5 gives

$$(1 + s)^{\bar{n}+1} = 1 + \frac{s}{2x_0 \mu N}. \tag{6}$$

Taking logarithms of both sides of Equation 6, we have

$$\bar{n} = \frac{\log\left(1 + \dfrac{s}{2x_0 \mu N}\right)}{\log(1 + s)} - 1, \tag{7}$$

which is Equation 10.12.

11

Human Evolution

11.1 Fossil Evidence on the Origin of Man

What is man? We have no difficulty in classifying a living being as human or nonhuman, because the gap between us and our nearest living relatives is large. A well-known paleontologist, G. G. Simpson (1949, 1966) lists twelve anatomical features by which he thinks a museum curator would always be able to recognize a specimen of *Homo sapiens*. None of them is, however, especially striking by itself. Simpson further lists nine behavioral traits that he believes are unique to man. Our slowly developing knowledge of anthropoid apes may well considerably reduce this list. Thus, recent observations on chimpanzees show that man is not the only tool-using primate (Goodall, 1965), even though the complexity of tools made by man is vastly more impressive than that of the tools used by chimpanzees (as far as we know). Similarly, the most important differentiating feature suggested by Simpson, language, is also not an absolute one. Recently Gardner and Gardner (1969) have described a chimpanzee who was able to learn a simple sign language and to construct sentences using this language. Our present incapacity to communicate with primates leaves some doubts as to the real extent of the discrepancies between us and our nonhuman relatives.

There is, however, a considerable evolutionary gap between men and apes which is, to some extent, confirmed by cytological and other observations. Man has 46 chromosomes; anthropoid apes have 48. The paleontological record suggests that the time available for the evolutionary divergence between man and apes is of the order of 10 million years, or more. There is no direct, fully satisfactory evidence that matings between man and apes are infertile, but *a priori* this would seem likely.

The differentiation between living organisms and their assignment to different species, even to different genera, are usually no problem. When we examine the paleontological record, however, we are confronted with an essentially continuous range of variation, which, inevitably, makes classification somewhat arbitrary. Thus, the recognition of *Homo*, and *Homo sapiens* in particular, among our progenitors in the fossil record is to some extent a matter of taste. Whatever discontinuity is observed may be mainly a result of the scarcity of available specimens. Discontinuity is magnified by the tendency of discoverers to claim uniqueness for their own finds. A great deal of systematic "splitting" has resulted. The creation of many new genera of fossil primates is probably unwarranted, whatever definition is adopted for a generic distinction, and has recently been the subject of much criticism. Attempts to clarify this taxonomic confusion and introduce synthetic points of view are constructive and welcome (Simons, 1963; Clark, 1964). Not being experts in the field, we shall give only a short summary of background information that might be useful for placing human evolution in genetic perspective, using known facts and theories of population genetics.

There is substantial general agreement that bipedal posture, tool making, and increase in brain size, accompanied by the development of higher intelligence, are the main features of human evolution. It seems most probable that the development of these features was interrelated. Size of the brain is one, very indirect, measure of intelligence which has the advantage that it can be measured in fossil specimens. Though it is far from perfect as an index of intelligence, it is, nevertheless, perhaps the most striking parameter of human evolution. Though many evolutionary developments have taken place in the formation of man, none is as important as intelligence for the development of a complex society, and with it, of a rich culture. Cultural evolution and biological evolution have inevitably interacted strongly, with results that seem, to us at least, quite remarkable. That man is unique in nature has been a creed for most men for a long time, and the suggestion, made about a century ago, that men are cousins of the apes, raised violent protest. Even now, it arouses in most people a strong negative emotional reaction. The uniqueness of man cannot, of course, be questioned, and is in some respects indeed striking. All the scientific observations that have been carried out on our nearest nonhuman cousins in recent times have, however, reduced rather than increased the gap between man and ape.

11.2 Primates, Hominoids, and Hominids

The primates can be subdivided as indicated in Table 11.1. The inclusion of extinct forms in a classification inevitably introduces complications. One of the main aims of a "natural" classification is to be consistent with the fossil record, or

TABLE 11.1
Major Subdivisions of Primates

Suborder	*Infraorder*	*Superfamily*	*Family*	*Some Important Genera*
Prosimii (lower primates, or prosimians)	Lemuriformes (lemurs) Lorisiformes (lorises) Tarsiiformes (tarsiers)			*Galago* *Lemur* *Tarsius* *Tupaia*
Anthropoidea (higher primates or simians)	Catarrhini (Old World higher primates)	Cercopithecoidea (Old World monkeys, or cercopithecoids)		*Macaca* *Papio* *Cercocebus* *Cercopythecus*
		Hominoidea (hominoids)	Pongidae (apes, pongines, dryopithecines)	*Pan* (chimpanzee) *Gorilla* *Pongo* (orangutan) *Dryopithecus* (extinct)
			Hominidae (hominids, or humanoids, men and allied forms)	*Homo* *Australopithecus* (extinct)
	Platyrrhini (New World higher primates)	Ceboidea (New World monkeys or ceboids)		*Cebus* *Ateles*

Source: From Simons (1960).

more accurately, with its most acceptable interpretation. The inevitable gaps in the fossil record create unavoidable uncertainties, and even *a priori* it might be argued that ancestors need not fall exactly into any of the categories created for the classification of living specimens. A classification, however, may be so tailored as to accommodate extinct forms. Fossil primates of the last seventy-five million years have thus been given a place in this classification. A summary of the information available for fossil primates up to the beginning of the Pleistocene (2-3 million years ago) is given in Table 11.2. The genus *Homo* itself is so defined as first to appear in the late Pleistocene. Some of the presumed "missing links" have been located in fossils dated in the early Pleistocene that have been given the generic name of *Australopithecus* (as well as a variety of alternative names). The data available at the present time are scanty enough that further discoveries may well change the picture.

It is sometimes stated (e.g., Simons, 1963) that the dryopithecines close the gap between man and the great living apes of Africa and Borneo (gorillas, chimpanzees, orangutans). The dryopithecines comprise several extinct genera *Dryopithecus*, *Sivapithecus*, *Proconsul*, and *Ramapithecus*, and include several other extinct hominoids of the Pliocene and Miocene that lived in a large area extending from Africa and Europe to India. If the dryopithecines were the progenitors of both apes and man, then the split between apes and man must have occurred after the period in which the dryopithecines lived, which was some time between 25 and 10 million years ago. Finds dated from 10 to 25 million years old in the evolutionary

TABLE 11.2
Partial or Complete Skulls of Tertiary Fossil Primates. Specimens identified, described or reassigned since 1950 are marked with an asterisk.

Epoch and Approx. Years Since Beginning	*Genus*	*Group*	*Continent*
Pliocene	*Dolichopithecus*	cercopithecoid	Europe
1.2×10^7	*Libypithecus*	cercopithecoid	Africa
	Mesopithecus	cercopithecoid	Europe, Africa
	*Oreopithecus**	hominoid	Europe
Miocene	*Cebupithecia**	ceboid	S. America
2.6×10^7	*Pliopithecus**		
	[*Epipliopithecus*]	hominoid	Europe
	Dryopithecus		
	[*Proconsul*]	hominoid	Africa
	*Progalago**	lorisiform	
		prosimian	Africa
	*Komba**	lorisiform	
		prosimian	Africa
	*Mioeuoticus**	lorisiform	
		prosimian	Africa
Oligocene	*Fayum frontal**		
4.0×10^7	[*?Apidium*]	catarrhine	Africa
	*Rooneyia**	prosimian	N. America
	*Aegyptopithecus**	hominoid	Africa

TABLE 11.2 (*continued*)

Eocene 6.0×10^7	*Adapis*	prosimian	Europe
	Anchomomys	prosimian	Europe
	*Cynodontomys**	prosimian	N. America
	*Hemiacodon**	prosimian	N. America
	Microchoerus	prosimian	Europe
	*Microsyops**	prosimian	N. America
	*Nannopithex** [*Pseudoloris*]		Europe
	Necrolemur	advanced prosimian	Europe
	Notharctus	prosimian	N. America
	*Phenacolemur**	prosimian	N. America
	Pronycticebus	prosimian	Europe
	*Protoadapis** [*Megatarsius*]	prosimian	Europe
	Pseudoloris	advanced prosimian	Europe
	*Smilodectes** [*Aphanolemur*]	prosimian	N. America
	Tetonius	adv. prosimian	N. America
Paleocene 7.5×10^7	*Plesiadapis**	prosimian	Europe
	*Palaechthon** (undescribed)	prosimian	N. America

Source: From Simons (1960).

branch called hominids which leads to *Homo*, include *Oreopithecus*, a genus dated to some 10 million years ago that is believed by most to be a side branch with respect to the main branch leading directly to man.

The australopithecines are the immediate antecedents of the genus Homo.

Miocene and Pliocene apes are usually believed to have been arboreal creatures. Some of them, for example *Proconsul*, were large enough that they must have spent most of their arboreal life, as present-day gorillas do, in the lower parts of the trees, where branches could more easily support their weight, leading otherwise a terrestrial life. It is not clear when the higher primates made the transition to a completely terrestrial life. Gradual deforestation, which left wooded areas separated by open land may have favored the transition. Clark is responsible for the interesting speculation that "the evolution of the ground living forms in the ancestry of the Hominidae was the result of adaptations primarily concerned not with the abandonment of arboreal life, but (paradoxically) with an attempt to retain it. For in regions undergoing gradual deforestation they would make it possible to cross intervening grasslands in order to pass from one restricted and shrinking wooded area to another."

Australopithecines comprise a number of different fossil finds, made over a large area even larger than that already mentioned for the dryopithecines: from China and Indonesia to Asia Minor, Central, East, and South Africa. They have been given, probably on insufficient grounds, a number of different generic names, *Australopithecus* being that of the first find (the Taung skull from Bechuanaland). Other generic names suggested by independent discoverers include: *Telanthropus*, *Plesianthropus*, *Paranthropus*, and *Zinjanthropus*. The most interesting finds are probably those made by the Leakeys (see, e.g., Leakey, 1967) in the Olduvai Gorge in Tanzania, which have been dated by the potassium-argon technique to some 1.75 million years ago. The inference that the creature whose remains were found in the Olduvai Gorge was indeed a tool maker is based on the finding of numerous primitive stone implements in the gorge and the observation that many baboon skulls associated with the hominid remains carry evidence of fractures, most probably caused by well-aimed blows. The name *Homo habilis* has been suggested for some of the Olduvai hominids, but there are considerable differences between the views of different authors (see, e.g., Simons, 1968) on the validity of the classification of these hominids.

Many anatomical features contribute to making *Australopithecus* nearly human. Perhaps the most important are those—skull, pelvis, and limb structures—that suggest a bipedal posture. The size of the brain of *Australopithecus*, however, was far from that of *Homo*, certainly less than 850 cc, its range extending from 450 to more than 600 cc. This, as we shall see, is less than half the size of the brain of *Homo*, and not much larger than that of the living apes. It should be remembered, however, that brain weight cannot be dissociated from body weight, the relationship being a complex one.

The genus Homo *began at least* 500,000 *years ago.*

Specimens that are now almost universally given a position in man's genus, but classified as a distinct species, *Homo erectus* (formerly called *Pithecanthropus* and *Sinanthropus*), first appeared approximately 500,000 years ago. Their distribution is widespread as is that of their forebears: from Java to China, to Africa (in a later layer of the Olduvai Gorge and also in Ternifine in Algeria), to Europe (Heidelberg). *Homo erectus* has almost double the brain size of *Australopithecus*, having a cranial capacity of about 1000 cc. Several other characteristics of the skull, the limb skeleton, and the dentition still differentiate it, however, from modern man.

The first specimens classified as *Homo sapiens* (according to Clark) date to the beginning of the second (Mindel-Riss) interglacial period, namely 200,000 years ago. Fossils from the last 50—100 thousand years classified as *Homo sapiens* show no marked difference from present-day man.

It is believed that Homo neanderthalensis *is an extinct species.*

During the last period of hominid evolution, another type appeared, called *Homo neanderthalensis*, a name derived from the valley of Neanderthal near Düsseldorf, West Germany, where the first find was located. *Homo neanderthalensis* differs from *Homo sapiens* mainly in the facial skeleton, especially the upper rim of the orbits, which is much larger than that of *Homo sapiens*, and in its protrusion (see Figure 11.1). Cranial capacity is large (1300–1600 cc), perhaps even larger than that of *Homo sapiens*. Several other differences are noticeable. In general, all the bones are more coarsely built, giving reconstructions of *Homo neanderthalensis* a superficially "simian" look.

Remains that can be assigned to *Homo neanderthalensis* have been found over a vast area. Apart from many finds in different parts of Europe (England, France, Spain, Italy, Germany, Croatia) and in Israel (Mount Carmel), there are also finds from Java and Rhodesia that proved to be rather similar to the European neanderthaloids. It should be added, however, that the differences between neanderthaloid remains and those assigned to *Homo sapiens* are not very consistent. Some finds show only some neanderthaloid features, other finds show other features. Many people consider the two species to be distinct, mostly owing to the fact that the more recent *Homo neanderthalensis* specimens, dating from 30,000–40,000 years ago, differ from *Homo sapiens* more than the earlier ones. Considerable speculation was raised by the finding in one of two caves near each other on Mount Carmel of remains assigned to *Homo sapiens* and in the other remains assigned to *Homo neanderthalensis*. The caves may have been inhabited during somewhat different periods. In another place, however, Krapina in Northern Croatia, remains from a single site show great variation in shape, ranging almost all the way from bone structures characteristic of *Homo sapiens* to ones characteristic of *Homo neanderthalensis*. Perhaps, the specific barrier was really not present and intercrosses were fertile. The problem of delineating the exact relationship between *Homo neanderthalensis* and *Homo sapiens* is not completely solved. An illuminating discussion can be found in Clark's (1964) book.

11.3 Paleolithic and Neolithic Man

The time during which the genus *Homo* evolved has been divided into three periods, Paleolithic, Mesolithic, Neolithic, which are distinguished on the basis of the sophistication of the stone implements (see Table 11.10) found with fossil

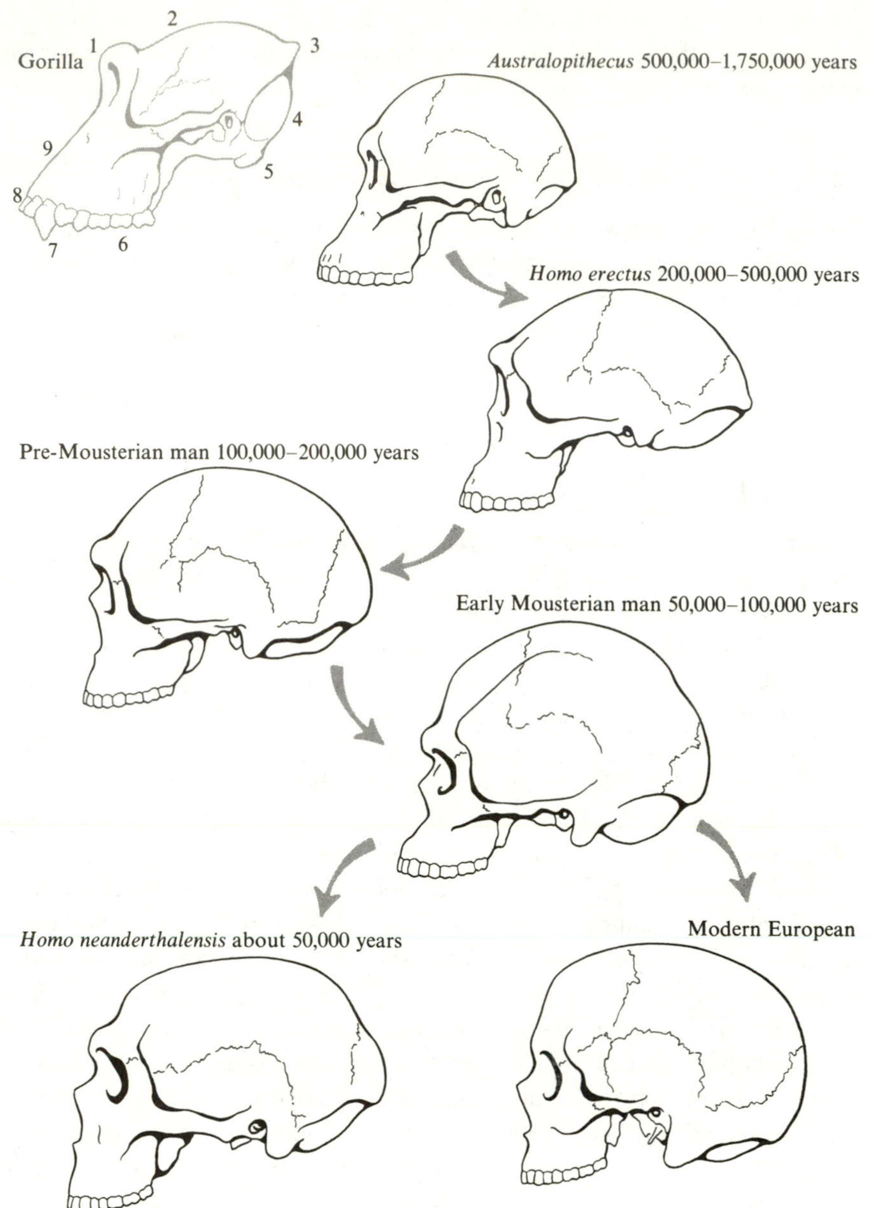

FIGURE 11.1
Appearance of the skull in a series of fossil hominid types arranged in their temporal sequence. The antiquity of each type has been estimated by methods of relative or absolute dating, but should be regarded as no more than approximate within very broad limits. It is also to be noted that some of these types over-lapped in time in different geographical regions. The representation of the genus *Australopithecus* is based on a skull found at Sterkfontein in South Africa, *Homo erectus* on a skull cap and portion of jaws found in Java, pre-Mousterian man on the Steinheim skull, early Mousterian man on one of the skulls found at Mount Carmel in

remains. These, in turn, precede the periods in which metal tools came into use. The end of the Paleolithic coincides approximately with the end of the fourth (and last) glaciation, some 20,000 years ago. It thus corresponds with the end of the Pleistocene, the last geological period before the present one. The origin of agriculture is associated with the beginning of the Neolithic period. Present data indicate that the first signs of human life centering on agriculture date back 10,000 years (in southwest Asia), but future discoveries may move this time back in some parts of the world. The transition cannot have been very abrupt, although agriculture may have spread quickly. It has not even today, however, spread over the whole world, as there are still some human communities whose acquisition of food is almost entirely based on hunting and gathering, as it was for Paleolithic man.

The changing shapes of stone tools during the Paleolithic show an increasing speed of cultural evolution. Communication between the different parts of the world was probably not totally absent, considering the similarity of the rates of biological and cultural evolution in different parts of the world. Some degree of geographic mobility for man, and even for his remote ancestors, may be the most likely explanation for the fact that evolution seems to have proceeded almost in parallel over a very wide area, namely most of the habitable parts of Europe, Asia, and Africa. It is not necessary to postulate a very high mobility to achieve this relative uniformity in evolutionary rates.

The increases in the size of the human skull and brain are among the fastest evolutionary processes known, but they were, nevertheless, very slow.

Figure 11.1 summarizes the evolution of the shape of the skull and the state of our paleoanthropological knowledge. We would like to be able to follow more closely this evolution, and especially that of the capacity of the skull, which is probably the preservable character most closely related to intelligence. It is known from data on living individuals, however, that even the correlation between cranial capacity and intelligence is not an extremely close or unvarying correlation. It is easy to list reasons why this should be so: (1) Cranial capacity is not the same as brain size; the latter is smaller. (2) There is a correlation between brain weight and body

---

Israel, and Neanderthal man on a skull found at Monte Circeo in Italy. Inset on the left is shown, for comparison and contrast, the skull of an adult female gorilla; the numerals indicate a few of the fundamental characters that, taken in combination, constitute a total morphological pattern distinguishing the anthropoid ape type of skull from the hominid type of skull. The characters indicated are (1) The forward projection of the brow ridges well beyond the front end of the braincase; (2) The low level of the cranial roof in relation to the upper border of the orbital aperture; (3) The high position of the external occipital protuberance; (4) The steep slope and great extent of the nuchal area of the occiput for the attachment of the neck muscles; (5) The relatively backward position of the occipital condyles; (6) Except in advanced stages of attrition the teeth are not worn down to a flat, even surface; (7) The canines form conical, projecting, and sharply pointed "tusks;" (8) The large size of the incisor teeth; (9) The massive upper jaw. All the skulls have been drawn to the same scale. (From Clark, 1964.)

weight. A larger body requires a larger number of nerve cells, or larger cells, or both, for vegetative life. (3) The folding of the cortex may provide an increased number of cortical cells without substantially altering brain weight. (4) The interconnections between cells, and the ability to form interconnections, as well as a variety of other poorly understood behavioral traits, can considerably influence brain performance quite independently of brain size.

Nevertheless, the fact remains that human intelligence is in practically all respects higher than that of other primates and that brain weight is also higher in absolute measures, and even more so if properly correlated with body weight. Thus the gorilla's body is much heavier than that of a human, but the size of the gorilla's brain is only half that of man's. It is approximately the same as that of *Australopithecus*, which was a much smaller creature than the gorilla, somewhat smaller than man himself.

If we divide the evolution of man into two periods, the first between the first dated *Australopithecus* and the first dated *Homo erectus* and the second between the first *Homo erectus* and the present, we find that in each period there was an increase of about 50 percent in skull capacity. The first period, however, lasted for more than a million years, and the second lasted only half a million years. The evolution of brain size in man turns out to be among the most rapid, if not the most rapid, of known evolutionary processes.

It is of interest to consider the evolutionary rate of the human skull in terms of selection differentials (see Chapter 9). Using a standard deviation of 100 grams for brain weight, the change in the second period (500 grams over 500,000 years, or 25,000 generations), gives a rate of change in units of standard deviations per generation of $500/(100 \times 25000) = 0.0002$. With a heritability of 50 percent, this is an average selection differential of 0.0004 per generation, which is very small indeed when compared with selection differentials achieved with artificial selection. The figure of 25,000 for the number of generations in the computation deserves some comment. For present-day man, the average generation time is slightly more than 30 years. In hunters and gatherers it is probably less, perhaps about 25 years. It may have been still less in earlier times, considering that the female chimpanzee, for instance, has a reproductive period beginning at approximately ten years of age and ending at 35 (the human female's reproductive period spans, on the average, ages 15–45 years). The generation time used in the computation is 20 years.

It is impossible to exclude the possibility that the evolution of the brain was not gradual and that periods of rapid increase were followed by periods with little or no change. The fossil record is too limited to exclude this interpretation. The suggestion that the change was gradual is based on the fact that if a graph is made, plotting the increasing brain size in the three available observations against the time at which the bearers of the brains lived, an approximately straight line can be drawn. Paleontology also provides other evidence, such as in the evolution of horses, of a gradual increase in some character (for example, overall size) over a very long period of time.

It is not inconceivable that some important change in body structure that took place at the beginning of "hominization" paved the way for the long period during which the human brain gradually evolved. This may have been erect posture, which permanently freed hands for specialization in toolmaking. It may also have been a change in the larynx, permitting vocalization and the development of a more complex language than any other primate has been able to acquire. Probably both of these factors, and also other, still unknown factors each contributed to give a small but roughly constant selective advantage to increasing brain size and complexity. As factors such as the two just mentioned developed, a great deal of later evolution was made possible, which proceeded at the slow pace at which genetic changes usually take place. Natural changes almost always occur more slowly than those due to artificial selection. Probably because of this, the final "plateau," namely, the end of response to selection, seems to be less important. The plateau probably changes during the process itself, because there is time for the appearance of new variation and the accompanying removal of barriers against the phenotypic increase under selection. Since such barriers seem to arise fairly quickly in most artificial-selection processes, a plateau is reached after only a few generations of relatively rapid increase.

There are various possible models for hominid evolution.

Systematic analysis of the kinetics of the changes in the evolution of man that are preserved in fossils is difficult. The genetic basis of the differences is not known. In any case genetic differences are likely to be mostly polygenic, and therefore less easily amenable to the more informative types of Mendelian analysis. Little more can be done than the simple calculations already shown for the rate of evolution of the brain. The genetics of bone variation, as judged from knowledge about mice (Grüneberg, 1963) is likely to be especially complex.

There is, however, an important lesson to be learned from the paleoanthropological data. It would seem that most of our ancestral lines (the dryopithecines, the australopithecines as well as *Homo erectus*, *Homo neanderthalensis*, and fossil specimens of *Homo sapiens*) included individuals living over a vast geographical area. They must, therefore, have been represented by a large number of individuals at any one time, say of the order of hundreds of thousands, or even millions. The fossil record has very large gaps, and the number of specimens on record is very small. The possibility that there have at certain times been significant bottlenecks in population sizes is not, therefore, excluded. What models of hominid evolution can be visualized, in terms of total population sizes and population movement, over the large area that we know was occupied by our ancestors?

It would seem that several models are compatible with the available fossil evidence. We shall examine some possibilities with a view to gauging the relative roles of selection and drift.

1. In accord with a continuous distribution model, man's ancestors may have occupied more-or-less uniformly the whole area which at one time was covered by forest. They may then have adapted to gradual deforestation, thus maintaining a fairly uniform density. This model assumes a rather low mobility to allow for the development of local differentiation and adaptation, but, nevertheless, the group evolves as a whole.

2. Another possible model assumes that a population develops in a relatively isolated part of the available area, undergoes a demographic explosion, and invades the rest of the area, mixing with the local inhabitants or supplanting them. This may have happened repeatedly, at different times and places.

3. According to a third model, the nomadic habits of hunters and their need to follow game may have been expressed by the high mobility of relatively small groups. Such nomadic groups may occasionally have left splinter groups in various localities they visited, which then remained relatively isolated.

TABLE 11.3
The ABO Groups in Primates

Species	*Blood Groups* O	A	B	AB	*Totals*	*Reactions of of Blood*	*Reactions of Saliva*
Chimpanzee *Pan satyrus*	17	113	0	0	130	As in man	All secretors of A–B–H substances
Orangutans *Pongo pygmaeus*	0	22	1	3	26	As in man	Among 26 treated, 25 secretors and 1 non-secretor of A–B–H substancess
Gibbons *Hylobates lar*	0	2	4	4	10	As in man	All secretors of A–B–H substances
Hylobates lar pileatus	0	0	1	0	1		
Gorillas *Gorilla gorilla gorilla*	0	0	2	0	2	B-like[a]	Secretor of B and H substances
Baboons *Papio cynocephalus and anubis*	0	33	14	29	76	Red cells do not react[b]	All secretors of A–B–H substances
Inbred hybrids	0	0	35	2	37		
Back-cross *Papio anubis* × *cynocephalus*	0	0	4	7	11		
Papio anubis and cynocephalus	0	3	3	5	11		

TABLE 11.3 (*continued*)

Pigtail monkeys *Macaca nemestrina*	0	0	5	0	5	Red cells do not react[b]	All secretors of H and very weak B
Rhesus monkeys *Macaca mulatta*	0	0	10	0	10	Red cells do not react[b]	Secretors of H; secretors of B of various strengths
Celebes black apes *Cynopithecus niger*	17	7	2	0	10	Red cells do not react[b]	All secretors of A–B–H substances
Java macaques *Macaca irus*	1	8	1	3	13	Red cells do not react[b]	All secretors of A–B–H substances
Squirrel monkeys	1	3	0	0	4	Red cells B-like[c]	All secretors of A and/ or H substances
Cebus monkeys *Cebus albifrans*	1	0	3	0	4	Red cells B-like[c]	All secretors of B and/ or H substances

Source: Adapted from Moor-Jankowski et al. (1964).

[a] Red cells only feebly clumped by anti-B reagents.

[b] Blood group classification based on reactions of saliva and serum.

[c] Blood group classification based entirely on reactions of saliva. The B-like reactivity of red cells is characteristic of all New World monkeys and is apparently independent of the saliva reactions.

People who like to think that man originated at a single place (the "garden of Eden") would find their viewpoint expressed by the second model. There must be at least one population explosion and subsequent spread in this model, or if there were several waves in the spread of the original population, Eden must have been the origin of the last wave. This view seems to be losing support today. The recent high concentration of new fossil finds in East Africa has moved the attention of the Eden searchers from Mesopotamia to East Africa.

It seems more plausible to assume, however, that the concentration of finds in East Africa is the result of the area's having conditions favorable to early human life or to preservation of fossil specimens, rather than evidence of the location of Eden. In any case, the statistically very small sample of fossil specimens makes it impossible to choose between these models at the present time.

The attempt to answer the question of whether human polymorphisms persist throughout the primates encounters technical difficulties, especially for immunological characters.

An ABO-like polymorphism has been detected in several primates (and even in other mammals and vertebrates). Table 11.3 shows a summary of some relevant data. There are technical difficulties in using sera prepared for human red-cell typing for testing the red cells of other primates. Special absorptions of nonspecific antibodies are usually necessary, and the reactions are sometimes not clearly comparable to those in man. It does seem, however, that ABO antigens in hominoids can be tested successfully. Chimpanzees seem to be almost uniformly negative for Lewis (a), and orangutans to be all positive. All chimpanzees, apparently, have the Rh allele *Ro* (*cDe*), but it is also clear that the associated Rh antigens detected are not identical to those in humans. Polymorphism for M-like or N-like antigens is also claimed. As discussed in Chapter 5, of all the Old World higher primates studied so far only the chimpanzee and to some extent the orangutan have shown detectable polymorphism for human Gm factors. Some of the human Gm reagents give reactions with primate sera (see Table 11.4) that somewhat parallel the evolutionary similarity between man and the other primates. Also, as mentioned in Chapter 5, the human Gm(a) peptide is found only in chimpanzees and gorillas, and not in the Old World monkeys.

Absent or incomplete chemical knowledge about blood-group antigens limits the validity of these conclusions. When differences between humans and other primates are not found at the immunological level, the doubt remains that there may still be important undetected differences, and that the apparent similarity may be superficial. The omnipresence of A, B, and H-like antigens in living organisms, including bacteria, suggests that variations in this system may be highly recurrent.

A very distinguished group of British scientists, R. A. Fisher, E. B. Ford, and J. H. Huxley, once went together to the London Zoo in order to decide whether the polymorphism for tasting phenylthiocarbamide was present in the primates. The main problem was how to recognize the taste reaction of an ape. The problem was solved by the chimpanzee to which a small amount of PTC was offered. The ape manifested its strong disgust for the substance he tasted by spitting in Fisher's face (Fisher et al., 1939). A polymorphism for tasting PTC seems to exist among chimpanzees. Later observations by Chiarelli have extended evidence for the polymorphism to several other primates. The fact remains, however, that when PTC sensitivity is properly tested for in humans by determining individual thresholds, most groups show extensive variation with some bimodality of the distribution. Some groups, however, do not show a bimodal distribution, but rather, for them the variation seems to be continuous. When tests are carried out with a single concentration of PTC, as is usual in testing primates, it is not even possible to verify whether there is true bimodality of the threshold. It is difficult, therefore, to reach a definitive conclusion concerning the validity of these observations, except to say that there are individual, and species differences in the ability to taste PTC.

Such relatively dubious conclusions can be eliminated when it is possible to analyze proteins whose full amino acid sequence is known. Differences between primate and human proteins (hemoglobin and a few others) have been described and usually amount to only a few amino acid substitutions. The main conclusions are that the differences between primates and man at the protein level fall in line with other observations on molecular evolution.

TABLE 11.4

Gm Factors in Nonhuman Primates

Species	Number of Sera Tested	Percent Frequency of Gm Factors at Human γ_{2c} Locus											Percent Frequency of Gm Factors at Human γ_{2b} Locus					
		s	c^5	b^0	b^3	b^4	c^3(B)	c^3(v)	$b^0_{(m)}$	b_1	t	g	z	a	$a(m)$	x	f	Ad
Hominoidea (Man and Apes)																		
Gorilla gorilla	14	100	0	100	100	0	0	0	0	0	0	0	100	100	100	0	0	100
Pan troglodytes (Chimpanzee)	78	82	82	82	14	14	94	0	0	0	0	0	100	100	0	0	0	45
Pongo pygmaeus (Orangutan)	25	100	100	70	70	0	0	0	0	0	0	0	0	0	0	0	0	0
Hylobates lar (Gibbon)	14	100	100	100	0	0	0	0	0	0	0	0	100	100	0	0	0	0
Hylobates pillcatus (Gibbon)	3	100	100	100	0	0	0	0	0	0	0	0	100	100	0	0	0	0
Cercopithecoidea (Old World Monkeys)																		
Papio doguera (Anubis baboon)	1	100	100	100	0	0	0	0	0	0	0	0	100	0	0	0	0	0

Source: From Van Loghem et al. (1968).

Molecular analyses should show the persistence of polymorphisms over periods as long as those involved in primate evolution much more clearly than immunological or any other types of tests can. This is still an underdeveloped field, however. A review by Barnicot, Jolly, and Wade (1967) shows that some polymorphisms for blood proteins have been detected in primates, none of which, however, bears any simple relationship to those found in man.

11.4 Racial Differentiation in Man

A race is considered, in common usage, to be a subdivision of a species formed by a group of individuals sharing common *biological* characteristics that distinguish them from other groups. In addition to the biological, there are often also important *cultural* differences, which probably develop through complex interactions with biological differences. The characters used to delineate the traditional subdivisions of man are mainly pigmentation, especially of the skin; facial characters; and body build. These characters are, to a large extent, inherited and thus stable; but they may also be somewhat arbitrary since they have certainly been chosen as criteria simply because they are conspicuous. The number of groups distinguished varies greatly. Many traditional classifications give five races corresponding more-or-less to the earliest inhabitants of the five continents. Some prefer to split these groups further—for example, European from extra-European Caucasians, and Eskimos from American Indians (see Schwidetzky, 1962). These large groups are still heterogeneous, and so can be split further. Thus, for example, Dobzhansky (1962) defines 14 races and Garn (1961) defines 32 races of man. No human race thus defined is completely isolated from all others and "hybrid" groups are occasionally formed. Difficulties in classification are encountered, especially because of the existence of these "hybrid" groups. Sometimes the origin of a group is known, on historical evidence, to be mixed. Very old mixtures, however, can only be inferred on the basis of biological evidence (see Chapter 8), and difficulties in interpretation may arise.

The existence of many different racial groups in man has been interpreted by some as a sign that the human species is, or was before the recent increase in communication and migration, heading towards splitting into several different species.

The differences that exist between the major racial groups are such that races could be called subspecies if we adopted for man a criterion suggested by Mayr (1963) for systematic zoology (see Coon, 1962; Clark, 1964). Mayr's criterion is that two or more groups become subspecies when 75 percent or more of all the individuals constituting the groups can be unequivocally classified as belonging to a particular group. As a matter of fact, when human races are defined fairly broadly, we could achieve a much lower error of classification than 25 percent, implying, according to Mayr, the existence of human subspecies. Most of the error would, in fact, come from hybrids of recent origin. These are today sufficiently numerous, however, to obscure the validity of this approach, and they are probably increasing in frequency.

The fossil record of human races is practically nonexistent.

The fossil record is inevitably limited to observation on bones, which like most anthropometric data, reveal nothing about their genetic determination. To make things worse, there does not seem to exist a really systematic investigation of differences between the skeletal systems of the living races. As Clark (1964) has commented:

> The determination of the evolutionary differentiation of the major races of *H. sapiens* presents exceptionally difficult problems for the paleontologist. These races may be distinctive enough in the flesh, but the anatomical distinction between one and another is not reflected to anything like the same degree in the bones. While they are recognizable by a number of external (and seemingly rather superficial) characters, such as skin color, hair texture, and nose form, they are by no means so easily distinguishable by reference to skeletal characters alone, at least not in the case of individual and isolated specimens (which are usually all that paleoanthropology has to offer). It may, indeed, be possible to identify a skull of a modern Negro, an Australian aboriginal, or a European, in individual cases where the racial characters are exceptionally well marked; but the variation within each group is so great that skulls of each type may be found which are impossible of racial diagnosis.

It would seem therefore that it is at present impossible to accept or to reject the views that Negroes (and especially Pygmies) originate from Rhodesian man (this was offered as a tentative hypothesis by Coon); that the caves near Grimaldi on the northern shores of the Mediterranean were inhabited by a Negroid race, or that Peking Man (*Homo erectus*) is the progenitor of Orientals, or many other suggestions of this kind. Moreover, the problem of antiquity of races cannot be solved, today, on the basis of paleontological evidence alone.

Coon has suggested, mostly on the basis of dentition, that the development of human races was a very long process, taking place over the last million years, and that, in fact, *Homo erectus* evolved into *Homo sapiens*, independently though similarly, in each of his five human "subspecies": Mongoloid, Australoid, Negroid, Capoid, and Caucasoid. The evidence adduced is, however, very scanty. It is considered by many others dangerous to base phylogenetic analysis on a small group of related characters. It is, moreover, unlikely that two species would evolve in entirely parallel ways, both because selective conditions are likely to be different in different areas and because drift, which must always accompany differentiation, is more likely to cause divergence than parallelism. It is even more unlikely that evolution of a new species could have happened five times with parallel results. A "polyphyletic" origin, is therefore, always less likely than a "monophyletic" one. Moreover, as long as there does not exist a specific barrier making gene exchange between groups impossible, a polyphyletic origin must be discarded *a priori*, unless it is reasonably certain for geographic reasons that no gene exchange occurred at all between the groups. It is unlikely that this was the case for man. It is certain that gene diffusion could have taken place, even though it may have been a very rare event. Provided migration to a given place is physically possible there is no reason why it should take place only once. Geographic changes have not been so large or rapid in the period of our evolution. The most important changes were glaciations, which affected mostly the northern regions, and even these did not completely bar human life. Deserts may form in a relatively short time. Thus, the Sahara is believed to have originated at some time during the last three or four thousand years and may have increased the separation between the inhabitants of Europe and those of Africa. The case for a polyphyletic origin of *Homo* does not seem well founded. Even if the persistence of some traits in a given region over a very long period can be really well documented, models of evolution other than a strictly polyphyletic one seem more acceptable. Among possible alternative explanations are the selective advantage of some characters in some areas, or, more probably, the absorption of a local isolate by an invading group (as in the second model described in Section 11.2).

The main determinant of race formation is relative isolation in different environments, which allows drift and differential natural selection to bring about divergence.

Often enough in human history, for one reason or another, a colony from one group has penetrated a previously uninhabited region. Some degree of isolation, if only because of physical separation of the two places by distance, has thus obtained between a parental and a filial population. In addition, when such separations have taken place, there have probably been differences between the old and the new environment. Isolation (which need not be absolute) and environmental differences result in differentiation between the two groups. The biological forces at play are natural selection, causing differential adaptation because environments differ, and drift, causing nonadaptive—that is, random—differentiation. The more different the two environments, the more important adaptive divergence is. The stricter the isolation, the more scope there is for differential selection. However, even if there are frequent migratory exchanges, as we have seen, differences in environmental conditions still foster divergence between two groups. Some degree of isolation is essential for the development of differences due to drift, which is, of course, larger the smaller are the groups concerned.

The split of a human population into two or more fragments, one of which emigrates, must have been a common event, and must account for the occupation of Australia, Melanesia, the Pacific Islands, and Pre-Columbian America. The occupation of many parts of Europe must have taken place repeatedly, perhaps after every glaciation. Even the invasion of Asia from Africa, or vice versa, may have been the outcome of such emigrations. Such splits are most probably, therefore, not the origin of present divergence. At least some of the splits may not have been abrupt and may have been followed by considerable cross exchange. Nevertheless, it is convenient for the purpose of model building, to assume that the history of human differentiation was a series of abrupt splits. The reconstruction of the phylogeny of human races becomes, under this assumption, the reconstruction of the history of these splits. A problem does exist of whether it is possible to do this in the almost complete absence of paleontological information, using only evidence from present-day differences.

Differences in gene frequencies between populations are expected to exist.

We expect to find differences in gene frequencies among different groups of individuals of a given species. Genes for which selective differences exist in the various environments should show the greatest variation among groups. On the other hand, drift should affect all genes equally. Thus, in principle, groups that have been isolated for some time should show differences for *all* genes. Stabilizing selection for some genes may, however, reduce the variation between groups for those genes. Almost all genetic loci should therefore be found to vary in different populations, but they should vary to different extents.

The analysis of gene frequencies in different populations confirms these expectations. It also shows that human groups considered to be distinct on anthropological evidence are also distinct in terms of gene frequencies (see, e.g., Boyd, 1952).

We would like, on the basis of these considerations, to construct a genetic definition of a race. A commonly repeated definition is that a race is a group showing differences in gene frequencies from other groups. Unfortunately such a definition is, in practice, essentially worthless. Most human populations show sufficient local variation that almost every town or city would have to be considered a race on its own if this definition were not further qualified. In order to make it useful, the definition should therefore be augmented by an indication of the amount of difference to be considered as interracial.

Such an amount would have to be set arbitrarily. However, certain kinds of information might assist in the choice of a reasonable standard. For example, if any clear-cut discontinuities are found between populations, clearly they indicate the need for a classification that, at least, separates the groups among which the discontinuities are observed. This may also help in setting thresholds for the differences in gene frequencies that are to be considered as interracial. Inevitably, however, a whole hierarchy of racial groups will be delineated. We must expect this because the "splits" leading to separation into isolated groups, which were mentioned before, must have happened a number of times. The extent of the separation between groups is roughly proportional to the differences in gene frequencies between them. Thus, depending on the time at which given splits occurred, and depending on factors affecting the subsequent differentiation, we can expect to find a whole gradation of differences between groups.

It is worth adding that the same factors that cause genetic divergence, also inevitably cause cultural divergence. Both tend to increase with time, and thus a correlation is to be expected between the amounts of cultural and genetic differences between two groups that were formerly a single group. This will be true, irrespective of interactions between biological and cultural factors.

When the aims of an analysis of racial differentiation are carefully defined, the best characters to work with are genetic polymorphisms.

The analysis of racial differences may be aimed simply at finding a hierarchy in race differences that allows them to be grouped in a rational way. It may, on the other hand, be aimed at the more ambitious goal of reconstructing racial phylogeny. Some might argue that a "rational" grouping is the only one that leads to a reconstruction of the phylogeny. This is not, however, true. There can be a great variety of "rational" classifications, each of which answers a specified purpose. Thus, we might wish to classify dogs according to their uses; hunting, guarding, sniffing truffles, etc. A classification of dogs using these criteria was prepared by Dr. John Caius of Cambridge in 1576, and happens to have no relationship whatsoever with the phylogenetic classification.

It has been suggested that most of the methods that have recently been introduced for use in "numerical taxonomy" are good for "all purpose" classification (see Sokal and Sneath, 1963). Inevitably, some purposes are less well served than others by any given method. These methods, in general, do not follow any specific evolutionary model, and it is not, therefore, clear whether they can give a classification that is meaningful from the phylogenetic point of view. In practice, they group objects according to their similarity, which is measured by some arbitrary criterion. If the chosen criterion is a satisfactory measure of evolutionary divergence then the method may give a fair representation of phylogenetic history. However, it is possible to develop methods in which specific evolutionary models are introduced and tested. These seem preferable, at least for phylogenetic analysis, and may also be useful for general purpose classification, since there can hardly be any other reason for classifying men into races than to understand their origins. (For reference to the work on the construction of evolutionary trees see the papers by Cavalli-Sforza and Edwards, 1963, 1967; Cavalli-Sforza, Barrai, and Edwards, 1964; Edwards and Cavalli-Sforza, 1964; and Cavalli-Sforza, 1966, 1969a.)

Which characters should be chosen for analysis? In numerical taxonomy the following recommendations are usually given: (1) Characters are sought that are *uncorrelated*: a character that is highly correlated with another does not add, at least superficially, anything to what the other already says. (2) The *weight* of each character should be known. In the absence of this knowledge, equal weight is usually given to each character. (3) As many characters as possible are needed for accuracy. (4) Characters of many different types are desirable; if possible, all meaningful characters should be used to ensure completeness.

The list of these desirable qualities is good for a classification that is all purpose in the sense that its aims have not been clearly specified. The aim of such a general classification might, perhaps, be that of being able to place an individual taken at random in the correct group with the smallest possible error. In such a classification, we would probably be interested in distinguishing Pygmies from non-Pygmies and would therefore use height, even if it happened to be known, for instance, that the difference in height were exclusively due to a different diet (which is not believed to be true). The mere existence of Pygmies would force us to create a separate group for them and to be able to recognize them. This does not, however, seem to be an essential requirement for a phylogenetic analysis, nor for a classification based on it. Clearly, if some of the characters used could be changed by the environment in the course of one or a few generations, a classification based on them would be of little use for the study of biological evolution. With this aim in mind, we should add to the list of the desirable qualities of character, the following essential one: that characters chosen for analysis are not subject to significant change by the environment in the course of a short time period.

We should also modify, for a phylogenetic analysis, the first of the above recommendations. The search for noncorrelated characters, or the use of analyses that remove such correlations, is valid when the purpose is pure discrimination. In the reconstruction of phylogenies, however, characters may happen to be correlated simply because of evolutionary history. Such correlations should not be eliminated. Those that have a physiological basis, on the other hand, are more likely to represent mere duplication and thus contribute undue weight. When independent genes are used as characters, it is probably true that they are physiologically independent. Correlations found between them are likely to have an evolutionary origin and thus should not be neglected. In fact, the correlation between two populations for their inherited traits is a direct estimate of the relative length of their common history (see Chapter 8).

The biological traits that can be studied in man can be roughly classified as follows: anthropometric (measurements, external or internal); physiological, behavioral, immunological, and biochemical (especially protein differences). Only among the latter two are many characters encountered that show discontinuous variation—that is, characters whose different expressions can be attributed to the presence of different alleles. Segregation in families can be studied for characters showing discontinuous variation, most of which conform to simple Mendelian expectations. It is not surprising that this is so, because biochemical variations depending on protein structure or immunological differences are much nearer to the origin of the long chain of cause and effect that starts with the gene and ends with a measurable trait. There is, thus, much less opportunity for other gene differences and environmental effects to obscure the picture.

All anthropometric characters are usually genetically complex and also subject to strong environmental influences. Even when heritability is relatively high, as for example for height, it is always dangerous to use the character for comparative observations between races, because there can be unsuspected environmental effects. For height we know, for instance, that even though the degree of genetic determination is high, between-family differences within a population are not adequately taken into account, and environmental differences between populations are certainly not considered. Even within a population, significant environmental effects are often observed, for example with respect to socioeconomic status, and family size. Stature and most other anthropometric traits are thus difficult to use for phylogenetic analysis because of the unknown short-term effects on them of environmental changes. In addition, they are influenced by an unknown number of genes, and it is difficult to evaluate their weight in comparison with single-gene differences. It would seem therefore, that characters for which a simple Mendelian basis has been ascertained are the obvious ones to choose. In fact, they do satisfy all the requirements that have been listed, and, in particular, satisfy the all-important last one: that a character chosen for analysis not be subject to short-term changes provoked by environmental influences.

The genetic traits must show variation between populations in order to be informative. This is true of almost all polymorphic genes. Rare mutant alleles, maintained at equilibrium by mutation and selection, can vary somewhat from one population to another due to drift effects, but the variation is likely to be small and due to the more recent history of the population. They cannot, therefore, be of much help for a probe into the distant past. Polymorphic genes, on the other hand, as long as they show variation between populations, are likely to tell more about the history of the remote past.

11.5 Measurement of Genetic Similarity and Distance Between Populations, Using Polymorphic Genes

The simplest way of computing a **genetic distance** between two populations A and B that takes into account several loci might seem to be that of computing the difference between the gene frequencies of A and B and adding their absolute values for different loci. The use of absolute values always, however, creates analytical difficulties that can, in part, be avoided by using sums of squares instead of sums of absolute values. A constant difficulty in working with gene frequencies is the dependence of their sampling error on the gene frequencies themselves. This can be solved, for a wide range of frequencies, by using the angular transformation (see Appendix I and Chapter 8), which has been generalized in the following way, suggested by Fisher, so as to take account of multiallelic loci. The quantity

$$\cos\theta = \sum_i \sqrt{p_{iA}\, p_{iB}} \tag{11.1}$$

can be used to measure genetic similarity between two populations A and B, where p_{iA} and p_{iB} are the gene frequencies of an allele at a given locus in populations A and B and the summation is taken over all alleles at each locus. In this way populations can be represented graphically, with respect to one locus, as shown in Figure 11.2. The coordinates are the square roots of the gene frequencies. A triallelic locus

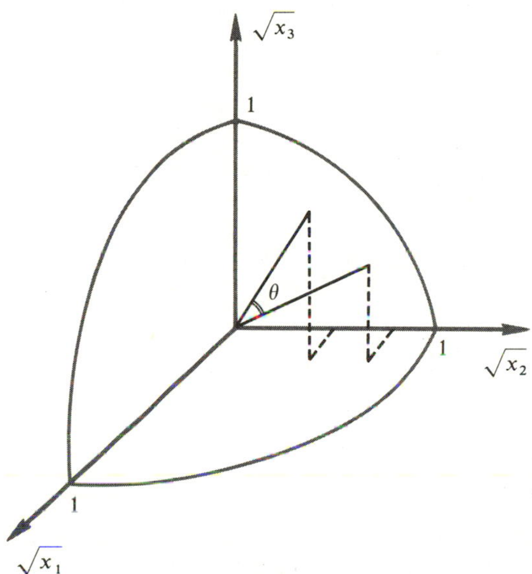

FIGURE 11.2
The concept of the genetic distance for three alleles at one locus. The gene frequencies of the three alleles are represented by x_1, x_2, and x_3. (Redrawn from Cavalli-Sforza and Conterio, 1960, with a slight modification.)

is shown in the figure. Because the sum of the frequencies of all alleles at a locus in one population is 1, every population will be represented by a point on the surface of an octant of a sphere (a quarter of a circle for a two-allele locus, and in general, a 2^{-n} section of an $(n-1)$ dimensional hypersurface for n alleles). It can be shown that the angle formed between the radii corresponding to the two populations is θ, as defined by Equation 11.1.

Genetic distance can conveniently be defined by d, where

$$d \propto \sqrt{1 - \cos\theta}, \tag{11.2}$$

which in geometrical terms is proportional to the chord subtended by the angle θ. The coefficient of proportionality can be chosen according to the purpose of the analysis. A quantity having a value of one for a complete gene substitution has been used by Cavalli-Sforza and Edwards. The methods of computing genetic distances used by these authors, although basically the same and always making use of the angular transformation, differed slightly in successive investigations. (Other procedures have been suggested by Balakrishnan and Sanghvi (1968) which do not, however, have the advantages offered by the angular transformation.) In later work it was found useful to use the relationship derived in Chapter 8 between genetic distance and the kinship coefficient f, namely,

$$f = \frac{\sigma_p^2}{\bar{p}(1 - \bar{p})}. \tag{11.3}$$

When only two populations are considered σ_θ^2, the variance between angular values, is equal to

$$\frac{(\theta_A - \theta_B)^2}{2} \sim \frac{\theta^2}{2} \sim 1 - \cos\theta.$$

Here θ_A and θ_B are the respective angles between each population and an arbitrary origin, and θ is the angle between the two populations. As $\sigma_p^2/\bar{p}(1 - \bar{p}) \sim 4\sigma_\theta^2$, it follows that

$$f \sim 4(1 - \cos\theta). \tag{11.4}$$

Thus, four times the square of the genetic distance d defined by $d = (1 - \cos\theta)^{1/2}$ (Equation 11.2) gives the value of f between two populations. The relationship 11.4 is valid for a biallelic locus. For a multi-allelic locus, an average of the f values over the k alleles is given by

$$f_\theta = \frac{4(1 - \cos\theta)}{k - 1}. \tag{11.5}$$

Summing over various loci, after weighting by the number of alleles at each locus, we obtain

$$f_\theta = \frac{4\sum(1 - \cos\theta)}{\sum(k - 1)}. \tag{11.6}$$

A standard error based on the variation of f values between loci can be computed.

Some simple approximate expectations for the relationship between f and evolutionary time show why it is useful, for some applications, to consider f, rather than d. Populations separated for a time T and undergoing drift subject to (effective) population size N have an expected f value given (see Chapter 8) by

$$1 - e^{-T/2N} \tag{11.7}$$

which, for small T, is simply

$$f = \frac{T}{2N}. \tag{11.8}$$

Thus, it is the *square* of the genetic distance that is proportional to time for divergence under a simple drift model. The use of squares of distances or f values as variates for evolutionary analyses thus depends on the assumption that drift is the main cause of differentiation. If the time available for divergence is large, it is not f itself but $-\log(1-f)$ that is proportional to evolutionary time (more precisely, to $T/2N$).

When local selection differences are believed to be responsible for differentiation, genetic distance itself should be used. In this case, gene frequencies diverge in two independently evolving populations according to the simple equation for selection with additive fitnesses (see Chapter 4):

$$\log \frac{p}{q} = sT + \log \frac{p_0}{q_0},$$

where p_0 is the gene frequency at the time of separation, and p is that of a population having been subjected to selection of intensity s for time T. It can be shown (see Cavalli-Sforza, 1969) by computing the variance of the quantity $\log p/q$ (assuming p_0 is constant but s varies from population to population) and then, from it, the f value, that f is in this case proportional to the square of T, and therefore that a quantity like genetic distance, varying according to the square root of f, is proportional to evolutionary time.

Depending on the model, therefore, $\log(1-f)$, f, or $f^{1/2}$ (the latter proportional to genetic distance) will be proportional to the time scale. The first two are valid for pure drift (the second being an approximation of the first for small values of $T/2N$) and the third is valid for differentiation due to local differences in selection intensity. The real situation, on the average, lies between the two extremes of only drift or only differential selection. The various expressions, therefore, give upper and lower bounds for the time of divergence T. Should information on T and N be available, then the comparison between observed and expected f values, and thus a choice between the two models, becomes possible.

Genetic distance can also be useful for the evaluation of racial mixtures (Cavalli-Sforza et al., 1969). It can easily be shown that a mixed population M, made up of a proportion m of population A and $(1-m)$ of population B, is at a genetic distance d_{MB} from B, where d_{MB} approximately equals md_{AB} and d_{AB} is the distance between A and B. The distance between M and A, d_{MA}, similarly, is approximately equal to $(1-m)d_{AB}$. Thus,

$$d_{MA} + d_{MB} = d_{AB}. \tag{11.9}$$

This equation is only an approximation, using the genetic distances given, because the distances are only approximately linear in the gene frequencies. A more exact method for evaluating whether a population corresponds exactly to the mixture of two given populations, and of estimating m, was given in Chapter 8.

11.6 Analysis of Racial Differences on the Basis of Polymorphic Genes

It was already clear to Hirszfeld and Hirszfeld in 1918 that ABO blood groups showed ethnic differences and might therefore be informative for the study of evolution. However, the ABO system alone is a poor indicator of racial differences. The major variations are the almost complete absence of B and of A in American Indians, and the higher frequency of B among Orientals than among Caucasians, forming a regular, though modest, cline across Eurasia. The Rh system is much more informative. There is a high prevalence of an allele Ro (*cDe*) among African Negroes that is rare elsewhere, and a high frequency of an allele *r* (*cde*) among European Caucasians that is rare or absent elsewhere. Boyd in 1949 could already show a qualitative agreement between the classical major subdivisions of mankind and blood-group frequency data. The accumulation of knowledge concerning new markers has provided the opportunity for a more refined analysis. The Diego blood group has been found to be present only in populations of Oriental origin, though not in all, establishing more clearly the link between Mongolians and American Indians. The Duffy (Fy) blood group has reinforced the difference between persons of African descent and non-Africans. Other markers discovered more recently, such as HL-A, are also very promising as indicators of racial differentiation.

A quantitative analysis of racial differences was attempted by Cavalli-Sforza and Edwards in 1963. The measure of genetic distance already explained was made the basis of the analysis, and methods were developed for the "*post*diction" (as opposed to *pre*diction, to use a neologism suggested to us by D. Gomberg) of the evolutionary history of the human races. These will be summarized briefly in the next section. The basic data are the frequencies of alleles at polymorphic loci in the various racial groups. A great deal of relevant information has accumulated in the last 50 years. Since the monumental work of Mourant, which was published in 1954, no critical and complete collection of data has been undertaken. The task of collecting extremely widely scattered material is, itself, a job of major proportions. It is unfortunate that there are very many gaps in our knowledge. Investigations made before the discovery of the more recent polymorphisms inevitably contain no data on them. The variety of techniques now necessary for a complete survey, and the cost and limited availability of some reagents, are the reasons for the incompleteness of many investigations. Today it is, however, possible to keep samples of blood refrigerated at very low temperatures, awaiting future analysis by still-to-be-discovered techniques or tests with reagents unavailable at the time of the investigation. This also, however, is expensive.

Incompleteness of information forces the selection of populations and markers for which more material is available, or the averaging of relatively heterogeneous groups. Later in this chapter, a table of polymorphisms with a summary of available information on alleles and their frequencies is given and discussed briefly. Some of the analyses to be described in the text of this chapter, however, were carried out earlier than those mentioned in the table and used somewhat different bodies of information. Some discrepancies have therefore, inevitably, arisen. For detailed conclusions, the presently available information may still be inadequate, and some of the statements are bound to change with advancing knowledge.

There are two basic ways of choosing data for the calculation of genetic distances between human races. One is to select a sample of racial groups, that may be representative of larger "racial" groups and carry out the analysis on these. The choice is then limited by the availability of suitable groups that have been analyzed for all the informative markers. The groups that are selected may inevitably bias the results. This method has, however, the advantage of permitting the use of investigations selected for reliability and completeness. It also introduces discontinuity, which may be artificial, but helps in assuming the validity of the methods of phylogenetic analysis to be described later. These, to some extent, rely on the effective isolation of the populations after the initial splitting.

An alternative method is that of grouping all known human racial groups into larger "racial" groups and taking averages for these. There are practical difficulties in considering all possible groups at once, and pooling makes the analysis much easier. Moreover, in the averaging process, gaps in the information on rare markers tend to disappear, even though this may be at the expense of making the reliability of the average estimates rather variable. This method is more comprehensive, but inevitably depends on the validity of the pooling process and of the groups formed. Serious problems may arise due to intercrossing and miscegenation.

An analysis of five loci gave a fairly clear picture of racial similarities and differences.

Both methods were tried. For the first, fifteen indigenous populations were chosen, three from each continent. They are all fairly widely separated. For Europe, the representatives were English, Lapps, and Turks; for Asia, Indians (Veddahs from Ceylon), Gurkhas (from Nepal, believed to derive from mixed Indian and Oriental ancestry), and Koreans; for America, Eskimos, and a North American and a South American aboriginal Indian tribe; for Africa, Ethiopians (East Africa), Ghanaians (West Africa) and Bantus (Central Africa); for Oceania, Australian aborigenes, Melanesians (New Guinea), and the Maoris (the aborigenes of New Zealand of mysterious origin).

The matrix of genetic distances between the 15 groups, two by two, was computed on the basis of five loci (ABO, 4 alleles; MNS 4 alleles; Rh, 6 alleles; Diego (Di) and Duffy (Fy), 2 alleles each).

The reconstruction of the phylogenesis of these groups by two independent methods by Cavalli-Sforza and Edwards produced the tree given in Figure 11.3.

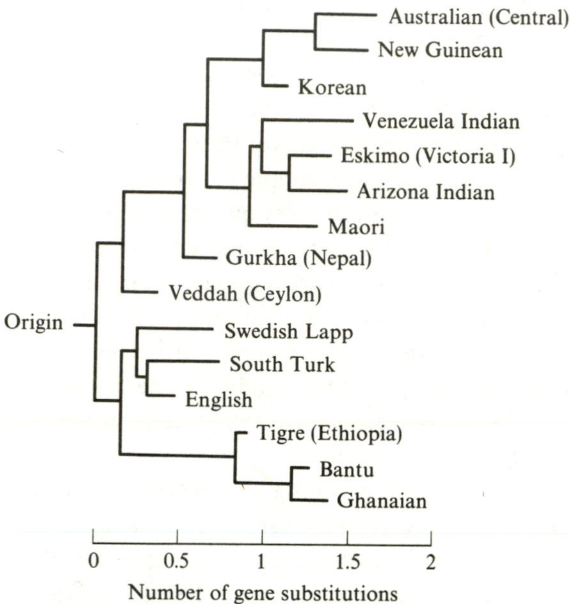

FIGURE 11.3
Evolutionary tree computed from blood-group gene frequencies. (From Cavalli-Sforza and Edwards, 1963.)

It will be noted that all three African groups came together, as do the three European populations, in spite of the considerable differences between them. The two groups from South Asia, Veddahs and Gurkhas, are intermediate between Europeans and other Asiatic groups, American Indian groups, and Oceania groups. American aborigenes also stay together, as do Australians and Melanesians.

The number of possible trees that can be constructed with 15 groups is enormous, and the tree shown is one among several that all give an almost equally good fit to the data. All trees that give a good fit retain the above mentioned relationships, which are expected from what is known of the similarities between these groups from other evidence. Another tree obtained from this material, giving an equally good fit, is shown in Figure 11.4, projected on a world map. No attempt was made, in this figure, to preserve the correct relative lengths of the various branches, the only feature retained being the sequence of splits. It is remarkable that the tree fits so closely the probable routes of mankind's migration.

Another independent way of analyzing the same data may be mentioned. It has the advantage of permitting the graphical presentation of multidimensional data in a reduced number of dimensions, for example, two or three, with minimum loss of information. It is called **principal-components analysis**. The basis of this method is illustrated in Figure 11.5. In Figure 11.5,A the four points plotted correspond to four observations with two characters. The straight oblique line passing through them is so drawn as to minimize the sum of the squares of the distances of each point from the line itself. These distances are indicated as dotted segments in the diagram, and the projection of each point on the line allows a simplified representation of each original point using one dimension only, as indicated in the right-hand part of the figure. The loss of information is measured by the distances themselves (or more precisely, their sums of squares) which have been minimized, as already mentioned, to provide the best-fitting line. Figure 11.5,B shows how this principle can be extended to three characters. The process can be repeated using a second principal-component orthogonal (at right angles) to the first.

The original gene-frequency data on eighteen alleles, which, therefore, required eighteen dimensions, can by this procedure be reduced to two variables, losing only half of the total information, or can be reduced to three, losing one-third. The two-dimensional representation is given in Figure 11.6. The clustering of Africans, of Europeans, and of Americans, is clear. Using a third dimension, the Maoris are seen to be nearer to Americans than to Europeans, and the Veddahs to Europeans than to other groups.

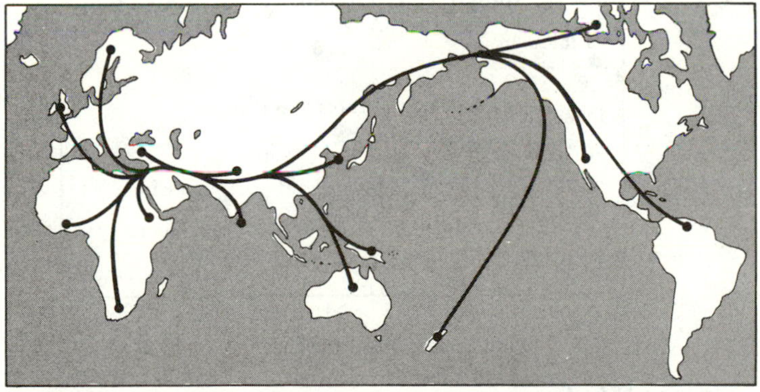

FIGURE 11.4
A phylogenetic tree of man projected on a world map to show correspondence with possible routes of migration. (From Edwards and Cavalli-Sforza, 1964.)

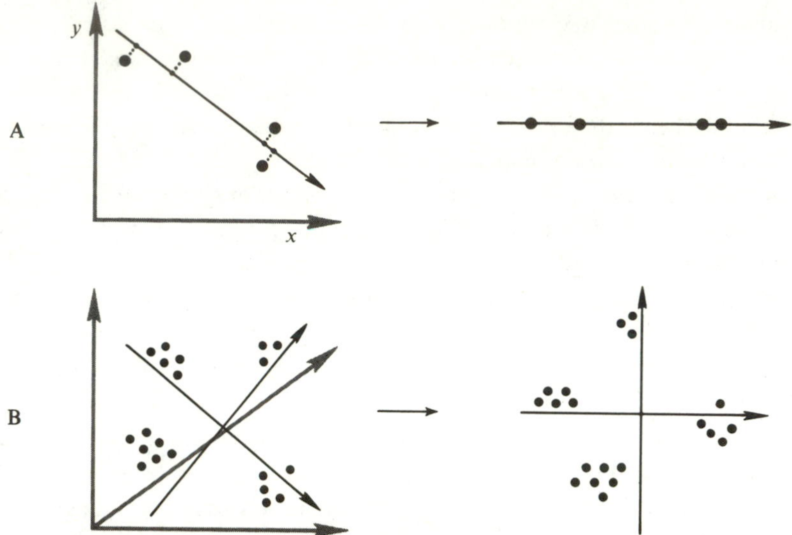

FIGURE 11.5
Illustration of the reduction in dimensions made possible by principal components analysis.

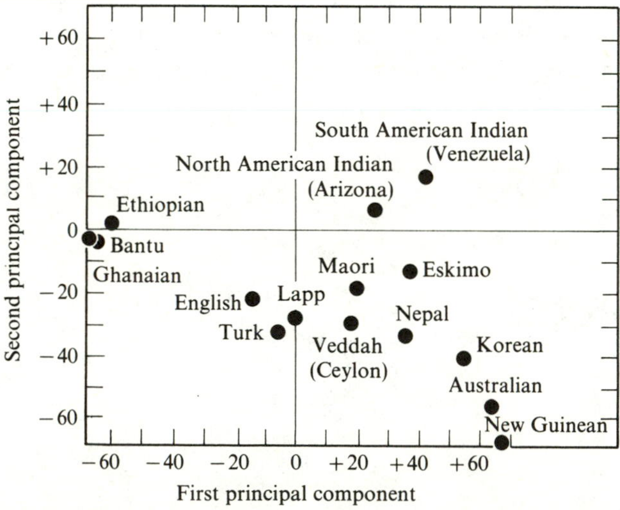

FIGURE 11.6
Chart of first two principal components, synthesizing frequencies of five blood-group systems (ABO, Rh, MNS, Fy, Di). (From Cavalli-Sforza and Edwards, 1963.)

Extending the analysis to more loci confirmed the general picture.

A tabulation of a large amount of gene-frequency data by Schwidetzky (1962) was used for an analysis based on an alternative approach—namely, pooling all known ethnic groups into larger racial groups. The "races" suggested by Schwidetzky were provisionally accepted. Pooling over many groups permitted a considerable increase in the number of loci. Only a few populations of uncertain and very probably mixed origins were left out (Ainus, Polynesians). The matrix of genetic distances computed between the seven "races" suggested by Schwidetzky is given in Table 11.5 and the best tree in Figure 11.7.

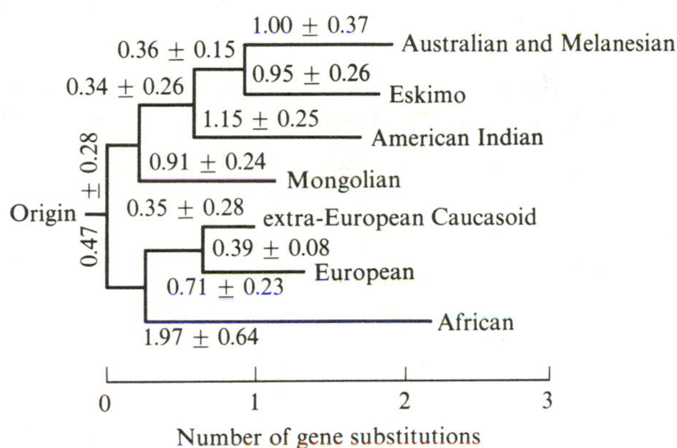

FIGURE 11.7
Tree of human evolution reconstructed from the matrix of distances given in Table 11.5. (From Cavalli-Sforza, 1966.)

Genetic distances range from 1.1 (between Europeans and extra-European Caucasoids) to 4.6 (between Africans and Eskimos). It is clear that the separation into Europeans and extra-European Caucasoids (Arabs, Jews, Indians) encompasses a smaller distance than other differences. These data agree with the former analysis and suggest three major groups: Americans, Caucasians (including extra-Europeans), and Mongoloids (including all the Pacific populations). The last group is the most heterogeneous. Comparisons of genetic distances within this group, using the data of Table 11.5, range from 1.9 to 2.7. Intergroup comparisons are all more than 2.8, ranging up to 4.6, the only exception being comparisons between extra-European Caucasoid and Mongoloid groups. History confirms the existence of some intermixture between these. Most of the subdivisions in the Mongoloid group must have taken place at some fairly early time. With the surprising exception of Australians and Eskimos, who show a somewhat lower difference than the others, this indicates a centrifugal tendency from an early group, located perhaps in South

East Asia (which is approximately the center of diffusion) at some relatively early time.

A rough estimate of the separation times between the major racial groups can be obtained on the assumption of constant evolutionary rates.

We have seen that it is possible to predict the evolutionary rate of divergence if differentiation is due to drift, unchecked by migration, and population sizes are known. At the microgeographic level, the assumption that variation is mostly due to drift can usually be accepted, but its extrapolation to an analysis covering a much greater geographical area is not generally acceptable. In such an analysis, f values for individual loci, which are expected to be equal if only drift is acting, are found not to be homogeneous. The variation between major racial groups, measured by f, is very low for the ABO and Kell blood groups. It is somewhat larger for MNS, and for the majority of other markers. It reaches its peak with Gm, Fy, and the Ro allele of the Rhesus system (see Table 11.6). If we know T, the time of divergence,

TABLE 11.5

Genetic Distances Between Racial Groups (Sum of Chords in Angular Transformation—See Equation 11.2) Computed from Alleles Given Below

	B	C	D	E	F	G
A	1.103	2.795	3.095	2.977	2.773	3.262
B	—	2.111	2.866	2.595	2.967	2.782
C	—	—	2.359	2.583	2.989	2.633
D	—	—	—	2.427	3.743	2.554
E	—	—	—	—	4.600	1.950
F	—	—	—	—	—	4.184

Racial groups. A=Europeans; B=extra-European Caucasoids: C=Mongoloids; D=Amer. Indians; E=Eskimos; F=Africans; G=Australians and Melanesians.

Alleles. A, B, O; MS, Ms, NS, Ns; P, p; CDE, CDe, Cde, cDE, cdE, cDe, cde; K, k; Lu^a, Lu^b, Le^a, Le^b; Fy^a, Fy^b; Di^a, Di^b, Hp^{1S}, Hp^{1F}; Hp^2; T, t; Gm^a, Gm^{ax}, Gm^b, Gm^{ab}; Ge^1, Ge^2, Gc^{ab}.

Source: From Cavalli-Sforza (1966).

and N, the average population size, we can predict the f value expected under drift alone. Unfortunately, we cannot say whether this value is higher or lower than the observed average. Selection could, in fact, give rise to observed values of f both above and below that expected under drift. Thus, the evidence for the existence of the ABO polymorphism in primates might suggest that it is balanced by heterosis,

TABLE 11.6
World Variation of Gene Frequencies

Genetic System	*Allele*	*Number of Racial Groups*[a]	$f=\sigma^2/\bar{p}\bar{q}$[b]
ABO	A	125	0.070
	B	125	0.055
	O	125	0.081
MN	MS	45	0.071
	Ns	45	0.094
Rhesus	R_o	75	0.382
	R_1	75	0.297
	R_2	75	0.141
	r	75	0.172
Duffy	Fy	62	0.358
Diego	Di	64	0.093
Kell	k	64	0.029
Haptoglobin	Hp^1	60	0.096
Gm	Gm^a	25	0.226
Gc	Gc^1	42	0.051

Source: From Cavalli-Sforza (1966).

[a] This is the number of racial groups used for the calculation of f.

[b] σ^2 is the variance in gene frequency between the groups, $\bar{p}$ is the mean gene frequency, $\bar{q}=1-\bar{p}$.

in which case the ABO f value should be lower than that expected under drift alone. If differential selection exists in the various environments occupied by the main racial groups (or existed at an earlier time), then f values should be greater than those expected under drift. It seems likely, though it is difficult to prove, that drift gives rise to the intermediate f values observed.

A simple empirical approach is to estimate the average observed f value between two groups that once were a single group and for which the separation date is known, and to use this as an estimated rate of evolutionary divergence. If the same average rate applies between two other groups that were once a single group but for which the separation date is not known, an estimate of the separation time can then be made on the basis of the observed f. Actually, the considerations of Section 11.4 show that the scale to be used for time estimation should be $-\log(1-f)$ under drift alone (the "drift" scale) and $f^{1/2}$ under differential selection alone (the "selection" scale). A time estimate intermediate between these two is probably the most satisfactory one at present.

The following tentative estimates are based on the best date known in human evolution, which is the first settling of the American continent some 15,000 years ago. The $\bar{f}$ value, computed according to Equation 11.6, on the basis of 14 loci, using data from American Indians on the one hand and from a large number of tribes from Australia and New Guinea representing the best known S. Pacific populations on the other is $\bar{f} = 0.176 \pm 0.053$. This implies an increase in $-\log(1-\bar{f})$ of $-\log(1-0.176)/1.5 = 0.13$ every 10,000 years and in $\bar{f}^{1/2}$ of $0.176^{1/2}/1.5 = 0.28$ in 10,000 years. The estimate of T is probably an underestimate, as it is likely that the separation between American Indians and Australians took place even before that between American Indians and their nearest neighbors, the Mongolians (about whom less is known).

The drift estimate of the rate of change of f is not in disagreement with other data. The f value between the New Guinean tribes, said to have differentiated, on the basis of the rate of linguistic changes, during the last 3000–4000 years is 0.045, so that $-\log(1-f) \sim 0.05$. The separation between Australians and New Guineans is larger than this—namely, $\bar{f} = 0.103$. These two f values may, however, be biased by the fact that different loci were used in the two studies. Variation between loci makes it advisable always to use the same loci if possible.

In general, related racial groups occupying the same continent give f values lower than 0.1. Thus the main African groups (Pygmies, Bushmen, Bantus, West Africans, and Southern Nilotes) give an $\bar{f}$ value of only 0.045; Australian tribes, 0.040; and South American Indian tribes, 0.082.

The three major human groups (using 16 loci) gave the following $\bar{f}$ values:

Negroid–Caucasoid	0.352 ± 0.078
Negroid–Mongoloid	0.416 ± 0.140
Caucasoid–Mongoloid	0.242 ± 0.062

On the basis of the estimate of the rate of change in f given, using the drift time scale (0.13 per 10,000 years), the greatest difference, that between Negroid and Mongoloid, would correspond to a separation time of $-\log(1-0.416)/0.13 = 4.1 \times 10{,}000$ years. If, instead of the drift scale, the selection scale is used, then the estimated time is $0.416^{1/2}/0.28 = 2.3 \times 10{,}000$ years. If 15,000 years for time of the separation between American Indians and their neighbors is an underestimate by, say, a factor of 2, then the estimates of the time of separation of Negroids and Mongoloids would be doubled. The estimates given do not permit a satisfactory distinction to be made between two hypotheses concerning the time of the major racial splits. Thus it is not clear whether the first split was between the branch that became the indigenous populations of Africa and the branch that populated Europe and the East, which then split later, or if the first separation divided the branch that populated Africa and Europe and the branch that populated the East. The greater similarity between Caucasoid and Mongoloid than between Caucasoid and Negroid populations on the basis of the most comprehensive analysis so far made using 16 loci, favors the first hypothesis, but the difference is not significant.

Analyses based on only anthropometric characters are less satisfactory.

The methods developed for phylogenetic analysis, which have been successfully used with genetic markers, can also be applied to anthropometric measurements. An analysis was made using fifteen populations chosen with criteria as similar as possible to those that had been used for choosing the group of fifteen populations that were analyzed using genetic markers (see Cavalli-Sforza and Edwards, 1963). Twenty-six anthropometric characters were used, including height, skin color, and hair shape. Unfortunately, estimates of the variation in these characters and especially of the correlations between them are not usually available, in the references from which the data were taken. Standard deviations were assumed to be proportional to $(\bar{x})^{-0.8}$, that is, to the overall mean of the character to the power 0.8. All deviations from the mean were standardized on the basis of this standard deviation, squared, and summed. Correlations were neglected. The outcome of this analysis (Cavalli-Sforza and Edwards, 1963), was somewhat unsatisfactory. Many groups known to be related were correctly placed together, but important exceptions were found. Thus North and South American Indians were separated, the former being associated with indigenous European populations, and the latter remaining in the group of populations of the East. Australians and Africans were fairly close, while genetic markers separate them sharply.

It may be expected that an enlarged list of anthropometric characters would give better results. In principle, there is no reason why they should give different results from genetic markers, though they are likely to be affected by serious errors, for the reasons mentioned earlier. Considering the inconsistencies in the results of the analysis using anthropometric characters, especially with respect to the separation of North and South American Indians and the association of Africans and Australians, conclusions based on genetic markers seem to be more reliable.

11.7 Methods of Phylogenetic Analysis

A short summary of the methods used in reconstructing the phylogenies given in Figures 11.4 to 11.6, is given in this section. For more details the original papers should be consulted. We are concerned here only with the general principles and main assumptions.

The model of evolution used for reconstructing phylogenies is one of successive branchings, or the successive splitting of a single population into two or more subpopulations, in which independent evolutionary changes then proceed in the various branches. The genetic changes are assumed to take place under drift or under selection, as specified.

The evolutionary model can be given a simple mathematical formulation.

If $q_{g,0}$ is the frequency of gene g in a population before the population splits into two or more subpopulations (at time zero), its frequency at time T, in a subpopulation labeled r, $q_{g,T,r}$, is

$$q_{g,T,r} = q_{g,0} + \Delta_{g,T} + \Delta_{g,T,r} + \varepsilon_{T,r}(1 - b_{g,T,r}). \tag{11.10}$$

The Δ values represent selection effects, the first of which, $\Delta_{g,T}$, is independent of local environmental conditions, while the second depends on them. In the comparison of different populations, only the second, $\Delta_{g,T,r}$, is involved. The quantity $\varepsilon_{T,r}$ is the contribution due to drift and is a random variate that has expectation zero and a variance that is a function of the population size and of the time elapsed. The coefficient b is a sort of "recall coefficient" representing heterotic effects that, if present, can reduce the magnitude of the variation of ε.

Under the simplest conditions, only drift is operative and then $\Delta_{g,T}$ and $\Delta_{g,T,r}$ have zero mean and variance, and can be neglected. If b varies only with respect to g, appropriate scaling of the gene frequencies can eliminate its effects. Various assumptions can be made for $\Delta_{g,T,r}$. If it is considered to be a random variate, we have a situation of selective drift superimposed on a situation of random drift, and scaling of the gene frequencies (for example, by equalizing the variation between populations for all genes) can eliminate its effects. If $\Delta_{g,T,r}$ varies in a more systematic way for one or a few genes then the averaging out allowed by the simultaneous consideration of many genes will still make some of the procedures developed later applicable. Random, or semirandom, variation of $\Delta_{g,T,r}$ is not unrealistic, as the adoption of new conditions of life, made necessary by a split, makes the resulting change in selective coefficients an almost random one.

The model of successive branching applied to observations on differences in gene frequencies makes it possible to reconstruct a presumptive genealogy ("tree") of populations examined.

Various methods have been developed for the construction of evolutionary trees and they fall, in general, into two categories: methods producing trees with an apex (or "root") and methods producing trees without an apex (unrooted). The main difference between the two types of methods lies in the use of the assumption of constant evolutionary rates in the various branches.

Suppose we have four populations, A, B, C, and D. There are altogether fifteen different ways in which these populations may have arisen from a single population by successive dichotomous splits, which are shown in Figure 11.8. Ignoring the apex, the number of trees is reduced to three, as shown in Figure 11.9. With one population less, the number of rooted trees is three and there is only one unrooted

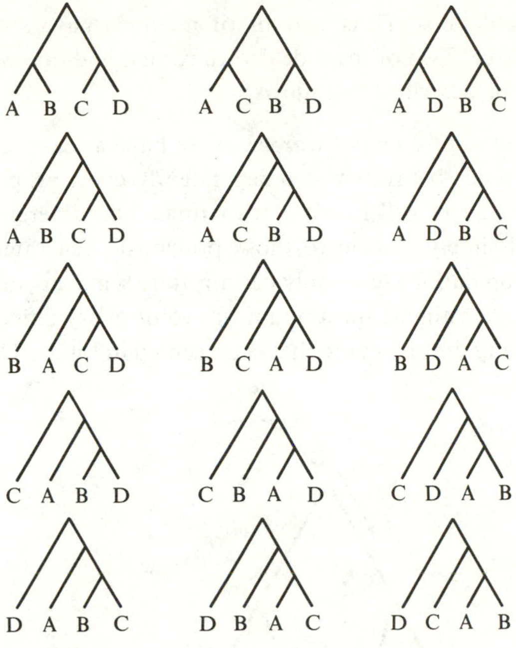

FIGURE 11.8
15 possible "rooted" trees for four populations.

tree. With one population more (five in all) these numbers are 90 and 15, respectively. The number of different possible trees increases very rapidly with the number of populations.

The method of analysis has to be divided into two parts; the choice of a tree and the estimation of its parameters, such as the length of its branches. The choice of the tree depends on the goodness-of-fit of the model chosen. In general, every possible tree should be fitted, and the best fitting tree chosen. At the moment there are no shortcuts to this procedure that are fully reliable. The complete analysis cannot, however, be carried out when the number of populations is more than seven or eight—even with the aid of a computer—because of the very large number of possible trees. For a larger number of populations, therefore, the investigator has to give up a full analysis and be satisfied with considering a relatively small number of the more likely trees. The differences between them are often, however, trivial. This shows that it is best to use a small number of well-studied

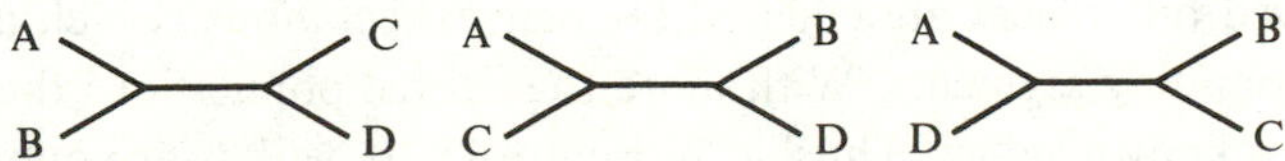

FIGURE 11.9
The 3 possible "unrooted" trees for four populations.

representative populations. Three groups of methods have so far been used for choosing the best tree. Two of them deal with rooted and one with unrooted trees. The methods can be described as follows:

1. *The additive model for unrooted trees.* If we have a reasonable measure of the accumulated difference between two independently evolving populations, we can assume that the measure will increase with time. The differences generated in a given time should simply add on to those previously generated. In a given tree, various pairs of populations have only certain time segments of evolution in common. It is possible to estimate the amount of evolutionary difference accumulating in each segment using the procedure that is presented in Figure 11.10. The difference

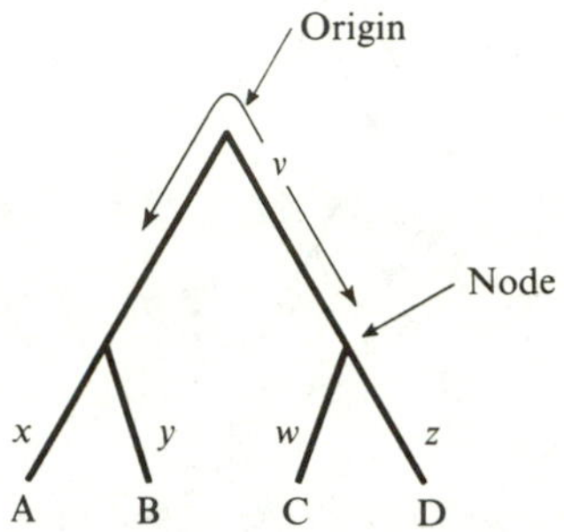

FIGURE 11.10
The additive model for estimating evolution. The letters A–D stand for four populations.
Distance between
A–B $= x + y$
A–C $= x + v + w$
A–D $= x + v + z$
B–C $= y + v + w$
B–D $= y + v + z$
C–D $= w + z$
Distances A–B etc. are observed: x, y, v, w and z are estimated by minimizing the sum of squares
$[(A - B - (x + y)]^2 + [A - C - (x + v + w)]^2 + \cdots + [(C - D - (w + z)]^2$

between A and B has accumulated over segments x and y, that between A and D over segments x, v, and z, etc. A system of equations can thus be constructed in which observed differences are equated to their expectations, which are the sums of the corresponding segments. With more than three populations there are fewer unknown than known values. Thus, as in Figure 11.10, with four populations there are six equations with five unknowns. The method of "least squares" can be used to provide a solution for the five unknowns, by minimizing the sum of the squares of the differences between observed and expected distances for each pair of populations. The resulting minimum sum of squares gives an estimate of the goodness-of-fit of the hypothesis of additivity for the particular tree chosen. In this model the assumption that evolutionary rates are constant is not made. However, unless this assumption is introduced later, the position of the apex cannot be given. Distances must be measured on a scale believed to be proportional to elapsed time.

2. *The minimum-path model.* The additive model gives no "spatial representation" of the tree (only the distance between its nodes) and to this extent, it is not realistic. It does not, in fact, provide an estimate of the gene frequencies of the common ancestors. If these gene frequencies are postulated and if they are represented in a space of as many dimensions as the analyzed gene frequencies, it is possible to predict them by another method, which has been called the minimum path. In Figure 11.11 the gene frequencies for three loci are given for four populations (the circles) together with the "minimum-path net." The possible positions

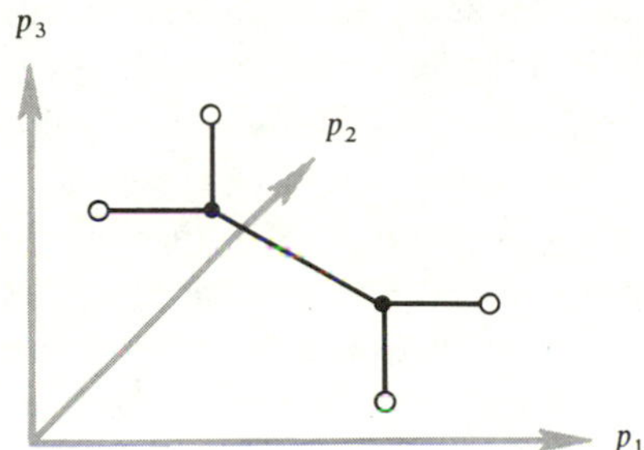

FIGURE 11.11
The minimum-path model for estimating evolution for four populations (circles). Coordinates p_1, p_2, and p_3 are gene frequencies for three loci (or three alleles at a locus). Solid dots mark positions of common ancestors.

of common ancestors are marked with solid dots. The "topology" of a tree is defined by the branch points (or nodes) and their interconnections, without specifying the actual positions of the nodes (or, therefore, the distances between them). An algorithm is available for obtaining the positions of the nodes for a given tree that minimize the total length of all the branches of the tree, called the "minimum net." This is, in simple terms, equivalent to minimizing the amount of evolution needed to lead to the observed populations. There is no simple reason for assuming that the amount of evolution that has actually taken place is the minimum possible. This method does however, almost always produce the best trees, which are very closely related to those given by the additive method. Gene frequencies are in this case the starting point for computations and thus genetic distances need not be computed. They can, however, be used to decrease the number of dimensions to the number of groups minus one, when the number of alleles for which frequencies are available is more than the number of groups minus one. The technique for doing this is described in the original papers (see, e.g., Cavalli-Sforza and Edwards, 1967).

The two methods suggested by these first two models require comparable amounts of computer time. Neither assumes constant evolutionary rates and without this assumption the apex cannot be located.

3. *Model based on assumption of constant evolutionary rates.* Gene frequencies at time T are assumed to vary according to a normal distribution that has, for any given branch, the same expected value as before the branching and a variance that increases in proportion to the time since the branching. A maximum likelihood method for estimation of the time between any two branchings—including the first one, which, of course, locates the apex—can be developed. This assumes a constant evolutionary rate, at least for a given gene. Differences in evolutionary rate for different genes can be considered to equalize the variances (for example, by suitable change of scale). This model corresponds closely to one in which Equation 11.10 is used, with the assumption that the expected value of $\varepsilon(1 - b)$ is proportional to the time length of each branch and, for a given gene, is constant throughout the tree. Selective drift can be accommodated with similar restrictions. Nonrandom variation of $\Delta_{g,T,r}$ in various parts of the tree for a given gene cannot, however, be accommodated. A lack of fit of this model therefore suggests that evolutionary rates along the various branches are not constant for one or more genes.

Each of the three methods just outlined can give a "most likely tree" for a given group of populations. The choice of populations must, however, be made carefully. There is as yet no good information on how hybridization between two distant branches may affect the procedure, but it is certainly likely to confuse the estimation. When results do not fit any tree well, the occurrence of hybridization is a possible interpretation. If, however, hybridization due to cross migration takes place between branches that have only recently separated (that is, the split is not as abrupt as implied by the theory) the estimation difficulties are less serious. In fact, intuitively, we can see that separation between two branches followed by some exchange of genes, namely incomplete separation, only decreases the rate of differentiation between the two branches and thus shortens the apparent length of the separation.

11.8 The Frequency of Polymorphisms and Its Meaning

Genetic variation can be partitioned, for simplicity, into continuous and discontinuous variation. We have seen in Chapter 9 the difficulties encountered in the study of continuous variation. Discontinuous variation is more directly amenable to genetic analysis and is usefully subdivided into that due to rare alleles and that to frequent (polymorphic) alleles. Many of the rare alleles are rare because they are maintained in the population by the balance between mutation and selection (as discussed in Chapter 3). They may also be polymorphisms in their initial or terminal stages.

Variation for frequent alleles (polymorphism proper) may be transient or balanced and can be due to a variety of conditions (see Chapters 4, 5). In the long run, however, all polymorphisms—other than sex!—are bound to be transient, and there probably are few, if any, that survive long enough to be present in different species. ABO may be one case in point. The possibility exists, however, that the ABO system is produced by a recurrent type of variation, which can be regenerated by high mutation rates or by mutation to similar, but not identical, alleles or by the fact that conditions favoring this polymorphism recur in nature, perhaps at long time intervals.

The major polymorphisms known in man are given in Table 11.7. The table gives the gene frequencies of most of the known polymorphic alleles for representative Caucasoid, Negroid, and Mongoloid populations. Many of these polymorphisms have already been discussed to some extent. The data given, however, should, we hope, provide a useful summary indication of the present knowledge concerning human polymorphisms and their variation among racial groups. The number of well-studied polymorphisms is still relatively small (roughly thirty), but is rapidly increasing. At the moment the data on polymorphisms are not quite enough to mark all the 23 pairs of human chromosomes.

The observed proportion of all genes that are polymorphic, based on the prevailing methods used to detect polymorphic differences, is roughly 30 *percent.*

Two independent lines of evidence suggest that polymorphic genes may actually be so frequent that almost all loci may be polymorphic.

The analysis of an arbitrarily chosen set of red-cell or serum enzymes by the technique of starch-gel electrophoresis showed that a substantial fraction of the loci specifying the enzymes were polymorphic (Harris, 1966). Of ten enzymes studied up to 1966, three were polymorphic. At present, out of eighteen (including the earlier ones) six are polymorphic, with one of them (phosphoglucomutase) showing polymorphism at two out of the three loci known to participate in its determination (see Table 11.8). The enzymes were chosen for study independently of their being polymorphic, and can thus probably be considered as a random sample. The detection of electrophoretic polymorphism in one-third of the enzymes studied probably implies a greater total frequency of polymorphism, since only a certain fraction of possible protein differences are picked up by this technique. Generally only amino acid substitutions that result in a change in the electric charge in the protein are

TABLE 11.7

Gene Frequencies of Human Polymorphisms in Three Racial Groups. The largest sample in existence in each group, showing Hardy-Weinberg equilibrium, was chosen, selecting whenever possible populations from Northern Europe for Caucasoid, Central or South Africa for Negroid, and China for Mongoloid. N is the sample size. (Data on the Gm polymorphism and limited data on HL–A and other white-cell and platelet antigen polymorphisms were given in Chapter 5.)

ABO

	Gene frequencies			
Frequent alleles	*Caucasoid* (N=3459)	*Negroid* (N=858)	*Mongoloid* (N=817)	*Notes*
A_1	0.2090	0.0969	0.1864	Weak *A* and *B* alleles exist. For a full survey up to 1958, see Mourant, Kopec, and Domaniewska (1958).
A_2	0.0696	0.0811	0	
B	0.0612	0.1143	0.1700	
O	0.6602	0.7077	0.6436	

MNS

	Gene frequencies			
Frequent alleles	*Caucasoid* (N=1000)	*Negroid* (N=205)	*Mongoloid* (N=103)	*Notes*
MS	0.2371	0.0925	0.0405	S^u (no S or s produced) only in Africans (up to 35%). M^g in Switzerland. Hunter, Henshaw, antigens associated with M (in Africans). *Mi*, *Vw*, and other satellite private antigens closely associated with MNS. For reviews, see Mourant (1954) and Race and Sanger (1968).
Ms	0.3054	0.4880	0.5663	
NS	0.0709	0.0436	0.0144	
Ns	0.3866	0.3760	0.3788	

P

	Gene frequencies			
Frequent alleles	*Caucasoid* (N=1166)	*Negroid* (N=900)	*Mongoloid* (N=293)	*Notes*
P_1	0.5161	0.8911	0.1677	Formerly, P_1 and P_2 were designated P and p. It is probable that there is a rare silent allele n for this locus and also an apparently recessive rare allele p^k. See Race and Sanger (1968).
P_2	0.4839	0.1089	0.8323	

TABLE 11.7 (*Continued*)

RHESUS

Frequent alleles	Gene frequencies: Caucasoid (N=8297)	Negroid (N=644)	Mongoloid (N=250)	Notes
R_0 (cDe)	0.0186	0.7395	0.0409	Many other rare alleles and associated antigens have been described (see Chapter 5). The Negroid R_0 (cDe) frequency includes 9% cD^ue. For reviews, see Mourant (1954) and Race and Sanger (1968).
R_1 (CDe)	0.4036	0.0256	0.7591	
R_2 (cDE)	0.1670	0.0427	0.1951	
R_z (CDE)	0.0008	0	0.0049	
R_1^w(C_wDe)	0.0198	0.0031		
r (cde)	0.3820	0.1184	0	
r' (or R')(Cde)	0.0049	0.0707	0	
r'' (or R'')(cdE)	0.0029	0	0	
r'^w (C^wde)	0.0003	0		

LUTHERAN

Frequent alleles	Gene frequencies: Caucasoid (N=1446)	Negroid (N=205)	Mongoloid	Notes
Lu^a	0.0354	0.0272	0	Rare alleles are Lu($a-$b) – a dominant and a recessive form, both very rare. For reviews see Mourant (1954) and Race and Sanger (1968).
Lu^b	0.9646	0.9728	1.0	

KELL

Frequent alleles	Gene frequencies: Caucasoid (N=9875)	Negroid (N=1205)	Mongoloid (N=7200)	Notes
K	0.0462	0.0029	0	Closely associated loci are Kp and Js. Kp^a antigen is found in about 2% of all whites but only 1% of whites who are K+. Kp^a is not found in blacks. Kp^b is antithetical to Kp^a. Rare K–k– Kp (a–b–) individuals are found. Js^a is found only in Africans. See Race and Sanger (1968).
k	0.9538	0.9971	1.0	

TABLE 11.7 (*Continued*)

SECRETOR

Frequent alleles	Gene frequencies: Caucasoid ($N=1118$)	Negroid ($N=115$)	Notes
Se	0.5233	0.5727	*Se* is dominant over *se* for presence of ABH substances in saliva. Interaction with *Le* for secretion in saliva (see Lewis in this table).
se	0.4767	0.4273	

LEWIS

Frequent alleles	Gene frequencies: Caucasoid ($N=1000$)	Negroid ($N=125$)	Mongoloid ($N=85$)	Notes
Le	0.8156	0.3188	0.7575	There is interaction with secretor locus, *SeLe* are Le(a − b+) on red cells, *seseLe* are Le(a+b−), *Se·le le* and *se se le le* are Le (a−b−). *Le* is dominant over *le* for presence of Le[a] substances in saliva. See Race and Sanger (1968).
le	0.1844	0.6812	0.2425	

DUFFY

Frequent alleles	Gene frequencies: Caucasoid ($N=1944$)	Negroid ($N=365$)	Mongoloid ($N=103$)	Notes
Fy	0.0300	0.9393	0.0985	The Caucasoid *Fy* frequency is an indirect and approximate measurement. See Mourant (1954) and Race and Sanger (1968).
Fy^a	0.4208	0.0607	0.9015	
Fy^b	0.5492			

KIDD

Frequent alleles	Gene frequencies: Caucasoid ($N=4275$)	Negroid ($N=105$)	Mongoloid ($N=103$)	Notes
Jk^a	0.4858	0.7818	0.3103	*Jk* is a rare silent allele. See Mourant (1954) and Race and Sanger (1968).
Jk^b	0.5142	0.2182	0.6897	

TABLE 11.7 (*Continued*)

DIEGO

Frequent alleles	Gene frequencies: Caucasoid (*N* = 2600)	Negroid	Mongoloid	Notes
Di^a	0	0	0–30%	No Di(a−b−) yet reported. Di^a most frequent in South American Indians, present in Mongoloids at a lower frequency, absent in Eskimos. See Race and Sanger (1968).
Di^b	1	1		

YT

Frequent alleles	Gene frequencies: Caucasoid (*N* = 2568)	Negroid (*N* = 69)	Notes
Yt^a	0.9564		See Race and Sanger (1968).
Yt^b	0.0413	0.07 (1 out of 69)	

DOMBROCK

Frequent alleles	Gene frequencies: Caucasoid (*N* = 755)	Negroid (*N* = 237)	Mongoloid	Notes
Do^a	0.4200	0.3064	0.07–0.34	Do^a has a frequency of 7% in Thais and 34% in American Indians. See Race and Sanger (1968).
Do	0.5800	0.6936		

AUBERGER

Frequent alleles	Gene frequencies: Caucasoid (*N* = 544)	Negroid (*N* = 39)	Notes
Au^a	0.6213	0.6419	This polymorphism is probably associated with the Lutheran polymorphism. See Race and Sanger (1968).
Au	0.3787	0.3581	

TABLE 11.7 (*Continued*)

XG

Frequent alleles	Gene frequencies: Caucasoid (N=2082)	Negroid (N=219)	Mongoloid (N=549)	Notes
Xg^a	0.675	0.55	0.54	This is the only sex-linked blood group. See Race and Sanger (1968).
Xg	0.325	0.45	0.46	

ACID PHOSPHATASE

Frequent alleles	Gene frequencies: Caucasoid (N=367)	Negroid (N=317)	Mongoloid (N=620)	Notes
P^a	0.36	0.16	0.22	A rare allele P^r is found only in Negroids, and another rare allele P^d is found only in Mongoloids. See Giblett (1969).
P^b	0.60	0.81	0.78	
P^c	0.04	0.001	0	

GLUCOSE-6-PHOSPHATE DEHYDROGENASE

Frequent allele	Gene frequencies	Notes
GdA+, *A*−, *B*	See Table 4.11 for some frequencies	Many *GdA*− alleles are known. Their geographic distribution is highly variable. See Giblett (1969).

6-PHOSPHOGLUCONATE DEHYDROGENASE

Frequent alleles	Gene frequencies: Caucasoid (N=4558)	Negroid (N=209)	Mongoloid (N=228)	Notes
PGD^A	0.979	0.943	0.934	A rare silent allele has also been found. See Giblett (1969).
PGD^C	0.021	0.057	0.066	

TABLE 11.7 (*Continued*)

PHOSPHOGLUCOMUTASE

Frequent alleles	Gene frequencies: Caucasoid (N=2115)	Negroid (N=318)	Mongoloid (N=417)	Notes
PGM_1^1	0.764	0.781	0.750	Other variants and loci are known. PGM_1 designates the first locus. For PGM_2 and PGM_3 see Table 11.8 and Harris (1969.) See Giblett (1969).
PGM_1^2	0.235	0.219	0.242	

ADENYLATE KINASE

Frequent alleles	Gene frequencies: Caucasoid (N=1887)	Negroid (N=800)	Mongoloid (N=227)	Notes
AK^1	0.955	1		See Giblett (1969).
AK^2	0.045	0	0.002	

HAPTOGLOBINS

Frequent alleles	Gene frequencies: Caucasoid (N=5811)	Negroid (N=483)	Mongoloid (N=339)	Notes
Hp^1	0.376	0.559	0.280	Hp^1 can be subdivided into Hp^{1s} and Hp^{1f}. The third line of the table gives ratios of these two subtypes. An Hp^{2-1} modified phenotype occurs in 13% of Negroids. It is probably due to a modified Hp^2 allele. A null type, Hp_0, showing no haptoglobins occurs in 3–5% of adults. It may, in part, be due to a silent allele. At least 12 rarer phenotypes have also been described. For a review see Kirk (1968).
Hp^2	0.623	0.441	0.720	
Hp^{1s}/Hp^{1f}	0.18–0.65	0.36	1	

INV

Frequent alleles	Gene frequencies: Caucasoid (N=402)	Negroid (N=399)	Mongoloid (N=270)	Notes
Inv^a	0.1022	0.3190	0.2774	Inv is an antigenic marker (see Chapter 5) on gamma globulin IgG light chain. See Ropartz et al. (1961).
Inv^b	0.8978	0.6810	0.7226	

TABLE 11.7 (*Continued*)

TRANSFERRIN

Frequent alleles	Gene frequencies: Caucasoid (N=2845)	Negroid (N=1143)	Mongoloid (N=868)	Notes
Tf^c	0.995	0.973	0.991	There are two common *D* types, D_1 and *D*-chi. The first is present in Africa, Australia, and Melanesia; and the second in Orientals, southeast Asians and American Indians. The parentheses around Tf^D indicate that the identification of *D* is uncertain. See Giblett (1969).
(Tf^D)	0.0005	0.027	0.0070	
Tf^{B2}	0.0037	0	0.0022	
Tf^{B1}	0.0009	0		

GROUP SPECIFIC COMPONENT

Frequent alleles	Gene frequencies: Caucasoid	Negroid	Mongoloid	Notes
Gc^1				Other phenotypes have been described. See Schultze and Heremans (1966).
Gc^2	~0.26	<0.10	0.30–0.35	

BETA LIPOPROTEINS

Frequent alleles	Gene frequencies: Caucasoid (N=248)	Negroid (N=54)	Mongoloid (N=1205)	Notes
Ag^x	0.23	0.69	0.73	See Giblett (1969).
Ag^y	0.77	0.31	0.27	

LIPOPROTEINS

Frequent alleles	Gene frequencies: Caucasoid (N=1109)	Negroid (N=107)	Mongoloid (N=234)	Notes
Lp^a	0.195	0.191	0.009	Only the product of one allele, Lp^a, can be identified. See Giblett (1969).

TABLE 11.7 (*Continued*)

PSEUDOCHOLINESTERASE

Frequent alleles	Gene frequencies: Caucasoid (N=8314)	Negroid (N=460)	Mongoloid (N=371)	Notes
E_1^a	0.016	0.001	0	Only the product of the E_1^a allele can be identified. See Giblett (1969).

detected using this technique. These constitute about one-third of all possible amino acid substitutions. The frequency of polymorphic proteins may thus be as high as 100 percent. An unknown correction factor reducing this estimate may still be necessary, because many proteins are made of subunits coded for by different, and usually independent, loci. Man is certainly not unique in respect to the frequency of polymorphism; Lewontin and Hubby (1966) working with *Drosophila pseudoobscura* and using a similar technique found that about 30 percent of all proteins they examined were polymorphic and that about 12 percent of the respective loci were heterozygous, on the average, in any given individual.

TABLE 11.8

Results of an Electrophoretic Survey of Eighteen Arbitrarily Chosen Enzymes in Caucasian and Negro Populations. Only common alleles (frequency > 0.01) are listed. Many "rare" alleles were also detected. None of the twelve enzymes unlisted in the body of the table showed common electrophoretic polymorphism.

	Caucasians			Negroes		
Enzymes[a]	Allele 1	Allele 2	Allele 3	Allele 1	Allele 2	Allele 3
Red-cell acid phosphatase	0.36	0.60	0.04	0.17	0.83	—
Phosphoglucomutase						
Locus PGM_1	0.76	0.24	—	0.79	0.21	—
Locus PGM_2	1.00	—	—	0.99	0.01	—
Locus PGM_3	0.74	0.26	—	0.34	0.66	—
Adenylate kinase	0.95	0.05	—	1.00	—	—
Peptidase A	1.00	—	—	0.90	0.10	—
Peptidase D (prolidase)	0.99	0.01	—	0.95	0.03	0.02
Adenosine deaminase	0.94	0.06	—	0.97	0.03	—

Source: From Harris (1969).

[a] The twelve other enzymes studied were: phosphohexoseisomerase, malate dehydrogenase, isocitrate dehydrogenase, red-cell hexokinase, lactate dehydrogenase, methaemoglobin, red-cell pyrophosphatase, pyruvate kinase, placental acid phosphatase, peptidases B and C, and a red-cell "oxidase."

The observed frequency of polymorphism estimated from loci whose products are detected by immunological techniques is also about 30 *percent.*

Genes must be picked at random if the estimate of what proportion of them are polymorphic is to be valid. This may be difficult to accept for immunologically detected polymorphisms. For these, the probability of detecting a polymorphism is a function of the gene frequencies, being higher the more alleles there are and the nearer their frequencies are to an equal distribution. With n equally frequent (and equally antigenic) alleles, it can be shown that the total probability that a mother has a child with an antigen different from any of her own is $(n-1)^2/n^2$. The probability that a donor used for an immunization has an antigen not present in a randomly chosen recipient is also a maximum with equal allele frequencies (see Chapter 5). In both cases, an antibody may arise, given such a difference, and so the polymorphism would be detected. For two alleles with gene frequencies p and $q = 1 - p$ the probability of detection is $p(1-p)$. Thus, the higher the level of heterozygosis, the more probable it is that a polymorphism will be detected. It is not surprising, therefore, that all the blood-group systems found earlier were highly polymorphic. (See Table 11.9.) In more recent years, however, more and more "private" and "public" blood groups have been found—that is, those present in very few persons and those present in almost all. The proportion of polymorphic blood-group systems has thus been decreasing constantly, as can be seen from the last column of Table 11.9. It now seems to be reaching an asymptote with about 1/3 of the red-cell antigens being polymorphic. Lewontin (1967) believes that with the accumulation of information, the ascertainment bias described above may have disappeared.

The proportion of polymorphic loci among those detected immunologically is high, but may be smaller than among enzymes. This is not surprising, since incompatibility may select against polymorphisms of this kind and thus make them somewhat rarer.

Why are there so many polymorphisms?

That such a large proportion of all loci are polymorphic was undoubtedly a surprise to many geneticists. We turn out to be much more different, one person from another, than we once thought we were. This is a consequence of the fact that more powerful techniques have become available for detecting genetic differences, but is in no way in contradiction with expectations based on relatively simple population-genetics models, as we shall see below.

It is likely that the observed polymorphisms belong to several classes. The most important ones must be the following:

1. *Classical Balanced Polymorphisms, in Which the Heterozygote is at a Constant Advantage, at Least in a Certain Environment.* They are mentioned first because

TABLE 11.9

Human Blood Groups with Their Dates of Discovery, Gene Frequency of the Most Common Allele, Heterozygosity per Locus, and Cumulative Proportion of Loci Polymorphic in the English Population

Blood Group	*Year Discovered*	*Frequency of Most Common Allele*	*Heterozygosity at Locus*	*Cumulative Heterozygosity*	*Proportion Polymorphic*
ABO	1900	0.437	0.512	0.512	1.00
MNS	1927	0.389	0.700	0.606	1.00
P	1927	0.540	0.497	0.569	1.00
Se	1930	0.523	0.499	0.552	1.00
Rh	1940	0.407	0.662	0.574	1.00
Lu	1945	0.961	0.075	0.491	1.00
K	1946	0.936	0.122	0.438	1.00
Le	1946	0.815	0.301	0.421	1.00
Levay	1946	~1.00	~0	0.374	0.889
Jobbins	1947	~1.00	~0	0.337	0.800
Fy	1950	0.549	0.520	0.353	0.818
Jk	1951	0.514	0.500	0.366	0.833
Becker	1951	~1.00	~0	0.337	0.769
Ven	1952	~1.00	~0	0.313	0.714
Vel	1952	~1.00	~0	0.292	0.667
H	1952	~1.00	~0	0.274	0.625
Wr	1953	0.999	0.002	0.258	0.588
Bc	1953	~1.00	~0	0.244	0.556
Rm	1954	~1.00	~0	0.231	0.526
By	1955	~1.00	~0	0.219	0.500
Chr	1955	0.999	0.002	0.209	0.476
Di	1955	~1.00	~0	0.199	0.454
Yt	1956	0.995	0.010	0.191	0.434
Js	1958	~1.00	~0	0.183	0.417
Sw	1959	0.999	0.002	0.176	0.400
Ge	1960	~1.00	~0	0.169	0.384
Good	1960	~1.00	~0	0.163	0.370
Au	1961	0.576	0.489	0.175	0.393
Lan	1961	~1.00	~0	0.168	0.380
Bi	1961	~1.00	~0	0.163	0.366
Xg	1962	0.644	0.458	0.173	0.387
Sm	1962	~1.00	~0	0.167	0.375
Tr	1962	~1.00	~0	0.162	0.364

Source: From Lewontin (1967).

they have been known to exist for a long time, not because they are believed to be especially frequent. In fact, there is only one proved case of this class in man, hemoglobin S. Others are based on claims that have not been confirmed (ABO, MN) or on only weak evidence (see Chapters 4 and 5).

It seems from theoretical considerations that this class can form only a relatively small fraction of all polymorphisms observed. The maintenance of polymorphisms of this class involves a "cost" or "load" as discussed in Chapter 7 whose value for each locus is approximately the loss of homozygotes incurred for the maintenance of equilibrium. We know that this is of order s/2, where s is the mean selective disadvantage of homozygotes. Assuming all such polymorphic loci act *independently*, the mean fitness of the population relative to a fitness of one for multiple heterozygotes is, for n loci, $(1 - s/2)^n$. If $s = 0.001$ and $n = 500$, the proportion of zygotes lost due to the heterotic polymorphisms is 0.92. Total prereproductive mortality, in man, is, at most, 70 percent or 80 percent of the conceptions in primitive populations, and much less in economically advanced ones. Even if much of this loss were due to genetic causes, which cannot be true, the total load implied by only 500 such loci acting independently would exceed observed levels of prereproductive mortality. Sved, Reed, and Bodmer (1967), King (1967), and Milkman (1967) have pointed out that the assumption of independent action of polymorphic loci is arbitrary and severely limits the number of heterotic loci that can be maintained polymorphic for a given level of selective elimination. Specifically, they showed that if there is an upper limit, or threshold, to the fitness, a much larger number of loci can be held polymorphic. For example, if the upper fitness limit for multiple heterozygotes is 10, relative to a mean population fitness of unity, and $s = 0.01$ for each of $n = 10{,}000$ loci, it can be shown that the average heterozygote advantage at each locus is only reduced from one percent to about 0.99999 percent, although the maximal individual fitness is only ten times the population mean. In fact, in this case only about 1 in 10^6 individuals in the population would have the maximum fitness of 10. This simple modification of the standard model thus increases by at least an order of magnitude the estimate of the number of heterotic loci that can be held polymorphic for a given level of elimination. The effects of inbreeding with this model, may, however, be quite severe. Thus, it can be shown that with an average population inbreeding coefficient, F, the reduction in mean fitness due to inbreeding is approximately given by a factor of $[1 - (Fs/2)]^n$. When $n = 10{,}000$ and $s = 0.01$, this factor is 1/20 when F is as low as 6 percent. Such extreme reductions in fitness with inbreeding, have not been observed (see Chapter 7). Though this effect can be reduced somewhat if there is heterogeneity in the inbreeding levels, it still imposes a severe limit on the number of loci that can be maintained polymorphic. Thus, though the threshold model would readily allow for, say, 1000 heterotic polymorphic loci, taking into account the inbreeding restriction, this can only be a small proportion of the total number of polymorphic loci.

2. *Mutant Alleles That Have in the Past Been Subject to Strong Selection Pres-*

sures, Either Directional or Balancing, and Have Now Disappeared. The selective agent may have been an infectious disease, whose pathogenic agent has evolved to become harmless. Other environmental and cultural changes may be important. For example, clothing and housing may have radically changed selective values of genes connected with cold resistance. If selective values are reduced to zero, these alleles are now drifting around gene frequencies that were arrived at under different conditions. It is very difficult to estimate what the frequency of such loci may be. It could be very large, if the rate of loss of such polymorphisms is less than their rate of formation, as they would then increase in number almost without limit (Bodmer, 1968).

3. *Mutant Alleles That Have Practically No Selective Effect, and so May Accumulate under Mutation Pressure, or Simply under Drift.* As Kimura (1968) has noted, genes whose selective values are less than $1/2N_e$ behave essentially as if they were neutral with respect to drift.

4. *Mutant Alleles That Have a Constant Absolute Selective Advantage and Are Caught in the Process of Actual Evolutionary Substitution, at Least in Those Populations in Which They Are at an Advantage.* Only these are true "transient" polymorphisms. Those of type 2 and 3 are also transient except that they, on the average, remain in the population for a very much longer period of time. Even polymorphisms of the first type are, in practice, transient, if there is a change of environmental conditions or in the genotypic background. Thus, the ABO polymorphism, believed to be the most stable, is not found among South American Indians. The evidence that it was probably present in their ancestors and subsequently disappeared is the following: (a) most closely related populations (Orientals and Eskimos), and all other world populations have the polymorphism; (b) the ABO polymorphism is found in some North American Indian tribes; (c) the *B* gene is found at very low frequencies in some South American Indian tribes that are relatively isolated from gene flow; (d) presence of ABO antigens has been claimed in pre-Columbian mummies (Boyd and Boyd, 1937).

The hemoglobin S polymorphism itself—the only thoroughly documented example of a classical balanced polymorphism in man—is limited to very few populations and is likely to disappear with the eradication of malaria.

Observed heterozygosity is at approximately the expected level.

We have seen in Chapter 8 that neutral mutations accumulate in a population, and reach an equilibrium at which

$$F = \frac{1}{1 + 4N_e\mu}, \tag{11.11}$$

where F is the proportion of homozygotes, N_e is the effective population size, and μ is the average mutation rate. From Table 11.9 the overall proportion of homozygotes for blood groups is $1 - 0.162$ or 84 percent. Since for enzymes, including those that are not detectable electrophoretically, heterozygosity must be even higher than 0.16, this value of F is an upper limit.

The effective population size is not known exactly. The estimate of F based on the data given in Table 11.9 is from the English population, but a similar estimate would be valid for indigenous European populations in general. The population size to be considered is not, however, that of today, but the one that prevailed for most of the period of human evolution. Population sizes must have started to increase rapidly with the agricultural revolution (the Neolithic period), which began some 10,000 years ago. Only minor changes in gene frequencies have probably occurred since that time and whatever fluctuations drift may have caused before then were most likely frozen by the increase in population size. Thus, the population sizes before the Neolithic, which may have been almost constant or

TABLE 11.10
The Increase in World Population

Years Ago	*Cultural Stage*	*Area Populated*	*Assumed Density per Square Kilometer*	*Total Population (Millions)*
1,000,000	Lower paleolithic		 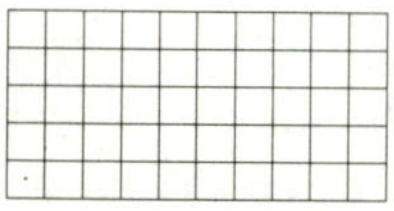 0.00425	0.125
300,000	Middle paleolithic		 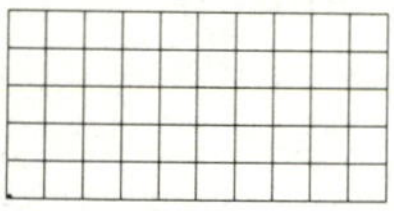 0.012	1
25,000	Upper paleolithic		 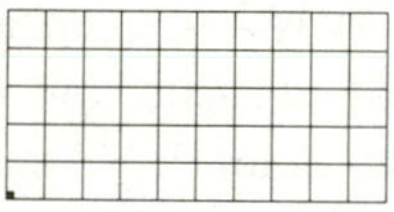 0.04	3.34
10,000	Mesolithic		 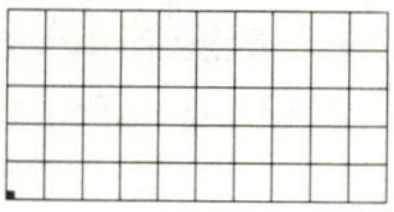 0.04	5.32
6,000	Village farming and early urban		 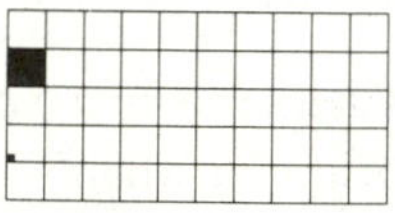 1.0[a] 0.04	86.5
2,000	Village farming and urban		 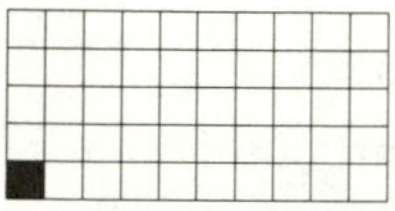 1.0	133

[a] The higher density refers to the area where farming has replaced the earlier type of economy (hunting and gathering).

growing only slowly for perhaps a million years, are those to be used in the formula given. Within major racial groups there is likely to have been sufficient genetic exchange to consider the whole group as being unitary from the point of view of drift. In Table 11.10 the increase in population size during Paleolithic times is based mostly on the assumption of an entirely African origin of mankind, which can be disputed. The earliest sizes given in Table 11.10 are thus probably underestimates. There is *today* about that number of hunters and gatherers still living in Africa, in conditions not very different from those of the Paleolithic. These people have been pushed into a few restricted areas regarded as being uninhabitable by most other groups. They now occupy a very small fraction of the African continent. Pygmies, for example, who may today number 100,000 or slightly less, now occupy a total area of the order of 500,000 square kilometers (1/60 of all Africa). It would

TABLE 11.10 (*continued*)

Year A.D.	*Cultural Stage*	*Area Populated*	*Assumed Density per Square Kilometer*	*Total Population (Millions)*
1650	Farming and industrial		 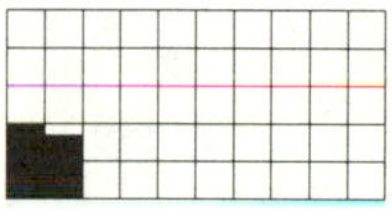 3.7	545
1750	Farming and industrial		 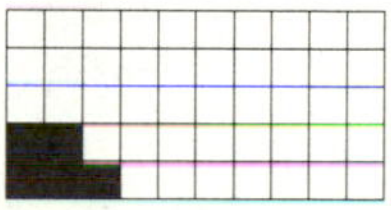 4.9	728
1800	Farming and industrial		 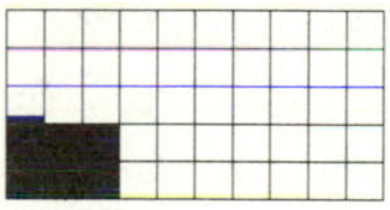 6.2	906
1900	Farming and industrial		 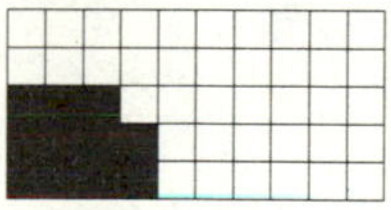 11.0	1,610
1950	Farming and industrial		 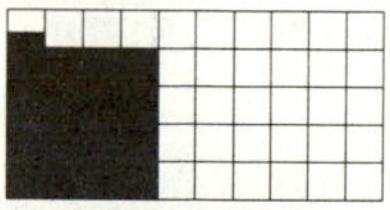 16.4	2,400
2000	Farming and industrial		 46.0	6,270

Source: From Deevey (1960).

seem more reasonable, therefore, to assume that in the Paleolithic, the number of humans must have been about 2 or 3 million for the whole species, and so, of course, somewhat less for just Europe.

The observed figure for F based on Table 11.9—namely, 0.84—corresponds from Equation 11.11 to $N_e\mu = \frac{1}{4}(1/F - 1) = 0.045$. With a population size less than 10^6, μ must be larger than 0.45×10^{-7}. Our unbiased estimate for average mutation rate per locus (Chapter 3), 0.3×10^{-7}, includes only mutation to "visible," and usually deleterious, characters. It does not include phenotypically "invisible" mutations, which are nevertheless detectable at the molecular level and which will increase the estimate of mutation rate by an unknown amount. The comparison between the two values is quite satisfactory on the assumption that almost all mutations are effectively neutral. With present information, it would be difficult to ask for more than finding that orders of magnitude correspond.

It should be emphasized that these are estimates for drift only, and are thus compatible only with polymorphic genes of class 3, and, possibly, of class 2. Heterotic genes in class 1, are, as already discussed, probably less common. They decrease the F value by a disproportionately larger amount than genes of classes 2 and 3 but the change in the number of alleles maintained by heterosis has been shown to be small (Kimura and Crow, 1963) as was mentioned in Chapter 8. Genes in class 4, whose transition through the polymorphic state, as discussed above, is only a fraction of their total life, tend to counterbalance the effects of type 1 genes. In fact, they contribute less to heterozygosity than genes that are subject to the effect of drift only. The main deficiency of the estimate of F based on assuming that drift is predominantly responsible for the observed level of polymorphism is that this does not adequately take into account the "relic" polymorphisms in class 2. Even if most mutations were not neutral, as required by this theory, accumulation of relic polymorphisms could readily explain observed levels of heterozygosity.

Further distinction between the various types of polymorphic genes requires other considerations. The existence of estimates of evolutionary rates from molecular genetics may help in making such distinctions.

11.9 Molecular Evolution and the Fate of Amino Acid Substitutions in Proteins

A comparison of the same (functional) protein in different organisms usually shows that during evolution a number of differences have accumulated, most of which are amino acid substitutions, but some of which are amino acid deletions and additions. The number of proteins, or polypeptides, whose amino acid sequence is completely known is still small though it is increasing rapidly. Some conclusions from the known sequences have, however, already clearly emerged.

The number of amino acid substitutions for any given protein is roughly proportional to the time available for divergence.

The paleontological record is especially good for vertebrates, and it is among these that the correlation of the number of amino acid substitutions with time can be best studied. An example is given in Figure 11.12. Proportionality to time is

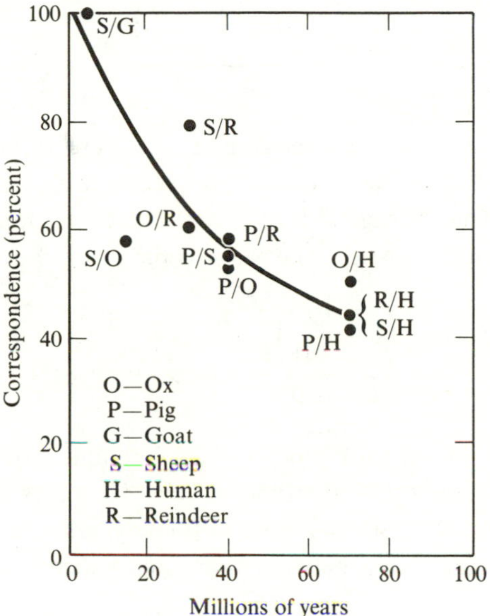

FIGURE 11.12
Amino-acid sequence correspondence between fibrinopeptides of various pairs of mammals plotted against the time that has elapsed since the last common ancestor lived. Percentage correspondence is based on number of identical amino acids in the same sequential position in both fibrinopeptides A and B. (From Doolittle and Blomback, 1964.)

actually an approximation that is only valid for relatively short periods. For longer periods the following treatment (Zuckerkandl and Pauling, 1965) is more satisfactory. Assume that the probability of a mutation occurring in the codon (the nucleotide triplet of DNA) corresponding to a given amino acid in a protein is α per unit of time. If the occurrence of mutations follows a Poisson distribution in time, the probability that no mutation occurs in a given codon in time t is $e^{-\alpha t}$, and the probability that at least one mutation occurs is $1-e^{-\alpha t}$ (see Appendix I for a description of the Poisson distribution). Assume that the rate α is the same for all codons in the cistron corresponding to a given protein. Then, if C is the number of amino acids in the protein, the mean number of amino acid differences expected in time t is

$$d = C(1 - e^{-\alpha t}). \tag{11.12}$$

Given d, C, and t, we can obtain α, the probability of a mutation per unit time, or $\tau = 1/\alpha$, the average time for one amino acid substitution, from

$$\alpha = \frac{-\log_e\left(1 - \frac{d}{C}\right)}{t}. \tag{11.13}$$

Thus, for the α chains of the hemoglobins, there are 18 amino acids that differ (out of 141) between man and horse, 17 between man and mouse, and 23 between horse and mouse. The average $\bar{d}$ is 19.3 and the corresponding time separation is believed to be 75 million years. The rate α is then from Equation 11.13

$$\frac{\log_e\left(1 - \frac{19.3}{141}\right)}{7.5 \times 10^7 \text{ years}} = 1.98 \times 10^{-9} \text{ years.}$$

Considering that evolution took place in *both* branches since the time when they split from a common ancestor, the time of 7.5×10^7 years must be doubled, and thus α must be halved. This gives $\alpha = 9.9 \times 10^{-10}$ years, corresponding to $\tau = 1/\alpha = 10^9$ years for the average time between amino acid substitutions at a given position in the protein.

Different proteins show different rates of change.

Table 11.11 shows similar estimates of rates of amino acid substitutions in mammalian evolution for other proteins and polypeptides. It is clear that there is a great range in evolutionary rates for different proteins, with the fastest of them being ten times as great as the slowest of them. The median value of τ is that of hemoglobins—namely, 10^9 years. The reasons for the differences in rates between proteins are not immediately clear. There are hints that at the beginning of their evolutionary history proteins are more variable. The improvement of function by natural selection may be accompanied by a decreasing frequency of advantageous changes. This may, therefore, account for the fact that older proteins seem to undergo, on an average, a smaller number of changes. It is also clear that since proteins like cytochrome C that are widespread and occur in almost all organisms are quite essential, change in them must be under strict selective control.

The data given in Table 11.11 are based on only a fraction of the possible evolutionary comparisons that could be made and are therefore not very accurate.

TABLE 11.11
Rates of Amino Acid Substitutions in Mammalian Evolution

Protein	*Total Number of Comparisons of Amino Acids*	*Observed Number of Amino Acid Differences*	*Observed Number of Differences per Codon*	*Estimated Number of Substitutions per Codon*	*10^{-10} Substitutions per Codon per Year*[a]
Insulin A and B	510	24	0.047	0.049	3.3
Cytochrome C	1040	63	0.061	0.063	4.2
Hemoglobin α-chain	432	58	0.137	0.149	9.9
Hemoglobin β-chain	438	63	0.144	0.155	10.3
Ribonuclease	124	40	0.323	0.390	25.3
Immunoglobulin light chain (constant half)	102	40	0.392	0.498	33.2
Fibrinopeptide A	160	76	0.475	0.644	42.9

Source: From King and Jukes (1969).
[a] The estimate for time elapsed since the divergence of euplacental mammalian orders is 75 million years. The average rate of evolution for the seven protein species in the table is 16×10^{-10} substitutions per codon per year.

More precise determinations, taking into account a larger number of species comparisons, indicate that the hemoglobin chain β is more variable than the α chain. The data do, however, give a valid background for the main propositions we have discussed.

Not all amino acids in a protein change at the same rate.

Some amino acids show no change whatsoever, even during long evolutionary time periods. An analysis of the number of changes observed per amino acid site shows that the variation in evolutionary stability is significant. However, a model in which a few amino acids (perhaps 5 percent of the total) are absolutely stable, while the others show a random variability (that is, all have the same probability of change) seems to fit the observed data fairly well (King and Jukes, 1969). Many of the amino acids that are stable during evolution probably have a key role in the function of the protein, and cannot, therefore, be changed without a loss of function.

Gene duplication must play a key role in evolution.

The existence of several related proteins, or polypeptide chains, in an organism immediately suggests the importance of gene duplication, especially if the genes determining the related proteins are closely linked. This fact was well known in *Drosophila*, both on the basis of theoretical considerations and of cytogenetic data. Molecular analysis has emphasized gene duplication in a striking way. Hemoglobins are made of several chains α, γ, β, and δ, which evolved one from the other probably in this order (Ingram, 1963). The chains β, δ, and probably γ are closely linked, suggesting an origin by "tandem" duplication. Translocation may have separated originally contiguous duplicates, as the α and β chain genes are not linked. Table 11.12 shows the probable history of the hemoglobin chains, based on the assumption that evolutionary rates were constant. Further variation can take place by

TABLE 11.12
The Approximate Time of Derivation of Different Hemoglobin Chains from Their Common Ancestor

Chains being compared	*Number of differences*[a]	*Estimated time of derivation from common chain ancestor* ($\times 10^6$ *years*)		*Corresponding geological period*
β and δ	~6	44		Eocene
β and γ	~36	260		Beginning of Carboniferous
α and β	78	565		Toward end of Precambrian
α and γ	~83	600		Toward end of Precambrian
Gorilla α and human α	2	14.5	Mean 11	Pliocene
Gorilla β and human β	1	7 3		

Source: From Zuckerkandl and Pauling (1965).
[a] The presence or absence of one or more contiguous amino acid residues in one of the chains is counted as one mutational change.

unequal crossing over between duplicates, giving rise to aberrant forms such as the Lepore hemoglobins (see Chapter 4). Other interesting examples of duplications, such as in the haptoglobins and immunoglobulins, have been discussed before (see Chapters 4 and 5).

The direct analysis of nucleic acids has recently posed an important problem connected with duplication, which is so far unsolved. It is possible to dissociate, for instance by treatment at high temperature, the two strands of a double-stranded DNA chain. Allowing the separated strands to remain together under appropriate conditions of temperature and ionic environment, reassociation between the complementary strands can take place. The rate of reassociation is also a function of the chain size. A similar phenomenon occurs on mixing DNA and messenger RNA. This latter nucleic acid is complementary to one of the two strands of the DNA and can, under certain conditions, associate with it to form a RNA-DNA hybrid.

A large fraction of the DNA in higher organisms seems to be highly repetitious.

The analysis of the reassociation of dissociated DNA duplexes from many higher organisms including man has shown that there exists a fraction that reassociates more readily (Britten and Kohne, 1968; Walker, 1968). The rapidity of reassociation is very probably due to the fact that a substantial fraction of the DNA segments are repeated a large number of times and thus have a much higher probability of reassociating. The fraction of the total DNA that is repetitious varies from one species to another, but can be as high as 40 percent. Many degrees of repetition probably exist. Some sequences, each made of some hundreds of nucleotides, may be repeated up to 100,000 times, while others are repeated less. It is known that the less highly repetitive DNA is functional, at least in the sense that RNA is produced from it. The significance of the repetitious DNA is not known. Some possible sources of repetition are, however, known. Thus, ribosomal RNA is known to be formed from DNA sections that may be repeated hundreds or even thousands of times. The same is true of mitochondrial DNA. These two together, however, on an average account for less than 1 percent of the total DNA in general. Other sources could be genes synthesizing different immunoglobulins (see Chapter 5). Such genes plus ribosomal and mitochondrial RNA cannot explain all of the repetitious DNA—for example, not accounted for is that found in plants and other organisms not endowed with the ability to form antibodies. It has been suggested that repetitious DNA may serve some important function, in cellular or chromosomal structure, perhaps in relation to the control of protein synthesis (Britten and Davidson, 1969).

Homology of DNA of different species parallels the evolutionary relationship.

Studies using DNA and RNA from different species have shown a close parallelism between the evolutionary separation between two species and the degree of divergence between their DNA's, as measured by the extent and rate of association of separated DNA strands from the different species. Earlier studies (Hoyer et al., 1965) dealt with the DNA now known to be repetitious. This fraction also shows, in general, the expected evolutionary behavior. More recent work has been aimed at studying the similarities of the nonrepetitious DNA, believed to be made of unique, or mostly unique sequences. The techniques are based on the thermal stability of reassociated duplexes and were calibrated on the basis of results with synthetic or artificially altered polynucleotides differing by a known percentage of base pairs. These can supply estimates of the percentage of nucleotides that differ in two DNA's. From then on, in a similar way to that used for amino acid substitutions, the mean time for a nucleotide substitution can be estimated (Laird et al., 1969). The percentage of mispaired bases has been found to change at different rates for the unique and the repetitious DNA. The former varies almost at the rate of fibrinopeptide A; the latter changes less—approximately at the rate of change of hemoglobin chain β.

A further category of DNA investigated by these authors was that synthesizing ribosomal RNA, which they did by studying the stability of DNA-rRNA hybrids. This DNA shows a considerably higher level of evolutionary stability, somewhat higher than that of cytochrome C and insulin. It seems reasonable that the protein-making machinery, which includes the ribosomes, must be kept rather constant, or else the functioning of cells and organisms may become impossible. Presumably, requirements for large amounts of protein bring about duplication of the DNA coding for the protein-making machinery. Some mechanism must prevent this DNA from diverging too rapidly under mutation pressure.

11.10 Theoretical Considerations on the Mean Time for Gene Substitution

The mean rate of gene substitution must depend on the frequency, μ, with which new mutant genes appear, on their selective advantage s, if any, in the living environment, which we assume to be constant, for simplicity, and on the population size N. This last quantity is important not only in determining the magnitude of statistical fluctuations, but also the rate of formation of new mutant genes at every generation.

In a population of N diploid individuals, the number of new mutant genes per generation is $2N\mu$. Many of these are doomed to extinction, irrespective of their selection coefficients, because of chance fluctuations. If the probability of ultimate fixation is P, then the number of new mutant genes appearing per generation that are destined to be fixed is

$$M = 2N\mu P. \tag{11.14}$$

The quantity M is the expectation of the number of mutant genes produced per generation that will be fixed, and is therefore proportional to α, the mean number of substitutions at one amino acid site per year. When account is taken of the duration of a generation in years, g, and the mutation rate is measured per amino acid site, then

$$M = g\alpha. \tag{11.15}$$

The mean evolutionary time for one gene substitution is

$$\tau = \frac{1}{\alpha} = \frac{g}{M}. \tag{11.16}$$

Similar considerations apply to the mutation rate per nucleotide substitution, which, however, is less accurately determined but must be comparable. From present knowledge of mammalian evolution, as we have already seen, $\alpha \sim 10^{-9}$. Assuming an average generation time of four years, this gives $M = 4.10^{-9}$, from Equation 11.15.

The quantity P in Equation 11.14 depends on the selection coefficients of the mutations. For a mutant allele whose selective advantage is s, it has been shown by Kimura (1962) and others that

$$P = \frac{1 - e^{-4Nsp_0}}{1 - e^{-4Ns}}, \tag{11.17}$$

where p_0 is the initial gene frequency. For a new mutant gene $p_0 = 1/2N$ and, therefore,

$$P = \frac{1 - e^{-2s}}{1 - e^{-4Ns}}. \tag{11.18}$$

The limit of this expression when s approaches zero, is $P = (1/2N)$, which is the probability of fixation of new neutral genes. For neutral genes, therefore, from Equation 11.14,

$$M = \mu. \tag{11.19}$$

If s is greater than zero, P is greater than $1/(2N)$ and, therefore, M is greater than μ.

Probably every new mutant allele has a fitness different from that of the normal allele at the same locus. Some new mutant alleles have lower fitnesses than the normal alleles, some have higher, and some have fitnesses so nearly equal to those of the normal type that they can be considered as neutral ($|s| < 1/2N$, as already mentioned). Mutant genes having fitnesses lower than those of the normal alleles have a negligible chance of being fixed and can be neglected in the computation of M.

If p_A is the proportion of mutant genes that are advantageous, for which the probability of fixation is greater than $1/2N$, p_N is the proportion that are neutral, for which the fixation probability is $1/2N$, and the others have a fixation probability of zero, then

$$M = p_N\mu + p_A\mu k, \tag{11.20}$$

where $p_N + p_A < 1$ and $k > 1$. The quantity μk is the average rate of substitution for the advantageous mutant genes. Thus, M can be greater, equal to, or smaller than μ, depending on p_A and k—that is, depending on the distribution of fitness values among the advantageous mutant genes. If assumptions concerning the form of this distribution could be made, then Equation 11.20 could be expressed more precisely.

With an average mutation rate per locus of 10^{-7}, accounting for "nonvisible" mutations by an arbitrary increase of the observed unbiased mutation rate, the mutation rate per amino acid (assuming 100 or more amino acids per locus) is 10^{-9} or less. It is thus somewhat lower than the observed value of M of 4×10^{-9}. There are, however, various factors that are set arbitrarily in this calculation. It seems reasonable to conclude that the estimates of mutation rate and of observed substitution rate are quite similar. They seem to be at least of the same order of magnitude as would be expected, from Equation 11.19, if all mutations were neutral. This does not mean that all mutations are in fact neutral, but merely that disadvantageous mutations are balanced, approximately, in their effects on M, by the greater probability of fixation of advantageous mutant genes. In other words, from Equation 11.20, $p_N + p_A k$ is approximately unity.

Kimura (1968) has suggested that most mutations are neutral because the load that would be demanded by substitution through selection of as many mutant genes as are actually substituted is too large. His computations are based, however, on the assumption of independent action of different loci at the level of fitness. This assumption is arbitrary. If a threshold model is assumed, as in the heterotic balanced polymorphisms already discussed, then it has been shown by Sved (1968) that the excessive substitutional load disappears.

The rate of gene substitution inferred from molecular evolution may be compared with that obtained from interracial comparisons. The time for an amino acid substitution is $1/\alpha = 10^9$ years. With more than 100 amino acids per locus, the time for substitution per locus is less than 10^7 years. From the estimate of the variation in $\log(1-f)$ between racial groups already given, 0.13 per 10^4 years, the rate of divergence in one of the two branches is $0.13/2 = 0.065$ per 10^4 years, or one unit of $-\log(1-f)$ per 150,000 years. At this rate 10^7 years of change would correspond to an f value of 0.9986. A complete gene substitution corresponds to $f = 1$, but this is, of course, only approached asymptotically. It is likely that interracial comparisons give an evolutionary rate higher than that expected for the whole species, both on account of the fact that smaller, and hence more rapidly evolving, groups are compared, and the fact that they live in different environments.

11.11 The Relative Roles of Natural Selection and Drift in Human Evolution

The role of natural selection in shaping man must have been extremely important. The same is true for any other living organism. Natural selection is an immediate consequence of the existence of genetic variation. It is also the only force capable of bringing about long-term—that is, inherited—adaptation to the environment. There is no question, therefore, of the possible importance of natural selection. The problem is determining its intensity and the magnitude of nonadaptive changes due to drift, relative to those due to natural selection. Even if the role of natural selection is clear, from theoretical considerations, the measurement of its intensity is not so simple, as we have seen.

It is often very difficult to discover the detailed mechanism by which natural selection acts on a given character.

An example in point is skin color. To say that a dark skin color has a selective advantage in the tropics seems trivial. Protection against the sun *may* be the right explanation, but how does the selective process take place? Several explanations have been advanced. It may be that erythematous responses in white-skinned people, especially children, exposed to the tropical sun are sufficiently dangerous to be the basis for selective response. However, in several African populations albinos survive and their frequency is perhaps not too different from that in white populations. Thus, a more extreme lack of pigmentation than found in the ordinary Caucasian does not seem to be very dangerous even in the tropics. It has been suggested that black skin affords protection against skin tumors induced by ultraviolet radiation, which seem to be rarer among black than among white people. But the frequency of such tumors, at least that known today, hardly seems large enough to account for a selective effect of the necessary magnitude. It has also been suggested that the selective factor may be vitamin D deficiency or excess. This vitamin is produced in the skin, usually from a precursor supplied by the diet. Dark skin would prevent the formation of excessive amounts of the vitamin, which can be dangerous, where the amount of sunlight is great per unit area, as in the tropics. A light skin would allow the formation of sufficient amounts where sunlight is not sufficient. All of these mechanisms, and others, may have taken part in the process of skin-color change, but it is very difficult to assess their relative importance. It is worth mentioning, however, that skin-color variation may be rapid. Descendants of Jews who went to Iraq and Iran some 2000 years ago have a much darker skin color than descendants of those who migrated to less desert-like areas at a similar latitude or who migrated to the north. There are as yet, unfortunately, no data available to solve the problem of what contribution was made to this change by gene flow.

Even for the best-known polymorphism, sickle-cell trait and anemia, many important physiological details of the selective mechanism are still missing, and many demographic details are not fully understood. It is difficult, for example, to assess the relative contributions to selection from differential mortality as compared to fertility. The limits to experimentation in man, the impossibility of reconstructing in all its aspects the environment of the past, and the difficulties of considering all possible relevant aspects even of the present-day environment, make this kind of problem unusually difficult.

Ecological variation is not always easy to interpret.

The existence of several conspicuous differences between human races, which are considered *a priori* likely to represent local adaptations to different environments, has stimulated the search for "explanations." Much variation accompanies the change in climate with latitude (or altitude) and thus much attention has rightly been paid to the climate. "Ecological rules" derived from observations in animals have been partially confirmed in man.

Schreider (1964) has studied the weight-to-surface ratio, which is expected to be higher in areas having a cold climate and lower in those having a hot climate, to facilitate regulation of body temperature. In Europe the ratio, in units of kilograms per square meter, goes from 39 in Germans or Finns to 37 in the inhabitants of Calabria (Southern Italy) and 35.4–36 in Sahara Berbers and in Yemenites. There is an approximate gradient with temperature, though it is somewhat different, for Africans, ranging from 37 in Mali and the Upper Volta to 33–36 in Central Africa and 30–31 among Bushmen and Pygmies. Among populations of the Pacific, inhabitants of cold countries again have a high ratio (Eskimos 38–39, Mapuches of the Andes 39.2) while Mexicans have 34.9–36.4 and hotter countries give lower figures (31–32 for the Philippines and 30.9 for Malaya).

With some exceptions, and subject to the reasonable limitation that different racial groups show different gradients, the rule on the variation of weight with temperature surface seems to hold for men. It should be noted, however, that it does not hold for women. This may reflect the tendency toward division of labor between men and women, men being more exposed to conditions of extreme stress, in which selection due to climatic conditions may be more important.

Another classical ecological rule, which is valid in animals and which also implies a dependence on the climate because of its importance for heat regulation, is based on the ratio of the length of the limbs to that of the trunk. The greater the ratio, the greater the loss of heat. The ratio is therefore expected to be lower in colder climates and higher in hotter ones. It is found to be lowest in Northern Europe and largest in Pygmies, with a fairly regular gradient in between. No major exceptions to this rule are known.

As for any other anthropometric measurement, there is very little information on the short-term effects of environment on these traits. An unknown fraction of the variation could be under direct environmental control, through nutrition or other variables.

Other possible adaptations to the climate have been mentioned. These include the decrease in nostril size in colder climates, and the mongolian shape of the eye, thought to be a protection of the conjunctive against glare (on snow and ice) and against cold. Empirically observed correlations, for example of Diego and Gm factors with climate, may also be mentioned (see Wilson and Franklin, 1968). For the morphological features it is easier to imagine what the selective mechanisms may be. Like all correlations, these also may be spurious and reflect chance historical accidents, or nonadaptive (for example, sexual) selection.

It is difficult at the present time to assess the relative importance of adaptive and nonadaptive changes.

The problem of the relative importance of adaptive and nonadaptive changes is of great theoretical interest. It can be answered only tentatively, at this stage of knowledge. It may be useful for this discussion to consider drift as the only nonadaptive force, combining in it all chance factors that are, by definition, nonadaptive. It is easy to absorb migration into drift, for it is only the combined effect of drift and migration that is observed. Consideration of population size without migration is meaningless with respect to the prediction of drift, unless we consider the total size of a species that cannot, by definition, receive immigrants and for which, therefore, migration is zero. Even mutation has a strong element of chance, both because of the random nature of its occurrence, and the unpredictability of its result. Total mutation pressure, however, remains a factor to be considered separately.

In some organisms mutability is known to be under genetic control. It has already been mentioned that the DNA which controls ribosomal RNA shows a smaller than average range of evolutionary variation. This may be the consequence of stricter stabilizing selection for these genes. It may also be that this is, in part, achieved by lower mutability. It is not impossible that different segments of the genome have different mutability, for example because of duplication by different DNA polymerases. This suggestion is, however, entirely speculative at the present time.

It is still impossible to answer quantitatively the question of the relative roles of drift and selection. The following facts seem, however, to be fairly well established.

1. When microgeographic variation is analyzed, it is often of the same order of magnitude as that expected under drift. In areas of recent immigration, however, such variation may also be the result of incomplete admixture.

2. When variation over a larger area is considered, some genes show selective effects. A few of these may be somewhat buffered by balancing pressures (for example, ABO and haptoglobins). A few may be, or may have been, under differential pressures in different parts of the world. Fairly large local selection differentials probably apply for skin color and perhaps for other racial traits connected with body build. They may also apply to some polymorphic markers such as Fy, Gm, and Diego. The bulk of polymorphic markers, however, show variation from one racial group to another that is not far from the amount expected under random drift. The information available is, however, too scanty to permit a valid comparison between the observed variation and that expected under drift alone. More knowledge on separation times, population sizes, and migration is needed.

3. The data on molecular evolution indicate mean evolutionary times that are not far from those expected for drift alone, though once again, this does not, of course, rule out a major contribution from advantageous mutations.

4. Selection for continuous variation, as, for example, for skin color, body build, or behavioral characters, usually affects several genes at once. For these characters drift is likely to be of less importance. When several genes are at play, the amount of random variation (in terms of variance) is decreased, as is true of an arithmetic mean, by a factor equal to the number of components—that is, the number of genes—affecting the character. Selection, when gauged from the frequencies of the individual genes affecting the character, will, however, be less intense as more genes contribute to the phenotype, and the effect of a particular gene will be correspondingly less important.

These considerations suggest that the selective differential of the average gene may be small, for many genes so small as to be within the range that can be considered as neutral. Changes in the environment, and therefore in selection intensities, however, complicate the situation.

Intergroup selection may be responsible for some changes in the frequencies of polymorphic genes.

Natural selection is commonly defined in terms of the individual's "competition" with the environment. Sometimes, however, natural selection may take the form of direct competition between individuals. The more general definition of selection, namely that which uses the concept of fitness as measured by the number of descendants, includes both the struggle for existence between individuals of the same species and the struggle against the environment.

Groups are often sufficiently isolated that it is meaningful to consider them as separate entities, even though they may, of course, have some degree of gene exchange. Differential rates of growth of different groups may then be a cause of change in the gene frequencies, when these are considered for the whole species. This is referred to as **intergroup** selection, as contrasted with **intragroup** selection, which is that between individuals.

Intergroup selection may take the form of direct competition between the groups. Wars are an example of such competition. By causing excessive mortality in one group as compared to another, by forcing gene flow, miscegenation, forced migrations, and other similar phenomena, wars have undoubtedly played a role in evolutionary changes. Intergroup selection, however, like intragroup (or individual) selection, need not be entirely competitive. The mere fact that one population grows at a different rate than another one, perhaps because of more favorable environmental conditions or of some technological advance increasing food production or reducing mortality, causes a change in the world gene frequencies. The concept of fitness can be applied to groups in the same way as it applies to individuals for intragroup selection. Group selection is now, however, determined mostly by cultural rather than by biological factors. It was noted by Hulse (1963) that the human species today has a much lighter skin color than it must have had some time ago, simply because Caucasians, who happened to have a lighter complexion, first adopted agriculture and thus increased disproportionately in numbers. Many other genes followed a fate similar to that of skin-color determining genes. The Rhesus allele, R_0, which may have been the original Rh allele (see Chapter 5) now has a frequency of less than 10 percent when estimated for the whole population of the world. This change has probably taken some 50,000 years. Intergroup selection may be the chief factor bringing about this change, which would under intragroup selection have required a selective coefficient of the order of 0.006, which is relatively large.

A picture of the evolution of our species can at present only be speculative.

Consideration of the potential importance of intergroup selection brings us back to the models of hominid evolution discussed earlier. Our incomplete knowledge of the geography, climates, fauna, and flora in the periods during which our genus was evolving, and the poverty of the paleontological record of human evolution, make any tentative reconstruction highly speculative. Probably, much more information will be made available in the future. At present, however, it does not seem inconceivable that the model of hominid evolution in which a relatively isolated population "explodes" and then invades, possibly supplanting neighboring populations (model 2 on page 694), is the nearest to the actual evolution of the genus *Homo*. Barriers between continents and within them have arisen and disappeared during this time. Large areas have become habitable and inhabitable at various times under the spread and the retreat of glaciers and with the increase or decrease in humidity. Long periods of hunger may thus have alternated with periods of plenty. It seems likely that this may have favored the formation of large isolates over long periods, one of which may have been situated in Africa and another in Asia sometime approximately 50,000 years ago. A proto-African type, of which Bushmen and Pygmies are perhaps the most direct representatives, may have been prevalent in Africa during approximately the same period in which Rhodesian man evolved. The connections between these proto-Africans and Rhodesian man are unknown. The group located in Asia, perhaps in the southeast, started spreading in all directions some 20 or 30 thousand years ago. Thus future Orientals went to the north, from where future American Indians and Eskimos were to go farther east across the Bering Strait, and Australians and Melanesians to the southeast. Earlier still, from this group or from Africans, or from both, the Caucasian was formed and occupied the Middle East and was the main actor of the Neolithic revolution, as far as we now know. Before the last glaciation the three main groups—Negroids, Caucasoids, and Mongoloids—had probably already been formed.

At the end of the last glaciation an important cultural and biological wave originated in the Middle East, and from there spread to the northwest towards Europe, perhaps absorbing—culturally and biologically—hunters who had migrated there earlier or who may have remained there during the whole period of the glaciation. This wave also spread to the southwest, and mixed with proto-Africans, perhaps introducing there the Neolithic revolution as well as genes of Caucasian type found mostly, but not exclusively, in East Africa. The wave also spread eastwards to occupy, or reoccupy, India.

Part of this picture is in agreement with known facts, part is hypothetical or based on very approximate time estimates. It is offered as a speculation to fill the gap in our actual knowledge. It does, however, help to visualize how a theory of shifting centers of cultural and biological success, created by environmental changes and technological innovations, may have shaped man's early history, the record of which is still inadequate.

General References

Cavalli-Sforza, L. L., "Population structure and human evolution." *Proceedings of the Royal Society, Series B* **164:** 362–379, 1966.
(A review of approaches to the establishment of the phylogeny of the major human racial groups.)

Clark, W. E. LeGros, *The Fossil Evidence for Human Evolution* (second edition). Chicago: University of Chicago Press, 1964. (A very useful review.)

Coon, C. S., *The Origin of Races.* New York: Alfred A. Knopf, 1962. (A useful survey.)

Washburn, S. L., and P. C. Jay, *Perspectives on Human Evolution.* New York: Holt, Rinehart and Winston, 1968. (A stimulating collection of papers.)

12

Eugenics, Euphenics, and Human Welfare

Most of the potential applications to human welfare of the knowledge we have surveyed so far in this book fall under two headings: *eugenics* and *euphenics*. Because of the social importance of such applications it is difficult to remain completely objective (in the sense of being free from all emotional factors) but we have at least made the attempt to be so. Clearly, the interaction between genetics and the social sciences can be a very important part of this underdeveloped subject, which is sometimes called humanics.

12.1 The History of Eugenics and Some General Problems of Its Practice

The aim of **eugenics** is the improvement of the human species by decreasing the propagation of the physically and mentally handicapped (*negative* eugenics) and by increasing that of the "more desirable" types (*positive* eugenics). It is, in other words, the application to man of the methods developed by breeders for improving their stocks by artificial selection. As we shall see in this chapter, we now know that the scope for eugenics is severely limited for both theoretical and practical reasons.

Most objections are avoided by an approach that is euphenical, rather than eugenical. **Euphenics** is the improvement of the phenotype by manipulation of the environment. Its definition could be broadened so as to include all of education, psychology, and medicine but such a broadening would dilute its significance. It seems best to limit the use of the word to applications for which genetic knowledge is essential. This greatly restricts present applications, though those of the future are likely to be very important. Psychologists and educationists have shown until now very little, if any, interest in the importance of considering innate individual differences. It seems inevitable that education can be more easily perfected if it can be better adapted to individual needs and moulded to fit, as much as is compatible with practical considerations, the potential of each child.

The term **genetic engineering** is commonly used to designate all applications of genetic techniques, from artificial selection to the direct chemical manipulation of genetic structure made possible by recently acquired knowledge of the mechanisms of life. A few of these techniques are potentially of enormous social significance and must be applied with great caution. The development of methods of controlling the sex of human offspring or of "cloning" individuals (Section 12.5), for example, must be accompanied by the realization that their indiscriminate use may have unwelcome social consequences. On the other hand, techniques such as those now used for prenatal diagnosis of fetal disease can have essentially beneficial effects when they have been perfected for use with humans. Such distinctions must be made, for although no scientific discovery is in itself dangerous, man's inability, or refusal, to control its use may have disastrous results.

Eugenics is an old concept and practice.

Many primitive societies practice negative eugenics. When infanticide is not regarded by a society as necessarily being a crime, and is used as a means of birth control, it is easily extended to the elimination of the malformed. Negative eugenics reappears over and over again throughout human history. In early Roman times, the Tarpeian rock was said to be the place for the disposal of handicapped children. Positive eugenics was advocated by philosophers—Plato, for example—but was practiced only rarely. Frederick II of Prussia is said to have given the best girls to his best soldiers, to promote the formation of an "elite."

In modern European culture, eugenic thought was given great emphasis by Francis Galton. He advocated the study of twins for the understanding of certain aspects of inheritance (see Chapter 9) and created correlation and regression methods for measuring the resemblance between parents and their children (Galton, 1894). He worked mostly during the second half of the nineteenth century and did not have the benefit of knowledge of Mendelian theory. In fact, Galton quantitated the "blending" theory of inheritance. Galton's theory predicted, at least approximately, the observed correlations between relatives for quantitative characters, which he himself was the first to measure. It was only following Weinberg's work in 1909–10 and Fisher's famous 1918 paper that it could be proved that the correlations between relatives for quantitative characters, the last stronghold of the anti-Mendelian currents of thought, were fully and more adequately explained by Mendelian theory.

Eugenic thought was not greatly disturbed by the fall of the Galtonian theory of inheritance. It was, in fact, largely independent of it. The practical application of artificial selection did not require the support of a precise theory of inheritance. The very considerable evidence accumulated by the success of empirical breeding for better animals and plants was sufficient to convince people of its applicability to man.

In the United States, the beginning of the twentieth century saw the flourishing of eugenics, especially under the leadership of Davenport, who tried to capitalize on the new Mendelian theory of inheritance. Unfortunately, the application of Mendelian rules to human genetics was conducted in a highly uncritical way by the group of workers connected with Davenport. Every character was assumed to mendelize, on the basis of insufficient evidence, and grotesque conclusions were reached (for a review of this history see Haller, 1963). The reaction of scientific circles to the poor scientific performance of the eugenists of that time might well have suffocated the new wave of eugenics. However, the growth of human genetics during the thirties revived eugenic aspirations. In more recent times, such distinguished geneticists as H. J. Muller have taken a strong stand in favor of positive eugenics. The development of new techniques and knowledge, and the realization of the genetic dangers connected with the use of atomic energy and the increase in radiation backgr und, were powerful elements in restoring a balanced interest in eugenics and putting the field on a firmer scientific basis.

There is always a racist danger in eugenics.

We consider **racism** to be the belief that some races are inherently superior to others and that this gives them a right to dominate or eliminate their supposed inferiors.

One danger to which eugenics has been constantly exposed during its development is its association with racism. At the time when eugenics was developing in the mind of Galton, racism was put forward as a scientific theory by a French diplomat, the Count de Gobineau, who was a great propagandist of the myth of the superiority of the German "race." In his book *Sur les inegalités des races humaines* (1853) he presented a history of the world which, in his view, proved that most of the good things on earth were due to the congenital superior ability of people of Germanic origin and that "degeneration" is the consequence of racial admixture of the original founders of a country. Racists of other countries take different views on which particular "race" is superior (usually, of course, their own), but still accept the principle that some "races" are superior to others. Gobineau was unusual in that he gave the supremacy to a "race" other than his own.

If the idea that some races are superior to others is accepted, eugenists are likely to become racists simply by asking the question, should not the "superior" races be expanded and the propagation of the "inferior" ones be limited? A most serious scientific objection is that superior and inferior are not at all easily defined, as we shall discuss later, and that the comparison of people from different countries and cultures provokes an almost inextricable confounding of biological and cultural factors. The fact that biological evolution is slow and cultural evolution may be fast makes it, *a priori*, more likely that differences in socially important traits between races are cultural rather than biological.

Many eugenists of the nineteenth and early twentieth century were racists (including Galton). In more recent times, the American Eugenic Society has openly condemned racism. A poll of opinion among geneticists would show that they share—practically unanimously—the same point of view. Almost the only evidence of exception to this point of view is a recent book collecting some genetic articles that are largely inspired by racism. A review of it written by L. C. Dunn for *Eugenics Quarterly* in 1968 provides an eloquent and illuminating answer.

Eugenic proposals at the intragroup level escape, in part, the criticisms that can be levelled against racism.

It is difficult or impossible to say, for many loci, which allele is "good" and which is "bad." The judgment depends inevitably on the environmental, genotypic, and even social background. Today, the sickle-cell gene is "bad" in the European or North American environment. But it is still "good" in Africa, where it helps to fight malaria, and was also "good" elsewhere until a short while ago for the same reason. Similarly, the G6PD-deficiency gene is good in a malarial environment. It is bad if its carriers need some relatively common drugs, which may, in some cases, cause serious hemolytic crises. It is also bad when it is associated with another, less well-known, gene. For, if the carriers of *both* genes inhale the pollen or ingest the raw beans of the plant *Vicia faba*, they suffer from serious and often lethal attacks of the hemolytic disease called favism. Epistasis, heterosis, and genotype-environment interactions may make it impossible to state whether a given gene is good or bad without a series of specifications that often render useless any such judgment. In the vast majority of cases, moreover, our ignorance paralyzes our judgment. We know that the large store of genetic variation accumulated in every species serves two very important purposes. One of them, common to all species, is that of meeting possible and unpredictable changes in the environment. Variation plays the role of an insurance policy against possible changes in the environment. If we could decrease at will the present amount of genetic variation by selecting certain "desirable" types, we might later have to pay very dearly for the resulting temporary improvement. A type defined as desirable might, for example, lack resistance to a new pathogen that suddenly arises; a type that had been selected against, and effectively removed from the population, might have had resistance to it. In our present state of ignorance it seems reasonable to subscribe to the plea of Neel (1970) and others, that it is dangerous to take steps that may grossly alter existing levels of genetic variation.

The second purpose of accumulated genetic variation is probably important mainly for human societies. Division of labor requires different abilities. Variation supplies them. A trend towards greater homogeneity may be seriously counterproductive in this respect.

The fact remains that some genes or inherited conditions, such as Down's syndrome, are unambiguously disadvantageous and that eugenic measures that could decrease their incidence would be universally welcome. Some traits such as, for example, inventiveness, artistic talents, and resistance to disease are almost unambiguously favorable and their development should be encouraged, if at all possible. They may, to some extent, be inherited. Thus, though in a more limited way than was conceived a century ago, eugenics could play an important role in shaping man's future.

Some very serious difficulties remain. The right of the individual to reproduce is considered an important part of individual freedom. Very few people are, at least at the present time, willing to abandon it, and probably with good reason. There are, however, two considerations that mitigate against this attitude. Man is confronted today with another motive for controlling his reproduction, which is much more urgent than eugenic aspirations for improvement of the species.

The explosive rate of growth of the human species is making it more and more urgent to generate quickly the right attitude toward birth control, which is at the moment restricted to a very small fraction of the world population. It is possible that the solution of this urgent problem may help to facilitate eugenic programs. We still do not know how to get the whole species to adopt measures of birth control. If we can learn how to do this we may in the process also learn to introduce eugenic considerations, for similar human attitudes are required for both.

Another kind of reason for limiting human reproduction pertains to the individual family rather than to the species. **Genetic counseling** is a service for married or engaged couples to help them understand and solve problems related to the risk of having genetically abnormal progeny. The basis for genetic counseling is a knowledge of the genetics of the abnormalities, which makes possible the prediction of the risk of abnormal progeny in a given mating together with an unequivocal clinical diagnosis of the abnormality. Prediction may be based on Mendelian expectations for simple recessive and dominant conditions, or on empirically determined risks for diseases whose inheritance is less well understood, such as diabetes or schizophrenia. The principles of genetic counseling are necessarily those of Mendelian inheritance and population genetics, as discussed throughout this book. In practice, genetic counseling is, to a large extent, really a branch of psychological medicine (see WHO report on Genetic Counselling, 1969, also Reed, 1955).

Spreading and increasing knowledge of human genetics will undoubtedly strengthen genetic counseling. There are no obvious moral objections to genetic counseling. People who now seek it do so on a voluntary basis, and so are likely to follow the advice they receive. Thus, eugenic programs are, in fact, being carried out, although so far only to a very limited extent.

Among the major difficulties facing eugenics is the fact that most of the means of eugenic action, such as abortion of abnormal fetuses, are often barred by religious prejudice and legal restriction. For instance, Roman Catholic obstetricians tend to deny their patients this treatment. The very slow rate of change observed so far in Roman Catholic ethics suggests that a significant fraction of the world population will not, or will not easily, have legal access to this powerful remedy.

12.2 Prospects for Negative Eugenics

Negative eugenics, which is the elimination of genetic defects, can, in theory, be practiced in two main ways: by avoidance of reproduction by individuals whose progeny is likely to be handicapped (which in some cases may be achieved simply by assortative mating) and by elimination of the defective progeny, which, of course, poses moral and social problems.

The avoidance of reproduction in cases of genetic risk can be applied before a potentially "dangerous" marriage takes place, or after. In practice, this includes the following possibilities.

Avoidance of Marriages Having High Genetic Risk. This method, practiced by means of counseling rather than legal restriction, is accepted by almost all societies. There are two main types of genetic risks: those that obtain even if only one of the parents is a carrier; and those that obtain only if both are. The latter type of genetic risk can be avoided by suitable choice of marriage partners (negative assortative mating).

Avoidance of Pregnancy in Marriages Having High Genetic Risk. This practice also encounters no moral opposition in almost any circles, at least as long as it is not compulsory but is based, rather, on counseling, and carried out by approved means of birth control.

Sterilization of Carriers of Genetic Handicaps. Technically this is an easy surgical procedure in men. It is less easy in women. Formerly sterilization of women was accomplished by irradiation of the ovaries with X-rays; today it is done by tying the fallopian tubes or by removing the uterus. Sterilization by these procedures seems to have no significant effect on the hormonal balance or the personality of the sterilized persons, but may obviously affect their lives in other ways. Sterilization does encounter enough moral and social opposition in some societies and countries for it to be illegal, while in some other countries is a legalized procedure, though not necessarily widely practiced. Institutionalization of mental defectives also, of course, effectively prevents reproduction.

When pregnancies that might give rise, or will surely give rise, to a severely handicapped offspring have occurred, the only eugenic measure left is the elimination of the offspring. There are widely different moral reactions to this measure, especially depending on whether it is the zygote or the newborn child that is eliminated. These two possibilities must therefore be considered separately.

Induced Abortion. During the first two months of pregnancy it is a medically simple matter to effect an abortion by minor surgery or minor nonsurgical procedures. Aborting the fetus is still surgically possible, without serious danger to the mother, until about the twentieth week of pregnancy. Early abortion is practiced as a means of birth control, legally in a few countries and illegally in almost all others. Moral constraints are therefore relatively ineffective. It would seem that the use of abortion to prevent the birth of a handicapped individual should have a stronger moral backing than its use for birth control, which can be achieved more simply in other ways. There may be psychological contraindications against abortion that have to be considered on an individual basis. It seems, however, that, in general, voluntary abortion for the elimination of abnormal fetuses should be legalized in all countries. Indications for abortion may, of course, vary. In almost no human mating is the probability of genetically handicapped progeny higher than 50 percent. In high risk cases, it is usually of the order of 25 percent or less. The problem of making a decision is greatly eased where there is the possibility of diagnosis of genetic defects by **amniocentesis**, which is the aspiration of amniotic fluid and subsequent examination of cells of fetal origin (see Steele and Breg, 1966). All readily detectable chromosome aberrations can in principle, be diagnosed prenatally using this technique. In Table 12.1 a list is given of some diseases that can at present be diagnosed in the fetus by amniocentesis before the nineteenth week, and thus in time for therapeutic abortion. It is also possible to obtain blood from the fetus, and to detect fetal cells circulating in the mother's blood, both of which procedures might increase diagnostic possibilities. The direct collection of fetal blood is not as widely practiced as amniocentesis because of the risk of damage to the fetus. It should be stressed that the proportion of pregnancies in which therapeutic abortion for eugenic purposes might be indicated, at this time, is small. With increasing knowledge it will rise, but almost certainly not to more than 5 percent of all pregnancies.

TABLE 12.1
Familial Metabolic Disorders Demonstrable in Tissue Culture

Disorder	*Deficient Enzyme or Mutant Phenotype*
Acatalasemia	Catalase
Branched-chain ketonuria (maple syrup urine disease)	Branched-chain α-ketoiso-caproate decarboxylase
Chediak-Higashi syndrome	Cytoplasmic inclusions
Citrullinemia	Argininosuccinate synthetase
Cystathionuria	Cystathionase
Cystic fibrosis	Metachromatic granules
Cystinosis	Increased free-cystine
Galactosemia	Galactose-1-phosphate uridyl transferase
Gaucher's disease	Glucocerebrosidase
Glucose-6-phosphate dehydrogenase deficiency	Glucose-6-phosphate dehydrogenase
Glycogen storage disease type II (Pompe's disease)	α_1-4 glucosidase
Homocystinuria	Cystathionine synthetase
Marfan's syndrome	Metachromatic granules
Mucopolysaccharidosis	Metachromatic granules
Orotic aciduria	Orotidylic pyrophosphorylase and orotidylic decarboxylase
Phytanic acid storage disease (Refsum's disease)	Phytanic acid α-hydroxylase
Sphingomyelinosis (Niemann-Pick variants)	Sphingomyelinase and/or sphingomyelin and cholesterol accumulation
X-linked uric aciduria (Lesch-Nyhan syndrome)	Hypoxanthine guanine phosphoribosyl transferase

Source: From Nadler (1969).

Physical Elimination of the Handicapped Individual (*a Form of Euthanasia*). This measure seems to be incompatible with the great majority of moral attitudes in socially advanced countries. It is cited here because it is practiced, mostly in the form of infanticide, in a number of cultures. Some economically advanced societies have practiced it in recent times even on adults. For instance, in Nazi Germany a large number of patients of mental hospitals were subjected to euthanasia. It seems clear that the only socially acceptable procedure is the development of special educational techniques for the handicapped and, if necessary, institutionalization.

It should be noted that when eugenic measures are being considered not only short-term but also long-term genetic effects should be taken into account.

We shall treat separately the following: dominant genetic defects; recessive genetic defects; defects with complex (or unknown) genetic causation; and possibilities, and consequences, of assortative mating. For eugenic purposes, the use of the terms dominant and recessive will be restricted to the genetic defect itself. Thus a *dominant* condition is one in which the heterozygote has the physical handicap. As these defects are usually rare, the homozygous condition is usually unknown, or is sometimes known to be lethal. A *recessive* condition is a pathological condition appearing only in the homozygote, while the heterozygote is normal or almost so from a medical point of view, even though he may be readily distinguished from the homozygote for the "normal" allele.

Negative eugenics for dominant defects: if the fitness of the bearer of a genetic defect caused by a dominant allele is low, as is usually the case, most affected individuals carry the gene as the result of recent mutation.

The frequency of most "dominant" defects at birth is very low because they usually impose a severe handicap and therefore are exposed to strong selection. Equilibrium between mutation and selection is established fairly rapidly. The loss at each generation is equal to the frequency of diseased D multiplied by the selection coefficient s, and this product must equal the rate of mutant gene formation per generation, which is 2μ (see Chapter 3). If selection takes place by mortality after birth or by infertility (of which celibacy is one component), the frequency of diseased at birth, D, will be $2\mu/s$. The fitness $f = 1 - s$ is low for most of these defects. It is, for instance, 20 percent for chondrodystrophy, and thus 80 percent of all affected births are the result of new mutations (or phenocopies) and only 20 percent are due to reproduction of already existing mutant genes. By abolishing completely the reproduction of chondrodystrophic individuals, the frequency at birth would decrease by 20 percent in the first generation but would remain constant thereafter, unless means were found of decreasing the mutation rate.

Down's syndrome (mongolism) is another example of an effectively dominant defect. A few instances of reproduction by individuals with Down's syndrome are known, and among their offspring the expected 1 : 1 proportion of affected is found. Their fitness is so low, however, that practically all cases, as least for the 47-chromosome type, are the result of fresh "mutation," more exactly, of nondisjunction in one of the parents, almost always the mother (Penrose and Smith, 1966). It is known that older women have a much higher chance (nearly 5 percent for mothers older than 45 years) of producing nondisjunctional gametes. Therefore, avoidance of reproduction at a late age for women seems highly recommendable. In pregnancies of older women, analysis of the chromosome set of the fetus by amniocentesis should be recommended. Such a recommendation is even more important for pregnancies in which either parent, whatever his or her age, is known to be the carrier of a translocation involving chromosome-21. The carrier of such a translocation has a 25 percent chance of producing defective progeny. Normal carriers are usually detected only when they, or their sibs, or other relatives properly located in the line of descent, do produce offspring afflicted with Down's syndrome.

The great majority of sex-chromosome aberrations cause infertility but do not severely impair longevity. They thus impose a significant load on the individual and on society because the affected are likely to spend most of their lives in institutions. Their detection by amniocentesis would, however, require the monitoring of *all* pregnancies, which seems, at present, to be an impossible proposition. For some time to come, the application of amniocentesis, considering the difficulties of the procedure and its cost, is likely to be limited to pregnancies carrying a high risk of giving rise to a recognizable genetic defect.

At the beginning of the century, eugenic propaganda exploited a few instances of idiocy, probably having genetic components but not clearly characterized, in which some mentally defective or antisocial persons were found to have a large number of "affected" descendants. The statistics used were largely biased, and it is possible that a substantial part of the unproductive or antisocial behavior in the incriminated families had environmental causes. In general, eugenic measures for dominant defects can decrease only to an insignificant extent the incidence of the defects.

Negative eugenics for recessive defects: it is almost impossible to eliminate disadvantageous alleles.

If a defect is recessive, its frequency at birth, D, depends on the mutation rate and the fitness, as for dominants, but is equal to μ/s (half as much as for dominants—see Chapter 3). The formula is only valid for genes that are fully recessive with respect to fitness—that is, genes for which the normal homozygote and the heterozygote have exactly the same fitness. This may be true even if the two genotypes are phenotypically distinguishable. The differences that we are able to detect may be of no importance from the point of view of natural selection. Vice versa, natural selection may be sensitive to differences that we have not detected. Even small departures from the equality of fitness of the normal homozygote and the heterozygote can affect the equilibrium frequency. Moreover, inbreeding and drift may also affect the frequencies of the affected homozygotes, as was discussed in Chapters 7 and 8.

The frequency of the disadvantageous allele is, of course, much higher than that of carriers of the defect. The latter will be of the order of mutation rates, and higher only if the fitness of the recessive homozygote is high—that is, if the defect has a small effect. The proportion of heterozygotes, however, will always be much higher, being approximately twice the square root of the frequency of recessive homozygotes at birth. Thus, for a defect like phenylketonuria, which has a frequency at birth of 1/10,000, the frequency of heterozygous carriers is about 2 percent.

POSSIBLE APPLICATIONS OF NEGATIVE EUGENICS TO RECESSIVE DEFECTS

Eliminate All (or as Many as Possible) Defectives Before Birth. In the list (Table 12.1) of those defects that can be diagnosed by amniocentesis, some are recessive. This approach would, of course, eliminate the problems for families and society of handicapped children. Gene frequencies would, however, be only slightly affected. The equilibrium gene frequency would change from μ/s to μ, and would not change at all for a gene, such as that for PKU, for which the fitness of the recessive homozygote is almost zero. Moreover, the change would be very slow, as in all selection against a recessive.

Eliminate All Progeny of Marriages Between Heterozygotes When These Are Recognizable. Such a rule would probably meet great moral opposition, in that heterozygotes are normal and in that three-quarters of their progeny are expected to be normal. The change in gene frequency would, in any case, be small. An equilibrium at one-half the gene frequency valid under natural selection would be approached slowly.

Bar All Heterozygotes From Reproduction. This measure would make acceptable for reproduction only a *very small* fraction of humans, if any. Considering only recessive genes for which the heterozygotes are at present distinguishable (see Table 12.1), only about 30 percent of the human population would be eligible for reproduction. These genes are, however, only a small minority of the recessive deleterious genes in existence. The overall estimate of lethal equivalents is one or more per gamete (see Chapter 7). This would give two or more, say four, fully penetrant ($s = 1$) deleterious recessives per zygote. The probability that an individual would be free of such deleterious recessives would be e^{-4} or only 1/55. It might even be much less than this figure, which is computed on the assumption that all recessive deleterious genes have full penetrance, as the expected number of deleterious genes per zygote would be higher if they had incomplete penetrance.

With recessive defects, therefore, negative eugenics is essentially powerless, and the only hope of decreasing the cost to society of the recessive heterozygotes is prenatal diagnosis and therapeutic abortion. This procedure would, of course, decrease the load, at the expense of another social cost: that of diagnosis and abortion. It is, however, very likely that the latter cost would be smaller than the former.

Sex-linked recessive genes behave more-or-less like dominants.

Sex-linked deleterious recessives, like the hemophilia gene, are effectively dominants in the male and hence respond to selection as dominants. The fitness of hemophilia has, however, changed greatly in recent decades and should be considered in the context of selection relaxation (see Section 12.6).

Some important diseases, such as schizophrenia and diabetes, have an incompletely known, probably complex, mode of inheritance, but the genetic component of their determination is likely to be large. Similar considerations, perhaps intermediate to those given above for dominant and recessive defects, apply in this case. For genetic counseling, statistics of risks, compiled from observations on the incidence of the diseases in families, can satisfactorily substitute for Mendelian probabilities.

12.3 Eugenic Potentialities of Consanguineous and Assortative Mating

Premarital genetic counseling is becoming increasingly common. Its influence is still small when judged for a whole population, but is bound to increase. The question arises then of its possible consequences. When only one member of a potentially "dangerous" mating is the known carrier of a genetic defect, the effects of genetic counseling, if any, are similar to those of sterilization. In a number of instances, however, the danger is tied to the specific union, and either person should have little problem in finding a suitable alternative mate. It is useful to distinguish four categories of potentially dangerous matings.

Consanguineous Matings. The empirical risks of consanguineous matings are fairly well known and are not, in general, very high. The risk that the closest consanguineous mating accepted by most societies (uncle-niece, aunt-nephew, or double first cousins, which all have $F = 1/8$) will produce progeny dead or defective at birth is perhaps five- to tenfold higher than that for nonconsanguineous unions. Each halving of the inbreeding coefficient approximately halves this value. The risk relates only to damage from recessive genes, and consanguinity has practically no effect on the risk that damage will be caused by dominant genes. When, however, a recessive defect is known to be present in a specific family, the increased risk may be the cause of greater concern. Thus, if it is known that one or more sibs of an individual are affected by PKU, and he wants to marry his first cousin, whose sibs do not have the disease, the chance that the union can potentially produce PKU progeny is about 1/6. A test for heterozygosis of the potential mates may, in part, relieve them of the worry.

From the point of view of population projection, avoidance of consanguineous marriages, due to genetic counseling or any other reason, does not solve the problem of recessive deleterious genes. It merely delays it. Genetic deaths that do not occur now will occur later. A high level of consanguinity helps to keep *down* the gene frequency of deleterious recessives. In fact, if the population is increasing in numbers, the avoidance of consanguinity *increases* the absolute number of deaths, when these are counted over a long time period.

Blood-group Incompatible Matings. The most common type of Rh incompatibility in Caucasian populations would be eliminated if Rh-negative women avoided marriage with Rh-positive men. As only about one out of six Caucasoids are Rh-negative, this would impose a serious limitation on the range of possible choices of a husband for these women. Moreover, as the numbers of Rh-negative men and women are approximately equal, almost all Rh-negative men would have to marry Rh-negative women. The human species would thus be split, at least among Caucasoids, into a larger Rh-positive "race" and a smaller Rh-negative "race", with cross marriage being dangerous for Rh-negative women. This drastic measure is not necessary, however, since Rh disease has been brought under control by recent therapeutic procedures (see Chapter 5). There are still, however, major difficulties in producing adequate amounts of the anti-Rh antibody used for the prophylactic treatment.

Matings Between Heterozygotes for Frequent Deleterious Recessives. Heterozygous advantage due to resistance to malaria has created several polymorphisms in which one allele is usually inferior to the "normal" type. For G6PD deficiency, the disadvantage is small. For sickle-cell anemia it is fairly high, the ensuing reduction of fitness being almost 80 percent. In thalassemia, the homozygous condition is practically lethal. In places that had, or still have, a high incidence of malaria, these genetic diseases represent a real social problem. It is estimated that sickle-cell anemia causes some 100,000 deaths a year in the whole world. Frequencies of the relevant genes are so high in some areas that thalassemia and sickle-cell anemia may affect 1–2 percent of all the newborn. The diagnosis of heterozygotes is easy. There is at present no effective therapy, but this was true for Rh and several other genetic diseases until a short time ago. As a therapy will probably be developed in a relatively short time, the suggestion of assortative mating to avoid the disease may be valid for only a relatively short period. The fact, however, remains that for the present, heterozygotes for either sickle-cell or thalassemic anemia would do better if they avoided marriage with a person heterozygous for the same gene. The problem of such heterozygotes finding a genetically suitable mate is much less difficult than the same problem would be for the Rh-negative woman. Heterozygotes for these anemias are rarely more than twenty percent of the population. Thus, the necessity of choosing, for a mate, a normal homozygote would not be an impossible limitation. In theory, assortative mating of this type could be continued indefinitely, eliminating the disease and keeping gene frequencies almost unaltered. Thus with a heterozygote frequency of 20 percent, 40 percent of marriages would be between homozygotes and heterozygotes, and 60 percent between normal homozygotes. Probably, in practice, the minority heterozygote group would be at a slight handicap in finding a marriage partner, as it would be limited to finding a mate from the majority group, while the majority group would have essentially no limitations. The smaller fertility of the minority group that might thus ensue could slowly lead to the disappearance of the deleterious gene in nonmalarial areas, perhaps at a faster rate than its decrease due to natural selection against homozygotes. Campaigning in favor of such assortative mating is certainly no easy task. A therapy for the disease may solve the problem before long, but the cost of the therapy may be high and resort to assortative mating may still be worthwhile.

Matings Between Heterozygotes for Rare Deleterious Recessives. By far the great majority of deleterious recessive genes are rare. The principle of assortative mating would operate on them in the same way as on the frequent recessives. In fact, it would create even less difficulty for the heterozygotes initially, because relatively few marriages would be precluded for them. This, however, is in part a consequence of our present incomplete knowledge; for, if all heterozygotes for all deleterious genes were detectable, a larger number of matings would be precluded.

The number would not, however, in any case be very large. Suppose every individual is heterozygous for four deleterious recessives. If for each of these, the heterozygote frequency is around one percent, the frequency of individuals not carrying any one of the four genes is, roughly, $(0.99)^4$ or 96 percent, so that only 4 percent of potential mates would be excluded by the assortative mating. The task of detecting all these deleterious recessives would, however, be enormous, as thousands of genes would have to be examined in each person. At the moment, in any case, we can carry out such an examination for only a very small proportion of recessives.

Barker (1966) has examined more fully the consequences of assortative mating for a recessive lethal gene, considering the effects of partial assortative mating and of the reproductivity of marriages between heterozygotes. Both factors are important for making predictions.

Mutation pressure not (or not fully) checked by natural selection will increase the frequency of deleterious genes. As we have already seen, however, and shall see again in more detail later, the increase under mutation pressure is so slow that it cannot be a cause for much concern.

In conclusion, negative eugenics, using sterilization or related means, can achieve only very limited success for defined genetic diseases because a large fraction of genetic defects are the direct outcome of mutation, the more so, the more serious the defect. Hence, the only hope would be the reduction of mutation rates, at present only a theoretical possibility. When handicapped individuals can be detected before birth, therapeutic abortion may be indicated. This procedure will, however, only slightly improve the future genetic picture of the population. Its effects are comparable to those obtained by sterilization. Therapeutic abortion, however, can be justified because it may considerably relieve social and individual burdens. Genetic counseling may lead to assortative mating. This can prevent some serious genetic diseases. Its application should be seriously considered, especially for diseases for which there is as yet no known therapy.

12.4 Methods of Positive Eugenics

Elimination of genetic defects by negative eugenics is severely limited in its scope. In theory, however, breeding for improvement of desirable traits might give a faster response. Practically all socially important traits have a polygenic type of inheritance. The results of artificial selection depend upon the following factors:

1. The rate of improvement by selection depends directly on the extent of genetic variability with respect to the trait to be improved. The gain expected in one generation of selection is equal to the product of the heritability and the selection differential, which is the difference between the mean of the individuals selected as parents of the next generation and the mean of the whole population (see Chapter 9).
2. Possible side effects on superficially unrelated characters may accompany selection. Selection for one trait usually brings about correlated responses in other characters. One prediction that often turns out to be true is that selection toward an extreme type brings about a reduction in fertility. It also leads to inbreeding, which itself may be a partial cause of infertility. Pleiotropic effects and linkage of the selected genes with other polymorphic genes may also be responsible for other less predictable types of side effects.
3. The achieved selection differential (that is, the difference between the mean of the offspring of the selected parents and that of the general population) may differ from the expected values based on heritability estimates if the genetic determination of the trait selected is at least in part due to complex types of interactions between genes, or between the genotype and the environment.
4. For many socially desirable traits, the relative importance of nature and nurture is unknown.

Given these limitations, selection for desirable traits in theory could be carried out in humans and the results predicted with some accuracy. The methods of selection could be based on **eutelegenesis**, the use of sperm from selected donors for voluntary artificial insemination. In mammals, there is a great excess of germ cells, especially from the male. The semen produced by one man during his life could fertilize perhaps tens of thousands of ova. There is also a wastage of female gametes, which is, however, not so large. Only about one percent of human ova actually give rise to progeny. The use in artificial selection of the other 99 percent would require either the transplantation of ova into the ovaries or oviducts of sterile women or *in vitro* fertilization followed by uterine implantation. Transplantation and implantation of ova are not beyond the reach of modern techniques (see R. G. Edwards et al., 1969) but are far less acceptable at the social level than the utilization of sperm, which can be stored almost indefinitely. Techniques that were developed by cattle breeders have now been adapted to man. Artificial impregnation of women is currently used for the treatment of certain kinds of sterility, in some cases using the sperm of the husband, and in others that of an unrelated volunteer donor. In fact, some degree of eutelegenesis is already practiced. Human sperm donors for artificial fertilization are undoubtedly chosen with a minimum of eugenic considerations in mind. The incidence of artificial insemination is, however, still extremely low.

Muller (1966) has been one of the major advocates of eutelegenesis. The storage of the semen of "excellent men" would, in his view, ensure the possibility of a sustained eugenic advance in "excellence." Screening the stored sperm against radiation would, in addition, insure against possible harmful effects of future atomic disasters.

EXPECTATIONS OF EUTELEGENESIS

It is easy to compute the expected gain in a desirable trait due to eutelegenesis. The most favorable model is that of culling selection (that is, the choice of the best extreme deviants) as practiced in plant and animal breeding. If only sperm is selected, only one sex, and thus half of the population, participates, and the possible gain is, accordingly, halved. Some numerical results are given in Table 12.2 for a desirable trait scaled as IQ is: namely, with a mean of 100 and a standard deviation of 15. The results are given for various fractions, p, of females participating in the eutelegenesis program, the remainder of the females reproducing at random. The participating females are assumed to be chosen at random with respect to IQ. The proportion, P, of males used as sperm donors is the upper P percent in the population with IQ's higher than S. The fraction of males with IQ's above the threshold S for selection as a sperm donor, is computed from the equation $P = z\sigma/(\bar{x} - M)$ where M is the mean and σ the standard deviation in the general population, $\bar{x}$ the mean IQ in the selected sample of sperm donors, and z the ordinate of the normal curve at the threshold for selection S (of sperm donors). This is a standard formula in selection theory, as discussed in Chapter 9. The mean IQ expected in the first generation $\bar{x}_1$ is computed from

$$\bar{x}_1 = p\,\frac{\bar{x} - M}{2}\,h^2 + M,$$

TABLE 12.2

Expectations under Eutelegenesis (Sperm Selection) as a Function of the Proportion of Females Participating in the Program, and the Proportion of Males Used as Sperm Donors: Mean IQ in First Generation Assuming 50 Percent Heritability

Sperm Donors			*Percentage of Females Participating* (*p*)					
IQ								
Mean Trait Value ($\bar{x}$)	*Selection Threshold* (*S*)	*Percentage Selected* (*P*)	1	5	10	20	50	100
115	105	38	100	100.2	100.4	100.8	100.9	103.8
130	122	6	100.1	100.4	100.8	101.5	103.8	107.5
145	140	0.36	100.1	100.6	101.1	102.2	105.6	111.2
160	156	0.008	100.1	100.8	101.5	103.0	107.5	115.0
175	171	0.0001	100.2	101.0	101.9	103.8	109.4	118.8

where the divisor 2 is due to the fact that selection only takes place in one sex, and the factor h^2 is the heritability. The values given in the body of the table are the gains, $\bar{x}_1$ for one generation, assuming a heritability of 50 percent, as is approximately true for IQ in a human Caucasian population. In later generations, the gain would decrease slightly, and would eventually level off after a large number of generations. From the data given in Table 12.2, it is clear that in order to obtain significant effects in a short period of time, the participation of females must be massive. With smaller participation, as noted by Maynard Smith, the increase to be expected for IQ might easily be matched by that obtainable, in the same period of time, with improved schooling. If selection were carried out also on females, the gain (that is, the excess over 100 of the figures in Table 12.2) would be at most double.

It may be added, by way of extrapolation, that if we could alter environmentally intelligence to the same extent as stature has been, we could obtain an amount of change that would only be possible with 100 percent participation of the females in the eutelegenesis program. The change in stature was in no way planned, and we are still uncertain about its causes, but we have seen (Chapter 9) that its origin must have been mostly environmental. The amount of the change was of the order of one standard deviation per generation. To obtain a similar effect for IQ, in addition to having 100 percent of females participating in the eutelegenesis program, it would be necessary to apply approximately the greatest strength of selection represented in Table 12.2. We do not yet really know what improvements in IQ could be obtained by improvements in education. Obviously, however, environmental improvement is possible, is potentially less dangerous with respect to disturbing the social structure, and is more readily accepted by everybody.

There is significant social resistance to eutelegenesis.

For a rapid effect of eutelegenesis, as noted above, females must participate on a large scale. The social and emotional problems created by such a policy could be enormous. How many males are prepared to be fathers of children who are not reared in their own families? What would be the consequences of the lack of a father figure, which psychologists consider so important, or worse, of the existence of many hostile fathers? How many people are interested in being outwitted by their children? It is not too surprising that these proposals have so far met with no substantial response in our present-day society. But there is an even greater difficulty in accepting eutelegenesis: agreement on what are desirable traits. Most people would agree on the desirability of some traits, like intelligence, social responsibility, artistic talent, generosity, and beauty. It is, however, very unlikely that all these traits could be selected for at the same time. Some of them, at least, may be to some extent mutually incompatible. In any case, simultaneous selection for several desirable traits is much more difficult than selection for a single trait. When a choice for the character to be selected has to be made, there may be much disagreement on which is the most important trait and whose sperm should be stored. As L. C. Dunn (1962) and others noted, even H. J. Muller contradicted himself on this point. In his first papers on the subject, he gave a list of famous thinkers, mainly scholars and scientists, whose progeny, he thought, should be increased by eutelegenesis. After a sojourn in Russia, during which he decided to withdraw his support of communist theories, he removed Marx and Lenin from the list of the men whose sperm should have been stored for future generations.

In addition, since insurance against unpredictable changes of the environment requires genetic variation, selection that fixes only one type is potentially dangerous.

12.5 Euphenics and Genetic Engineering

There is undoubted evidence that socially important traits like intelligence and mental balance are to a fair extent genetically determined. But there is almost equally impressive data showing that the environment is also important. Manipulation of the environment is at present an easier task than genetic manipulation. It is also clear that some genotypes can do very well in certain environments, while they do less well, or worse than average, in others. It would be a terrible waste to force a potential Bach to become a bricklayer or an engineer, or a potential Einstein to become an accountant. One difficulty is, of course, that we cannot accurately predict the potential performance of an individual. Even if we could, society has limited space for most types of activities. Ideally, we would like to be able to explore, as early as possible, the potential abilities of each child, and provide him or her with the best possible environment for the development of his or her abilities both from an individual and a social point of view.

We are still very far from this ideal. Even if we had the necessary know-how to create the right environments for the development of desired talents and to prevent unwanted physical and psychological tendencies, there would still remain, at least in many cases, the problem of deciding the most favorable course of action. As we have seen concerning eugenics, it is usually easier to decide which defects are to be prevented than which positive tendencies are to be fostered.

In a similar way, certain techniques that are considered a part of genetic engineering may prove to be a mixed blessing. These techniques essentially involve the manipulation of the genotype. If they alter the genotype of *somatic* cells or utilize the transplantation of somatic cells, they affect only the phenotype of the individual and will not carry over to future generations. They should then be considered as a part of euphenics. There are, however, genetic engineering techniques that affect germinal cells and therefore might be considered in eugenic programs.

Genetic engineering in man is, at the moment, a program for the future.

Among the techniques of genetic engineering are ones that have been tested successfully in other organisms but never yet applied to man. Some of these are successful in organisms sufficiently similar to man that their carryover to our species does not look like a formidable scientific program. Others have not been worked out yet in mammals but their *a priori* chances of success do not seem to be too low, and certainly will increase in the near future. (See the general references at the end of this chapter.) Among the possible techniques of euphenics and genetic engineering, are the following:

The Transplantation of Cell Nuclei from One Cell to Another. A major application could be the formation of clones of genetically identical individuals, by transplantation of a nucleus from a somatic cell into a fertilized egg whose nucleus has previously been inactivated, for example, by X-rays. This technique was developed for frogs by King and Briggs (1956), further developed by Gurdon and others (see, e.g., Gurdon, 1963), and is now being worked out for mice.

The Formation of Artificial Chimeras by Cell and Organ Transplantation. Kidney transplantation has been used in attempts to cure genetic diseases affecting the kidney, notably cystinosis (see, e.g., Lucas et al., 1969). This use of transplantation is certain to be extended to other genetic diseases affecting the kidney and other organs, especially the liver. Many genetic diseases affect enzymes specific to the liver. So far, liver transplantation has met with only limited success, mainly for technical reasons, though synthesis of donor specific proteins following a liver transplant has been reported by Kashiwagi and co-workers (1968). Transplantation of bone marrow in man following extensive irradiation with X-rays has been attempted, so far unsuccessfully, as a cure for leukemia (Mathé et al., 1965). If it becomes successful, it may prove to be an important way of curing many genetic diseases of blood cells such as sickle-cell anemia, thalassemia, and hemophilia (Motulsky et al., 1962). Bone marrow transplantation differs from the transplantation of such organs as the kidney, liver, or heart, in that the transplanted cells recolonize the bone marrow and divide to produce new differentiated blood cells of the donor genotype.

As discussed in Chapter 5, the main barrier at the present time to the success of transplantation is the recipient's immune reaction to "foreign" histocompatibility antigens present in the donor tissue. A variety of techniques directed at solving this problem, including specific matching of donor and recipient and specific inhibition of the immune response, are either known to work in animals, especially mice, or are being worked out. One approach, known to work in animals, is the use of very young recipients whose immune response is not yet fully developed. This could work for diseases that can be detected immediately at birth or before birth using amniocentesis. The dangers of transfusion for, or surgery on, fetuses are still, of course, relatively large and serious long-term consequences could result from the transfer of too many cells. This is because these cells may later produce antibodies directed against the recipient, leading to a serious disease, known as "runt disease," because of the resulting thwarted growth. There is no doubt, however, that a solution to the immunological problem will sooner or later be found for man, making transplantation available as a euphenic cure for genetic diseases.

An extension of this principle would follow from the development of *in vitro* long-term cultures of somatic cells that express normal *differentiated* functions. Such cultured cells could be transplanted into individuals genetically deficient for such functions. (Bone marrow transplantation for hemoglobinopathies is a special case of this approach.) Given suitable techniques for genetic manipulation of somatic cells in culture, it is even conceivable that cells from an affected individual could be genetically changed *in vitro* so as to function normally, and then transplanted back into the same individual. Such an approach would overcome the immunological barrier to transplantation. A serious problem in the use of somatic-cell cultures for transplantation is always the risk of inducing cancers by viruses carried by, or contaminating, the *in vitro* cultures.

The result of cell transplantation is an **artificial chimera**, in which tissues of different genotypes coexist. Chimeras are known to arise spontaneously in human dizygotic twins, as mentioned in Chapters 1 and 9, formed by tissue exchange between the fetuses. Artificial chimeras have been successfully used in a model case in mice, involving hereditary spherocytosis. In this case transplantation of the bone marrow into newborns has given rise to chimeras in a relatively large number of individuals (Motulsky et al., 1962). The extension to humans with related diseases is, as already discussed, an obvious possibility, though it is still fraught with problems.

Transduction and Transformation. These techniques have their origins in microbial genetics (for a review see Hayes, 1968). They involve the incorporation of a piece of foreign DNA into the chromosome of an organism, usually a bacterium. In transformation, the foreign DNA from the donor can penetrate into the bacterium because of the existence of a mechanism that permits the introduction of DNA fragments into the cell and their pairing with the bacterial chromosome. In transduction, the DNA from a foreign bacterium enters as part of a viral chromosome formed by reproduction of the virus in the donor bacterium. In either case, the existence of extensive homology between the DNA fragment and the recipient's chromosome is essential for the incorporation to take place. Homology assures pairing and exchange by crossing over. The DNA segment incorporated is that included between two regions of genetic exchange. The existence of phenomena similar to transduction in higher organisms, including presumably man, is made extremely likely by observations on tumor-producing viruses such as polyoma or SV40. The transformation to malignancy is believed to be similar to a transducduction. Evidence has been accumulated that part of the virus DNA enters a host cell chromosome and is responsible for the transformation in cell behavior that leads to malignancy (see Westphal and Dulbecco, 1968). Claims of bacterial-type DNA transformation in somatic cells (in the absence of known viral transmitters) exist in published reports, both for mammalian cell cultures and whole organisms, but they so far belong to the class of nonreproducible experiments. Similar techniques, if they could be made to work for man, might provide the best hopes of curing genetic diseases, perhaps even at the germinal level. These are, however, possibilities that still require a great deal of experimental work before they can be applied in practice.

Control of the Sex of Children To Be Born. Known differences between Y- and X-carrying spermatozoa could be exploited to separate the two types of cells and thus effect artificial fertilizations producing one or other of the two sexes at will. Partially successful experiments have been claimed in cattle and rabbits (see Gordon, 1957). Another approach to sex control that is already feasible is the selective abortion of fetuses whose sex has been determined following amniocentesis. These techniques, if applied to man, might have profound social influences. Etzioni (1968) was impressed by the danger of overproduction of one sex (presumably the male sex, at least to start with) that could result from the application of the techniques. A large excess of one sex could lead to dangerous competition between its individuals. Time would be required to balance out the initial excess and oscillations would probably be inevitable, unless a strict control were imposed. Pohlmann (1967) is not so pessimistic. He believes that sex control might decrease the conflicts arising in families when children of the undesired sex are born, and that it may reduce the sizes of families. The latter consideration is in agreement with the interpretation of sex-sequence variation within families that was discussed in Chapter 10. If couples kept having children until they had one of each sex (certainly an oversimplification of actual behavior), families would have an average of three children (see Chapter 10). This would be reduced to two if sex control were available.

It seems possible that if a suitable convenient form of sex control became available, it would be used on a large scale. The possibility of having children of desired sex may be a powerful help in making a reduction and standardization of progeny size acceptable to many people. If every family had two children of different sexes, then the numerical equality of the sexes, the constancy of genetic variation, and a trend toward reduction of the total world population, all desirable goals at the present time, would be assured.

Parthenogenesis—the Development of Eggs, Which Are Haploid, Without Fertilization. Artificial stimulation of eggs to undergo parthenogenesis can be done with some organisms and may also become feasible in man. Social problems that might be raised are related to those cited for the manipulation of the sex ratio, but are exaggerated by the fact that only the female sex is susceptible to parthenogenesis, which would therefore give rise only to female progeny.

12.6 Cultural Evolution and Its Effect on Natural Selection

The rate of cultural and social evolution can be very fast.

This book has mostly been concerned with the *biological* evolution of man, which clearly is a very slow process. Its tempo can be gauged by the average time taken for a nucleotide substitution, which is of the order of once every thousand million years (see Chapter 11). For organisms whose chromosome set contains some five billion nucleotide pairs, as does ours, differences accumulate at the rate of five substitutions per year. This is a high absolute, but a low relative, frequency. Thus, for a time of separation between man and gorillas or chimpanzees of at least ten million years, the total number of expected nucleotide differences is of the order of fifty million (in absolute terms) but only one in every hundred nucleotides will be changed. We can thus see how we can be so different, and yet still so similar to the more human-like apes.

The rate of biological evolution may have varied from time to time, and may differ from one organism to another. It must, however, have been small at all times, which is not true for another type of evolution affecting man (and possibly to some extent all other animals), namely, social or cultural evolution. An *increasing rate* is characteristic of this type of change. There may, of course, be times of crisis involving cataclysms and radical changes, as in biological evolution. However, from the history of man it is clear that the tempo of social or cultural evolution is continuously increasing at a rate that may be even faster than an exponential. For example, agriculture was invented about ten thousand years ago and has taken about all of the time up to the present to spread across the whole earth. There are, even today, a few groups in highly isolated areas, such as the Pygmies in Africa, that have not yet changed to this mode of life. Their first contact with it may, however, date from only a few hundred years ago, and acculturation is in progress even among those people. Other inventions have taken progressively less time to spread. The industrial revolution has spread across perhaps half of the earth in less than three hundred years. The atomic revolution has, in twenty-five years, taken a firm grip in at least a dozen countries. The main cause of the increasing rate of cultural evolution is the increase in the speed and the amount of communication. Even now, however, there are people who are economically ten thousand years behind the most advanced, and a whole range of others that are between two thousand years and ten years behind.

There are similarities between cultural and biological evolution. The latter has mutation as its element of change, the former has invention. Both of these are rare phenomena, sometimes recurrent. A fitness value can be attached to either type of change and a sort of natural selection theory could be conceived for inventions, which is not entirely dissimilar from that which is valid for mutations. The spread of inventions around the world could also be governed by migration laws that are similar to those for genetic mutations. One basic difference, however, between the two types of evolution is in the mechanism of transmission. Biological inheritance has strict laws of carry-over from generation to generation. These laws are not readily violated, are extremely conservative, and are responsible for making biological evolution such a slow process. Cultural modes, on the other hand, are transmitted *infectively*. All the progeny from type A and type B is usually of one type only. Progeny of mixed marriages acquire most, if not all, of their rights and customs from one of the two parents. This by itself gives greater flexibility to social evolution. Often, however, there is not even a need to wait for the next generation. A new habit or new ability may be acquired almost at once, or in a few years at most. Thus the rate of cultural evolution may be such that complete changes may take place in a population in the course of a few years. Naturally, many social customs have a strong inertia, and may carry over unchanged or almost unchanged for many generations. The force of tradition may be great, and for some traits at least, it may equal or even surpass the stability of biological characters.

We can therefore expect a whole variety of results in social and cultural evolution, ranging from great stability to very rapid change. We will be concerned here only with those social phenomena whose changes are likely to affect biological evolution.

Changes in mortality patterns have led to selection relaxation, which has in some cases eugenic and in others dysgenic effects.

The most striking social changes are perhaps those that have affected mortality patterns (see Chapter 6). The last three or four hundred years have witnessed the development of modern medicine, with a characteristic rate of change that increases exponentially, or even faster. Mortality between birth and reproductive age has consequently decreased by a factor of twenty, mostly in the last fifty years. This change has affected some diseases more than others. New diseases and new causes of mortality have been created by the change of environment. In any case, the effects of natural selection due to mortality have been considerably altered. Some genetic trends may have been fully reversed.

It is possible that many diseases which have disappeared had no clearcut genetic effects. Some diseases are known to have genetic effects, however. Malaria is among them, and is by far the most important example. Practically all other known cases are still debatable or are far less important.

A major eugenic preoccupation has been the increase in the number of cases of genetic disease due to relaxation of natural selection. This preoccupation is not always justified. In fact it is quite unjustified for malaria-dependent polymorphisms. Almost all of these seem to have been maintained chiefly as balanced polymorphisms, one homozygote being more sensitive to malaria, the other to a gene-determined disease. The disappearance of malaria releases the balance, and the elimination of the genetic disease then begins and proceeds to completion. Naturally, if the genetic defect is a recessive, its elimination requires a large number of generations. Exactly how many depends to some extent on the unknown small selective disadvantage of the heterozygote in the nonmalarial environment. Table 12.3 gives some computations of the gene frequencies and the frequency of homozygotes, and therefore of overt disease, on the basis of various hypotheses. Selective disadvantages of 0, 1 percent, and 5 percent have been assumed for the heterozygote. Remember that disadvantages of 1–5 percent are sufficiently small to be difficult to measure. The disadvantage of the homozygote has been assumed to be 100 percent (the present case for thalassemia), or 67 percent (a partially cured sickle-cell anemia) or 0, the case in which these diseases can be fully cured by a satisfactory therapy. The case in which the heterozygote disadvantage remains after a complete cure of the homozygote is probably rather artificial. In any case, it is clear that the individual and social burdens of these diseases are there to stay for some time. Sickle-cell anemia will gradually disappear, except in the limiting case in which the heterozygote has no disadvantage and the disease is completely cured phenotypically. In the other cases Table 12.3 shows that selection relaxation has slow eugenic effects.

Today a few genetic diseases are cured but only phenotypically as, for example, are phenylketonuria, galactosemia, fructosemia, and several other defects that can be corrected by appropriate dietary treatment of the infants. Diabetes and several other endocrine disorders can be cured by lifelong substitutional therapy. Tendency to dental caries, sensory defects, like vision acuity loss, congenital deafness, and accompanying mutism, all require expensive apparatus or rehabilitation programs. All this poses a financial burden on the affected person and his relatives. Society should undoubtedly, and sometimes does, take on itself this burden as completely

TABLE 12.3

Selection Relaxation for Recessive Traits, Like Anemias of the Sickle-cell and Thalassemia Type, Which Are Kept in Balance by Heterozygote Advantage. It is assumed that the heterozygote advantage ceases at time 0 and is replaced by a disadvantage of 0, 1 percent or 5 percent. Computations are carried out for the case in which the defective homozygote has 100 percent selective disadvantage, as for thalassemia, 67 percent as for sickle-cell anemia, 0 as if either of these diseases could be completely cured. Figures in the body of the table indicate the frequencies (in percent) of the heterozygote and of the homozygote at birth. The initial gene frequency of the recessive is 10 percent in all cases.

Generation	*Homozygote Disadvantage*																	
	100 *Percent (Thalassemia)*						67 *Percent (Sickle-cell Anemia)*						0 *Percent (Homozygote Cured)*					
	Heterozygote Disadvantage						*Heterozygote Disadvantage*						*Heterozygote Disadvantage*					
	0 *Percent*		1 *Percent*		5 *Percent*		0 *Percent*		1 *Percent*		5 *Percent*		0 *Percent*		1 *Percent*		5 *Percent*	
	het	*hom*	*het*	*hom*	*het*	*hom*	*het*	*hom*	*het*	*hom*	*het*	*hom*	*het*	*hom*	*het*	*hom*	*het*	*hom*
0	18.0	1.00	18.0	1.00	18.0	1.00	18.0	1.00	18.0	1.00	18.0	1.00	18.0	1.00	18.0	1.00	18.0	1.00
1	16.5	0.83	16.4	0.81	15.9	0.76	17.0	0.88	16.9	0.87	16.4	0.81	18.0	1.00	17.9	0.99	17.4	0.93
3	14.2	0.59	13.9	0.56	12.7	0.47	15.3	0.70	15.0	0.67	13.8	0.55	18.0	1.00	17.7	0.96	16.3	0.80
10	9.5	0.25	8.9	0.22	6.7	0.12	11.4	0.37	10.7	0.32	8.0	0.18	18.0	1.00	16.8	0.86	12.6	0.46
30	4.9	0.06	4.1	0.04	1.8	0.01	6.5	0.11	5.4	0.08	2.3	0.01	18.0	1.00	14.7	0.64	5.4	0.08
100	1.8	0.01	1.0	0.00	0.04	0.00	2.6	0.02	1.4	0.01	0.1	0.00	18.0	1.00	8.5	0.20	0.2	0.00

as possible. How is this load going to increase in the future because of selection relaxation? We will try to take into account in our discussion the possible effects of a usually unknown factor, the fitness of the heterozygote.

For some relatively frequent diseases such as PKU, there is some doubt about whether its fairly high frequency is maintained only by mutation pressure. A slight selective advantage of the heterozygote may be a contributing factor. For other diseases, the heterozygote may be at a slight disadvantage. If mutation rate is the only determinant, we can then be sure that the increase due to selection relaxation will be extremely slow. Small differences in the fitness of the heterozygote may, however, affect the rate of increase. The cases represented in Table 12.4 indicate, as usual, that the increase is never fast.

It should be added that some special cases may be more complicated. A disease like diabetes is very sensitive to the environment. In countries in which sugar consumption is low and there is extensive social stratification, diabetes is mostly a disease of the upper social strata, probably because the consumption of sugar is higher there. The improvement of medical treatment is usually accompanied by an

TABLE 12.4
Selection Relaxation for a Formerly Lethal Recessive that from Time Zero Onwards Is Cured, so that Its Fitness Is Restored to 100 Percent. Prediction of the frequency of the defect, with various combinations of mutation rates and selection in favor of the heterozygote, assuming that gene frequencies are at equilibrium at time zero and the heterozygote advantage remains unchanged.

	Initial Frequency of Defect at Birth and Mutation Rate				
	1/10,000			1/100,000	
	10^{-4}	10^{-5}	10^{-6}	10^{-5}	10^{-6}
Time, in Generations	*Heterozygote Selective Advantage*				
	0%	0.92%	1.01%	0%	0.3%
	Frequencies of Defect				
	$\times 10^{-4}$			$\times 10^{-5}$	
0	1	1	1	1	1
1	1.02	1.02	1.02	1.01	1.01
3	1.06	1.06	1.06	1.02	1.02
10	1.21	1.22	1.22	1.06	1.06
30	1.69	1.79	1.80	1.20	1.21
100	4.00	6.47	6.84	1.73	1.86

increase in general welfare, and thus an increase in the penetrance and the incidence of the disease. The increase in survival due to therapy is thus correlated in time with a phenotypic increase in frequency of the disease. The two effects may, to some extent, balance each other from a selective point of view. In any case, an increase in disease incidence need not be genetic. If it is fast, it is more likely to be of environmental origin.

We shall now examine the situation with dominant defects. Suppose that retinoblastoma, Huntington's chorea, or chondrodystrophy could be cured (presumably by an expensive treatment) so that patients could be restored to full health and genetic fitness. What would be the frequency of these diseases in future generations, and hence their social burden? The data are given in Table 12.5. Again, the increase is slow, being determined by mutation pressure.

In conclusion, selection relaxation may be dysgenic—that is, it may bring about increases in the incidence of disease in future generations. Such increases, however, occur very slowly. In such balanced polymorphisms as malaria-correlated anemias, selection relaxation is eugenic—that is, it effects changes in the desired direction. The changes are still, however, rather slow.

TABLE 12.5

Prediction of the Frequency (in percent) of Dominant Defects in Future Generations Following Selection Relaxation. Initial frequencies are taken as 1/500, 1/10,000, 1/100,000. Various combinations of mutation rates and selective disadvantages of the heterozygotes are considered, all giving equilibrium at the above frequencies of the defect at birth, before selection. It is assumed that after time zero the defect is completely cured, so that its fitness (namely the fitness of heterozygotes) becomes normal, while the homozygotes for the mutant gene remain lethal.

	Initial Frequency of Defect at Birth and Mutation Rate						
	1/500	1/10,000				1/100,000	
Time, in Generations	0.001	0.00005	0.00004	0.000025	0.00001	0.00005	0.00001
	Fitness of Heterozygote prior to Cure						
	0	0	20%	50%	80%	0	80%
0	0.2	0.01	0.01	0.01	0.01	0.001	0.001
1	0.4	0.02	0.02	0.015	0.01	0.002	0.0011
3	0.8	0.04	0.03	0.025	0.01	0.004	0.0016
10	2.1	0.11	0.09	0.06	0.03	0.011	0.003
30	4.7	0.30	0.24	0.16	0.07	0.031	0.007
100	6.2	0.86	0.71	0.47	0.20	0.099	0.021

12.7 Other Interactions of Cultural and Biological Evolution in Man

Cultural evolution of man has deeply affected the world around him. Almost all other living species have been changed in numbers, often very drastically, with many even becoming extinct, as a result of man's actions. Conditions for natural selection have changed. Industrial melanism, the mimicry of butterflies in an environment made darker by particles deposited by industrial smoke at the time when coal was the major source of power, is of no major social significance, but is an outstandingly clear example (see Ford, 1964, for a review). Today, the use of insecticides is an even more powerful agent of evolutionary change. We have even started to pollute the moon, both chemically and perhaps biologically.

Cultural evolution also affects the biological evolution of man. The reduction in mortality has decreased almost to nil the impact of natural selection through differential mortality. Large changes in fertility have also been induced by cultural changes.

Several other aspects of cultural evolution have genetic consequences. We will limit our consideration to those for which some facts are already known and which have greater social significance. Unfortunately, the interactions between cultural and genetic evolution have not been investigated in any depth. Very often it is difficult, on the basis of present knowledge, to separate strictly environmental and cultural changes from genetic changes. The importance of these interactions should not, however, be underestimated. We need to increase rapidly our knowledge of these factors. We urgently need to know much more about man, if we wish to avoid destruction under the impact of forces that we have ourselves unleashed.

A fundamental principle to be learned from population genetics in general, and from the genetics of human populations in particular, is that the amount of individual variation is enormous. Therefore, while learning more about man, in particular that which is essential for social welfare and the avoidance of our own destruction, we should remember that the picture of stereotyped man is fallacious. We should not only learn about averages but also about individual variation, without which our knowledge is incomplete and misleading. The sources, extent, and nature of this variation should always be present in our minds if we want to understand more about human nature, and make humanics a useful science.

12.8 Patterns of Fertility

Changes in marriage and reproduction patterns have been very important and will probably continue to be so. The almost complete disappearance of natural selection due to differential mortality follows the decrease of mortality. This has brought about parallel changes in fertility. Differential fertility, however, has not yet disappeared, and natural selection continues to act through individual differences in fertility.

Patterns of age at marriage, which have been changing in several countries, can have selective consequences, for instance with respect to deleterious traits that have an age of onset that is approximately the same as the marriage age, such as schizophrenia. Onset of these diseases considerably decreases the chance of marriage. A decrease in the age of marriage, therefore, increases the fitness of persons destined to have such diseases and an increase in age at marriage decreases their fitness. Naturally, it is also possible that a changed environment affects the pattern of age of onset, thus complicating the picture. Various countries have quite different patterns of age at marriage. In Ireland a late age at marriage became a popular means of birth control after the famine of the middle of the last century, and this pattern has survived until the present. The proportion of unmarried people has also probably changed with time. In the past centuries in Europe a considerable proportion of people did not marry. It is possible that this was part of the birth control mechanism operating at that time. There are various ways in which this fact may have had selective consequences, but at the moment it would be highly speculative to consider them in detail.

We have considered in various places the possible consequences of assortative mating. It seems likely that present trends are towards an increase in assortative mating, producing an increased variance.

Fertility is in many western societies strongly negatively correlated today with socioeconomic status (see Figure 12.1). The negative correlation between socioeconomic status and fertility may, however, be transient, as was pointed out in Chapter 6. It is probable that this is merely a reflection, at the intragroup level, of the demographic transition from high to low mortality, which is accompanied, after a lag, by a transition from high to low fertility. The transition is a consequence of increased hygiene and nutritional welfare, and thus is bound to show intergroup as well as intragroup variation. As the transition is still in progress this negative correlation is not surprising. If there were other causes for this correlation—that is, if certain stages of civilization were accompanied for other (psychological or social) reasons by a lower fertility in the higher classes—a phenomenon of "involution" might take place. Fisher (1930) thought that the fall of several civilizations of the past might have been a consequence of the lower fertility of the upper classes. Naturally, if there is enough social selection and social mobility to bring into the upper classes the most active and intelligent portion of the population and if this upper class of the population reproduces less there will be a dysgenic effect. The effect will have an intensity that is proportional to the extent of the negative correlation between social class and fertility and the positive correlation between social class and desirable qualities, and the heritability of these qualities. It would seem, however, that the present time is exceptional and, that whatever was true in the past, may not apply directly to the presently observed situation.

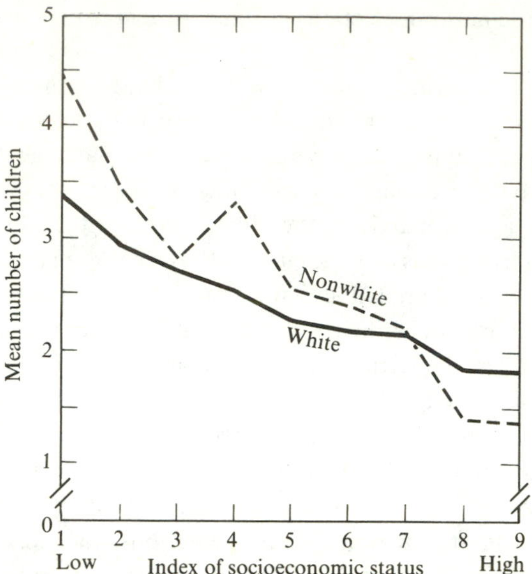

FIGURE 12.1
Average number of children per woman 25–29 years old, married once, with husband present, for whites and nonwhites, by socioeconomic status. From 1960 U.S. Census. (After Mitra, 1966; from Jensen, 1969.)

Several primitive societies do not show a similar trend. When polygamy (usually in the form of polygyny) is accepted, it is normally a measure of high socioeconomic status. As polygyny entails, on the average, a fertility which is roughly proportional to the number of wives, the correlation between socioeconomic status and fertility must be positive.

Another possible dysgenic change is the likely decrease in relative fecundity (potential fertility) due to the spreading of obstetrical help. At one time postpartum mortality of women was high. The stillbirth rate also was high and showed a socioeconomic gradient. Increased care at childbirth is minimizing these rates and increasing, presumably, the frequency of women who have difficulties at childbirth, on the assumption that these difficulties may, in part, have a genetic basis.

Birth control may have eugenic implications.

One major current problem for human society is the need to decrease birth rates. The precipitous decreases in mortality have not been accompanied by equal and concomitant decreases in fertility. The discrepancy between the theoretically unlimited geometric increase of population numbers and the limited increase of resources clearly, though naively, enunciated by Malthus has not—and cannot—readily be removed. Birth control is now more necessary than ever. Probably no species has ever been confronted with as serious a problem of population explosion as we are facing today. It is therefore unlikely that means of natural self-regulation of numbers can apply an adequate brake to our inordinate increase, before very serious suffering and perhaps major catastrophies intervene.

Even primitive societies have, and use, methods of birth control. But at no other time has the need for it been so seriously felt, and the dangers of population increase been so real. The underfed proportion of the species may have been even higher in the past than it is now. The availability of the daily bread may then have been an even more important consideration for an even larger part of the world population than it is today. Now, however, pressures other than just that of getting the daily bread are also being felt in modern industrial society, such as those of environmental pollution and overcrowding. In spite of this, birth control measures are still very unevenly used. Only recently have really effective techniques become available, and even these are not perfect. We are unable to convince some nations and population groups that value high fertility of the urgent need for birth control. Really efficient means of persuasion must be found, and more relevant research on birth control and its dissemination is greatly needed.

On the assumption that birth control will adequately spread and be accepted in time without coercion (which at present is a repulsive idea to most people), it is possible that not only the average size of families, but also the relative variation in progeny size will be reduced. If everyone married and all families had two children, natural selection due to differential fertility would be eliminated and no further systematic changes in the human population's genetic composition would take place, except by mutation. The discovery of ways to convince people that they should limit the number of children they produce may also be used to exercise eugenic influences.

Intergroup differences in reproductive rates have evolutionary consequences.

Intergroup differences in net reproductive rates, mainly due to cultural innovations, have led to genetic changes in the species as discussed in the previous chapter. To give one example of this, no longer than 100,000 years ago, most people may have carried the Rh blood-group allele *cDe* (R_0). Now the common allele is not *cDe* but, for example, *CDe* (R_1) or *cDE* (R_2) among Orientals and *r* (*cde*) among Caucasians. The disproportionate increase in frequency of the non-*cDe* alleles is mainly a consequence of the fact that Orientals and Caucasians adopted an agricultural way of life a few thousand years before Africans did, and thus were able greatly to increase in numbers. Similarly, as emphasized by Hulse (1963), the average skin color of the species, on the assumption that it originally was as dark as that of Negroes, has changed rapidly to a much lighter color simply because of the greater increase in those groups with lighter skin color, probably mostly due to technological, rather than selective, changes. The rate of change of skin color is, however, unknown. If the change in skin color shown by Iranian and Iraqi Jews is not due to gene flow but to direct genetic adaptation—that is, by means of selection—then the rate of genetic change may be high.

12.9 Migration

From the point of view of evolution, emigration is comparable to mortality and may have similar selective effects if emigrants are not a random sample of the population. The analysis of the selective effects of migration has been given very limited attention. Migrants are usually dissatisfied with their place of origin, owing, perhaps, to poverty or to certain local disadvantages. Usually, however, only a fraction of a population emigrates. This suggests that migration is selective with respect to behavioral factors that may be, in part, under genetic control, such as initiative, courage, restlessness, taste for adventure, and related traits. There have been studies on the mental balance of migrants, stimulated by the observation of a higher incidence of mental disorders among recent immigrants. It is very difficult, however, to distinguish, at present, a possible genetic basis for this effect from the adverse pressures of a new, and possibly, hostile, environment.

Apart from migration to remote countries, there have been important changes in the mobility of individuals within a country, usually connected with the search for jobs. The very rapid improvement in transportation has been a major factor in increasing mobility. More than three thousand years ago the domestication of horses and cattle and the invention of boats started the increase in mobility. During the last hundred and fifty years, however, the invention of railways and motorized vehicles has greatly increased mobility. This has had the effect of widening the group among which the choice of mates is made and of destroying or limiting previous isolation. Among the genetic consequences of the increase in mobility and decrease in isolation are the following.

1. On a wide geographic scale, several "hybrid" groups have been created between populations that were formerly widely separated.

2. On a smaller geographic scale, the "breakdown of isolates," in Dahlberg's sense, has followed in many countries. Inbreeding pockets have tended to disappear. The average inbreeding coefficient is, as a result, decreasing. This is not always the case, however, because the frequency of consanguineous matings, for example, also depends on economic and religious factors. It is believed, for instance, that the sharp increase in first-cousin matings at about the beginning of the nineteenth century in many European countries was the consequence of the introduction of napoleonic legislature. By equalizing the inheritance of land for all progeny, these laws tended to disrupt the unity of land ownership. Cousin marriages became

popular when it was realized that they could, in part, counteract the undesired effects of this legislation. The breakdown of isolates may also cause a moderate increase in the frequency of heterozygotes. It should be noted that this effect has occasionally been exaggerated, as was discussed in Chapter 9 with respect to the increase in height.

3. If the choice of a mate can take place from a larger group of people, it may be more selective, and thus increase the chances for assortative mating. It would be interesting to follow this hypothesis further, since it leads to an expected increase in phenotypic variance.

Urbanization involves selective migration and differential fertility.

One important change in the last ten thousand years has been the growth of towns and cities. All social studies show differences between urban and rural environments. One of potential genetic importance is the differential rate of reproduction of the population in these two environments. It was known to Galton that reproductivity is low in an urban environment. If there is differential migration of different genotypes from rural to urban environments this may create a powerful selection pressure, which may be further complicated by different selective conditions in the two environments.

12.10 Environmental Pollution and Mutagenicity

Environmental pollution has always been a problem. Today industrial pollution is reaching levels that are sufficiently alarming that concerned citizens are determined to make great efforts to change the situation. One effect of pollution especially relevant to human population genetics is the possible increase of mutation rates due to inadvertent introduction of mutagens into the environment. Under the impact of modern medicine, diseases are gradually being reduced to the most resistant core: those of genetic origin. These, we have seen, are ultimately the outcome of mutation.

We do not know the causes of spontaneous mutation in man. We do know, however, that a fraction of them must be due to radiation from cosmic rays and radioactivity from natural radioisotopes. In recent times man-made sources of radiation have been added to these natural sources, mainly through the medical use of X-rays and other radiation sources, radioactive waste from atomic bomb testing and atomic power plants, and other uses of atomic energy and radioactive isotopes. The "natural" radioactive background reaching the gonads, and thus of genetic significance, has been computed to be, on an average, 100 millirad per year, or about 3 rad (radiation units) per generation (see U.N. Scientific Committee Report on the Effects of Atomic Radiation, 1962). The rad measures radiation in terms of the energy absorbed by the irradiated material. Another commonly used unit, which is limited to X or gamma rays, is the roentgen, or r, which is based on the ionizing properties of the radiation. It is believed that about 28 percent of the natural radioactive background is due to cosmic rays, and 50 percent to gamma rays from the disintegration of the radioactive uranium, thorium, and potassium contained in the earth's crust. These two components vary greatly from place to place, even inside and outside homes. The remainder comes from radiation due to radioactive natural elements that are present in the body, mostly (20 percent of the total) potassium-40, and a much smaller proportion from carbon-14 or from ingested and inhaled radionuclides of the radium and thorium series.

Medical use of X-rays has increased the overall radiation reaching the gonads by an amount that varies from individual to individual and country to country. It is estimated to be at the most 100 percent, and on an average 30 percent, of that due to the natural radiation background.

The total radiation from atomic bomb testing between 1956 and 1965 that will reach the gonads by the year 2000 is estimated to be 76 millirads. By that time, all the short-lived radioactive elements (caesium-137 and strontium-90) will practically have finished their life, but the long-lived carbon-14, which is responsible for 13 of these 76 millirads, will still contribute an additional 167 millirads after the year 2000. Strontium-90 does not make an important contribution to the irradiation of the gonads. It affects mostly tissues such as the bone marrow, and therefore might induce leukemia or bone tumors. Altogether, it is believed that radioactive fallout up to December 1965 has added about 2 percent to the genetically significant dose due to the natural background, and about 8 percent to the dose significant for leukemia, or bone tumors. A limited period of atomic bomb testing has thus raised appreciably the levels of obnoxious radiation.

All radiation is harmful though the dose dependence is still not properly established.

Animal experiments, especially early work on *Drosophila*, and also work on bacteria, have shown that there is no threshold below which radiation is harmless. The effect, in terms of induced mutation, increases with the dose, proportionately at first, and then with a somewhat diminishing return for higher doses. Thus, no radiation dose is completely safe and all increases in radiation should be avoided if possible.

The only existing data on the relation between radiation dose and mutation rate for a mammal are those obtained using mice. The most extensive are those given by W. L. Russell and others (1963) who have measured the average mutational effects of irradiation on seven specific genetic loci. There are differences in the sensitivity of the gametes to radiation that depend on sex, maturity, and the intensity at which the dose is delivered. Chronic irradiation is less effective than acute irradiation by a factor of about five, for a 10,000-fold difference in radiation dose-rate per minute. The type of ionizing radiation (gamma, X-rays, or other), is of less importance, as long as the dose reaching the gonads is the same. The dose required to induce as many mutations as those produced spontaneously in one generation (the so-called "doubling dose") is believed to be, in mice, approximately 30–80 rad. This quantity is influenced by the level of the spontaneous mutation rate and is less satisfactory than a direct estimate of the induced mutation rate, when this is available. In the mouse the induced mutation rate for acute irradiation of spermatogonia is around 10^{-7} mutations per locus per r unit with great variation between loci (see United Nations 1966 report). X-rays cause, in addition to gene mutations, chromosomal breaks and, consequently, aberrations. These effects can also be studied in human somatic cells cultivated *in vitro*. The number of breaks induced has been shown to respond linearly to the dose, as is also true for gene mutation (Chu et al., 1961).

Our knowledge of X-ray induced gene mutation in man is still very limited. The analysis of the United Nations Scientific Committee on the Effects of Atomic Radiation (1962) had set as likely limits for the doubling dose in man values of 10 and 100 rads. It is still difficult to improve on this estimate. If the lower value is correct, this means that the natural background radiation contributes 30 percent of the observed mutations. Thus, if 10 additional rads reaching the gonads give rise to as many mutations as does the spontaneous mutation rate, then the natural background (3 rad per generation) must be responsible for $3/10 = 0.3$ of the total spontaneous mutation rate. If the upper doubling dose (100 rad) is right, then the natural background is responsible only for about 3 percent of the total spontaneous mutation rate. A later approximate estimate by the ad hoc U.N. Committee (1966) indicated that the addition of one acute (that is, administered at high intensity in a short time) rad per generation would add one seventieth to the total spontaneous mutation rate.

Radiation is only one type of mutagen. It is the one whose level we know will undoubtedly increase in the environment. A nuclear war would cause a dramatic increase, perhaps by more than one order of magnitude. In the case of a nuclear war, however, the immediate problems are likely to be so great that those due to the ensuing increase in mutation rates may appear to be of a second order. It is clearly not only because of the potential increase in mutation rates that nuclear war should at all costs be avoided, though its genetic consequences should not be forgotten.

If we could learn more about other causes of spontaneous mutation, and learn how to reduce them, we could ensure considerable progress in public health problems related to genetic diseases. The recognition of the increased incidence of Down's syndrome with maternal age was an important step in this direction. Because Down's syndrome is due to a chromosomal aberration rather than a specific gene mutation, however, this information is not as useful as it might otherwise be.

Point mutation is much less sensitive, if it is at all, to parental age. Other factors must be looked for.

At the moment, rather than augmenting our knowledge of how to decrease spontaneous mutation rates, we may be inadvertently increasing them. Several public health, ecological, and esthetic consequences of environmental pollution are well known. Are we covering this world with pollutants that are also capable of increasing mutation rates? The question is, of course, aimed especially at chemical pollutants, because we know that radioactive mutagens are already doing so, though probably not at an alarming rate. The answer is that we do not know. Tests of mutagenicity are laborious and often uncertain. They can only be carried out on animals, and extrapolation of the conclusions to man involves many uncertainties. They are, however, very useful indicators, and at the moment the only ones we have. Tests on *in vitro* cell cultures may prove very useful in this respect. The recently created Society of Environmental Mutagens aims to coordinate investigations of this problem and collect available information. Of course, it is also necessary that accumulated knowledge, when available, is actually put to use.

12.11 Segregation and Amalgamation

We know very little about intergroup competition in other animals, including those nearest to man, the primates. But we do know that intergroup competition in man is very important, and consumes, directly or indirectly, a good part of our time. Even in the intervals between wars, much money and effort goes for "defense," namely, the preparation for the more acute phases of intergroup competition. The social organization of man into separate nations is designed to defend the interests of a relatively large, but still limited, group of people, occupying the same territory. It helps to unite individuals of the same country but also strengthens the segregation between countries. There still seems to be much change in attitudes needed before national barriers can effectively be dropped. The increase in communication has helped to broaden the domain of individual languages, thus reducing the difficulties of communication. But we are still far from a global interchange of thought.

Even when different groups come to live in the same territory and share a common language and way of life, barriers may persist. Some traditions die hard and help to keep segregation alive. Religious heterogeneity is a powerful bar to admixture. When visible differences exist in facial appearance and skin color, psychological separation may obtain, which prevents amalgamation. Thus, intergroup competition also appears at the intranational level and is frequently the cause of considerable social discomfort and danger.

In most cases, amalgamation (miscegenation) does occur, sooner or later. However, for historical and cultural reasons, resistance to amalgamation differs greatly in different places. In Brazil, miscegenation is believed to be the rule. An investigation of a rural Brazilian area by Krieger and co-workers (1965) has given the picture shown in Table 12.6. Mates were classified into "racial types" mostly on the basis of skin color, as indicated in the table. The expected frequencies, on the assumption of random mating, for the matings in which both partners are of the same racial type (those on the main diagonal) are given in the last column. Observed matings between individuals of the same type tend, on the average, to be somewhat more frequent than expected under random mating. Thus, even in Brazil, mating is not

TABLE 12.6
Racial-mating-type Frequencies in North Eastern Brazil

Husbands by Racial Types[a]	Wives by Racial Type[a] 0	1	2	3	4	5	6	Total Husbands	Expected Number on Main Diagonal
0	**269**	24	4	95	41	14	5	452	183.27
1	8	**4**	0	7	3	1	2	25	13.88
2	2	1	**1**	5	1	2	0	12	0.19
3	94	18	7	**82**	39	10	1	251	62.81
4	39	4	3	44	**46**	11	11	158	24.38
5	14	5	1	23	26	**19**	7	95	7.33
6	5	3	1	10	8	25	**18**	70	2.90
Total Wives	431	59	17	266	164	82	44	1063	294.76[b]

Source: From Krieger et al. (1965).
[a] Racial types are: 0, White (branco); 1, Light mestizo (amarelo claro); 2, Dark mestizo (amarelo escuro); 3, Light mulatto (mulato claro); 4, Medium mulatto (mulato medio); 5, Dark mulatto (mulato escuro); 6, Negro (preto).
[b] Observed total on main diagonal is 439.

entirely at random. Nevertheless, the total frequency of matings in the main diagonal is 439, which is much nearer to the expected number for randomness (the sum of the expectations for the main diagonal is 294.8) than to that expected for complete assortative mating (1063). This is a maximum expected if all matings could be between persons of the same class—that is, on the main diagonal.

The quantity

$$\frac{439 - 294.8}{1063 - 294.8} \sim 0.2$$

gives an approximate measure of the deviation from random mating. On the basis of a genetic analysis by Krieger and co-workers, the ancestry of 30 percent of the persons in this Brazilian population is Negro, 11 percent is Indian, and 58 percent is Caucasian. A more elaborate measurement of the deviation from random mating based on the correlation between uniting gametes, gives an even lower estimate of the deviation from random mating.

The main settlers of Brazil, the Portuguese, have shown here as elsewhere less resistance to miscegenation than other white settlers. In the southern United States where Africans were brought as slaves, some gene flow has taken place. The proportion of births per generation in which the mother was black and the father was white may occasionally be or have been as high as 5 percent, but resistance to amalgamation is still extremely strong, and a relatively small proportion of black ancestry causes a person to be classified as black in the United States.

As Stern (1960) noted, black skin color would practically disappear in the United States after a few generations of random mating. It would be substituted by a color designated as " amarelo " in Table 12.6, with no great individual variation. This does not seem to be an imminent possibility, however. Data on interracial mating frequencies in some areas of the United States are given in Table 12.7. Conspicuous differences exist between states. The ratio indicated in the last column, which would be one under random mating, shows that Hawaii is almost half-way towards random mating. The average value for the United States, however, indicates very little tendency for cross marriages. Only 2.3 percent of the marriages between whites and nonwhites that would be expected assuming random mating occurred in 1960.

Nevertheless, a trend toward an increase of mixed marriages can be noted, although it is slow. Heer (1967) has computed the expected rate of amalgamation of the black and white populations, on the assumption that presently perceptible trends might continue. Several projections made using more-or-less conservative hypotheses indicate that, at least in the next hundred years, no major change in the racial composition of the population will occur. The two extreme projections give, however, widely different results as to the time needed for complete amalgamation: from more than 1000 to only 12 generations. Whichever projection turns out to be right, it seems probable that the presently acute racial problem in the United States will not quickly be solved by amalgamation.

Resistance to miscegenation also shows up in subtle ways. The north-south gradient of blood groups found in the British Isles (Mourant, 1954) is accurately reproduced in Australia when people are classified according to their ethnic origin by means of their surname (Hatt and Parsons, 1965). Random mating should have modified this situation, and the persistence of the gradient indicates strong assortative mating according to place of origin. To test this further, Hatt and Parsons investigated, still using surnames as an indicator of ethnic origin, the distribution of marriages in Victoria, Australia. Their data are given in Table 12.8, and show a considerable excess of marriages between people having Scottish surnames and an almost equal excess for Irish surnames. Traditions die hard.

TABLE 12.7

Actual and Expected Percentages of Whites Marrying Blacks in California, Hawaii, Michigan, and Nebraska, During Recent Years and of Whites Marrying Nonwhites, for the United States, 1960

State and Year	*Actual Percentage of Whites Marrying Nonwhites*	*Ratio of Actual Percentage to Expected Percentage*[a]
California		
1955	0.14	0.024
1957	0.17	0.024
1958	0.17	0.024
1959	0.21	0.029
Hawaii		
1956	0.13	0.232
1957	0.20	0.333
1958	0.10	0.147
1959	0.22	0.310
1960	0.24	0.444
1961	0.29	0.426
1962	0.24	0.393
1963	0.27	0.370
1964	0.38	0.458
Michigan		
1953	0.07	0.008
1954	0.06	0.008
1955	0.08	0.009
1956	0.09	0.010
1957	0.07	0.008
1958	0.09	0.010
1959	0.10	0.011
1960	0.10	0.011
1961	0.11	0.012
1962	0.12	0.013
1963	0.15	0.017
Nebraska		
1961	0.00	0.000
1962	0.00	0.000
1963	0.01	0.003
1964	0.02	0.006
United States marriage registration area		
1960	0.28	0.023

Source: From Heer (1967).

[a] The expected percentage was calculated on the basis of random mating.

TABLE 12.8

Frequencies of the Ten Possible Marriage Classes from English (E), Scottish (S), Irish (I) and All Other (X) Surnames, with Expectations Based on Random Mating

Marriage class	*Observed frequencies*	*Probability assuming random mating*	*Expected frequencies*	*Expected/Observed*
E × E	141	p^2	110.231	0.782
E × S	148	$2pq$	115.481	0.780
E × I	100	$2pr$	82.895	0.829
E × X	1,087	$2ps$	1,198.162	1.102
S × S	50	q^2	30.245	0.605
S × I	41	$2qr$	43.421	1.059
S × X	558	$2qs$	627.608	1.125
I × I	26	r^2	15.584	0.599
I × X	415	$2rs$	450.515	1.086
X × X	3,364	s^2	3,255.858	0.968

Source: From Hatt and Parsons (1965).

12.12 Race and Society

The existence of reproductively isolated subgroups that are culturally, and often racially, distinct, tends to elicit social tensions that are the seeds of racism.

This has held, as far as we can tell, throughout the history of mankind and is by no means unique to the present-day tensions between different racial groups of the United States. The African farmers of the equatorial rain forest treat the Pygmies almost as their slaves and hardly even consider them as fellow human beings. Their attitude undoubtedly predates the contact of whites with these parts of Africa, which has, in any case, been relatively slight in many regions where Pygmies are found. Also, from various places around the world numerous examples of conflicts between different religious groups could be cited as demonstrating a form of inter-group competition between populations inhabiting the same geographic region.

Racial differences, as we have already emphasized, are inevitably accompanied by cultural differences. Genetic divergence and cultural divergence between populations each accumulate, according to much the same principles, following effective reproductive isolation. Cultural divergence is often accompanied by relative economic deprivation in one or the other of the subgroups in a population, which, of course, only serves to aggravate the tensions between them. This is surely an important contributing factor, for example, to the conflicts between French and English Canadians, and between the Roman Catholics and Protestants of Ireland, as well as to those between black and white Americans.

The striking outward differences between blacks and whites, mainly, of course, the color of their skin, must be a major additional factor contributing to the racial tensions between them. A disappearance of the cultural differences between, for example, the Protestants and Roman Catholics of Ireland, would leave them essentially indistinguishable. The same is not true for black and white Americans, as we have already emphasized. In order to even out their skin-color difference random mating would be necessary. The data on the frequency of mixed marriages, given in Table 12.7, show clearly that the movement in recent years toward a strong legal stand in favor of desegregation has had so far very little effect on the reproductive isolation between these two groups of people.

The relative socioeconomic deprivation of one racial group inevitably raises the question of whether the difference in socioeconomic strata has a significant genetic component, and, for blacks and whites in the United States, this question has recently been focused on their average difference in IQ.

Many studies have shown the existence of substantial differences between the distribution of IQ in blacks and whites in the United States. An example of data obtained in one such study, which are based on IQ tests given to 1800 black elementary school children in the southern United States, is shown in Figure 12.2. The distribution found is compared with a 1960 based "normative" sample of the

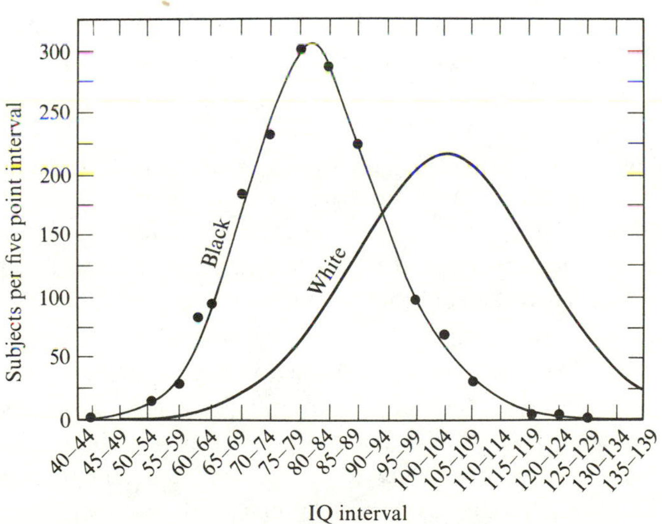

FIGURE 12.2
Distribution of IQ in black and white Americans. The distribution on the left was obtained from tests on 1800 black school children from the South; that on the right is based on a "normative" sample of white Americans. (From Kennedy et al., 1963.)

United States white population. The mean difference in IQ is 21.1, and the standard deviation of the distribution among the blacks is some 25 percent less (12.4 versus 16.4) than that of the normative distribution. Thus, while there is considerable overlap between the two distributions, nevertheless, 95.5 percent of the blacks have IQ's below the white mean of 101.8 and 18.4 percent have IQ's lower than 70, whereas only 2 percent of the normative sample have IQ's below 70. Since reported differences between the mean IQ's of blacks and whites generally lie between 10 and 20 points, the value found in this particular study is one of the most extreme reported. The difference is usually less for the northern than for the southern states, and clearly depends very much on the particular populations tested and on many other factors. A well-known study of Army intelligence-test results showed that some groups of northern blacks made higher averages than some groups of southern whites, though in any given region whites scored higher than blacks (Yerkes, 1921, quoted by Haldane, 1946). Though there are indeed many uncertainties and uncontrollable variables that influence the outcome of IQ tests, the observed mean differences between blacks and whites are undoubtedly more or less reproducible and quite striking.

The argument in favor of this mean difference having a substantial genetic component is based on the premise that existing estimates of the heritability of IQ can reasonably be applied to the racial difference.

As discussed in Chapter 9, most estimates of the heritability of IQ are quite high, ranging from about 40–80 percent, depending on the investigation and whether the data refer to heritability (ratio of additive genetic variance, V_A, to total phenotypic variance, V_P) or to the degree of genetic determination (ratio of *total* genetic variance to V_P). It will be recalled that V_A/V_P is a measure of that part of the genetic variance that could, in principle, be fixed in a selection experiment. Undoubtedly, substantial innate differences with respect to IQ exist within most human populations. A systematic breeding program aimed at selecting for higher IQ should, thus, systematically increase IQ, though it could surely do so only slowly. We have emphasized repeatedly that heritability estimates are only valid for the particular population being studied and for the range of environmental variation to which it is subjected. Essentially all published heritability estimates pertain to Caucasian populations, and usually to the middle socioeconomic stratum. The major problem confronting the extrapolation of such heritability estimates to the differences between blacks and whites is whether the range of environmental variation within a middle-class Caucasian population is really comparable to the average environmental difference between black and white Americans. Those arguing that there is a significant genetic component to the difference (see especially Jensen, 1969) do so on the basis of (1) comparison between blacks and whites matched, formally, with respect to socioeconomic status, and (2) evaluation of possible environmental factors contributing to the IQ difference.

There are significant differences in mean IQ between the various socioeconomic strata.

One of the most comprehensive and widely quoted studies on IQ differences between social classes and the reasons for their apparent stability is that published by Burt in 1961. His data come from school children in a typical London borough, and from their parents. Classification was made on the basis of the head of the household's occupation, into six classes. These range from class I, including "university teachers, those of similar standing in law, medicine, education or the church and top people in commerce, industry or civil service," to class VI including "unskilled laborers, casual laborers and those employed in coarse manual work." The means and standard deviations of the IQ distributions of parents and their offspring, according to this occupational classification are given in Table 12.9 together with the relative proportions of individuals in each class. There are four main features of these data.

TABLE 12.9
Distribution of Intelligence According to Occupational Class

	Occupational Class[a]					
	I *Higher Professional*	II *Lower Professional*	III *Clerical*	IV *Skilled*	V *Semi-skilled*	VI *Unskilled*
Proportion in class (percent)	0.3	3.1	12.2	25.8	32.5	26.1
Mean IQ's ± Standard deviations						
Parents	139.7 ± 4.7	130.6 ± 6.7	115.9 ± 9.3	108.2 ± 9.9	97.8 ± 9.9	84.9 ± 10.9
Offspring	120.8 ± 12.5	114.7 ± 11.2	107.8 ± 13.6	104.6 ± 14.3	98.9 ± 13.8	92.6 ± 13.8

Source: Data from London, based on Burt (1961).
[a] See the text for a further discussion of the definition of occupational class and the origin of this data.

1. There is a very large and very significant direct dependence of mean IQ on occupational class. The mean difference between the highest and lowest classes is more than 50, which is at least $2\frac{1}{2}$ times the largest reported mean IQ difference between blacks and whites.

2. In spite of the significant variation between the means, the residual variation in IQ of parents *within* each class is still remarkably large. The mean standard deviation of the parental IQ's for the different classes is 8.6, or almost $\frac{2}{3}$ the standard deviation for the whole group.

3. The offspring means for each class are between the parental and the overall population means. This so called "regression to the population mean" is expected for directional selection with respect to a quantitative character that is, at least in part, genetically determined. This, as discussed in Chapter 9, is both because parents in each class must be selected on the basis of their phenotype, which will therefore tend to be toward the extreme for the corresponding genotype, and because of genetic segregation in matings which are clearly not perfectly assortative with respect to class or IQ. The offspring means, in fact, are almost exactly between the parental means and the population mean of 100, as expected for 100 percent heritability. The resulting very high parent-offspring correlation must, to a fair extent, be inflated by the effects of assortative mating and of environmental correlations *within* families.

4. The last important feature of the data shown in Table 12.9 is that the standard deviations of the offspring IQ distributions are almost the same as that of the general population (15), averaging to 13.2. This is more-or-less expected on the basis of segregation for genes determining IQ.

Undoubtedly the most straightforward interpretation of these data is that IQ is itself a major determinant of occupational class and is, to an appreciable extent, genetically determined. Burt argues that because of the fairly wide distribution of IQ within each class, appreciable mobility between classes is needed each generation to maintain the class differences. He estimates that to maintain a stable distribution of IQ differences between classes, a minimum of 22 percent of offspring would have to change class with respect to their parents each generation, which is well below the observed intergenerational social mobility in Britain of about 30 percent. Burt attempted, further, to assess the psychological factors influencing occupational mobility by conducting interviews with the families to determine the parental and offspring attitudes towards social advancement and the general home background. He derived the following regression for predicting mobility

$$S = 0.35I + 0.27M + 0.16H + 0.15E,$$

where S is social mobility, I is intelligence, M is motivation, H is home background, and E is educational achievement. This suggests that intelligence and motivation are the most important factors influencing social advancement, though home background and educational achievement have a substantial influence, as might be expected.

There are two main features that clearly distinguish the IQ differences between social classes found by Burt from those between blacks and whites.

In the first place, the differences relate to the environmental variation within the relatively homogeneous British population. As emphasized before, it cannot be assumed that this range of environmental variation is comparable to the average difference between black and white Americans. Secondly, and even more important, these differences are maintained by *mobility* between occupational classes based to a significant extent on selection for higher IQ in the higher occupational classes. There is clearly no counterpart to this mobility with respect to the differences between blacks and whites in the United States. Skin color is an effective bar to mobility between races. Thus, while the IQ differences between classes emphasize the stratification of occupations with respect to IQ differences that may be, to a considerable extent, genetically determined (and this is why we have devoted a fair amount of space to their discussion), they do not seem to bear any direct comparison to the mean IQ difference between blacks and whites.

Socioeconomic status, measured mainly in terms of schooling, occupation, and income, is necessarily in part a measure of the environment. A number of studies, quoted by Jensen, have shown that in the United States even at the same socioeconomic status, blacks still have a substantially lower mean IQ than whites. Taken at face value—that is, under the assumption that socioeconomic status is truly a measure of the total environment—these data would, of course, indicate that the IQ difference was genetically determined. It is, however, difficult to see how the socioeconomic status of blacks and whites can be compared. The predominantly black schools of the United States are notoriously less adequate than the white schools, so that equal numbers of years of schooling certainly do not mean equal education. Wide variation in level of occupation must exist within each occupational class. Thus, one would certainly expect even for equivalent occupational classes that the black level was on average lower than that of the white. No amount of money can readily buy a black person's way into a privileged upper class white community, or buy away more than two hundred years of accumulated racial prejudice on the part of the whites, or remove the relative disruption of the black family, in part culturally inherited from the days of slavery. Thus, we consider it impossible to accept the idea that matching for socioeconomic status provides an adequate, or even a substantial, control over the most important environmental differences between blacks and whites.

What then can be said concerning environmental differences known, or suspected, to affect IQ? First of all it should be emphasized that, in spite of high estimates of IQ heritability, the mean intrapair IQ difference, for example, between the monozygotic twins reared apart studied by Newman and co-workers (1937) was 8, and the range was from 1 to 24. Thus there is clearly, even within the white population, substantial environmental variation with respect to IQ. The following four important environmental effects were already discussed in Chapter 9.

1. There is a systematic difference of as much as five IQ points between twins and their nontwin sibs (with the twins having the lower IQ's), irrespective of socioeconomic and other variables. This reduction in the IQ of twins, as discussed before,

could be due either to the effects of the *in utero* maternal environment on twins, or to the reduced attention parents are able to give to each of two very young children born at the same time.

2. It has been reported that the IQ of blacks tested by blacks was at least 2–3 points higher than when the same persons were tested by whites (see Pettigrew, 1964). This simply serves to emphasize the intrinsic error of IQ measurements and the fact that such measurements devised for a white population by white people could, because of the cultural differences, easily be unsuitable to measure level of performance among black people.

3. Maternal malnutrition during pregnancy is known to be significantly associated with impaired mental development of the offspring. Studies in rats by Zamenhof and co-workers (1968) on the effects of the mother rat's having protein-deficient diets before and during pregnancy, have shown a substantial reduction in total brain DNA content of the resulting offspring and so, also, presumably, a reduction in brain cell number. The reductions were correlated with behavioral deficiencies, and this experiment clearly could point to the basis, in man, for substantial IQ differences. There can be no doubt that in many areas, dietary deficiency is a part of poor socioeconomic conditions.

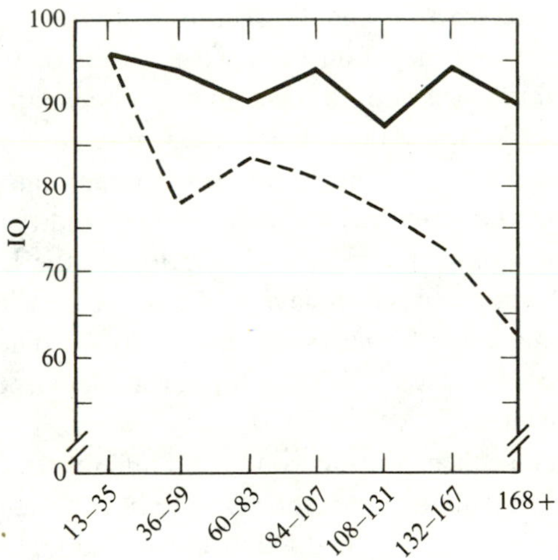

FIGURE 12.3
Mean IQ's of 596 children of 88 mothers as a function of age of children. The solid line plots data on children whose mothers have IQ's of 80 or higher; the dashed line plots that of children whose mothers have IQ's lower than 80. (Data from Heber, Dever, and Conry, 1968; graph from Jensen, 1969.)

4. The very early home environment has long been suspected to be of substantial importance for intellectual development. Clear-cut data exist that demonstrate the detrimental effects of severe sensory deprivation very early in life (see Freedman, in Washburn and Jay, 1968). It is no great extrapolation from the known effects of severe sensory deprivation, to likely effects of lesser deficiencies in the early home environment.

While Jensen (1969) chooses to minimize environmental effects such as we have discussed above, we believe that there is no evidence against the notion that these, among others, could explain essentially all the difference in IQ between blacks and whites in the United States. As an example of a difference in interpretation of the same information, consider the data shown in Figure 12.3 on IQ of children as a function of age and of mother's IQ. Jensen interprets the decline in the IQ of offspring of mothers whose IQ is less than 80 as evidence for "genetic factors involved in the growth rate of intelligence." It is not clear to us, however, why this decline could not equally be attributed to the relatively detrimental environment likely to be associated with a mother whose IQ is below 80. Early IQ may well be a better representation of the innate potentialities of the offspring. The decline with age could then be due to an inferior environment's preventing the full development of this potential.

In summary, therefore, we do not exclude the possibility that there could be a genetic component to the mean difference in IQ between black and white Americans, but simply maintain that presently available data are inadequate to resolve this question in either direction.

The physicist Shockley, who received a Nobel prize for his part in the invention of the transistor, has in recent years strongly advocated that more research effort should be devoted to understanding the basis for the black-white IQ difference. He has coupled his stance with statements widely quoted in the press that he believes the difference is, most probably, largely genetic and that welfare programs may, to some extent, be responsible for an average decline in the social adaptation of some segments of the population resulting, apparently, from selection for a decreased IQ. Welfare programs have been in operation for at most 30–40 years, or equivalent to between one and two generations. We have seen in Chapter 9 that even the most rapid response to directional selection for a quantitative character is relatively slow in terms of generations, let alone years. There is, therefore, little theoretical basis for believing that welfare programs could already have had detectable genetic effects through selection for low IQ. Moreover, all that we now know about demographic patterns (see Chapter 6) suggests that the most rapid way to equalize reproductive differences is to improve socioeconomic conditions, which at least should be possible within at most one or two generations.

In this context, it is worth mentioning that Jensen states that because the gene pools of whites and blacks are known to differ and "these genetic differences are manifested in virtually every anatomical, physiological, and biochemical comparison one can make between representative samples of identifiable racial groups" therefore "there is no reason to suppose that the brain should be exempt from this generalization." There is, however, no reason why genes affecting IQ which differ in frequency in the gene pools of blacks and whites, should be such that, on the average, whites would have significantly higher frequencies of genes increasing IQ than would blacks. On the contrary, one should expect, assuming no tendency for high IQ genes to accumulate by selection in one or other race, that the more polymorphic genes there are that affect IQ and that differ in frequency in blacks and whites, the less likely it is that there is an average genetic difference in IQ between the races. This follows from that most basic law of statistics, the law of large numbers, which predicts increasing accuracy of a mean based on increasing numbers of observations (See Appendix I).

The main questions to be answered are: (1) Is it, in fact, feasible at the present time to do studies that could clearly determine the extent to which the black-white IQ difference is genetically determined? (2) Why is it important to do such studies? The only really satisfactory approaches to taking account of intrafamilial environmental correlations in the study of the inheritance of IQ are to work with identical twins reared apart or with adopted children (see Chapter 9). The only approach applicable to the study of the race IQ difference is, therefore, that of working with black children adopted into white homes and vice versa. Adoptions would, of course, have to be at a very early age to be sure of taking into account any possible effects of the very early home environment. To our knowledge no scientifically adequate studies of this nature have ever been done. The IQ's of say, the black children adopted into white homes would, presumably, have to be compared with those of white children adopted into comparable white homes. It is questionable whether such studies could be done in a reasonably controlled way at the present time. Even if they could, they would not remove the effects of prejudice almost inevitably directed to some extent against black people in most white communities. We, therefore, suggest that the question of whether there is a genetic basis for the race IQ difference will be almost impossible to answer satisfactorily before the environmental differences between blacks and whites in the United States have been substantially reduced.

Perhaps the only argument in favor of research on the racial IQ difference is that, since the question of whether difference is genetic has been raised, an attempt should be made to provide an answer. Otherwise, those who now believe, we think on quite inadequate evidence, that the difference is genetic, will be left to continue their campaigns for an adjustment of our educational and economic systems to take account of innate racial differences. Hopefully, a demonstration that the difference is not primarily genetic would counter these campaigns. An answer in the other direction clearly would not, though in a really democratic society free of racial prejudice, the answer should make no difference.

Apart from the intrinsic difficulties in answering the question of whether the racial IQ difference is mainly genetic, it seems to us, that there is no good case for encouraging the support of such studies on either theoretical or practical grounds. From a theoretical point of view, it seems unlikely that they would throw much light on the general problem of the genetic control of IQ. Any racial difference would be a small fraction of the total variation in IQ. The mere fact that even the relatively crude studies on the inheritance of IQ done so far have not taken advantage of racial differences suggests that these are not the most convenient differences to study. Much basic work on the biology and biochemistry of mental development under controlled conditions, and making use of known genetic differences, is needed before a fuller understanding of the inheritance of IQ can be achieved.

It is hard to see what direct practical applications in a democratic society could follow from knowledge that there is an average genetic difference in IQ between the races.

A democratic society such as that in the United States believes that there should be no discrimination against an individual on the basis of race, religion, or other *a priori* categorization (including sex). Each individual should be given equal and maximum opportunity, according to his or her needs, to develop to his or her fullest potential. Surely, as we have previously emphasized, innate differences in ability and other individual variations should be taken into account by our educational system. These must, however, be judged on the basis of the individual and not the race to which he or she belongs. To claim otherwise, shows a lack of appreciation of the distinction between differences between individuals as compared to those between populations.

We are, of course, aware of the dangers of either overt or implicit political control over scientific inquiry. The suppression of Galileo and the success of Lysenko are two notorious examples of the evils of such control. Most scientists, however, do submit to certain controls over research on human beings such as, for example, the right of an individual not to be experimented on, and the confidentiality of the information collected by the census bureau. These controls are imposed to protect the individual from possible direct detrimental effects of scientific investigations. The treatment of the Jews in Nazi concentration camps is a testimonial never to be forgotten to the needs for such controls. There can be no doubt that in the present racial climate of the United States, studies on racial differences in IQ, however well intentioned, could easily be misinterpreted as a form of racism and lead to unnecessary accentuation of racial tensions. Since we believe that no good case can, at present, be made for such studies on scientific or practical grounds, it follows naturally that we do not see the point in particularly encouraging the use of government or other funds for their support as Shockley has advocated. There are many more challenging biological problems for the scientist to attack, while one of the most urgent social problems is to equalize the *environment* of black and white Americans, and so remove this and other similar racial imbalances.

Hopefully, equalization of the environment will help to eliminate racial prejudice. This, in turn, should lead to the breakdown of the cultural barriers against intermarriage, and so, eventually, to the reproductive merging of racial groups, which is the only ultimate, even if remote, solution to the problem of racism.

12.13 Changes in the Complexity of Society

The Pygmies represent one of the simplest types of human society. Specialization and division of labor are extremely limited: they exist to a limited extent only between the sexes, and between age groups. There are individuals who are better than others at medicine, hunting, singing, and dancing, but they never entirely specialize (see Turnbull, 1962, 1965). The main individual activity is always the provision of food. In contrast, in the economically most advanced societies, there are thousands of specializations. Many require years, some a lifetime, of application for proficiency. Many require talents so highly specialized that most of an individual's time is spent in just one highly specialized activity. The degree of specialization is at the opposite extreme to that of more primitive human societies.

In a complex society, the need for a variety of individual talents has undoubtedly increased tremendously. Even if some extreme educationists believe that anybody, properly educated, could be made proficient in any job, art, or science, few educationists maintain that the cost would be the same for every individual. If a great deal of research were devoted to genetic differences in response to education, we might arrive at diagrams, such as the hypothetical one given in Figure 12.4.

Our present capacity to predict the potentials of a child are still very limited. Our knowledge as to the relative efficiencies of different teaching techniques directed, for a given individual, to a given end, is essentially nonexistent. There is therefore very little hope (or fear) for some time at least that a good (or bad) social planner might make use of such data. Should they become available, their social uses will be extremely important. As educationists are now becoming aware of the existence of individual differences, and of the potential importance of the different susceptibilities of various individuals to different methods of teaching, it is possible that such a development will sooner or later take place.

When such knowledge is available, it would seem that the time for a Utopia has come. Knowledge can, however, always be put to bad or good use. The specter of a dictator exploiting this knowledge for his own power is, and will be for a long time, around to haunt us. Experience has taught us to beware. A significant part of the educational effort should go into building up adequate critical judgment, desire to participate in public decisions, and resistance to political demagogery. Only then will the danger from dictators, or political puppets in the hands of power groups, decrease.

The major problem in social planning will remain that of the choice of quantities to be maximized or minimized. The total cost to society may be a quantity to be

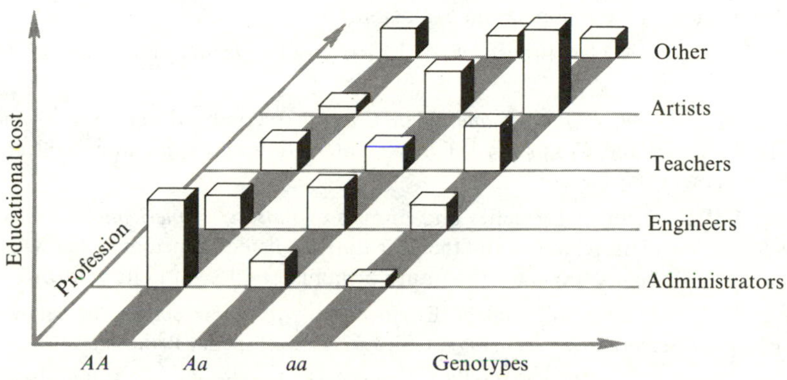

FIGURE 12.4
A hypothetical model of the effect of a diallelic locus on the average educational cost of training individuals of each genotype for different professions. Only one pair of alleles at a locus is considered for simplicity, though the number of significant genotypes is likely to be very large. In each case, the height of the bar represents the cost to society per individual.

minimized. Once it was determined how many administrators, engineers, teachers, artists, and so on were required, their education might be obtained at minimum cost. But an approach aimed towards efficiency alone can hardly create a stable society. For social stability, individual satisfaction also must be maximized. Today, we are entirely unable to assess, except very roughly, motives and inhibitions, feelings of achievement or frustration, satisfaction and discontent, and in general the basic forces underlying our behavior. Perhaps the most sensible approach will be to use such knowledge, when available, in the endeavor to maximize the correlation between usefulness to society and individual satisfaction. Individual differences, genetically, or environmentally determined, will inevitably be of the greatest importance, here as elsewhere. No study that ignores individual variation can fully contribute to the problems of society, especially now that it is becoming more and more necessary to meet the needs of increasing complexity with increasing specialization, and still respect and encourage the basic unity of human nature.

General References

Ehrlich, P. R., and A. H. Ehrlich, *Population, Resources, Environment*. San Francisco: W. H. Freeman and Company, 1970. (A stimulating elementary textbook on the problems posed by the population explosion.)

Haller, M. H., *Eugenics*. New Brunswick, N.J.: Rutgers University Press, 1963. (A history of the subject.)

Hardin, G., *Population, Evolution, and Birth Control* (second edition). San Francisco: W. H. Freeman and Company, 1969. (An interesting and stimulating collection of miscellaneous essays.)

Lederberg, J., "Experimental genetics and human evolution." *American Naturalist* **100:** 519–531, 1966. (This reference and the four that are listed immediately below contain discussions of the prospects for the future of human biology and its control.)

Lederberg, J., "Humanics and Genetic Engineering." In *Encyclopedia Britannica, 1970 Yearbook of Science and Technology*. Chicago: Encyclopedia Britannica.

Osborn, F., *The Future of Human Heredity*. New York: Weybright and Talley, 1968.

Sonneborn, T. M., ed., *Control of Human Heredity and Evolution*. New York: MacMillan, 1965.

Wolstenholme, G., ed., *Man and His Future*. Ciba Foundation Symposium. Boston: Little, Brown, 1963.

APPENDIX I

Statistics and Probability

I.1 Introduction

An understanding of elementary statistics and probability theory is essential to a full appreciation of the problems of human population genetics. Such an appreciation, however, requires more proficiency with algebra and calculus than we have assumed in this text. Certainly, the models we have discussed require some knowledge of probability and probability distributions, and to test whether observed data adequately fit the models requires some statistical analysis. But, basically, little demand is made on the reader in these two areas. In this appendix we do not intend a thorough coverage of probability theory and statistics for that would fill another book. Instead, we provide the following survey as a refresher for those who already have some background in this area and also some knowledge of calculus. It is intended to be little more than an annotated list of useful definitions and formulas, with special reference to applications in human population genetics.

I.2 Distribution

Given a set of *discrete* states i, $i = 0, \ldots n$, the relative expected frequency p_i defines the discrete frequency distribution of the states. Note that $\sum_{i=0}^{n} p_i = 1$ for such a distribution. The quantity p_i is the relative frequency with which the state

i would be found in a *very large* number of samples, for example, the proportion of families of size n with i males out of a very large number of families. Thus p_i can be thought of as the probability of observing state i in a single sample.

Suppose that the states are intervals of a continuous measurement x, such as height or IQ, where the ith state is the interval x_i to x_{i+1}. As the interval becomes smaller, taking the form x to $x + dx$ where dx is small, the frequency distribution takes the form $f(x)\,dx$, where $f(x)$ is a function of the continuous measurement x. The function $f(x)$ is then the continuous frequency distribution of x. Now

$$\int_A^B f(x)\,dx = 1,$$

where (A, B) is the interval over which $f(x)$ is defined. For continuous distributions, integrals replace sums.

I.3 Expected Values

The average or *expected* value of i for a discrete distribution is defined by

$$E(i) = \sum_i ip_i, \tag{1}$$

and for a continuous distribution, the expected value of x is

$$E(x) = \int_A^B xf(x)\,dx. \tag{2}$$

In general, the expected value of any function of i or x, $\phi(i)$ or $\phi(x)$ (for example, i^2 or x^2), is given by

$$E[\phi(i)] = \sum_i \phi(i)p_i \tag{3}$$

or

$$E[\phi(x)] = \int_A^B \phi(x)f(x)\,dx, \tag{4}$$

where the E stands for expectation. The expectation of a sum is the sum of the expectations. If c is a constant,

$$E(cx) = cE(x).$$

I.4 The Normal Distribution

The normal distribution is the best known distribution, both because it can be used to fit many types of observed distributions and because it turns out to be central to statistical theory. The normal distribution is defined by the function

$$f(x) = \frac{1}{\sigma\sqrt{2\pi}} e^{-(x-\mu)^2/2\sigma^2}. \tag{5}$$

It is a bell-shaped curve, symmetrical about its peak at $x = \mu$ and tending to zero when x is very large and either positive or negative. The mean of the distribution is μ and is equal to the expected value of x: $\mu = E(x)$.

I.5 Variance

The expectation (expected value) of $(x - \mu)^2$, which is the mean of the squared deviations from the mean, is an important general measure of the spread of a distribution called the **variance** and often written $V(x)$ or σ^2. The variance of the normal distribution given in Equation 5 is σ^2. Note that

$$E(x - \mu)^2 = E(x^2 - 2\mu x + \mu^2) = E(x^2) - 2\mu E(x) + \mu^2, \tag{6}$$

and therefore

$$E(x - \mu)^2 = E(x^2) - \mu^2 \tag{7}$$

since μ is a constant, and the expectation of a sum is the sum of the expectations. Equations 6 and 7 are very useful formulas for calculating the expected or theoretical variance. (For estimating the variance from an observed sample see Section I.21.)

I.6 Parameters

The mean, μ, and the variance, σ^2, of the normal distribution are called its **parameters**. They define the position and spread of the curve. When $\mu = 0$ and $\sigma^2 = 1$, the resulting distribution

$$\frac{1}{\sqrt{2\pi}} e^{-x^2/2}$$

is called the **standard normal distribution**. In general, parameters are constant values which define a distribution function.

The normal distribution is of great theoretical importance because the distribution of the sum of a large number of variables, whatever their individual distributions, always tends to be normal. Even the distribution of the sum of any two normally distributed variables is itself precisely normal.

A normal distribution with a mean μ and a variance σ^2 is often represented by the symbol $N(\mu, \sigma^2)$.

I.7 Change of Scale and Location—Mean and Variance of a Function of the Original Variable

The expectation of $ax + b$ is $aE(x) + b$. The variance of $ax + b$ is $a^2V(x)$. Given an x that is $N(\mu, \sigma^2)$, then

$$E\left(\frac{x - \mu}{\sigma}\right) = \frac{1}{\sigma} E(x - \mu) = 0, \tag{8}$$

and

$$V\left(\frac{x - \mu}{\sigma}\right) = \frac{1}{\sigma^2} V(x) = 1. \tag{9}$$

Therefore, $x - \mu/\sigma$ is $N(0, 1)$. More generally it can be shown that if x has a distribution with a mean m and a variance σ^2, then $f(x)$ has a distribution with, approximately, a mean $f(m)$ and a variance

$$v = \sigma^2[f'(m)]^2, \tag{10}$$

where $f'(m)$ is the value of df/dx at $x = m$.

Examples:

The variance of the square of a variable x is

$$V(x)^2 = (2x)^2_{x=m}\sigma^2 = 4m^2\sigma^2. \tag{11}$$

According to the fact that $f(x)$ has a distribution with approximate mean $f(m)$, the mean of x^2 is approximately m^2. It is given exactly by (see Equation 7)

$$E(x^2) = m^2 + \sigma^2. \tag{12}$$

The mean and variance of $\sqrt{x}$ are

$$E(\sqrt{x}) \sim \sqrt{m};\ V(\sqrt{x}) \sim \left(\frac{1}{2\sqrt{x}}\right)^2_{x=m} \times \sigma^2 = \frac{\sigma^2}{4m}. \tag{13}$$

The approximate variance of $\log x$ is

$$V(\log x) = \left(\frac{1}{x}\right)^2_{x=m} \times \sigma^2 = \frac{\sigma^2}{m^2}. \tag{14}$$

I.8 Moments

Moments are expectations of the rth powers of a variable or of its deviations from the mean. The first types of moments (also called moments about zero) are thus defined by

$$\mu'_r = E(x)^r,$$

and the second type of moments (moments about the mean) by

$$\mu_r = E[x - \mu]^r,$$

where $\mu'_1 = \mu$, the mean, is the first moment about zero. The second moment about the mean is the variance. (Note that $\mu_1 = 0$ and $\mu_0 = \mu'_0 = 1$.) Moments are useful for defining distributions. The first four moments are the ones most widely used. The third moment about the mean, if different from zero, indicates skewness, and the fourth moment about the mean, if different from $3\mu_2^2$, indicates departure from normality; that is, **leptokurtosis** if $(\mu_4/\mu_2^2) - 3 > 0$; **platykurtosis** if $\mu_4/\mu_2^2 - 3 > 0$.

The following formulas are useful for converting moments about the mean into those about zero, and vice versa:

$$\begin{aligned} \mu'_2 &= \mu_2 + (\mu'_1)^2 & \mu_2 &= \mu'_2 - \mu'^2_1 \\ \mu'_3 &= \mu_3 + 3\mu'_1\mu_2 + (\mu'_1)^3 & \mu_3 &= \mu'_3 - 3\mu'_2 + 2(\mu'_1)^3 \\ \mu'_4 &= \mu_4 + 4\mu'_1\mu_3 + 6(\mu'_1)^2\mu_2 + (\mu'_1)^4 & \mu_4 &= \mu'_4 - 4\mu'_1\mu'_3 + 6(\mu'_2)^2\mu'_2 - 3(\mu'_1)^4. \end{aligned} \tag{15}$$

I.9 The Combination of Probabilities

The probability of state *i or j* is $p_i + p_j$ (for example, the probability of either i or j males in families of size n).

Two events are said to be *independent* if the probability of their occurring together is the *product* of their separate probabilities. For example, if two families are observed independently—that is, there is no connection between them in any way—then the probability that the first has i males *and* the second j males is $p_i \times p_j$.

I.10 The Binomial Distribution

If the probability of a male birth is p and of a female birth is q, where $p + q = 1$, then the probability that the first r out of n births are male and the remainder female is $p^r q^{n-r}$. This is also the probability for *any* particular sequence of n births including r males. For the total probability of r males, we need to multiply $p^r q^{n-r}$ by the number of ways in which r male births can occur with $n - r$ female births. This is the same as the total number of ways of arranging r things out of n, which is the binomial coefficient

$$\frac{n(n-1)\ldots(n-r+1)}{1\times 2\times \ldots \times r} = \binom{n}{r}. \tag{16}$$

The total probability of r males in n births is therefore

$$p_r = \binom{n}{r} p^r q^{n-r}, \tag{17}$$

which is called the **binomial distribution**. Note that

$$\sum_{r=0}^{n} p_r = q^n + npq^{n-1} + \cdots + \binom{n}{r} p^r q^{n-r} + \cdots + p^n = (p+q)^n = 1. \tag{18}$$

The mean of the binomial distribution is given by

$$\begin{aligned} m &= \sum_{r=0}^{n} r p_r = \sum_{r=0}^{n} (r) \frac{n(n-1)\ldots(n-r+1)}{1\times 2\times \ldots \times r} p^r q^{n-r} \\ &= np \sum_{r=0}^{n} \frac{(n-1)(n-2)\ldots(n-r+1)}{1\times 2\times \ldots (r-1)} p^{r-1} q^{(n-1)-(r-1)} \\ &= np(p+q)^{n-1} \end{aligned}$$

or

$$m = np. \tag{19}$$

The variance can be shown to be

$$v = np(1-p). \tag{20}$$

Example:

In a mating $Aa \times Aa$, involving heterozygotes for the albino gene a, what is the probability that 3 out of 8 children are albinos? Use the binomial with $p = \frac{1}{4}$, $q = 1 - p = \frac{3}{4}$, $n = 8$, and $r = 3$ to give

$$\frac{8\times 7\times 6}{1\times 2\times 3} (\tfrac{1}{4})^3 (\tfrac{3}{4})^5 = 0.2076. \tag{21}$$

I.11 The Poisson Distribution

If in the binomial, p becomes very small and n very large in such a way that the mean $np = m$ remains constant, then it can be shown that the binomial distribution tends to the limit

$$p_r = \frac{m^r}{r!} e^{-m}. \tag{22}$$

(Note $r! = 1 \times 2 \times 3 \times \cdots \times r$, $0! = 1$, and $r!$ is r factorial.)

This is called the **Poisson distribution**. If events occur at random with respect to time, then the number of events occurring in a given time interval, corresponding to a mean of m, has a Poisson distribution. The mean is given by

$$\sum_{r=0}^{\infty} rp_r = e^{-m} \sum_{r=1}^{\infty} m \times \frac{m^r}{(r-1)!} = m \times e^{-m} \times e^m = m. \tag{23}$$

This distribution is sometimes used to approximate the distribution of family sizes.

The variance of a Poisson distribution is equal to its mean m. The binomial for large n and the Poisson distribution for large m both tend to be normally distributed.

I.12 The Geometric Distribution

This distribution is defined by

$$p_r = (1-p)^r p, \tag{24}$$

where $r = 0, 1, 2, \ldots,$ to ∞.

This is, for example, the probability of having r female births before having a male birth (see Chapter 10). The mean total births (expectation of $r + 1$) is $1/p$ and the variance $1/p^2 + 1/p$.

I.13 The Negative Binomial Distribution

This distribution is given by the terms of $(q - p)^{-n}$, where $q = 1 + p$ and p is positive. Thus,

$$p_r = q^{-n} \frac{(n+r-1)!}{r!(n-1)!} \left(\frac{p}{q}\right)^r, \tag{25}$$

where $r = 0, 1, 2, \ldots.$ For example,

$$p_0 = q^{-n}, \qquad p_1 = q^{-n} \frac{np}{q}, \qquad p_2 = q^{-n} \frac{n(n+1)}{2} \left(\frac{p}{q}\right)^2, \text{ etc.}$$

This distribution has two parameters, n and p (or q). It can be shown to be equivalent to a particular type of mixture of Poisson distributions with different means. More exactly, the means must be distributed as in a gamma distribution (see Section I.15). It is often used to give a good approximation to the distribution of family sizes (see Chapter 6). The mean of the distribution is np, and the variance is $np(1 + p)$. For $p' = 1/q$ and $n = 1$ this is the same as the geometric distribution.

I.14 The Multinomial Distribution

Suppose that there are more than two categories of events, for example, three genotypes from an intercross $Aa \times Aa$. Then, given the probability that the ith category out of k is p_i, $i = 1, \ldots, k$, and given respective observations $r_1, r_2, \ldots, r_i, \ldots, r_k$ in the k categories out of a total of $\sum_{i=1}^{k} r_i = n$, the probability of this set of observations is

$$\frac{n!}{r_1!r_2!\ldots r_k!} p_1^{r_1} p_2^{r_2} \ldots p_i^{r_i} \ldots p_k^{r_k}. \tag{26}$$

This is called the **multinomial distribution**, being given by the terms of $(p_1 + p_2 + \cdots + p_i + \cdots + p_k)^n$. It is a straight generalization of the binomial distribution. The mean of the ith category is np_i, and its variance is $np_i(1 - p_i)$.

Example:

In a mating $MN \times MN$, what is the probability that out of 5 offspring 2 are MN, 2 are MM, and 1 is NN? Use the multinomial distribution with $n = 5$, $r_1 = 2$, $r_2 = 2$, and $r_3 = 1$, where $p_1 = \frac{1}{2}$, $p_2 = p_3 = \frac{1}{4}$ to give

$$\frac{1 \times 2 \times 3 \times 4 \times 5}{(1 \times 2) \times (1 \times 2) \times 1} (\tfrac{1}{2})^2 (\tfrac{1}{4})^2 (\tfrac{1}{4}) = \frac{15}{128}.$$

I.15 The Gamma Distribution

This one-parameter continuous distribution is defined by

$$f(x) = \frac{1}{\Gamma(n)} e^{-x} x^{n-1}, \tag{27}$$

where the gamma function, $\Gamma(n)$, is defined by

$$\Gamma(n) = \int_0^\infty e^{-x} x^{n-1}\, dx. \tag{28}$$

When n is an integer $\Gamma(n) = (n-1)!$. The gamma function is thus a generalization of the factorial. The mean and variance of the gamma distribution are both n. It is especially important in the theory of statistical tests, many of which are based on functions that have distributions related to the gamma distribution (t distribution, F-test, and chi-square—see Sections I.23 to I.30).

I.16 Distributions of Sums of Variables in Some Special Cases

If two variables x_1 and x_2 are normally distributed with means μ_1 and μ_2 and standard deviations σ_1 and σ_2, the distribution of the new variable formed by their sum, $y = (x_1 + x_2)$, is normal with a mean $\mu_1 + \mu_2$ and a variance $\sigma_1^2 + \sigma_2^2$. If the two variables are Poisson, the results are the same. If x_1 and x_2 are gamma variables with parameters n_1 and n_2, their sum is also a gamma variate with a parameter $n_1 + n_2$.

I.17 The Exponential Distribution

This distribution takes the form

$$f(x) = \lambda e^{-\lambda x} \tag{29}$$

and has a mean $1/\lambda$ and a variance $1/\lambda^2$. It is the continuous limit of the geometric distribution $(1-p)^r p$. [If $p = \lambda\, dx$ and $r\, dx = x$, then as $r \to \infty$, $dx \to 0$, and $(1-p)^r p = (1 - \lambda x/r)^r \lambda\, dx \to \lambda\, e^{-\lambda x}\, dx$ as $r \to \infty$.]

The **exponential distribution** is the distribution of the time interval between two random events. For example, if the distribution of the number of births in a family is Poisson, then the distribution of the time interval between the births is exponential. When $\lambda = 1$, the exponential distribution is a special case of the gamma distribution, namely, with $n = 1$.

I.18 Beta Distributions

There are two kinds of beta distributions. The first kind is given by

$$f(x) = \frac{x^{l-1}(1-x)^{m-1}}{B(l, m)} \tag{30}$$

in the range $0 < x < 1$, where $B(l, m)$ is the beta function defined by

$$B(l, m) = \int_0^1 x^{l-1}(1-x)^{m-1}\, dx. \tag{31}$$

It can be shown that

$$B(m, n) = \frac{\Gamma(m)\Gamma(n)}{\Gamma(m + n)}, \tag{32}$$

where the gamma function is defined by Equation 28. The mean and variance of this distribution are $l/(l + m)$ and $lm/(l + m)^2(l + m + 1)$, respectively. Some of the results of the analysis of genetic drift (see Chapter 8) can be interpreted in terms of the beta distributions of gene frequencies at equilibrium. When $l, m > 1$, $f(0) = f(1) = 0$, and there is a mode at $x - (l - 1)/(l + m - 2)$. However, when l or $m < 1, f(0)$ or $f(1) \to \infty$.

The second kind of beta distribution is defined by

$$f(x) = \frac{x^{l-1}}{B(l, m)(1 + x)^{l+m}} \tag{33}$$

in the range $0 < x < \infty$ and has a mean $l/(m - 1)$ and a variance $l(l + m - 1)/(m - 1)^2(m - 2)$. It is the distribution of the ratio of two gamma variates with parameters l and m and also the distribution of the variance ratio F (see Section I.25).

I.19 The Lognormal Distribution

If x is $N(0, 1)$ and is equal to $\gamma + \sigma \log y$, then y is said to have a **lognormal distribution**. As x varies from $-\infty$ to $+\infty$, y varies from 0 to ∞. The moments of y about zero are

$$\mu_r' = e^{r/2\sigma^2 - r\gamma/\sigma}.$$

I.20 Analysis of Data

There are two important aspects to the statistical analysis of observed data.

1. *Goodness of Fit and Tests of Significance.* We may wish to test whether an observed body of data adequately fits a particular expected distribution. For example, does the distribution of IQ's obtained from a random sample of people fit a normal distribution with a mean of 100 and a variance of $15^2 = 225$? Or does the number of males in a sample of families of size 8 fit a binomial distribution with parameters $p = \frac{1}{2}$ and $n = 8$?

2. *Estimation.* On the assumption that a body of data follows some given expected theoretical distribution, what are the values of the parameters of the distribution that give rise to the best fit to the observed data? For example, given

the distribution of the height of a random sample of people and assuming they fit a normal distribution, what are the estimates for the mean and variance of the normal distribution that give the best fit to the observed data?

The two problems of estimation and significance testing are closely interrelated. For example, if we wish to test the fit of a distribution without specifying its parameters, we first estimate the parameters which give the best fit to the data and then assess whether this resulting fit is adequate. (See, for example, Chapter 10 in which we considered the fit of the binomial distribution to sex ratio distribution in families of a given size.) On the other hand, we may wish to test whether a certain parameter estimated from the data, such as the proportion of recessives in an intercross mating $Aa \times Aa$, is compatible with an *a priori* expected value, in this case $\frac{1}{4}$.

I.21 Statistics and Estimation

We need criteria for constructing functions of observed data which provide estimates of desired parameters. Functions of the observations are called **statistics**. For example, given a set of quantitative measurements (for example, an IQ) $x_1, x_2, \ldots, x_n$, the *arithmetic mean*

$$\bar{x} = \frac{1}{n}(x_1 + x_2 + \cdots + x_n) \tag{34}$$

is a statistic which, on the assumption that the measurements are a random sample from a normal distribution with a mean μ and a variance σ^2, is an estimate of μ. An appropriate estimate of σ^2 is given by

$$s^2 = \frac{1}{n-1}\sum_{i=1}^{n}(x_i - \bar{x})^2 = \frac{1}{n-1}\left(\sum_{i=1}^{n} x_i^2 - \bar{x}\sum_{i=1}^{n} x_i\right), \tag{35}$$

using Equation 6, where $\bar{x}$ is given by Equation 34. As we shall see in Section I.22, the estimates of μ and σ^2 given by Equations 34 and 35 are in some sense the best that can be constructed from the data.

In these calculations, the concept of a *random sample* is a very important one. In random sampling, an individual is chosen without reference to his value, and so all individuals have an equal chance of being chosen. Only when sampling randomly is it correct to assume that the distribution which applies to the whole population is also appropriate for the sample. As we discussed in Chapter 6, biases in sampling human populations can be very hard to avoid. We shall see in Appendix II in discussing pedigree analysis how many problems arise as a result of sampling biases.

Another important statistic is the *observed proportion*. For example, if we observe

r homozygotes *aa* among a total of n offspring from an $Aa \times Aa$ intercross, then the observed proportion r/n is an obvious estimate of the expected proportion p of *aa* offspring. The expected variance of r is $np(1-p)$ from the binomial distribution—Equation 21. Thus, since

$$V\left(\frac{r}{n}\right) = \frac{1}{n^2} V(r), \tag{36}$$

the expected variance of the observed proportion is given by

$$V\left(\frac{r}{n}\right) = \frac{p(1-p)}{n}, \tag{37}$$

and the estimated variance, substituting $p = r/n$, is

$$\frac{r}{n} \frac{1 - \dfrac{r}{n}}{n} \tag{38}$$

so long as r is not equal to zero. This variance is, of course, a function of the proportion p or r/n.

It is sometimes very useful to consider a transformation of r/n which is such that its expected variance is at least approximately independent of p. (See the discussion on migration matrices in Chapter 8.) The *angular transformation* defined by

$$\frac{r}{n} = \sin^2 \theta \tag{39}$$

or

$$\theta = \sin^{-1} \sqrt{\frac{r}{n}} \tag{40}$$

has this property of independence from p. Thus, from Equation 10

$$V\left(\frac{r}{n}\right) \sim (2 \sin \bar{\theta} \cos \bar{\theta})^2 V(\bar{\theta}), \tag{41}$$

where $\bar{\theta}$ is the expected value of θ, defined approximately by

$$p = \sin^2 \bar{\theta}. \tag{42}$$

Thus, since $\cos^2 \bar{\theta} = 1 - \sin^2 \bar{\theta} = 1 - p$, Equation 41 gives

$$V(\bar{\theta}) = \frac{1}{4p(1-p)} V\left(\frac{r}{n}\right) = \tfrac{1}{4} n. \tag{43}$$

Since a statistic is a function of observations which vary according to some assumed distribution, it must itself have an expected frequency distribution. Thus, different random samples (for example, of the IQ of individuals from a given population) will lead to different estimates of the observed mean $\bar{x}$. These give an observed frequency distribution that tends more and more nearly to the expected distribution as the number of samples on which it is based increases. For example, given a set of independent measurements x_i, $i = 1 \ldots n$, from an $N(\mu, \sigma^2)$, the statistic

$$\bar{x} = \sum_{i=1}^{n} \frac{x_i}{n}$$

can be shown to have a normal distribution with a mean μ and a variance σ^2/n. Clearly, as n becomes larger, the variance of $\bar{x}$, σ^2/n, becomes smaller so that a single observed $\bar{x}$ approaches more and more closely its expected value μ.

In order for a statistic to be a valid estimate of a parameter of an expected distribution, it must be *consistent*. This means that as the number of observations on which the statistic is based increases without limit, its value must tend to the parameter it is supposed to be estimating, exactly as in the example just mentioned of the mean of a set of observations from a normal distribution. One way to assure consistency is to equate a statistic with its expected value. Thus, for a sample x_i, $i = 1 \ldots n$, from an $N(\mu,\sigma^2)$

$$E(\bar{x}) = E\left(\frac{1}{n} \sum_{i=1}^{n} x_i\right) = \frac{1}{n} \sum_{i=1}^{n} E(x_i) = \frac{1}{n} n\mu = \mu \tag{44}$$

since the expectation of a sum is the sum of the expectations, and by the definition of the sample, $E(x_i) = \mu$ for all values of i. One useful method of estimation, originally advocated by the English statistician Karl Pearson, is to equate the observed moments of a distribution to their expected values, which is, effectively, what we are doing in Equation 44.

Another consistent estimate of μ in a normal distribution is the **median**, namely, that value of x which when all values are placed in order of magnitude splits the sample exactly in half. Still another estimate is the **mode**, which is the value of x_i that occurs most frequently.

I.22 Maximum Likelihood Estimation

Given a variety of consistent statistics for the estimation of a parameter, the question naturally arises, Which is the best? Since the mean of the frequency distribution of an estimate must be very close to the value of the parameter being estimated, the narrower the frequency distribution of the statistic, the closer a single observation of the statistic is likely to be to the parameter being estimated. Thus, the statistic with the smallest variance is, in general, the best one to choose. For example, the variance of the median of a sample from an $N(\mu, \sigma^2)$ is $\pi\sigma^2/2n$, which is $\pi/2 = 1.57$ times as large as the variance of the mean $\bar{x}$. As we shall see, $\bar{x}$ is in fact the estimate of μ that has the lowest possible variance.

The binomial probability

$$\binom{n}{r} p^r(1-p)^{n-r}$$

of observing r events out of n with probability p can, for a given observed proportion r/n, be thought of as a function of p. Considered in this way, it gives the *relative* probability of observing r/n for different values of p, which is called the **likelihood**. The likelihood is not a frequency distribution with respect to p since its sum over all values of p is not unity. An intuitively rational approach to choosing the value of p that best fits the observation is to choose that value which maximizes the likelihood. This method of estimation by **maximum likelihood** was shown by its originator R. A. Fisher to give—almost always—consistent statistics which have the minimal possible variance. It therefore also has a theoretical basis as the optimum method for estimating parameters from observed data. For mathematical reasons it is often convenient to consider maximizing the logarithm of the likelihood instead of the likelihood itself, which, of course, gives the same result. In the case of a binomial observation the logarithm of the likelihood is given by

$$L(p) = \text{const} + r \log p + (n - r)\log(1 - p). \tag{45}$$

The maximum value of $L(p)$ is given by the value of p which satisfies the equation

$$\frac{dL}{dp} = 0 = \frac{r}{p} - \frac{n-r}{1-p}, \tag{46}$$

giving $p = r/n$, the observed proportion.

It can be shown quite generally that

$$\frac{1}{E\left(\dfrac{-d^2L}{dp^2}\right)}$$

is approximately the variance of the maximum likelihood estimate of p. Fisher called the quantity

$$E\left(-\frac{d^2L}{dp^2}\right),$$

which is the inverse of the variance, the *expected amount of information* provided by the sample with respect to the parameter p that is being estimated. The intuitive basis for this is clear since the larger the amount of information, the smaller the variance, and so the closer the estimate to its true expected value. From Equation 46

$$\frac{d^2L}{dp^2} = \frac{-r}{p^2} - \frac{n-r}{(1-p)^2}; \tag{47}$$

therefore the expected information is

$$E\left(-\frac{d^2L}{dp^2}\right) = \frac{np}{p^2} + \frac{n(1-p)}{(1-p)^2} = \frac{n}{p} + \frac{n}{1-p} = \frac{n}{p(1-p)} \tag{48}$$

and the variance $p(1-p)/n$, where $p = r/n$ as given by Equation 37. We shall discuss in Sections I.37 and I.38 special procedures for solving maximum likelihood equations numerically, when they cannot readily be solved analytically as for a binomial observation.

The probability of making a series of independent observations $x_1, x_2, \ldots, x_n$ from an $N(\mu, \sigma^2)$ is

$$\left(\frac{1}{\sigma\sqrt{2\pi}}e^{-(x_1-\mu)^2/2\sigma^2}\right) \times \left(\frac{1}{\sigma\sqrt{2\pi}}e^{-(x_2-\mu)^2/2\sigma^2}\right) \cdots \times \left(\frac{1}{\sigma\sqrt{2\pi}}e^{-(x_n-\mu)^2/2\sigma^2}\right). \tag{49}$$

The likelihood with respect to μ and σ^2 is therefore

$$\frac{1}{\sigma^n(2\pi)^{n/2}}\exp\left(-\frac{1}{2\sigma^2}\sum_{i=1}^{n}(x_i-\mu)^2\right) \tag{50}$$

and is a function of the two parameters μ and σ^2, given the x_i's. The logarithm of the likelihood is

$$L = \text{const} - \frac{1}{2\sigma^2}\sum_{i=1}^{n}(x_i-\mu)^2 - n\log\sigma. \tag{51}$$

To obtain the maximum of L with respect to μ, we differentiate L partially with respect to μ and equate to zero. Thus

$$\frac{\partial L}{\partial \mu} = 0 = \frac{1}{\sigma^2}\sum_{i=1}^{n}(x_i-\mu) = 0, \tag{52}$$

which gives

$$\mu = \frac{1}{n}\sum_{i=1}^{n}x_i,$$

the observed arithmetic mean, as the maximum likelihood estimate of μ. For the estimation of σ we have

$$\frac{\partial L}{\partial \sigma} = \frac{1}{\sigma^3}\sum_{i=1}^{n}(x_i-\mu)^2 - \frac{n}{\sigma} = 0,$$

giving the estimate

$$\hat{\sigma}^2 = \frac{1}{n} \sum_{i=1}^{n} (x_i - \mu)^2. \tag{53}$$

This is the estimate of the variance appropriate for the situation when μ is assumed known. The estimate

$$s^2 = \frac{1}{n-1} \sum_{i=1}^{n} (x_i - \bar{x})^2$$

given in Equation 35 is the one to be used if μ is not known, and we then consider what is the best estimate of σ^2 given the observed mean $\bar{x}$.

I.23 Significance Tests and Confidence Limits

Given observations x_i, $i = 1 \ldots n$, from an $N(\mu, \sigma^2)$ and assuming σ^2 is known, the observed arithmetic mean $\bar{x}$ is distributed as $N(\mu, \sigma^2/n)$. Thus, from equations 8 and 9

$$\frac{\bar{x} - \mu}{\sqrt{\sigma^2/n}} = \frac{(\bar{x} - \mu)\sqrt{n}}{\sigma}$$

is distributed as an $N(0, 1)$. From tables of the $N(0, 1)$ (for example, in Fisher and Yates, Table I or VIII) we find that 95 percent of the distribution lies in the range $+1.96$ to -1.96 and thus 5 percent lies outside this range. Knowing μ, we could therefore say that 95 percent of the samples of size n would give values of $\bar{x}$ lying in the range defined by

$$\frac{\sqrt{n}(\bar{x} - \mu)}{\sigma} = +1.96 \text{ to } \frac{(\bar{x} - \mu)\sqrt{n}}{\sigma} = -1.96,$$

or

$$\mu + 1.96 \frac{\sigma}{\sqrt{n}} \text{ to } \mu - 1.96 \frac{\sigma}{\sqrt{n}}. \tag{54}$$

An observed value of $\bar{x}$ outside this range would be expected in only 5 percent of random samples. Accepting this probability level as unlikely, we then say that such an $\bar{x}$ differs from μ at the 5 percent significance level. More generally if for an observed $\bar{x}$ and an assumed μ and σ^2

$$\frac{\sqrt{n}(\bar{x} - \mu)}{\sigma} = d_\alpha, \tag{55}$$

where from tables of the $N(0, 1)$ we find that α percent of values lie outside the range $\pm d_\alpha$, then $\bar{x}$ differs from μ at the α percent significance level. This is one of the simplest forms of a test of significance. Here we are testing whether an observed mean $\bar{x}$ is compatible with an assumed mean μ, given that σ^2 is known. We express the result in terms of the probability of finding an $\bar{x}$ that differs from μ by as much as that observed or more.

Assume now that μ is unknown and we are estimating it by $\bar{x}$. We can then ask a complementary question to that posed above, namely, what is the range of values of μ that, if known, would not lead to $\bar{x}$ being significantly different from μ at a given significance level, α. This range is defined by the limits

$$\bar{x} + d_\alpha \frac{\sigma}{\sqrt{n}} \qquad \text{and} \qquad \bar{x} - d_\alpha \frac{\sigma}{\sqrt{n}}, \tag{56}$$

known as $(100 - \alpha)$ percent *confidence limits*. So long as μ lies inside these limits, it will be compatible with the observed $\bar{x}$ at the α percent significance level. In general terms we can say that given an observed $\bar{x}$, the probability of μ being in the range defined by Equation 56 is effectively $(100 - \alpha)$ percent. Equations 54 and 56 clearly express the complementarity between the problems of significance testing and estimation.

In most real situations σ^2 is not known but is estimated from the data by s^2, as in Equation 35. So long as the sample is large enough, the difference between the true value of σ^2 and its estimate s^2 will be sufficiently small so that substituting s^2 for σ^2 in Equations 54 and 56 will give reasonably accurate results. When, however, the sample is not large, the variation in s^2 must be taken into account in setting confidence limits and performing tests of significance. It can be shown that the ratio

$$t = \frac{\bar{x} - \mu}{s/\sqrt{n}}, \tag{57}$$

which is a function of $\bar{x}$ and s, both of which are themselves functions of the observations, has a beta distribution of the second kind (see Equation 33) with parameters $l = \frac{1}{2}$ and $m = \frac{1}{2}(n - 1)$, which are independent of the unknown mean and variance, μ and σ^2. This distribution is called **student's t distribution** after the statistician W. S. Gosset who used "student" as his pseudonym. The single parameter, $n - 1$, of the t distribution is called the **number of degrees of freedom**. Tabulations of this distribution for various $n - 1$ values give the values t_α, such that α percent of the distribution lies outside the range $-t_\alpha$ to $+t_\alpha$. We can now replace Equations 55 and 56, giving significance levels and confidence intervals by the equations

$$\frac{\sqrt{n}(\bar{x} - \mu)}{s} = t_\alpha \tag{58}$$

and

$$\bar{x} \pm t_\alpha \frac{s}{\sqrt{n}}, \tag{59}$$

which take into account the sampling variation of s^2.

I.24 Significance of the Difference Between Observed Means

Instead of testing the agreement between an observed and an expected mean, it is often necessary to test for the significance of the difference between two observed means. We assume two independent sets of observations $x_i^{(1)}$ $i = 1 \ldots n_1$, and $x_i^{(2)}$ $i = 1 \ldots n_2$ giving means $\bar{x}_1$ and $\bar{x}_2$ and variance estimates s_1^2 and s_2^2. We wish to test effectively whether the difference $\bar{x}_1 - \bar{x}_2$ is significantly different from zero. Assuming both sets of observations have the same expected variance σ^2, we find that the combined variance estimate is given by

$$s^2 = [(n_1 - 1)s_1^2 + (n_2 - 1)s_2^2]/(n_1 + n_2 - 2). \tag{60}$$

The expected variance of the difference between the means can be shown to be

$$V(\bar{x}_1 - \bar{x}_2) = \sigma^2\left(\frac{1}{n_1} + \frac{1}{n_2}\right). \tag{61}$$

Thus, assuming all observations are normally distributed,

$$t = \frac{\bar{x}_1 - \bar{x}_2}{s\sqrt{1/n_1 + 1/n_2}}, \tag{62}$$

where s is given by Equation 60, has a t distribution with $n_1 + n_2 - 2$ degrees of freedom, and provides the basis for testing whether $\bar{x}_1 - \bar{x}_2$ differs significantly from zero. Thus, if the t value given by Equation 62 corresponds to the α percentage point of the t distribution for $n_1 + n_2 - 2$ degrees of freedom, $\bar{x}_1 - \bar{x}_2$ differs significantly from zero at the α percent significance level.

I.25 One-way Analysis of Variance and the F or Variance Ratio Distribution

If instead of just two sets of values we have several sets x_{ij}, $i = 1 \ldots k, j = 1 \ldots n_i$ (for example, IQ measurements in different stratified population samples), then we may wish to test whether all the means

$$\bar{x}_i = \frac{1}{n_i} \sum_{j=1}^{n_i} x_{ij} \tag{63}$$

are equal. An alternative equivalent question is to ask whether the variation between the $\bar{x}_i$ is that expected if all have the same expected mean μ. We assume that the expected variance σ^2 of the observations within each set is the same. Then an overall estimate of σ^2 is given by the average of estimates from all the data sets, namely,

$$s^2 = \frac{1}{N-k} \sum_{i=1}^{k} \sum_{j=1}^{n_i} (x_{ij} - \bar{x}_i)^2, \tag{64}$$

where

$$N = \sum_{i=1}^{k} n_i \tag{65}$$

is the total number of observations. Assuming that all the $\bar{x}_i$ have expectations μ, the observed variance estimate derived from them, namely,

$$B = \frac{1}{k-1} \sum_{i=1}^{k} n_i(\bar{x}_i - \bar{x})^2, \tag{66}$$

where

$$\bar{x} = \frac{1}{N} \sum_{i=1}^{k} n_. \bar{x}_i = \frac{1}{N} \sum_{i=1}^{k} \sum_{j=1}^{n_i} x_{ij} \tag{67}$$

is the overall mean, should also be an estimate of σ^2. If B is significantly bigger than s^2, then the $\bar{x}_i$ presumably have different expected values since the variation in these expected values would inflate the expected variance of the $\bar{x}_i$.

The ratio of two independent variance estimates $F = s_1^2/s_2^2$, say, where s_1^2 and s_2^2 are based on n_1 and n_2 observations, respectively, was shown by Fisher to have a beta distribution of the second kind (see Equation 23) with parameters $l = n_1 - 1$ and $m = n_2 - 1$. This is called the F distribution with degrees of freedom $n_1 - 1$ and $n_2 - 1$ and is independent of the expected means and variances of the observations used to calculate s_1^2 and s_2^2. The F values, which, for given $n_1 - 1$ and $n_2 - 1$, are exceeded with any given probability, have been tabulated (see Fisher and Yates, Table V). Conventionally, F is always calculated and tabulated for values greater than 1, and so we always assume $s_1^2 > s_2^2$. Thus, for example, if $n_1 = 6$, $n_2 = 7$, and $F = 4.4$, this being approximately the 5 percent limit of the F distribution for 5 and 6 degrees of freedom, respectively, we say that the two variances in the ratio differ at a 5 percent significance level. Since it can be shown that B and s^2 of Equations 64 and 66 are independent variance estimates, the ratio

$$F = \frac{B}{s^2} \tag{68}$$

provides an F test for the significance of the differences between the expected means of the $\bar{x}_i$. Note that when $k = 2$, $F = t^2$. In other words, the t distribution is a special case of the F distribution, namely, t^2 has an F distribution for degrees of freedom 1 and $N - 2$.

I.26 Analysis of Variance

Another way of looking at the F test for the variation between the sets of measurements, x_{ij}, is in terms of an **analysis of variance**. This procedure was originated by Fisher and is the cornerstone of much of the analysis of experimental data. It depends on breaking down the *total* variation in the set of measurements x_{ij} into two components, one within sets x_{ij} for given i, and the other between sets, or effectively between the means $\bar{x}_i$. Thus, considering the data as a whole and ignoring its subdivisions into different sets, we estimate the overall variance as

$$\sum_{i=1}^{k} \sum_{j=1}^{n_i} (x_{ij} - \bar{x})^2. \tag{69}$$

Considering first the terms for a given value of i, we have

$$\begin{aligned} \sum_{j=1}^{n_i} (x_{ij} - \bar{x})^2 &= \sum_{j=1}^{n_i} (x_{ij} - \bar{x}_i + \bar{x}_i - \bar{x})^2 \\ &= \sum_{j=1}^{n_i} (x_{ij} - \bar{x}_i)^2 + n_i(\bar{x}_i - \bar{x})^2 \end{aligned} \tag{70}$$

since the cross product term

$$2 \sum_{j=1}^{n_i} (x_{ij} - \bar{x}_i)(\bar{x}_i - \bar{x}) = 2(\bar{x}_i - \bar{x}) \sum_{j=1}^{n_i} (x_{ij} - \bar{x}_i) = 0 \tag{71}$$

from the definition of $\bar{x}_i$ (Equation 63). Thus the *total sum of squares* given by Equation 69 can be broken down into two components, as follows,

$$\sum_{i=1}^{k} \sum_{j=1}^{n_i} (x_{ij} - \bar{x})^2 = \sum_{i=1}^{k} \sum_{j=1}^{n_i} (x_{ij} - \bar{x}_i)^2 + \sum_{i=1}^{k} n_i(\bar{x}_i - \bar{x})^2. \tag{72}$$

The first of these is $(N - k)s^2$, from Equation 64 and is called the **within** sum of squares while the second is $(k - 1)B$ from Equation 66 and is called the **between** sum of squares. The results of this so-called *one-way analysis of variance* are usually presented in the form of a table, as follows:

Source of variation	*Sum of squares*	*Degrees of freedom*	*Mean squares*
Between	$\sum_{i=1}^{k} n_i(\bar{x}_i - \bar{x})^2$	$k-1$	B
Within (or residual)	$\sum_{i=1}^{k} \sum_{j=1}^{n_i} (x_{ij} - \bar{x}_i)^2$	$N-k$	s^2
Total	$\sum_{i=1}^{k} \sum_{j=1}^{n_i} (x_{ij} - \bar{x})^2$	$N-1$	

The ratio of the mean squares, B/s^2 following Equation 68 provides the basis for testing the significance of the differences between the $\bar{x}_i$. The total sum of the squares is usually computed as

$$\sum_i \sum_j x_{ij}^2 - N\bar{x}^2, \tag{73}$$

the between sum of squares as

$$\sum_i n_i \bar{x}_i^2 - N\bar{x}^2, \tag{74}$$

and the within, or residual, from the difference between Expressions 73 and 74.

One of the main genetic applications of the analysis of variance is in the analysis of quantitative characters, where it is used for assessing the significance of differences in a measurement between different types of relatives. Much more complex subdivisions of a body of data into multiple cross tabulations—for example, with respect to relationship and socioeconomic status—can be treated in an analogous way. In each case, the total variance is broken down into a series of subcomponents, measuring differences between category means of various sorts, and a residual for the estimate of variation within a cross classification. Genetic applications of the analysis of variance are discussed in detail by Kempthorne (1957).

I.27 Goodness of Fit and the Chi-square Test

So far we have discussed the problems of estimation and significance testing mainly in terms of continuous distributions, in particular the normal distribution. Given an observed proportion r/n, we have, however, shown that r/n itself is the best estimate of the expected proportion p, assuming a binomial distribution. If we assume that a binomial distribution is, for large n, nearly a normal distribution with a mean p and a variance $p(1-p)/n$, then

$$\frac{r}{n} \pm d_\alpha \sqrt{\frac{r}{n} \times \frac{\frac{1-r}{n}}{n}}$$

are α percent confidence limits for the estimate of p (see Equations 38 and 56) and provide the basis for testing whether the observed proportion differs significantly from a given expected proportion. However, in applying these limits, we ignore the question as to whether the binomial distribution is in fact a good fit to the data. We have also ignored this question in our discussion of fitting means and variances to a normal distribution. Consider, for example, the question raised in Chapter 10 concerning sex ratio distributions in families. Here, given a body of data consisting of the numbers of males and females in a sample of families of given size, we wished to test whether the data conformed to the expected binomial distribution. Assuming a binomial, we can estimate p from the overall proportion of males for all the families and then calculate the expected numbers of families with a particular number of males and females from the formula for the binomial distribution. How then do we decide whether these expected numbers differ significantly from the observed numbers? The most convenient and widely used approach is the **Chi-square test for goodness of fit**. If for each data compartment, such as that defined by a given number of males and females, we observe a number O and calculate on the basis of some assumed model, such as the binomial distribution, an expected number E, then the **Chi-square criterion** is defined by

$$\chi^2 = \sum \frac{(O - E)^2}{E}, \tag{75}$$

where summation is over all data compartments. It can be shown generally that this criterion has approximately a distribution known as the *chi-square distribution*, which depends only on one parameter n, *the number of degrees of freedom*. The number of degrees of freedom is generally the number of observations minus one minus the number of independent parameters fitted in order to calculate the expectation. Actually, $\frac{1}{2}\chi^2$ has a gamma distribution (see Equation 27) with parameter $\frac{1}{2}n$. The chi-square distribution is effectively the distribution of the observed variance derived from a set of data from an $N(\mu, \sigma^2)$. In particular, if we wish to test whether a set of $N(\mu, \sigma^2)$ observations, x_i, are compatible with an expected variance σ^2, then we use the fact that

$$\sum_i \frac{(x_i - \bar{x})^2}{\sigma^2} \tag{76}$$

has a chi-square distribution with $n - 1$ degrees of freedom. Thus chi-square is actually a special case of the F distribution, namely, for degrees of freedom $n_1 = n$ and $n_2 \to \infty$ (second variance known exactly). Tables of the chi-square distribution give, for different values of n, the chi-square value, which would be exceeded by chance only α percent of the time. The expected value of chi-square for n degrees of freedom is n. High chi-square values indicate discrepancies between the observed and the expected.

The simplest application of the chi-square test is in testing the fit of an observed proportion with an expected proportion. Suppose we have observed A of one type and B of another type, where $A + B = N$, and we expect that there are p of the first type and q of the second where $p + q = 1$; then we have

	First type	*Second type*	*Total*
Observed O	A	B	N
Expected E	Np	Nq	N
Contribution to χ^2: $(O - E)^2/E$	$\frac{(A - Np)^2}{Np}$	$\frac{(B - Nq)^2}{Nq}$	

Thus

$$\chi^2 = \frac{(A - Np)^2}{Np} + \frac{(B - Nq)^2}{Nq} = \frac{(Aq - Bp)^2}{Npq}, \tag{77}$$

since

$$A - Np = Aq - Bp = B - Nq \quad \text{and} \quad (1/p) + (1/q) = 1/pq.$$

This chi-square has one degree of freedom since there are two observations and no parameters are estimated from the data. Note that when $p = q = 1/2$, Equation 77 takes the form

$$\chi^2 = (A - B)^2/N, \tag{78}$$

a very useful, quick formula for testing for equality of two observed numbers. It is useful to remember that the 5 percent point for a chi-square with one degree of freedom is 3.84 or nearly 4. The result given by Equation 77 can be written in the form

$$\frac{(A - Np)^2}{Npq} = \chi^2 \qquad \text{or} \qquad \frac{A}{N} - p = \pm\chi\sqrt{\frac{pq}{N}},$$

which, since it can be shown that χ is an $N(0, 1)$, shows that the use of chi-square is, in this case, equivalent to approximating the binomial distribution by a normal distribution (see the expression given above for the confidence limits of an observed proportion).

I.28 2 × 2 Contingency Tables

A very important application of chi-square is in testing the equality of two observed proportions, for example, the proportion of concordant monozygotic versus dyzygotic twins for some given disease (see Chapter 9). The observed data can be set out in the form of a 2 × 2 contingency table.

	First type	*Second type*	
Data 1	a	b	$a+b$
Data 2	c	d	$c+d$
	$a+c$	$b+d$	$N=a+b+c+d$

We assume that the proportion of the first type versus the second type is $a + c : b + d$, as observed. Then, for example, the expected number of the first type in Data 1 is $(a + c)(a + b)/N$ and its contribution to χ^2 is

$$\left[a - \frac{(a+c)(a+b)}{N}\right]^2 \Big/ \frac{(a+c)(a+b)}{N}.$$

The other contributions can be similarly obtained to give, after a little algebraic manipulation,

$$\chi_1^2 = \frac{N(ad - bc)^2}{(a+c)(a+b)(b+d)(c+d)}. \tag{79}$$

This chi-square has one degree of freedom only since there are four observations, but two parameters are effectively estimated from the data, namely, the proportion of first to second type, and of Data 1 to Data 2. Looked at another way, the marginal totals are assumed given in calculating the expectations, and so as soon as one of the four data cells is fixed, the remainder are determined. When expected numbers are small, the chi-square with "Yates' continuity correction" is generally used as it is thought to give a better fit to the expected chi-square distribution, which is effectively based on normality assumptions, as indicated above. The corrected χ^2 is defined by

$$\chi_1^2 = \frac{N(|ad - bc| - \frac{1}{2}N)^2}{(a+c)(a+b)(b+d)(c+d)}, \tag{80}$$

where $|ad - bc|$ indicates the positive value of $ad - bc$. An exact test for agreement of observed proportions in a 2×2 table was devised by Fisher and is tabulated by Finney and co-workers (1963).

I.29 $2 \times k$ and $l \times k$ Contingency Tables

More generally, we may wish to test for the equality of a set of k proportions $a_i : b_i$, $a_i + b_i = N_i$, for $i = 1 \ldots k$. Following the same procedures as above, assuming the marginal totals N_i and $\sum_i a_i$, $\sum_i b_i$ given, it can be shown that the appropriate chi-square takes the form

$$\chi^2_{k-1} = \frac{N^2}{AB}\left(\sum_{i=1}^{k} \frac{a_i^2}{N_i} - \frac{A^2}{N}\right), \tag{81}$$

where

$$A = \sum_{i=1}^{k} a_i, \qquad B = \sum_{i=1}^{k} b_i, \qquad \text{and} \qquad N = A + B.$$

This chi-square has $k - 1$ degrees of freedom and is known as the $2 \times k$ contingency chi-square. The chi-square appropriate for a general $l \times k$ contingency table, having l classifications along rows and k along the columns, can be calculated in an analogous way. It takes the form

$$\chi^2_{(l-1)(k-1)} = \sum_{ij} \frac{\left(n_{ij} - \frac{n_{i.}n_{.j}}{N}\right)^2}{\frac{(n_{i.}n_{.j})}{N}}$$

$$= N\left(\sum \frac{n_{ij}^2}{n_{i.}n_{.j}} - 1\right) \tag{82}$$

for $(l-1)(k-1)$ degrees of freedom, where n_{ij} is the observation in the ith row and the jth column of the table,

$$n_{i.} = \sum_{j=1}^{k} n_{ij}, \; n_{.j} = \sum_{i=1}^{l} n_{ij},$$

and

$$N = \sum_{ij} n_{ij}$$

is the total number of observations.

I.30 Goodness of Fit of the Poisson Distribution

The chi-square distribution has a special application in testing whether a set of observations x_i, $i = 1 \ldots n$, conforms to the Poisson distribution. The test is based on the fact that the mean and variance of a Poisson distribution are equal. Thus, if $\bar{x}$ is the observed mean, it is also the expected value of the variance, and so

$$\chi^2 = \sum_{i=1}^{n} \frac{(x_i - \bar{x})^2}{\bar{x}} \tag{83}$$

should be distributed as a chi-square with $n - 1$ degrees of freedom. A variance that is significantly higher than the mean, as caused, for example, by variation in the parameter of a Poisson distribution leading in certain cases to a negative binomial as discussed above, will be indicated by a significantly high chi-square.

I.31 Pairs of Measurements, Correlation, and Regression

We shall now consider distributions of pairs of measurements, such as, for example, the height and IQ measured on a series of individuals. Suppose we observe pairs x_i and y_i, $i = 1 \ldots n$, with respective frequency distributions $f(x)$ and $g(y)$. If the measurements are *independent*, then by definition (see Section I.9) the probability of observing a particular pair x and y is $f(x)\, g(y)\, dx\, dy$. Thus

$$\begin{aligned} E(xy) &= \int_x \int_y xyf(x)g(y)\, dx\, dy \\ &= \left[\int_x xf(x)\, dx\right]\left[\int_y yg(y)\, dy\right] = E(x)E(y). \end{aligned} \tag{84}$$

In other words, if x and y are independent observations, then the expectation of their product is equal to the product of their expectations. If x and y are not independent, then the probability of the pair x and y is some function of x and y, say, $F(x, y)\, dx\, dy$, called their joint distribution, which cannot be expressed as the product of a function of x only and a function of y only. In this case

$$E(xy) \neq E(x)E(y), \tag{85}$$

and we define the *covariance* of x and y by

$$\begin{aligned} \operatorname{cov}(x, y) &= E(x - \bar{x})(y - \bar{y}) \\ &= E(xy + \bar{x}\bar{y} - \bar{x}y - \bar{y}x) \\ &= E(xy) - \bar{x}\bar{y}. \end{aligned} \tag{86}$$

The covariance is a measure of the interdependence of x and y, which is of course, zero when they are independent. The **correlation coefficient** which is effectively a standardized covariance, is defined by

$$r(x, y) = \frac{\operatorname{cov}(x, y)}{\sqrt{V(x) \times V(y)}}, \tag{87}$$

where $V(x)$ and $V(y)$ are the variances of the distributions of x and y when they are considered separately. The correlation coefficient varies between $+1$ and -1 and is a widely used standardized symmetric measure of the association between a pair of random variables. Note that

$$V(x + y) = E(x + y - \bar{x} - \bar{y})^2 = E[(x - \bar{x})^2 + 2(x - \bar{x})(y - \bar{y}) + (y - \bar{y})^2]$$

so that

$$V(x + y) = V(x) + 2 \operatorname{cov}(x, y) + V(y). \tag{88}$$

Thus the variance of the sum of two variables is only equal to the sum of the variances if they are independent. This is the basis for the fact that given a set of independent observations x_i, all drawn from the same distribution with variance σ^2,

$$V(\bar{x}) = \frac{1}{n^2} V\left(\sum_{i=1}^{n} x_i\right) = \frac{n\sigma^2}{n^2} = \frac{\sigma^2}{n}. \tag{89}$$

I.32 The Bivariate Normal Distribution

This generalization of the one-dimensional normal distribution (see Equation 5) is defined by

$$f(x, y) = \frac{1}{2\pi\sigma_x\sigma_y\sqrt{1-\rho^2}} \exp\left[\left[-\frac{1}{2(1-\rho^2)}\right]\right. \\ \left. \times \left[\frac{(x-\mu_x)^2}{\sigma_x^2} - \frac{2\rho(x-\mu_x)(y-\mu_y)}{\sigma_x\sigma_y} + \frac{(y-\mu_y)^2}{\sigma_y^2}\right]\right], \tag{90}$$

where μ_x, σ_x^2, and μ_y, σ_y^2 are the respective means and variances of the separate distributions of x and y. The correlation coefficient between x and y can be shown to be ρ. Note that when $\rho = 0$, Equation 90 simply breaks down into the product of two normal distributions, as expected. Given an observed value of x, the frequency distribution of y can be shown to be

$$f(y) = \frac{1}{\sigma_y\sqrt{2\pi(1-\rho)^2}} \exp\left[-\frac{1}{2(1-\rho)^2\sigma_y^2}\left[y - \mu_y - \rho\frac{\sigma_y(x-\mu_x)^2}{\sigma_x}\right]\right], \tag{91}$$

where x is considered a constant. This is a normal distribution with mean

$$\mu_y + \frac{\rho\sigma_y}{\sigma_x}(x - \mu_x)$$

and variance

$$\sigma_y\sqrt{1-\rho^2}.$$

This distribution of y for given values of x is called the **regression** of y on x. By symmetry the regression of x on y is given by a normal distribution with mean

$$\mu_x + \frac{\rho\sigma_x}{\sigma_y}(y - \mu_y)$$

and variance

$$\sigma_x\sqrt{1-\rho^2}.$$

Each of these regressions can be thought of as depicting a distribution of observed values about an expected straight line. The most common and useful genetic application of these concepts is to the study of quantitative inheritance (see Chapter 9).

I.33 Multinomial Distributions

Suppose r_i and r_j are the numbers of observations in the ith and jth categories of a multinomial distribution as defined by Equation 26. Then the covariance between r_i and r_j is given by

$$\operatorname{cov}(r_i, r_j) = -np_ip_j \tag{92}$$

while

$$V(r_i) = np_i(1 - p_i). \tag{93}$$

Restricting the sum of the numbers of observations to n, or equivalently the sum of the proportions to unity, imposes a negative association on pairs of observations.

I.34 Analysis of Observed Pairs of Measurements

Given a set of observed pairs of measurements x_i and y_i, $i = 1 \ldots n$, we can calculate variance, covariance, and correlation coefficient estimates as follows:

$$s_x^2 = \frac{1}{n-1}\sum_i (x_i - \bar{x})^2 = \frac{1}{n-1}\left(\sum_i x_i^2 - \bar{x}\sum_i x_i\right), \tag{94}$$

$$s_y^2 = \frac{1}{n-1}\sum_i (y_i - \bar{y})^2 = \frac{1}{n-1}\left(\sum_i y_i^2 - \bar{y}\sum_i y_i\right), \tag{95}$$

$$\operatorname{cov}(x, y) = \frac{1}{n-1}\sum (x_i - \bar{y})(y_i - \bar{y}) = \frac{1}{n-1}\left(\sum_i x_i y_i - \bar{x}\sum_i y_i\right), \tag{96}$$

and

$$r = \operatorname{cov}(x, y)/s_x s_y. \tag{97}$$

Note that only the sums $\sum_i x_i$, $\sum_i y_i$ and $\sum_i x^2$, $\sum_i y^2$, $\sum_i x_i y_i$ need to be computed. A simple test of independence of the pairs is whether r is significantly different from zero. Fisher showed that for observations from a bivariate normal distribution (Equation 90) the distribution of

$$\zeta = \tfrac{1}{2}\log_e \frac{1+r}{1-r} \tag{98}$$

is approximately normal with mean

$$\zeta = \tfrac{1}{2}\log_e \frac{1+\rho}{1-\rho}$$

$$= 0 \text{ if } \rho = 0, \tag{99}$$

where ρ is the expected correlation coefficient and $1/(n-3)$ is the variance, which is *independent of* ρ. This provides a simple approximate test for the significance of any observed r value (see Fisher and Yates, 1953).

I.35 Linear Regression Analysis

In many cases, given pairs of observations, it is appropriate to think of one of them, y, as depending on the other, x, for example, birth weight as a function of parity. Birth weight is then the observation subject to statistical error and is to be considered, potentially, as a function of parity. A linear relation of the form

$$y = \alpha + \beta(x - \bar{x}), \tag{100}$$

where $\bar{x}$ is the mean of the x values, is the simplest relation that can be tested for. We assume that for a given value of x, y is $N(\alpha + \beta(x - \bar{x}), \sigma^2)$. Suppose now that we are given a series of paired observations x_i and y_i, $i = 1 \ldots n$. Then, following our discussion of regression with respect to the bivariate normal distribution (see Equation 91), estimates of α and β are given, respectively, by

$$a = \bar{y} \tag{101}$$

and

$$b = \frac{rs_y}{s_x} = \frac{\text{cov}(x, y)}{s_x^2}, \tag{102}$$

where $\text{cov}(x, y)$, r, s_x, and s_y are as given by Equations 94 to 97. Another way of obtaining these estimates is by minimizing the expression

$$\sum_i [y_i - \alpha - \beta(x_i - \bar{x})]^2 \tag{103}$$

with respect to α and β. This procedure is known as **least squares** estimation, and is often used for fitting linear relationships. It is equivalent to maximum likelihood as long as y is normally distributed, as assumed above. The results of a regression analysis can be expressed in the form of an analysis of variance.

The total sum of squares for y is $\sum_i (y_i - \bar{y})^2$, ignoring the potential linear relationship. The sum of squares about the fitted line is

$$\sum_i (y_i - \bar{y} - b(x_i - \bar{x}))^2, \tag{104}$$

where b is given by Equation 102. The sum of squares "removed" by fitting the line is the difference between the total and Expression 104. This can easily be shown to be

$$b^2 \sum_i (x_i - \bar{x})^2 = [\text{cov}(x, y)]^2 / s_x. \tag{105}$$

In the form of an analysis-of-variance table, we have, therefore,

Source of variation	*Sum of squares*	*Degrees of freedom*	*Mean squares*
Regression of y on x	$[\text{cov}(x, y)]^2/s_x = A$	1	A
Residual about fitted line	$\sum_i [y_i - \bar{y} - b(x_i - \bar{x})]^2 = B$	$n - 2$	$s^2 = \dfrac{B}{n-2}$
Total	$\sum_i (y_i - \bar{y})^2$	$n - 1$	

The F ratio, A/s^2, provides a test for the significance of the amount of variation removed by fitting the line, that is, for the significance of a linear relationship. The value of B is usually computed by subtraction. It can be shown that an estimate of the variance of the regression coefficient b (defined by Equation 102) is given by

$$V(b) = \frac{s^2}{s^2_x}. \tag{106}$$

Assuming b is normally distributed, this allows one to test for the significance between the observed and an expected regression coefficient. Thus, if β is the expected value of b,

$$t = \frac{b - \beta}{\dfrac{s}{s_x}} \tag{107}$$

follows the t distribution for $n - 2$ degress of freedom. Confidence limits at the α percent level are given by

$$b \pm t_\alpha \frac{s}{s_x}. \tag{108}$$

The concept of regression is often generalized in two ways:

1. The quantity y may be a function of several different types of variables, say, $x_1, x_2, \ldots, x_k$, leading to a predicted multiple linear relationship

$$y = \alpha + \beta_1(x_1 - \bar{x}_1) + \cdots + \beta_k(x_k - \bar{x}_k). \tag{109}$$

Multiple regression is often used to "remove" biases due to stratification effects as in the relation between fertility or mortality and inbreeding (see Chapters 7 and 9).

2. The relation between y and x may not be linear. Generally it can be assumed to take the form of a polynomial

$$y = \alpha + \beta_1(x - \bar{x}) + \beta_2(x - \bar{x})^2 + \cdots + \beta_k(x - \bar{x})^k. \tag{110}$$

Thus if it is found that a linear relation does not give an adequate fit, we can next test for a quadratic relationship, then a cubic, and so on. (Recall the relation between survival and birth weight discussed in Chapter 9.) In all cases the coefficients α, β_1, etc., are fitted by the method of least squares. Mather's book *Statistical Analysis in Biology* gives a good elementary discussion and derivation of these procedures.

I.36 Intraclass Correlation Coefficient

In some cases involving pairs of measurements there is no way to distinguish between the pairs. For example, given IQ measurements of a pair of identical twins, how can it be decided which measurement is x and which is y? In such cases a different type of correlation coefficient, called the **intraclass correlation coefficient** which treats the pairs of measurements symmetrically, is calculated to assess the relation between them. This is defined by

$$r = \frac{2\sum_i (x_i - \bar{x})(x_i' - \bar{x})}{[\sum_i (x_i - \bar{x})^2 + \sum_i (x_i' - \bar{x})^2]}, \tag{111}$$

where x_i and x_i', $i = 1 \ldots n$, are the pairs of measurements taken in an arbitrary order, and

$$\bar{x} = \frac{1}{2n}\left(\sum_i x_i + \sum_i x_i'\right). \tag{112}$$

Now

$$\begin{aligned}\sum_i (x_i - x_i')^2 &= \sum_i [(x_i - \bar{x}) - (x_i' - \bar{x})]^2 \\ &= \sum_i (x_i - \bar{x})^2 + \sum_i (x_i' - \bar{x})^2 - 2\sum_i (x_i - \bar{x})(x_i' - \bar{x})\end{aligned} \tag{113}$$

so that from Equation 111

$$\frac{\sum_i (x_i - x_i')^2}{\sum_i (x_i - \bar{x})^2 + \sum_i (x_i' - \bar{x})^2} = 1 - r. \tag{114}$$

This justifies the relationship between the intraclass correlation coefficient and the variance of the differences, which is $\sum_i (x_i - x_i)^2/n$, that was used in the construction of Holzinger's twin heritability index (see Chapter 9).

I.37 Fisher's Scoring Method for the Solution of Maximum Likelihood Equations: One Parameter

In many, if not most situations, maximum likelihood equations for the estimation of parameters cannot be solved analytically. The equations can then be solved numerically by iterative methods that is, by repeated correction methods. Fisher devised a procedure, which he called **scoring** for such numerical solutions, which is extremely convenient and provides in a readily available form variances of the estimates and tests of significance. These procedures are widely used in the statistical analysis of human pedigrees, which we shall discuss in Appendix II.

We will consider first the general multinomial one-parameter distribution. We are given a sample of observations a_i, $i = 1 \ldots k$, in k categories with the likelihood

$$l = \frac{n!}{a_1!\,a_2!\ldots a_k!}\, p_1^{a_1} p_2^{a_2} \ldots p_k^{a_k},$$

where $a_1 + a_2 + \cdots + a_k = n$ and each p_i is a function of some parameter θ, which we wish to estimate.

An appropriate genetic example would be the estimation of the recombination fraction from double intercross data, using the mating $AB/ab \times AB/ab$. The logarithm of the likelihood is given by

$$L = \text{const} + a_1 \log p_1 + a_2 \log p_2 + \cdots + a_k \log p_k. \tag{115}$$

We wish to solve the equation

$$\frac{dL}{d\theta} = 0 = \sum_i a_i \frac{d}{d\theta}(\log p_i) = \sum_i a_i \frac{1}{p_i}\frac{dp_i}{d\theta}. \tag{116}$$

If we write $S(\theta) = dL/d\theta$, then the Newton-Raphson formula for the approximate solution of an equation states that if θ_0 is a trial solution for the equation $S(\theta) = 0$, an improved solution is given by

$$\theta_1 = \theta_0 - \frac{S(\theta_0)}{\left(\frac{dS}{d\theta}\right)_{\theta=\theta_0}}. \tag{117}$$

Now

$$\frac{dS}{d\theta} = \frac{d^2L}{d\theta^2} = -\tilde{I}(\theta), \tag{118}$$

where for given θ, $\tilde{I}(\theta)$ is the information realized with respect to the estimate θ (see Section I.22). Thus Equation 117 can also be written in the form

$$\theta_1 = \theta_0 + \left[\frac{S(\theta_0)}{\tilde{I}(\theta_0)}\right]. \tag{119}$$

The expected information is given by

$$E\left(\frac{dS}{d\theta}\right) = E\left(\frac{d^2L}{d\theta^2}\right) = -I(\theta). \tag{120}$$

Another form of correction for an initial estimate θ_0 is given by

$$\theta_0 + \left[\frac{S(\theta_0)}{I(\theta_0)}\right], \tag{121}$$

where the expected information is substituted for the realized information in Equation 119.

The quantity $S(\theta_0)$ is called the score with respect to θ at the value θ_0. The expected information can be evaluated as follows:

$$\begin{aligned} I(\theta) &= -E\left(\frac{d^2L}{d\theta^2}\right) \\ &= -E\left[\sum_i a_i\left(\frac{1}{p_i}\frac{d^2p_i}{d\theta^2} - \frac{1}{p_i^2}\left(\frac{dp_i}{d\theta}\right)^2\right)\right] \\ &= -n\left[\sum_i \frac{p_i}{p_i}\frac{d^2p_i}{d\theta^2} - \sum_i \frac{p_i}{p_i^2}\left(\frac{dp_i}{d\theta}\right)^2\right] \\ &= n\sum_i \frac{1}{p_i}\left(\frac{dp_i}{d\theta}\right) \end{aligned} \tag{122}$$

since

$$\sum_i \frac{d^2p_i}{d\theta^2} = \frac{d^2}{d\theta^2}\sum_i p_i = 0, \qquad \text{for} \qquad \sum_i p_i = 1.$$

The amount of information expected per observation is

$$i(\theta) = \frac{I(\theta)}{n} = \sum_i \frac{1}{p_i}\left(\frac{dp_i}{d\theta}\right)^2. \tag{123}$$

The expectation of the score is given by

$$E[S(\theta)] = E\sum_i a_i \frac{1}{p_i}\frac{dp_i}{d\theta} = n\sum_i \frac{p_i}{p_i}\frac{dp_i}{d\theta} = 0 \tag{124}$$

since

$$\sum_i p_i = 1.$$

The variance of the score is, therefore,

$$\begin{aligned} E[S(\theta)]^2 &= E\left[\sum_i a_i^2 \frac{1}{p_i^2}\left(\frac{dp_i}{d\theta}\right)^2 + \sum_{ij} \frac{a_i a_j}{p_i p_j}\frac{dp_i}{d\theta}\frac{dp_j}{d\theta}\right] \\ &= \sum_i np_i \frac{1}{p_i^2}\left(\frac{dp_i}{d\theta}\right)^2 - n\left(\sum_i \frac{dp_i}{d\theta}\right)^2. \end{aligned} \tag{125}$$

In other words,

$$V[S(\theta)] = I(\theta) \tag{126}$$

from Equation 122. Thus if $\hat{\theta}$ is an assumed true value,

$$\chi_1^2 = \frac{S^2(\hat{\theta})}{I(\hat{\theta})} \tag{127}$$

has a chi-square distribution for one degree of freedom for testing the agreement between observed and expected values of θ. Recall that $1/I(\theta)$ is approximately the variance of the maximum likelihood estimate $\hat{\theta}$. It is often useful to calculate the approximate information realized numerically from the fact that

$$\left(\frac{dS}{d\theta}\right)_{\theta=\theta_0} = \frac{S(\theta_0) - S(\theta_0')}{\theta_0 - \theta_0'} \sim -\tilde{I}(\theta_0), \tag{128}$$

provided θ_0' is close to θ_0. In fact, all the calculations can be done numerically by starting with the logarithm likelihood L and calculating derivatives approximately as in Equation 128.

As an illustration of the use of the various scoring formulas, consider the application to the estimation of the binomial parameter discussed earlier. For observations a and b, where $a + b = n$, the logarithm of the likelihood is (see Equation 45)

$$L = \text{const} + a \log p + b \log (1 - p), \tag{129}$$

where p is the parameter to be estimated. The score for p is

$$S(p) = \frac{dL}{dp} = \frac{a}{p} - \frac{b}{1 - p}. \tag{130}$$

The information realized is

$$\tilde{I}(p) = \frac{-dS}{dp} = \frac{a}{p^2} + \frac{b}{(1-p)^2}$$

while the information expected is, from Equation 122

$$I(p) = n\left(\frac{1}{p} + \frac{1}{1-p}\right) = \frac{n}{p(1-p)}, \tag{132}$$

as given in Equation 48. The chi-square for testing agreement between the observed and the expected is, from Equation 127,

$$\chi_1^2 = \frac{\left(\dfrac{a}{p} - \dfrac{b}{1-p}\right)^2}{\dfrac{n}{p(1-p)}} = \frac{[bp - a(1-p)]^2}{np(1-p)}, \tag{133}$$

as given before in Equation 77. In this case of course, since the scoring Equation 130 can be solved analytically, numerical procedures are not needed.

The method of scoring can be applied very simply to the situation in which there are a number of different *independent* bodies of information with respect to a parameter θ—for example, data on a particular recombination fraction from both a backcross and a double intercross. Since the bodies of data are independent, the joint likelihood is the product of the separate likelihoods, and so the joint logarithm of the likelihood is the sum of the logarithms of the separate likelihoods. From this it follows that the total score and information with respect to θ are the sums of the separate scores and information:

$$\begin{aligned} &\text{Total score} = S = S_1 + S_2 + \cdots \\ &\text{Total information} = I = I_1 + I_2 + \cdots. \end{aligned} \tag{134}$$

The combined correction for a trial value θ is, therefore, simply $-S/I$ (see Equation 121). The test for agreement with an expected value is $\chi_1^2 = S^2/I$, as before. Heterogeneity between the k sources of information with respect to the parameter θ is tested for by a heterogeneity chi-square,

$$\chi_{k-1}^2 = \frac{S_1^2}{I_1} + \cdots + \frac{S_k^2}{I_k} - \frac{S^2}{I} \tag{135}$$

for k degrees of freedom. When θ is at its maximum likelihood estimate, then $S = 0$, and so the last term in Equation 135 is zero.

I.38 Scoring for Several Parameters

This section requires a knowledge of matrix theory.

When there are several parameters θ_i, $i = 1 \ldots s$, there is a score for each parameter, which is given by the partial derivative of the logarithm of the likelihood with respect to each parameter. Thus

$$S_i = \frac{\partial L}{\partial \theta_i} = \sum_{j=1}^{k} a_j \frac{1}{p_j} \frac{\partial p_j}{\partial \theta_i} \tag{136}$$

from Equation 115, where S_i is itself still a function of all the parameters θ_i, $i = 1 \ldots s$. The maximum likelihood estimates are now the solutions of a set of s simultaneous equations,

$$S_i = 0, \qquad i = 1 \ldots s. \tag{137}$$

The set of informations is now an $s \times s$ symmetric matrix. Thus, for example, the expected information with respect to the ith and jth parameters is defined by

$$I_{ij} = -E\left(\frac{\partial^2 L}{\partial \theta_i \, \partial \theta_j}\right). \tag{138}$$

Following the derivation of Equation 122, it can be shown that

$$I_{ij} = n \sum_{l=1}^{k} \frac{1}{p_l} \frac{\partial p_l}{\partial \theta_i} \frac{\partial p_l}{\partial \theta_j}, \tag{139}$$

where summation, as before, is over the k categories of observations. The inverse of the information matrix I_{ij} is, approximately, the variance-covariance matrix V_{ij}, where V_{ii} is the variance of the estimate of θ_i and V_{ij} is the covariance between the estimates of θ_i and θ_j. The corrections $d\theta_i$ to a set of trial values θ_i are now given by

$$d\theta_i = -\sum_{j=1}^{s} V_{ij} S_j, \tag{140}$$

or, in matrix notation,

$$\boldsymbol{\theta}_1 = \boldsymbol{\theta}_0 + (\boldsymbol{I}^{-1} \boldsymbol{S}_{\theta=\theta_0}), \tag{141}$$

where $\boldsymbol{\theta}_0$ are the trial values and $\boldsymbol{\theta}_1$ the corrected values. In matrix notation the test for the goodness of fit is given by

$$\chi_s^2 = \boldsymbol{S}'\boldsymbol{I}^{-1}\boldsymbol{S}, \tag{142}$$

where S' is the transpose of the vector of scores S. As before, given a series of independent bodies of data, the various scores and informations are additive, and all the same formulas now apply to the total scores and information. The use of these formulas will be illustrated in Appendix II by application to pedigree analysis and gene-frequency estimation.

I.39 An Example of a Computer Program

Most computations in genetics require the use of a fairly large number of significant figures to get sufficiently accurate results. Moreover, the amount of computing, except in the simplest instances, is sufficiently large to make the use of a computer almost essential. Using a very simple example of a simulation program, we will illustrate the structure of a common computer language, PL1, for readers unfamiliar with computers.

The computer can carry out arithmetic operations and can follow certain logical, conditional instructions such as, "if $x = 0$, then do..., otherwise do...." where the ellipses indicate some other set of instructions. The essence of writing a program is to break an operation down into a sequence of simple operations or instructions and then to translate these instructions into a language that the computer can interpret.

The program we shall write simulates genetic drift in a population of size n, beginning with a gene frequency of p_0 (which the computer writes as *p*0) and continuing for t (nonoverlapping) generations. We will run concurrently m such populations, and every i generations the mean and variance of the gene frequencies of the m populations will be computed and printed.

The flow chart of the program is as follows:

1. Reserve space in the computer for the m populations.
2. Decide on the values of n, m, p_0, i, and also the total duration of the experiment. This is designated by another variable, which we call *ttot*.
3. Give all m populations the initial gene frequency p_0. This is considered as time t equal to zero.
4. Advance time by one unit.
5. Carry out a random sampling of m genes according to a binomial scheme for each of the m populations, using as a probability the gene frequency of a given population at the last generation. The random sample forms the gene frequency of the same population at the next generation. The gene frequency of population j at time t is $p_{j,t}$. Thus, at time $t = 0$, $p_{j,t} = p_{j,0} = p_0$ for all j. The term $p_{j,t+1}$ is the frequency of a random sample of size n from population $p_{j,t}$.
6. Repeat procedures 4 and 5 for the desired number of generations (*ttot*).

7. If t is equal to i or a multiple of it, the mean and variance of the m $p_{j,t}$ values should be computed and printed.

8. If t is equal to *ttot*, the computer must stop.

We will now describe how these operations are carried out.

1. *Reserve space for the m populations.* We need to keep a record of the gene frequency p of all m populations at the last generation. A vector of m values is sufficient, and it is labeled p. We must also, at this stage, declare the maximum number of populations that we want to store. This is done by the instruction:

declare $p(100)$;

the instruction is preceded by a number and followed by a semicolon. First, however, we must name the program; this will be instruction 1. Thus, the first two lines of the program will read:

```
1    program DRIFT;
2    declare p(100);
```

2. *Decision on n, p_0, m, i, and ttot.* This is communicated to the computer by writing in the next instruction, which will be the third one:

```
3    get data (n, p0, m, i, ttot);
```

During execution, the computer asks for the numerical values of the five variables given, and then proceeds to the following instructions.

3. *Put time $t = 0$ and $p = p_0$.* There is a variable to indicate time, which we may call t. Rules for giving names to variables are simple. In some computers, only capital letters can be used, but in others small letters can also be used. The maximum number of letters (or digits) in the name given to a variable is usually six or eight. The setting of initial time equal to zero can be done in the iteration process to be described below. The initial setting of all gene frequencies equal to p_0 can be done by the fourth instruction:

```
4    p = p0;
```

It should be noted that in computer language the equal sign corresponds to an *order*. The order is that of making the quantity indicated on the left equal to the quantity indicated on the right. When, as here, p is a matrix (a vector), the operation is iterated automatically for all the elements of the matrix. When these have to be treated individually, as happens in the later part of this program, each element is referred to by its index, given in parentheses after the matrix symbol. Thus, $p(1)$ is the first element of the vector p; $p(22)$ is the twenty-second element, etc.

4. 6. and 8. *Iteration for the desired number of generations.* Iteration can be achieved by a "do" statement, namely, by putting all the instructions to be repeated (indicated by dots in what follows) between two instructions, which are, in the present case:

```
do t = 1 to ttot;
    ...
end;
```

The dots represent, in this case, the operations that are described below as "sample binomially from all populations." The execution of the "do" loop is carried out by the computer by initially setting the variable t equal to 1 and performing all the instructions indicated by the dots until the word "end"; then going back to the beginning of the "do" loop, setting t equal to 2, and performing again all the operations indicated by the dots, now with $t = 2$, and so on, until t has been increased, one by one, and $t = ttot$. When this last run has been made, the computer leaves the "do" loop and proceeds to the next instruction after the word "end."

5. *Sample binomially from all populations.*

a. *We will first consider population 1* and then show how this can be generalized to all populations. We want to sample n genes from population 1 so that the n genes are a random sample with probability equal to the gene frequency $p_{1,t}$. This sample will then give us the gene frequency at the next time interval, that is, $p_{1,t+1}$. In this case it is not necessary to specify all the genes of the individual population. In fact, we can imagine that the gene frequency at time t in population 1, $p_{1,t}$, indicates that at time t population 1 contains a proportion $p_{1,t}$ of type A among all its n genes, and therefore a proportion of $1 - p_{1,t}$ of genes of type a. If we want to simulate a population of diploid individuals, we must remember that the number of genes is equal to twice the number of individuals. It is not necessary, however, in this simple case to simulate matings and give symbols to genes. It is enough to sample n genes, each with probability $p_{1,t}$ of being A, and to compute $p_{1,t+1}$ as the proportion (out of n) of A genes thus found. To take an A gene at random with probability $p_{1,t}$, we take a random number between 0 and 1 and compare it with $p_{1,t}$. A random number is usually obtained by an instruction using the *function* random (x) where x is a dummy variable. It must be given a variable name, for example, "chance." Thus the instruction

```
chance = random (x);
```

generates a random number defined to lie between 0 and 1, to have eight decimal digits, and to have equal probability of being one of the 10^8 numbers from 0 to 0.99999999. The random number thus extracted has been given the name "chance," and every time this function is used, a new random number is generated and given the name "chance."

The procedure for obtaining a gene *A*, or *a* with probability $p_{j,t}$, is thus simulated by comparing the numerical value of "chance" with $p_{j,t}$. If "chance" is smaller than $p_{j,t}$, we consider that we have sampled an *A* gene. If "chance" is greater or equal to $p_{j,t}$, we consider that we have sampled an *a* gene. The comparison is carried out by an "if" instruction:

```
if chance < p j, t then... ;
```

where the dots indicate that we have sampled an *A* gene; but we must still decide what to do if we have had an *A* gene. The simplest way to follow the process is to count the number of *A* genes produced in *n* consecutive samplings of this kind. If we call "*A* gene" the number of *A* genes and if we set this initially equal to zero, before starting each operation of sampling our *n* genes, then counting an *A* gene is equivalent to setting

```
A gene = A gene + 1;
```

which can be put in place of the dots in the "if" instruction. To understand this method of counting *A* genes, we must realize that to the computer the equal sign means that we replace the quantity *A* gene with the same quantity increased by one. It can easily be seen that if we take exactly *n* samples from population 1 at time *t*, we do not need also to count the *a* genes because all genes that are not *A* are *a*. It is, therefore, enough to repeat the sampling procedure *n* times, thus:

```
do samples = 1 to n;
    chance = random (x);
    if chance <p j, t  then A gene = A gene + 1;
end;
```

This "do" loop makes sure that we carry out *n* samplings, by repeating the indented part of the "do" loop exactly *n* times, and counts the number of *A* genes thus produced.

We still have to do two things. First, we must make sure that the variable *A* gene is zero at the beginning, that is, before we enter the "do" loop; this is accomplished by the instruction: "*A* gene = 0" immediately before the "do" instruction. We must also, eventually, compute the proportion of *A* genes out of *n* and call this the proportion of *A* genes in the next generation, that is, $p_{j,t+1} = A$ gene/n. We do not have, however, a subscript of time in our *p* values as we store them in the computer because we need to keep track only of the last value. As we shall not need the former $p_{1,t}$ value anymore, we can just write the new $p_{j,t+1}$ values in place of the old ones. This value is indicated for the computer by $p(1)$, for we are dealing with the gene frequency of the first population. The whole procedure is therefore:

```
A gene = 0;
do samples = 1 to n;
     chance = random (x);
     if chance <p(1) then A gene = A gene + 1;
end;
p(1) = A gene/n;
```

b. We now have to extend this procedure to all m populations. We can do this by including the above sampling procedure in another "do" loop, which carries out the same thing m times, the first time for population 1, the second for population 2, etc., until all m have been done. This constitutes the work to be done in one generation. It is enough to put the symbol j instead of 1 for the element of p in the sampling procedure just described and make the "do" loop go through all j values from 1 to m. Thus

```
do j = 1 to m;
  A gene = 0;
  do samples = 1 to n;
       chance = random (x);
       if chance <p(j) then A gene = A gene + 1;
  end;
  p(j) = A gene/n;
end;
```

At the beginning, $j = 1$ and the first population is sampled, and n samplings done; the $p(1)$ value is replaced by the new value $p(1)$ obtained by sampling n genes from it and counting the number that are A. Then the outside "do" loop increases j to 2, and the inside loop is repeated, generating the new $p(2)$ value for the second population, and so on, until the last $p(m)$ value is obtained. All m populations have then undergone one generation of random drift, and the cycle can be repeated for another generation.

The above eight instructions form the content of the outside "do" loop, which we have described as points 4, 6, and 8. They thus take the place of the dots in the outside "do" loop. It inevitably takes some practice to be able to follow the meaning of nested "do" loops, such as those we have described here. Each "end" always closes the "do" loop that comes immediately before it.

7. *Compute mean and variance of gene frequencies when $t = i$ or its multiple.* We must now show how to compute the mean and variance, how to print them, and how to do it at the desired time. We will first consider the last step, which is done by the simple instruction:

if mod $(t, i) = 0$ then do;... ; end;

The function "mod" computes the residual of the quotient between time t and the present variable i. If the residual is zero, then t is a multiple of i, and then the instructions between "do" and "end" following "then" are executed. We therefore put in the place of the dots the instructions for computing the mean and variance of the p values, which we will now describe.

To compute the mean and variance of the p values, we must compute the sums of the p values and of their squares. From these, standard formulas give us the mean and variance (see, for example, Equations 34 and 35). To obtain the sum of the p's and that of their squares, we use two new variables, which we may call Sp and $Sp2$ and which may be zeroed at the beginning:

$$Sp, Sp2 = 0,$$

We then carry out the two sums by a "do" loop which cover all the m populations. We need an index that varies between 1 and m, and to avoid confusion with those already used, we shall call this k. Summing the p values is done by

$$Sp = Sp + p(k).$$

The "do" covers all k values from 1 to m. This part of the program is, therefore,

```
do k = 1 to m;
    Sp = Sp + p(k);
end;
```

At the beginning, $Sp = 0$; then $Sp = Sp + p(1) = p(1)$ when $k = 1$; $Sp = Sp + p(2) = p(1) + p(2)$ when $k = 2$; and so on, until all m values of p have been summed. The *mean* is then Sp/m. At the same time the sum of squares can be computed by a similar procedure, namely,

$$Sp2 = Sp2 + p(k)*p(k);$$

the asterisk being the conventional symbol for a product. Alternatively, we could write,

$$Sp2 = Sp2 + p(k)**2;$$

a double asterisk separating the base from the exponent in calculating a power.

For the variance, we use the formula

$$\text{variance} = (\sum x^2 - (\sum x)^2/m)/(m-1);$$

which is written for the computer as follows:

variance = (*Sp*2 − (*Sp***2)/*m*)/(*m* − 1);

All brackets must be round ones, but can be nested *ad infinitum*, the internal ones being computed first. Actually, the internal ones in this case are not necessary but have been inserted for clarity. The computer executes powers first, then multiplications and divisions (usually from left to right), and finally additions and subtractions, exactly as we would do.

We still have to ask the computer to print the results of the computations. This can be done by the simple instruction:

put data (*t*, mean, variance);

and the numerical values of the three variables are then printed in that order. This instruction must follow the computation of the mean and variance and like them, be nested in the "if" instruction specified at the beginning of step 7.

Finally, this completed series of instructions must be inserted *inside* the "do" loop, specifying the advance of time, and *outside* the "do" loops, internal to this one, which compute the *p* values one generation after the other. It must follow the internal "do" loops because we first want to compute the new *p* values, then their mean and variance.

The end of the program must be indicated to the computer by the word "end" followed by the name of the program:

end DRIFT;

The whole program is then:

```
program DRIFT;
declare p(100);
get data (n, p0, m, i, ttot);
p = p0;
   do t = 1 to ttot;
         do j = 1 to m;
                  A gene = 0;
                  do samples = 1 to n;
                           chance = random(x);
                           if chance <p(j) then
                              A gene = A gene + 1;
                  end;
                  p(j) = A gene/n;
         end;
```

(Lines 6–13: Simulation of drift by generating a new population from the old one by random sampling of its genes)

```
            if mod (t, i) = 0 then do;
               Sp, Sp2 = 0;
               do k = 1 to m;
                    Sp = Sp + p(k);
                    Sp2 = Sp2 + p(k)**2;
               end;
               mean = Sp/m;
               variance = (Sp2 - (Sp**2/m))/(m - 1);
               put data (t, mean, variance);
            end;
      end;
      end DRIFT;
```

Computation and printout of mean and variance of gene frequencies every i generations

A typical run was as follows:

$n =$? 50,
$p0 =$? 0.5
$m =$? 20
$i =$? 10
$ttot =$? 100

that is, a run for 100 generations, 20 populations each of 50 genes, with initial gene frequency $p_0 = 0.5$, printing mean and variance every 10 generations. The printout of one such run is tabulated below.

t	*Mean*	*Variance*
10	0.487999	0.034964
20	0.595000	0.058984
30	0.609000	0.091725
40	0.625000	0.098984
50	0.650000	0.150042
60	0.648000	0.163217
70	0.641000	0.171052
80	0.648000	0.172859
90	0.607000	0.217443
100	0.611000	0.224062

General References

Bailey, N. J., *Statistical Methods in Biology*. New York: John Wiley and Sons, 1959. (An elementary text.)

Fisher, R. A., *Statistical Methods for Research Workers*, (thirteenth edition). New York: Hafner 1954. (A classic text whose first edition was published in 1925 in London and Edinburgh by Oliver and Boyd.)

Fisher, R. A., and F. Yates, *Statistical Tables for Biological, Agricultural, and Medical Research*. London and Edinburgh: Oliver and Boyd, 1953.

Kendall, M. G., and A. Stuart, *The Advanced Theory of Statistics*. New York: Hafner, Vol. I—1958; Vol. II—1961. (An advanced comprehensive survey.)

Mather, K., *Statistical Analysis in Biology*. New York: Interscience, 1947. (Another elementary text.)

Rao, C. R., *Advanced Statistical Methods in Biometric Research*. New York: John Wiley and Sons, 1952. (An advanced text.)

Sokal, R. R., F. J. Rohlf, *Biometry*. San Francisco: W. H. Freeman and Company, 1969. (A comprehensive elementary text.)

Rohlf, F. J., and R. R. Sokal, *Statistical Tables*. San Francisco: W. H. Freeman and Company, 1969.

Weatherburn, C. E., *A First Course in Mathematical Statistics*. Cambridge: Cambridge University Press, 1946. (A useful source of concise information on statistical distributions and tests.)

APPENDIX II

Segregation and Linkage Analysis in Human Pedigrees and the Estimation of Gene Frequencies

II.1 Pedigrees and Segregation Analysis

Conventional genetic analysis in human populations is complicated by the fact that instead of being able to make controlled crosses, as can be done with experimental animals, appropriate matings have to be sought in the population that is already available. Genetic hypotheses concerning the inheritance of a given attribute can therefore be tested only indirectly, by fitting probability models to family data, and such models must take into account the sampling problems involved in finding appropriate pedigrees. This indirect approach to the study of human genetics has been called **segregation analysis**. It necessarily involves some knowledge of population genetics and statistics. Our aim here is to outline various approaches of segregation analysis and to illustrate their application with a few key examples. We then follow this outline with a discussion of the special problems in detecting linkage in human pedigrees and some of the methodology used in estimating gene frequencies in human populations.

The search for appropriate pedigrees in the human population has three important implications:

1. The observed distribution of pedigrees is a function of the frequency of relevant genotypes in the population. This is, of course, why population genetics is so relevant to segregation analysis. The Hardy-Weinberg law, for example, implies that almost all the matings that give rise to a rare homozygous phenotype, such as PKU, are between heterozygotes (see Chapter 2).

2. The pedigrees found may be biased by the way in which they have been sought, that is, by the method of **ascertainment**. Biases may be introduced simply by the fact that genotypes cannot be directly determined; then the only way to detect a mating between normal heterozygotes is by their production of affected homozygous offspring. Especially for rare traits, sampling biases may depend on whether information is collected by screening comprehensively a large population or by

selecting those carriers of a particular disease that come to light in a specialty clinic.

3. Since human families are now generally small, no single family is ever large enough to test a given genetic hypothesis. Thus data from several pedigrees must be pooled and statistical variations within and between families taken into consideration.

Further problems often arise because of missing individuals in a pedigree, incomplete diagnoses, and sometimes inaccurate diagnoses. The last is especially relevant in relatively broadly defined diseases such as diabetes and schizophrenia, for which, of course, no adequate simple genetic models are available (see Chapter 9).

In constructing the models of segregation analysis we generally consider the contrast between a normal and an "affected" state and are interested in determining the pattern of inheritance of the affected state. The probability p of an affected offspring in matings of a given type, which is called the **segregation frequency**, is one of the key parameters that must be estimated. For uncomplicated simple Mendelian characters, p is usually $\frac{1}{2}$ or $\frac{1}{4}$ depending on whether matings are backcrosses or intercrosses. Incomplete penetrance and viability effects cause departures from these expected values. The distribution of the number of affected persons, r, in sibships of size s is generally binomial. Individuals through whom a family is ascertained are called **probands**. A specialty clinic, for example, might ascertain a family of hemophiliacs through a primary individual who came for treatment and provided information on his relatives both normal and affected. Even for rare diseases, a family may be ascertained twice through two affected members coming independently to a clinic. Each of these is then a proband.

The basic information available from families is their size s, the number of affected individuals in the sibship r, and the number of these which are probands a. On the basis of such information from a pool of sibships, our primary goal is to obtain a valid estimate of the segregation frequency p, keeping in mind the biases inherent in the way the data were collected.

Weinberg (1921, 1927) was the first to realize the importance of taking into account the method of ascertainment in analyzing human pedigrees. His methods were elaborated first by Haldane (1932b, 1938, 1949b) and Fisher (1934), and later, more extensively, by Morton and his colleagues (see, for example, Morton, 1959, 1962, 1964, Morton and Chung, 1959a, Barrai et al., 1965, Dewey et al., 1965).

II.2 Modes of Ascertainment

There are two primary ways to ascertain families that are segregating affected offspring.

1. **Complete selection.** Ascertainment through random sampling of parents

without regard to the children. This is actually possible only for dominant or for very frequent—that is, polymorphic—traits.

2. **Incomplete selection.** Ascertainment through one or more affected individuals in a sibship. This method can be divided into three categories according to the relative probabilities with which affected individuals are ascertained.

   a. **Truncate selection.** Every affected individual is ascertained by an exhaustive survey of an entire population, either for a rare recessive trait or for a common recessive polymorphic trait, such as the inability to taste PTC. All families, independent of size and number of affected sibs, are then equally likely to be included in the sample.

   b. **Single selection.** The probability of an affected individual being a proband is so low that there is never more than one proband per family. Here the probability of a family being ascertained depends directly on the number of affected individuals in the sibship.

   c. **Multiple selection.** A category intermediate between truncate and single selection in which there is often more than one proband per sibship, but not all affected individuals are probands.

In selection through affected offspring, families potentially able to segregate (for example, intercrosses $Aa \times Aa$) that have no affected offspring are not included in the sample.

II.3 Truncate Selection

We consider now a recessive condition with ascertainment of potentially segregating families of normal parents with affected offspring. The probability that none of the s offspring of an intercross $Aa \times Aa$ are affected is q^s, where as usual $q = 1 - p$. The relative probability of observing r out of s affected, given that only families with at least one affected individual are ascertained, is the "truncated" binomial distribution,

$$p_r = \binom{s}{r} p^r q^{s-r}/(1 - q^s), \quad r = 1, \ldots, s. \tag{1}$$

The mean number of affected individuals in families of size s is

$$\bar{r} = \frac{ps}{1 - q^s}. \tag{2}$$

Suppose that in a sample of families ascertained independently in this way, a_r had r offspring. Then the log likelihood of the observations is

$$
\begin{aligned}
L &= \text{const} + \sum_{r=1}^{s} \log\left(\frac{p^r q^{s-r}}{1-q^s}\right)^{a_r} \\
&= \text{const} - \log(1-q^s)\sum_{r=1}^{s} a_r + \log p\left(\sum_{r=1}^{s} ra_r\right) + \log q\left[\sum_{r=1}^{s}(s-r)a_r\right] \\
&= \text{const} - n\log(1-q^s) + A\log p + (B-A)\log q, \qquad (3)
\end{aligned}
$$

where

$$A = \sum_{r=1}^{s} ra_r, \qquad n = \sum_{r=1}^{s} a_r, \qquad B = ns.$$

Thus A is the total number of affected offspring and n the number of families, with the result that A/n is the mean number of affected per sibship. The maximum likelihood estimate of p is given by

$$S(p) = \frac{dL}{dp} = \frac{A}{p} - \frac{B-A}{q} - \frac{nsq^{s-1}}{1-q^s} = 0$$

or

$$\frac{A}{pq} - \frac{B}{q(1-q^s)} = 0 \qquad (4)$$

since $B = ns$.

This can be written in the form

$$\frac{A}{n} = \frac{ps}{1-q^s}, \qquad (5)$$

which, from Equation 2, is equivalent to equating the observed mean, A/n, to its expected value. The information with respect to p can be shown to be

$$I(p) = -E\left(\frac{d^2L}{dp^2}\right) = \frac{B(1-q^s-spq^{s-1})}{pq(1-q^s)^2}, \qquad (6)$$

noting from Equation 2 that

$$E(A) = Bp/(1-q^s) \qquad (7)$$

(see Haldane, 1932, and Fisher, 1934). Usually we have observations for various family sizes. Then, following the scoring procedure outlined in Appendix I (see Equations 115 to 128), we can calculate a total score and information for trial p values by evaluating the expressions given in Equations 4 and 6 for each value of s, using the relevant, observed values A, B, and n, and then summing with respect to s. Tables of $p/1-q^s$ and $(1-q^s)-spq^{s-1}/pq(1-q^s)^2$ for various values of s and p were set up by Maynard-Smith, Penrose, and Smith (1961) to facilitate the scoring procedure. A corrected p value can be calculated using Equation 117 of Appendix I, with a variance given by $1/I(p)$ evaluated at the corrected value. Heterogeneity between the families with respect to values of p can be calculated using Equation 23 from Appendix I.

II.4 Single Selection

Truncate selection generally applies only to common traits or very exhaustive surveys of well-defined populations since these are the only conditions under which families are selected at random with respect to their composition. When a trait is very rare and families carrying a recessive condition are ascertained through a single affected offspring, the simplest model for a sampling bias is that the relative probability of ascertaining a family is directly proportional to the number of affected offspring in the family. In this case, the probability of observing a family of size s with r affected individuals is

$$p_r = kr\binom{s}{r}p^r q^{s-r}, \tag{8}$$

where k, a constant of proportionality, is given by

$$\sum_{r=1}^{s} p_r = 1 = k\sum_{r=1}^{s} r\frac{s!}{r!(s-r)!}p^r q^{s-r} = kps. \tag{9}$$

that is, $k = 1/ps$. Thus p_r takes the form

$$p_r = \frac{r}{ps}\frac{s!}{r!(s-r)!}p^r q^{s-r} = \frac{(s-1)!}{(r-1)!(s-r)!}p^{r-1}q^{s-r}, \tag{10}$$

which is the binomial probability of observing $r - 1$ out of $s - 1$. This is equivalent to estimating p from the sibs of the proband and ignoring the proband itself. This procedure for allowing for ascertainment bias was first suggested in 1912 by the German physician Weinberg, who contributed so much to human population genetics in its early days.

Alcaptonuria is a rare recessive condition resulting from accumulation of homogentisic acid. It was one of Garrod's original "inborn errors of metabolism." A sample of $N = 37$ sibships with affected individuals contained a total of $A = 66$ alcaptonurics, including probands, out of a total of $T = 181$ offspring. Ignoring ascertainment bias, the estimate of the segregation frequency would be

$$p = A/T = 66/181 = 0.364,$$

with a binomial standard error of

$$\sqrt{\frac{p(1-p)}{T}} = 0.035.$$

The estimate is significantly higher than the expected $\frac{1}{4}$. When the sib method outlined above is used, the estimate of p is

$$p = \frac{A - N}{T - N} = 0.2, \tag{11}$$

with a binomial standard error

$$\sqrt{\frac{A - N}{T - N} \times \frac{T - A}{T - N}} = 0.03, \tag{12}$$

and is not significantly different from $\frac{1}{4}$. It is, however, somewhat on the low side as might be expected for a deleterious trait because of viability losses before the time of observation. The difference between the estimate 0.2 and the expected value of 0.25 could be taken as an indication that the relative viability of alcaptonurics is 0.75, though other factors, such as the inadequacy of the ascertainment model, could also account for this difference.

II.5 Multiple Selection

This is the situation intermediate between truncate and single selection. Here more than one proband per sibship may be ascertained, though not all affected individuals are probands. Weinberg (1927) gave an extension of the method of estimation just discussed for single selection. He suggested that if a family contained a probands, it should be counted a times in estimating p, each time leaving out a single proband. The estimate of p then takes the simple form

$$p = \frac{\sum a(r-1)}{\sum a(s-1)}, \tag{13}$$

where as usual r is the number of affected and s the total number of sibships and the summations are over all families. This is sometimes called **Weinberg's proband method**. When $a = r$, so that all affected are probands, Equation 13 gives an approximate formula for estimating p under truncate selection.

In order to construct a model for ascertainment bias in this intermediate situation, we introduce the parameter π, which is the probability that an affected individual is a proband. The probability that a out of r affected are probands is then the binomial

$$\binom{r}{a}\pi^a(1-\pi)^{r-a}. \tag{14}$$

Note that the probability of observing at least one proband out of r is

$$1-(1-\pi)^r \approx r\pi, \tag{15}$$

when π is small. This corresponds to the model used for single ascertainment. Extending Weinberg's proband method, Fisher (1934) suggested that a simple estimate of π is given by

$$\pi = \frac{\sum a(a-1)}{\sum a(r-1)}, \tag{16}$$

where again summation is over all families. The variance of the estimate of p for given values of s was shown by Fisher (1934) to take the form

$$V(p_s) = \frac{p(1-p)[1+\pi+\pi p(s-3)]}{(s-1)\sum a}, \tag{17}$$

where p and π are estimated by Equations 13 and 16, respectively, and $\sum a$ is the total number of probands in families of size s. Estimates of p calculated separately for different values of s can be combined by calculating a weighted mean, using $1/V(p_s)$ as weights. This leads to a slightly different estimate than that given by Equation 13 since it takes into account differences in the relative amount of information with respect to the parameter p provided by families of differing size.

In line with our treatment of truncate and single selection, we can set up a model for multiple selection using the ascertainment probability π and then estimate p and π by the method of maximum likelihood scoring. The probability that a family of size s contains r affected offspring of which a are probands is

$$\binom{r}{a}\pi^a(1-\pi)^{r-a}\binom{s}{r}p^s(1-p)^{s-r}. \tag{18}$$

Ignoring the distribution of the number of probands per family, the probability that a family of size s with r affected is ascertained is the sum of expression 18 over all values of a from 1 to r. This is

$$[1-(1-\pi)^r]\binom{s}{r}p^r(1-p)^{s-r}, \tag{19}$$

where $(1-(1-\pi)^r$ is the probability that $a \geqslant 1$, namely that the family is ascertained at least once (see Equation 15). The proportion of families of size s that are not ascertained is

$$\sum_{r=0}^{s} \binom{s}{r} p^r(1-p)^{s-r}(1-\pi)^r = \sum_{r=0}^{s} \binom{s}{r} [p(1-\pi)]^r(1-p)^{s-r}$$
$$= [1-p+p(1-\pi)]^s = (1-p\pi)^s. \qquad (20)$$

Thus, among families of size s that are ascertained at least once, the proportion with r affected is (see Morton, 1959)

$$p_r = \frac{\binom{s}{r} p^r(1-p)^{s-r}[1-(1-\pi)^r]}{1-(1-p\pi)^s}. \qquad (21)$$

Note that when $\pi = 1$, this is the same as Equation 1. The log likelihood of a sample of n families with a_r having r affected individuals, ignoring the distribution of probands per family, is

$$L = \text{const} - n\log[1-(1-p\pi)^s] + \sum_{r=1}^{s} a_r \log[1-(1-\pi)^r]$$
$$+ A\log p + (ns - A)\log(1-p), \qquad (22)$$

where

$$n = \sum_{r=1}^{s} a_r, \quad \text{and} \quad A = \sum_{r=1}^{s} ra_r$$

is the total observed number of affected offspring. Maximum likelihood scores dL/dp and $dL/d\pi$ can be calculated from Equation 22 for the parameters p and π. The scoring procedure outlined in Appendix I can thus be used for the joint estimation of p and π and their variances.

In practice it appears that Weinberg's proband method and its extension by Fisher to the estimation of π as given by Equations 13 and 16 provide a simple and generally satisfactory estimate for multiple selection. The overall variance of the estimate is approximately

$$\frac{1}{V} = \sum_{s=1}^{r} \frac{1}{V(p_s)}, \qquad (23)$$

where $V(p_s)$ is given as a function of s by Equation 17, using the additive property of the information and the fact that the variance is approximately the inverse of the information (see Crow, 1965). Numerical examples indicate that the proband method gives estimates that are almost fully efficient in the sense that their variance is little higher than that of the maximum likelihood estimates. Data on families segregating for cystic fibrosis that illustrate the numerical application of the proband method are shown in Table II.1.

TABLE II.1

Summary, by Family Size, of Data on Sibships with Normal Parents, Segregating for Cystic Fibrosis. For each family: a = number of probands, r = number of affected, s = family size.

s	$S = \sum a(s-1)$	$R = \sum a(r-1)$	$A = \sum a(a-1)$	$1/C$	CS	CR
10	9	2	0	2.08	4.33	0.963
9	8	2	0	1.98	4.04	1.010
8	7	3	0	1.88	3.73	1.597
7	30	7	2	1.78	16.88	3.940
6	10	1	0	1.68	5.97	0.597
5	48	15	8	1.58	30.40	9.530
4	42	12	2	1.47	28.45	8.130
3	38	9	6	1.37	27.65	6.550
2	26	8	4	1.27	20.40	6.290
Total	218	59	22		141.85	38.607

Source: From Crow (1965).

Note: See the text for details on the estimation of the segregation frequency p and the ascertainment probability π.

If relative weighting is ignored and Equation 13 is used, an initial estimate of p is given by

$$p_0 = \frac{\sum R}{\sum S} = \frac{59}{218} = 0.271, \tag{24}$$

where summation is now with respect to family size. The corresponding estimate of π from Equation 16 is

$$\pi = \frac{\sum A}{\sum R} = \frac{22}{59} = 0.373. \tag{25}$$

The weights for the estimates of p from families of size s are, from equation 17,

$$\frac{1}{V(p_s)} = \frac{S}{p(1-p)[1+\pi+\pi p(s-3)]} = \frac{SC}{p(1-p)}, \tag{26}$$

where

$$1/C = 1 + \pi + \pi p(s-3). \tag{27}$$

Using the initial estimates of π and p given by Equations 25 and 28, we can write the weighted estimate of p in the form

$$\hat{p} = \frac{\sum \frac{R}{S} \times \frac{SC}{p_0(1-p_0)}}{\sum \frac{SC}{p_0(1-p_0)}} = \frac{\sum RC}{\sum SC}. \tag{28}$$

The variance is given by

$$V(p) = \frac{1}{\sum V(p_s)} = \frac{\hat{p}(1-\hat{p})}{\sum SC}, \tag{29}$$

using now the corrected estimate of p.

For the data in Table II.1, we have

$$\hat{p} = \frac{38.607}{141.85} = 0.272$$

and

$$V(\hat{p}) = \frac{\hat{p}(1-\hat{p})}{141.85} = 0.001396,$$

giving a standard error of $\sqrt{0.001396} = 0.037$. In this case, the weighting has little effect on the estimate of p, which is not significantly different from the expected value of $\frac{1}{4}$.

As we shall see later, Weinberg's proband method is almost the only procedure for estimating segregation frequencies that does not require excessive numerical computation. It is for this reason that we felt it was worthwhile to discuss a numerical example in some detail. Maximum likelihood solutions of models such as that given by Equation 21 and the more general models we will now discuss can be reasonably contemplated only by using an electronic computer. Our aim in the remaining sections of this chapter is therefore to show how the models can be constructed and what conclusions can be reached by using them rather than to illustrate their detailed numerical application.

II.6 Sporadic Cases

It was first emphasized by Haldane in 1949 that sibships containing only one affected individual should receive separate treatment from those with two or more affected individuals because single, or as he called it, "sporadic" cases of a given condition could occur, which were not the result of Mendelian segregation for the major gene known to cause the condition (Haldane, 1949b). These sporadic cases could have a variety of causes, such as mutations, **phenocopies** resulting from environmental causes (for example, thalidomide babies), and phenocopies determined by a mimicking polygenic complex rather than the major gene. (Such a situation, for example, is known to exist with respect to polydactyly in the house mouse.) All these aı likely to be rare events; thus the probability of a sporadic case in any given family is in general low enough to make it very unlikely that two such cases occur in one family. Thus families with one affected offspring, called **simplex** families, are a mixture of sporadic cases and instances of Mendelian segregation for the relevant gene. Families with two or more affected offspring, called **multiplex** families, are, however, almost certainly the result of Mendelian segregation for the major gene. We can allow for sporadic cases in the estimation of segregation frequencies by extending the concept of truncate selection to fitting our model only to data on multiplex families. Thus, for example, assuming the probability of ascertainment $\pi = 1$, the relative frequency of families of size s with r affected, among those with at least two affected, is

$$p_1(r \mid r > 1) = \binom{s}{r} \frac{p^r(1-p)^{s-r}}{1 - q^s - spq^{s-1}}, \tag{30}$$

$$r = 2, 3, \ldots, s,$$

since $1 - q^s - spq^{s-1}$ is the total frequency of all families with $r \geqslant 2$, where the vertical bar in the parenthesis of the left-hand side means "given that." For single selection, when π is very small and the probability of ascertaining a family is proportional to the number of affected in the sibship,

$$p_0(r \mid r > 1) = \binom{s-1}{r-1} \frac{p^{r-1}q^{s-r}}{1 - q^{s-1}}. \tag{31}$$

This is obtained by combining Equation 10 for single selection with the concept of truncate selection. Haldane (1949b) showed that if A is the total number of abnormals and n the total number of families of size s, then for given values of s,

$$I(p_1) = \frac{1}{V(p_1)} = \frac{A}{p^2 q} + \frac{1}{q^2} \frac{[s(s-1)(1 - sp - q^s)q^{s-1}n]}{(1 - spq^{s-1} - q^s)^2}, \tag{32}$$

and

$$I(p_0) = \frac{1}{V(p_o)} = \frac{A-n}{p^2 q} - \frac{1}{q^2} \frac{(s-1)^2 nq^{s-1}}{(1 - q^{s-1})^2}, \tag{33}$$

Allowing for multiple selection gives

$$p_\pi(r \mid r > 1) = \frac{\binom{s}{r} p^r q^{s-r}[1 - (1-\pi)^r]}{1 - (1 - p\pi)^s - \pi spq^{s-1}}. \tag{34}$$

This follows from Equations 20 and 21 and from the fact that the probability of observing a family with just one affected, ignoring sporadic cases, is πspq^{s-1} (putting $r = a = 1$ in Equation 18).

Morton (1959, and later) generalized the concept of sporadic cases to the estimation of the proportion x of all cases that are sporadic. This can be related to the proportion, w, of simplex sporadic cases among families of size s with at least one affected, as follows. Among segregating families with at least one affected, the mean number of affected is $\bar{r} = ps/(1 - q^s)$ from Equation 2. The proportion of sporadic cases among all affected offspring in families of size s is, therefore,

$$x = \frac{w}{w + (1 - w)\bar{r}} \tag{35}$$

since a proportion $1 - w$ of families are segregating with mean affected $\bar{r}$ and the remaining w are the sporadics, each with just one affected offspring. The estimate of x should in general be the same for all family sizes so long as x is not a function of parity or parental age. Substituting for $\bar{r}$ in Equation 35 gives

$$w = \frac{xsp}{1 - q^s - x(1 - sp - q^s)}. \tag{36}$$

We now calculate the expected proportion of ascertained simplex families of size s, assuming these have at least one affected offspring. This is given by

$$m_1 = C\left[\pi w + \frac{\pi spq^{s-1}}{1 - q^s}(1 - w)\right] \tag{37}$$

since $spq^{s-1}/(1 - q^s)$ is the proportion of segregating nonsporadic families that have only one affected offspring; therefore, the second term in Equation 37 is the probability of ascertaining a nonsporadic simplex family. C is a constant of proportionality to be determined later. The proportion of ascertained families which are multiplex is

$$m_2 = C(1 - w)\left[\frac{1 - (1 - p\pi)^s}{1 - q^s} - \frac{s\pi pq^{s-1}}{1 - q^s}\right], \tag{38}$$

where the first term is, from Equation 20, the total proportion of families of size s which are ascertained and the second term is, as before, the proportion of ascertained simplex families. The constant C is given by setting $m_1 + m_2 = 1$ so that

$$C = \frac{1 - q^s}{\pi w(1 - q^s) + (1 - w)[1 - (1 - p\pi)^s]}. \tag{39}$$

Substituting this into Equation 37 and then for w from Equation 36 gives

$$m_1 = pr(r = 1 | r \geqslant 1) = \frac{sp\pi[x + (1 - x)q^{s-1}]}{xsp\pi + (1 - x)[1 - (1 - p\pi)^s]}. \tag{40}$$

For single selection, when $\pi \to 0$, this probability takes the form

$$pr(r = 1 \mid r \geqslant 1) = x + (1 - s)q^{s-1}. \tag{41}$$

Maximum likelihood scoring can be used to fit jointly the parameters x, p, and π to (a) the relative proportions of simplex to multiplex families from Equation 40 and (b) the distribution of the number of affected among multiplex families from Equation 34.

Initial estimates of p and π can be obtained by the proband method, ignoring sporadic cases. An initial estimate of π can also be obtained from the distribution of the number of probands in families with r affected. There must be at least one proband per family ascertained, and so this is equivalent to truncate selection with a replacing r and r replacing s. Thus the relative probability of observing a probands among r affected is

$$p(a \mid a > 0) = \frac{\binom{r}{a}\pi^a(1 - \pi)^{r-a}}{1 - (1 - \pi)^r} \tag{42}$$

An estimate of π can now be obtained by maximum likelihood scoring as discussed for truncate selection.

The general procedure followed for maximum likelihood scoring to estimate parameters such as x, p, and π is to calculate scoring coefficients, which are the relevant partial derivatives of the logarithm of the expectations m_1, etc., and hence the contribution of each family to the score for each parameter. Thus the contribution of a simplex family with s children to the score for x is, from Equation 40,

$$\frac{\partial}{\partial x}(\log m_1) = \frac{1 - q^{s-1}}{x + (1 - x)q^{s-1}} - \frac{\pi s p - [1 - (1 - p\pi)^s]}{\pi x s p + (1 - x)[1 - (1 - p\pi)^s]}, \tag{43}$$

calculated at the trial values for x, p, and π. The total score is always the sum of the individual contributions from different families, which may come from different types of data. Thus, scores for p and π from the relative proportion of simplex to multiplex families can be added to those for the distribution of r among multiplex families. The information is calculated in the same way from the second derivatives, and then the corrections to the trial values and their variances are calculated as indicated in Appendix I (Sections I.37 and I.38). Morton and his colleagues have given the appropriate scoring and information coefficients for most of the segregation models that we have discussed so far, and those that will be mentioned in subsequent sections. (See references given at bottom of page 852.) The scoring and information coefficients for the models given by Equations 34 and 40 can be calculated as follows (Morton, 1959). First define:

$$
\begin{aligned}
A &= x + (1 - x)q^{s-1}, \\
B &= xsp\pi + (1 - x)[1 - (1 - p\pi)^s], \\
W &= 1 - q^{s-1}, \\
Y &= sp\pi - 1 + (1 - p\pi)^s \\
X &= x + (1 - x)q^{s-1} - p(1 - x)(s - 1)q^{s-2}, \\
Z &= x + (1 - x)(1 - p\pi)^{s-1}, \\
J &= (1 - p\pi)^{s-1} - q^{s-1}, \text{ and} \\
K &= J + p(s - 1)q^{s-2}.
\end{aligned} \tag{44}
$$

Then, from Equation 40 and its complement, scores for simplex families are

$$S(x) = \frac{BW - AY}{AB}, \qquad S(\pi) = \frac{B - sp\pi Z}{\pi B}, \qquad \text{and} \qquad S(p) = \frac{Bx - sp\pi AZ}{pAB}. \tag{45}$$

Scores for multiplex families are

$$S(X) = -\frac{1 - x)Y + B}{(1 - x)B}, \qquad S(\pi) = \frac{sp(BJ - DZ)}{BD}, \qquad \text{and} \qquad S(p) = \frac{s\pi(BK - DZ)}{BD}. \tag{46}$$

Information contributions are calculated from

$$I_{xx} = \sum pS^2(x), \ I_{xp} = \sum pS(x)S(p), \text{ etc.}, \tag{47}$$

where summation is over all families and p and $S(x)$ refer to the probability and score coefficients for each family.

Using Equation 34, we obtain that the scores for multiplex families with r affected are

$$S(p) = \frac{r}{pq} - e_p, \ S(\pi) = \frac{r(1 - \pi)^{r-1}}{1 - (1 - \pi)^r} - e_\pi, \tag{48}$$

where

$$e_p = \frac{s(D + q\pi K)}{qD} \quad \text{and} \quad e_\pi = \frac{spJ}{D}. \tag{49}$$

The information coefficients are

$$I_{pp} = \frac{s(s - 1)[1 - (1 - \pi)^2(1 - p\pi)^{s-2}]}{q^2D + (e_p/pq) - e_p^2}, \tag{50}$$

$$I_{\pi\pi} = \sum_{r=2}^{s} p_r[S(\pi)]^2, \tag{51}$$

where p_r is given by Equation 34, and

$$I_{\pi p} = s(s-1)p(1-\pi)(1-p\pi)^{s-2}/qD + (e_\pi/pq) - e_p e_\pi . \tag{52}$$

Each family contributes to the total score by an amount calculated from the coefficients given in Equations 45 to 52 for appropriate trial values of p, π, and x. The calculations are tedious but straightforward. Anyone wishing to use these methods should of course take advantage of computer programs for carrying out the numerical calculations.

Consider as an example the data on cystic fibrosis given in Table II.1. Scoring multiplex versus simplex families at $p = 0.25$, $\pi = 0.35$, and $x = 0$, using Equations 45, 46, and 47, and, for the sake of simplicity, ignoring further adjustments to π gives accumulated values over all families: $S_p = 16.03$, $S_x = -7.85$, and $I_{pp} = 642.90$, $I_{xx} = 125.95$, and $I_{px} = -260.49$. The elements of the inverse of the matrix

$$\begin{pmatrix} I_{pp} & I_{px} \\ I_{px} & I_{xx} \end{pmatrix} = \begin{pmatrix} 642.90 & -260.49 \\ -260.49 & 125.95 \end{pmatrix}$$

are given by

$$\begin{pmatrix} V_{pp} & V_{px} \\ V_{px} & V_{xx} \end{pmatrix} = \frac{1}{\Delta}\begin{pmatrix} I_{xx} & -I_{px} \\ -I_{px} & I_{pp} \end{pmatrix} = \begin{pmatrix} 0.0096 & 0.0199 \\ 0.0199 & 0.0490 \end{pmatrix}, \tag{53}$$

where

$$\Delta = I_{pp} I_{xx} - (I_{px})^2.$$

The corrected values of p and x are therefore given by (see Appendix I, Equations 140 and 141)

$$\begin{aligned} p_1 &= p + S_p V_{pp} + S_x V_{px} = 0.25 + (16.03)(0.0096) + (-7.85)(0.0199) \\ &= 0.248 \end{aligned} \tag{54}$$

and

$$x_1 = x + S_x V_{xx} + S_p V_{px} = 0 + (-7.85)(0.0490) + (16.03)(0.0199) = -0.066. \tag{55}$$

The corresponding variances and standard errors are $V_{pp} = 0.0096$, SE $(p) = 0.098$ and $V_{xx} = 0.0490$, SE $(x) = 0.221$. There is hardly any change in the value of p, which certainly does not differ significantly from 0.25. The standard error is high because only information from the relative proportions of simplex and multiplex families has been used. When the distribution of r among multiplex families is also used, the estimate of p becomes 0.2485 with a standard error of 0.05. This is higher than the standard error 0.037 obtained using the proband method (Table II.1) because we have taken into account the possibility of sporadic cases. The estimate of x is negative and very small, clearly giving no indication for the existence of sporadic cases, although its standard error is very high. For a fully penetrant, relatively common, clear-cut recessive such as cystic fibrosis, it is not surprising to find that x is negligible. This is not the case for many other diseases for which the estimation of x is one of the major contributions of segregation analysis.

II.7 Inbreeding in Sporadic Cases

As in most of the cases we have so far discussed, if the disease under analysis is caused by one or more recessive genes, the inbreeding coefficient of the affected familial as opposed to sporadic cases should be appreciably higher than that in the general population. On the other hand, the inbreeding coefficient of the sporadic cases is probably not increased. This, as Morton (1962) pointed out, provides an independent estimate of the incidence of sporadic cases. Thus let F_f = inbreeding coefficient of the familial cases, F_i = the inbreeding coefficient of isolated cases (affected in simplex sibships), α = the inbreeding coefficient in the general population, and y = the proportion of isolated cases which are sporadic.

Since isolated cases are a mixture of sporadic and chance familial single cases, their inbreeding coefficient must be the weighted sum of the respective inbreeding coefficients. Thus

$$F_i = y\alpha + (1 - y)F_f \tag{56}$$

or

$$y = \frac{F_f - F_i}{F_f - \alpha}, \tag{57}$$

where F_f, F_i, and α are observables. If now c is the proportion of probands which occur in simplex families, then the overall proportion of sporadic cases is given by

$$x = cy. \tag{58}$$

This can be compared with the estimates of x obtained by maximum likelihood scoring of the proportion of simplex families in order to test the hypothesis that there is no increase in the inbreeding coefficient of sporadic cases. For some data on severe mental defect, excluding Down's syndrome and other cases with a known environmental origin, Dewey and co-workers (1965) found $c = 354/379$ (354 simplex families and 25 probands in multiplex families), $F_f = 12{,}500 \times 10^{-6}$, $F_i = 1015 \times 10^{-6}$, and $\alpha = 406 \times 10^{-6}$. As expected, assuming that mental defect is largely recessively determined, F_f is very much larger than α. The estimate of x using Equations 57 and 58 was 0.887 ± 0.046. This compared with a maximum likelihood scoring estimate from the familial segregation data of 0.881 ± 0.026. The agreement between these estimates shows that in marked contrast to the familial cases the sporadic cases are not associated with an increased level of inbreeding. Similar results have been obtained with respect to deaf-mutism and limb-girdle muscular dystrophy. The proportion of sporadic cases is very high, indicating that the majority of cases of severe mental defect are not simple recessive traits.

II.8 Estimation of Prevalence with Incomplete Selection

The prevalence of a trait is the number of cases existing in a given population at a given time. The incidence is the proportion of individuals who have the trait at birth.

If A is the total number of probands observed in a sample of families, and π, as before, is the ascertainment probability estimated from these same families, then from the definition of π, the prevalence n is given by

$$n = \frac{A}{\pi}. \tag{59}$$

The standard error of this estimate of the prevalence is $A\sigma_\pi/\pi^2$, where σ_π is the standard error of the estimate of π. Thus, for the data on mental defectives discussed above (see Dewey *et al.*, 1965), $A = 379$, $\pi = 0.2565$, and so $n = 379/0.2565 = 1478$. Only $0.12 \times 1478 = 176 \pm 66$ of these are "high-risk" familial cases. These cases were drawn from a population of approximately two million so that the incidence of high-risk familial cases is about 88 per million, with a standard error of 33.

II.9 The Separation of Segregating and Nonsegregating Families with Complete Selection

With complete selection, usually in the case of relatively frequent polymorphic differences, useful information on segregation can be obtained from the proportion of segregating to nonsegregating families (see Morton, 1959, and later). This proportion is available with complete selection since ascertainment is through the parents. Consider, for example, the PTC taster–nontaster polymorphism. Ability to taste PTC is determined by a dominant gene T. Matings of tasters by nontasters can be either $Tt \times tt$, which will segregate, or $TT \times tt$, which will not segregate. The proportion of matings which cannot segregate is

$$h = \frac{P^2}{P^2 + 2PQ} = \frac{P}{P + 2Q}, \tag{60}$$

where $P = 1 - Q$ is the frequency of the gene T. The expected proportion of segregating families of size s (namely, those with at least one affected offspring) is, therefore,

$$p(r > 0) = (1 - h)(1 - q^s); \tag{61}$$

thus the expected proportion of nonsegregating families is

$$p(r = 0) = h + (1 - h)q^s, \tag{62}$$

where h is given by Equation 60. The proportion of segregating families, among all families of size s with r affected offspring, is

$$p(r) = (1 - h)\binom{s}{r} p^r q^{s-r}. \tag{63}$$

These probabilities can be scored for estimates of the gene frequency P and the segregation frequency p. Independent scores for P may be obtained from Hardy-Weinberg fits to the population phenotype frequencies (see below). Snyder's (1932) ratios for the expected proportion of dominant to recessive offspring in dominant × dominant and dominant × recessive matings are another approach to solving the problem of dealing with pedigree data for dominant traits for which the heterozygote cannot be distinguished (see Chapter 2, Example 1). The probability of a segregating mating involving two dominant parents is

$$p(r > 0) = (1 - h)^2(1 - q)^s, \tag{64}$$

where as before h is given by Equation 60. Scores and information coefficients are calculated from these equations in the usual way. Morton and his colleagues have, for example, used segregation models which are simple generalizations of Equations 60, 61, 62, and 63 to test for segregation distortion in the ABO blood group system (Chung and Morton, 1961). They have also emphasized the use of this form of segregation analysis in searching for exceptional "bivalent" alleles, such as, for example, the Rh haplotypes carrying the genetic determinants for both C and c (Morton, Mi, and Yasuda, 1966; see Chapter 5). The existence of Cc as well as C and c alleles modifies, of course, the expected segregation patterns and is indicated by a significant estimate of the frequency of Cc haplotypes resulting from the maximum likelihood fit. Using this approach, Morton, Mi, and Yasuda (1966) found that about half the apparent *Cde* haplotypes present in the Brazilian population they were studying were in fact *Ccde*.

II.10 Incorporation of Family-size Distributions for the Estimation of Prevalence

In cases where only multiplex families have been recorded, the assumption of explicit family size distributions, such as the Poisson or negative binomial, can provide an estimate of the expected number of simplex families (Barrai *et al.*, 1965). This assumption is needed to obtain an estimate of the prevalence. Thus if A^*

is the number of probands in multiplex families and θ the expected proportion of probands occurring in multiplex families, then the prevalence is estimated by

$$n = \frac{A^*}{\pi\theta}. \tag{65}$$

The parameter θ is estimated as follows. The probability that a proband comes from a family of size s is

$$\frac{sf(s)}{\sum sf(s)}, \tag{66}$$

where $f(s)$ is the relative frequency of families of size s and summation is over all values of s. The probability that none of the $s - 1$ sibs of a proband is affected and so the proband comes from a simplex sibship is q^{s-1}. Thus the mean probability that a proband comes from a simplex sibship is the mean of q^{s-1} weighted by the probabilities given by Equation 66, that is,

$$\frac{sf(s)q^{s-1}}{\sum sf(s)}. \tag{67}$$

Thus θ, the mean probability that a proband comes from a multiplex sibship, is the complement of Equation 67, namely,

$$\theta = 1 - \frac{\sum sf(s)q^{s-1}}{\sum sf(s)}. \tag{68}$$

For example, if $f(s)$ is the truncated Poisson distribution (only families with at least one child are observed), then

$$f(s) = \frac{m^s e^{-m}}{s!(1 - e^{-m})}, \tag{69}$$

$s = 1, 2, \ldots,$ where m is the mean family size. Now

$$\sum sf(s)q^{s-1} = \frac{m}{1 - e^{-m}} \sum \frac{(mq)^{s-1}}{(s-1)!} = \frac{m}{1 - e^{-m}} \times e^{mq}, \tag{70}$$

and

$$\sum sf(s) = \frac{m}{1 - e^{-m}} \sum \frac{m^{s-1}}{(s-1)!} = \frac{m}{1 - e^{-m}} e^m. \tag{71}$$

Thus,

$$\theta = 1 - \frac{e^{mq}}{e^m} = 1 - e^{-mp} \tag{72}$$

and can be estimated from a knowledge of p, obtained by fitting a segregation distribution to multiplex families, and of m, by fitting Equation 69 to the family sizes. The probability of family size s, given at least one ascertained and affected individual, can be shown to be

$$p(s \mid r > 0) = \frac{m^s[xsp\pi + (1 - x)(1 - (1 - p\pi)^s)]}{[xmp\pi + (1 - x)(1 - e^{-mp\pi})]s!\, e^m} \tag{73}$$

Expressions similar to Equations 72 and 73 can be derived for the negative binomial distribution (Barrai, Mi, Morton, and Yasuda, 1965).

The segregation models we have discussed so far are those which are most generally used and which apply to a majority of observed situations. Morton and his colleagues have suggested some further ramifications which may be useful in certain special circumstances.

II.11 Analysis of Linkage in Human Pedigrees

The analysis of linkage is a cornerstone of genetic analysis in experimental animals. In man, however, it has met with considerable difficulty. While many other organisms, among the plants, animals, and microorganisms, have well developed chromosome maps, in man at the present time only the X chromosome is mapped to some extent and only a few autosomal linkages are known. Only recently has knowledge about linkage been accumulating more rapidly than in the past. A summary of the main aspects of the problem and the difficulties follows.

1. In most cases, analysis is limited for practical reasons to the study of two genes (loci) at a time as it is unlikely, except for some X-linked conditions, that more than two genes belonging to one chromosome are available for study in the same pedigree. Sex linkage, of course, is determined from the segregation behavior in relation to sex. When two autosomal genes are tested for linkage, both may be polymorphic, or one may be polymorphic and the other rare, or both rare. About thirty polymorphic genes are now available, and almost all the approximately 600 two-by-two combinations have been examined. Some linkages, as expected, have been found, for example between the Lutheran blood groups and Secretor.

When pedigrees of a rare gene (or, more precisely, phenotype) are available and individuals in the pedigree have been tested for polymorphic genes, it is possible to test for linkage between these and the rare phenotype. If enough polymorphic genes were known, the genetic mapping of the rare phenotype would always be possible. We are not yet in this position, but the increasing knowledge about polymorphisms will soon make this possible. Meanwhile, some such linkages have been found. Occasionally, the geneticist may also look for linkage between the rare traits, but the chance that two such traits occur together in the same family is very low.

2. *Only few of the possible matings are informative*. Consider the case of a gene difference *G*, *g*, whose linkage with a polymorphic marker *T*, *t* is under examination. For simplicity, and without loss of generality, we will only consider two alleles at each locus. *The only informative matings are those in which one of the parents is heterozygous for both genes*, that is, *GgTt*. The other parent may be homozygous at both (double backcross or BX), heterozygous at both (double intercross or IX), or homozygous for one and heterozygous for the other (single backcross). Potentially informative and uninformative matings are indicated in Table II.2. Not all of the potentially informative matings will be so, however, depending on dominance. Thus, if *G* is dominant to *g* and *T* to *t*, only the double backcrosses labeled 2, 28, 35, and 36 in the table will be informative since in the other two, one parent is *GT*/*GT*, and therefore all offspring will be phenotypically *GT*. There will be similar limitations on the single backcrosses and intercrosses while the double intercrosses give a very limited amount of information on the recombination fraction.

The expected proportion of informative matings depends on the gene frequencies.

TABLE II.2

Informative and Noninformative Matings for Two Loci G and T Which Are on the Same Chromosome. Genotypes of parents are given on the first line and first column.

Female Parent \ *Male Parent*	*GT/GT*	*GT/Gt*	*GT/gT*	*Gt/Gt*	*GT/gt* (*C*)	*Gt/gT* (*R*)	*gT/gT*	*gt/Gt*	*gt/gT*	*gt/gt*
GT/GT					1	2				
GT/Gt		Noninformative			3	4		Noninformative		
GT/gT					5	6				
Gt/Gt					7	8				
GT/gt(*C*)	9	11	13	15	17	19	21	23	25	27
Gt/gT(*R*)	10	12	14	16	18	20	22	24	26	28
gT/gT		Noninformative			29	30				
gt/Gt					31	32		Noninformative		
gt/gT					33	34				
gt/gt					35	36				

C = *coupling double heterozygote.* *R* = *repulsion double heterozygote. Double backcrosses are matings* 1, 2, 9, 10, 27, 28, 35, and 36. *Single backcrosses and single intercrosses are* 3–8, 11–16, 21–26, and 29–36. *Double intercrosses are* 17–20.

It is greater with codominance (when all three genotypes are recognizable), and also the nearer the gene frequencies are to 50 percent.

3. An important problem concerns the linkage *phase* of the double heterozygote, that is, whether it is in coupling or repulsion. The expected progeny frequencies are reversed in the two cases (see Chapter 1). If the population is in equilibrium, the two phases are expected to be equally frequent. However, if linkage is close and the population has undergone relatively recent admixture, one of the two phases may be less frequent than the other (see Chapter 5 and 10). Moreover, very often few pedigrees are available for testing, and there may be a substantial random deviation from the 1 : 1 expectation for *C* and *R* phases. The only way in which information on the phase can be obtained is by examination of the parents of the double heterozygote, and even that may not always be enough.

4. Sex linkage is easier to study than autosomal linkage. The peculiar segregation of X-linked markers makes their assignment to the X chromosome almost foolproof. In addition, the haploid (hemizygous) state of the X in males makes it especially easy to follow recombination in heterozygous mothers (see Chapter 10). In practice, therefore, linkage between X-chromosome markers is studied only in females, whose sons give the same information formally, as does a **testcross** in autosomal inheritance—that is, the backcross of the multiple heterozygote to the multiple recessive homozygote for all the genes involved. This is the most informative type of cross for the study of linkage because all the gametes that are formed can be scored directly for their genotype, without the complications introduced by dominance. When information is available on the parents, and in particular the father of the heterozygous females whose progeny is being studied, it is possible to give the *phase* of linkage.

5. There are probably at least slight differences in crossing over between males and females. Therefore pedigrees should be separated whenever possible into those in which the double heterozygote is the male and into those in which it is the female. Most analyses have, however, been carried out so far with the assumption that it is the same in males and females.

6. In theory, it is possible to study linkage even with data from only one generation. Penrose (1935, 1953b) has developed methods according to which sibs are grouped in pairs, and pairs are assigned to one of four categories: alike for trait *A* and *B*, alike for *A* but unlike for *B*, alike for *B* but unlike for *A*, and unlike for both. This 2×2 contingency table can then be tested by chi-square for independence, this being the same as the test for absence of linkage. It will be clear that in the absence of knowledge about the parents of the sibs, informative and uninformative matings cannot be separated, and the former may be a small fraction of the total. Linkage can lead only to a small deviation from independence under these conditions, and this deviation can easily be simulated by totally unrelated effects, for example, heterogeneity or stratifications in the population. It is very time-consuming to accumulate data on human linkage, and therefore informative pedigrees

should be examined by the most efficient method. Two-generation data are preferable, and, when possible, three (or more) generation pedigrees should be collected, for they can give information on the linkage phase.

The first systematic attempts to develop procedures for detecting and estimating genetic linkage in human pedigrees were made by Bernstein (1931b), Hogben (1934), Haldane (1934, 1936), Fisher (1935, 1936a, 1936b, and later), and Penrose (1935). Fisher developed an elaborate maximum likelihood scoring procedure for two-generation data, which was later considerably extended by Finney (see Bailey, 1961, for a review of this approach). Haldane and Smith (1947) introduced a more direct approach to the problem that was based on the calculation of the likelihood of a set of pedigrees for different values of the recombination fraction. Following further elaboration of this approach by C. A. B. Smith (1953), and especially Morton (1955, 1956a and b, 1957, 1962), it has become the standard approach for the analysis of linkage in human pedigrees. A comparative review of the various approaches to linkage detection was given by Morton (1955). The problem is a highly complex one, and it is clearly beyond our scope to give a detailed treatment here. We shall, therefore, concentrate on outlining only the likelihood approach, leaving the reader to consult the references at the end of this appendix for further information (see especially Smith et al., 1961).

II.12 Linkage Analysis by Likelihood

We will consider as an example of linkage estimation a double backcross, using data on Rh and on an autosomally inherited dominant red-cell defect called **elliptocytosis**. We will use the family F_1, part of a pedigree given by Goodall *et al.* in 1954 (see Figure II.1). Black circles indicate affected progeny. The mother was not examined but must have been the carrier of the trait. Her Rh genotpye is not directly known but can be inferred from those of her progeny to be $R_1r(DCe/dce)$. It is not known whether the dominant gene for elliptocytosis is on the R_1 or the r chromosome. We will therefore give equal probability to each hypothesis and

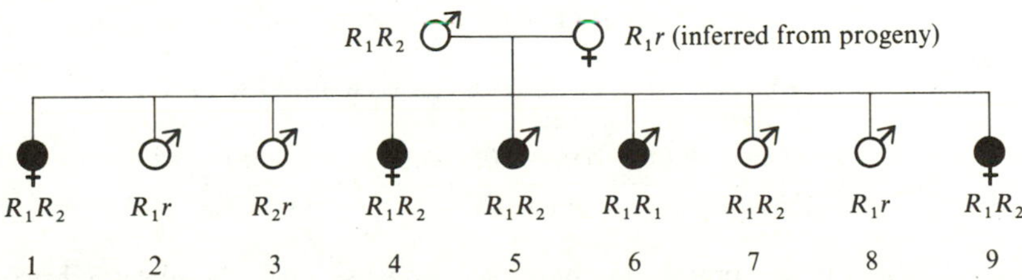

FIGURE II.1
$r = dce$; $R_1 = DCe$; $R_2 = DcE$. (Based on Goodall et al., 1954.)

consider them separately, computing for each, the probability of obtaining the pedigree.

1. If the *El* gene (*El* for elliptocytosis) is on the same maternal chromosome as R_1, the mother has the genotype $R_1El/r\,el$. Then, in the absence of recombination, progeny $R_1R_1(DCe/DCe)$ and $R_1R_2(DCe/DcE)$ should be *El*, and R_1r, R_2r should be normal (*el*/*el*). Eight out of the nine progeny are thus *non*recombinants. Only one (individual, the seventh) who is R_1R_2 and *el*/*el* is a recombinant, on the hypothesis that *El* is on the maternal R_1 chromosome. If θ is the probability of recombination, namely that R_1 and *El* are separated by crossing over at meiosis, then $1-\theta$ is the probability that recombination does not occur and that R_1 and *El* stayed together, and so the probability of observing this pedigree is $\theta(1-\theta)^8$.

2. If the elliptocytosis gene were on the same chromosome as r, the maternal genotype would be R_1el/rEl. Then if no recombination occurred, all progeny R_1r, R_2r would be *El*, and all R_1R_1,R_1R_2 would be normal. All individuals except one (again the seventh) are now recombinants on this hypothesis, and the probability of the pedigree is $\theta^8(1-\theta)$.

The two hypotheses are a *priori* equally probable, assuming equilibrium of the coupling and repulsion linkage phases, and so the total probability of pedigree F_1 is

$$P(F_1\,|\,\theta) = \tfrac{1}{2}[\theta(1-\theta)^8 + \theta^8(1-\theta)]. \tag{74}$$

Choosing the value of θ for which this quantity is largest is the same as maximizing the likelihood of the observations. The log likelihood of the pedigree is in fact, apart from a constant, simply the log of $P(F_1\,|\,\theta)$. The procedure used for maximizing these expressions is entirely numerical, for they usually give rise to unwieldy expressions after differentiation and setting the derivative equal to zero.

It can be shown by trial and error (inserting different numerical values of θ into Equation 74 and computing the likelihood) that in the present case the likelihood, and so also its logarithm, is a maximum, approximately, at $\theta = 0.11$. This is thus the maximum likelihood estimate of θ. In the present case, the second hypothesis (R_1el/rEl) for the maternal genotype) contributes negligibly to the likelihood, and the result is almost the same as if the phase were known exactly and corresponded to the first hypothesis. If this were the case, the log likelihood would reduce to

$$L = \log[\theta(1-\theta)^8] + \text{const}, \tag{75}$$

and its derivative with respect to θ, equated to zero, would be

$$\frac{dL}{d\theta} = 0 = \frac{(1-\theta)^8 - 8\theta(1-\theta)^7}{(1-\theta)^8} = \frac{1-9\theta}{\theta(1-\theta)}, \tag{76}$$

giving $\hat{\theta} = 1/9$ as the maximum likelihood solution, which is the observed proportion of recombinants in the pedigree. This is very close to the solution obtained assuming *both* phases were possible.

The problem as to whether this pedigree is sufficient to "prove" linkage can be answered in the following way. The probability of obtaining a pedigree if the hypothesis of linkage is true is compared with that of obtaining it under the hypothesis of no linkage, that is, independent segregation. In the latter case, $\theta = 1/2$, and the probability of the pedigree F_1 becomes

$$P(F_1 \mid \tfrac{1}{2}) = \tfrac{1}{2}[(\tfrac{1}{2})^9 + (\tfrac{1}{2})^9] = (\tfrac{1}{2})^9. \tag{77}$$

The logarithm (to base 10) of the ratio of the two probabilities is called the *z* or *lod* (from logarithm of the odds) score and is

$$z = \log \frac{P(F_1 \mid \theta)}{P(F_1 \mid \frac{1}{2})} = \log 2^8 + \log\{\theta(1-\theta)^8 + \theta^8(1-\theta)\}. \tag{78}$$

The lod score for $\hat{\theta} = 0.11$ is 1.044, corresponding to the logarithm of the ratio of odds of 11 : 1. Thus, this pedigree lends considerable weight to the hypothesis of linkage between Rh and elliptocytosis by showing that there are 11 to 1 odds (or a probability of $1/(1+11) = 1/12$) in favor of the hypothesis of linkage versus independence.

Whether this is sufficient evidence for establishing linkage is a difficult question to answer in a completely satisfactory way, as it is for all problems involving a statistical decision. What the odds ratio tells us is that it is eleven times more probable to have the observed pedigree if there is linkage with $\theta = 0.11$ than if there is no linkage. Usually one looks for higher ratios, say, 1/20 (or 19 : 1) to be reasonably sure of linkage.

We may sometimes want to "feed" into this conclusion some knowledge concerning the "prior probability" of the two hypotheses. *A priori*, the probability that two genes are found to be linked depends on the number of chromosomes or, more precisely, the mean number of crossovers among them. The smaller the number of chromosomes, or more exactly, the fewer crossovers that occur, the more probable *a priori* that any two genes will be found to be linked. This modifies the lod approximately by adding to it a constant quantity. However, the problem does not fundamentally change since the level at which we are prepared to accept the lod as evidence of linkage is itself arbitrary, being the probability of error in a significance test. Further discussion of the Bayesian approach, as this is called, and the non-Bayesian approaches will be found in Morton (1962) and Smith and co-workers (1968).

The pedigree F_1 just discussed has been called somewhat loosely a double backcross. Here actually the three alleles at the Rh locus make the situation slightly different, but the main feature of a double backcross, namely a $1-\theta : \theta : \theta : 1-\theta$ segregation of the four types (parental: recombinant: recombinant: parental), is preserved.

It is useful to show how the standard error of the maximum likelihood estimate of the recombination fraction θ can be evaluated. In the estimation of a single parameter, the error variance of the estimates is minus the inverse of the second derivative of the likelihood taken at the value of θ corresponding to the maximum likelihood. The square root of this quantity is the standard error of the estimate (see Appendix I, Sections I.22, I.37, and I.38). Given several pedigrees, the lods, z, have to be cumulated over all pedigrees, and the sum $Z = \sum z_i$ over the pedigrees is used to obtain the likelihood estimate. The correction factor 2.303 ($= \log_e 10$) is necessary, as the lods are usually computed as logarithms to base 10 while the theory is valid for logarithms to base e. The computation of the second derivative can be done numerically and will be illustrated here, using the example just given even though it is limited to a single pedigree.

Values of the lods must be computed at or near the maximum Z values and also in the neighborhood of the maximum. In this case, the maximum is near $\theta = 0.11$, and the computed values are as shown in the following table.

θ	Z	*First difference*	*Second difference*
0.10	1.04218		
		$+0.00258 = 1.04475 - 1.04218$	
0.11	1.04475		$-0.00406 = -0.00148 - (+0.00258)$
		$-0.00148 = 1.04328 - 1.04475$	
0.12	1.04328		

First and second differences are computed between the three values. The second difference, -0.00406, taken with a positive sign, multiplied by $\log_e 10 = 2.303$ and divided by the square of the interval between successive values of θ (which is 0.01), is approximately the second derivative of the log likelihood. This is

$$2.303 \times (0.00406)/0.01^2 = 93.7.$$

The inverse of the square root of this quantity, taken with a positive sign, is $1/\sqrt{93.7} = 0.1034$, and this is an estimate of the standard error of the linkage estimate, $\hat{\theta} = 0.11$. With such small numbers, the sampling distribution of the statistic can hardly be normal. This estimate of the standard error can be used only with larger numbers on the assumption that its distribution is approximately normal.

General Treatment for the Double Backcross. The cross can be symbolized by $GgTt \times ggtt$, and the progeny expectations are shown in the following table.

		GT	*Gt*	*gT*	*gt*
Progeny phenotypes		*GT*	*Gt*	*gT*	*gt*
Number observed		*a*	*b*	*c*	*d*
Expectations for					
Coupling	(*GT*/*gt*)	$(1-\theta)/2$	$\theta/2$	$\theta/2$	$(1-\theta)/2$
Repulsion	(*Gt*/*gT*)	$\theta/2$	$(1-\theta)/2$	$(1-\theta)/2$	$\theta/2$

The lod score z is thus

$$z = \log\ 2^{s-1} + \log[(1-\theta)^{a+d}\theta^{b+c} + (1-\theta)^{b+c}\theta^{a+d}]$$

when the phase is not known, where $s = a + b + c + d$ and the first term relates to the probability of the pedigree for $\theta = \frac{1}{2}$. If the phase is known and is, for example, coupling, then

$$z = \log 2^s + \log[(1-\theta)^{a-d}\theta^{b-c}] = s \log 2 + (a+d)\log(1-\theta) + (b+c)\log \theta.$$

General Treatment for Single Backcross. The expectations for the cross in which the double heterozygote is known to be in coupling are given in the table below.

Gametes from single heterozygote parent

Gametes from GT/gt parent	1/2 *Gt*	1/2 *gt*
$(1-\theta)/2$ *GT*	*GT* $(1-\theta)/4$	*GT* $(1-\theta)/4$
$\theta/2$ *Gt*	*Gt* $\theta/4$	*Gt* $\theta/4$
$\theta/2$ *gT*	*GT* $\theta/4$	*gT* $\theta/4$
$(1-\theta)/2$ *gt*	*Gt* $(1-\theta)/4$	*gt* $(1-\theta)/4$

Summing over the four phenotypes, one obtains the expectations given in the coupling column in the table that follows. The repulsion case can be analyzed in the same way.

Progeny phenotypes	*Coupling* $(\frac{1}{2})$	*Repulsion* $(\frac{1}{2})$	*Weighted total*
GT	$\frac{1-\theta}{2} + \frac{\theta}{4} = \frac{2-\theta}{4}$	$\frac{1+\theta}{4}$	$\frac{3}{8}$
Gt	$\frac{\theta}{2} + \frac{1-\theta}{4} = \frac{1+\theta}{4}$	$\frac{2-\theta}{4}$	$\frac{3}{8}$
gT	$\frac{\theta}{4}$	$\frac{1-\theta}{4}$	$\frac{1}{8}$
gt	$\frac{1-\theta}{4}$	$\frac{\theta}{4}$	$\frac{1}{8}$

The probability of the pedigree F if there is linkage θ of unknown phase is then

$$P(F|\theta) = \tfrac{1}{2}[(2-\theta)^a(1+\theta)^b\theta^c(1-\theta)^d + (1+\theta)^a(2-\theta)^b(1-\theta)^c\theta^d] \times 4^{-s}. \tag{79}$$

If there is no linkage, $\theta = \frac{1}{2}$, and the above reduces to

$$P(F|\tfrac{1}{2}) = 3^{a+b} \times 2^{-s} \times 4^{-s}, \tag{80}$$

and the lod score is

$$z = \log \frac{P(F|\theta)}{P(F|\frac{1}{2})} = \log 2^{s-1}3^{-a-b} + \log[(2-\theta)^a(1+\theta)^b\theta^c(1-\theta)^d + (1+\theta)^a(2-\theta)^b(1-\theta)^c\theta^d].$$

A slightly more complicated formula can be similarly obtained for the double intercross.

II.13 Heterogeneity of Linkage Estimates

Being logarithms of probabilities, the z scores obtained from independent pedigrees for the same value of θ can be added together. It is thus possible to give a curve of z as a function of θ, showing for what value of θ z is maximum. This is then the combined maximum likelihood estimate of linkage if θ is significantly less than $\frac{1}{2}$.

The test for heterogeneity between pedigrees is important because it can disclose sources of disturbance and also the possibility that the same trait may be due to different genes, some linked and some unlinked to the markers being used.

Figure II.2 shows how the lods vary as a function of θ in the seven most informative pedigrees for Rh and elliptocytosis. Four of them (3, 4, 5, and *R*) have a maximum value of z at or near zero. The remaining three (*J*, *P*, *N*, *Ae*, and *Z*), on the other hand, show a maximum at $\frac{1}{2}$. It is quite clear from this picture that there is heterogeneity between these seven pedigrees. They fall into two groups, one showing close linkage and the other no linkage with Rh, suggesting that there exist two loci for elliptocytosis, one linked to Rh and the other unlinked. Summing the z scores of those showing linkage, the maximum z value is found at a θ value of about 0.04, which represents the maximum likelihood estimate of recombination between Rh and the Rh-linked elliptocytosis locus.

Morton (1956b) has suggested the following method for testing for heterogeneity. Let $\hat{Z}$ be the maximum value of the sum of the z values from all pedigrees, $\sum z_i$, and let the maximum value of z for the ith pedigree be $\hat{z}_i$. It can then be shown that $(\sum\hat{z}_i - \hat{Z}) \times 4.605$ is distributed approximately as a chi-square with degrees of freedom equal to the number of independent pedigrees minus one. The fourteen pedigrees examined were very significantly heterogeneous. Restricting the analysis

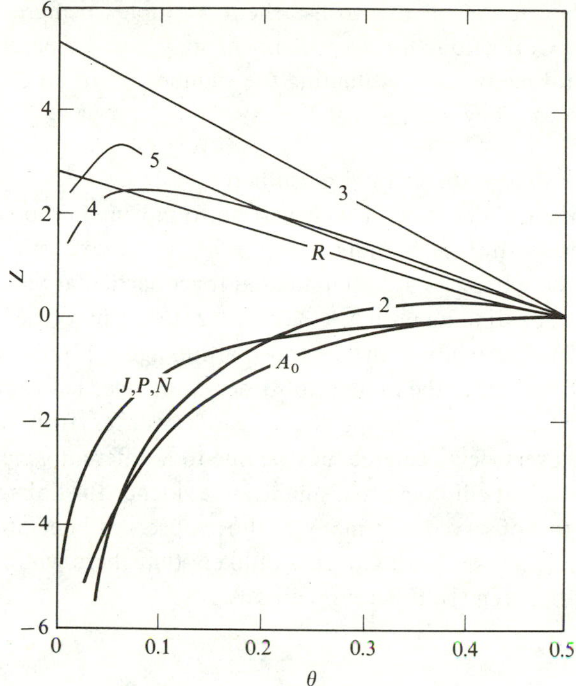

FIGURE II.2
Scores for pedigrees, 2, 3, 4, 5, R, Ae, and J, P, N, showing linkage between Rh and elliptocytosis in four of them (4, 5, 3, R) and absence of linkage in the other three (J, P, N, Ae, 2). (From Morton, 1956b.)

to the nine most informative ones, seven of which are represented in Figure II.2, there is no significant heterogeneity between sections of the same pedigree, between pedigrees of the Rh-linked locus, or between pedigrees of the Rh-unlinked locus.

The possibility of such heterogeneity as a result of "mimic" genes should always be kept in mind; therefore, testing should always be done for heterogeneity between pedigrees. Other cases of mimics separable by linkage analysis are known. Thus, the the two loci for colorblindness and the two types of hemophilia (*A* and *B*) are located at different positions on the X chromosome (see Chapter 10).

II.14 General Methods and Tables for Linkage Analysis

The presentation given here is intended to be a brief introduction to the general methodology. Tables of lod scores were published by Morton (1955 and later). They have been extended and republished by Smith, Penrose, and Smith (1961) with a detailed description of how to use them. Complex pedigrees may require special treatment. Corrections for ascertainment may also be necessary.

The three general methods of evaluating the significance of an observed linkage all use the lod scores. They are:

1. *Maximum likelihood*, the method given here.
2. *Bayesian inference*, details of which will be found in the books and articles cited in the references to this appendix.
3. *Sequential*, where z scores are accumulated for a particular value of θ, usually near the maximum, continuing until $\hat{Z} > 3$ or < -2. In the first case, the conclusion is that linkage is present while in the second that linkage is absent. The sequential method is claimed to reduce the number of observations needed to reach a conclusion. It does not, however, exempt the research worker from testing further pedigrees because every new pedigree may be due to a different gene. In testing for linkage between Rh and elliptocytosis, significant evidence for linkage was reached by the third pedigree observed, but many of those observed later proved to be due to the other locus, unlinked with Rh. It is unlikely that there will be any substantial discrepancies between the different methods.

II.15 Partial Sex Linkage

Partial homology of X and Y chromosomes has been demonstrated in some animals and in cytogenetic observations of chiasmata between X and Y in man, thus supporting the possibility that if genes can be found in this region, they could undergo crossing over and therefore would, in some pedigrees or parts of them, behave as Y-linked genes and in others as X-linked genes (see also Chapter 10). It was claimed by Haldane (1936) that the five diseases: xeroderma pigmentosum, achromatopsia, retinitis pigmentosa, Oguchi's disease, and epidermolysis bullosa, showed partial sex linkage. The basis for this claim was essentially the contrast between the sex ratio among trait carriers that inherited the condition through their fathers from their paternal grandfathers (in which case the gene was probably on the paternal Y chromosome and transmitted mostly as Y-linked) or from their paternal grandmothers (in which case it should be transmitted as X-linked). This of course could be simulated by a tendency of the genes to be manifest in one sex in some pedigrees and in the other sex in others because of variable sex limitation.

Analysis by the methods just described, adjusted so as to be independent of one possible type of sex-biased manifestation, has not confirmed Haldane's conclusions (Morton, 1957). At the moment there seems to be no significant evidence for partial sex linkage in man.

II.16 Map Length and Physical Length

All estimates of genetic distance by recombination are of course in relative units since they are measured as recombination frequencies. These are not directly proportional to the number of crossovers that have taken place between two genes which lie on the same chromosome. The number of crossovers can increase indefinitely, but the upper limit to the recombination frequency is, under assumptions which are at least approximately correct, 50 percent (it can be somewhat higher if there is interference).

Under some assumptions it is possible to transform an observed recombination frequency into the number of crossovers that gave rise to it. **Interference** is the interaction between crossovers occurring near to each other in the same chromosome. **Positive** interference is a reduction in the probability of occurrence of a second crossover near the place where another first one occurred. It is an almost general rule that genes which are on the same chromosome arm show positive interference while there is probably no interference across the centromere. **Negative** interference is the opposite phenomenon, namely, an increased probability of a second crossover. It is rarer, and probably only occurs at the intracistronic level, and in some special situations involving, for example, incomplete pairing.

In the absence of interference, the relationship between the number of crossovers x and the frequency of recombination y between two loci is given by Haldane's (1919) formula:

$$y = \tfrac{1}{2}(1 - e^{-2x}), \tag{81}$$

a simple derivation of which follows.

Suppose A_1, A_2, and A_3 are three loci on the same chromosome and that y_1 and y_2 are the respective recombination fractions between A_1 and A_2 and between A_2 and A_3.

$$A_1 \quad \overset{y_1}{\rule{3em}{0.4pt}} \quad A_2 \quad \overset{y_2}{\rule{3em}{0.4pt}} \quad A_3$$

Assuming recombination in the interval A_1A_2 is independent of that in A_2A_3 the recombination fraction between A_1 and A_3 must be

$$y_{1+2} = \underbrace{y_1(1 - y_2)}_{\text{Probability of recombination in } A_1A_2 \text{ but not } A_2A_3} + \underbrace{y_2(1 - y)}_{\text{Probability of recombination in } A_2A_3 \text{ but not } A_1A_2}$$

This can be rewritten

$$1 - 2y_{1+2} = (1 - 2y_1)(1 - 2y_2). \tag{82}$$

In general, therefore, for a series of loci $A_1, A_2, \ldots, A_r$ such that the recombination fraction between A_i and A_{i+1} is y_i

$$(1 - 2y_{1+2+\cdots+r}) = (1 - 2y_1)(1 - 2y_2)(1 - 2y_3)\cdots(1 - 2y_r). \tag{83}$$

Since for loci close together $y_i < \frac{1}{2}$ so that $1 - 2y_i < 1$, as the number of intervening loci increases $1 - 2y_{1+2+\cdots+r} \to 0$, and so $y_{1+2+\cdots+r} \to \frac{1}{2}$. Thus the recombination fraction between two loci tends to 1/2 as the distance between them on the chromosome increases. Actually when all the y_i are small and equal to y

$$(1 - 2y_1)(1 - 2y_2)\cdots(1 - 2y_r) = (1 - 2y)^r = \left(1 - 2\frac{x}{r}\right)^r \sim e^{-2x}, \tag{84}$$

where $x = y_1 + y_2 + \cdots + y_r = ry$. Thus, approximately, $y(x) = y_{1+2+\cdots+r}$, the recombination fraction between loci distance x apart, where distance is measured by the number of intervening closely linked loci, is given by

$$y(x) = \tfrac{1}{2}(1 - e^{-2x}), \tag{85}$$

which is known as Haldane's formula (1919). Note that when x is small,

$$y(x) \sim \tfrac{1}{2}[1 - (1 - 2x)] \sim x. \tag{86}$$

Haldane's equation represents in a simple way the relation between the recombination fraction y and the map length x measured effectively by the average number of crossover events which occur in the relevant interval.

The unit of map length is the morgan or, more usually, the centimorgan, which corresponds to 0.01 in the x scale and therefore approximately to a one percent recombination fraction.

Map length x is measured on a scale defined by the constant probability of a crossover per unit length. It need not, and usually is not, proportional to physical length, as the frequency of crossing-over may vary from point to point on the chromosome. The demonstration of differences in crossover frequencies in various chromosome sections, or in different parts of the life cycle (for example, "somatic" crossing-over versus the conventional "meiotic" crossing-over, an event shown to occur, though rarely, in several organisms) requires additional studies by other techniques. An empirical formula for the relation between map distance and recombination fractions which takes account of the effects of positive interference was given by Kosambi (1944):

$$y = \frac{1}{2}\frac{e^{2x} - e^{-2x}}{e^{2x} + e^{-2x}} = \tfrac{1}{2}\tanh 2x. \tag{87}$$

Its inverse is

$$x = \tfrac{1}{4}\log(1 + 2y)/(1 - 2y) \tag{88}$$

and has been found to be useful in several organisms. Other related formulas have been suggested. A theoretical treatment of the general problem was initiated by Fisher, Lyon, and Owen (1947) and can be found in some detail in Owen's (1950) and Bailey's (1961) discussions. However, knowledge of human linkage is much too slight at this point to benefit from such theories.

The problem of the total amount of recombination going on in human meiosis can be tackled from a cytological point of view. Examination of first meiotic divisions in human gametogenesis has given counts of the number of "chiasmata" per nucleus.

Chiasmata are the visible counterparts of crossovers in meiotic chromosomes although the exact relationship is not clear (see Chapter 1). One chiasma corresponds to one crossover, which involves only two of the four strands of which a chromosome pair is made (a crossover when homologues have duplicated) and therefore to 0.5 of a morgan. It is estimated from cytological observations of chiasmata in human meiosis that the total number of crossovers occurring is on an average, such that the total map length of the human chromosome set should be about 26 morgans.

II.17 Assignment of Genes to Chromosomes

Sex-linked genes can be assigned to the sex chromosome, which is distinguished on the basis of comparison of male and female nuclei. For autosomes, assignment is sometimes made possible by: (1) cytological peculiarities of the chromosomes, (2) chromosome aberrations, and more recently (3) somatic cell genetic studies using somatic cell hybrids.

The first category of methods depends on the availability of cytologically marked chromosomes. Thus, there is one rare variant of chromosome 1 (population frequency of about 0.5 percent) in which one of the two members of the pair shows an attenuation of an otherwise constant secondary constriction near the centromere. This is believed to be due to an *uncoiler* locus. Linkage of *uncoiler* with the blood group Duffy (gene symbol, *fy*) is highly significant and fairly close (5 map units or centimorgans). Duffy is known to be linked to a gene that causes cataracts (the congenital zonular pulverulent types). The relevant locus, *Cae*, is therefore also believed to be on chromosome 1. Another uncoiler locus, with the same heterozygous frequency, has been reported for chromosome 17, but no linkage with it has yet been established (for a recent review of this data and the following discussion see Renwick, 1969).

Chromosome aberrations of various kinds can help in assigning loci to specific chromosomes. The first example published in man indicates, however, some of the difficulties occasionally encountered in this approach. The progeny of a father of phenotype *Hp* 2-2 was found to have a ring chromosome of group D, presumed to be the thirteenth, and a phenotype *Hp* 1-1. If there were a terminal deletion in the father's gamete associated with ring-chromosome formation, and the *Hp* locus were located in the deleted section, this would explain the loss of the paternal Hp_2 allele. Later work (Cook et al., 1969; Robson et al., 1969) proved, however, that this result must have been due to the segregation in this family of the rare silent allele Hp_0. On the basis of further data, essentially of pedigrees for translocations between chromosome 16 and other chromosomes, the *Hp* locus is now assigned to chromosome 16. Another case exists in which a deletion helped to assign a gene (this time for the rare disease pyknodysostosis to a chromosome (of Group G).

We have discussed in Chapter 2 the possibility of using aneuploidy (in particular for the case of trisomy 21) to assign genes to chromosomes. Monosomics can also be helpful. The greater karyotypic variability of somatic cells cultured *in vitro* may be of help in this respect, at least for those markers, especially biochemical ones, which can be recognized in cultured cells. Alkaline phosphatase mutations arising in association with segregation of chromosome 21 suggest that the corresponding gene is located in this chromosome. This agrees with the fact that leukocytes from leukemic patients with the "Philadelphia" chromosome (characterized by a strong heterochromatic zone in one member of pair 21, probably because of inactivation) show a smaller concentration, and leukocytes from trisomics for 21 a greater concentration of this enzyme. These quantitative data on enzyme concentration are obviously less satisfactory than direct cytological observations (De Carli, 1964).

From studies on somatic cell hybrids between mouse cells and human cells that had lost all but one of their human chromosomes, a thymidine kinase gene has been shown to be associated with chromosome 17 (or 18) (Weiss and Green, 1967;

TABLE II.3

Unassigned Autosomal Linkage Groups in Man. Other linkages and assignments to specific chromosomes are given in the text.

Odds on Linkage	*Loci*	*Locus Symbols*	*Estimated Map Intervals (f = female: m = male) (Centimorgans)*	*95 Percent Probability Limits (Centimorgans)*
$>10^6$:1	*ABO*:*Nail-patella syndrome*	*ABO*:*Npl*	11 (*f*, 14; *m*, 8)	6–19
$>10^6$:1	*Nail-patella syndrome*:*AK*	*Npl*:*AK*	0	0–7
200:1	*Adenylate kinase*:*ABO*	*AK*:*ABO*	16 (*f*, 24; *m*, 8)	5–30
$>10^6$:1	*Haemoglobin*$_\beta$:*Haemoglobin*$_\delta$	Hb_β:Hb_δ	0.1	0–6
$>10^6$:1	*Albumin*:*Gc protein*	*Alb*:*Gc*	2	1–6
$>10^6$:1	*Elliptocytosis 1*:*Rhesus*	*EL1*:*Rh*	3	2–7
186:1	*Sclerotylosis*:*MNS*	*Tys*:*MNS*	4	0.3–19
$>10^6$:1	*Lutheran*:*Secretor*	*Lu*:*Se*	13 (*f*, 16; *m*, 10)	7–19
75:1	*Transferrin*:*Cholinesterase 1*	*Tf*:E_1	16 (*f*, 19; *m*, 12)	7–28

Source: From Renwick (1969).

Migeon and Miller, 1968; Matsuya *et al.*, 1968). Similar studies of cell cultures have confirmed the linkage between the X-linked marker *G6PD* and the enzyme hypoxanthine-guanine-phosphorybosil transferase (Nabholz *et al.*, 1969). Extensions of this approach will undoubtedly soon lead to a rapid increase in our knowledge of the human linkage map. A list of other autosomal linkages is given in Table II.3.

II.18 Estimation of Gene Frequencies

The estimation of gene frequencies is straightforward when no genes are recessive and the sample is a truly random sample from the original population. Under such conditions, genes can be counted directly. The estimates thus obtained are also maximum likelihood estimates, and so is the usual error variance. This error variance is $p(1-p)/2N$ for a gene frequency p, where N is the number of individuals in the sample.

We have already seen how to deal with some of the complications resulting from recessivity. The basis for computations is the assumption of the Hardy-Weinberg equilibrium. Approximate estimates can always be obtained by short-cut methods, such as equating observed phenotype frequencies with those expected under Hardy-Weinberg. There are, however, with more than two alleles usually more equations than unknowns, and thus conflicting estimates of the gene frequencies may be obtained depending on the equations used for estimation. A method for correcting the approximate estimates for ABO genes (or, in general, for three alleles of which one only is recessive to the others) was given by Bernstein and was disscused in Chapter 2, Section 2.5 (see also Neel and Schull, 1954). It can be proved that the estimates thus obtained, although different from those obtained by maximum likelihood, are fully efficient and that the variance estimates obtained from a maximum likelihood approach can be applied to them.

A complete treatment of the ABO system by maximum likelihood was given by Neel and Schull (1954). We will illustrate here only the general methodology.

If A, B, AB, and O are the observed (absolute) frequencies of the four phenotypes (A + B + AB + O = N) and p, q, and r are the gene frequencies of A,B, and O $(p+q+r=1)$, then the log likelihood is the log of the multinomial probability, namely,

$$L = \log\left[\frac{N!}{\text{A!B!AB!O!}}(p^2+2pr)^A(q^2+2qr)^B(2pq)^{AB}(r^2)^O\right]. \tag{89}$$

It is easier to express this by letting f_i be the observed frequency and E_i the expected relative frequency in each phenotypic class; then

$$L = \sum f_i \log E_i + \text{const.} \tag{90}$$

The sum extends over all phenotypic classes. We have to differentiate L with respect to two unknown parameters, p and q (since $r = 1 - p - q$).

The derivatives of the expectations are computed keeping in mind that $r = 1 - p - q$, giving the results in the Table following

Class	E_i	$\partial E_i/\partial p$	$\partial E_i/\partial q$
A	$p^2 + 2pr$	$2r$	$-2p$
B	$q^2 + 2qr$	$-2q$	$2r$
AB	$2pq$	$2q$	$2p$
O	r^2	$-2r$	$-2r$

The derivatives of the likelihood are given in Equation 91.

$$\frac{\partial L}{\partial p} = \sum \frac{f_i}{E_i}\frac{\partial E_i}{\partial p}, \qquad \frac{\partial L}{\partial q} = \sum \frac{f_i}{E_i}\frac{\partial E_i}{\partial q_i}, \tag{91}$$

giving rise, when set equal to zero, to very complicated equations, which can however, be solved by successive approximations. Following the scoring system, (Appendix I.37 and I.38), the trial values p_0 and q_0 are first obtained by some approximate method, and then the scores $s_p = (\partial L/\partial p)_{p=po}$ and $s_q = (\partial L/\partial q)_{q=q_0}$ are computed. The second derivatives (giving the information matrix) allow one to compute corrections to p_0 and q_0. Recycling is done if necessary.

The elements of the information matrix I can be obtained by differentiating the $\partial L/\partial \theta$ values (for parameters θ) and by putting the observed values equal to their expectations, or more simply, following Appendix I, Equation 129,

$$I_{\theta, \theta_k} = N \sum_i \frac{1}{E_i}\left(\frac{\partial E}{\partial \theta_j}\right)\left(\frac{\partial E}{\partial \theta_k}\right), \tag{92}$$

where the sum is extended to all the observed classes, and θ_j and θ_k are the parameters to be estimated (here p and q), taken in all possible pairs. There are thus three elements of the information matrix to be computed: I_{pp}, I_{pq}, and I_{qq} (since the matrix is symmetrical and $I_{pq} = I_{qp}$).

As an example of an application, let us again take the data from Chapter 2:

A	B	AB	O	N
44	27	4	88	163

from which $p = 0.165$, $q = 0.1$, $r = 0.735$ were estimated by approximate square-root formulas. From the first derivative equations (combining the results given in the table at top of this page and Equation 91) the scores can be shown to be:

$$s_p = \frac{\partial L}{\partial p} = -9.853$$

and

$$s_q = \frac{\partial L}{\partial p} = -0.476,$$

and the information matrix elements are

$$I_{pp} = 2196.73,$$
$$I_{pq} = 379.66,$$

and

$$I_{qq} = 3499.18.$$

Corrections to the trial values $p_0 = 0.165$ and $q_0 = 0.1$ are obtained by multiplying the vector of scores, s, by the inverse of the information matrix, $\mathbf{V} = \mathbf{I}^{-1}$ which can be shown to be

$$\mathbf{I}^{-1} = \begin{Bmatrix} 2196.73 & 379.66 \\ 379.66 & 3499.18 \end{Bmatrix}^{-1} = \begin{Bmatrix} 0.0004639 & 0.0000503 \\ 0.0000503 & 0.0002912 \end{Bmatrix} = \mathbf{V}.$$

Then, multiplying the scores vector by the inverse of the information matrix gives corrections to the trial values:

$$\delta p_0 = s_p V_{pp} + s_q V_{pq} = -9.853 \times 0.0004639 - 0.476 \times 0.0000503 = -0.00459$$

and

$$\delta q_0 = s_p V_{pq} + s_q V_{qq} = -9.853 \times 0.0000503 - 0.476 \times 0.0002912 = -0.00063,$$

giving $p_1 = p_0 + \delta p_0 = 0.16041$, $q_1 = q_0 + \delta q_0 = 0.09937$. After recomputation of the scores with the new p and q values, they become

$$s_p = 0.4822 \qquad s_q = 3.4949,$$

and therefore on average nearer to zero. New approximations p_2 and q_2 can be computed and the cycle repeated until the scores are as near to zero as desired.

Successive cycles gave the following p and q values, where the standard errors are given for the first and last and are obtained as the square roots of the elements of the main diagonal of the inverted information matrix (for example, $\sigma(p_0) = \sqrt{0.0004639} = 0.0215$). They change slightly with changes in the information matrix:

$$p_0 = 0.165 \pm 0.0215 \qquad q_0 = 0.1 \pm 0.0171,$$
$$p_1 = 0.160405 \qquad q_1 = 0.09937,$$
$$p_2 = 0.160793 \qquad q_2 = 0.100401,$$

and

$$p_{10} = 0.160454 \pm 0.0213 \qquad q_{10} = 0.100357 \pm 0.0171.$$

The same result can be obtained numerically, without going through the tedious algebra which is sometimes necessary for obtaining the scores and the information matrix. By definition, the derivative (for example, of the likelihood L) is the following limit:

$$\frac{\partial L}{\partial \theta} = \lim_{\varepsilon \to 0} \frac{L(\theta + \varepsilon) - L(\theta)}{\varepsilon}. \tag{93}$$

Therefore the value of the derivative can be calculated approximately by computing the ratio

$$\frac{\partial L}{\partial \theta} \sim \frac{L(\theta + \varepsilon) - L(\theta)}{\varepsilon}, \tag{94}$$

provided ε is small. Computer accuracy is usually eight digits, therefore ε can be safely taken as 0.0001. With several parameters, "double precision" can be used, giving sixteen-digit accuracy, and then ε can be as small as 10^{-8}.

For use with the formulas just given, the derivatives of the expectations E_i are calculated. With two parameters, p and q, two derivatives computed for each phenotypic class are:

$$\frac{\partial E_i}{\partial p} \sim \frac{E_i(p + \varepsilon, q) - E_i(p - \varepsilon, q)}{2\varepsilon} \tag{95}$$

and

$$\frac{\partial E_i}{\partial q} \sim \frac{E_i(p, q + \varepsilon) - E_i(p, q - \varepsilon)}{2\varepsilon},$$

and the general procedure already given is used. The algebraic differentiation of the expected frequencies is thus avoided.

Several computer programs have been written which can carry out all-purpose estimation by numerical maximum likelihood. The problem thus solved is a general one in multinomial estimation and is not restricted to gene frequencies. The only functions to be specified are the phenotype frequencies of each class in terms of the gene frequencies. Programs are available for distribution (see, for example, Reed and Schull, 1968).

II.19 Samples of Related Individuals

Very often, populations are analyzed by family groups, and thus a peculiar type of *non*random sampling is practiced. Even when this is not done on purpose and a whole population is analyzed, the existence of close relationships between the individuals forming the population will create statistical difficulties. If the relationship is ignored, it is not so much the estimate of the gene frequency which is biased but rather that of its sampling variance, which will be artificially decreased. In fact, the same genes, reappearing for example, in the parents and in the progeny, are counted twice. When a whole population, which is made up on the average of almost three generations, is tested, each gene will be counted almost three times. This factor is probably the limit to the underestimation of the sampling variance. When, on the other hand, the sample is a small and random one of the general population, the relationship between individuals forming the sample is on the average low, and no untoward effects are expected. The bias is minimized, in general, by selecting for the sample individuals of the same age class, say 20–40 years old.

When the degree of relationship in a sample is high, a correction can be made in various ways.

1. We can discard invididuals who are closely related to others (for example, progeny when parents are given).

2. To minimize the loss of information, one may discard, at random, only an adequate fraction of genes or individuals. For example, if in a family, one parent and one child are available, one of the two genes can be discarded at random, the gene remaining in the sample representing, hopefully, the gene from the missing parent.

3. It is perhaps more satisfactory to keep all the genes in the estimation procedure, but we can correct the number of individuals by subtracting duplicated genes. Thus, if one parent and one child are present, they should count as 3 genes; one parent and two children, $3\frac{1}{2}$ genes; one parent and three children $3\frac{3}{4}$ genes; etc. Two parents and n children would therefore count as 4 genes; one parent and n children as $4 - \frac{1}{2}^{n-1}$.

4. A counting method applicable to pedigrees was proposed by Ceppellini, Siniscalco, and Smith (1955). It is based on an iteration procedure which assumes that gene frequencies are known initially and then corrects them. Parents in the pedigree whose genotype is unknown are assumed to take all acceptable genotypes with a frequency equal to that expected in the population (on the basis of trial values for the gene frequencies). Gene frequencies are then computed and the trial values substituted by the new values. Recycling may be necessary, and heavy numerical work can be expected, which can as usual be minimized by use of the computer.

General References

Bailey, N. T. J., *Mathematical Theory of Genetic Linkage*. Oxford: The Clarendon Press, 1961. (Contains a useful review of some relevant statistical methodology and the theory relating recombination fractions to map distances.)

Haldane, J. B. S., "The combination of linkage values and the calculation of distance between the loci of linked factors." *Journal of Genetics* **8**: 299–309, 1919. (The first attempt to relate recombination fractions to map distances.)

Mather, K., *The Measurement of Linkage in Heredity*. London: Methuen, 1951. (A useful review of some specialized statistical methods.)

Morton, N. E., "Sequential tests for the detection of linkage." *American Journal of Human Genetics* **7**: 277–318, 1956. (A comprehensive paper on the maximum likelihood and sequential approaches.)

Morton, N. E., "Genetic tests under incomplete ascertainment." *American Journal of Human Genetics* **11** (1): 1–16, 1959. (Morton's first major contribution to segregation analysis, with a compendium of methods and formulas.)

Smith, S. M., L. S. Penrose, and C. A. B. Smith, *Mathematical Tables for Research Workers in Human Genetics*. London: Churchill, 1961. (A useful collection of methods, formulas, and tables.)

APPENDIX III
Sample Problems

These problems are based primarily on questions given in the final examination of a preclinical medical genetics course for students at Stanford Medical School. They are arranged to correspond, approximately, to the sequence of topics treated in this book.

1. a Give the frequencies of all the genotypes that arise from crossing two individuals, each of which is heterozygous (*Aa Bb*) at a pair of unlinked loci.
   b. If the genes *A* and *B* are dominant but AB and A phenotypes are not distinguishable, what are the frequencies of the observable phenotypic categories?
   c. How would these frequencies be affected if the two loci were linked?

2. Given the following data, construct a linkage map for the three loci involved in the backcross of a heterozygote *Aa Bb Cc* to the multiple recessive *aa bb cc*. Lowercase letters indicate the mutant phenotype, plus (+) indicates normal phenotype.

	Percent of total
+ + +	2
a + +	6
+ b +	37
+ + c	11
a b +	9
a + c	30
+ b c	4
a b c	1
	100

3. A man with a skeletal abnormality determined by the dominant gene *B* marries a woman who has a minor defect determined by the recessive gene *c*. They have one normal child, four children who have the recessive defect only, one child with both defects, and five children who only have the skeletal abnormality. Give an interpretation of these data, assuming that the genes under consideration are fully penetrant, and indicate the most probable genotypes of all members of the family.
4. Two parents have three normal and one galactosaemic offspring. What are the chances of a union between two normal offspring (each coming from such a family), giving rise to a galactosaemic offspring? How could you make your prediction more accurate?
   b. How many different types of gametes could arise from an individual known to be heterozygous at 15 loci, all on different chromosomes? How would your answer be affected if you were told that some of the loci were linked?
   c. Albinism is generally considered to be a simple recessive defect. How would you explain a mating between two albinos that gave rise only to normal offspring?
5. Four babies were born in a hospital on the same night, and their blood groups were later found to be O, A, B and AB. The four pairs of parents were:

   O and O

   AB and O

   A and B

   B and B

   Assign the four babies to their correct parents.
6. The frequency of ABO types in a group of about 300 American Indians was 81 percent O, 19 percent A, and no B or AB. Estimate the frequencies of the A, B, and O alleles in this population.
7. Consider the following pedigree for phenylketonuria.

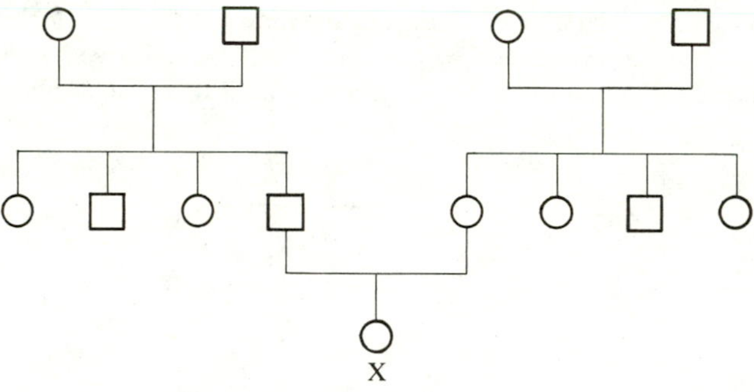

   a. On the basis of this information, what is the predictive possibility that X will be a phenylketonuric?
   b. What measures could you take that would improve the accuracy of your prediction?
   c. If X has a sib who is a phenylketonuric, does this affect the prediction for X? If so, how?

8. Assume that albinism is determined by a fully penetrant recessive gene *a* with no effects on fitness or fertility and that the frequency of albinos in a population is 1/20,000.
   a. What is the probability of a mating between an albino and a normal individual (randomly chosen from the population), producing an albino offspring?
   b. What proportion of the albinos in the population have parents neither of whom is an albino?
9. Ten thousand people were typed for two characters in which the homozygotes could be phenotypically distinguished from one another and from the heterozygotes. The percentage frequencies of the nine possible classes were as follows:

AABB: 0.36	*AaBB*: 29.16	*aaBB*: 6.48
AABb: 0.48	*AaBb*: 38.88	*aaBb*: 8.64
AAbb: 0.16	*Aabb*: 12.96	*aabb*: 2.88

   a. Is this population in genetic equilibrium? How can you tell? Do you need further information?
   b. Are the *A* and *B* loci linked? How can you tell from this information? Do you need further information?
10. Using galactose tolerance tests or enzyme assays, it is estimated that one of every 200 people carry the recessive gene *g* for galactosaemia. Assuming this has a mutation rate of 2.5×10^{-5}, estimate the relative fitness of carriers of the gene *g*, as compared with noncarriers.
11. A fully penetrant recessive defect has a population incidence of 4/10,000, and affected individuals have 50 percent of normal fitness. How would you explain the maintenance of this gene in the population?
12. The frequency of sickle-cell trait carriers at birth in a malarious, underdeveloped district in West Africa is 20 percent. Calculate the frequency of the sickle-cell trait carriers at maturity in this locality. Ignore any factors of differential fertility other than mortality. Assume that sickle-cell disease causes death before maturity and that the sickle-cell gene is maintained in this population through a balanced polymorphism.
13. Two alleles, *A* and *a*, are such that in a large random mating population the relative fitnesses of the genotypes are

AA	*Aa*	*aa*
0.9	1	0.2

   Describe what happens to the allele *a* in a population which is predominantly *A* (i.e., $>99.99\%$). What happens if the fitness of *aa* is 1.1 instead of 0.2? What happens if the fitness of *Aa* is 0.85 instead of 1?
14. a. Gene *A* mutates to *a* with a frequency 2×10^{-6} per generation. Mating is at random in the population, and there are no other forces acting on these two alleles. How many generations are needed to increase the frequency of gene *a* from 1 percent to 2 percent?
   b. Homozygotes *gg* die at an early age, but heterozygotes *Gg* are indistinguishable in fitness or fertility from normal *GG* individuals. How many generations are needed to reduce the frequency of the gene *g* from 1 percent to 0.5 percent?

c. Approximately how many generations are needed if *Gg* has a fitness of 0.9 relative to *GG*?

15. Ability to taste PTC is a dominantly inherited characteristic that has a frequency of 70 percent in the American population. What can you say about the selective values needed to maintain this balanced polymorphism?

16. The following data were obtained using a monospecific antiserum that detects a new blood-group antigen G (+ indicates presence; − indicates absence).

Pedigree data

Parents ♀	Parents ♂	Offspring ♂ G+	Offspring ♂ G−	Offspring ♀ G+	Offspring ♀ G−
G+	G+	23	12	29	0
G+	G−	9	4	10	7
G−	G+	0	3	4	0
G−	G−	0	2	0	1

Population data

	♂	♀
+	95	167
−	59	21

a. What is the mode of inheritance of the G+ : G− antigenic difference?
b. What is (are) the frequency (frequencies) of the relevant gene (genes) in the population? Give the number expected in each phenotypic category on the basis of your model. (Option—test the goodness-of-fit of your expected numbers.)

17. a. Data on the MN types of a very large number of offspring from M × MN and N × MN matings in the U.S. show proportions of 1 *MM* : 1.03 *MN* and 1 *NN* : 1.04 *MN*, respectively.
What genotype frequencies would be *predicted* by an explanation based on the above data for the *MN* polymorphism?
b. In 1900, 1000 *NN* individuals migrated into a country with approximately 40,000,000 inhabitants. In that country the *MM* × *MN* matings were found to produce 52 percent *MM* offspring whereas the *NN* × *MN* matings behaved as described in 3a. What are the expected ultimate gene frequencies of *M* and *N*, respectively, in this population? (State any additional premises you specify in producing an answer to this question.)
c. If you were to premise that the mutation rates per generation of $M \rightarrow N$ and of $N \rightarrow M$ were both 5×10^{-6}, how would this affect your answers to a and b?

18. A gene *g*, when homozygous, causes a relatively benign disease, as a result of a deficiency of carbohydrate metabolism. The fitness of affected individuals (*gg*) is estimated to be 0.8 and of carriers (*Gg*) 0.95, relative to a fitness of 1 for normal individuals (*GG*). The frequency of affected individuals in the population is 1/10,000.
a. Assuming the gene *g* is maintained at equilibrium by mutation-selection balance, what is the approximate mutation rate of $G \rightarrow g$?
b. What proportion of the genetic load caused by this gene is due to the selective disadvantage of the heterozygous carriers?

c. Give *one* plausible alternative explanation to that suggested in a for the present gene frequencies in the population.

d. A change in the general dietary intake of the population changes the relative fitnesses so they now have the values 1, 1.05, and 0.9 for the three genotypes *GG*, *Gg*, and *gg*, respectively. Assuming these new conditions are maintained indefinitely, what will be the ultimate fate of the *g* gene?

19. Inability to taste phenylthiocarbamide (PTC) is determined by a recessive gene, *t*. The frequency of tasters of PTC in a typical population sample is 64 percent. Suppose, further, that in those matings between tasters and nontasters that give rise to nontaster offspring, the observed proportion of this latter type is 5/11.

a. Assuming the departure from 1:1 ratio in these families is due solely to a slight selective disadvantage of the nontasters, what is the estimated fitness of nontasters relative to heterozygous tasters?

b. Briefly, how would you account for this situation as a balanced polymorphism? On the basis of your hypothesis, what are the relative fitnesses of the three relevant genotypes?

20. In a given part of West Africa, the frequency of sickle-cell homozygotes (*SS*) at birth is 0.25. The proportion among people without sickle-cell anemia who have the sickle-cell trait increases by a factor 1.25 from birth to maturity. Calculate the contribution to the fitness of the trait carriers relative to the normal homozygotes (*AA*) arising from this increased chance of survival.

21. Huntington's chorea is a dominant defect with a relatively late age of onset. Its frequency in the population is about 1/100,000. Individuals homozygous for the defective gene have never been detected but are presumed to die during an early stage of development.

a. Assuming random mating and a mutation rate for the gene for Huntington's chorea of 10^{-6}, estimate the fitness of affected individuals.

b. Fertility studies suggest that affected individuals tend to have appreciably more children than average during their fertile period, which is before the onset of the disease. The current trend toward a lowering of the average age at which reproduction is terminated could, therefore, conceivably result in the affected individuals having a higher than average fitness. Assuming this trend in fact increases the previously estimated fitness by an amount corresponding to 30 percent of its original value, what ultimately will be the new equilibrium frequency of individuals with Huntington's chorea?

22. Given two alleles *A* and *a* at a single locus in a random mating population, and given that the population contains predominantly *A*, describe the fate of the allele *a* for each of the following set of fitness values:

	AA	*Aa*	*aa*
(i)	0.99	1.0	1.01
(ii)	0.9	1.0	0.9
(iii)	0.9	0.8	1.0
(iv)	0.9	0.7	0.9

How would your conclusion be altered if the gene frequencies of *A* and *a* were both 1/2?

23. Mutant allele h is produced with a mutation rate of 1 in 10,000. (The normal allele is H.) Of those people heterozygous for this gene, half contract a certain disease and die before age 10. The other half of the heterozygotes and all HH homozygotes never get the disease. (Ignore hh homozygotes, since they are very rare.) Suppose a treatment is found that cures half of those that get the disease. Assuming random mating, what is the mutation load for this locus: (1) before the cure is discovered? (2) in the first generation of people born after the cure is put into practice? (3) after the population has reached a new stable equilibrium?

24. Individuals in the following pedigree were classified for ABO and MN red-cell types and also for two white-cell antigens, HLA1 and HLA2. O, A, M and MN refer to ABO and MN types. HLA1 + or − and HLA2 + or − refer to the presence and absence of HLA1 and HLA2 antigens on the white cells

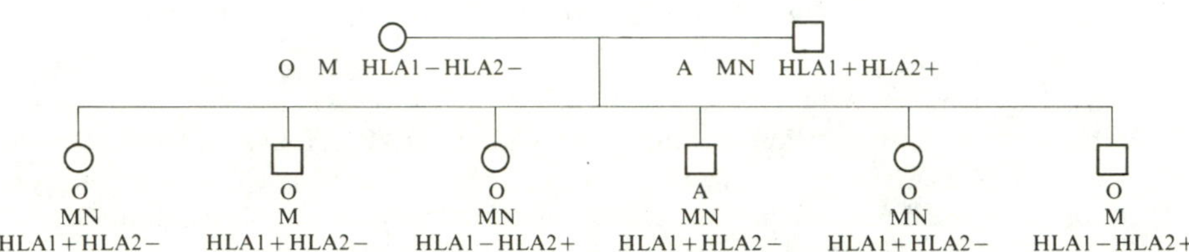

a. Give the probable genotypes of all the individuals in the pedigree for both red-cell and white-cell systems.
b. What can you say about the linkage relationships between the red-cell and white-cell systems?

25. Three children known to have been born to three mothers in a certain hospital at approximately the same time were mixed up soon after birth so that it was not known which child belonged to which mother. Given the blood groups of the children were OM, NB, AMN and the blood groups of the three pairs of parents were AMN × ABN, AN × BN, and AM × AMN, assign each child to its correct parents.

26. Arrange the following transplantations from the fastest to the slowest expected time of rejection of the transplant. If you expect two transplantations to have the same survival time, indicate this by an equality sign. The format of your reply might be like "$a = d > c > b = e$." A, B, and O, refer to the ABO blood groups in man.
a. Type A red blood cells injected into an unrelated individual of type O.
b. Type A red cells injected into an unrelated type AB individual.
c. Type A red cells injected into a sib of type O.
d. HLA1 white cells injected into an HLA2 male who had not previously received a blood transfusion.
e. Type AB red cells stored and reinjected into the donor individual.

Make a similar arrangement of the following transplantations.
f. Skin transplanted from a donor to his dizygotic male cotwin.
g. Skin transplanted from a donor to her monozygotic cotwin.
h. Skin transplanted from a donor to his first cousin.
i. Skin transplanted from a donor to her son.
j. Skin transplanted from a donor to his brother.

27. The information given concerns the phenotype of a mother and of her baby. What are the possible phenotypes of the father? List all the possibilities, which may include " no possibility," determined by the data. For each possibility, also list the possible genotypes of the child. Consider only those factors for which the data have predictive value, that is, narrow the range of choices from the scope of the whole population. Note that O means no reaction with anti-A or anti-B and H-positive unless specified.

Mother	*Child*	*Possible father's phenotype(s)*	*Child's genotype(s)*
AB	O		
AB	O (H-negative)		
O	A		
O Se	O se		
Lewis^a (red cells)	Lewis^b (red cells)		
Lewis negative	Lewis^b		

28. a. A population contains predominantly the rhesus negative allele, *r*. Ten percent of fetuses that are incompatible with the mother are lost because of hemolytic disease. Assuming the frequency of the allele *R* is small enough for the frequency of the genotype *RR* to be negligible, what is the required lower limit for the fitness of the heterozygote *Rr* to allow *R* to increase in frequency?
    b. What is this limit when *R* is the common allele and *r* the rare allele?
    c. In a given population, the frequencies of the ABO blood group alleles *O*, *A*, and *B* are 0.6, 0.3, and 0.1; and of the rhesus alleles *R* and *r* are 0.6 and 0.4. A woman is *ORh*−. What is the probability that she will marry a man whose genotype is such that all her offspring are potentially capable of immunizing her to produce Rh+ antibodies? (Assume random mating.)

29. a. Two pure antisera anti-a and anti-b react only with 7S gamma globulin and not with 19S gamma globulin. On which polypeptide chains would you predict the antigens a and b are located?
    b. When 1000 individuals were tested for a and b, the following results were obtained:

a + b−	a + b+	a − b+
40	330	630

    What is the most plausible hypothesis to account for this distribution in terms of: (1) how many sites, (2) how many alleles, and (3) the frequency of each allele?
    c. A third antibody is found which reacts with 7S gamma globulin. This reagent, anti-c, reacts with the 1000 sera as follows:

	a + b−	a + b+	a − b+
c+	20	175	305
c−	25	160	315

Which of the following statements are inconsistent with the above data?

(1) The antigen is controlled by an allele of the gene for *a*.

(2) The antigen is controlled by a gene at a site closely linked to the site for b.

(3) The antigen is controlled by a gene at a site unlinked to the site(s) for a or b.

(4) We have presumptive evidence for c's being controlled at a new site but do not know whether it is linked or unlinked to the site(s) for a or b.

(5) c is an allele of a but not of b.

(6) The evidence can not be used to support any genetic hypothesis.

d. It is found that anti-c also reacts with β_{2A} as well as β_{2M} (19S) globulins. Which of the above statements is most plausible, taking into account this fact along with the previously given information?

30. The frequencies of the 7 commonest Rh alleles in a Caucasian population are:

cDe	*CDe*	*cDE*	*cde*	*Cde*	*CDE*	*cdE*
0.053	0.415	0.110	0.404	0.008	0.004	0.005

An individual from this population was typed as *CcDe* using sera that identifies the antigens C, c, D, e, and E. What is this individual's most probable genotype?

31. For testing the following Caucasoid mating, only the *Gm* factors indicated were scored: Gm (a+ g+ f+ n+) × Gm (x+ a+ g+). What are the probable combinations of these Gm factors in offspring of this mating?

32. Put in the best order that you can, indicating ambiguities when appropriate, the following prospective donor-recipient pairs for transplantation according to their *a priori* chances of being matched with respect to histocompatibility types: (1) husband (O) to wife (AB), (2) brother to sister, (3) sister to sister, (4) woman (B) to sister-in-law (A), (5) father to daughter, (6) mother to daughter, (7) Joan to Cousin Mary, (8) husband to wife (identical in ABO and Rh groups), and (9) sibs, one AB cde, the other CDE. Write the sequence in decreasing order of matching.

33. R and S are two antigens of a red-cell blood-group locus. Given the following pedigree, what are the most probable genotypes of all the individuals in the pedigree? Define any gene symbols used.

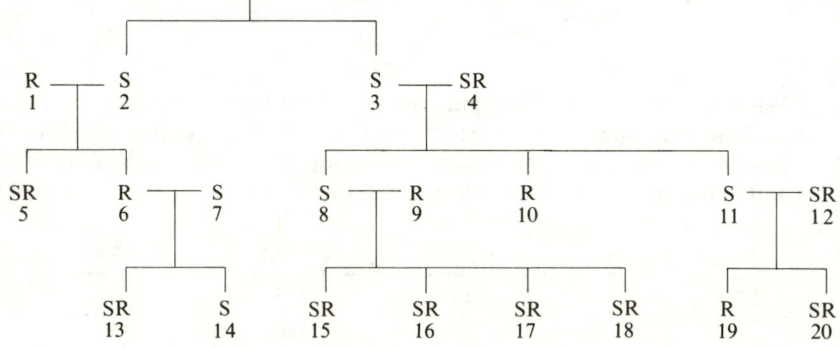

34. a. Name two genes that are presumed to have arisen by a nonhomologous crossover in man.
Indicate which of the following are true and which are false. In the ABO blood group system:
b. An AB father could have an O child.
c. An AB mother could have an O child.
d. An H (Bombay) child cannot be distinguished serologically from an O child.
e. All SeSe individuals are Lewis-positive.

35. Are the following statements true or false?
a. All immunoglobulin molecules in an individual have a common polypeptide chain.
b. All γG immunoglobulin molecules in an individual have the same H chain and differ from each other by the L chain type they contain.
c. The offspring from all matings of phenotypically Gm a+b+ individuals are all Gm a+b+.
d. A myeloma protein from an individual who received the gene for Gm a from his father and the gene for Gm b from his mother may have neither Gm a nor Gm b determinants (specificities).
e. In matings of Gm (a+ x+) by Gm (b+) individuals, all offspring are Gm (a+ x+) or Gm (b+) or Gm (a+ b+ x+). This indicates that Gm (a+) and Gm (x+) are here controlled by linked genes in repulsion.

36. a. Two sera X and Y were discovered that have the following reactions with a random sample of individuals (X+ means an individual reacts to serum X; X− means he does not).

	X+	X−
Y+	30	21
Y−	45	0

From this information what can you infer concerning: (1) Number of loci: one or two? (2) Number of alleles at each locus? (3) Frequency of each allele?
b. A third serum, Z, was subsequently discovered that reacts as follows:

	X+Y+	X+Y−	X−Y+	X−Y−
Z+	0	20	12	4
Z−	30	25	9	0

Set up a specific genetic hypothesis to explain all the data. Illustrate your hypothesis by showing the expected outcome of two kinds of matings. These should be carefully chosen so as to bring out the salient features of the system.

37. A serum that reacted with white cells from about 70 percent of the population was analyzed by absorption, using a fairly large number of positive cell donors. Only two types of cells were found:
a. 30 percent of the cell samples used for absorption left no detectable activity behind when after absorption the serum was tested on all cells positive to it originally.

b. The remaining 70 per cent left behind an antibody that reacted only with the 30 percent of cells described in a.
   (1) What is the likely minimum number of antibodies you would suggest is present in the serum?
   (2) How would you describe the antigenic constitution of the two types of positive cells?
   (3) Can you give a simple *genetic* interpretation for the pattern of results obtained with this serum?

38. The average Rh(−) frequency in the U.S. is 16 percent. Assuming random mating with respect to Rhesus blood type,
   a. What is the frequency of the *r* gene?
   b. How many genetic types of Rh(+) × Rh(−) matings are there?
   c. What are the frequencies of the various types of Rh(+) × Rh(−) matings?
   d. What is the expected proportion of Rh(+) offspring from a random sample of Rh(+) × Rh(−) matings?

39. The table shows the HLA white-cell antigen phenotypes of a family of two parents and six children. Data are given only for the HLA1, HLA2, HLA3, HLA9, 4a, and 4c antigens (+ indicates presence of the antigen, − absence).
   a. Identify the genotypes of parents and offspring with respect to the given HL-A antigens.
   b. Child number 1 is in need of a kidney transplant. The father, mother, and children 2, 3 and 4 are all ready to donate their kidneys. Assuming they are all equally suitable from a physical and psychological viewpoint, which would you choose as a potential donor, given the HL-A antigen data?

	HLA1	HLA2	HLA3	HLA9	4a	4c
Father	−	+	+	−	+	+
Mother	+	−	−	+	+	+
Children						
1	+	+	−	−	+	+
2	−	−	+	+	−	+
3	+	+	−	−	+	+
4	+	−	+	−	+	−
5	−	−	+	+	−	+
6	+	+	−	−	+	+

40. a. Given are the following frequencies of ABO types:

A	B	O	AB	Total
4500	1300	3600	600	10,000

   Assuming random mating, what are the ABO allele frequencies? What is the frequency of *AA* × *BO* matings?
   b. An O Rh− woman is married to a man who is an AB Rh+. They have had three children, all Rh+, and are planning to have a fourth child. What would you advise them concerning the likelihood of hemolytic disease of the newborn in this fourth child, knowing that no previous children were affected?

c. Hemolytic disease of the newborn in ABO incompatible matings cannot be the selective mechanism maintaining the ABO polymorphism because:
   (1) It is not sufficiently frequent.
   (2) It only affects certain types of matings.
   (3) The heterozygote advantage is not a factor in maintaining polymorphisms.
   (4) The heterozygote disadvantage gives an unstable equilibrium.
   (3) There is interaction with the Rh system.
   (6) The ABO types are more frequent in certain diseases.

41. a. Give a genetic interpretation of the following data, representing the successive stages in the development of a red-cell blood-group system.
   (1) Two anti-sera, X and Y gave the following reactions on 100 individuals, none of whom were X−Y−.

X+Y−	X+Y+	X−Y+
9	42	49

   (2) A third serum, Z, was found which reacted as follows with a group of 1000 people:

	X+Y−	X+Y+	X−Y+
Z+	88	365	187
Z−	3	55	302

   (3) A further serum, U, was found which was such that all U− individuals were Z+.

b. A new serum, T, gave rise to the following pedigree:

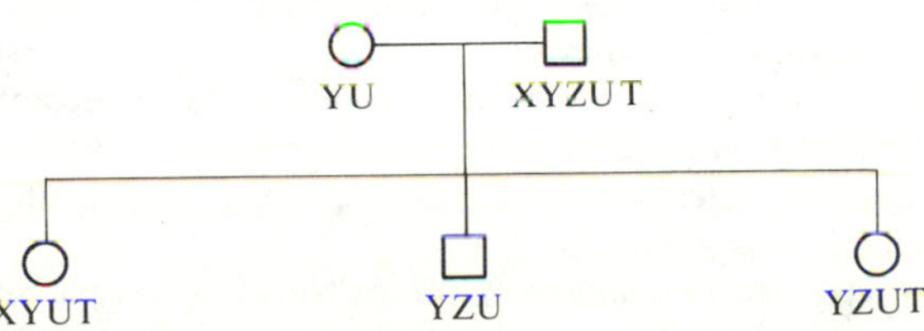

(Only positive reactions are indicated.) What can you say about the segregation of T in relation to the XYZU system?

42. a. Phenylketonuria is a recessive defect which, in the absence of appropriate treatment that has only recently become available, results in severe mental deficiency. It now occurs in the U.S. population with a frequency of 1/10,000. Assuming that phenylketonurics do not have any offspring calculate the genetic load resulting from this trait:
   (1) Under the assumption that normal homozygotes and carrier heterozygotes have equal fitness.
   (2) Under the assumption that carriers are slightly fitter than normal homozygotes.

b. Given a frequency of 0.005 for first cousin marriages in the general population what proportion of normal parents of phenylketonurics would you expect to be first cousins?

c. What on the basis of the two hypotheses in a, would be the ultimate effect on the frequency of phenylketonuria of increasing the fitness of phenylketonurics to 0.9 of the normal?

43. a. What is the probability that a normal man marrying at random will have an albino offspring if he has an albino sister? Given that the frequency of the albino gene is 1/140.
    b. How would this probability be altered if the man married a first cousin, given that none of his uncles and aunts were albinos?

44. Two people who are first cousins wish to marry. They are related through one of their grandmothers who was normal and had six normal children but whose brother was an albino. Assuming the frequency of the recessive gene determining albinism is 1/140, what is the increased risk (expressed as a ratio of probabilities) that a child of theirs will be albino over what it would be if they marry at random?

45. Galactosaemia is a rare recessive disease that is effectively lethal since, at least until recently, affected individuals rarely, if ever, live to have children. Assays for the affected enzyme can be used to detect heterozygotes and indicate a frequency of 1/200. Assuming the population has an inbreeding coefficient of 1 percent, what proportion of galactosaemics are the offspring of consanguineous unions? If the heterozygous carriers have a fitness of 0.99 relative to homozygous "normal" individuals, what proportion of the genetic load resulting from this locus is caused by the inbreeding?

46. A woman, who had a brother who died in infancy of phenylketonuria, marries her nephew. The nephew (who is an only child) and both his parents are clinically unaffected.
    a. What is the chance that their first child will have phenylketonuria?
    b. If you were told that the nephew had a brother who also suffered from phenylketonuria, how would this affect your answer to a?

47. A survey of marriages in a sample of rural communities in the U.S. gave the following frequencies for different types of consanguineous unions: first cousins 0.5 percent, first cousins once removed 1 percent, second cousins 3 percent, second cousins once removed 2 percent, and third cousins 1 per cent.
    a. Neglecting all other forms of consanguinity, what is the mean coefficient of inbreeding in this population?
    b. The frequency of first cousin marriages amongst the parents of offspring with a rare recessive trait found in the same population sample indicated a gene frequency of $q = 0.0025$. What would you estimate is the expected frequency of the trait in the population?
    c. Migration to the cities reduces the average level of inbreeding to one-fifth of its former value. How would this affect the frequency of the recessive trait among the offspring of a representative group of such migrants?

48. Consider two loci C and D, such that the relative fitnesses of CC and DD are 1 and the relative fitnesses of cc and dd are 0. If the relative fitness of Cc is 0.8 and that of Dd is 0.2, and if the mutation rate at both loci is 1 in 10,000, what is the relationship between the mutational loads at the two loci? Explain why this should be so.

49. Consider two loci A and B, such that the mutation rates for $A \rightarrow a$ and $B \rightarrow b$ are both equal to 1 in 10,000. The relative fitnesses of the various genotypes are $AA = 1$, $Aa = 0.5$, $aa = 0$, $BB = 1$, $Bb = 1$, $bb = 0.5$. What are the mutation loads for the two loci?

50. Assume 10,000 mutable loci and an increased mutation rate of 2.5×10^{-7} per locus per r (roentgen), and an average frequency of recessive alleles of 10^{-3}. What would be

the effect of an exposure of a population to an average dose of 25r on the frequency of homozygous recessive offspring in the generation after exposure?

51. What would be the increased mutation load if a population were exposed to a radiation dose which doubled the spontaneous mutation rate? Should you be able to calculate the total genetic load from this information?

52. Which has a greater effect on the genetic load of a panmictic population, a single acute exposure of:
    a. 100r to each of 1,000 children
    or
    b. 1r to each of 100,000 children?

53. A man who had an albino brother marries his own niece. If neither uncle nor niece are albinos, what is the chance that an offspring of theirs will be an albino?
    How would this probability be affected if you were told that the niece had an albino brother?

54. Cystic fibrosis of the pancreas has been shown to be an inherited disease. Analysis of data from pooled sibships from unaffected parents indicate that about 25 percent of offspring are affected. An exhaustive study of death certificates and hospital records in the state of Ohio suggests that the disease has an incidence of 1 per 37,000 births.
    A child in the pediatrics ward of the Stanford-Palo Alto Hospital has this disease. His parents, uncles, and aunts are unaffected. He is one of seven children. Two older brothers and a younger sister have already died of the disease, a younger sister has been diagnosed as affected and two younger sisters as unaffected.
    a. "A" is an unaffected sister of the proband. She is not related to her husband. What is the risk that A's first child will be affected?
    b. Until recently cystic fibrosis has been considered to be a lethal disease of childhood. Which of the following is the most likely explanation for the relatively high frequency of this deleterious gene:
        (1) There has been a mutation-selection balance involving only homozygous affected individuals. The heterozygote has the same fitness as the homozygous normal individual.
        (2) The heterozygote shows an advantage in fitness of approximately 5 percent over the normal homozygous individual.
        (3) The heterozygote shows an advantage in fitness of almost 2 percent over the the normal homozygous individual.
        (4) The mutational and segregational loads are so minimal with respect to this particular locus that the gene frequency can attain this magnitude.
    c. Although the gene determining cystic fibrosis occurs with a relatively high frequency in Caucasian populations, its frequency in Negro populations is essentially zero. Which one of the following statements most likely explains this difference:
        (1) The rate of mutation to the deleterious gene is higher in Caucasians than in Negroes.
        (2) There is a significantly higher proportion of consanguineous matings in Caucasian populations than in Negro populations.
        (3) With respect to this particular locus, the segregational load is higher in Caucasian populations than in Negro populations.
        (4) Special environmental conditions of Negro populations are such that individuals heterozygous for the deleterious gene are less fit than are normal homozygous individuals.
        (5) In Negro populations, affected individuals die *in utero*.

55. Assume that hemophilia and colorblindness are each determined by a single, sex-linked locus. The recombination frequency between the two loci is about 10 percent. If a hemophilic, color-blind man and a normal woman produce twin daughters who are hemophilic and color-blind, what is the probability that these twins are dizygotic?

56. Two unlinked loci, each with two alleles, *A* and *a*, *B* and *b*, affect diastolic arterial pressure in such a way that the genotype *aabb* has a pressure of 85, each *A* gene adds 2 units, and each *B* gene adds 4 units. The *A* and *B* gene frequencies are respectively 0.5 and 0.25. Assume random mating and that arterial pressures are corrected for age.
    a. Give the frequencies with which all possible pressures are found in the population.
    b. What is the mean pressure in the population?
    c. What is the mean pressure of the children of a mating between individuals with pressures 95 and 91, respectively?
    d. The next generation of individuals is formed only from the offspring of those members of the population whose arterial pressure is less than 92. Give the gene frequencies, phenotype frequencies, and mean arterial pressure for this next generation.

57. Genes *a*, *b*, and *c* are fully penetrant, nonlinked autosomal recessives, and gene *d* is a sex-linked recessive. Twin boys marry twin girls. None of the twins show any of the traits *a*, *b*, *c*, or *d*. Each marriage produces a son, one of which is *a*, *b*, *d*, and the other of which is *a*, *b*, *c*, *d*. From this data, what is the probability that the boy twins were dizygotic? The girls? Assume that all of these recessive genes are very rare and that there was no inbreeding in either of the families.

58. You are interning in the ObGyn clinic, and a patient tells you that twins run in her husband's family. She wants to know if she is more likely to have twins than is her sister (whose husband's family does not have any twins). What would you tell her?

59. You are confronted with two mothers who have three children between them. One of the mothers has one child, but feels that one of the children of the other mother is really an identical twin of her own child. What might be the easiest way to settle the question? How might you show that they are (or are not) dizygotic twins, if your first test shows that they are not monozygotic twins?

60. If you were interested in the conditions favoring the occurrence of monozygotic twins in humans would the *Weinberg Differential Method* be of any use to you? (Assume that a good friend works in the U.S. Bureau of Statistics.) Why would an accurate value for the secondary sex ratio be important?

61. Suppose that an egg undergoes abnormal meiotic cleavage, resulting in two identical functional eggs, and that each functional egg is fertilized by a separate sperm, resulting in twins. What would be the genetic nature of these twins, and how might this situation be verified?

62. Colorblindness is due to a sex-linked recessive gene. A color-blind but otherwise healthy father and a normal mother have a color-blind child with gonadal agenesis, no sex-chromatin bodies in interphase nuclei, and 45 chromosomes. How many X and Y chromosomes, respectively, do you expect in
    a. The father?
    b. The mother?
    c. The child?

    What are the likely chromosomal events that occurred in this family?
    Be sure to indicate when and in which individual these events took place.

63. An infant born of a normal pregnancy is found to have a negative buccal smear and an XO chromosome pattern.
   a. What is the most likely phenotypic characteristic of her external and internal sexual organs, and why?
   b. What would her phenotype most likely be at 16 years of age, assuming that no treatment has been given?
   c. Suppose the same infant instead of an XO karyotype has one normal X chromosome and a chromosome consisting only of the short arms of the X. What is the most likely phenotype at birth (sexual organs) and at puberty?

64. An enzyme GOOP-transvestase (GOTVase) is found in liver, red cells, and tissue cultures of skin explants. Deficiency of this enzyme is found to be controlled by a rare recessive gene linked to colorblindness and hemophilia.
   a. On which chromosome is the locus for GOTVase?
   b. If the normal activity of the enzyme is one unit/mg tissue (liver or skin) and the clinically affected individuals have no detectable activity, how much activity would you expect in:
      (1) A liver biopsy of the clinically normal mother of an affected boy?
      (2) A liver biopsy of the clinically normal mother of an affected girl?
      (3) Individual clones of cells from explanted skin of the clinically normal mother of an affected boy?
      (4) Individual clones of cells from explanted skin of the clinically normal father of an affected boy?

65. The following pedigree was obtained for a patient (III.4) with Klinefelter's syndrome, who was color-blind. The + and − phenotype refer to the Xg^a blood-group types. The type Xg(a+) is determined by a dominant X-linked gene.

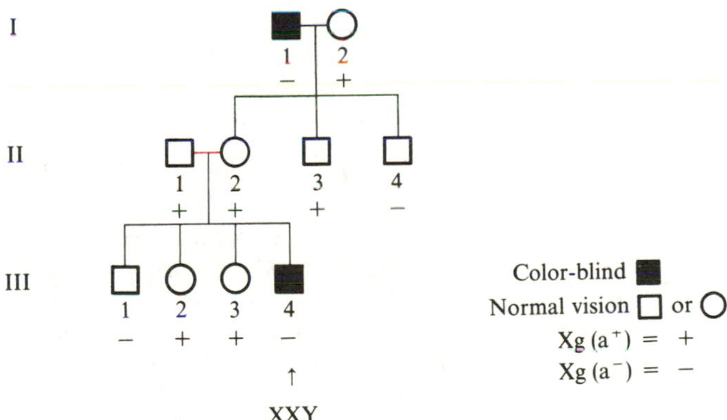

   a. What is the most probable genotypic constitution of the affected individual III.4?
   b. Assuming that the nondisjunctional event giving rise to the Klinefelter's syndrome occurred during meiosis, you realize that it could have occurred at the first or second meiotic division in either the mother or the father. Which of these is most likely, and why? Ignore the hypothesis of X-inactivation.

c. How might your answer to b be affected by assuming the hypothesis of mosaic X-inactivation applies to individuals with Klinefelter's syndrome?

d. Give the most probable genotype(s) of *all* the individuals in the pedigree. Can you say anything, from this information, concerning the linkage between colorblindness and Xg^a?

66. a. Indicate where possible, which parent must have produced a nondisjunctional gamete, given the following sex-chromosome types:

(1) XO (3) XXX

(2) XYY (4) XXXX

b. Given that the gene frequency for sex-linked recessive colorblindness is 10 percent, estimate the frequency of colorblindness among females with Turner's syndrome.

67. Twelve cases of a rare disease, each from a different family, were ascertained through a survey of hospital records. Further inquiries revealed that the twelve families included a total of 77 offspring of which a total of 27 had the disease. None of the parents had the disease, but 5 sets were first cousins. Knowing that the frequency of first cousin marriages in the population is 0.5 percent, what can you say

a. About the mode of inheritance, if any, of the disease?

b. The expected segregation frequency of the disease, if you conclude it is inherited?

c. The frequency of the gene (or genes) that determine it?

68. A geneticist in a small country whose government provides almost all the medical care for the population reviewed the records of the only three hospitals in the country and of all government clinics for cases of phenylketonuria. A review of the family history of each of the 125 sibships which he ascertained showed that none of the parents were affected and that a high proportion of the sibships issued from related parents. The pooled data from families were as follows:

Size of sibship	*Number of sibships*	*Number affected*	*Number unaffected*
2	70	80	60
3	37	48	63
4	18	26	46

Do the pooled data support the hypothesis that the disease is inherited as an autosomal recessive? If so why?

69. Gaucher's disease is a rare metabolic disorder characterized by the accumulation of cerebrosides in the reticuloendothelial system, presumably resulting from a defect in cerebrosidase activity. The familial nature of this disease has been repeatedly observed, many cases having been reported in siblings and twins. Cases have been noted among offspring of consanguineous matings. There is an equal sex distribution among affected persons. For 80 sibships containing one or more affected individuals collected from the medical literature, parents were not affected. Two generation involvement (that is, one parent and one or more offspring) has been reported in 6 additional cases found in the medical literature. Involvement in more than 2 generations of a kindred has never been observed.

a. What is the *most likely* mode of inheritance of Gaucher's disease?

b. The following data were taken from 80 sibships containing one or more affected individuals. For each sibship the parents were unaffected. Compute for these families the segregation frequency *without* a correction for ascertainment. Then compute the segregation frequency, assuming single selection. What is the mode of inheritance suggested by this value? Does this value differ significantly from 0.25?

Number in family	*Number of families*	*Number affected*	*Number unaffected*
2	26	36	16
3	16	29	19
4	9	17	19
5	9	16	29
6	5	8	22
7	2	3	11
8	3	5	19
9	5	12	33
10	1	1	9
11	2	7	15
12	1	2	10
13	1	2	11
	80	138	213

70. The following pedigree is an extract from a larger study on a family segregating a variant albumin in addition to the polymorphic "group-specific component" types Gc 1–1, Gc 2–1, and Gc 2–2. These three are electrophoretically distinguishable forms of a serum protein of unknown function.

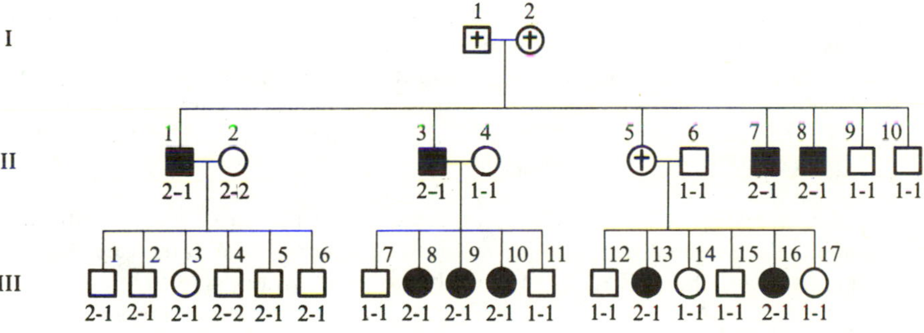

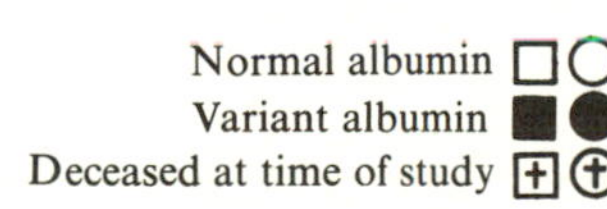

a. What are the likely modes of inheritance of the Gc types and the albumin variation?
b. What can you say, if anything, about the possibility of linkage between the genes controlling albumin and Gc?
c. In terms of your answers to a and b indicate, wherever possible, the most probable genotypes of all the individuals in the pedigree.

71. The following pedigree illustrating joint segregation of muscle dystrophy and the Pelger-Huët anomaly was obtained by Dr. L. Schneiderman in the Stanford clinics. The Pelger-Huët anomaly is recognized by the abnormal shape of the nuclei of the

affected individual's white blood cells and occurs with a frequency of about 1/5000. It is a benign anomaly with no obvious clinical consequences. Muscular dystrophy is a degenerative disease of the muscle with a late age of onset (around middle age), which often results in total crippling. The form of muscular dystrophy observed in this pedigree seems to be different from that reported in other familial studies.

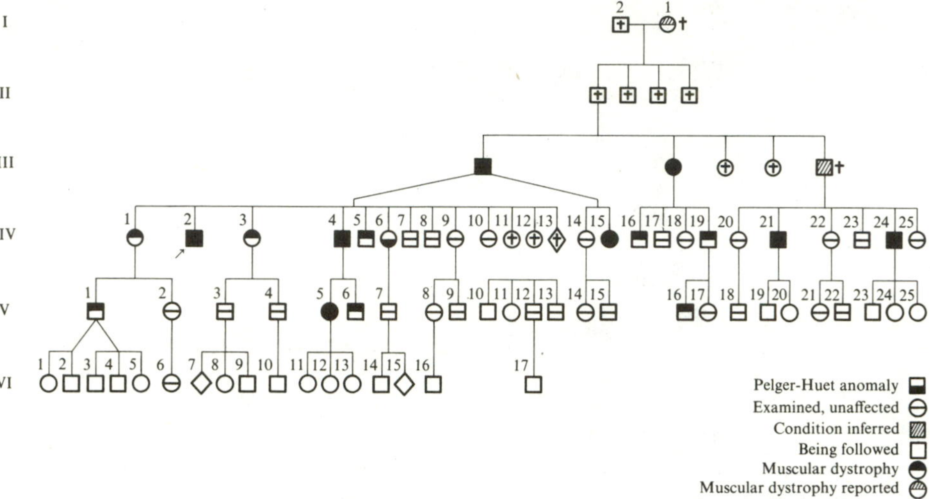

a. What is the most likely mode of inheritance of the two anomalies segregating in this pedigree?
b. In terms of your answer to a indicate, wherever possible, the most probable genotypes of the individuals in the pedigree.
c. Does the pedigree show any evidence for linkage between the genes determining these two anomalies, and, if so, what is the indicated recombination fraction?
d. Would demonstration of linkage in this way be of any use in clinical practice? If so, indicate how in *one sentence*.

72. In the following pedigree, an "X" indicates that a person is affected with the rare dominantly inherited defect elliptocytosis. The letters A and B indicate ABO blood type, and Rh(+) and Rh(−) indicate Rhesus type with respect to D.

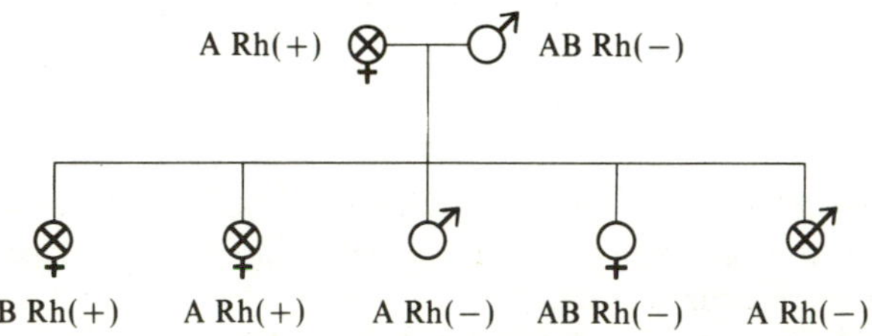

Given that Rh type and elliptocytosis are fairly closely linked, wherever possible give the most probable genotypes of all the individuals in the pedigree.

73. A 20-year-old mother has given birth to an infant with Down's syndrome. Her health is excellent, and her family history as well as that of her husband is negative for any congenital defect. This young couple wants to know if the risk for recurrence of the same problem is greater for them than that of the general population of the same maternal age.
   a. What is your answer?
   b. If your answer to a is yes or maybe, they now want to know how much greater is the risk. The baby's chromosomes are analyzed. He has 47 chromosomes in the analyzed cells with trisomy of chromosome 21. What is the recurrence risk for this couple? How would your answer be changed if the mother were 40 years old rather than 20?
   c. If the infant's cells had 46 chromosomes with a D/21 translocation and the mother had 45 chromosomes with the same translocation chromosome, what would the recurrence risk be now?
   d. Suppose the infant and the mother have a G/21 translocation, the mother with 45 and the infant with 46 chromosomes. What is the recurrence risk in this case?
   e. What would the recurrence risk be if the mother had Down's syndrome herself (a rare but not an impossible event) with trisomy 21?

74. What type of pedigree could establish whether the gene for the β chain of hemoglobin is or is not linked to the gene for the γ chain? State which type of mutant genes would be appropriate and give a plausible pedigree, indicating how a relevant family might be discovered.

75. Indicate which of the following statements is correct and which is incorrect. If incorrect, state why briefly.
   a. Mendelian inheritance is superior to blending inheritance, from an evolutionary point of view, because it does not conserve genetic variability.
   b. Inbreeding increases homozygosity.
   c. It is usually possible to detect linkage between two loci from population data.
   d. Ignoring mutation and selection for two alleles at a single locus in a random mating population, a Hardy-Weinberg equilibrium is always achieved after one generation of random mating.

76. Indicate which of the following statements is correct and which is incorrect. If incorrect state in *one* sentence why
   a. Linkage between two loci, alleles *A* and *a* and *B* and *b*, where *a* and *b* refer to recessive characters, can only be detected in those families which produce all the four observable phenotypes (*AB*, *ab*, *Ab*, *aB*).
   b. A new allele can increase in a population only if it has a selective advantage in the heterozygote.
   c. Unions between first cousins have a higher chance of producing genetically abnormal offspring than unions between unrelated individuals.
   d. In a random sample of families that produced offspring with a perfectly penetrating and fully viable recessive characteristic, both parents being normal, the overall frequency of such offspring will, in general, be greater than 1/4.

77. The following questions are based on case histories from the Stanford Hospital:
   a. The first three children born to a Caucasian couple are normal healthy girls. The fourth child was a boy who died at 10 months of age of a rare progressive, degenerative neurological disorder, Krabbe's disease. This disease is so rare that only about

30–40 cases have been reported in the world medical literature. Some of the published cases had affected siblings, but in none of the cases have parents been affected. At least one sibship, with affected and unaffected individuals, has issued from parents who were first cousins.

Review of the family trees of the parents of the Stanford case reveals no evidence of Krabbe's disease in previous generations nor evidence of parental consanguinity. Which of the following do you expect to be true:

(1) In this family the disease is not inherited because the parents are unrelated.

(2) In this family the disease can be inherited or the patient was a sporadic case. Thus the risk to the parents ranges from a negligible one to 25 percent for each future pregnancy.

(3) The risk to these parents is that one out of four children will be affected. Since they have had four children and one has been affected, their next child will be normal.

(4) In this family the disease is inherited as a sex-linked recessive because the mother is unaffected, her son is affected, and her daughters are unaffected.

b. Achondroplasia is a rare disease so affecting long bone development that an affected individual will be dwarfed, showing short extremities and relatively large head and trunk. The disease has been shown to be inherited as an autosomal dominant. Recently an unaffected child was born to achondroplastic parents (both affected) at the Stanford Hospital. This was the mother's fifth pregnancy. Her first two pregnancies terminated in miscarriages. Each of the next two pregnancies resulted in an affected child, who died soon after birth. Indicate whether the following statements are true or false.

(1) The facts prove that the unaffected child is a sporadic case.

(2) The parents are affected because they carry deleterious genes at different loci; otherwise they would have only affected children.

(3) Inheritance of achondroplasia because of a recessive gene is ruled out in this family.

(4) The risk to these parents of having an affected child at each pregnancy is 0.5.

(5) Both parents are heterozygous for the deleterious gene.

(6) At least one maternal grandparent and one paternal grandparent of the unaffected child must be affected.

78. For each of the following questions indicate which answer(s) you consider to be appropriate.

a. An increased incidence of consanguinity among the parents of patients with an apparently genetically determined trait

(1) Proves that its inheritance is not mediated by a dominant gene.

(2) Is suggestive evidence that the trait is determined by a recessive gene.

(3) Indicates the importance of inbreeding as a factor maintaining the relevant gene in the population.

(4) Suggests that the inheritance of the trait may be due to a dominant gene with low penetrance.

b. The estimation of mutation rates for recessive genes is subject to considerable uncertainty because

(1) Recessive traits generally exhibit variable expression and low penetrance.

(2) The incidence of inbreeding complicates the estimation of recessive gene frequencies.

(3) The fitness of heterozygote carriers of the gene is hard to determine.

(4) Recessive traits generally have low fitness values.

(5) Mutation rates for recessive genes are generally higher than those of dominant genes.

c. The correlation between sibs with respect to a quantitative character, which is at least in part genetically determined, is expected to be higher than that between parent and offspring because

(1) Sibs tend to be reared in an environment that is more favorable for the expression of quantitative characters.

(2) Most genes affecting such traits have approximately additive effects.

(3) The environment dominates the expression of quantitative characters.

(4) Sibs may be alike with respect to different dominant genes inherited from both parents.

(5) Interactions between environment and genotype can be ignored when comparing parents with their offspring.

79. a. Two inbred lines of corn (whose variances in height are the same) are crossed, and the variance in height of the resulting F_1 is measured. Which of the following do you expect to be true?

(1) The F_1 variance is equal to the sum of the two parental (inbred line) variances.

(2) The F_1 variance is the same as the variance of the original outbred population.

(3) The F_1 variance is the same (or less than) the parental variance.

(4) The F_1 variance is zero.

(5) The F_1 variance is equal to the square root of the sum of the parental variances.

b. Human twin studies are important because:

(1) They provide a unique opportunity for studying single gene differences.

(2) They allow a clear-cut estimation of the genetic component in behavioral characteristics, but not for difference in susceptibility to infectious diseases.

(3) They provide the best available control over genetic and environmental variables with a view to obtaining some idea of the genetic components of various characteristics.

(4) They allow an estimation of the genetic load.

c. If one assumes that differences in handedness between individuals are entirely due to an accidental developmental circumstance, selective breeding of left-handed individuals would:

(1) Increase rapidly the frequency of left-handed people in the population.

(2) Balance the frequency of left- and right-handed people in the population.

(3) Have no effect on the frequency of left-handed people.

(4) Decrease the frequency of left-handed people.

80. a. What is the inbreeding coefficient for offspring from a marriage between second cousins once removed?

b. Blood-group antigens Jk(a+) and Jk(b+) are determined by a pair of codominant alleles Jk^a and Jk^b. In a sample of 120 people the numbers which were Jk(a + b−), Jk(a+b+) and Jk(a−b+) were 20, 56, and 44, respectively. Estimate the frequency of the Jk^b allele.

c. The observed correlations for fingerprint ridge counts on a group of parent-offspring, mother-father, and sib-sib pairs were 0.49, 0.05, and 0.48, respectively. Assuming the simplest possible model of quantitative inheritance is valid for this case, which of the following intervals would you judge most likely to include the estimated heritability for ridge counts: 0.2–0.4, 0.4–0.6, 0.6–0.8, or 0.8–1?

d. The identity tags of two babies in a maternity ward were lost. The ABO and MN blood types of the babies were ON and AM. The only two women who could have been their mothers had blood types ABM and AMN. Which baby belongs to which mother?

81. What is the probability that at least two of thirty people chosen at random have birthdays on the same day? First calculate the probability that there are no birthdays on the same day.

Literature Cited

Adams, M. S., and J. V. Neel

1967. "Children of incest." *Pediatrics* **40**(1): 55–62.

Aird, I., H. H. Bentall and J. A. Fraser Roberts

1953. "A relationship between cancer of stomach and the ABO blood groups." *Brit. Med. J.* **1**: 799–801.

Allan, T. M.

1959. "ABO blood groups and sex ratio at birth." *Brit. Med. J.* **1**: 553–554.

Allen, G.

1957. "The mathematical relationships among plural births." *Amer. J. Hum. Genet.* **9**: 181–190.

Allison, A. C.

1954a. "The distribution of the sickle-cell trait in East Africa and elsewhere, and its apparent relationship to the incidence of subtertian malaria." *Trans. Roy. Soc. Tropi. Med. Hyg.* **48**: 312–318.

1954b. "Protection afforded by sickle-cell trait against subtertian malarial infection." *Brit. Med. J.* **1**: 290–294. (Reprinted *in* S. H. Boyer, ed., *Papers on Human Genetics*. Englewood Cliffs, N.J: Prentice-Hall, 1963.)

1961. "Genetic factors in resistance to malaria" *Ann. N.Y. Acad. Sci.* **91** (3): 710–729.

1964. "Polymorphism and natural selection in human populations." *Cold Spring Harbor Symp. Quant. Biol.* **29**: 137–149.

Arner, G. B. L.

1908. "Consanguineous marriage in the American population." *Columbia Univ. Studies Hist. Econ. Public Law* **31**(3): 1–99.

Azevedo, E., H. Krieger, M. P. Mi, and N. E. Morton
1965. "PTC taste sensitivity and endemic goiter in Brazil." *Amer. J. Hum. Genet.* **17**(1): 87–90.
Azevedo, E. H. Krieger, and N. E. Morton
1964. "Smallpox and the ABO blood groups in Brazil." *Amer. J. Hum. Genet.* **16**(4): 451–454.
Azevedo, E., N. E. Morton, C. Miki, and S. Yee,
1969. "Distance and kinship in Northeastern Brazil." *Amer. J. Hum. Genet.* **21**(1): 1–22.
Bach, F. H., and D. B. Amos
1967. "Hu-1—major histocompatibility locus in man." *Science* **156:** 1506–1508.
Bailey, N. T. J.
1959. *Statistical Methods in Biology.* New York: John Wiley and Sons.
1961. *Mathematical Theory of Genetic Linkage.* Oxford: The Clarendon Press.
Bailit, H. L., S. T. Damon, and A. Amon
1966. "Consanguinity on Tristan da Cunha in 1938." *Eugen. Quart.* **13**(1): 30–33.
Bajema, C. J.
1963. "Estimation of the direction and intensity of natural selection in relation to human intelligence by means of the intrinsic rate of natural increase." *Eugen. Quart.* **10**(4): 175–187.
Balakrishnan, W., and L. D. Sanghvi
1968. "Distance between populations on the basis of attribute data." *Biometrics* **23**(4): 859.
Balner, H., F. J. Cleton, and J. G. Ernisse, eds.
1965. *Histocompatibility Testing, 1965.* Copenhagen: Munksgaard.
Balner, H., A. Van Leeuwen, H. Dersjant, H. and J. J. Van Rood
1967. "Chimpanzee iso-antisera in relation to human leukocyte antigens." *In* E. S. Curtoni, P. L. Mattiuz, and R. M. Tosi, eds. *Histocompatibility Testing, 1967.* Copenhagen: Munksgaard.
Barker, J. S. F.
1966. "The effect of partial exclusion of certain matings and restriction of their average family size on the genetic composition of a population." *Ann. Hum. Genet.* **30:** 7–11.
Barnicot, N. A., C. J. Jolly, and P. T. Wade
1967. "Protein variations and primatology." *Amer. J. Phys. Anthropol.* **27**(3): 343–355.
Barr, M. L. and E. G. Bertram
1949. "A morphological distinction between neurones of the male and female, and the behaviour of the nucleolar satellite during accelerated nucleoprotein synthesis." *Nature* **163:** 676–677.
Barrai, I. and H. M. Cann
1965. "Segregation analysis of juvenile diabetes mellitus." *J. Med. Genet.* **2,** 8–11.
Barrai, I., H. M. Cann, L. L. Cavalli-Sforza, and P. de Nicola
1968. "The effect of parental age on rates of mutation for hemophilia and evidence for differing mutation rates for hemophilia A and B." *Amer. J. Hum. Genet.* **20:** 175–196.
Barrai, I., L. L. Cavalli-Sforza, and M. Mainardi
1964. "Testing a model of dominant inheritance for metric traits in man." *Heredity* **19**(4): 651–668.
Barrai, I., L. L. Cavalli-Sforza, and A. Moroni
1962. "Frequencies of pedigrees of consanguineous marriages and mating structure of the population." *Ann. Hum. Genet.* **25:** 347–376.
1965. "Record linkage from parish books." *In* Medical Research Council, *Mathematics and Computer Science in Biology and Medicine.* London: H.M.S.O.
Barrai, I., M. P. Mi, N. E. Morton, and N. Yasuda
1965. "Estimation of prevalence under incomplete selection." *Amer. J. Hum. Genet.* **17**(3): 221.
Bateman, A. J.
1960. "Blood-group distribution to be expected in persons trisomic for the ABO gene." *Lancet* **1:** 1293–1294.
Beckman, L., and R. Elston
1962. "Assortive mating and fertility." *Acta Genet. Statist. Med.* **12:** 117–122.

Bemiss, S. M.
1858. "Report on influence of marriages of consanguinity upon offspring." *Trans. Amer. Med. Ass.* **11**: 319–425.

Bennett, J. H.
1958. "The existence and stability of selectively balanced polymorphism at a sex-linked locus." *Australian J. Biol. Sci.* **11**: 598–602.

Benzer, S.
1961. "Genetic Fine Structure." *In Harvey Lectures* 56, New York: Academic Press.

Berg, K., and A. G. Bearn
1968. "Human serum protein polymorphisms." *Ann. Rev. Genet.* **2**: 341–362.

Bernstein, F.
1924. "Ergebnisse einer biostatischen zusammenfassenden Betrachtung uber die Erblichen Blutstrukturen des Menschen." *Klini. Stochensch.* **3**: 1496.
1925. "Zusammenfassende Betrachtungen über die erblichen Blutstrukturen des Menschen." *Z. Ind. Abst. Vererb.-Lehre* **37**: 237-270.
1930. "Fortgesetzte Untersuchungen aus der Theorie der Blutgruppen." *Z. Ind. Abst. Vereb.-Lehre* **56**: 233–73.
1931a. "Die geographische Verteilung der Blutgruppen und ihre anthropologische Bedeutung." In *Comitato Italiano per lo Studio dei Problemi della Popülazione.* Rome: Istituto Poligrafico dello Stato.
1931b. "Zur Grundlegung der Chromosomentheorie der Vererbung beim Menschen mit besondere Berücksichtung der Blutgruppen." *Z. Ind. Abst. Vereb.-Lehre* **57**: 113–138.

Beutler, E., R. J. Dern, and C. L. Flanagan
1955. "Effect of sickle-cell trait on resistance to malaria." *Brit. Med. J.* **1**: 1189–1191.

Blank, C. E.
1960. "Alpert's syndrome (a type of acrocephalosyndactyly), Observations on a British series of 39 cases." *Ann. Hum. Genet.* **24**: 151.

Bodmer, J. G., and W. F. Bodmer
1970. "Studies on African Pygmies IV: a comparative study of the HL-A polymorphism in the Babinga Pygmies and other African and Caucasian populations." *Amer. J. Hum. Genet.* **22**: 396–411.

Bodmer, W. F.
1960. "Discrete stochastic processes in population genetics." *Roy. Statist. Soc. B* **22**: 218–236.
1965. "Differential fertility in population genetics models." *Genetics* **51**: 411–424.
1967. "Models for DNA mediated bacterial transformation." *In* Lucien Le Cam and Jerzey Neyman, eds. *Proceedings of the Fifth Berkeley Symposium on Mathematical Statistics and Probability.* Berkeley: University of California Press.
1968. "Demographic approaches to the measurement of differential selection in human populations." *Proc. Nat. Acad. Sci.* **59**(3): 690–699.
1970. "The evolutionary significance of recombination in prokaryotes." *In* H. P. Charles and B. C. J. G. Knight, eds. *Twentieth symposium Soc. Gen. Microbiol.* Cambridge: Cambridge University Press.

Bodmer, W. F., J. Bodmer, S. Adler, R. Payne, and J. Bialek
1966. "Genetics of '4' and 'LA' human leukocyte groups." *Ann N. Y. Acad. Sci.* **129**: 473–489

Bodmer, W. F., J. Bodmer, D. Ihde, and S. Adler
1969. "Genetic and serological association analysis of the HL-A leukocyte system." *In* N. E. Morton, ed. *Computer Applications in Genetics.* Honolulu: University of Hawaii Press.

Bodmer, W. F., and L. L. Cavalli-Sforza
1966. "Perspectives in genetic demography." *Proc. 2nd World Population Conf.* (Belgrade, Yugoslavia) p. 455–459.
1968. "A migration matrix model for the study of random genetic drift." *Genetics* **59**: 565–592.

Bodmer, W. F., and A. W. F. Edwards
1960. "Natural selection and the sex ratio." *Ann. Hum. Genet.* **24**: 289–244.

Bodmer, W. F., and J. Felsenstein

1967. "Linkage and selection: theoretical analysis of the deterministic two locus random mating model." *Genetics* **57**(2): 237–265.

Bodmer, W. F., and A. Jacquard

1968. "La Variance de la dimension des familles." *Population* **23**: 870–878.

Bodmer, W. F., and J. Lederberg

1967. "Census data for studies of genetic demography." *Proc. third Int. Congr. Hum. Genet.* Baltimore: The Johns Hopkins Press.

Bodmer, W. F., and P. A. Parsons

1960. "The initial progress of new genes with various genetic systems." *Heredity* **15**: 283–299.

1962. "Linkage and recombination in evolution." *Advance. Genet.* **11**: 1–100.

Bodmer, W. F. and R. Payne

1965. "Theoretical consideration of leukocyte grouping using multispecific sera." *In* H. Balner, F. J. Cleton, and J. G. Eernisse, eds. *Histocompatibility Testing, 1965*. Copenhagen: Munksgaard.

Boettcher, B.

1964. "The Rh 'deletion' phenotypes and the information they provide about the Rh genes." *Vox Sang.* **9**: 641–652.

1965. "The Rh antigens of anthropoid apes in relation to Rh 'deletion' phenotypes." *Amer. J. Hum. Genet.* **17** (4): 308–310.

1966. "Modification of Bernstein's multiple allele theory for the inheritance of the ABO blood groups in the light of modern genetical concepts." *Vox Sang.* **11**: 129–136.

Bonné, B.

1963. "The Samaritans: a demographic study." *Hum. Biol.* **35**: 61–89.

Böök, J. A.

1956. Genetical investigations in a north Swedish population." *Ann. Hum. Genet.* **20**(3): 239–250.

1957. "Genetical investigations in a north Swedish population: the offspring of first-cousin marriages." *Ann. Hum. Genet.* **21**: 191–221.

Borberg, A.

1951. *Clinical and Genetic Investigations into Tuberous Sclerosis and Recklinghausen's Neurofibromatosis*. Copenhagen: Munksgaard.

Bourgeois-Pichat, J.

1951. "Evolution de la population française depuis le XVIIIe siècle." *Population* **4**: 635–662 An English translation of this paper has been published *in* D. V. Glass and D. E. C. Eversley, eds. *Population in History*. Chicago: Aldine, 1965.

Boyce, A. J., C. F. Kuchemann, and G. A. Harrison

1968. "The reconstruction of historical movement patterns." *In* E. D. Acheson, ed. *Proceedings of Oxford Record Linkage Symposium*. London: Oxford University Press.

Boyd, M. F., ed.

1949. *Malariology*. Philadelphia: Saunders.

Boyd, W. C.

1952. *Genetics and the Races of Man*. Boston: Little, Brown.

Boyd, W. C., and L. G. Boyd

1937. "Blood grouping tests on 300 mummies." *J. Immunol.* **32**(4): 307–319.

Boyer, S. H.

1963. *Papers on Human Genetics*. Englewood Cliffs, New Jersey: Prentice-Hall.

Braidwood, R. J., and C. Reed

1957. "The achievement and early consequences of food-production: a consideration of the archeological and natural-historical evidence." *Cold Spring Harbor Symp. Quant. Biol.* **22**: 19–31.

Brass, W.

1958. "Models of birth distributions in human populations." *Bull. Inst. Int. Statist.* **36**: 165–178.

British Medical Bulletin,

1969. Vol. 25(1). London: Medical Department, the British Council.

Britten, R. J., and E. H. Davidson

1969. "Gene regulation for higher cells: a theory." *Science* **165:** 349.

Britten, R. J., and D. E. Kohne

1968. "Repeated sequences in DNA." *Science* **161:** 529–540.

Brues, A. M.

1954. "Selection and polymorphism in the ABO Blood groups." *Amer. J. Phys. Anthropol.* **12:** 559–597.

1963. "Stochastic tests of selection in the ABO blood groups." *Amer. J. Phys. Anthropol.* **21:** 287–299.

Bulmer, M. G.

1958. "The effect of parental age, parity and duration of marriage on the twinning rate. "*Ann. Hum. Genet.* **23:** 454–458.

1960. "The familial incidence of twinning." *Ann. Hum. Genet.* **24:** 1–3.

Burdette, W. J., ed.

1962. *Methodology in Human Genetics.* San Francisco: Holden-Day.

Burks, B. S.

1928. "The relative influence of nature and nurture upon mental development; a comparative study of foster parent-foster child resemblance and true parent-true child resemblance." *27th Yearbook Nat. Soc. for Study of Ed.* **1:** 219–316.

Burnet, F. M.

1962. *The Integrity of the Body.* Cambridge: Harvard University Press.

Burt, C.

1961. "Intelligence and social mobility." *Brit. J. Statist. Psychol.* **14:** 3–24.

1966. "The genetic determination of differences in intelligence: a study of monozygotic twins reared together and apart." *Brit. J. Psychol.* **57** (1 and 2): 137–153.

Burt, E., and M. Howard

1956. "The multifactorial theory of inheritance and its application to intelligence." *Brit. J. Statist. Psychol.* **8** (2): 95–131.

Cann, H. M. and L. L. Cavalli-Sforza

1968. "Effects of grandparental and parental age, birth order, and geographic variation on the sex ratio of live-born and stillborn infants." *Amer. J. Hum. Genet.* **20** (4): 381–391.

Cantrelle, P., and M. Dupire

1964. "L'endogamie des Peul's du Fouta-Djallon." *Population* **19**(3): 529–558.

Carr, D. H.

1965. "Chromosome studies in spontaneous abortions." *Obstet Gynecol.* **26**(3): 308–326.

1967. "Chromosome anomalies as a cause of spontaneous abortion." *Amer. J. Obstet. Gynecol.* **97**(3): 283–293.

Carter, C. O.

1965. "The inheritance of common congenital malformations." *In* A. G. Steinberg and A. G. Bearn, eds. *Progress in Medical Genetics*, Vol. 4. New York: Grune and Stratton.

Cavalli-Sforza, L. L.

1958. "Some data on the genetic structure of human populations." *Proc. Tenth Int. Congr. Genet.* **1:** 389–407. Toronto: University of Toronto Press.

1960. "Demographic attacks on genetic problems." In *The Use of Vital and Health Statistics for Genetic and Radiation Studies* (Proceedings of the seminar sponsored by the United Nations and the World Health Organization). New York.: United Nations

1962a. "The distribution of migration distances: models and applications to genetics." In *Human Displacements* (Entriens de Monaco en sciences humaines, Première session 1962).

1962b. *Indagine Speciale Su Alcune Caratteristiche Genetiche Della Popolazione Italiana.* Note e Relazioni, No. 17. Rome: Istituto Centrale di Statistica.

1962c. "Demographic attacks on genetic problems. Some possibilities and results." In *The Use of Vital and Health Statistics for Genetic and Radiation Studies* (Proceedings of the seminar sponsored by the United Nations and the World Health Organization). New York: United Nations.

1963. "Genetic drift for blood groups." *In* E. Goldschmidt, ed. *Genetics of Migrants and*

Isolated Populations. Baltimore: Williams and Wilkins.

1966. "Population structure and human evolution." *Proc Roy. Soc. B* **164:** 362–379.

1967. "Human populations." *In* A. Brink, ed. *Heritage from Mendel*. Madison: The University of Wisconsin Press.

1968. "Studi sulla struttura genetica di una popolazione italiana." *Le Scienze* **1:** 7–19.

1969a. "Human diversity." *Proc. Twelfth Int. Congr. Genet.* **3:** 405–416. Tokyo: Science Council of Japan.

1969b. "Genetic drift in an Italian population." *Sci. Amer.* **221**(2): 30–37.

Cavalli-Sforza, L. L., I. Barrai, and A. W. F. Edwards

1964. "Analysis of human evolution under random genetic drift." *Cold Spring Harbor Symp. Quant. Biol.* **29:** 29.

Cavalli-Sforza, L. L., and F. Conterio

1960. "Analisi della fluttuazione di frequenze geniche nella popolazione della val Parma." *Atti Ass. Genet. Ital.* **5:** 335–344.

Cavalli-Sforza, L. L., and A. W. F. Edwards

1963. "Analysis of human evolution." *In* S. J. Geerts, ed. *Genetics Today*, Vol. 3 (Proceedings of the Eleventh International Congress of Genetics. The Hague. September 1963). New York: Pergamon Press, 1965.

1967. "Phylogenetic analysis: models and estimation procedures." *Amer. J. Hum. Genet.* **19**(3): 233.

Cavalli-Sforza, L. L., M. Kimura and I. Barrai

1966. "The probability of consanguineous marriages." *Genetics* **54:** 37–60.

Cavalli-Sforza, L. L., and G. Zei

1967. "Experiments with an artificial population." *Proc. Third Int. Congr. Hum. Genet.* 473–478. Baltimore: The Johns Hopkins Press.

Cavalli-Sforza, L. L., L. A. Zonta, F. Nuzzo, L. Bernini, W. W. W. De Jong, P. Meera Kyan, A. K. Ray, L. N. Went, M. Siniscalco, L. E. Nijenhius, E. van Loghem, and G. Modiano

1969. "Studies on African Pygmies. I. A pilot investigation of Babinga Pygmies in the Central African Republic (with an analysis of genetic distances)." *Amer. J. Hum. Genet.* **21:** 252–274.

Ceppellini, R.

1952. *In* R. Ceppellini, S. Nasso, and F. Tecilazich, eds. *La Malattia Emolitica del Neonato*. Milano.

1959. "Physiological genetics of human factors." *In* G. E. W. Wolstenholme and C. M. O'Connor, eds. *Ciba Foundation Symposium on Biochemistry of Human Genetics*. London: Churchill.

1966. "Genetica delle immunoglobuline." (XII Annual Meeting of Associazione Genetica Italiana. Parma.) *Atti Ass. Genet. Ital.* **12:** 3.

1968. "The genetic basis of transplantation." *In* F. T. Rapaport and J. Dausset, eds. *Human Transplantation*. New York: Grune and Stratton.

Ceppellini, R., A. Arata, R. Scarzella, G. Ceccarelli, A. Zanalda, and M. Zaccala

1963. "Aspetti genetici delle malattie psichiatriche dell'età senile.", *Il Lavoro Neuropsichiatrico*, **32:** 1–21.

Ceppellini, R., E. S. Curtoni, P. L. Mattiuz, G. Leigheb, M. Visetti, and A. Colombi

1966. "Survival of test skin grafts in man: Effect of genetic relationship and of blood groups incompatibility.", *Ann. N. Y. Acad. Sci.* **129:** 421–445. (Seventh International Transplantation Conference. F. T. Rapaport, ed.).

Ceppellini, R., E. S. Curtoni, P. L. Mattiuz, V. Miggiano, G. Scudeller, and A. Serra

1967. "Genetics of leukocyte antigens. A family study of segregation and linkage." *In* E. S. Curtoni, P. L. Mattiuz, and R. M. Tosi, eds. *Histocompatibility Testing, 1967*. Copenhagen: Munksgaard.

Ceppellini, R., P. L. Mattiuz, and E. S. Curtoni

1965. "Characteristics and evolution of leukoagglutinins developed in a patient receiving multiple transfusions from selected donors." *In*. P. S. Russell, H. J. Winn, and D. B.

Amos, eds. *Histocompatibility Testing* (Nat. Acad. Sci. Pub. 1229: 71).

Ceppellini, R., M. Siniscalco, and C. A. B. Smith
1955. "The estimation of gene frequencies in a random-mating population." *Ann. Hum. Genet.* **20:** 97–115.

Chiarelli, B.
1963. "Sensitivity to P.T.C. (phenyl-thio-carbamide) in primates." *Folia Primatologica*, **1:** 103–107.

Childs, B. and V. Der Kaloustian
1968. "Genetic heterogeneity." *New Engl. J. Med.* **279:** 1205–1212.

Chu, E. H. Y., N. H. Giles, and K. Passano
1961. "Types and frequencies of human chromosome aberrations induced by X-rays." *Proc. Nat. Acad. Sci.* **47:** 830–839.

Chung, C. S. and N. E. Morton
1961. "Selection at the ABO locus." *Amer. J. Hum. Genet.* **13:** 9–27.

Cisternas, J. P. and A. Moroni
1967. "Estudio sobre la consanguinidad en España." *Biologica* **40:** 3–20.

Clark, W. E. LeGros
1964. *The Fossil Evidence for Human Evolution.* Chicago: University of Chicago Press.

Clarke, C. A.
1961. "Blood groups and disease." *In* A. G. Steinberg, ed. *Progress in Medical Genetics*, Vol. 1. New York: Grune and Stratton.

Coale, A. J.
1957. "How the age distribution of a human population is determined." *Cold Spring Harbor Symp. Quant. Biol.* **22:** 83–89.

Cohen, B. H., and B. Glass
1956. "The ABO blood groups and the sex ratio." *Hum. Biol.* **28**(1): 20–42.
1959. "Further observations on the ABO blood groups and the sex ratio." *Amer. J. Hum. Genet.* **11**(3): 274–278.

Cohen, B., and J. E. Sayre
1968. "Further observations on the relationship of maternal ABO and Rh types to fetal death." *Amer. J. Hum. Genet.* **20**(4): 310–360.

Colombo, B.
1957. "On the sex ratio in man." *Cold Spring Harbor Symp. Quant. Biol.* **22:** 193–202.

Conterio, F., and I. Barrai
1966. "Effetti della consanguineità sulla mortalità e sulla morbilità nella popolazione della diocesi di Parma." *Atti Ass. Genet. Ital.* **11:** 378–391.

Conterio, F., and L. L. Cavalli-Sforza
1957. "Evolution of the human constitutional phenotype: an analysis of mortality effects." *Convegno Genet.* 3–14.
1960. "Selezione per caratteri quantitativi nell'uomo." *Atti Ass. Genet. Ital.* **5:** 295–304.

Conterio, F., and B. Chiarelli
1962. "Study of the inheritance of some daily life habits." *Heredity* **17:** 347–359.

Cook, P. J. L., J. E. Gray, R. A. Brack, E. B. Robson, and R. M. Howlett
1969. "Data on the haptoglobin and the D group Chromosomes." *Ann. Hum. Genet.* **33:** 125–138.

Coon, C. S.
1962. *The Origin of Races.* New York: Alfred A. Knopf.

Cooper, A. J., B. S. Blumberg, P. L. Workman, and J. R. McDonough
1963. "Biochemical polymorphic traits in a U.S. White and Negro population." *Amer. J. Hum. Genet.* **15:** 420–428.

Cooper, R., and J. Zubek
1958. "Effects of enriched and restricted early environments on the learning ability of bright and dull rats." *Canad. J. Psychol.* **12:** 159–164.

Court Brown, W. M.
1967. *Human Population Cytogenetics.* New York: John Wiley and Sons.
1968. "Males with an XYY sex chromosome complement." *J. Med. Genet.* **5**(4): 341.
Crew, F. A. E.
1927. *The Genetics of Sexuality in Animals.* Cambridge: Cambridge University Press.
1933. *Sex-Determination.* London: Methuen.
Crick, Francis
1966. *Of Molecules and Men.* Seattle: University of Washington Press.
Crow, J. F.
1950. *Genetics Notes.* Minneapolis: Burgess.
1954. "Breeding structure of populations. II. Effective population number." *In* O. Kempthorne, T. A. Bancroft, J. W. Gowen, and J. L. Lush, eds. *Statistics and Mathematics in Biology.* Ames: Iowa State College Press.
1958. "Some possibilities for measuring selection intensities in man." *Hum. Biol.* **30:** 1–13.
1961a. "Population genetics." *Amer. J. Hum. Genet.* **13**(1): 137–150.
1961b. "Mutation in man." *In* A. G. Steinberg, ed. *Progress in Medical Genetics,* Vol. 1. New York: Grune and Stratton.
1965. "Problems of ascertainment in the analysis of family data." *In* J. V. Neel, M. W. Shaw, and W. J. Schull, eds. *Epidemiology and Genetics of Chronic Diseases* (Public Health Service Publication No. 1163). Washington, D.C.: U.S. Dept. of Health, Education, and Welfare.
Crow, J. F., and J. Felsenstein
1968. "The effect of assortative mating on the genetic composition of a population." *Eugen. Quart.* **15**(2): 85–97.
Crow, J. F., and M. Kimura
1956. "Some genetic problems in natural populations." *In* Jerzy Neyman, ed. *Proceedings of the Third Berkeley Symposium on Mathematical Statistics and Probability.* (Vol. 4). Berkeley: University of California Press.
1965. "Evolution in sexual and asexual populations." *Amer. Natur.* **99:** 439–450.
1969. "Evolution in sexual and asexual populations: a reply." *Amer. Natur.* **103:** 89.
1970. *An Introduction to Population Genetics Theory.* New York: Harper and Row.
Crow, J. F., and A. P. Mange
1965. "Measurement of inbreeding from the frequency of marriages between persons of the same surname." *Eugen. Quart.* **12**(4): 199–203.
Curtoni, E. S., P. L. Mattiuz, and R. M. Tosi, eds.
1967. *Histocompatibility Testing, 1967.* Copenhagen: Munksgaard.
Dahlberg, G.
1926. *Twin Births and Twins from a Hereditary Point of View.* Stockholm: Bokforlags-A. -B. Tidens Tryckeri.
1948. *Mathematical Methods for Population Genetics.* Basel: S. Karger; New York: Interscience.
Damon, A.
1969. "Race, ethnic groups, and disease." *Soc. Biol.* **16**(2): 69–80.
Danforth, C. H.
1921. "The frequency of mutation and the incidence of hereditary traits in man." *Second Int. Congr. Eugen.* **1:** 120–128.
Darlington, C. D.
1958. *Evolution of Genetic Systems.* London: Oliver and Boyd.
Dausset, J., J. Colombani, L. Legrand, and N. Feingold
1968. "Le deuxieme sub-locus du systeme HL-A: le sub-locus 4-5-8." *Nou. Rev. Francaise Hematol.* **8:** 841.
Dausset, J., P. Ivanyi, and D. Ivanyi
1965. "Tissue alloantigens in humans: Identification of a complex system (Hu-1)." *In* H. Balner, F. J. Cleton, and J. G. Eernisse, eds. *Histocompatibility Testing, 1965.* Copenhagen: Munksgaard.

Dausset, J., and F. T. Rapaport
1966. "The role of blood group antigens in human histocompatibility." *Ann. N. Y. Acad. Sci.* **129:** 408. (Seventh International Transplantation Conference. F. T. Rapaport, ed.).

Dayhoff, M. O.
1969. *Atlas of Protein Sequence and Structure, 1969.* Silver Spring, Maryland: National Biomedical Research Foundation.

De Carli, L., J. J. Maio, F. Nuzzo, and A. S. Benerecetti
1964. "Cytogenetic studies with alkaline phosphatase in human heteroploid cells." *Cold Spring Harbor Symp. Quant. Biol.* **29:** 223–231.

Deevey, E. S.
1960. "The human population," *Sci. Amer.* **203:** 194–204.

Dempster, E. R.
1955. "Maintenance of genetic heteorogeneity." *Cold Spring Harbor Symp. Quant. Biol.* **20:** 25–32.

Dewey, W. J., I. Barrai, N. E. Morton, and M. P. Mi
1965. "Recessive genes in severe mental defect." *Amer. J. Hum. Genet.* **17:** 237–256.

Dobzhansky, T.
1951. *Genetics and the Origin of Species.* New York: Columbia University Press.
1962. *Mankind Evolving.* New Haven: Yale University Press.

Doolittle, R. F., and B. Blomback
1964. "Amino-acid sequence investigations of fibrino-peptides from various mammals: evolutionary implications." *Nature* **202:** 147.

Dronamraju, K. R.
1960. "Hypertrichosis of the pinna of the human ear, Y-linked pedigrees." *J. Genet.* **57:** 230–243.
1964. "Mating system of the Andhra Pradesh people." *Cold Spring Harbor Symp. Quant. Biol.* **29:** 81–84.

Dronamraju, K. R., and P. M. Khan
1960. "Inbreeding in Andhra Pradesh." *J. Hered.* **51**: 239–242.

Dunham, H. W., P. Phillips, and B. Srinivasan
1966. "A research note on diagnosed mental illness and social class." *Amer. Sociol. Rev.* **31:** 223.

Dunn, L. C.
1956. "Analysis of a complex gene in the house mouse." *Cold Spring Harbor Symp. Quant. Biol.* **21:** 187–195.
1962. "Cross currents in the history of human genetics." *Amer. J. Hum. Genet.* **14**(1): 1–13.
1968. "Review of *Race and Western Science* (R. E. Kultner, ed.)" *Eugen. Quart.* **15:** 298–301.

East, E. M.
1916. "Studies on size inheritance in Nicotiana." *Genetics* **1:** 164–176.

Eaton, J., and A. Mayer,
1954. "Man's capacity to reproduce." *Hum. Biol.* **25**(3): 1–58.

Edington, G. M.
1959. "Some observations on the abnormal haemoglobin diseases in Ghana." *In* J. H. P. Jonxis and J. F. Delafresnaye, eds. *Abnormal Haemoglobins.* Oxford: Blackwell.

Edwards, A. W. F.
1958. "An analysis of Geissler's data on the human sex ratio." *Ann. Hum. Genet.* **23**(1): 6–15.
1960a. "The meaning of binomial distribution." *Nature* **186:** 1074.
1906b. "Natural selection and the sex ratio." *Nature* **188:** 960–961.
1961. "The population genetics of 'sex ratio' in *Drosophila pseudoobscura.*" *Heredity* **16**(3): 291–304.
1962. "Genetics and the human sex ratio." *Advanc. Genet.* **11:** 239–272.
1963. "Natural selection and the sex ratio: the approach to equilibrium." *Amer. Natur.* **97:** 397–400.
1966. "Sex-ratio data analysed independently of family limitation." *Ann. Hum. Genet,* **29:** 337–347.

Edwards, A. W. F., and L. L. Cavalli-Sforza

1964. "Reconstruction of evolutionary trees." *In* V. E. Heywood and J. McNeill, eds. *Phenetic and Phylogenetic Classification* (Systematics Association Publication No. 6). London: The Systematics Association.

1965. "A method for cluster analysis." *Biometrics* **21**(2): 362–375.

Edwards, A. W. F., and M. Fraccaro

1958. "The sex distribution in the offspring of 5477 Swedish ministers of religion, 1585-1920." *Hereditas* **44:** 447.

Edwards, J. H.

1958. "Congenital malformations of the central nervous system in Scotland." *Brit. J. Prev. Soc. Med.* **12**(3): 115–130.

1960. "The simulation of mendelism." *Acta Genet. Statist. Med.* **10**(1–3): 63–70.

1965. "The meaning of the associations between blood groups and disease." *Ann. Hum. Genet* **29:** 77–83.

1969. "Familial predisposition in man." *Brit. Med. Bull.* **25**(1): 58–64.

Edwards, R. G., B. D. Bavister, and P. C. Steptoe

1969. "Early stages of fertilization *in vitro* of human oocytes matured *in vitro*." *Nature* **221:** 632–635.

Ehrlich, P. R., and A. H. Ehrlich

1790. *Population, Resources, Environment*. San Francisco: W. H. Freeman and Company.

Eichner, E. R., R. Finn, and J. R. Krevans

1963. " Relationship between serum antibody-levels and the ABO blood group polymorphism" *Nature* **198:** 164–165.

Elandt-Johnson, R. C.

1969. "Survey of histocompatibility testing." *Biometrics* **25**(2): 208–277.

Erlenmeyer-Kimling, L.

1968. "Studies on the offspring of two schizophrenic parents." *In* D. Rosenthal and S. S. Kety, eds. *The Transmission of Schizophrenia*. New York: Pergamon Press.

Erlenmeyer-Kimling, L., and L. F. Jarvik

1963. "Genetics and intelligence: a review." *Science.* **142:** 1477–1479. Copyright 1963 by the American Association for the Advancement of Science.

Erlenmeyer-Kimling, L., and W. Paradowski

1966. "Selection and schizophrenia." *Amer. Natur.* **100:** 651–665.

Etzioni, A. M.

1968. "Sex control, science, and society." *Science* **161:** 1107–1112.

Ewens, W. J.

1969. *Population Genetics*. London: Methuen.

Falconer, D. S.

1960. *Introduction to Quantitative Genetics.* Edinburgh: Oliver and Boyd.

1965. "The inheritance of liability to certain diseases, estimated from the incidence among relatives." *Ann. Hum. Genet.* **29:** 51.

1967. "The inheritance of liability to diseases with variable age of onset, with particular reference to diabetes mellitus." *Ann. Hum. Genet.* **31:** 1–20.

Falek, A.

1959. "Handedness: a family study." *Amer. J. Hum. Genet.* **11**(1): 52–62.

Feldman, M. W., M. Nabholz, and W. F. Bodmer

1969. "Evolution of the Rh polymorphism: a model for the interaction of incompatibility, reproductive compensation, and heterozygote advantage." *Amer. J. Hum. Genet.* **21**(2): 171–193.

Finney, D. J., R. L. Latscha, B. M. Bennett, and P. Hsu

1963. *Tables for Testing Significance in a* 2×2 *Contingency Table*. Cambridge: Cambridge University Press.

Firschein, I. L.

1961. "Population dynamics of the sickle-cell trait in the Black Caribs of British Honduras, Central America." *Amer. J. Hum. Genet.* **13:** 223–254.

Fisher, R. A.

1918. "The correlation between relatives on the supposition of Mendelian inheritance." *Trans. Roy. Soc.* (*Edinburgh*) **52:** 399–433.

1922. "On the dominance ratio." *Proc. Roy. Soc. Edinburgh* **42:** 321–341.

1925–1954. *Statistical Methods for Research Workers.* Edinburgh and London: Oliver and Boyd.

1930. [See 1958 entry.]

1934. "The effect of methods of ascertainment upon the estimation of frequencies." *Ann. Eugen.* **6:** 13–25.

1935. "The detection of linkage with recessive abnormalities." *Ann. Eugen.* **6:** 339–351.

1936a. "Tests of significance applied to Haldane's data on partial sex linkage." *Ann. Eugen.* **7:** 87–104.

1936b. "Heterogeneity of linkage data for Friedreich's ataxia and the spontaneous antigens." *Ann. Eugen.*, **7:** 1–21.

1937. "The wave of advance of advantageous genes." *Ann. Eugen.* **7:** 355–360.

1944. Cited by R. R. Race. *In* "An 'incomplete' antibody in human serum." *Nature* **153:** 771–772.

1947. "The Rhesus factor." *Amer. Sci.* **35:** 95–102, 113.

1950. "Gene frequencies in a cline determined by selection and diffusion." *Biometrics* **6**(4): 353–361.

1953. "Population Genetics." *Proc. Roy. Soc. B.* **141:** 510–523.

1958. *The Genetical Theory of Natural Selection*, 2nd Ed. New York: Dover. (The first edition of this book was published in 1930.)

Fisher, R. A., E. B. Ford, and J. Huxley

1939. "Taste-testing the anthropoid apes." *Nature* **144:** 750.

Fisher, R. A., M. F. Lyon, and A. R. G. Owen

1947. "The sex chromosome in the house mouse." *Heredity* **1:** 355–365.

Fisher, R. A., R. R. Race, and G. L. Taylor

1944. "Mutation and the Rhesus reaction." *Nature* **153:** 106.

Fisher, R. A., and F. Yates

1953. *Statistical Tables for Biological Agricultural and Medical Research*, London and Edinburgh: Oliver and Boyd.

Fitch, W. M., and E. Margoliash

1967. "Construction of phylogenetic trees." *Science* **155:** 279–284.

Ford, E. B.

1945. "Polymorphism." *Biol. Rev.* **20:** 73–88.

1964. *Ecological Genetics.* New York: John Wiley and Sons.

Fraser, G. R.

1965. "The role of Mendelian inheritance in the causation of childhood deafness and blindness." *In* R. R. Hoňcariv, ed. *Mutation in Population* (*Proceedings of the Symposium on the Mutational Process*). Prague: Academia.

Freda, V. J., and J. G. Gorman

1966. "Rh factor: prevention of isoimmunization and clinical trial on mothers." *Science* **151:** 828–830.

Freedman, D. G.

1968. "Personality development in infancy: a biological approach." *In* S. L. Washburn and P. C. Jay eds. *Perspectives on Human Evolution.* New York: Holt, Rinehart and Winston.

Freire–Maia, N.

1968. "Inbreeding levels in American and Canadian populations: a comparison with Latin America." *Eugen. Quart.* **15:** 22–27.

Fuhrmann, W., and F. Vogel

1969. *Genetic Counseling.* New York: Springer-Verlag.

Galton, F.

1894. *Natural Inheritance.* New York: Macmillan.

Gardner, R. A., and B. T. Gardner
1969. "Teaching sign language to a chimpanzee." *Science* **165:** 664.
Garn, S. M.
1961. *Human Races.* Springfield: Thomas.
Garrod, A. E.
1902. "The incidence of alkaptonuria: a study in chemical individuality." *Reprinted in* S. H. Boyer, ed. *Papers on Human Genetics.* Englewood Cliffs, N. J.: Prentice Hall.
Geissler, A.
1889. "Beiträge zur Frage des Geschlechtsverhältnisses der Geborenen." *Z. K. Sächsischen Statistischen Bureaus* **35:** 1.
Giblett, E. R.
1969. *Genetic Markers in Human Blood.* Philadelphia: F. A. Davis.
Glass, B.
1949. "The relation of Rh incompatibility to abortion." *Amer. J. Obstet. Gynecol.* **57:** 323–332.
1950. "The action of selection on the principal Rh alleles." *Amer. J. Hum. Genet.* **2:** 269–278.
Glass, B. and C. C. Li
1953. "The dynamics of racial intermixture, an analysis based on the American Negro." *Amer. J. Hum. Genet.* **5:** 1–20.
Glass, B., M. S. Sacks, E. F. Jahn, and C. Hess
1952. "Genetic drift in a religious isolate: an analysis of the causes of variation in blood group and other gene frequencies in a small population." *Amer. Natur.* **86:** 145–159.
Gobineau, Le Comte de.
1933. *Sur l'inégalité des Races Humaines.* Paris: Firmin-Didot.
Goldberg, E. M., and S. L. Morrison
1963. "Schizophrenia and social class." *Brit. J. Psychiat.* **109:** 785.
Goldfarb, C., and L. Erlenmeyer-Kimling
1962. "Mating and fertility trends in schizophrenia." *In* F. J. Kallmann, ed. *Expanding goals of genetics in psychiatry.* New York: Grune and Stratton.
Goldschmidt, E., ed.
1963. *The Genetics of Migrant and Isolate Populations.* Baltimore: Williams and Wilkins, on behalf of the Association for Crippled Children, New York City.
Goldschmidt, R. B.
1927. *Physiologische Theorie der Vererbung.* Berlin: J. Springer.
1958. *Theoretical Genetics.* Berkeley: University of California Press.
Goodall, J.
1965. "Chimpanzees of the Gombe stream reserve." *In* I. DeVore, ed., *Primate Behavior.* New York: Holt, Rinehart and Winston.
Goodall, H. B., D. W. W. Hendry, S. D. Lawler, and S. A. Stephen
1954. "Data on linkage in man: elliptocytosis and blood groups. III. Family 4." *Ann. Eugen.* **18:** 325–327.
Goodman, H. O., and J. J. Thomas
1968. "Kell types in Down's Syndrome." *Ann. Hum. Genet.* **31:** 369–372.
Gordon, M. J.
1957. "Control of sex ratio in rabbits by electrophoresis of spermatozoa." *Proc. Nat. Acad. Sci.* **43:** 913–918.
Gottesman, I. I.
1968. "Severity/concordance and diagnostic refinement in the Maudsley–Bethlem schizophrenic twin study." *J. Psychiat. Res.* **6:** 37–48.
Gottesman, I. I., and J. Shields
1967. "A polygenic theory of schizophrenia." *Proc. Nat. Acad. Sci.* **58.:** 199–205.
Greulich, W. W.
1934. "Heredity in human twinning." *Amer. J. Phys. Anthropol.* **19:** 391.
Grubb, R.
1965. "Agglutination of erythrocytes coated with 'incomplete' anti-Rh by certain rheumatoid

arthritic sera and some other sera. The existence of human serum groups." *Acta Pathol. Microbiol. Scand.* **39:** 195–197.

Gruneberg, H.
1963. *The Pathology of Development.* New York: John Wiley and Sons.

Gunther, M., and L. S. Penrose
1935. "The genetics of epiloia." *J. Genet.* **31:** 413.

Gurdon, J. B.
1963. "Nuclear transplantation in amphibia and the importance of stable nuclear changes in promoting cellular differentiation." *Quart. Rev. Biol.* **38:** 54–78.

Haga, H.
1959. "Studies on natural selection in ABO blood groups with special reference to the influence of environmental changes upon the selective pressure due to maternal-fetal incompatibility." *Jap. J. Hum. Genet.* **4:** 1–20.

Hajnal, J.
1963. "Random mating and the frequency of consanguineous marriages." *Proc. Roy. Soc. B* **159:** 125–177.

Haldane, J. B. S.
1919. "The combination of linkage values and the calculation of distance between the loci of linked factors." *J. Genet.* **8:** 299–309.
1924. "A mathematical theory of natural and artificial selection." *Proc. Cambridge Phil. Soc.* **23:** 19–41.
1926. "A mathematical theory of natural and artificial selection. Part III." *Proc. Cambridge Phil. Soc.* **23:** 363–372.
1927. "A mathematical theory of natural and artificial selection." Part V." *Proc. Cambridge Phil. Soc.* **23:** 838–844.
1932a. *The Causes of Evolution.* New York: Harper. (A classic.)
1932b. "A method for investigating recessive characters in man." *J. Genet.* **25:** 251–255.
1934. "Methods for the detection of autosomal linkage in man." *Ann. Eugen..* **6:** 26–65.
1935. "The rate of spontaneous mutation of a human gene." *J. Genet.* **31:** 317–326.
1936. "A search for incomplete sex-linkage in man." *Ann. Eugen.* **7:** 28–57.
1937. "The effect of variation on fitness." *Amer. Natur.* **71:** 337–349.
1938. "The estimation of the frequencies of recessive conditions in man." *Ann. Eugen.* **8:** 255–262.
1942. "Selection against heterozygosis in man." *Ann. Eugen.* **11:** 333–340.
1946. "The interaction of nature and nurture." *Ann. Eugen.* **13**(3): 197–205.
1948. "The theory of a cline." *J. Genet.* **48:** 277–284.
1949a. "Disease and evolution." *Ricerca Sci.* **19:** (Suppl. 1): 3–10.
1949b. "A test for homogeneity of records of familial abnormalities." *Ann. Eugen.* **14:** 339–341.
1954. *The Biochemistry of Genetics.* London: G. Allen and Unwin.
1957. "The conditions for co-adaptation in polymorphism for inversions." *J. Genet.* **55:** 218–225.

Haldane, J. B. S., and S. D. Jayakar
1963. "Polymorphism due to selection of varying direction." *J. Genet.* **58**(2): 237–242.
1964. "Equilibria under natural selection at a sex-linked locus." *J. Genet.* **59**(1): 29–36.
1965. "The nature of human genetic loads." *J. Genet.* **59**(2): 53–59.

Haldane, J. B. S., and C. A. B. Smith
1947. "A new estimate of the linkage between the genes for colour-blindness and haemophilia in man." *Ann. Eugen.* **14:** 10–31.

Haller, M. H.
1963. *Eugenics.* New Brunswick, N. J.: Rutgers University Press.

Hardin, G.
1966. *Biology Its Principles and Implications*, 2nd Ed. San Francisco: W. H. Freeman and Company.

1969. *Population, Evolution, and Birth Control*, 2nd Ed. San Francisco: W. H. Freeman and Company.

Hardy, G. H.

1908. "Mendelian Proportions in a Mixed Population." *Science* **28:** 49–50. *Reprinted in* J. H. Peters, ed. *Classic Papers in Genetics*. Englewood Cliffs, N. J.: Prentice-Hall, 1959.

Harris, H.

1959. *Human Biochemical Genetics*. Cambridge: Cambridge University Press.

1966. "Enzyme polymorphisms in man." *Proc. Roy. Soc. B* **164:** 298–310.

1969. "Enzyme and protein polymorphism in human populations." *Brit. Med. Bull.* **25:** 5–13.

Harrison, G. A., and J. J. T. Owen

1964. "Studies on the inheritance of human skin colour." *Ann. Hum. Genet.* **28:** 27–37.

Harrison, G. A., J. S. Weiner, J. M. Tanner, and N. A. Barnicot

1964. *Human Biology*. Oxford: The Clarendon Press.

Harvald, B., and M. Hauge

1965. "Hereditary factors elucidated by twin studies." *In* J. V. Neel, M. W. Shaw, and W. J. Schull, eds. *Genetics and the Epidemiology of Chronic Diseases* (Public Health Service Publication No. 1163). Washington, D.C.: U.S. Dept. of Health, Education, and Welfare.

Hatt, D., and P. A. Parsons

1965. "Association between surnames and blood groups in the Australian Population." *Acta Genet.* **15:** 309–318.

Hayes, W.

1968. *The Genetics of Bacteria and their Viruses*, 2nd Ed. New York: John Wiley and Sons.

Heber, R., R. Dever, and J. Conry

1968. "The influence of environmental and genetic variables on intellectual development." *In* H. J. Prehm, L. A. Hamerlynchk, and J. E. Crosson, eds. *Behavioral Research in Mental Retardation*. Eugene: University of Oregon Press.

Heer, D. M.

1967. "Intermarriage and racial amalgamation in the United States." *Eugen. Quart.* **14:** 112–120.

Heiken, A., and M. Rasmuson

1966. "Genetical studies on the Rh blood group system." *Hereditas Lund.* **55:** 192–212.

Herzenberg, L. A., H. O. McDevitt and L. A. Herzenberg

1968. "Genetics of antibodies." *Ann. Rev. Genet.* **2:** 209–244.

Heston, L. L.

1966. "Psychiatric disorders in foster home reared children of schizophrenic mothers." *Brit. J. Psychiat.* **112:** 819–825.

Hiernaux, J.

1968. *La diversité humaine en Afrique subsaharienne*. Editions de l'Institut de Sociologie, Universite Libre de Bruxelles.

Higgins, J. V., E. W. Reed, and S. C. Reed

1962. "Intelligence and family size: a paradox resolved." *Eugen. Quart.* **9:** 84–90.

Hiorns, R. W., G. A. Harrison, A. J. Boyce, and C. F. Kuchemann

1969. "A mathematical analysis of the effects of movement on the relatedness between populations." *Ann. Hum. Genet.* **32:** 237–250.

Hiraizumi, Y.

1964. "Prezygotic selection as a factor in the maintenance of variability. *Cold Spring Harbor Symp. Quant. Biol.* **29:** 51–60.

Hirszfeld, L., and H. Hirszfeld

1918–19. "Essai d'application des méthodes sérologiques au problème des races." *Anthropologie* **29:** 505–537.

Hirszfeld, L., and H. Zborowsky

1925. "Gruppenspezifische Beziehungen Zwischen Mutter und Frucht und elektive Durchlaessigkeit der Placenta." *Klini. Stochensch.* **4:** 1152–1157.

Histocompatibility Testing
1965. H. Balner, F. J. Cleton, and J. G. Eernisse, eds. Copenhagen: Munksgaard.
1967. E. S. Curtoni, P. L. Mattiuz, and R. M. Tosi, eds. Copenhagen: Munksgaard.
1970. Paul Terasaki, ed. Copenhagen: Munksgaard.

Hogben, L.
1934. "The detection of linkage in human families." *Proc. Roy. Soc. B* **114:** 340–363.

Holt, Sarah
1961. "Dermatoglyphic patterns." *In* G. A. Harrison, ed. *Genetical Variation in Human Populations*, Vol. 4. New York: Pergamon Press.

Holzinger, K. J.
1929. "The relative effect of nature and nurture on twin differences." *J. Educ. Psychol.* **20:** 241–248.

Hoyer, B. H., E. T. Bolton, B. J. McCarthy, and R. Roberts
1965. "The evolution of polynucleotides." *In* V. Bryson and H. J. Vogel, eds. *Evolving Genes and Proteins*. New York: Academic Press.

Huehns, E. R., and E. M. Shooter
1965. "Human haemoglobins." *J. Med. Genet.* **2:** 48–90.

Hulse, F. S.
1963. *The Human Species*. New York: Random House.
1968. "The breakdown of isolates and hybrid vigor among the Italian Swiss." *Proc. Twelfth Int. Congr. Genet.* **2:** 177. Tokyo: Science Council of Japan.

Huron, R. and J. Ruffié
1959. *Les methodes en génétique generale et en génétique humaine*. Paris: Masson.

Ingram, V. M.
1957. "Gene mutations in human haemoglobin: the chemical difference between normal and sickle cell haemoglobin." *Nature* **180:** 326–328.
1963. *The Hemoglobins in Genetics and Evolution*. New York: Columbia University Press.

Jacobs, P. A., M. Brunton, M. M. Melville, R. P. Brittain, and W. F. McClemont
1965. "Aggressive behaviour, mental subnormality, and the XYY male." *Nature* **208:** 1351.

Jacquard, A.
1967. "La reproduction humaine en régime malthusien." *Population* **22:** 897–920.

Jennings, H. S.
1916. "The numerical results of diverse systems of breeding." *Genetics* **1:** 53–89.
1917. "The numerical results of diverse systems of breeding with respect to two pairs of characters, linked or independent, with special relation to the effects of linkage." *Genetics* **2:** 97–154.

Jensen, A. R.
1969. How much can we boost IQ and scholastic achievement?" *Harvard Educ. Rev.* **39**(1): 1–123. (Copyright © 1969 by President and Fellows of Harvard College.)

Johannsen, W.
1909. *Elemente der exakten Erblichkditslehre*. Jena, Germany: Fisher.

Juel-Nielsen, N., and B. Harvald
1958. "The electroencephalogram in uniovular twins brought up apart." *Acta Genet.* **8:** 57–64.

Kahan, B. D., and R. A. Reisfeld
1969. "Transplantation antigens." *Science* **164:** 514–521.

Kaij, L.
1957. "Drinking habits in twins." *Acta Genet.* **7:** 437–441.

Kalmus, H., and C. A. B. Smith
1960. "Evolutionary origin of sexual differentiation and the sex-ratio." *Nature* **186:** 1004–1006.

Karlsson, J. L.
1968. "Genealogic studies of schizophrenia." *In* D. Rosenthal and S. S. Kety, eds. *The Transmission of Schizophrenia*. New York: Pergamon Press.

Karn, M. N., and L. S. Penrose
1951. "Birth weight and gestation time in relation to maternal age, parity, and infant survival." *Ann. Eugen.* **15:** 206–233.

Kashiwagi, N., C. G. Groth, and T. E. Starzl
1968. "Changes in serum haptoglobin and group specific component after orthotopic liver homotransplantation in humans." *Proc. Soc. Exp. Biol. Med.* **128:** 247–250.
Kempthorne, O.
1957. *An Introduction to Genetic Statistics.* New York: John Wiley and Sons.
Kendall, M. G., and A. Stuart
1958. *The Advanced Theory of Statistics,* Vols 1 and 2. New York: Hafner.
Kennedy, W. A., V. Van De Riet, and J. C. White, Jr.
1963. "A normative sample of intelligence and achievement of Negro elementary school children in the southeastern United States." *Monogr. Soc. Res. Child Devel.* **28**(6).
Kety, S. S., D. Rosenthal, P. H. Wender, and F. Schulsinger
1968. "The types and prevalence of mental illness in the biological and adoptive families of adopted schizophrenics." *In* D. Rosenthal and S. S. Kety eds. *The Transmission of Schizophrenia.* New York: Pergamon Press.
Keyfitz, N.
1966. "Finite approximations in demography." *Population Studies* **19**(3): 281–295.
1968. *Introduction to the Mathematics of Population.* Reading, Massachusetts: Addison-Wesley.
Kimura, M.
1954. "Process leading to quasi-fixation of genes in natural populations due to random fluctuation of selection intensities." *Genetics* **39:** 280–295.
1955a. "Stochastic processes and distribution of gene frequencies under natural selection." *Cold Spring Harbor Symp. Quant. Biol.* **20:** 33–53.
1955b. "Solution of a process of random genetic drift with a continuous model." *Proc. Nat. Acad. Sci.* **41:** 149.
1956. "Rules for testing stability of a selective polymorphism." *Proc. Nat. Acad. Sci.* **42:** 336–340.
1960. *Outline of Population Genetics.* Tokyo: Baifukan. (In Japanese.)
1961. "Some calculations on the mutational load." *Jap. J. Genet.* **36** (Suppl.): 179–190.
1962. "On the probability of fixation of mutant genes in a population." *Genetics* **47:** 713–719.
1964. *Diffusion Models in Population Genetics.* London: Methuen.
1968. "Evolutionary rate at the molecular level." *Nature* **217:** 624–626.
Kimura, M., and J. F. Crow
1963. "The measurement of effective population number." *Evolution* **17**(3): 279–288.
1964. "The number of alleles that can be maintained in a finite population." *Genetics* **49:** 725–738.
Kimura, M., and T. Ohta
1969. "The average number of generations until fixation of a mutant gene in a finite population." *Genetics* **61**(3): 763–771.
Kimura, M., and G. H. Weiss
1964. "The stepping stone model of population structure and the decrease of genetic correlation with distance." *Genetics* **49:** 561–576.
King, J. L.
1967. "Continuously distributed factors affecting fitness." *Genetics* **55:** 483–492.
King, J. L., and T. H. Jukes
1969. "Non-Darwinian evolution." *Science* **164:** 788–798.
King, T. J., and R. Briggs
1956. "Serial transplantation of embryonic nuclei." *Cold Spring Harbor Symp. Quant Biol.* **21:** 271–290.
Kirk, R. L.
1968a. *The Haptoglobin Groups in Man.* Basel: S. Karger.
1968b. "The world distribution of transferrin variants and some unsolved problems." *Acta Genet. Med. Gemell.* **17:** 613–640.
Kiser, C. V.
1968. "Assortative mating by educational attainment in relation to fertility." *Eugen. Quart.* **15**(2): 98–112.

Kissmeyer-Nielsen, F., A. Svejgaard, and M. Hauge
1968. "Genetics of the human HL-A transplantation system." *Nature* **29:** 1116-1119.
Kohn, M. L.
1968. "Social class and schizophrenia: a critical review." *In* D. Rosenthal and S. S. Kety, eds. *The Transmission of Schizophrenia*. New York: Pergamon Press.
Kolman. W. A.
1960. "The mechanism of natural selection for the sex ratio," *Amer. Natur.* **94:** 373-377.
Kosambi, D. D.
1944. "The estimation of map distance from recombination values." *Ann. Eugen.* **12:** 172–175.
Krehbiel, E. L.
1966. "An estimation of the cumulative mutation rate for sex-linked lethals in man which produce fetal deaths." *Amer. J. Hum. Genet.* **18:** 127–143.
Krieger, H., N. E. Morton, M. P. Mi, E. Azevedo, A. Freire-Maia, and N. Yasuda
1965. "Racial admixture in northeastern Brazil." *Ann. Hum. Genet.* **29:** 113–125.
Kultner, R. E. ed.
1967. *Race and Modern Science*. New York: Social Science Press.
Kunkel, H. G., J. B. Natvig, and F. G. Joslin
1969. "A "Lepore" type of hybrid γ globulin.", *Proc. Nat. Acad. Sci.* **62:** 144–149.
Laberge, C.
1967. "La consanguinité des Canadiens Français." *Population* **22**(5): 861–896.
Laird, C. D., B. L. McConaughy, and B. J. McCarthy
1969. "Rate of fixation of nucleotide substitutions in evolution." *Nature* **224:** 149–154.
Lalezari, P., and G. F. Bernard
1965. "Distribution of leukocyte antigens in various blood cells and body tissues." *In* H. Balner, F. J. Cleton, and J. G. Eernisse, eds. *Histocompatibility Testing, 1965*. Copenhagen: Munksgaard.
Landsteiner, K.
1900. "Zur Kenntnis der antifermentativen, lytischen und agglutinierenden Wirkungen des Blutserums und der Lymphe." *Zentralbe. Bakteriol.* **27:** 357–362.
Landsteiner, K., and A. S. Wiener
1940. "An agglutinable factor in human blood recognized by immune sera for rhesus blood." *Proc. Soc. Exp. Biol. N.Y.* **43:** 223.
Laughlin, W. S.
1950. "Blood groups, morphology and population size of the Eskimos." *Cold Spring Harbor Symp. Quant. Biol.* **15:** 165–173.
Leahy, A. M.
1935. "Nature-nurture and intelligence." *Genet. Psychol. Monogr.* **17**(4): 245–306.
Leakey, L. S. B.
1967. "An early miocene member of Hominidae." *Nature* **213:** 155–163.
Lederberg, J.
1966. "Experimental genetics and human evolution." *Amer. Natur.* **100:** 519–531.
1969. "Humanics and Genetic Engineering." In *Encyclopedia Britannica, 1970 Yearbook of Science and Technology*. Chicago: Encyclopedia Britannica.
Lerner, I. M.
1968. *Heredity, Evolution, and Society*. San Francisco: W. H. Freeman and Company.
Leslie, P. H.
1945. "On the use of matrices in certain population mathematics." *Biometrika* **33:** 183–212.
Levene, H.
1953. "Genetic equilibrium when more than one ecological niche is available." *Amer. Natur.* **87:** 331–333.
1963. "Inbred genetic loads and the determination of population structure." *Proc. Nat. Acad. Sci.* **50:** 587–592.
Levene, H., and R. E. Rosenfield
1961. "ABO incompatibility." *In* A. G. Steinberg, ed. *Progress in Medical Genetics*, Vol. 1. New York: Grune and Stratton.

Levine, P.
1943. "Serological factors as possible causes in spontaneous abortions." *J. Hered.* **34:** 71–80.
1958. "The influence of the ABO system on Rh hemolytic disease." *Hum. Biol.* **20:** 14-28.
Levine, P., and R. E. Stetson
1939. "An unusual case of intragroup agglutination." *J. Amer. Med. Ass.* **113:** 126–127.
Levine, P., and R. K. Waller
1946. "Erythroblastosis fetalis in the first born." *J. Hematol.* **1:** 143.
Lewontin, R. C.
1967. "An estimate of average heterozygosity in man." *Amer. J. Hum. Genet.* **19**(5): 681–685.
Lewontin, R. C., and L. C. Dunn
1959. "The evolutionary dynamics of a polymorphism in the house mouse." *Genetics* **45**(6): 705–722.
Lewontin, R. C., and J. L. Hubby
1966. "A molecular approach to the study of genetic heterozygosity in natural populations. II. Amount of variation and degree of heterozygosity in natural populations of *Drosophila pseudoobscura*." *Genetics* **54:** 595–609.
Li, C. C.
1955. *Population Genetics.* Chicago: University of Chicago Press.
Li, C. C., and D. G. Horvitz
1953. "Some methods of estimating the inbreeding coefficient." *Amer. J. Hum. Genet.* **5:** 107–117.
Livingstone, F. B.
1967. *Abnormal Hemoglobins in Human Populations.* Chicago: Aldine.
Lotka, A. J.
1956. *Elements of Mathematical Biology.* New York: Dover.
Lucas, Z. J., R. L. Kempson, J. Palmer, D. Korn, and R. B. Cohn
1969. "Renal allotransplantation in humans." *Amer. J. Surg.* **118:** 158.
Lunghi, C.
1965. "Teoria delle frequenze di incompatibilità da trapianto." *Atti Ass. Genet. Ital.* **10:** 311–319.
Luzzatto, L., E. A. Usanga, and S. Reddy
1969. "Glucose-6-phosphate dehydrogenase deficient red cells: resistance to infection by malarial parasites." *Science* **164:** 839–841.
Lyon, M. F.
1961. "Gene action in the X-chromosomes of the mouse (*Mus musculus L.*)." *Nature* **190:** 372–373.
Lyon, M. F., and A. R. G. Owen
1947. "The sex chromosomes in the house mouse." *Heredity* **1:** 355–365.
Mac Cluer, J. W.
1967. "Montecarlo methods in human population genetics; a computer model incorporating age specific birth and death rates." *Amer J. Hum. Genet.* **19**: 303–312.
MacCluer, J. W., R. Griffith, C. F. Sing, and W. J. Schull
1967. "Some genetic programs to supplement self-instruction in FORTRAN." *Amer. J. Hum. Genet.* **19**(3, Part 1): 189–221.
McConnell, R. B.
1966. "The prevention of Rh haemolytic disease." *Ann. Rev. Med.* **17:** 291–306.
McKeown, T., and C. R. Lowe
1951. "The sex ratio of stillbirths related to cause and duration of gestation." *Hum. Biol.* **23:** 41–60.
McKusick, V. A.
1964a. *Human Genetics.* Englewood Cliffs, N. J.: Prentice Hall.
1964b. *On the X Chromosome of Man.* Washington, D.C.: American Institute of Biological Sciences.

1968. *Mendelian Inheritance in Man*, 2nd Ed. Baltimore. The Johns Hopkins Press.

McLaren, A.

1962. "Does maternal immunity to male antigen affect the sex ratio of the young." *Nature* **195**: 1323–1324.

Mainardi, M., L. L. Cavalli-Sforza, and I. Barrai

1962. "The distribution of the number of collateral relatives." *Atti Ass. Genet. Ital.* **7**: 123–130.

Malécot, G.

1945. "La diffusion des gènes dans une population mendélienne." *Compt. Rend. Acad. Sci.* **221**: 340–342.

1948. *Les mathematiques de l'hérédité*. Paris: Masson.

1950. "Quelques schémas probabilistes sur la variabilité des populations naturelles." *Ann. Univ. Lyon Sci. A* **13**: 37–60.

1965–1966. "Les covariances dans un milieu en équilibre statistique." *Cahiers Rhodaniens* **14**: 1–29.

1966. *Probabilité et hérédité*. Paris: Presse Universitaire de France.

1967. "Identical loci and relationship." *Proc. Fifth Berkeley Symp. Math. Statist. Prob.* **4**: 317–332.

Mandel, S. P. H.

1959. "The stability of a multiple allelic system." *Heredity* **13**(3): 289–302.

Mange, A. P.

1964. "Growth and inbreeding of a human isolate." *Hum. Biol.* **36**: 104–133.

Mann, D. L., G. N. Rogentine, Jr., and J. L. Fahey

1969. "Molecular Heterogeneity of human lymphoid (HL-A) alloantigens." *Science* **163**: 1460–1462.

Mann, J. D., A. Cahan, A. G. Gelb, N. Fisher, J. Hamper, P. Tippett, R. Sanger, and R. R. Race

1962. "A sex-linked blood group." *Lancet*, **i**: 8–10.

Mathe, G., J. L. Amiel, L. Schwarzenberg, A. Cattan, M. Schneider, M. J. De Vries, M. Tubiana, C. Lalanne, J. L. Binet, M. Papiernik, G. Seman, M. Matsukura, A. M. Mery, V. Schwarzmann, and A. Flaisler

1965. "Successful allogenic bone marrow transplantation in man: chimerism, induced specific tolerance and possible anti-leukemic effects." *Blood* **25**(2): 179–196.

Mather, K.

1947. *Statistical Analysis in Biology*. New York: Interscience.

1949. *Biometrical Genetics*. London: Methuen.

1951. *The Measurement of Linkage in Heredity*. London: Methuen.

1955. "Polymorphism as an outcome of disruptive selection." *Evolution*. **9**: 52–61.

1963. "Genetical demography." *Proc. Roy. Soc. B* **159**: 106–125.

1964. *Human Diversity*. New York: The Free Press.

Matsunaga, E., and Y. Hiraizumi

1962. "Prezygotic selection in ABO blood groups." *Science* **135**: 432–434.

Matsunaga, E., Y. Hiraizumi, T. Furusho, and H. Izumiyama

1962. "Studies on selection in ABO blood groups." *Ann. Rep. Nat. Inst. Genet.* **13**: 103–106.

Matsunaga, E. and S. Itoh

1958. "Blood groups and fertility in a Japanese population, with special reference to intra-uterine selection due to maternal-foetal incompatibility." *Ann. Hum. Genet.* **22**: 111–131.

Matsuya, Y., H. Green, and C. Basilico

1968. "Properties and uses of human-mouse hybrid cell lines." *Nature* **220**: 1199.

1965. "Eugenics and Utopia." *Daedalus* **94**(2): 487–505.

Maynard-Smith, J.

1968. "Evolution in sexual and asexual populations." *Amer. Natur.* **102**: 469–473.

Mayr, E.
1963. *Animal Species and Evolution*. Cambridge: The Belknap Press of Harvard University Press.
Mi, M. P., and N. E. Morton
1966. "Allo-antigens and antigenic factors of human leukocytes. A hypothesis." *Vox Sang.* **11**: 326–331.
Miall, W. E., and P. D. Oldham
1963. "The hereditary factor in arterial blood pressure." *Brit. Med. J.* **19**: 75–80.
Mickey, M. R., D. P. Singal, and P. I. Terasaki
1969. "Serotyping for homotransplantation XXV. Evidence for three HL-A subloci." *Transp. Proc.* **1**: 347.
Migeon, B. R., and C. Miller
1968. "Human-mouse somatic cell hybrids with single human chromosome (group E) link with thymidine kinase activity." *Science* **162**: 1005.
Milkman, R. D.
1967. "Heterosis as a major cause of heterozygosity in nature." *Genetics* **55**(3): 493–495.
Mitra, S.
1966. "Income, socioeconomic status, and fertility in the United States." *Eugen Quart.* **13**: 223–230.
Mollison, P. L.
1967. *Blood Transfusion in Clinical Medicine*. Philadelphia: F. A. Davis.
Moor-Jankowski, J., A. S. Wiener, and C. M. Rogers
1964. "Human blood group factors in non-human primates." *Nature* **202**: 663–665.
Moran, P. A. P.
1962. *The Statistical Processes of Evolutionary Theory*. Oxford: The Clarendon Press.
Mørch, E. T.
1941. *Chondrodystrophic Dwarfs in Denmark*. Copenhagen: Munksgaard.
Moroni, A.
1961. "Sources, reliability and usefulness of consanguinity data with special reference to Catholic records." *World Health Organization Chron.* **15**: 465–472.
1964. "Evoluzione della frequenza dei matrimoni consanguinei in Italia negli ultimi cinquant'anni." *Atti Ass. Genet. Ital.* **9**: 207–223.
1966. "La consanguineità umana in Sardegna." *Ateneo Parmense* **87**: 3–28.
1967a. "Andamento della consanguineità nell'Italia settentrionale negli ultimi quattro secoli." *Atti Ass. Genet. Ital.* **12**: 202–222.
1967b. "Struttura ed evoluzione della consanguineità umana nelle isole Eolie (1860–1966)." *Archiv. Antrop. Etnol.* **98**(3): 135–150.
1969. *Historical Demography, Human Ecology, and Consanguinity*. London: International Union for the Scientific Study of Population, General Conference.
Morton, N. E.
1955. "The inheritance of human birth weight." *Ann. Hum. Genet.* **20**(2): 125–134.
1956a. "Sequential tests for the detection of linkage." *Amer. J. Hum. Genet.* **7**: 277–318.
1956b. "The detection and estimation of linkage between the genes for elliptocytosis and the Rh blood type." *Amer. J. Hum. Genet.* **8**(2): 80–96.
1957. "Further scoring types in sequential linkage tests, with a critical review of autosomal and partial sex linkage in man." *Amer. J. Hvm. Genet.* **9**: 55–75.
1959. "Genetic tests under incomplete ascertainment." *Amer. J. Hum. Genet.* **11**(1): 1–16.
1960. "The mutational load due to detrimental genes in man." *Amer. J. Hum. Genet.* **12**: 348–364.
1961. "Morbidity of children from consanguineous marriages." *In* A. G. Steinberg, ed. *Progress in Medical Genetics*, Vol. 1. New York: Grune and Stratton.
1962. "Segregation and linkage." *In* W. J. Burdette, ed. *Methodology in Human Genetics*. San Francisco: Holden-Day.
1964. "Models and evidence in human population genetics." *Proc. Eleventh Int. Congr. Genet.* **3**: 935–951. The Hague: Pergamon Press.

1967. "The detection of major genes under additive continuous variation." *Amer. J. Hum. Genet.* **19**(1): 23–34.

Morton, N. E., and C. S. Chung

1959a. "Formal genetics of muscular dystrophy." *Amer. J. Hum. Genet.* **11**(4): 360–379.

1959b. "Are the MN blood groups maintained by selection?" *Amer. J. Hum. Genet.* **11**(3): 237–251.

Morton, N. E., C. S. Chung, and M. P. Mi

1967. *Genetics of Interracial Crosses in Hawaii.* New York: S. Karger.

Morton, N. E., J. F. Crow, and H. J. Muller

1956. "An estimate of the mutational damage in man from data on consanguineous marriages." *Proc. Nat. Acad. Sci.* **42**: 855–863.

Morton, N. E., H. Krieger, and M. P. Mi

1966. "Natural selection on polymorphisms in northeastern Brazil." *Amer. J. Hum. Genet.* **18**(2): 153–171.

Morton, N. E., M. P. Mi, and N. Yasuda

1966. "Bivalent alleles." *Amer. J. Hum. Genet.* **18**(3): 233–242.

Morton, N. E., and C. Miki

1968. "Estimation of gene frequencies in the MN system." *Vox Sang.* **15**: 15–24.

Morton, N. E., C. Miki, and S. Yee

1968. "Bioassay of population structure under isolation by distance." *Amer. J. Hum. Genet.* **20**: 411–419.

Morton, N. E., N. Yasuda, C. Miki, and S. Yee

1968. "Population structure of the ABO blood groups in Switzerland." *Amer. J. Hum. Genet.* **20**: 420–429.

Motulsky, A. G.

1964. "Hereditary red cell traits and malaria." *Amer. J. Trop. Med.* **13**: 147–155.

Motulsky, A. G., R. Anderson, R. S. Sparkes, and R. H. Huestis

1962. "Marrow transplantation in newborn mice with hereditary spherocytosis: a model system." *Trans. Ass. Amer. Phys.* **75**: 64–71.

Mourant, A. E.

1954. *The Distribution of the Human Blood Groups.* Oxford: Blackwell.

Mourant, A. E., A. C. Kopeč, and K. Domaniewska

1958. *The ABO Blood Groups.* Oxford: Blackwell.

Muller, H. J.

1925. "Mental traits and heredity." *J. Hered.* **16**: 433–448.

1932. "Some genetic aspects of sex." *Amer. Natur.* **64**: 118–138.

1950. "Our load of mutations." *Amer. J. Hum. Genet.* **2**(2): 111–176.

1958. "Evolution by mutation" *Bull. Amer. Math. Soc.* **64**: 137-160.

1964. "The relation of recombination to mutational advance." *Mutation Res.* **1**: 2–9.

1966. "What genetic course will man steer?" *Proc. Third Int. Congr. Hum. Genet.* pp. 521–543. Baltimore: The Johns Hopkins Press.

Murdock, G. P.

1959. *Africa, Its Peoples and Their Culture and History.* New York: McGraw-Hill.

Nabholz, M., V. Miggiano, and W. F. Bodmer

1969. "Genetic analysis with human-mouse somatic cell hybrids." *Nature* **223**: 358–363.

Nadler, H. L.

1969. "Prenatal detection of genetic defects." *J. Pediat.* **74**: 132–143.

Natvig, J. B., H. G. Kunkel, and S. D. Litwin

1967. "Genetic markers of the heavy chain subgroups of human γG globulin." *Cold Spring Harbor Symp. Quant. Biol.* **32**: 173–180.

Neel, J. V.

1949. "The inheritance of sickle-cell anemia." *Science* **110**: 64–66. *In* S. H. Boyer, ed. *Papers on Human Genetics.* Englewood Cliffs, New Jersey: Prentice-Hall, 1963.

1970. "Lessons from a primitive people." *Science* **170**: 815–822.

Neel, J. V., S. S. Fajans, J. W. Conn, and R. Davidson
1965. "Diabetes mellitus." *In* J. V. Neel, M. Shaw and W. J. Schull, eds. *Genetics and the Epidemiology of Chronic Diseases* (Public Health Service Publication No. 1163). Washington, D.C.: U.S. Dept. of Health, Education, and Welfare.
Neel, J. V., M. Kodani, R. Brever, and R. C. Anderson
1949. "The incidence of consanguineous matings in Japan." *Am. J. Hum. Genet.* **1**(2): 156–178.
Neel, J. V., and R. H. Post
1963. "Transitory 'positive' selection for color-blindness?" *Eugen. Quart.* **10:** 33–35.
Neel, J. V., and W. J. Schull
1954. *Human Heredity.* Chicago: The University of Chicago Press.
1956. *The Effect of Exposure to the Atomic Bombs on Pregnancy Termination in Hiroshima and Nagasaki.* Washington, D.C.: National Academy of Sciences, National Research Council Publication.
Nei, M., and Y. Imaizumi
1966. "Genetic structure of human populations." *Heredity* **21:** 183–190.
Nei, M., and M. Murata
1966. "Effective population size when fertility is inherited." *Genet. Res.* **8:** 257–260.
Newcombe, H. B.
1964. [Remarks included in panel discussion of the session on epidemiologic studies.] *In* M. Fishbein, ed. *Papers and Discussions of the Second International Conference on Congenital Malformations.* New York: International Medical Congress.
1965. "Environmental versus genetic interpretations of birth-order effects." *Eugen. Quart.* **12**(2): 90.
1967. "Record linking: the design of efficient systems for linking records into individual and family histories." *Amer. J. Hum. Genet.* **19**(3): 335–359.
Newcombe, H. B., and J. M. Kennedy
1962. "Record linkage: Making maximum use of the discriminating power of identifying information." *Commun. Ass. Computing Machinery,* **5:** 563–566.
Newcombe, H. B., and P. O. W. Rhynas
1962. "Child spacing following stillbirth and infant death." *Eugen. Quart.* **9**(1): 25–35.
Newman,. H. H., F. N. Freeman, and K. J. Holzinger
1937. *Twins: A Study of Heredity and Environment.* Chicago: University of Chicago Press.
Newth, D. R.
1961. "Chance compatibility in homografting." *Transplant. Bull.,* **27:** 452.
Nichols, R. C., and W. C. Bilbro
1966. "The diagnosis of twin zygosity." *Acta Genet.* **16:** 265–275.
Novitzki, E., and A. W. Kimball
1958. "Birth order, parental ages, and sex of offspring." *Amer. J. Hum. Genet.* **10**(3): 268–275.
Novitski, E., and L. Sandler
1956. "The relationship between parental age, birth order and the secondary sex ratio in humans." *Ann. Hum. Genet.* **21**(2): 123–131.
Nuzzo, F., A. Bompiani, E. Moneta, F. Caviezel, and F. Mussinelli
1967. "Osservazioni su alcuni casi di variazioni strutturali del cromosoma Y nell' uomo." *Atti Ass. Genet. Ital.* **12:** 191–201.
O'Donald, P.
1969. "'Haldane's Dilemma' and the rate of natural selection." *Nature* **221:** 815–816.
Ohkubo, T.
1963. "Consanguineous marriages in West-Southern Sado." *Jap. J. Hum. Genet.* **8**(2): 128–135.
Ohno, S.
1967. *Sex Chromosomes and Sex-linked Genes.* Berlin: Springer-Verlag.
Osborn, F.
1968. *The Future of Human Heredity.* New York: Weybright and Talley.
Osborne, R. H., and F. V. De George
1959. *Genetic Basis of Morphological Variation.* Cambridge: Harvard University Press.
Owen, A. R. G.

1950. "The theory of genetical recombination." *Advanc. Genet.* **3:** 117–157.
1953. "A genetical system admitting of two distinct stable equilibria under natural selection." *Heredity* **7:** 97–102.
1954. "Balanced polymorphism of a multiple allelic series." *Caryologia* (Suppl.) **6:** 1240–1241.

Payne, R., and M. R. Rolfs
1958. "Fetomaternal leukocyte incompatibility." *J. Clin. Invest.* **37:** 1756–1763.

Payne, R., M. Tripp, J. Weigle, W. F. Bodmer, and J. Bodmer
1964. "A new leukocyte isoantigen system in man." *Cold Spring Harb. Symp. Quant. Biol.* **29:** 285.

Patel, R., M. R. Mickey, and P. I. Terasaki
1968. "Serotyping for homotransplantation." *New Engl. J. Med.* **279:** 501–506.

Pauling, L., H. A. Itano, S. J. Singer, and I. C. Wells
1949. "Sickle cell anemia, a molecular disease." *Science* **110:** 543–548. *In* S. H. Boyer, ed. *Papers on Human Genetics.* Englewood Cliffs, N. J.: Prentice-Hall, 1963.

Pearson, K., and A. Lee
1903. "On the laws of inheritance in man. I. Inheritance of physical characters." *Biometrika* **2:** 357–462.

Penrose, L. S.
1935. "The detection of autosomal linkage in data which consist of pairs of brothers and sisters of unspecified parentage." *Ann. Eugen.* **6:** 133–138.
1938. "A Clinical and Genetic Study of 1280 Cases of Mental Defect." *Special Report Series No. 229, Medical Research Council.* London: H.M.S.O.
1952. "Measurement of pleiotropic effects in phenylketonuria." *Ann. Eugen.* **16:** 134–141.
1953a. "The genetical background of common diseases." *Acta Genet.* 4: 257–265.
1953b. "The general purpose sib-pair linkage test." *Ann. Eugen.* **18:** 120–124.
1955. "Parental age and mutation." *Lancet* **11:** 312.
1963. *The Biology of Mental Defect*, 3rd Ed. New York: Grune and Stratton.

Penrose, L. S., and G. F. Smith
1966. *Down's Anomaly.* Boston: Little, Brown.

Perry, T. L., S. Hansen, B. Tischler, and R. Bunting
1967. "Determination of heterozygosity for phenylketonuria on the amino acid analyzer." *Clin. Chim. Acta* **18:** 51–56.

Peters, J. A., ed.
1959. *Classical Papers in Genetics.* Englewood Cliffs, N. J.: Prentice Hall.

Petri, E.
1935. "Untersuchungen zur Erbbedingtheit der Menarche." *Z. Morph. Anthrop.* **33:** 43–48.

Pettigrew, T. F.
1964. "Race, mental illness and intelligence: a social psychological view." *Eugen. Quart.* **11:** 189–215.

Pohlman, E.
1967. "Some effects of being able to control sex of offspring." *Eugen. Quart.* **14**(4): 274.

Polman, A.
1951. "Over consanguine huwelijken in Nederland." *Onderzoekingen en Mededelingen Uit Het Institut voor preventieve Geneeskunde, Leiden* **7:** 5–35.

Porter, I. H., J. Schulze, and V. A. McKusick
1962. "Genetic linkage between the loci for glucose-6-phosphate dehydrogenase deficiency and colour-blindness in American Negroes." *Ann. Hum. Genet.* **20:** 107–122.

Post, R. H.
1962. "Population differences in red and green color vision deficiency: a review, and a query on selection relaxation." *Eugen. Quart.* **9:** 131–146.

Potter, E. L.
1948. *Fundamentals of Human Reproduction.* New York: McGraw-Hill.

Powys, A. O.
1905. "Data for the problem of evolution in man. On fertility, duration of life and reproductive selection." *Biometrika*, **4:** 233.

Putnam, F. W., K. Titani, M. Wikler, and T. Shinoda
1967. "Structure and evolution of kappa and lambda light chains." *Cold Spring Harbor Symp. Quant. Biol.* **32:** 9–29.

Race, R. R., and R. Sanger
1968. *Blood Groups in Man.* Philadelphia: F. A. Davis.

Rao, C. R.
1952. *Advanced Statistical Methods in Biometric Research.* New York: John Wiley and Sons.

Reed, E. W., and S. C. Reed
1965. *Mental Retardation: a Family Study.* Philadelphia: W. B. Saunders.

Reed, S. C.
1955. *Counseling in Medical Genetics.* Philadelphia: W. B. Saunders.

Reed, T. E.
1961. "Polymorphisms and natural selection in blood groups." *In* B. S. Blumberg, ed. *Proceedings of the Conference on Genetic Polymorphisms.* New York: Grune and Stratton.
1969a. "Critical tests of hypotheses for race mixture using Gm data on American Caucasians and Negroes." *Amer. J. Hum. Genet.* **21**(1): 71–83.
1969b. "Caucasian genes in American Negroes." *Science* **165:** 762–768.

Reed, T. E., and J. H. Chandler
1958. "Huntington's chorea in Michigan." *Amer. J. Hum. Genet.* **10:** 201–206.

Reed, T. E., H. Gershowitz, A. Soni, J. Napier
1964. "A search for natural selection in six blood group systems and ABH secretion." *Amer. J. Hum. Genet.* **16:** 161–179.

Reed, T. E., and J. V. Neel
1959. "Huntington's chorea in Michigan." *Amer. J. Hum. Genet.* **11**(2): 107–136.

Reed, T. E., and W. J. Schull
1968. Letters to the Editor, "A general maximum likelihood estimation program." *Amer. J. Hum. Genet.* **20:** 579–580.

Renkonen, K. O., O. Mäkalä, and R. Lehtovaara
1962. "Factors affecting the human sex ratio." *Nature* **194:** 308.

Renwick, J. H.
1969. "Progress in mapping human autosomes." *Brit. Med. Bull.* **25**(1): 65–73.

Ridley, J. C., and M. C. Sheps
1966. "An analytical simulation model of human reproduction with demographic and biological components." *Population Studies* **19**(3): 297–310.

Rieger, R., A. Michaelis, and M. M. Green
1968. *A Glossary of Genetics and Cytogenetics.* New York: Springer-Verlag.

Roberts, D. F., and A. E. Boyo
1960. "On the stability of haemoglobin gene frequencies in West Africa." *Ann. Hum. Genet.* **24:** 375–387.

Roberts, D. F., and R. W. Hiorns
1962. "The dynamics of racial intermixture." *Amer. J. Hum. Genet.* **14:** 261–277.

Robertson, A.
1951. "The analysis of heterogeneity in the binomial distribution." *Ann. Eugen.* **16:** 1-15.
1961. "Inbreeding in artificial selection programmes." *Genet. Res.* **2:** 189–194.
1962. "Selection for heterozygotes in small populations." *Genetics* **47:** 1291–1300.

Robson, E. B.
1955. "Birth weight in cousins." *Ann. Hum. Genet.* **19:** 262–268.

Robson, E. B., P. E. Polani, S. J. Dart, P. A. Jacobs, and J. A. Renwick
1969. "Probable assignment of the alpha locus of haptoglobin to chromosome 16 in man." *Nature* **223:** 1163–1165.

Rohlf, F. J., and R. R. Sokal
1969. *Statistical Tables.* San Francisco: W. H. Freeman and Company.

Ropartz, C., P.-Y. Rousseau, L. Rivat, and J. Lenoir
1961. "Étude génétique du facteur sérique Inv; fréquence dans certaines populations." *Rev. Franc. Études Clin. Biol.* **6:** 374–377.

Rosenfield, R. E., F. H. Allen, S. N. Swisher, and S. Kochwa
1962. "A review of Rh serology and presentation of a new terminology." *Transfusion* **2:** 287–312.

Rosenthal, D.
1966. "The offspring of schizophrenic couples." *J. Psychiat. Res.* **4:** 169–188.

Rosenthal, D., and S. S. Kety, eds.
1968. *The Transmission of Schizophrenia*. New York: Pergamon Press.

Rucknagel, D. L., and J. V. Neel
1961. "The Hemoglobinophathies." *Prog. Med. Gen.*, **1:** 158–260.

Russell, L. B.
1963. "Mammalian X chromosome action: inactivation limited in spread and in region of origin." *Science* **140:** 976–978.

Russell, W. L.
1963. "Evidence from mice concerning the nature of the mutation process." *In* S. J. Geerts, ed. *Genetics Today*, Vol. 2 (Proceedings of the Eleventh International Congress of Genetics. The Hague. September 1963). New York: Pergamon Press, 1965.

Saldanha, P. H.
1957. "Gene flow from White into Negro population in Brazil." *Amer. J. Hum. Genet.* **9:** 299–309.

Salzano, F. M., J. V. Neel, and D. Maybury-Lewis
1967. "Further studies on the Xavante Indians. 1. Demographic data on two additional villages; genetic structure of the tribe." *Amer. J. Hum. Genet.* **19:** 463.

Sanghvi, L. D.
1966. "Inbreeding in India." *Eugen. Quart.* **13**(4): 291–301.

Schreider, E.
1964. "Ecological rules, body-heat regulation, and human evolution." *Evolution* **18**(1): 1–9.

Schreiner, W. E.
1959. "Über eine hereditäre Form von Pseudohermaphroditismus masculinus ("testiculäre Feminisierung"). *Geburtsch. Frauenheilk.* **19:** 1110–1118.

Schull, W. J., I. Komatsu, H. Nagano, and M. Yamamoto
1968. "HIRADO: temporal trends in inbreeding and fertility." *Proc. Nat. Acad. Sci.* **59**(3): 671–679.

Schull, W. J., and J. V. Neel
1965. *The Effects of Inbreeding on Japanese Children.* New York: Harper and Row.

Schultze, H., and J. F. Heremans
1966. *Molecular Biology of Human Proteins*, Vol. I. London: Elsevier.

Schwidetzky, I.
1962. *Die neue Rassenkunde.* Stuttgart: Gustav Fisher Verlag.

Scudo F. M.
1967. "The adaptive value of sexual dimorphism: I. Anisogamy." *Evolution* **21**(2): 285–291.

Serra, A.
1961. "La consanguineità e i suoi effetti nelle popolazioni umane." *In* L. Gedda, ed. *De Genetica Medica.* Rome: Apud Mendelianum Institutum.

Serra, A. and D. O'Mathuna
1966. "A theoretical approach to the study of genetic parameters of histocompatibility in man." *Ann. Hum. Genet.* **30:** 96–118.

Shaw, M. W., and H. Gershowitz
1962. "A search for autosomal linkage in a trisomic population: blood group frequencies in mongols." *Amer. J. Hum. Genet.* **14:** 317.

Shen, L., E. F. Grollman, and V. Ginsburg
1968. "An enzymatic basis for secretor status and blood group substance specificity in humans." *Proc. Nat. Acad. Sci.* **59**(1): 224–230.

Shields, J.
1962. *Monozygotic Twins Brought up Apart and Brought up Together*. London: Oxford University Press.

Shulman, N. W., V. J. Marder, M. C. Hillier, and E. M. Collier
1964. "Platelet and leukocyte isoantigens and their antibodies: Serologic physiologic and clinical studies." *In* L. M. Tocantins, ed. *Progress in Hematology*, Vol. 4. New York: Grune and Stratton.

Simons, E. L.
1960. "New fossil primates: a review of the past decade." *Amer. Sci.* **48:** (Reprinted *in* S. L. Washburn and P. C. Jay, eds. *Perspectives on Human Evolution.* New York: Holt, Rinehart, and Winston, 1968.)
1963. "Some fallacies in the study of hominid phylogeny." *Science* **141:** 879–889.

Simonsen, M.
1965. " Strong transplantation antigens in man." *Lancet* **228**(11): 415–418.

Simpson, G. G.
1949. *The Meaning of Evolution.* New Haven: Yale University Press.
1953. *The Major Features of Evolution.* New York: Columbia University Press.
1966. "The biological nature of man." *Science* **152:** 472–478.

Simpson, N. E.
1962. "The genetics of diabetes: A study of 233 families of juvenile diabetics." *Ann. Hum. Genet.* **26:** 1.

Singal, D. P., M. R. Mickey, and P. I. Terasaki
1969a. "Serotyping for homotransplantation." *Transplantation* **7**(4): 246–258.
1969b. "Serotyping for homotransplantation. XXXII HL-A haplotypes in Japanese families. *Transplantation* **8**(6): 829–836.

Siniscalco, M., L. Bernini, B. Latte, and A. G. Motulsky
1961. "Favism and Thalassaemia in Sardinia and their Relationship to Malaria." *Nature* **190:** 1179–1180.

Skellam, J. G.
1951. "Gene dispersion in heterogeneous populations." *Heredity* **5:** 433–435.

Slater, E.
1943. "A demographic study of a psychopathic population." *Ann. Eugen.* **12:** 121–137.
1968. "A review of earlier evidence on genetic factors in schizophrenia." *In* D. Rosenthal and S. S. Kety, eds. *The Transmission of Schizophrenia.* New York: Pergamon Press.

Slatis, H. M., R. H. Reis, and R. E. Hoene
1958. "Consanguineous Marriages in the Chicago Region." *Amer. J. Hum. Genet.* **10:** 446–464.

Smith, C. A. B.
1953. "The detection of linkage in human genetics." *J. Roy. Stat. Soc. B* **15:** 153–192.
1959. "Some comments on the statistical methods used in linkage investigations." *Amer. J. Hum. Genet.* **11:** 289.
1961. "Methodology in human genetics." *Amer. J. Hum. Genet.* **13:** 128.
1963. "Testing for heterogeneity of recombination values in human genetics." *Ann. Hum. Genet.* **27:** 175.
1968. "Linkage scores and corrections in simple two- and three-generation families." *Ann. Hum. Genet.* **32:** 127–150.
1969. "Local fluctuations in gene frequencies." *Ann. Hum. Genet.* **32:** 251–260.

Smith, S. M., L. S. Penrose, and C. A. B. Smith
1961. *Mathematical Tables for Research Workers in Human Genetics.* London: Churchill.

Smithies, O.
1964. "Chromosomal rearrangements and protein structure." *Cold Spring Harbor Symp. Quant. Biol.* **29:** 309–319.

Sneath, J. S., and P. H. A. Sneath
1955. "Transformation of the Lewis groups of human red cells." *Nature* **176:** 172.

Snell, G. D., and J. H. Stimpfling
1966. "Genetics of the tissue transplantation." *In* B. Roscoe, ed. Biology of the Laboratory Mouse. New York: McGraw-Hill.

Snyder, L. H.
1932. "Studies in human inheritance. IX. The inheritance of taste deficiency in man." *Ohio J. Sci.* **32:** 436–440.

Social Implications of the 1947 Scottish Mental Survey
1953. London: University of London Press.

Sokal, R. R., and F. J. Rohlf
1969. *Biometry*. San Francisco: W. H. Freeman and Company.

Sokal, R. R., and P. H. A. Sneath
1963. *Principles of Numerical Taxonomy*. San Francisco: W. H. Freeman and Company.

Sonneborn, T. M., ed.
1965. *Control of Human Heredity and Evolution*. New York: Macmillan.

Spuhler, J. N.
1968. "Assortative mating with respect to physical characteristics." *Eugen. Quart.* **15**(2): 128–140.

Spuhler, J. N., and C. Kluckhohn
1953. "Inbreeding coefficients of the Ramah Navaho population." *Hum. Biol.* **25**(4): 295–317.

Srb, A. M., R. D. Owen, and R. S. Edgar
1965. *General Genetics*, 2nd Ed. San Francisco: W. H. Freeman and Company.

Steele, M. W., and W. R. Breg
1966. "Chromosome analysis of human amniotic-fluid cells." *Lancet* **1:** 383–385.

Steinberg, A. G.
1965. "Evidence for a mutation or crossing-over at the Rh locus." *Vox Sang.* **10:** 721–724.

Steinberg, A. G., H. K. Bleibtreu, T. W. Kurczynski, A. O. Martin and E. M. Kurczynski
1966. "Genetic studies on an inbred human isolate." *In* J. F. Crow and J. V. Neel, eds. *Proceedings of the Third International Congress of Human Genetics*. Baltimore: The Johns Hopkins Press, 1967.

Stent, G. S.
1970. *Molecular Genetics*. San Francisco: W. H. Freeman and Company.

Stern, C.
1957. "The problems of complete Y linkage in man." *Amer. J. Hum. Genet.* **9:** 147–166.
1960. *Principles of Human Genetics*, 2nd Ed. San Francisco: W. H. Freeman and Company.

Stern, K., I. Davidsohn, and L. Masaitis
1956. "Experimental studies on Rh immunization." *Amer. J. Clin. Pathol.* **26:** 833–843.

Stevenson, A. C., and C. B. Kerr
1967. "On the distributions of frequencies of mutation to genes determining harmful traits in man." *Mutation Res.* **4:** 339–352.

Strickberger, M. W.
1968. *Genetics*. New York: Macmillan.

Sturtevant, A. H., and Th. Dobzhansky
1936. "Geographical distribution and cytology of 'sex ratio' in *Drosophila pseudoobscura* and related species." *Genetics* **21:** 473–490.

Sutter, J.
1958. "Recherches sur les effets de la consanguinité chez l'homme." *Biol. Med.* **47:** 463–660.

Sutter, J., and J.-M. Goux
1964. "Lethal equivalents and demographic measures of mortality." *Cold Spring Harbor Symp. Quant. Biol.* **29:** 41–50.

Sutter, J., and L. Tabah
1952. "Effets de la consanguinité et de l'endogamie." *Population* **7:** 249–266.
1953. "Structure de la mortalité dans les familles consanguines." *Population* **8:** 511–526.
1955. "L'evolution des isolates de deux departements francais: Loir-et-Cher, Finistere." *Population* **10**(4): 645–674.
1956. "Méthode mécanographique pour établir la généalogie d'une population." *Population* **3:** 515–520.

Sved, J. A.
1968. "Possible rates of gene substitution in evolution." *Amer. Natur.* **102:** 283–293.
Sved, J. A., T. E. Reed and W. F. Bodmer
1967. "The number of balanced polymorphisms which can be maintained in a natural population." *Genetics* **55:** 469–481.
Szeinberg, A.
1963. "G6PD deficiency among Jews—genetic and anthropological considerations." *In* E. Goldschmidt, ed. *Genetics of Migrant and Isolated Populations.* New York: Williams and Wilkins and the Association for Crippled Children.
Tanner, J. M.
1960. "Genetics of human growth." *In* J. M. Tanner, ed. *Symposia of the Society for the Study of Human Biology. Vol.* 3*: Human Growth.* New York: Pergamon.
Terasaki, P. I.
1969. "Human histocompatibility antigens of leukocytes." *Ann. Rev. Med.* **20:** 175–188.
Terman, L. M., and M. H. Oden, eds.
1959. *Genetic Studies of Genius. Vol.* 5*: The Gifted Group at Mid-Life.* Stanford, Calif.: Stanford University Press.
Thoday, J. M.
1960. "Effects of disruptive selection." *Heredity* **14**(1 and 2): 35–49.
1966. "New insights into continuous variation." *In* J. F. Crow and J. V. Neel, eds. *Proceedings of the Third International Congress of Human Genetics.* Baltimore: The Johns Hopkins Press, 1967.
Tietze, C.
1948. "A note on the sex ratio of abortions." *Hum. Biol.* **20:** 156.
Tisserand-Perrier, M.
1953. "Etude comparative de certains processus de croissance chez les jumeaux." *J. Genet. Hum.* **2:** 87–102.
Turnbull, C. M.
1962. *The Forest People.* New York: Doubleday.
1965. *Wayward Servants.* Garden City N.Y.: Natural History Press.
Twisselmann, F.
1961. "De l'évolution du taux de consanguinité en Belgique entre les années 1918 et 1959." *Proc. Second Int. Congr. Hum. Genet.* **1:** 142–150. Rome: Istituto G. Mendel.
Tyrrell, D. A. J., P. Sparrow, and A. S. Beare
1968. "Relation between blood groups and resistance to infection with influenza and some picornaviruses." *Nature* **220:** 819–820.
United Nations
1961. *Demographic Yearbook*, 13th Ed. New York: United Nations.
United Nations. Scientific Committee
1962. *Report on the Effects of Atomic Radiation.* General Assembly, 17th Session (1962), Official Records: Supplement No. 16 (A/5216). New York: United Nations.
1966. *Report on the Effects of Atomic Radiation.* General Assembly, 21st Session (1966), Official Records: Supplement No. 14 (A/6314). New York: United Nations.
United Nations Educational, Scientific, and Cultural Organization.
1956. *The Race Question in Modern Science.* Paris: UNESCO.
Van Der Weerdt, C. M.
1965. "The platelet agglutination test in platelet grouping." *In* H. Balner, F. J. Cleton, and J. G. Eernisse, eds. *Histocompatibility Testing, 1965.* Copenhagen:Munksgaard.
Van Loghem, E., J. Shuster, and H. H. Fudenberg
1968. "Gm factors in non-human primates." *Vox Sang.* **14:** 81–94.
Van Rood, J. J.
1962. *Leukocyte grouping. A Method and Its Application.* Unpublished Ph.D. Dissertation, University of Leyden.
Van Rood, J. J., J. G. Eernisse, and A. Van Leeuwen
1958. "Leukocyte antibodies in sera from pregnant women." *Nature* **181:** 1735–1736.

Van Rood, J. J., A. Van Leeuwen, A. M. J. Schippers, W. H. Vooys, E. Frederiks, H. Balner, and J. G. Eernisse

1965. "Leukocyte groups, the normal lymphocyte transfer test and homograft sensitivity." *In* H. Balner, F. J. Cleton, and J. G. Eernisse, eds. *Histocompatibility Testing, 1965*. Copenhagen: Munksgaard.

Vogel, F.

1961. *Lehrbuch der Allgemeinen Humangenetik*. Berlin: Springer-Verlag.

1963. "Mutations in man." *In* S. J. Geerts, ed. *Genetics Today*, Vol 3 (Proceedings of the Eleventh International Congress of Genetics. The Hague. September 1963). New York: Pergamon Press.

Vogel, F., H. J. Pettenkoffer, and W. Helmbold

1960. "Uber die Populationsgenetik der ABO-Blutgruppen." *Acta Genet. Statist. Med.* **10:** 267–294.

Vos, G. H.

1965. "The frequency of ABO-incompatible combinations in relation to maternal Rhesus antibody values in Rh immunized women." *Amer. J. Hum. Genet.* **17**(3): 202–211.

Wahlund, S.

1928. "Zusammensetzung von Populationen und Korrelationserscheinungen von Standpunkt der Vererbungslehre aus betrachtet." *Hereditas* **11:** 65–106.

Walker, P. M. B.

1968. "How different are the DNAs from related animals." *Nature* **219:** 228–232.

Walker, W., and S. Murray

1956. "Haemolytic disease of the newborn as a family problem." *Brit. Med. J.* **12:** 187.

Wang, A. C., J. Shuster, and H. H. Fudenberg

1969. "Evolutionary origin of the Gm 'a' peptide of immunoglobulins." *J. Mol. Biol.* **41:** 83–86.

Washburn, S. L., and P. C. Jay

1968. *Perspectives on Human Evolution*. New York: Holt, Rinehart and Winston.

Waterhouse, J. A. H.

1950. "Twinning in twin pedigrees." *Brit. J. Soc. Med.* **4:** 197.

Waterhouse, J. A. H., and L. Hogben

1947. "Incompatibility of mother and foetus with respect to the iso-agglutinogen A and its antibody." *Brit. J. Prev. Soc. Med.* **1:** 1–17.

Watkins, W. M.

1966. "Blood-group substances." *Science* **152:** 172–181.

Watson, J. D.

1965. *Molecular Biology of the Gene*. New York: W. A. Benjamin.

Weatherburn, C. E.

1946–1962. *A First Course in Mathematical Statistics*. Cambridge: Cambridge University Press.

Weinberg, W.

1901. "Beiträge zur Physiologie und Pathologie der Mehrlingsgeburten beim Menschen." *Arch. f. Ges. Physiol.* **88:** 346–430.

1908. "On the demonstration of heredity in man." *Reprinted in* S. H. Boyer, ed. *Papers on Human Genetics*. Englewood Cliffs, N.J.: Prentice-Hall, 1963.

1909. "Über Vererbungsgesetze beim Menschen." *Z. Ind. Abst. Vererb.-Lehre* **1:** 377–392, 440–460; **2:** 276–330.

1910. "Weitere Beiträge zur Theorie der Vererbung." *Arch. Rass. Ges. Biol.* **7:** 35–49; 169–173.

1921. "Methode und Fehlerquellen der Untersuchung auf Mendeleschen Zahlen beim Menschen." *Arch. Rass. Ges. Biol.* **9:** 165–174.

1927. "Mathematische Grundlagen der Probandenmethode." *Z. Ind. Abst. Vererb.-Lehre* **48:** 179–228.

Weiss, G. H., and M. Kimura

1965. "A mathematical analysis of the stepping stone model of genetic correlation." *J. Appl. Prob.* **2:** 129–149.

Weiss, M. C., and H. Green
1967. "Human-mouse hybrid cell lines containing partial complements of human chromosomes and functioning human genes." *Proc. Nat. Acad. Sci.* **58:** 1104–111.
Westphal, H., and R. Dulbecco
1968. "Viral DNA in polyoma- and SV40-transformed cell lines." *Proc. Nat. Acad. Sci.* **59:** 1158–1105.
Wiener, A. S., and J. Moor-Jankowski
1963. "Blood groups in anthropoid apes and baboons." *Science* **142:** 67–69.
Wiesenfeld, S. L.
1967. "Sickle cell trait in human biological and cultural evolution." *Science* **157:** 1134–1140.
Wilson, A. P., and I. R. Franklin
1968. "The distribution of the Diego blood group and its relationship to climate." *Caribbean J. Sci.* **8:** 1–13.
Wolstenholme, G., ed.
1963. *Man and His Future.* Boston: Little, Brown.
Woolf, C. M., F. E. Stephens, D. D. Mulaik, and R. E. Gilbert
1956. "An investigation of the frequency of consanguineous marriages among the Mormons or their relatives in the United States." *Amer. J. Hum. Genet.* **8:** 236–252.
Workman, P. L., B. S. Blumberg, and A. J. Cooper
1963. "Selection, gene migration and polymorphic stability in a U.S. White and Negro population." *Amer. J. Hum. Genet.* **15**(4): 429–437.
World Health Organization.
1969. *Genetic Counselling.* Technical Report Series, No. 416. Geneva: World Health Organization.
Wright, S.
1921. "Systems of mating, I, II, III, IV, and V. *Genetics* **6:** 111–1278.
1921. "Assortative mating based on somatic resemblance." *Genetics* **6:** 144–161.
1922. "Coefficients of inbreeding and relationship." *Amer. Natur.* **56:** 330–338.
1931. "Evolution in Mendelian Populations." *Genetics* **16**: 97–159.
1934. "The results of crosses between inbred strains of guinea pigs differing in number of digits." *Genetics* **19:** 537–551.
1937. "The distribution of gene frequencies in populations." *Proc. Nat. Acad. Sci.* **23:** 307–320.
1938. "The distribution of gene frequencies under irreversible mutation." *Proc. Nat. Acad. Sci.* **24:** 253–259.
1939a. "The distribution of self-sterility alleles in populations." *Genetics* **24:** 538–552.
1939b. "Statistical genetics in relation to evolution." 1–64, Paris: *In* G. Teissier, ed. *Exposés de biométrie et de statististique biologique*, Vol. 13. Paris: Hermann.
1943. "Isolation by distance." *Genetics* **28:** 114–138.
1946. "Isolation by distance under diverse systems of mating." *Genetics* **31:** 39–59.
1951. "The genetical structure of populations." *Ann. Eugen.* **15:** 322–354.
1955. "Classification of the Factors of Evolution." *Cold Spring Harbor Symp. Quant. Biol.* **20:** 16–24D.
1967. *Evolution and the Genetics of Populations, Vol. I.* Chicago: Genetic and Biometric Foundation.
Yasuda, N.
1968a. "An extension of Wahlund's principle to evaluate mating type frequency." *Amer. J. Hum. Genet.* **20:** 1–23.
1968b. "Distribution of matrimonial distance in the Mishima District." *Proc. Twelfth Int. Congr. Genet.* **2:** 178–179. Tokyo: Genetics Society of Japan.
Zamenhof, S., E. van Marthens, and F. L. Margolis
1968. "DNA (cell number) and protein in neonatal brain: alteration by maternal dietary protein restriction." *Science* **160:** 322–323.

Zipf, G. K.
1949. *Human Behavior and the Principle of Least Effort.* Cambridge, Mass.: Addison-Wesley.
Zuckerkandl, B., and L. Pauling
1962. "Molecular disease, evolution and genic heterogeneity." *In* M. Kasha and B. Pullman, eds. *Horizons in Biochemistry.* Chicago: Academic Press.
1965. "Evolutionary divergence and convergence in proteins." *In* V. Bryson and H. Vogel, eds. *Evolving Genes and Proteins.* New York: Academic Press.

Indexes

Author Index

Subject Index

Definitions are indicated by **boldface** type.